T0320772

Partially Observed Markov Decision Processes

Covering formulation, algorithms, and structural results, and linking theory to real-world applications in controlled sensing (including social learning, adaptive radars and sequential detection), this book focuses on the conceptual foundations of POMDPs. It emphasizes structural results in stochastic dynamic programming, enabling graduate students and researchers in engineering, operations research, and economics to understand the underlying unifying themes without getting weighed down by mathematical technicalities. Bringing together research from across the literature, the book provides an introduction to nonlinear filtering followed by a systematic development of stochastic dynamic programming, lattice programming, and reinforcement learning for POMDPs.

Questions addressed in the book include:

- When does a POMDP have a threshold optimal policy?
- When are myopic policies optimal?
- How do local and global decision-makers interact in adaptive decision making in multi-agent social learning where there is herding and data incest?
- How can sophisticated radars and sensors adapt their sensing in real time?
- How well can adaptive filtering algorithms track the state of a hidden Markov model?

Vikram Krishnamurthy is a Professor and Canada Research Chair in statistical signal processing in the Electrical & Computer Engineering department at the University of British Columbia. His research contributions focus on nonlinear filtering, stochastic approximation algorithms, and POMDPs. Dr Krishnamurthy is a Fellow of the IEEE and served as a distinguished lecturer for the IEEE Signal Processing Society. In 2013, he received an honorary doctorate from KTH, Royal Institute of Technology, Sweden.

Partially Observed Markov Decision Processes

From Filtering to Controlled Sensing

VIKRAM KRISHNAMURTHY
University of British Columbia

CAMBRIDGE
UNIVERSITY PRESS

University Printing House, Cambridge CB2 8BS, United Kingdom

One Liberty Plaza, 20th Floor, New York, NY 10006, USA

477 Williamstown Road, Port Melbourne, VIC 3207, Australia

314-321, 3rd Floor, Plot 3, Splendor Forum, Jasola District Centre, New Delhi - 110025, India

79 Anson Road, #06-04/06, Singapore 079906

Cambridge University Press is part of the University of Cambridge.

It furthers the University's mission by disseminating knowledge in the pursuit of education, learning and research at the highest international levels of excellence.

www.cambridge.org
Information on this title: www.cambridge.org/9781107134607

First published 2016

A catalogue record for this publication is available from the British Library

Library of Congress Cataloging in Publication data
Names: Krishnamurthy, V. (Vikram)
Title: Partially observed Markov decision processes : from filtering to controlled sensing / Vikram Krishnamurthy, University of British Columbia, Vancouver, Canada.
Description: Cambridge : Cambridge University Press, 2016. | Includes bibliographical references and index.
Identifiers: LCCN 2015047142 | ISBN 9781107134607
Subjects: LCSH: Markov processes–Textbooks. | Stochastic processes–Textbooks.
Classification: LCC QA274.7 .K75 2016 | DDC 519.2/33–dc23
LC record available at http://lccn.loc.gov/2015047142

ISBN 978-1-107-13460-7 Hardback

Additional resources for this title are available at www.cambridge.org/krishnamurthy

Contents

Preface

This book aims to provide an accessible treatment of partially observed Markov decision processes (POMDPs) to researchers and graduate students in electrical engineering, computer science and applied mathematics. "Accessible" means that, apart from certain parts in Part IV of the book, only an engineering version of probability theory is required as background. That is, measure theoretic probability is not required and statistical limit theorems are only used informally.

Contributions to POMDPs have been made by several communities: operations research, robotics, machine learning, speech recognition, artificial intelligence, control systems theory, and economics. POMDPs have numerous examples in controlled sensing, wireless communications, machine learning, control systems, social learning and sequential detection.

Stochastic models and Bayesian state estimation (filtering) are essential ingredients in POMDPs. As a result, this book is organized into four parts:

- *Part I: Stochastic models and Bayesian filtering*[1] is an introductory treatment of Bayesian filtering. The aim is to provide, in a concise manner, material essential for POMDPs.
- *Part II: Partially Observed Markov Decision Processes: models and algorithms* deals with the formulation and algorithms for solving POMDPs, together with examples in controlled sensing.
- *Part III: Partially Observed Markov Decision Processes: structural results* constitutes the core of this book. It consists of six chapters (Chapters 9 to 14) that deal with lattice programming methods to characterize the structure of the optimal policy without brute force computations.
- *Part IV: Stochastic approximation and reinforcement learning* deals with stochastic gradient algorithms, simulation-based gradient estimation and reinforcement learning algorithms for MDPs and POMDPs.

Appendix A is a self-contained elementary description of stochastic simulation. *Appendix B* gives a short description of continuous-time HMM filters and their link to discrete-time filtering algorithms.

[1] The term "Bayesian" here denotes using Bayes' rule for computation. This terminology is used widely in electrical engineering. It has no relation to Bayesian statistics, which deals with the interpretation of priors and evidence.

The abstraction of POMDPs becomes alive with applications. This book contains several examples starting from target tracking in Bayesian filtering to optimal search, risk measures, active sensing, adaptive radars and social learning.

The typical applications of POMDPs assume finite action spaces and the state (belief) space is the unit simplex. Such problems are not riddled with technicalities and sophisticated measurable selection theorems are not required.

Courses: This book has been taught in several types of graduate courses.

At UBC, Parts I and II, together with some topics in Part IV, are taught in a thirteen-week course with thirty-nine lecture hours. This is supplemented with basic definitions in random processes and simulation – Appendix A is a self-contained review of stochastic simulation. Parts III and IV can be taught as a second graduate-level course together with applications in controlled sensing and social networks.

I have also taught a more intensive graduate-level course at KTH comprised of parts of Chapter 2, 3, 6, 7 and a selection of topics in Chapters 11 to 14.

When the audience is assumed to have some background in target tracking (Bayesian filtering), an intensive course in POMDPs for active sensing can cover material from Chapters 3, 7, 8 and some material in Parts III and IV.

Problem sets for each chapter are at `www.cambridge.org/krishnamurthy`. The website will also include additional pedagogical material and an errata.

What this book is not

This book does not contain an encyclopedic coverage of filtering or POMDPs. The book focuses primarily on structural results in filtering and stochastic dynamic programming and shows how these structural results can be exploited to design efficient algorithms – such structural results are used widely in signal processing, control theory, economics and operations research. Parts II, III and IV constitute the core of this book. Part I (filtering) was written to provide the necessary background for this core material. Our main applications stem from controlled sensing in signal processing and, to a lesser extent, social networks.

Since filtering and POMDPs are studied by numerous research communities, it is inevitable that this book omits several areas that some readers might deem to be important. We apologize for this in advance. This book does not cover decentralized POMDPs (although the social learning models considered in the book are similar in structure), factored POMDPs or hierarchical POMDPs.

This book is not relevant to computational statisticians who deal with Markov chain Monte Carlo algorithms and their analysis, and have little interest in social learning, stochastic control, structural results in dynamic programming, gradient estimation or stochastic approximation algorithms (namely, Parts II–IV of the book). This book emphasizes the deeper structural aspects of Bayesian filters and stochastic control, and not brute-force Monte Carlo simulation.

Acknowledgments

I am grateful to collaborators, students and family for supporting this project.

I am grateful to Bo Wahlberg for hosting my visits to KTH where parts of the book were taught in short graduate courses.

Much of my research is funded by the Canada Research Chairs program, NSERC and SSHRC grants. Thanks to this block research funding and the supportive environment in the ECE department at UBC, I have had large amounts of unfettered time to work on this book.

This book is dedicated to the memory of my father E.V. Krishnamurthy. I wish we could spend more summer afternoons in Canberra, drink a few beers, and share wonderful times.

Vikram Krishnamurthy

1 Introduction

We start with some terminology.

- A Markov decision process (MDP) is obtained by controlling the transition probabilities of a Markov chain as it evolves over time.
- A hidden Markov model (HMM) is a noisily observed Markov chain.
- A partially observed Markov decision process (POMDP) is obtained by controlling the transition probabilities and/or observation probabilities of an HMM.

These relationships are illustrated in Figure 1.1.

A POMDP specializes to an MDP if the observations are noiseless and equal to the state of the Markov chain. A POMDP specializes to an HMM if the control is removed. Finally, an HMM specializes to a Markov chain if the observations are noiseless and equal to the state of the Markov chain.

The remainder of this introductory chapter is organized as follows:

- §1.1 to §1.4 contain a brief outline of the four parts of the book.
- §1.5 outlines some applications of controlled sensing and POMDPs.

1.1 Part I: Stochastic models and Bayesian filtering

Part I of this book contains an introductory treatment of *Bayesian filtering*, also called *optimal filtering*. Figure 1.2 illustrates the setup. A sensor provides noisy observations y_k of the evolving state x_k of a Markov stochastic system, where k denotes discrete time. The Markov system, together with the noisy sensor, constitutes a partially observed Markov model (also called a stochastic state space model or hidden Markov model[1]). The aim is to estimate the state x_k at each time instant k given the observations $y_1, \ldots, y_k$.

Part I of the book deals with *optimal filtering*. The optimal filter computes the posterior distribution π_k of the state at time k via the recursive algorithm

$$\boxed{\pi_k = T(\pi_{k-1}, y_k)} \tag{1.1}$$

[1] In this book, the term "hidden Markov model" is used for the special case when x_k is a finite state Markov chain that is observed via noisy observations y_k.

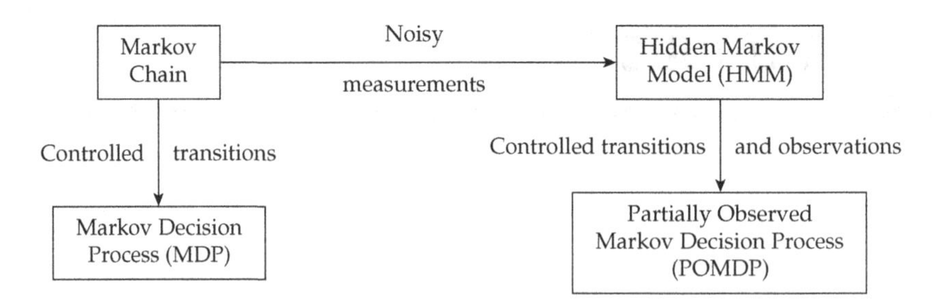

Figure 1.1 Terminology of HMMs, MDPs and POMDPs.

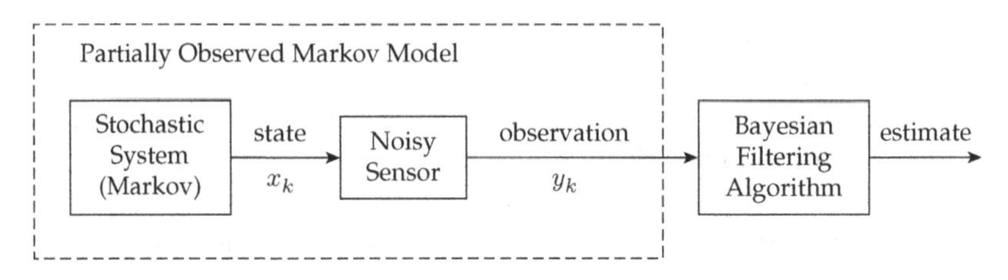

Figure 1.2 Part I deals with hidden Markov models and Bayesian filtering for state estimation. The framework is classical statistical signal processing.

where the operator T denotes Bayes' formula. Once the posterior π_k is evaluated, the optimal estimate (in the minimum mean square sense) of the state x_k given the noisy observations $y_1, \ldots, y_k$ can be computed by integration.

Part I of the book deals with the properties of the filtering recursion (1.1). The aim is to provide in a concise manner, material essential for POMDPs.

Chapters 2 and 3 cover classical topics including state space models, the Kalman filter, hidden Markov model filter and suboptimal filtering algorithms such as the particle filter.

Chapter 4 discusses how the Bayesian filters can be used to devise numerical algorithms (general purpose optimization algorithms and also expectation maximization algorithms) for maximum likelihood parameter estimation.

Chapter 5 discusses multi-agent filtering over a social network – social learning and data incest models are formulated; such models arise in applications such as online reputation systems and polling systems.

The material in Part I is classical (to a statistical signal processing audience). However, some nontraditional topics are discussed, including filtering of reciprocal processes; geometric ergodicity of the HMM filter; forward only filters for the expectation maximization algorithm; multi-agent filtering for social learning and data incest. Also, Appendix B discusses continuous-time HMM filters, Markov modulated Poisson filters, and their numerical implementation.

1.2 Part II: POMDPs: models and algorithms

Statistical signal processing (Part I) deals with extracting signals from noisy measurements. In Parts II, III and IV of the book, motivated by physical, communication and

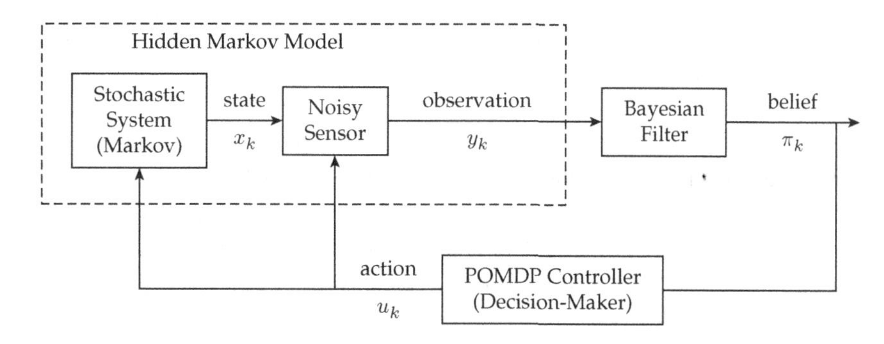

Figure 1.3 Schematic of partially observed Markov decision process (POMDP). Part II deals with algorithms and applications of POMDPs where the stochastic system (Markov Chain) and the sensor are controlled. Part III deals with determining the structure of a POMDP to ensure that the optimal action taken is a monotone function of the belief; this can result in numerically efficient algorithms to compute the optimal action (policy). Part IV deals with POMDPs when the stochastic system and sensor model are not known.

social constraints, we address the deeper issue of how to dynamically schedule and optimize signal processing resources to extract signals from noisy measurements. Such problems are formulated as POMDPs. Figure 1.3 displays the schematic setup.

Part II of the book deals with the formulation, algorithms and applications of POMDPs. As in the filtering problem, at each time k, a decision-maker has access to the noisy observations y_k of the state x_k of a Markov process. Given these noisy observations, the aim is to control the trajectory of the state and observation process by choosing actions u_k at each time k. The decision-maker knows ahead of time that if it chooses action u_k when the system is in state x_k, then a cost $c(x_k, u_k)$ will be incurred at time k. (Of course, the decision-maker does not know state x_k at time k but can estimate the cost based on the observations y_k.) The goal of the decision-maker is to choose the sequence of actions $u_0, \dots, u_{N-1}$ to minimize the expected *cumulative* cost $\mathbb{E}\{\sum_{k=0}^{N} c(x_k, u_k)\}$, where $\mathbb{E}$ denotes mathematical expectation.

It will be shown in Part II that the optimal action u_k at each time k is determined by a *policy* (strategy) as $u_k = \mu_k^*(\pi_k)$ where the optimal policy μ_k^* satisfies *Bellman's stochastic dynamic programming equation*:

$$\boxed{\begin{aligned} \mu_k^*(\pi) &= \operatorname*{argmin}_u Q_k(\pi, u), \quad J_k(\pi) = \min_u Q_k(\pi, u), \\ Q_k(\pi, u) &= \sum_x c(x, u)\pi(x) + \sum_y J_{k+1}\,(T(\pi, y, u))\sigma(\pi, y, u). \end{aligned}} \tag{1.2}$$

Here T is the optimal filter (1.1) used in Part I, and σ is a normalization term for the filter. Also π is the posterior computed via the optimal filter (1.1) and is called the *belief state*.

Part II of the book deals with algorithms for solving Bellman's equation (1.2) along with several applications in controlled sensing.

Chapter 6 is a concise presentation of stochastic dynamic programming for fully observed finite-state MDPs.

Chapter 7 starts our formal presentation of POMDPs. The POMDP model and stochastic dynamic programming recursion (1.2) are formulated in terms of the belief state computed by the Bayesian filter discussed in Part I. Several algorithms for solving POMDPs over a finite horizon are then presented. Optimal search theory for a moving target is used as an illustrative example of a POMDP.

Chapter 8 deals with the formulation and applications of POMDPs in controlled sensing. Several examples are discussed, including linear quadratic state and measurement control with applications in radar control, sensor scheduling for POMDPs with nonlinear costs and social learning.

1.3 Part III: POMDPs: structural results

In general, solving Bellman's dynamic programming equation (1.2) for a POMDP is computationally intractable. Part III of the book shows that by introducing assumptions on the POMDP model, important structural properties of the optimal policy can be determined without brute-force computations. These structural properties can then be exploited to compute the optimal policy.

The main idea behind Part III is to give conditions on the POMDP model so that the optimal policy $\mu_k^*(\pi)$ is monotone[2] in belief π. In simple terms, $\mu_k^*(\pi)$ is shown to be increasing in belief π (in terms of a suitable stocahastic ordering) by showing that $Q_k(\pi, u)$ in Bellman's equation (1.2) is *submodular*. The main result is:

$$\boxed{\underbrace{Q_k(\pi, u+1) - Q_k(\pi, u) \downarrow \pi}_{\text{submodular}} \implies \underbrace{\mu_k^*(\pi) \uparrow \pi.}_{\text{increasing policy}}} \tag{1.3}$$

Obtaining conditions for $Q_k(\pi, u)$ to be submodular involves powerful ideas in stochastic dominance and lattice programming.

Once the optimal policy of a POMDP is shown to be monotone, this structure can be exploited to devise efficient algorithms. Figure 1.4 illustrates an increasing optimal policy $\mu_k^*(\pi)$ in the belief π with two actions $u_k \in \{1, 2\}$. Note that any increasing function which takes on two possible values has to be a step function. So computing $\mu_k^*(\pi)$ boils down to determining the single belief π_1^* at which the step function jumps. Computing (estimating) π_1^* can be substantially easier than directly solving Bellman's equation (1.2) for $\mu_k^*(\pi)$ for all beliefs π, especially when $\mu_k^*(\pi)$ has no special structure.

Part III consists of six chapters (Chapters 9 to 14).

Chapter 9 gives sufficient conditions for a MDP to have a monotone (increasing) optimal policy. The explicit dependence of the MDPs optimal cumulative cost on transition probability is also discussed.

In order to give conditions for the optimal policy of a POMDP to be monotone, one first needs to show monotonicity of the underlying hidden Markov model filter. To this end, Chapter 10 discusses the monotonicity of Bayesian (hidden Markov model) filters.

[2] By monotone, we mean either increasing for all π or decreasing for all π. "Increasing" is used here in the weak sense; it means "non-decreasing". Similarly for decreasing.

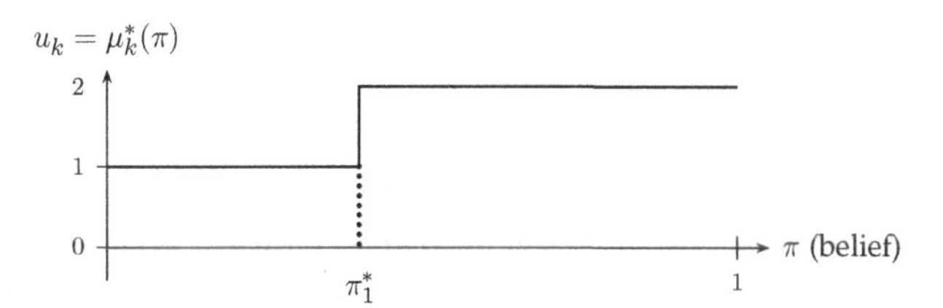

Figure 1.4 Example of optimal policy $\mu^*(\pi)$ that is monotone (increasing) in the belief π. The policy is a step function and completely characterized by the threshold state π_1^*.

This monotonicity of the optimal filter is used to construct reduced complexity filtering algorithms that provably lower- and upper-bound the optimal filter.

Chapters 11 to 14 give conditions on the POMDP model for the dynamic programming recursion to have a monotone solution. Chapter 11 discusses conditions for the value function in dynamic programming to be monotone. This is used to characterize the structure of two-state POMDPs and POMDP multi-armed bandits.

Chapter 12 gives conditions under which stopping time POMDPs have monotone optimal policies. As examples, Chapter 13 covers quickest change detection, controlled social learning and a variety of other applications. The structural results provide a unifying theme and insight to what might otherwise simply be a collection of examples.

Finally, Chapter 14 gives conditions under which the optimal policy of a general POMDP can be lower- and upper-bounded by judiciously chosen myopic policies. Bounds on the sensitivity of the optimal cumulative cost of POMDPs to the parameters are also discussed.

1.4 Part IV: Stochastic approximation and reinforcement learning

A major assumption in Parts I, II and III of the book is that the model of the stochastic system and noisy sensor is completely specified and known ahead of time. When this assumption does not hold, one needs to devise alternative methods. Part IV deals with *stochastic gradient algorithms* for estimating reasonable (locally optimal) strategies for POMDPs.

Suppose a decision-maker can observe the noisy response y_k of a controlled stochastic system to any action u_k that it chooses. Let $\mathcal{I}_k = \{u_0, y_1, \ldots, u_{k-1}, y_k\}$ denote the history of actions and observed responses up to time k. The decision-maker chooses its action as $u_k = \mu_\theta(\mathcal{I}_k)$ where μ_θ denote a parametrized policy (parametrized by a vector θ). Then, to optimize its choice of actions, the decision-maker needs to compute the optimal parameter θ^* which minimizes the expected cost criterion $\mathbb{E}\{C(\theta, \mathcal{I}_k)\}$. The decision-maker uses the following stochastic gradient algorithm to estimate θ^*:

$$\boxed{\theta_{k+1} = \theta_k - \epsilon \, \nabla_\theta C(\theta_k, \mathcal{I}_k), \quad k = 0, 1, \ldots.} \tag{1.4}$$

Here $\nabla_\theta C(\theta_k, \mathcal{I}_k)$ denotes the gradient (or estimate of gradient) of the instantaneous cost with respect to the parameter θ and ϵ denotes a small positive step size. Algorithms such as (1.4) lie within the class of reinforcement learning methods since the past experience $\mathcal{I}_k$ is used to adapt the parameter θ_k which in turn determines the actions; intuitively a good choice of θ would result in good performance which in turn reinforces this choice. Part IV deals with such stochastic gradient algorithms, including how to compute the gradient estimate and analyze the resulting algorithm.

Chapter 15 describes simulation-based gradient estimation methods that form the basis for gradient-based reinforcement learning. Chapter 16 deals with Q-learning and policy gradient algorithms for reinforcement learning. Chapter 17 presents stochastic approximation algorithms for estimating the parameters and states of a hidden Markov model (which can be used for adaptive control of a POMDP) and discrete policy search algorithms and mean field dynamics of large scale Markov chains that arise in social networks.

Remark

The three main equations described above, namely the filtering recursion (1.1), Bellman's dynamic programming equation (1.2), and the stochastic gradient algorithm (1.4), are ubiquitous in electrical engineering. Most algorithms in statistical signal processing and control boil down to these. The submodularity equation (1.3) is the basis for analysis of the optimal policy structure.

1.5 Examples of controlled (active) sensing

This section outlines some applications of controlled sensing formulated as a POMDP. Controlled sensing also known as "sensor adaptive signal processing" or "active sensing" is a special case of a POMDP where the decision-maker (controller) controls the observation noise distribution but not the dynamics of the stochastic system. The setup is as in Figure 1.3 with the link between the controller and stochastic system omitted.

In controlled sensing, the decision-maker controls the observation noise distribution by switching between various sensors or sensing modes. An accurate sensor yields less noisy measurements but is expensive to use. An inaccurate sensor yields more noisy measurements but is cheap to use. How should the decision-maker decide at each time which sensor or sensing mode to use? Equivalently, how can a sensor be made "smart" to adapt its behavior to its environment in real time? Such an active sensor uses *feedback* control. As shown in Figure 1.3, the estimates of the signal are fed to a controller/scheduler that decides the sensor should adapt so as to obtain improved measurements; or alternatively minimize a measurement cost. Design and analysis of such closed loop systems which deploy stochastic control is nontrivial. The estimates from the signal processing algorithm are uncertain (they are posterior probability distribution functions). So controlled sensing requires decision making under uncertainty.

We now highlight some examples in controlled sensing covered in this book.

Example 1: Adaptive radars

Adaptive multifunction radars are capable of switching between various measurement modes, e.g. radar transmit waveforms, beam pointing directions, etc. so that the tracking system is able to tell the radar which mode to use at the next measurement epoch. Instead of the operator continually changing the radar from mode to mode depending on the environment, the aim is to construct feedback control algorithms that dynamically adapt where the radar radiates its pulses to achieve the command operator objectives. This results in radars that autonomously switch beams, transmitted waveforms, target dwell and re-visit times. §8.4 and §12.7 deal with simplified examples of radar control.

Example 2: Social learning and data incest

A *social sensor* (human-based sensor) denotes an agent that provides information about its environment (state of nature) to a social network. Examples of such social sensors include Twitter posts, Facebook status updates, and ratings on online reputation systems like Yelp and TripAdvisor. Social sensors present unique challenges from a statistical estimation point of view since they interact with and influence other social sensors. Also, due to privacy concerns, they reveal their decisions (ratings, recommendations, votes) which can be viewed as a low resolution (quantized) function of their raw measurements.

In Chapter 5, the formalism of *social learning* [29, 56, 84] will be used for modeling the interaction and dynamics of social sensors. The setup is fundamentally different from classical signal processing in which sensors use noisy observations to compute estimates – in social learning agents use noisy observations together with decisions made by previous agents, to estimate the underlying state of nature. Also, in online reputation systems such as Yelp or TripAdvisor which maintain logs of votes (actions) by agents, social learning takes place with information exchange over a graph. Data incest (misinformation propagation) occurs due to unintentional reuse of identical actions in the formation of public belief in social learning; the information gathered by each agent is mistakenly considered to be independent. This results in overconfidence and bias in estimates of the state. How can automated protocols be designed to prevent data incest and thereby maintain a fair online reputation system?

Example 3: Quickest detection and optimal sampling

Suppose a decision-maker records measurements of a finite-state Markov chain corrupted by noise. The goal is to decide when the Markov chain hits a specific target state. The decision-maker can choose from a finite set of sampling intervals to pick the next time to look at the Markov chain. The aim is to optimize an objective comprising false alarm, delay cost and cumulative measurement sampling cost. Making more frequent measurements yields accurate estimates but incurs a higher measurement cost. Making an erroneous decision too soon incurs a false alarm penalty. Waiting too long to declare the target state incurs a delay penalty. What is the optimal sequential strategy for the decision-maker? It is shown in §13.5 that the optimal sampling problem results in a POMDP that has a monotone optimal strategy in the belief state.

Example 4: Interaction of local and global decision-makers

In a multi-agent network, how can agents use their noisy observations and decisions made by previous agents to estimate an underlying randomly evolving state? How do decisions made by previous agents affect decisions made by subsequent agents? In §13.4, these questions will be formulated as a multi-agent sequential detection problem involving social learning. Individual agents record noisy observations of an underlying state process, and perform social learning to estimate the underlying state. They make local decisions about whether a change has occurred that optimizes their individual utilities. Agents then broadcast their local decisions to subsequent agents. As these local decisions accumulate over time, a global decision-maker needs to decide (based on these local decisions) whether or not to declare a change has occurred. How can the global decision-maker achieve such change detection to minimize a cost function comprised of false alarm rate and delay penalty? The local and global decision-makers interact, since the local decisions determine the posterior distribution of subsequent agents which determines the global decision (stop or continue) which determines subsequent local decisions. We also discuss how a monopolist should optimally price their product when agents perform social learning.

Other applications of POMDPs

POMDPs are used in numerous other domains. Some applications include:

- Optimal search: see §7.7.
- Quickest detection and other sequential detection problems: see Chapter 12.
- Dialog systems: see [350] and references therein.
- Robot navigation and planning: see [194] and references therein.
- Cognitive radio dynamic spectrum sensing: see [353] and references therein.

Website repositories

Code for POMDP solvers is freely downloadable from:

- `www.pomdp.org`
- `bigbird.comp.nus.edu.sg/pmwiki/farm/appl/`

Part I

Stochastic models and Bayesian filtering

Part I of the book deals with the filtering problem.

Chapter 2 describes the partially observed Markov process model (stochastic state space model) and the Chapman–Kolmogorov equation for prediction.

Chapter 3 discusses the filtering problem and filtering algorithms – particularly the Kalman filter and HMM filter and their properties, and also a short discussion on suboptimal filtering algorithms such as the particle filter.

In Chapter 4, the filters are used to compute the maximum likelihood parameter estimate for HMMs via general purpose optimization algorithms and also the Expectation Maximization (EM) algorithm.

Chapter 5 discusses social learning and filtering on graphs. It deals with the structure of the filtering recursion rather than computations and is indicative of the rest of the book. We return to such structural results in Part III of the book.

Appendix B discusses filters for continuous-time HMMs and Markov modulated Poisson processes, and their numerical implementation.

The material of Part I forms the background to the rest of the book where *controlled* partially observed Markov models are considered. Please note that the term "Bayesian" here is consistent with the usage in signal processing. It means that Bayes' formula is used in *computation*.

Part I is an *introductory* treatment of Bayesian filtering and is not meant to be a comprehensive account. We mainly discuss filters for finite state Markov chains in noise (including social learning filters) and, to a lesser extent, filters for linear Gaussian state space processes and jump Markov linear systems.

In Parts II to IV of the book, we will deal with controlling the dynamics of the HMM filter. The reader who wishes to proceed rapidly to Part II only needs to read sections of Chapters 2 and 3 pertaining to finite state Markov chains and HMMs, and parts of Chapter 5 in social learning.

2 Stochastic state space models

Contents

This chapter discusses stochastic state space models and how optimal predictors can be constructed to predict the future state of a stochastic dynamic system. Finally, we examine how such predictors converge over large time horizons to a stationary predictor.

2.1 Stochastic state space model

The stochastic state space model is the main model used throughout this book. We will also use the phrase *partially observed Markov model* interchangeably with state space model.

We start by giving two equivalent definitions of a stochastic state space model. The first definition is in terms of a stochastic difference equation. The second definition is presented in terms of the transition kernel of a Markov process and the observation likelihood.

2.1.1 Difference equation form of stochastic state space model

Let $k = 0, 1, \dots,$ denote discrete time. A discrete-time stochastic state space model is comprised of two random processes $\{x_k\}$ and $\{y_k\}$:

$$x_{k+1} = \phi_k(x_k, w_k), \quad x_0 \sim \pi_0 \tag{2.1}$$

$$y_k = \psi_k(x_k, v_k). \tag{2.2}$$

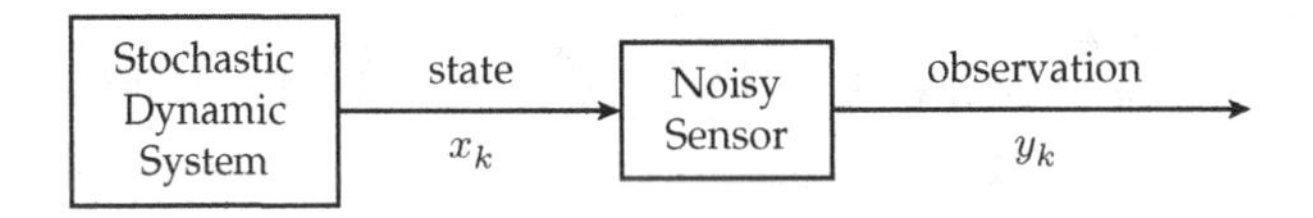

Figure 2.1 Block diagram of stochastic state space model.

The difference equation (2.1) is called the *state equation*. It models the evolution of the state x_k of a nonlinear stochastic system – *nonlinear* since $\phi_k(x, w)$ is any nonlinear[1] function; *stochastic* since the system is driven by the random process $\{w_k\}$ which denotes the "process noise" or "state noise". At each time k, the state x_k lies in the state space $\mathcal{X} = \mathbb{R}^X$. The initial state at time $k = 0$, x_0 is generated randomly according to prior distribution π_0. This is denoted symbolically as $x_0 \sim \pi_0$.

The *observation equation* (2.2) models a nonlinear noisy sensor that observes the state process $\{x_k\}$ corrupted by measurement noise $\{v_k\}$. At each time k, the observation y_k is a Y-dimensional vector valued random variable. Note that y_k defined by (2.2) is a doubly stochastic process. It is a random function of the state x_k which itself is a stochastic process evolving according to (2.1).

It is assumed that the state noise process $\{w_k\}$ is an X-dimensional *independent and identically distributed* (i.i.d.) sequence of random variables.[2] Also it is assumed that$\{w_k\}$, $\{v_k\}$ and x_0 are independent.

The observation noise process $\{v_k\}$ is assumed to be a Y-dimensional i.i.d. sequence of random variables. Denote the probability density function (pdf) or probability mass function (pmf) of w_k and v_k, respectively, as p_w and p_v. We denote this as

$$w_k \sim p_w, \quad v_k \sim p_v. \tag{2.3}$$

Additive noise state space model: Most models of engineering relevance involve additive noise and are the following special case of (2.1) and (2.2):

$$x_{k+1} = A_k(x_k) + \Gamma_k(x_k)w_k, \quad x_0 \sim \pi_0 \tag{2.4}$$

$$y_k = C_k(x_k) + D_k(x_k)v_k. \tag{2.5}$$

Examples such as linear Gaussian state space models and jump Markov linear systems are special cases of this additive noise model.

Figure 2.1 is a schematic representation of the stochastic state space model. The following preliminary example is useful in motivating the model. Consider a target (e.g. aircraft) whose kinematic coordinates (position and velocity) evolve according to the state vector x_k in three-dimensional Euclidean space. The process noise $\{w_k\}$ models our uncertainty about the target's dynamics. The noisy nonlinear sensor box in Figure 2.1 can be thought of as a radar which outputs noisy information about the

[1] Strictly speaking, ϕ and ψ need to be Borel measurable functions. Throughout this book we omit measure-theoretic considerations.

[2] $\{w_k\}$ is said to be i.i.d. if the random variables w_k, $k = 1, 2, \ldots$ are identically distributed and mutually independent. Mutual independence means that for any positive integer n and arbitrary times $k_1, k_2, \ldots, k_n$, $\mathbb{P}(w_{k_1} \in S_1, \ldots w_{k_n} \in S_n) = \prod_{l=1}^n \mathbb{P}(w_{k_l} \in S_l)$ for any (measurable) sets $S_1, S_2, \ldots, S_n$ where $\mathbb{P}$ denotes probability.

aircraft's position and velocity; this output is denoted as $\{y_k\}$. The observation noise $\{v_k\}$ models atmospheric effects and multi-path reflections that corrupt observations of the state process $\{x_k\}$.

2.1.2 Transition density form of state space model

The difference equation model (2.1), (2.2) is typically motivated by physics-based modeling. We now give an equivalent characterization for the evolution of the state and observations in terms of transition densities (or conditional probabilities). While less transparent from a modeling point of view, the transition density formulation streamlines notation.

State process

The state process $\{x_k\}$ is assumed to be Markov on the state space denoted as $\mathcal{X}$. This means: at each time k, x_k is a random variable defined on $\mathcal{X}$ such that:

1. At time $k = 0$, the initial state x_0 is generated according to probability density function (prior) π_0. Here, for any $S \subseteq \mathcal{X}$, the initial density π_0 is defined as

$$\mathbb{P}(x_0 \in S) = \int_S \pi_0(x)\,dx, \qquad \text{and } \mathbb{P}(x_0 \in \mathcal{X}) = 1,$$

   where $\mathbb{P}$ denotes probability.
2. Let n be an arbitrary positive integer. Then given any $n+1$ time instants $k_1 < k_2 < \ldots < k_n < k_{n+1}$, the following Markov property holds:

$$\mathbb{P}(x_{k_{n+1}} \in S | x_{k_1}, x_{k_2}, \ldots, x_{k_n}) = \mathbb{P}(x_{k_{n+1}} \in S | x_{k_n}).$$

   The Markov property says that the state at time k_{n+1} given the state at time k_n is conditionally independent of the past. The process $\{x_k\}$ is called a Markov process on state space $\mathcal{X}$ with *transition probabilities* $\mathbb{P}(x_{k+1} \in S|x_k)$. Also $\mathbb{P}(x_{k+1} \in \mathcal{X}|x_k) = 1$.

It is convenient to define the *transition density* $p(x_{k+1} = x|x_k)$ of the Markov process so that for any $S \subseteq \mathcal{X}$,

$$\mathbb{P}(x_{k+1} \in S|x_k) = \int_S p(x_{k+1} = x|x_k)\,dx, \qquad \int_{\mathcal{X}} p(x_{k+1} = x|x_k)\,dx = 1.$$

Observation process

The observation process is a probabilistic function of the state process. At each time k, the observation y_k recorded by a sensor is a random variable on the observation space $\mathcal{Y}$. For any subset S of $\mathcal{Y}$, the observation y_k is generated with conditional probability $\mathbb{P}(y_k \in S|x_k, x_{k-1}, \ldots, x_0)$. That is, the observation at time k depends probabilistically on the current and previous states. However, as an idealization of real sensors, we will assume that the following *conditional independence* property holds:

$$\mathbb{P}(y_k \in S|x_k, x_{k-1}, \ldots, x_0) = \mathbb{P}(y_k \in S|x_k), \quad S \subseteq \mathcal{Y}.$$

In other words, the observation y_k is a probabilistic function of state x_k only. Note that $\{y_k\}$ is a doubly stochastic process. It is a random function of the state x_k, which itself is a random process. Define the conditional density $p(y_k|x_k)$ as

$$\mathbb{P}(y_k \in S|x_k) = \int_S p(y_k = y|x_k)dy, \qquad \int_{\mathcal{Y}} p(y_k = y|x_k)dy = 1.$$

The conditional density $p(y_k|x_k)$ is called the *observation likelihood*.

Summary

The state space *model* consists of the initial state density π_0 and the following two sequences of conditional probability densities on the state space $\mathcal{X}$ and observation space $\mathcal{Y}$:

$$\text{State transition density: } p(x_{k+1}|x_k), \quad k = 0, 1 \ldots, \tag{2.6}$$

$$\text{Observation likelihood: } p(y_k|x_k), \quad k = 0, 1 \ldots. \tag{2.7}$$

The state space *process* is the observation process $\{y_k\}$ generated according to these probability laws. The stochastic state space model can also be viewed as a *partially observed Markov model* since the state is Markov and the observations provide partial (noisy) observations of the Markov state.

2.1.3 Stochastic difference equation to transition density

The stochastic difference equation representation (2.1), (2.2) and transition density representation (2.6), (2.7) are, of course, equivalent.

Consider the general state space model (2.1), (2.2). Assume $\phi_k(x_k, w)$ and $\psi(x_k, v)$ are invertible with respect to the variables w and v, respectively, so that there exist functions g_k and h_k such that

$$\begin{aligned} x_{k+1} = \phi_k(x_k, w) &\implies w = g_k(x_k, x_{k+1}) \\ y_k = \psi_k(x_k, v) &\implies v = h_k(x_k, y_k). \end{aligned}$$

Assuming these inverses are continuously differentiable then

$$p(x_{k+1}|x_k) = p_w\big(g_k(x_k, x_{k+1})\big) \, \|\nabla_{x_{k+1}} g_k(x_k, x_{k+1})\| \tag{2.8}$$

$$p(y_k|x_k) = p_v\big(h_k(x_k, y_k)\big) \, \|\nabla_{y_k} h_k(x_k, y_k)\| \tag{2.9}$$

where ∇ denotes the derivative.

These are shown as follows: Since $\{w_k\}$ is i.i.d. the transition probability

$$\mathbb{P}(x_{k+1} \leq x|x_k) = \mathbb{P}(\phi_k(x_k, w_k) \leq x) = \mathbb{P}(w_k \leq g_k(x_k, x)).$$

Taking derivatives with respect to x yields the transition density function (2.8). Similarly, since v_k is i.i.d., (2.9) holds because

$$\mathbb{P}(y_k \leq y|x_k) = \mathbb{P}(\psi_k(x_k, v_k) \leq y) = \mathbb{P}(v_k \leq h_k(x_k, y)).$$

As a concrete example, consider the state space model with additive noise, namely (2.4), (2.5). Assuming $\Gamma_k(x_k)$ and $D_k(x_k)$ are square invertible matrices, the transition density and observation likelihood form of (2.6), (2.7) are:

$$p(x_{k+1}|x_k) = |\Gamma_k^{-1}(x_k)|\, p_w\left(\Gamma_k^{-1}(x_k)\left[x_{k+1} - A_k(x_k)\right]\right) \tag{2.10}$$

$$p(y_k|x_k) = |D_k^{-1}(x_k)|\, p_v\left(D_k^{-1}(x_k)\left[y_k - C_k(x_k)\right]\right). \tag{2.11}$$

where $|\cdot|$ denotes determinant.

Equation (2.9) and (2.11) are called the *likelihood formula* – they model the input output behavior of the noisy nonlinear sensor box in Figure 2.1 in terms of the probability density function of the noise.

2.2 Optimal prediction: Chapman–Kolmogorov equation

Given the stochastic state space model for a partially observed Markov process, an important problem is to predict the state at time k, given the probability distribution of the state at time 0, namely, π_0. By prediction, we mean compute the probability density (or mass) function of the state at time k denoted as π_k. That is, for any subset $S \subseteq \mathcal{X}$, the density π_k is defined as

$$\mathbb{P}(x_k \in S) = \int_S \pi_k(x)dx, \qquad \int_{\mathcal{X}} \pi_k(x)dx = 1.$$

Once π_k is evaluated, one can compute the predicted mean, covariance, etc. at time k as will be explained below. We call π_k the *predicted* density.

The observation process $\{y_k\}$ is not used in this section since we are interested in predicting the state given the initial condition π_0 and the model. This section serves as a warm-up for the filtering problem considered in Chapter 3, where the observation process will be used to estimate the state process.

Given the transition density $p(x_{k+1}|x_k)$ for the state evolution in (2.6), a recursion for the predicted density of the state at each time k can be obtained as follows: from the total probability rule[3] it follows that

$$\pi_k(x) = \int_{\mathcal{X}} p(x_k = x|x_{k-1})\,\pi_{k-1}(x_{k-1})dx_{k-1}, \qquad \text{initialized by } \pi_0. \tag{2.12}$$

Equation 2.12 is called the *Chapman–Kolmogorov* equation. It is a recursion for computing the predicted density function π_k of the state at time k given the initial state probability π_0 and no observations. From π_k, one can then compute the predicted mean and covariance of the state at any time k as[4]

$$\begin{aligned} \hat{x}_k &= \mathbb{E}\{x_k\} = \int_{\mathcal{X}} x\,\pi_k(x)dx, \\ \operatorname{cov}(x_k) &= \mathbb{E}\{(x_k - \hat{x}_k)(x_k - \hat{x}_k)'\} = \mathbb{E}\{x_k x_k'\} - \hat{x}_k\hat{x}_k'. \end{aligned} \tag{2.13}$$

[3] In terms of probability density functions, the total probability rule is $p(x) = \int p(x|y)p(y)dy$.

[4] In this book, all vectors are column vectors. M' denotes transpose of a vector or matrix M.

The predicted state estimate $\hat{x}_k$ computed using π_k via (2.13) is optimal in the minimum mean square error sense. That is, if any other predictor $\phi(\pi_0)$ is used to predict state x_k, then[5]

$$\mathbb{E}\{(x_k - \hat{x}_k)^2\} \le \mathbb{E}\{(x_k - \phi(\pi_0))^2\}. \tag{2.14}$$

Hence this section is titled "optimal prediction".

Examples of stochastic state space models

The next three sections discuss the following stochastic state space models and their optimal predictors that arise in numerous applications.

- Linear Gaussian state space model
- Hidden Markov model
- Jump Markov linear system

2.3 Example 1: Linear Gaussian state space model

Linear Gaussian models are perhaps the most studied state space models. We use the following notation for an l-variate Gaussian (normal) pdf with mean μ and $l \times l$ covariance matrix Σ (assumed to be symmetric positive definite)

$$N(\zeta;\mu,\Sigma) = (2\pi)^{-l/2}\,|\Sigma|^{-1/2}\exp\left[-\frac{1}{2}(\zeta-\mu)'\Sigma^{-1}(\zeta-\mu)\right]. \tag{2.15}$$

Here $|\cdot|$ denotes the determinant of a matrix. When no confusion arises, the shorter notation $N(\mu,\Sigma)$ will be used.

2.3.1 The model

The linear Gaussian state space model is a special case of (2.4), (2.5):

$$x_{k+1} = A_k x_k + w_k, \quad x_0 \sim \pi_0 = N(\hat{x}_0, \Sigma_0),\ w_k \sim N(0, Q_k) \tag{2.16}$$

$$y_k = C_k x_k + v_k, \quad v_k \sim N(0, R_k). \tag{2.17}$$

Here the state $x_k \in \mathcal{X} = \mathbb{R}^X$ and measurement $y_k \in \mathcal{Y} = \mathbb{R}^Y$. At each time k, A_k is an $X \times X$ matrix and C_k is a $Y \times X$ matrix.

The processes $\{w_k\}$ $\{v_k\}$ are assumed to be i.i.d. Also $\{w_k\}$, $\{v_k\}$ and initial condition x_0 are statistically independent.

Using (2.10) and (2.11), we can express the linear Gaussian state space model in transition density form in terms of the following X-variate and Y-variate Gaussian densities (see notation in (2.15))

[5] By $(x_k - \hat{x}_k)^2$ we mean $(x_k - \hat{x}_k)'(x_k - \hat{x}_k)$. Equation (2.14) follows since $\mathbb{E}\{(x_k - \phi(\pi_0))^2\}$ is minimized by choosing $\phi(\pi_0) = \hat{x}_k$ because

$$\mathbb{E}\{(x_k - \phi(\pi_0))^2\} = \mathbb{E}\{(x_k - \hat{x}_k + (\hat{x}_k - \phi(\pi_0))^2\} = \mathbb{E}\{(\hat{x}_k - \phi(\pi_0))^2\} + \text{ constant}.$$

$$
\begin{aligned}
p(x_{k+1}|x_k) &= p_w\,(x_{k+1} - A_k(x_k)) = N(x_{k+1}; A_k x_k, Q_k) \\
p(y_k|x_k) &= p_v\,(y_k - C_k(x_k)) = N(y_k; C_k x_k, R_k).
\end{aligned}
\tag{2.18}
$$

2.3.2 Optimal predictor for linear dynamics

Using the Chapman–Kolmogorov equation (2.12), we can compute the predicted mean and covariance for the state of a linear Gaussian state space model.

THEOREM 2.3.1 *For a linear Gaussian state space model, the explicit solution to Chapman–Kolmogorov's equation (2.12) is* $\pi_{k+1} = N(\hat{x}_{k+1}, \Sigma_{k+1})$ *where the predicted mean* $\hat{x}_{k+1}$ *and predicted covariance* Σ_{k+1} *satisfy the following recursion:*

$$
\hat{x}_{k+1} = \mathbb{E}\{x_{k+1}\} = A_k \hat{x}_k \tag{2.19}
$$

$$
\Sigma_{k+1} = \operatorname{cov}\{x_{k+1}\} = A_k \Sigma_k A_k' + Q_k. \tag{2.20}
$$

In summary, if the initial density π_0 is Gaussian, then the predicted density π_k is Gaussian at each time k. The mean $\hat{x}_k$ is the predicted state at time k and covariance Σ_k measures the accuracy of this state estimate. These evolve according to (2.19) and (2.20). Equation (2.20) for the covariance is called the *Lyapunov equation* and arises in several places in this book.

Proof π_{k+1} is normal since a linear system preserves normal distributions. The above mean and covariance updates are obtained as follows: taking expectation of (2.16) yields (2.19). Also, evaluating $(x_{k+1} - \hat{x}_{k+1})(x_{k+1} - \hat{x}_{k+1})'$ yields

$$
A_k(x_k - \hat{x}_k)(x_k - \hat{x}_k)'A_k' + w_k w_k' + A_k(x_k - \hat{x}_k)w_k' + w_k(x_k - \hat{x}_k)'A_k'.
$$

Then evaluating $\mathbb{E}\{(x_{k+1} - \hat{x}_{k+1})(x_{k+1} - \hat{x}_{k+1})'\}$ and using the fact that w_k is zero mean and statistically independent of x_k yields (2.20). □

The expressions for the predicted mean (2.19) and covariance (2.20) hold for any i.i.d. zero mean state noise $\{w_k\}$ (not necessarily Gaussian). Actually (2.19) only needs zero mean state noise.

2.3.3 Stationary covariance for linear time invariant model

Here we discuss the asymptotic time behavior of the predictor in Theorem 2.3.1 when the linear state space model (2.16) is time invariant. Consider time invariant linear state space model

$$
x_{k+1} = A x_k + w_k, \quad x_0 \sim \pi_0,
$$

where $\mathbb{E}\{w_k\} = 0$ and $\operatorname{cov}\{w_k\} = Q$ for all k. Here $\{w_k\}$ and π_0 are not necessarily Gaussian. Suppose A is stable; that is, all its eigenvalues are smaller than 1 in magnitude. What can one say about the mean $\hat{x}_k$ and covariance Σ_k of the linear predictor (2.19), (2.20), as time $k \to \infty$?

From (2.19), (2.20) it follows that

$$\hat{x}_k = A^k x_0, \quad \Sigma_k = A^k \Sigma_0 A'^k + \sum_{n=0}^{k-1} A^n Q A'^n. \tag{2.21}$$

Clearly, from (2.21), since A is stable, $\hat{x}_k \to 0$ as $k \to \infty$. Next, if the covariance Σ_k converges, then, from (2.20), it is intuitively plausible that it converges to

$$\Sigma_\infty = A \Sigma_\infty A' + Q. \tag{2.22}$$

Equation (2.22) is called the *algebraic Lyapunov equation*. It is a linear equation in the elements of Σ_∞ and can be expressed as

$$(I - A \otimes A)\,\text{vec}(\Sigma_\infty) = \text{vec}(Q) \tag{2.23}$$

where $\otimes$ denotes the Kronecker product.[6] Solving (2.23) yields the asymptotic covariance of the linear predictor.

Under what conditions does the algebraic Lyapunov equation (2.22) admit a unique solution Σ_∞? First, let us express $Q = MM'$. Since Q is positive semi-definite, there exists an infinite number of matrices M such that $Q = MM'$. Next we introduce the following definition from linear systems theory [163]. Given a matrix M, the pair $[A, M]$ is said to be completely reachable if the matrix

$$[M, AM, A^2M, \ldots, A^{X-1}M] \tag{2.24}$$

has rank X. Equivalently, the matrix $\sum_{k=0}^{X-1} A^k MM' A'^k > 0$ (positive definite).

THEOREM 2.3.2 *Suppose A is stable (has all eigenvalues strictly smaller than 1 in magnitude) and $[A, M]$ is completely reachable (where $MM' = Q$). Then the algebraic Lyapunov equation (2.22) has a unique solution that is positive definite and satisfies $\Sigma_\infty = \sum_{k=0}^{\infty} A^k Q A'^k$.*

Proof Denote $\bar{\Sigma} = \sum_{k=0}^{\infty} A^k Q A'^k$. Since A is assumed stable, by the properties of convergent geometric series, $\bar{\Sigma}_\infty$ exists and is finite. It is trivially verified that $\bar{\Sigma}$ satisfies (2.22). Also clearly

$$\bar{\Sigma} \geq \sum_{k=0}^{X-1} A^k Q A'^k > 0$$

where the last inequality follows from the complete reachability assumption. Thus $\bar{\Sigma}$ is positive definite. To establish uniqueness of the solution, let $\tilde{\Sigma}$ be another solution of (2.22). Then since $\bar{\Sigma}$ and $\tilde{\Sigma}$ satisfy (2.22), it follows that

$$\bar{\Sigma} - \tilde{\Sigma} - A(\bar{\Sigma} - \tilde{\Sigma})A' = 0.$$

Therefore

$$A^{k-1}(\bar{\Sigma} - \tilde{\Sigma})A'^{k-1} - A^k(\bar{\Sigma} - \tilde{\Sigma})A'^k = 0$$

[6] For an arbitrary matrix M, vec(M) is a column vector constructed by stacking the columns of M below one another. It is easily shown that given three matrices A, X, B, then $\text{vec}(AXB) = (B' \otimes A)\,\text{vec}(X)$.

and adding such relations yields

$$\bar{\Sigma} - \tilde{\Sigma} - A^k(\bar{\Sigma} - \tilde{\Sigma})A'^k = 0.$$

Finally, since $A^k \to 0$ as $k \to \infty$, it follows that $\bar{\Sigma} = \tilde{\Sigma}$. □

The above theorem gives sufficient conditions for the covariance Σ_∞ of the infinite horizon predictor to be unique, positive definite, and bounded. It also shows that this predictor is asymptotically independent of the initial condition Σ_0. In Chapter 3, we will generalize this result by considering the Kalman filter which takes the noisy observations of the state into account.

2.4 Example 2: Finite-state hidden Markov model (HMM)

HMMs are used in numerous areas including speech recognition, neurobiology, protein alignment and telecommunications. Their simplicity and wide applicability stems from the fact that the underlying state is a finite-state Markov chain.

2.4.1 The finite-state HMM

An HMM is a finite-state Markov chain measured via a noisy observation process.

(i) *Markov chain*: Let the process $\{x_k\}$ denote an X-state Markov chain on the finite state space $\mathcal{X} = \{1, 2, \dots, X\}$. The dynamics of a homogeneous Markov chain $\{x_k\}$ are described by its $X \times X$ transition probability matrix P with elements P_{ij}, $i, j \in \{1, \dots, X\}$ such that

$$P_{ij} = \mathbb{P}(x_{k+1} = j | x_k = i), \quad 0 \le P_{ij} \le 1, \quad \sum_{j=1}^{X} P_{ij} = 1. \tag{2.25}$$

Each row of the transition probability matrix P adds to one. This can be written in matrix vector notation as $P\mathbf{1} = \mathbf{1}$, where $\mathbf{1}$ denotes the X dimensional column vector of ones. A matrix P with nonnegative elements such that $P\mathbf{1} = \mathbf{1}$ is called a *stochastic matrix*. The initial condition x_0 is drawn according to the probabilities $\mathbb{P}(x_0 = i) = \pi_0(i)$, $i = 1, \dots, X$ with $\sum_{i=1}^{X} \pi_0(i) = 1$.

Associated with the states $1, 2, \dots, X$, define the state level vector

$$C = [C(1), C(2), \dots, C(X)]'$$

where the element $C(i)$ denotes the physical state level of the Markov chain corresponding to state i. For example, in a binary communication system, if the transmitted signal is a two-state Markov chain with voltages 0 or 5, then $C = [0,\ 5]'$. The elements of C are called the "drift coefficients" or "state levels".

(ii) *Noisy observations*: For HMMs with continuous-valued observations, define the *observation probability* density function B as

$$B_{xy} = p(y_k = y | x_k = x), \quad x \in \mathcal{X}, y \in \mathcal{Y}. \tag{2.26}$$

For example, suppose, similar to (2.5), the HMM observations are obtained in additive noise as

$$y_k = C(x_k) + D(x_k)\, v_k, \quad x_k \in \{1, 2, \ldots, X\}, \tag{2.27}$$

where the measurement noise process $\{v_k\}$ is an i.i.d. sequence with density function p_v. Then the observation likelihood density $B_{xy} = p(y_k = y | x_k = x)$ is given by formula (2.11).

In many HMM applications, the observation space is finite: $\mathcal{Y} = \{1, \ldots, Y\}$. The observation probabilities $B_{xy} = \mathbb{P}(y_k = y | x_k = x)$ are then called "symbol" probabilities and B is an $X \times B$ dimensional matrix which satisfies $\sum_y B_{xy} = 1$.

2.4.2 Difference equation representation of HMM

The above HMM (2.25), (2.27) was in the transition probability form. An HMM can also be expressed in state space equation form similar to (2.4), (2.5). Let e_i denotes the X-dimensional unit vector with 1 in the i-th position and zeros elsewhere.[7] Define the state space

$$\mathcal{X} = \{e_1, e_2, \ldots, e_X\}.$$

So at each time k, $x_k \in \mathcal{X}$ is an $X-$dimensional indicator vector. For example, if the Markov chain is observed in additive i.i.d. measurement noise v_k, then the HMM can be expressed as the linear state space system

$$x_{k+1} = P' x_k + w_k, \tag{2.28}$$

$$y_k = C' x_k + v_k. \tag{2.29}$$

A key difference between (2.28) and (2.4) is that in (2.28), $\{w_k\}$ is no longer i.i.d. Instead $\{w_k\}$ is a martingale difference process. That is, the conditional expectation $\mathbb{E}\{w_k | x_0, x_1, \ldots, x_k\} = 0$.

To obtain the martingale representation (2.28) for a Markov chain, define the process

$$w_k = x_{k+1} - \mathbb{E}\{x_{k+1} | x_0, \ldots, x_k\}, \quad k = 0, 1, \ldots. \tag{2.30}$$

Then clearly $\mathbb{E}\{w_k | x_0, \ldots, x_k\} = 0$ implying w_k is a martingale difference. Since $\{x_k\}$ is Markov, it follows that

$$\mathbb{E}\{x_{k+1} | x_0, \ldots, x_k\} = \mathbb{E}\{x_{k+1} | x_k\} = \sum_j P_{x_k j} e_j = P' x_k.$$

Plugging this in (2.30) yields (2.28). Note also that (2.29) is equivalent to the measurement equation (2.27) with $D(x_k) = 1$.

[7] For example, for a 3-state Markov chain, $\mathcal{X} = \{e_1, e_2, e_3\}$ with $e_1 = [1, 0, 0]'$, $e_2 = [0, 1, 0]'$, $e_3 = [0, 0, 1]'$ where $'$ denotes transpose.

2.4.3 Optimal predictor for HMM and stationary distribution

For a finite-state Markov chain with state space $\mathcal{X} = \{1, \ldots, X\}$ and transition matrix P, define the X-dimensional state probability vector at time k as

$$\pi_k = \begin{bmatrix} \mathbb{P}(x_k = 1) & \ldots & \mathbb{P}(x_k = X) \end{bmatrix}'. \tag{2.31}$$

Then the Chapman–Kolmogorov equation (2.12) reads

$$\pi_{k+1} = P'\pi_k \quad \text{initialized by } \pi_0. \tag{2.32}$$

So (2.32) is the optimal predictor for an HMM. Note that the martingale representation (2.28) yields (2.32) by taking expectations on both sides of (2.28). Using π_{k+1} computed via (2.32), the predicted state level at time $k+1$ is

$$\mathbb{E}\{C(x_{k+1})\} = \sum_{i=1}^{X} C(i)\pi_{k+1}(i) = C'\pi_{k+1}.$$

Limiting distribution

The Chapman–Kolmogorov equation implies that $\pi_k = P'^k\pi_0$. The *limiting distribution* of a Markov chain is

$$\lim_{k\to\infty} \pi_k = \lim_{k\to\infty} P'^k\pi_0.$$

This limiting distribution may not exist. For example, if

$$P = \begin{bmatrix} 0 & 1 \\ 1 & 0 \end{bmatrix}, \quad \pi_0 = \begin{bmatrix} \pi_0(1) \\ \pi_0(2) \end{bmatrix}, \quad \text{then } \pi_k = \begin{cases} \begin{bmatrix} \pi_0(2) & \pi_0(1) \end{bmatrix}' & k \text{ odd} \\ \begin{bmatrix} \pi_0(1) & \pi_0(2) \end{bmatrix}' & k \text{ even} \end{cases} \tag{2.33}$$

and so $\lim_{k\to\infty} \pi_k$ does not exist unless $\pi_0(1) = \pi_0(2) = 1/2$.

Stationary distribution

The *stationary distribution* of a Markov chain is defined as any X-dimensional vector π_∞ that satisfies

$$\pi_\infty = P'\pi_\infty, \quad \mathbf{1}'\pi_\infty = 1, \tag{2.34}$$

where the second equality ensures that π_∞ is a valid probability mass function. So the stationary distribution π_∞ is the normalized right eigenvector of P' corresponding to the unit eigenvalue.[8] Equivalently, if a Markov chain is initialized at its stationary distribution, then from Chapman–Kolmogorov's equation, the state distribution at any time remains at the stationary distribution. That is, choosing $\pi_0 = \pi_\infty$ implies $\pi_k = \pi_\infty$ for all k. The stationary distribution is also called the *invariant, equilibrium or steady-state distribution*.

Clearly, the limiting distributions (if they exist) are a subset of stationary distributions. That is, any limiting distribution is a stationary distribution. The reverse implication is

[8] As a sanity check, since P is a stochastic matrix, $P\mathbf{1} = \mathbf{1}$. So P has an eigenvalue of 1 corresponding to the eigenvector $\mathbf{1}$. Therefore, P' also has an eigenvalue of 1.

not true in general. For example, given P in (2.33), $\pi_\infty = \begin{bmatrix} 0.5 & 0.5 \end{bmatrix}'$ is a stationary distribution but there is no limiting distribution.

DEFINITION 2.4.1 *A transition matrix P is said to be* irreducible *if every state i communicates with every state j in a finite amount of time. That is, for any two states $i, j \in \mathcal{X}$, there exists a positive integer k (which can depend on i, j) such that $P^k_{ij} > 0$. (Here P^k_{ij} denotes the (i, j) element of P raised to the power of k.)*

A transition matrix P is regular *(or primitive) if there exists a positive integer k such that $P^k_{ij} > 0$ for all i, j.*

Regular is a stronger condition than irreducible. For example, P in (2.33) is not regular since for each k, two elements of P^k are always zero – so there is no value of k for which all the elements are nonzero. However, P is irreducible.

THEOREM 2.4.2 *Consider a finite-state Markov chain with regular transition matrix P. Then:*

1. *The eigenvalue 1 has algebraic and geometric multiplicity of one.*
2. *All the remaining eigenvalues of P have modulus strictly smaller than 1.*
3. *The eigenvector of P' corresponding to eigenvalue of 1 can be chosen with strictly positive elements. (Indeed, the only eigenvectors with all strictly positive elements are associated with the eigenvalue 1.)*
4. *$P^k = \mathbf{1}\pi'_\infty + O(k^{m_2-1}|\lambda_2|^k)$ where λ_2 is the second largest eigenvalue modulus and m_2 is its algebraic multiplicity.*
5. *The limiting distribution and stationary distribution coincide.*

The proof of the above classical theorem is in [293] and will not be given here. Statements 1 to 3 of the above theorem constitute what is known as the *Perron–Frobenius theorem* for a regular stochastic matrix. (A more general Perron–Frobenius theorem holds for matrices with nonnegative elements; see [293].) Statement 4 asserts that if the transition matrix P is regular, then the state probability vector π_k forgets its initial condition geometrically fast. To see this, note that from the Chapman–Kolmogorov equation and Statement 4 of the above theorem,

$$\pi_k = {P'}^k \pi_0 = \pi_\infty \mathbf{1}' \pi_0 + O(k^{m_2-1}|\lambda_2|^k)\pi_0 = \pi_\infty + O(k^{m_2-1}|\lambda_2|^k)\pi_0. \tag{2.35}$$

Recall that π_k is a k-step ahead predictor of the state probabilities of the Markov chain given the initial probabilities π_0. So it follows that the k-step ahead predictor of a Markov chain forgets the initial condition geometrically fast in the second largest eigenvalue modulus, $|\lambda_2|$.

2.5 Example 3: Jump Markov linear systems (JMLS)

2.5.1 JMLS model

The linear state space model and HMM described above are special instances of the following more general stochastic state space model called a jump Markov linear system (JMLS):

$$z_{k+1} = A(r_{k+1})\, z_k + \Gamma(r_{k+1})\, w_{k+1} + f(r_{k+1})\, u_{k+1} \tag{2.36}$$

$$y_k = C(r_k)\, z_k + D(r_k)\, v_k + g(r_k) u_k. \tag{2.37}$$

Here the unobserved discrete-time Markov chain $r_k \in \{1, 2, \ldots, X\}$ with transition matrix P denotes the discrete component of the state of the JMLS, z_k denotes the continuous component of the state of the JMLS, w_k and v_k are i.i.d. noise processes, y_k denotes the noisy observation process and u_k is a known exogenous variable (input). $A(r_k)$, $\Gamma(r_k), f(r_k)$, $C(r_k)$, $D(r_k)$ and $g(r_k)$ are time-varying matrices (parameters) of the jump linear system; they evolve according to the realization of the unobserved finite-state Markov chain r_k.

Since they facilitate modeling continuous and discrete valued states (hybrid valued states), JMLS are one of the most widely used stochastic state space models in applications such as target tracking [32] (where they are called generalized HMMs) and deconvolution [238].

The above JMLS is a generalization of an HMM and linear state space model. For example, if $A = I$, $\Gamma = 0, f = 0$, then $\{r_k, y_k\}$, constitute the state and observations of an HMM. On the other hand, if the sample path $\{r_k\}$ is known, then the above JMLS reduces to the linear time varying stochastic state space model (2.16), (2.17).

It is important to note that in a JMLS, $\{z_k\}$ itself is not a Markov process and also y_k given z_k is not independent of past observations and states. The augmented state process $x_k = (z_k, r_k)$ is a Markov process. For example, if $z_k \in \mathbb{R}^z$, then x_k lies in the state space $\mathcal{X} = \mathbb{R}^z \times \{1, \ldots, X\}$.

The transition density (2.10) of the JMLS state x_k is

$$\begin{aligned} p\left(x_{k+1} = (\bar{z}, j) | x_k = (z, i)\right) &= p(r_{k+1} = j | r_k = i) p\left(z_{k+1} = \bar{z} | r_{k+1} = j, z_k = z\right) \\ &= P_{ij}\, |\Gamma^{-1}(j)|\, p_w \left(\Gamma^{-1}(j) \left[\bar{z} - A(j) z - f(j) u_{k+1}\right]\right). \end{aligned}$$

The observation likelihood (2.11) is evaluated as

$$p(y_k | x_k = (z, i)) = |D^{-1}(i)|\, p_v \left(D^{-1}(i) \left[y_k - C(i) z - g(i) u_k\right]\right).$$

2.5.2 Simulation-based optimal state predictor

By using the composition method for stochastic simulation described in Appendix A, one can simulate random samples from the predicted distribution. Consider the Chapman–Kolmogorov equation (2.12) which determines the predicted state density:

$$\pi_k(x) = \int_{\mathcal{X}} p(x_k = x | x_{k-1})\, \pi_{k-1}(x_{k-1})\, dx_{k-1}.$$

Suppose we can generate (or have) random samples from the predicted state density $\pi_{k-1}(x)$ at time $k-1$. How can samples be simulated from the predicted state density $\pi_k(x)$ at time k? The composition approach generates samples as follows:

1. Generate samples $x_{k-1}^{(l)}$, $l = 1, \ldots, L$ from $\pi_{k-1}(x)$.
2. Generate sample $x_k^{(l)} \sim p(x_k | x_{k-1} = x_{k-1}^{(l)})$, for $l = 1, 2, \ldots, L$.

It follows from the composition method that these L simulated samples are from the density $\pi_k(x)$. For large L, these samples can be used to approximate the density $\pi_k(x)$. From these simulated samples, the mean, variance, and other statistics can be estimated straightforwardly. This approach is useful in cases where the Chapman–Kolmogorov equation cannot be solved in closed form or is computationally intensive, such as in jump Markov linear systems.

2.6 Modeling moving and maneuvering targets

Tracking a moving target is of importance in surveillance and defence systems and serves as a useful example of the models described above. Below we give a bare-bones outline of a basic target tracking model.

The problem of estimating the kinematic state (position, velocity and heading) of a moving target (such as aircraft, ship or automobile) from noisy observations, is known as the target tracking problem [32, 60]. The noisy observations are obtained by a sensor such as a radar. A major issue in target tracking is the presence of false measurements: that is, spurious measurements that are unrelated to the target. (For example, measurements of a flock of birds while tracking an aircraft.) Maneuvering targets are often modeled as jump Markov linear systems (JMLS) where the maneuver of the target is modeled as a finite-state Markov chain. The difficulty in tracking a maneuvering target in the presence of false measurements arises from the uncertain origin of the measurements (as a result of the observation/detection process) and the uncertainty in the maneuvering command driving the state of the target.

2.6.1 Target dynamics

Consider a target moving in two-dimensional Euclidean space. Let t denote continuous time. Let $\Delta > 0$ denote the time discretization (sampling) interval. During each time interval $[k\Delta, (k+1)\Delta)$, assume that the target undergoes acceleration $q_k^{(1)}$ and $q_k^{(2)}$ in the x and y coordinates. At time t, let $p_t^{(1)}$ and $\dot{p}_t^{(1)}$ denote the position and velocity of the target in the x-direction, and $p_t^{(2)}$ and $\dot{p}_t^{(2)}$ denote the position and velocity in the y-direction. Then

$$\frac{d}{dt}\dot{p}_t^{(1)} = q_k^{(1)}, \quad \frac{d}{dt}\dot{p}_t^{(2)} = q_k^{(2)}, \quad t \in [k\Delta, (k+1)\Delta)$$

As a result, in discrete time for $i = 1, 2$,

$$\dot{p}_{k+1}^{(i)} = \dot{p}_k^{(i)} + \Delta\, q_k^{(i)}, \quad p_{k+1}^{(i)} = p_k^{(i)} + \Delta \dot{p}_k^{(i)} + \frac{\Delta^2}{2} q_k^{(i)}, \quad k = 1, 2, \ldots$$

Define the state of the target at time k as $z_k = (p_k^{(1)}, \dot{p}_k^{(1)}, p_k^{(2)}, \dot{p}_k^{(2)})'$. Then

$$z_{k+1} = A\, z_k + f\, r_{k+1}, \quad k = 0, 1, \ldots, N \tag{2.38}$$

$$A = \begin{bmatrix} 1 & \Delta & 0 & 0 \\ 0 & 1 & 0 & 0 \\ 0 & 0 & 1 & \Delta \\ 0 & 0 & 0 & 1 \end{bmatrix}, \quad f = \begin{bmatrix} \Delta^2/2 & 0 \\ \Delta & 0 \\ 0 & \Delta^2/2 \\ 0 & \Delta \end{bmatrix}, \quad r_{k+1} = \begin{bmatrix} q_k^{(1)} \\ q_k^{(2)} \end{bmatrix}. \tag{2.39}$$

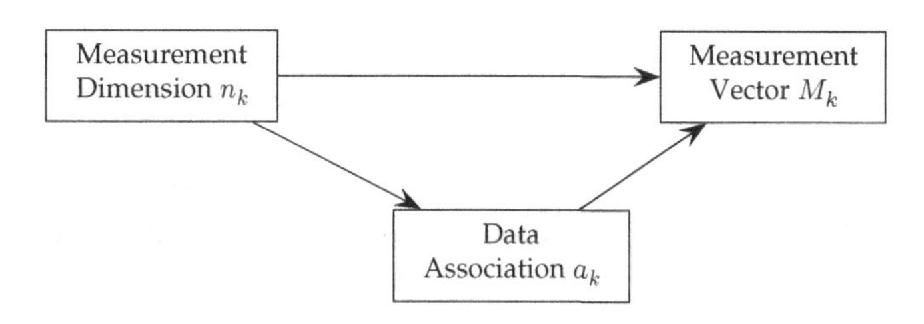

Figure 2.2 Dependency structure random variables in clutter model.

Here r_{k+1} denotes the constant acceleration (maneuvers) during the k-th sampling period. (We use r_{k+1} instead of r_k for notational convenience to match the JMLS model given in (2.37).) The maneuver process $\{r_k\}$ is modeled as a finite-state Markov chain. Finally, the motion of the target, as described above, is subject to small random perturbations $w_k \sim N(0, Q)$ occurring every Δ seconds. These perturbations are assumed to remain constant within the sampling interval, but change in magnitude and direction randomly from one interval to the next.

To summarize, the dynamics of the target are modeled as a jump Markov linear system where the four-dimensional kinematic state $\{z_k\}$ evolves as

$$z_{k+1} = Az_k + fr_{k+1} + w_k, \quad k = 0, 1, \ldots. \tag{2.40}$$

Here $\{r_k\}$ is a finite-state Markov chain, and $w_k \sim N(0, Q)$.

2.6.2 Measurement process and clutter

In the simplest setting, the target state is observed via a noisy linear sensor as[9]

$$y_k = Cz_k + v_k, \quad v_k \sim N(0, R). \tag{2.41}$$

For example, if only the position of the target is observed, then $y_k \in \mathbb{R}^2$, and C is a 2×4 matrix with elements $C_{1,1} = C_{2,3} = 1$ and the remaining elements are zero. If the position and velocity are observed, then C is an identity matrix.

In target tracking, the measurement model becomes more complicated due to presence of *clutter* – that is, false measurements of the target. The measurement at each time consists of true measurements (if the target is detected) and possibly false measurements due to random false alarms in the detection process.

The measurement process at each time k is modeled by the 3-tuple $\{n_k, a_k, M_k\}$. Here, the measurement vector M_k is an n_k dimensional vector comprising of possibly false measurements (clutter) and true measurement (the actual target in noise). a_k denotes the data association decision made at the tracker and specifies which measurement is the true measurement of the target. As shown in Figure 2.2, the dependency structure of $\{n_k, a_k, M_k\}$ is as follows: n_k is i.i.d.; a_k given n_k is i.i.d.; and finally, M_k depends on n_k, a_k and the underlying state z_k.

[9] In bearings-only target tracking, the angle of the target with reference to a fixed bearing is observed in noise: $y_k = \arctan(z_k(1)/z_k(3)) + v_k$. So the measurement equation is nonlinear in the state.

In more detail:

(i) *Number of false measurements*: $\{n_k\}$ is a nonnegative integer-valued i.i.d. random process denoting the number of false measurements due to clutter made at time k. If $n_k = 0$, then only the true measurement of the target is made at time k. Typically $\{n_k\}$ is modeled as a homogeneous Poisson process. That is, at each time k, n_k is a random variable with probability mass function

$$\mu_F(n) \stackrel{\text{defn}}{=} \mathbb{P}\{n_k = n \text{ false measurements detected}\} = e^{-\lambda V} \frac{(\lambda V)^n}{n!}. \tag{2.42}$$

Here λ is the *clutter density*, i.e. the spatial density of false measurements, and denotes the average number of false measurements per unit volume. Thus, λV is the expected number of false measurements in the measurement volume V.

(ii) *Data association*: The finite-state process $\{a_k\}$ models the data association step in the tracker. If $a_k = 0$, then all measurements obtained at time k are assumed to be false measurements due to clutter. If $a_k = i$ for $i \in \{1, 2, \ldots, n_k\}$ and $n_k > 0$, then it is assumed that the i-th measurement at time k is the true measurement. The standard assumption is that $\{a_k\}$ conditional on n_k is an i.i.d. discrete valued process [32] with

$$\mathbb{P}\{a_k = i | n_k\} = \begin{cases} \rho \mathrm{P_d} \mu_F(n_k - 1)/n_k, & \text{for } i \in \{1, 2, \ldots, n_k\},\ n_k > 0 \\ \rho(1 - \mathrm{P_d})\mu_F(n_k), & \text{for } i = 0. \end{cases}$$

where $\rho \stackrel{\text{defn}}{=} [\mathrm{P_d}\mu_F(n_k - 1) + (1 - \mathrm{P_d})\mu_F(n_k)]^{-1}$ is a normalizing constant, $\mu_F(n)$ is defined in (2.42) and $\mathrm{P_d}$ denotes the target detection probability.

(iii) *Measurement vector*: M_k is an n_k dimensional vector at each time k and denotes the measurements made at time k. The conditional probability density $p(M_k | z_k, a_k)$ is given by

$$p(M_k | z_k, a_k = i) = \begin{cases} V^{-n_k}, & \text{for } i = 0 \\ V^{-n_k+1} p(M_k^{(i)} | z_k), & \text{for } i \in \{1, 2, \ldots, n_k\}. \end{cases}$$

That is, the false measurements at time k are uniformly and independently distributed in the volume V. If $a_k = i$ for $i \in \{1, 2, \ldots, n_k\}$, then the i-th component of M_k is the true measurement; that is,

$$M_k^{(i)} = y_k = C z_k + v_k$$

so that $p(M_k^{(i)} | z_k) \propto p_v(y_k - C z_k)$.

Aim: Given the measurement process $\{n_k, a_k, M_k\}$, $k = 1, 2, \ldots$, the aim of the tracker is to estimate at each time k the target state z_k and maneuver r_k. This is a filtering problem. §3.10 deals with simplified versions of the above formulation in the context of particle filters. Also §8.3 discusses radar scheduling for tracking a maneuvering target.

2.7 Geometric ergodicity of HMM predictor: Dobrushin's coefficient

We now return to the HMM predictor discussed in §2.4.3. This section discusses bounds on how fast the HMM predictor forgets its initial conditions and converges to the stationary distribution. Such bounds determine the importance of the initial condition. If the initial condition is forgotten rapidly then the HMM predictor is relatively insensitive to the initial condition. In Chapter 3 we will extend these bounds to the optimal filter.

Below we will define the Dobrushin coefficient of ergodicity, denoted as $\rho(P)$, which maps the transition probability matrix P to a real number in the interval $[0, 1]$. Consider two arbitrary state probability vectors π and $\bar{\pi}$. How far apart do these probability vectors evolve after one time step of an HMM predictor with a transition matrix P? The Dobrushin coefficient gives the following useful bound on this distance:

$$\|P'\pi - P'\bar{\pi}\| \leq \rho(P)\, \|\pi - \bar{\pi}\|.$$

The above inequality says that one can replace the matrix P by the scalar $\rho(P) \in [0, 1]$ when analyzing the distance between the one-step evolution of two probability vectors. So if $\rho(P)$ is strictly smaller than 1, then by successively applying the above inequality, we obtain geometric convergence of the distance.

A related motivation of the Dobrushin coefficient stems from bounding the second largest eigenvalue of the transition matrix P. Recall from Statement 4 of Theorem 2.4.2 that the rate of convergence of π_k to the stationary distribution π_∞ is determined by the second largest eigenvalue modulus $|\lambda_2|$ of the transition matrix P. Computing $|\lambda_2|$ is difficult for large dimension transition matrices. The Dobrushin coefficient is an upper bound for the second largest eigenvalue modulus $|\lambda_2|$. This is established in Theorem 2.7.2, Property 7 below. It is often simpler to show that the Dobrushin coefficient is smaller than one, which implies that the Markov chain forgets its initial condition geometrically fast – this applies to nonlinear filtering problems that we will consider in Chapter 3; also for predicting Markov chains with time varying transition matrices, $\lambda_2(P_1 P_2 \cdots P_k) \neq \lambda_2(P_1)\lambda_2(P_2) \cdots \lambda_2(P_k)$. So knowing second largest eigenvalues for each transition matrix P_k is not useful in determining how the product converges, whereas with Dobrushin coefficients we can make use of its submultiplicative property detailed in Theorem 2.7.2, Property 5. We will also use the Dobrushin coefficient to analyze the statistical properties of simulation-based gradient estimators in Chapter 15.

2.7.1 Dobrushin coefficient and variational distance

Given two probability mass functions α and β on $\mathcal{X}$, define the total variational distance as

$$\|\alpha - \beta\|_{\text{TV}} = \frac{1}{2}\|\alpha - \beta\|_1 = \frac{1}{2}\sum_{i \in \mathcal{X}} |\alpha(i) - \beta(i)| \tag{2.43}$$

So the variational distance is just half the l_1 distance between two probability mass functions.[10]

[10] Suppose P and Q denote the cumulative distribution functions of the probability mass functions α and β. Let A denote any event, that is, any (measurable) subset of the reals. Then another interpretation of the variational distance is as follows:

$$\|\alpha - \beta\|_{\text{TV}} = \max_{A \subset \mathbb{R}} P(A) - Q(A). \tag{2.44}$$

DEFINITION 2.7.1 *For transition matrix P, the Dobrushin coefficient of ergodicity is defined as*

$$\rho(P) = \frac{1}{2} \max_{i,j} \sum_{l \in \mathcal{X}} |P_{il} - P_{jl}| = \max_{i,j} \|P' e_i - P' e_j\|_{TV}. \tag{2.45}$$

where e_i denotes the unit vector with 1 in the i-th position.

In words: the Dobrushin coefficient $\rho(P)$ is the maximum variational distance between two rows of the transition matrix P.

2.7.2 Properties of Dobrushin's coefficient

The Dobrushin coefficient $\rho(P)$ satisfies several useful properties that will be used in this book. Define a Markov chain with transition matrix P to be *geometrically ergodic* if the Dobrushin coefficient $\rho(P)$ is strictly smaller than 1. We summarize these properties in the following theorem:

THEOREM 2.7.2 *The Dobrushin coefficient $\rho(P)$ satisfies the following properties*

1. $0 \leq \rho(P) \leq 1$.
2. $\rho(P) = 0$ *if and only if* $P = \mathbf{1}\pi_\infty'$.
3. $\rho(P) = 1 - \min_{i,j} \sum_{l \in \mathcal{X}} \min\{P_{il}, P_{jl}\}$.
4. *Given two probability vectors π and $\bar{\pi}$, then*

$$\|P'\pi - P'\bar{\pi}\|_{TV} \leq \rho(P)\, \|\pi - \bar{\pi}\|_{TV}.$$

5. *$\rho(\cdot)$ is submultiplicative. Given two transition matrices P_1 and P_2, then $\rho(P_1 P_2) \leq \rho(P_1)\rho(P_2)$.*
6. *Let π_0 and $\bar{\pi}_0$ denote two initial possible distributions of a Markov chain. Denote the corresponding state probability vectors at time k as π_k and $\bar{\pi}_k$. Then for any stochastic matrix P,*

$$\|\pi_k - \bar{\pi}_k\|_{TV} \leq \|\pi_0 - \bar{\pi}_0\|_{TV}\, (\rho(P))^k, \quad k \geq 0.$$

   So for a geometrically ergodic Markov chain (i.e. when $\rho(P) < 1$), $\|\pi_k - \bar{\pi}_k\|_{TV}$ goes to zero geometrically fast.
7. *$|\lambda_2| \leq \rho(P)$. That is, the Dobrushin coefficient upper bounds the second largest eigenvalue modulus of a stochastic matrix P.*

That is, the variational distance is the largest possible difference between the probabilities that P and Q can assign to the same event A.

To prove (2.44), let B denote the set of indices i such that $\alpha(i) > \beta(i)$. Then the left-hand side of (2.44) is

$$\|\alpha - \beta\|_{\text{TV}} = \frac{1}{2}\|\alpha - \beta\|_1 = \frac{1}{2}\sum_{i \in B}(\alpha(i) - \beta(i)) + \frac{1}{2}\sum_{i \in B^c}(\beta(i) - \alpha(i)) = \sum_{i \in B}(\alpha(i) - \beta(i)).$$

Next consider the right-hand side of (2.44). Clearly, the choice of set $A \subset \mathbb{R}$ that maximizes $P(A) - Q(A)$ corresponds exactly to the set of indices B, since on set B, $\alpha(i) > \beta(i)$. (Terms $\alpha(i) - \beta(i)$ with indices i outside set B are negative and can only diminish $P(A) - Q(A)$.) So $\max_{A \in \mathbb{R}} P(A) - Q(A) = \sum_{i \in B}(\alpha(i) - \beta(i))$. Thus the left- and right-hand sides of (2.44) are identical.

Properties 1, 2 and 3 are straightforward to establish and are left as an exercise. Property 2 says that if every row of P is the same, i.e. $P = \mathbf{1}\pi'_\infty$ for some probability vector π_∞, then $\rho(P) = 0$. $P = \mathbf{1}\pi'_\infty$ represents the transition matrix of an independent and identically distributed process. As an immediate consequence of Property 3, if all the elements of P are nonzero, then $\rho(P)$ is strictly smaller than 1.

In the appendix, we prove Properties 4, 5, 6 and 7. These are crucial for establishing conditions for a Markov chain to forget its initial condition. The proof relies on the following key lemma, which will also be used in Chapter 3 for proving geometric ergodicity of the HMM filter.

LEMMA 2.7.3 *For any $f \in \mathbb{R}^X$ and X-dimensional probability vectors, $\pi, \bar{\pi}$,*

$$|f'(\pi - \bar{\pi})| \leq \max_{i,j} |f_i - f_j| \, \|\pi - \bar{\pi}\|_{TV}.$$

The proof of the lemma is in the appendix. Despite its simplicity, the above lemma is tighter than Holder's inequality which states

$$|f'(\pi - \bar{\pi})| \leq \|f\|_\infty \|\pi - \bar{\pi}\|_1 = 2 \max_i |f_i| \|\pi - \bar{\pi}\|_{\mathrm{TV}}.$$

Examples

Example 1: If $P = \begin{bmatrix} P_{11} & 1 - P_{11} \\ 1 - P_{22} & P_{22} \end{bmatrix}$, then clearly $\rho(P) = |1 - P_{11} - P_{22}|$ which coincides with the second largest eigenvalue modulus $|\lambda_2|$.

Example 2: If $P_{ij} \geq \epsilon$ for all i, j, then Property 3 of Theorem 2.7.2 implies that $\rho(P) \leq 1 - \epsilon$ which ties in with the intuition that a Markov chain with all non-zero transition probabilities is trivially irreducible and aperiodic.

Example 3: A weaker condition than Example 2 for $\rho(P) < 1$ is the *uniform Doeblin condition* also called the *minorization condition*:

THEOREM 2.7.4 *Suppose $\epsilon \in (0, 1]$ and there exists a probability mass function κ such that P satisfies the uniform Doeblin condition (minorization condition)*

$$P_{ij} \geq \epsilon \kappa_j \text{ where } \sum_{j \in \mathcal{X}} \kappa_j = 1, \quad \kappa_j \geq 0. \tag{2.46}$$

Then $\rho(P) \leq 1 - \epsilon$.

The proof follows from Property 3 of Theorem 2.7.2. From the Doeblin condition, $\min\{P_{il}, P_{jl}\} \geq \epsilon \kappa_l$. Therefore $\sum_l \min\{P_{il}, P_{jl}\} \geq \sum_l \epsilon \kappa_l = \epsilon$. Then using Property 3 of Theorem 2.7.2 that $\rho(P) = 1 - \min_{i,j} \sum_{l \in \mathcal{X}} \min\{P_{il}, P_{jl}\}$ yields $\rho(P) \leq 1 - \epsilon$.

The above examples show that the Dobrushin coefficient is useful when it is strictly less than one. For sparse transition matrices, $\rho(P)$ is typically equal to 1 and therefore not useful since it provides a trivial upper bound for $|\lambda_2|$. In such cases, it is useful to consider $\rho(P^n)$ for some positive integer $n > 1$. If $\rho(P^n)$ is strictly smaller than 1, then geometric ergodicity follows by considering blocks of length n, i.e. $\|P^{n\prime}\pi - P^{n\prime}\bar{\pi}\|_{\mathrm{TV}} \leq \rho(P^n)\|\pi - \bar{\pi}\|_{\mathrm{TV}}$.

2.7.3 Optimal prediction for inhomogeneous Markov chains

An inhomogeneous Markov chain has a time varying transition probability matrix $P(k)$, $k = 1, 2 \ldots$. For initial distribution π_0, the Chapman–Kolomogorov equation (optimal predictor) for the state probability vector at time k reads

$$\pi_k = P'(k)\pi_{k-1} = \Big(\prod_{n=1}^{k} P(n)\Big)'\pi_0 = P'(k)P'(k-1)\cdots P'(1)\pi_0. \tag{2.47}$$

The Dobrushin coefficient is useful in giving sufficient conditions for the predictor of an inhomogeneous Markov chain to forget the initial condition. Suppose the predicted distribution $\bar{\pi}_k$ is computed using (2.47) but with initial condition $\bar{\pi}_0$. Then applying Properties 4 and 5 of Theorem 2.7.2 to (2.47) yields

$$\|\pi_k - \bar{\pi}_k\|_{\mathrm{TV}} = \|\Big(\prod_{n=1}^{k} P(n)\Big)'\pi_0 - \Big(\prod_{n=1}^{k} P(n)\Big)'\bar{\pi}_0\|_{\mathrm{TV}} \leq \|\pi_0 - \bar{\pi}_0\|_{\mathrm{TV}}\ \rho\Big(\prod_{n=1}^{k} P(n)\Big). \tag{2.48}$$

We now give two types of conditions for the optimal predictor of an inhomogeneous Markov chain to forget its initial condition.[11]

1. *Weak ergodicity*: An inhomogeneous Markov chain is *weakly ergodic* if for all $l \geq 0$, the Dobrushin coefficients satisfy $\lim_{k\to\infty} \rho\big(\prod_{n=l}^{k} P(n)\big) = 0$. From (2.48), it is clear that for a weakly ergodic Markov chain, $\lim_{k\to\infty} \|\pi_k - \bar{\pi}_k\|_{\mathrm{TV}} = 0$, implying that the influence of the initial distribution dies out with k.

2. *Uniform Doeblin condition*: Assume the Doeblin condition (2.46), namely that the transition probability $P_{ij}(k)$ at each time k satisfies

$$P_{ij}(k) \geq \epsilon\kappa_j \text{ where } \sum_{j\in\mathcal{X}} \kappa_j = 1, \quad \kappa_j \geq 0 \tag{2.49}$$

where $\epsilon \in (0, 1]$ is larger than δ for some constant $\delta > 0$ for all k. Then Theorem 2.7.4 implies that the Dobrushin coefficient satisfies $\rho(P(k)) \leq 1 - \epsilon$. Using the sub-multiplicative Property 5 of Theorem 2.7.2, yields

$$\begin{aligned}\|\pi_k - \bar{\pi}_k\|_{\mathrm{TV}} &\leq \|\pi_0 - \bar{\pi}_0\|_{\mathrm{TV}}\ \rho\Big(\prod_{n=1}^{k} P(n)\Big) \leq \|\pi_0 - \bar{\pi}_0\|_{\mathrm{TV}} \prod_{n=1}^{k} \rho(P(n)) \\ &\leq \|\pi_0 - \bar{\pi}_0\|_{\mathrm{TV}}(1-\epsilon)^k, \quad \text{for } k \geq 0, \end{aligned}\tag{2.50}$$

implying that the difference in state probability vectors at time k with two different initial conditions π_0 and $\bar{\pi}_0$, dies away geometrically fast with k.

[11] Define $\Phi_{lk} = \prod_{n=l}^{k} P(n)$. Then weak ergodicity is equivalent to $\lim_{k\to\infty} \sup_{\pi,\bar{\pi}} \|\Phi'_{lk}\pi - \Phi'_{lk}\bar{\pi}\|_{\mathrm{TV}} = 0$ for all l and pmfs $\pi, \bar{\pi}$. An inhomogeneous Markov chain is strongly ergodic if there exists π such that $\lim_{k\to\infty} \sup_{\bar{\pi}} \|\Phi'_{lk}\bar{\pi} - \pi\|_{\mathrm{TV}} = 0$. Strong ergodicity implies weak ergodicity. In this book we will only use the uniform Doeblin condition. Clearly this implies weak ergodicity.

2.8 Complements and sources

For results on linear stochastic systems see [17, 74, 163, 354]. Accessible treatments of finite-state Markov chains include [69]. The seminal paper [274] in HMMs that sparked interest in electrical engineering. [293] is a seminal book on positive and stochastic matrices. Jump Markov linear systems (generalized hidden Markov models) are used widely in target tracking applications [60, 32]. The Dobrushin coefficient is a special case of a more general coefficient of ergodicity defined in terms of the Wasserstein metric [261] that applies to denumerable and continuous state Markov models. In particular, Properties 1, 4, 5 of Theorem 2.7.2 hold for this general ergodic coefficient.

2.8.1 Coupling inequality

The Doeblin (minorization) condition (2.46) has a simple interpretation in terms of coupling. Suppose $X \sim \pi$, $Z \sim \bar{\pi}$ are dependent random variables. The *coupling inequality* states that the variational distance (proof left as an exercise)

$$\|\pi - \bar{\pi}\|_{\text{TV}} \leq \mathbb{P}(X \neq Z). \tag{2.51}$$

Consider a Markov chain $\{x_n\}$ with transition probability P and initial distribution π_0. Using the above coupling inequality, we can prove that the sequence of state probability vectors $\{\pi_n\}$ of the Markov chain converges geometrically fast to the stationary distribution π_∞ under the Doeblin condition

$$P_{ij} \geq \epsilon \pi_\infty(j), \qquad \text{for constant } \epsilon \in (0, 1]. \tag{2.52}$$

To show this result, consider a fictitious Markov chain $\{z_n\}$ whose transition matrix is also P. Suppose z_0 is initialized at the stationary distribution π_∞ implying that $\{z_n\}$ remains at the stationary distribution forever. Clearly,

$$P_{ij} = \epsilon \pi_\infty(j) + (1 - \epsilon)\left[\frac{P_{ij} - \epsilon \pi_\infty(j)}{1 - \epsilon}\right]. \tag{2.53}$$

Because of the Doeblin condition, the term in square brackets is nonnegative. So (2.53) can be interpreted as follows: suppose the state at time $n - 1$ is i. At time n, flip a coin with probability of heads ϵ. If the coin comes up heads (probability ϵ), simulate x_n from π_∞. If the coin comes up tails (probability $(1 - \epsilon)$), simulate $x_n = j$ with probability $\frac{P_{ij} - \epsilon \pi_\infty(j)}{1 - \epsilon}$. Clearly, the first time that the coin comes up heads, the chain $\{x_n\}$ is stationary. Denote this time as $\tau = \inf\{n : x_n \sim \pi_\infty\}$ is the time the Markov chain x_n achieves the stationary distribution. Using the coupling inequality (2.51) with $x_n \sim P'^n \pi_0$ and $z_n \sim \pi_\infty$ yields

$$\|P'^n \pi_0 - \pi_\infty\|_{\text{TV}} \leq \mathbb{P}(x_n \neq z_n) = \mathbb{P}(n < \tau) = \mathbb{P}(\tau > n) = (1 - \epsilon)^n.$$

To summarize, under the Doeblin condition (2.52), π_n converges geometrically fast to π_∞; see [207] for details of the coupling method. From an algorithmic point of view, [66] formulates minimization of the second largest eigenvalue modulus as a convex optimization problem; see Theorem A.3.1 on page 438.

Appendix 2.A Proof of Theorem 2.7.2 and Lemma 2.7.3

Proof of Lemma 2.7.3

Given two probability vectors π and $\bar{\pi}$, define $\phi(i) = \min\{\pi(i), \bar{\pi}(i)\}$. That is, ϕ is the elementwise minimum of π and $\bar{\pi}$. Then it follows that

$$\pi - \bar{\pi} = \sum_{i,j} \gamma_{ij}(e_i - e_j), \qquad \text{where } \gamma_{ij} = \frac{(\pi(i) - \phi(i))\,(\pi(j) - \phi(j))}{\|\pi - \bar{\pi}\|_{\text{TV}}} \geq 0$$

and e_i is the unit vector with 1 in the i-th position. Therefore,

$$\sum_{i,j} \gamma_{ij} = \|\pi - \bar{\pi}\|_{\text{TV}}, \text{ and} \tag{2.54}$$

$$|f'(\pi - \bar{\pi})| = |\sum_{i,j} \gamma_{ij}(f_i - f_j)| \leq \sum_{i,j} \gamma_{ij}|f_i - f_j| \leq \max_{ij} \gamma_{ij} \max_{ij} |f_i - f_j|.$$

Proof of Theorem 2.7.2

Property 4: We first establish Property 4 for the case when π and $\bar{\pi}$ are unit vectors. Suppose $\pi = e_i$ and $\bar{\pi} = e_j$ and $i \neq j$, where e_i denotes the unit vector with 1 in the i-th position. Then the left-hand side is $\|P'e_i - P'e_j\|_{\text{TV}}$. Since $\|e_i - e_j\|_{\text{TV}} = 1$, the right-hand side is

$$\rho(P)\,\|e_i - e_j\|_{\text{TV}} = \max_{ij} \|P'e_i - P'e_j\|_{\text{TV}}.$$

So for unit vectors, Property 4 reads $\|P'e_i - P'e_j\|_{\text{TV}} \leq \max_{ij} \|P'e_i - P'e_j\|_{\text{TV}}$ which is obviously true.

To prove Property 4 for general probability vectors, the trick is to express the difference of two probability vectors in terms of unit vectors:

$$\pi - \bar{\pi} = \sum_{i,j} \gamma_{ij}(e_i - e_j), \qquad \text{where } \sum_{i,j} \gamma_{ij} = \|\pi - \bar{\pi}\|_{\text{TV}}; \text{ see (2.54).}$$

$$\text{Then, } \|P'\pi - P'\bar{\pi}\|_1 = \|\sum_{i,j} \gamma_{ij}P'(e_i - e_j)\|_1 \leq \sum_{i,j} \gamma_{ij} \max_{i,j} \|P'(e_i - e_j)\|_1.$$

Using the definition $\rho(P) = \frac{1}{2}\max_{i,j} \|P'(e_i - e_j)\|_1$ and $\sum_{i,j} \gamma_{ij} = \|\pi - \bar{\pi}\|_{\text{TV}}$,

$$\|P'\pi - P'\bar{\pi}\|_{\text{TV}} \leq \rho(P)\,\|\pi - \bar{\pi}\|_{\text{TV}}.$$

Property 5: By Definition 2.7.1, for the stochastic matrix $P_1 P_2$

$$\begin{aligned}\rho(P_1P_2) &= \max_{ij} \|P_2'P_1'e_i - P_2'P_1'e_j\|_{\text{TV}} \\ &\leq \max_{ij} \rho(P_2)\,\|P_1'e_i - P_1'e_j\|_{\text{TV}} \qquad \text{(by Property 4)} \\ &= \rho(P_2)\,\rho(P_1) \qquad \text{(by Definition 2.7.1).}\end{aligned}$$

Property 6:

$$\|\pi_k - \bar{\pi}_k\|_{\text{TV}} = \|P'^k\pi_0 - P'^k\bar{\pi}_0\|_{\text{TV}} \leq \|\pi_0 - \bar{\pi}_0\|_{\text{TV}}\rho(P^k) \leq \|\pi_0 - \bar{\pi}_0\|_{\text{TV}}\,(\rho(P))^k, \quad k \geq 0.$$

The last inequality follows from the sub-multiplicative Property 5.

Property 7: Suppose the possibly complex valued z is the eigenvector corresponding to the possibly complex valued eigenvalue λ_2, i.e. $P'z = \lambda_2 z$. Then

$$|\lambda_2| \max_{ij} |z_i - z_j| = \max_{ij} |\lambda_2(z_i - z_j)| = \max_{i,j} |\lambda_2(e_i - e_j)'z| = \max_{i,j} |(e_i - e_j)'P'z|$$
$$= \max_{i,j} |(P_i - P_j)z| \quad (2.55)$$

where the i-th row of P is denoted as P_i. Next, Lemma 2.7.3 yields

$$|(P_i - P_j)z| \le \max_{i,j} |z_i - z_j| \|P_i - P_j\|_{\mathrm{TV}}$$
$$\Longrightarrow \max_{i,j} |(P_i - P_j)z| \le \max_{i,j} |z_i - z_j| \|P_i - P_j\|_{\mathrm{TV}}$$
$$\le \max_{i,j} |z_i - z_j| \max_{i,j} \|P_i - P_j\|_{\mathrm{TV}} = \max_{i,j} |z_i - z_j| \, \rho(P). \quad (2.56)$$

Using (2.55) and (2.56) yields $|\lambda_2| \max_{ij} |z_i - z_j| \le \rho(P) \max_{i,j} |z_i - z_j|$.

3 Optimal filtering

Contents

The recursive state estimation problem described in this chapter is also called *optimal filtering*, *Bayesian filtering*, *stochastic filtering* or *recursive Bayesian estimation*. The setup is displayed in Figure 3.1. The aim is to estimate the underlying state x_k of a stochastic dynamic system at each time k given measurements $y_1, \dots, y_k$ from a sensor.

Optimal filtering is used in tracking moving objects, navigation, data fusion, computer vision applications and economics. The material in this chapter constitutes the background to Parts II and III of the book, where we will "close the loop" by using the filtered estimates to control (reconfigure) the sensor.

Recall from Chapter 2 that a partially observed Markov model consists of the initial state density π_0, and the following two sequences of conditional probability densities on the state space $\mathcal{X}$ and observation space $\mathcal{Y}$:

$$\text{State transition density: } p(x_{k+1}|x_k), \quad \text{where } x_k \in \mathcal{X}, \quad k = 0, 1 \dots, \tag{3.1}$$

$$\text{Observation likelihood: } p(y_k|x_k), \quad \text{where } y_k \in \mathcal{Y}, \quad k = 1, 2, \dots. \tag{3.2}$$

In this chapter it is assumed that the above probability densities are known.

3.1 Optimal state estimation

The framework is displayed in Figure 3.1. Suppose that the sample paths $\{x_k\}$ of the state process and $\{y_k\}$ of the observation process are generated according to the partially observed Markov model (3.1), (3.2). A sensor records the observation process up to time N. Denote this *observation history*[1] as

$$y_{1:N} = (y_1, \dots, y_N). \tag{3.3}$$

The aim is to estimate the underlying state x_k at time k using the observation history $y_{1:N}$. That is, construct an estimator (algorithm) $\kappa(y_{1:N})$ that processes the observations $y_{1:N}$ to yield an estimate of the state x_k.

Let $\boldsymbol{\kappa}$ denotes the set of all possible estimators that map $y_{1:N}$ to an estimate of the state x_k. By an *optimal state estimator* for the state x_k, we mean an estimator $\kappa^* \in \boldsymbol{\kappa}$ that minimizes the *mean square state estimation error*, i.e.

$$\kappa^* = \underset{\kappa \in \boldsymbol{\kappa}}{\operatorname{argmin}} \, \mathbb{E}\{\left(x_k - \kappa(y_{1:N})\right)' \left(x_k - \kappa(y_{1:N})\right)\}. \tag{3.4}$$

In words, the optimal estimator is a function κ^* that yields an estimate $\kappa^*(y_{1:N})$ of the state x_k with the smallest mean square error.

The main result is as follows. "Optimal" below is in the sense of (3.4).

THEOREM 3.1.1 *The optimal state estimate $\hat{x}_{k|N}$ at time k given observations $y_{1:N}$ is given by the conditional expectation:*

$$\hat{x}_{k|N} \overset{defn}{=} \kappa^*(y_{1:N}) = \mathbb{E}\{x_k | y_{1:N}\}. \tag{3.5}$$

On account of (3.4) and (3.5), the optimal state estimate is called the *minimum mean square error (MMSE) state estimate* or *conditional mean estimate.*

The above theorem is a special case of a more general result (Theorem 3.2.1 on page 38) that the conditional expectation minimizes the Bregman loss function.

Before proceeding we make two further remarks:

1. Since $\mathbb{E}\{x_n|y_{1:n}\} = \int_{\mathcal{X}} x_k p(x_k|y_{1:n}) dx_k$, evaluating the conditional expectation involves computing the posterior distribution $p(x_k|y_{1:n})$ via Bayes' formula. So optimal state estimation is also called *Bayesian state estimation.*

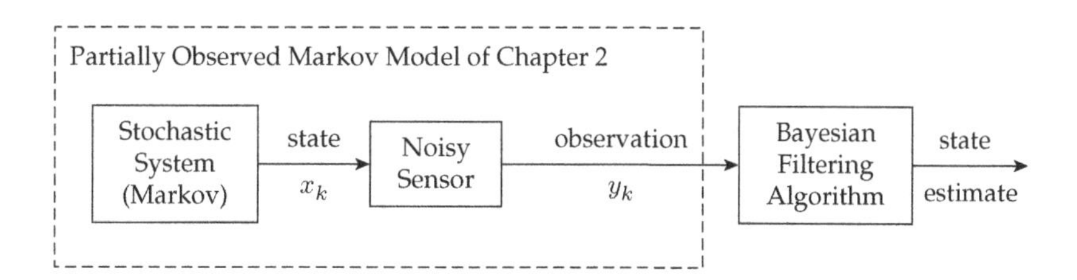

Figure 3.1 Bayesian filtering for state estimation.

[1] It is assumed that the observations start from time $k = 1$. If the observations start from time $k = 0$, then simply replace the prior $\pi_0(x)$ with $\frac{p(y_0|x)\pi_0(x)}{\int_{\mathcal{X}} p(y_0|x)\pi_0(x)dx}$.

2. The conditional mean (optimal state estimate) $\hat{x}_{k|N}$ is an unbiased estimate, i.e. $\mathbb{E}\{\hat{x}_{k|N}\} = \mathbb{E}\{x_k\}$. This follows from the smoothing property of conditional expectations (3.11) described below. As a result, the right-hand side of (3.4) is the minimum variance of an unbiased estimator. So the optimal state estimate is also the minimum variance estimate.

Prediction, filtering and smoothing

Consider the optimal state estimate $\hat{x}_{k|N}$ defined in (3.5). Depending on the choice of k and N there are three optimal state estimation problems:

- *Optimal prediction*: Choosing $k > N$, the optimal prediction problem is to compute the state estimate $\hat{x}_{k|N} = \mathbb{E}\{x_k|y_{1:N}\}$ of the state x_k at a future time k given measurements until time N.
- *Optimal filtering*: Choosing $k = N$, the optimal filtering problem is to compute the state estimate $\hat{x}_{k|k} = \mathbb{E}\{x_k|y_{1:k}\}$ of the state at current time k given past and current measurements until time k.
- *Optimal smoothing*: Choosing $k < N$, the optimal smoothing problem is to compute the state estimate $\hat{x}_{k|N} = \mathbb{E}\{x_k|y_{1:N}\}$ given past, present and future measurements until time N. Naturally a smoother is non-causal and will require storing future observations from time $k + 1$ to N in memory.

3.2 Conditional expectation and Bregman loss functions

Since the optimal filter aims to compute the conditional expectation of the state given the observations, we pause here to review conditional expectations. We will also show that the conditional expectation is optimal with respect to the Bregman loss function.

3.2.1 Conditional expectation

Given random variables[2] X and Y, the conditional expectation $\mathbb{E}\{X|Y\}$ is a random variable. Below we present three definitions of the conditional expectation, starting with special cases and ending in a more abstract definition.

(i) Conditioning on a discrete valued random variable: The conditional distribution of X given $Y = y$ is defined by

$$\mathbb{P}(X \leq x|Y = y) = \frac{\mathbb{P}(X \leq x, Y = y)}{\mathbb{P}(Y = y)},$$

providing $\mathbb{P}(Y = y) > 0$. From this, depending on whether X is discrete or continuous valued, the conditional pmf $\mathbb{P}(X = x|Y = y)$ or pdf $p(x|Y = y)$ can be obtained and the conditional expectation $\mathbb{E}\{X|Y = y\}$ evaluated. The above definition is useful for

[2] The notation in this book does not distinguish between random variables and their outcomes. However, in this section we make an exception and denote random variables by upper-case letters and their outcomes by lower-case letters.

discrete random variables Y that take on a finite number of values with $P(Y = y) > 0$ for some y.

(ii) Conditioning on a continuous valued random variable: The above definition is useless if Y is a continuous valued random variable since $\mathbb{P}(Y = y) = 0$. In such cases, the definition can be generalized if (X, Y) has a joint probability density function $p(x, y)$. Then Y has marginal density $p_Y(y) = \int_{\mathcal{X}} p(x, y)dx$ and the conditional density is defined as

$$p(x|y) = \frac{p(x, y)}{p_Y(y)} \tag{3.6}$$

at any point y where $p_Y(y) > 0$.

The conditional expectation of X given $Y = y$ is then defined as

$$\mathbb{E}\{X|Y = y\} = \int_{\mathcal{X}} x\, p(x|y)dx. \tag{3.7}$$

The conditional density and conditional expectation in (3.6), (3.7) are defined only at points y where $p_Y(y) > 0$. They can be defined arbitrarily on the set $\{y : p_Y(y) = 0\}$. Obviously, the set $\{y : p_Y(y) = 0\}$ has zero probability under the distribution of Y, so both $p(x|y)$ and $\mathbb{E}\{X|Y = y\}$ are defined uniquely with probability one.

(iii) General definition: The joint density function $p(x, y)$ may not exist. A more general definition of conditional expectation involving sigma-algebras is as follows. Let X be an integrable random variable on probability space $(\Omega, \mathcal{F})$. That is, $\mathbb{E}\{|X|\} < \infty$. Suppose $\mathcal{G} \subseteq \mathcal{F}$ is a sigma-algebra. Then the conditional expectation $\mathbb{E}\{X|\mathcal{G}\}$ is any $\mathcal{G}$ measurable random variable that satisfies[3]

$$\mathbb{E}\{X\, I(X \in B)\} = \mathbb{E}\left\{\mathbb{E}\{X|\mathcal{G}\}I(X \in B)\right\}, \qquad \text{for any set } B \in \mathcal{G},$$

where $I(\cdot)$ denotes the indicator function. By $\mathbb{E}\{X|Y\}$ is meant $\mathbb{E}\{X|\mathcal{G}\}$ with $\mathcal{G} = \sigma(Y)$ denoting the sigma-algebra generated by random variable Y.

Properties of conditional expectation

Three important properties of conditional expectations we will use are:

1. *Unconditional expectation*: The unconditional expectation is simply a special case obtained by choosing $\mathcal{G} = \{\emptyset, \Omega\}$ as the trivial sigma-algebra. Then
$$\mathbb{E}\{X|\mathcal{G}\} = \mathbb{E}\{X\}.$$
2. If Y is $\mathcal{G}$ measurable, then $\mathbb{E}\{XY|\mathcal{G}\} = Y\,\mathbb{E}\{X|\mathcal{G}\}$. This means that if $\mathcal{G}$ contains complete information about Y, then given $\mathcal{G}$, Y is a constant. In more familiar engineering notation, for any integrable function $\phi(\cdot)$,
$$\mathbb{E}\{X\phi(Y)|Y\} = \phi(Y)\,\mathbb{E}\{X|Y\}.$$
3. *Smoothing property*: If $\mathcal{G}_1 \subset \mathcal{G}_2$ are sigma-algebras, then
$$\mathbb{E}\{\mathbb{E}\{X|\mathcal{G}_2\}|\mathcal{G}_1\} = \mathbb{E}\{X|\mathcal{G}_1\}. \tag{3.8}$$

[3] The existence of the conditional expectation follows from the Radon–Nikodym theorem [59]. This general definition of conditional expectation will not be used in the book.

Note that $\mathcal{G}_1 \subset \mathcal{G}_2$ means that $\mathcal{G}_1$ contains less information than $\mathcal{G}_2$. In simplified engineering notation,

$$\mathbb{E}\{\mathbb{E}\{X|Y_1, Y_2\} \mid Y_1\} = \mathbb{E}\{X|Y_1\}. \tag{3.9}$$

The smoothing property (3.8) or (3.9) says: conditioning on more information and then conditioning on less information is equivalent to conditioning on less information. The higher resolution information $\mathcal{G}_2$ is smoothed out.

Choosing $\mathcal{G}_1 = \{\emptyset, \Omega\}$ as the trivial sigma-algebra yields the following important special case called the *Law of Iterated Expectations*:

$$\mathbb{E}\{\mathbb{E}\{X|\mathcal{G}_2\}\} = \mathbb{E}\{X\}. \tag{3.10}$$

This says that the unconditional expectation of a conditional expectation is the unconditional expectation. In other words, the conditional expectation $\mathbb{E}\{X|Y\}$ is an unbiased estimator of X since its expectation is $\mathbb{E}\{X\}$.
In simplified engineering notation, the law of iterated expectations (3.10) reads

$$\mathbb{E}\{X\} = \mathbb{E}\{\mathbb{E}\{X|Y\}\} = \int \mathbb{E}\{X|Y = y\}\, p(y)dy. \tag{3.11}$$

3.2.2 Conditional expectation, minimizes Bregman loss

Theorem 3.1.1 says that for the quadratic loss function, the conditional expectation $\mathbb{E}\{x|y\}$ is the optimal estimator for x given observation y. Are there other loss functions for which the conditional mean is the optimal estimator? Below, it is shown that the optimality property of the conditional mean holds for all *Bregman loss functions* [30]. Also, examples of Bregman loss functions are given.

Let $\phi : \mathbb{R}^X \to \mathbb{R}$ denote any strictly convex differentiable function. Then the *Bregman loss function* $\mathbf{L}_\phi : \mathbb{R}^X \times \mathbb{R}^X \to \mathbb{R}$ is defined as

$$\mathbf{L}_\phi(x, \bar{x}) = \phi(x) - \phi(\bar{x}) - (x - \bar{x})' \nabla \phi(\bar{x}). \tag{3.12}$$

Due to the strict convexity assumption, clearly for all $x, \bar{x} \in \mathbb{R}^X$, it follows that

$$\mathbf{L}_\phi(x, \bar{x}) \geq 0, \quad \text{and } \mathbf{L}_\phi(x, \bar{x}) = 0 \text{ iff } x = \bar{x}. \tag{3.13}$$

In terms of state estimation, given a sequence of measurements y, we are interested in estimators $\kappa(y)$ of state x that minimize the expected Bregman loss function $\mathbb{E}\{\mathbf{L}_\phi(x, \kappa(y))\}$. The main result is as follows:

THEOREM 3.2.1 *[30, Theorem 1] Let $\phi : \mathbb{R}^X \to \mathbb{R}$ be a strictly convex and differentiable function and $\mathbf{L}_\phi$ denote the corresponding Bregman loss function. Let y denote a sequence of noisy observations of a random variable x. Then the optimal estimator $\kappa^*(y)$ of x is the conditional mean. That is,*

$$\kappa^*(y) = \mathbb{E}\{x|y\} = \operatorname*{argmin}_{\kappa \in \mathcal{K}} \mathbb{E}\{\mathbf{L}_\phi(x, \kappa\,(y))\}.$$

Proof Denote $\hat{x} = \mathbb{E}\{x|y\}$. Below it is shown that

$$\mathbb{E}\{\mathbf{L}_\phi(x,\kappa)\} - \mathbb{E}\{\mathbf{L}_\phi(x,\hat{x})\} = \mathbb{E}\{\mathbf{L}_\phi(\hat{x},\kappa)\}. \tag{3.14}$$

The result then follows from (3.13) which says that $\mathbb{E}\{\mathbf{L}_\phi(\hat{x},\kappa)\}$ is minimized if $\kappa(y) = \hat{x} = \mathbb{E}\{x|y\}$.

So it only remains to show that (3.14) holds. From the definition (3.12)

$$\mathbb{E}\{\mathbf{L}_\phi(x,\kappa)\} - \mathbb{E}\{\mathbf{L}_\phi(x,\hat{x})\} = \mathbb{E}\{\phi(\hat{x}) - \phi(\kappa) - (x-\kappa)'\nabla\phi(\kappa) + (x-\hat{x})'\nabla\phi(\hat{x})\}.$$

The last term is zero via the smoothing property of conditional expectation:

$$\mathbb{E}\{(x-\hat{x})'\nabla\phi(\hat{x})\} = \mathbb{E}\{\mathbb{E}\{(x-\hat{x})'\nabla\phi(\hat{x}) \mid y\}\} = \mathbb{E}\{(\hat{x}-\hat{x})'\nabla\phi(\hat{x})\} = 0.$$

Using the smoothing property of conditional expectations also yields

$$\mathbb{E}\{(x-\kappa)'\nabla\phi(\kappa)\} = \mathbb{E}\{\mathbb{E}\{(x-\kappa)'\nabla\phi(\kappa)|y\}\} = \mathbb{E}\{(\hat{x}-\kappa)'\nabla\phi(\kappa)\}.$$

So

$$\mathbb{E}\{\mathbf{L}_\phi(x,\kappa)\} - \mathbb{E}\{\mathbf{L}_\phi(x,\hat{x})\} = \mathbb{E}\{\phi(\hat{x}) - \phi(\kappa) - (\hat{x}-\kappa)'\nabla\phi(\kappa)\} = \mathbb{E}\{\mathbf{L}_\phi(\hat{x},\kappa)\}$$

□

It is also shown in [30] that if the conditional expectation is the optimal estimator for a nonnegative continuously differentiable loss function, then the loss function has to be a Bregman loss function. In this sense, Bregman loss functions are both necessary and sufficient for the conditional mean to be optimal.

Examples of Bregman loss functions

(i) *Quadratic loss (Mahalanobis distance)*: Suppose the state $x \in \mathbb{R}^X$ and $\phi(x) = x'Sx$ where S is a symmetric positive definite matrix. The resulting loss function is called the Mahalanobis distance. Then from (3.12), the associated Bregman loss function is

$$\mathbf{L}_\phi(x,\bar{x}) = x'Sx - x'S\bar{x} - 2(x-\bar{x})'S\bar{x} = (x-\bar{x})'S(x-\bar{x}).$$

Therefore, the conditional mean is the optimal estimator in terms of the Mahalanobis distance:

$$\kappa^*(y) = \operatorname*{argmin}_{\kappa\in\mathcal{K}} \mathbb{E}\{(x-\kappa(y))'\, S\,(x-\kappa(y))\} = \mathbb{E}\{x|y\}.$$

This is what Theorem 3.1.1 asserted when $S = I$.

(ii) *Kullback–Leibler (KL) divergence*: Suppose that the state x is a probability mass function $x = (x_1,\dots,x_X)$ with $\sum_{i=1}^X x_i = 1$. In other words, the state x lives in a $X-1$ dimensional unit simplex. Define the negative Shannon entropy as $\phi(x) = \sum_{i=1}^X x_i \log_2 x_i$. This is strictly convex on the $X-1$ dimensional unit simplex. Then given another probability mass function $\bar{x} = (\bar{x}_1,\dots,\bar{x}_X)$, the associated Bregman loss function (3.12) is

$$\begin{aligned}\mathbf{L}_\phi(x,\bar{x}) &= \sum_{i=1}^X x_i \log_2 x_i - \sum_{i=1}^X \bar{x}_i \log_2 \bar{x}_i - (x-\bar{x})'\nabla\phi(\bar{x}) \\ &= \sum_{i=1}^X x_i \log_2 x_i - \sum_{i=1}^X x_i \log_2 \bar{x}_i\end{aligned}$$

Table 3.1 Examples of Bregman loss functions. In all these cases, the conditional mean $\mathbb{E}\{x|y\} = \text{argmin}_\kappa \mathbb{E}\{\mathbf{L}_\phi(x, \kappa(y))\}$ is the optimal estimate.

Domain	$\phi(x)$	$\mathbf{L}_\phi(x,\bar{x})$	Loss
$\mathbb{R}_+$	$-\log(x)$	$\frac{x}{\bar{x}} - \log\frac{x}{\bar{x}} - 1$	Itakura–Saito
(0,1)	$x\log x$ $+(1-x)\log(1-x)$	$x\log(\frac{x}{\bar{x}})$ $+(1-x)\log\left(\frac{1-x}{1-\bar{x}}\right)$	Logistic Loss
$\mathbb{R}$	e^x	$e^x - e^{\bar{x}} - (x-\bar{x})e^{\bar{x}}$	Exponential
$\mathbb{R}^X$	$x'Sx$	$(x-\bar{x})'S(x-\bar{x})$	Mahalanobis
Simplex	$\sum_{i=1}^X x_i \log x_i$	$\sum_{i=1}^X x_i \log\frac{x_i}{\bar{x}_i}$	KL divergence
$\mathbb{R}_+^X$	$\sum_{i=1}^X x_i \log x_i$	$\sum_{i=1}^X x_i \log\frac{x_i}{\bar{x}_i}$ $-\mathbf{1}'x + \mathbf{1}'\bar{x}$	Generalized I-divergence

which is the Kullback–Leibler divergence between x and $\bar{x}$. Therefore, with $\kappa_i(y)$ denoting the estimate of the i-th component of $\bar{x}$, the conditional mean minimizes the Kullback–Leibler divergence:

$$\kappa^*(y) = \underset{\kappa\in\mathcal{K}}{\text{argmin}}\, \mathbb{E}\left\{\sum_{i=1}^X x_i \log_2 \kappa_i(y) - \sum_{i=1}^X \kappa_i(y)\log_2 \kappa_i(y)\right\} = \mathbb{E}\{x|y\}.$$

(iii) Table 3.1 gives examples of Bregman loss functions $\mathbf{L}_\phi(x,\bar{x})$.

3.3 Optimal filtering, prediction and smoothing formulas

We are now ready to derive the formula for the optimal filtering recursion. Once the optimal filtering recursion is obtained, formulas for optimal prediction and smoothing can be derived straightforwardly.

3.3.1 Optimal filtering recursion

Consider the partially observed Markov model (3.1), (3.2). As discussed in §3.1, the aim is to compute the filtered state estimate $\hat{x}_k = \mathbb{E}\{x_k|y_{1:k}\}$ (to simplify notation, $\hat{x}_k$ is used instead of $\hat{x}_{k|k}$). How can one compute the filtered state estimate $\hat{x}_k$ recursively. From (3.7), $\hat{x}_k$ is obtained by integrating the posterior distribution over the state space $\mathcal{X}$:

$$\hat{x}_k = \mathbb{E}\{x_k|y_{1:k}\} = \int_{\mathcal{X}} x_k\, p(x_k|y_{1:k})dx_k, \quad k = 1,2\ldots.$$

Here $p(x_k|y_{1:k})$ is the posterior density of the state at time k given observations until time k. For notational convenience denote the posterior density as

$$\pi_k(x) = p(x_k = x|y_{1:k}), \quad x \in \mathcal{X}.$$

The optimal filtering problem is this: devise an algorithm to compute the posterior density $\pi_k = p(x_k|y_{1:k})$ recursively at each time k. That is, construct a formula (algorithm) T that operates on the observation y_k and π_{k-1}, so that

$$\pi_k = T(\pi_{k-1}, y_k), \quad k = 1, 2, \ldots, \quad \text{initialized by } \pi_0.$$

The main result is as follows:

THEOREM 3.3.1 *Consider the partially observed Markov model (3.1), (3.2). Then the posterior* $\pi_{k+1}(x_{k+1}) = p(x_{k+1}|y_{1:k+1})$ *satisfies the optimal filtering recursion:*

$$\pi_{k+1}(x_{k+1}) = T(\pi_k, y_{k+1}) \stackrel{defn}{=} \frac{p(y_{k+1}|x_{k+1}) \int_{\mathcal{X}} p(x_{k+1}|x_k)\, \pi_k(x_k)\, dx_k}{\int_{\mathcal{X}} p(y_{k+1}|x_{k+1}) \int_{\mathcal{X}} p(x_{k+1}|x_k)\, \pi_k(x_k)\, dx_k dx_{k+1}} \tag{3.15}$$

initialized by $\pi_0(x)$. *Thus the filtered state estimate (conditional mean estimate) at time* $k+1$ *is*

$$\hat{x}_{k+1} = \mathbb{E}\{x_{k+1}|y_{1:k+1}\} = \int_{\mathcal{X}} x\, \pi_{k+1}(x) dx. \tag{3.16}$$

The filtering recursion (3.15) is simply a version of Bayes' formula. Bayes' formula says that the posterior $p(x|y)$ of a random variable is proportional to the likelihood $p(y|x)$ times the prior $\pi(x)$. In comparison, the numerator of (3.15) has an additional term $p(x_{k+1}|x_k)$ which models the evolution of the state and is integrated out with respect to x_k.

Proof The posterior (conditional density) given the observations $y_{1:k+1}$ is

$$\frac{p(x_{k+1}, y_{1:k+1})}{p(y_{1:k+1})} = \frac{p(y_{k+1}|x_{k+1}, y_{1:k}) \int_{\mathcal{X}} p(x_{k+1}|x_k, y_{1:k}) p(x_k, y_{1:k}) dx_k}{\int_{\mathcal{X}} p(y_{k+1}|x_{k+1}, y_{1:k}) \int_{\mathcal{X}} p(x_{k+1}|x_k, y_{1:k}) p(x_k, y_{1:k}) dx_k dx_{k+1}}. \tag{3.17}$$

Since the state noise and observation noise are assumed to be mutually independent i.i.d. processes, they satisfy the following conditional independence properties:

$$p(y_{k+1}|x_{k+1}, y_{1:k}) = p(y_{k+1}|x_{k+1}), \quad p(x_{k+1}|x_k, y_{1:k}) = p(x_{k+1}|x_k).$$

Substituting these expressions into (3.17) yields the recursion (3.15). □

The filtering recursion (3.15) is often rewritten in two steps: a prediction step (3.18) (namely, Chapman Kolmogorov's equation of §2.2), and a measurement update (3.19) which is Bayes' rule:

$$\pi_{k+1|k}(x_{k+1}) \stackrel{\text{defn}}{=} p(x_{k+1}|y_{1:k}) = \int_{\mathcal{X}} p(x_{k+1}|x_k)\, \pi_k(x_k)\, dx_k \tag{3.18}$$

$$\pi_{k+1}(x_{k+1}) = \frac{p(y_{k+1}|x_{k+1}) \pi_{k+1|k}(x_{k+1})}{\int_{\mathcal{X}} p(y_{k+1}|x_{k+1}) \pi_{k+1|k}(x_{k+1}) dx_{k+1}}. \tag{3.19}$$

Note the denominator of (3.15) or (3.19) is a normalization term since it is the integral of the numerator. This denominator is the likelihood of the observation sequence $y_{1:k+1}$ and will play a key role in maximum likelihood parameter estimation in Chapter 4.

3.3.2 Un-normalized filtering recursion

Since the denominator of (3.15) is merely a normalization term, to derive filtering recursions it is often convenient to work with only the numerator. Denote the un-normalized filtered density of the state given observations $y_{1:k}$ as the joint density

$$q_k(x) = p(x_k = x, y_{1:k}). \tag{3.20}$$

Clearly the normalized (conditional) filtered density is $\pi_k(x) = q_k(x)/\int_{\mathcal{X}} q_k(x)dx$. A similar derivation to that of (3.15) shows that the update for the un-normalized filtered recursion is

$$q_{k+1}(x) = p(y_{k+1}|x_{k+1} = x) \int_{\mathcal{X}} p(x_{k+1} = x|x_k) q_k(x_k) dx_k. \tag{3.21}$$

This is initialized as $q_0 = \pi_0$. The filtered state estimate can be expressed as

$$\hat{x}_{k+1} = \frac{\int_{\mathcal{X}} x\, q_{k+1}(x)dx}{\int_{\mathcal{X}} q_{k+1}(x)dx}. \tag{3.22}$$

3.3.3 Finite dimensional optimal filters

Recursive computation of the filtered density $\pi_k(x)$ in (3.15) is the critical step in evaluating the filtered state estimate $\hat{x}_k$. Since (3.15) is an integral equation, it does not directly translate to a practical algorithm for computing $\pi_k(x)$. Indeed, one would need to compute $\pi_k(x)$ for each $x \in \mathcal{X}$, which is a hopeless task if the state space $\mathcal{X}$ is a continuum. However, if the density function $\pi_k(x)$, $x \in \mathcal{X}$ can be completely described by a *finite* dimensional statistic (parameter) at each time k, then from (3.15) we can obtain a recursion for this finite dimensional sufficient statistic at each time. Such a "finite dimensional optimal filter" is of key importance since it can be implemented exactly[4] by updating a finite dimensional vector, namely the sufficient statistic.

For what structure of the partially observed Markov model (3.1), (3.2) does the filtered density update $T(\pi, y)$ in (3.15) result in a finite dimensional statistic? It turns out that so far only two cases have been found where the filtered density $\pi(x)$ in (3.15) can be parametrized by a finite dimensional statistic: these are the celebrated *Kalman filter* and *hidden Markov model filter*. Since these two cases are the only two known finite dimensional filters, they are exceedingly important and are discussed in §3.4 to §3.7, In all other cases, the recursion (3.15) can only be approximated. Such sub-optimal filters are discussed in §3.8 and §3.9. Figure 3.2 depicts the various models and filters that are widely used.

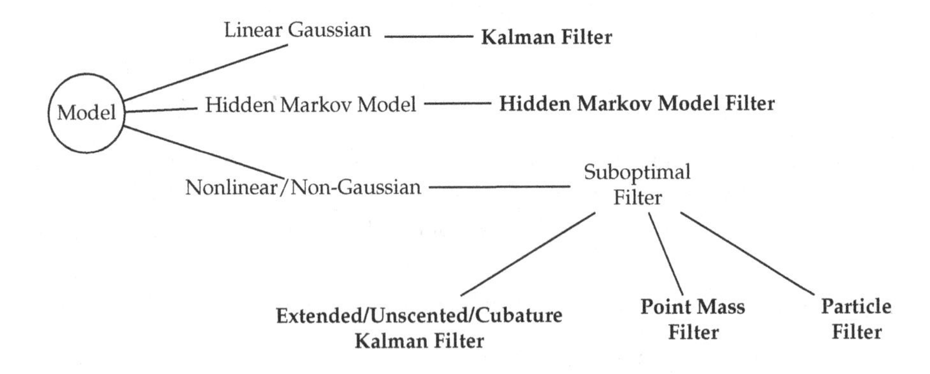

Figure 3.2 Various filters for Bayesian state estimation. The filter names are listed in boldface.

[4] "Exact" here means that there is no approximation involved in the filtering algorithm. However, the filtering algorithm is still subject to numerical round-off and finite precision effects.

Remark. Conjugate Priors: Consider the special case where the state x is a random variable (instead of random process) with prior π_0. Then finite dimensional filters can be characterized in terms of conjugate priors. In the area of statistics, π_0 is said to be a conjugate prior to the likelihood $p(y|x)$ if the posterior $\pi_1 \propto \pi_0 p(y|x)$ belongs to the same family as π_0. For example, the Gaussian is the conjugate prior to the Gaussian likelihood, the Beta distribution is the conjugate prior to the Bernoulli likelihood, and the Gamma distribution is the conjugate prior to the Poisson likelihood.

3.3.4 Optimal state predictor

The optimal predicted state estimate at time $k + \Delta$ given observations up to time k is $\hat{x}_{k+\Delta|k} = \mathbb{E}\{x_{k+\Delta}|y_{1:k}\}$. This can be computed from the Δ-step ahead predicted density $\pi_{k+\Delta|k} = p(x_{k+\Delta}|y_{1:k})$ which is obtained from the Chapman–Kolmogorov equation (2.12) initialized at π_k. Equivalently, the predictor is obtained from the filtering recursion (3.15) by assuming that observations are non-informative, i.e. choose $p(y_{k+1}|x_{k+1} = x)$ independent of the state x. The one-step ahead predictor was already presented in (3.18). The Δ-step optimal predictor in terms of the filtered density π_k is

$$\pi_{k+\Delta|k} = \int_{\mathcal{X}} \pi_k(x_k) \left(\prod_{t=1}^{\Delta} \int_{\mathcal{X}} p(x_{k+t}|x_{k+t-1}) dx_{k+t-1} \right). \tag{3.23}$$

3.3.5 Optimal smoothers

The optimal smoother aims to compute the state estimate

$$\hat{x}_{k|N} = \mathbb{E}\{x_k|y_{1:N}\} = \int_{\mathcal{X}} x\, \pi_{k|N}(x) dx \tag{3.24}$$

where $k < N$ and $\pi_{k|N} = p(x_k|y_{1:N})$. Below, formulas are given for evaluating the posterior $\pi_{k|N}$. Depending on the choice of N, three types of smoothers are of interest:
(i) *Fixed-interval smoother*: Here N is chosen as a fixed constant. Given a fixed batch of measurements $y_1, \dots, y_N$, the aim is to evaluate $\pi_{k|N}$ at each time $k = 0, \dots, N$. To derive the fixed-interval smoother, start with

$$\pi_{k|N}(x) = p(x_k = x|y_{1:N}) = \frac{p(y_{k+1:N}|x_k = x, y_{1:k})\, p(x_k = x|y_{1:k})}{\int_{\mathcal{X}} p(y_{k+1:N}|x_k = x, y_{1:k})\, p(x_k = x|y_{1:k}) dx}. \tag{3.25}$$

Note that conditional on state x_k, $y_{k+1:N}$ are independent of $y_{1:k}$. So

$$\pi_{k|N}(x) = p(x_k = x|y_{1:N}) = \frac{\pi_k(x)\beta_{k|N}(x)}{\int_{\mathcal{X}} \pi_k(x)\beta_{k|N}(x) dx} \tag{3.26}$$

where

$$\pi_k(x) = p(x_k = x|y_{1:k}), \quad \beta_{k|N}(x) = p(y_{k+1:N}|x_k = x). \tag{3.27}$$

Recall $\pi_k(x)$ is the filtered density evaluated according to (3.15). Since (3.15) is a forward recursion, π_k is called the *forward* variable. The *backward* variable $\beta_{k|N}$ in (3.27) is evaluated according to the backward time recursion

$$\beta_{k|N}(x) = \int_{\mathcal{X}} p(x_{k+1} = z|x_k = x)\, p(y_{k+1}|x_{k+1} = z)\, \beta_{k+1|N}(z) dz, \quad k = N-1, \ldots, 0$$

$$\text{initialized with } \beta_{N|N}(x) = 1. \tag{3.28}$$

To summarize, implementation of the fixed-interval smoother requires computing and storing the backward variables $\beta_{0|N}, \ldots, \beta_{N|N}$ using the backward recursion (3.28). Then the forward recursion (3.15) is implemented to evaluate the forward variables π_k, $k = 1, \ldots, N$. Finally the forward and backward variables are combined using (3.26) resulting in the fixed-interval smoothed estimate $\pi_{k|N}(x)$, $0 \leq k \leq N$.
(ii) *Fixed-lag smoother*: Here we choose $N = k + \Delta$ in (3.24) where Δ is a positive integer denoting the fixed lag. So at each time k, given the past, present and Δ future measurements, the aim is to compute the state estimate $\mathbb{E}\{x_k|y_{k+\Delta}\}$.

Define the fixed lag smoothed density $\pi_{k|k+\Delta} = p(x_k|y_{1:k+\Delta})$ for $k = 1, 2, \ldots$. The fixed-lag smoother is implemented as follows. At each time k, run the backward recursion (3.28) to evaluate $\beta_{k+t|k+\Delta}$ for $t = \Delta - 1, \ldots, 0$ initialized by $\beta_{k+\Delta|k+\Delta}(x) = 1$. Finally, combine $\beta_{k|k+\Delta}$ with the filtered density π_k using (3.26) to evaluate the fixed lag smoothed density $\pi_{k|k+\Delta}$.
(iii) *Fixed-point smoother*: Here $N = k, k+1, \ldots,$ denotes an increasing sequence of time instants. The aim is to refine the state estimate at fixed time k as additional future observations become available. The fixed-point smoother is implemented as a filter by augmenting the state. Suppose we wish to compute the fixed point estimate at time 0, namely, $\mathbb{E}\{x_0|y_1, \ldots, y_k\}$ as time k increases. Define the augmented state process $\bar{x}_k = (x_k', x_0')'$. With $\delta(x)$ denoting the Dirac delta function, clearly the state transition density and observation likelihoods of the augmented process $\bar{x}_k$ are

$$p(x_{k+1}, x_0 = z|x_k, x_0 = x) = p(x_{k+1}|x_k)\, \delta(z - x), \quad p(y_k|x_k, x_0) = p(y_k|x_k).$$

The filtered posterior $p(x_k, x_0|y_1, \ldots, y_k)$ can then be evaluated as in (3.15). Finally, marginalizing this over x_k yields $\pi_{0|k} = p(x_0|y_1, \ldots, y_k)$.

To summarize, the optimal smoother is obtained from the forward filtering recursion and a backward recursion.

3.4 The Kalman filter

The Kalman filter is the first finite dimensional filter that we will now discuss. A complete treatment of the Kalman filter would take a whole book [17]. Our goal here is to illustrate the Kalman filter as an example of the filtering recursion derived above and summarize some of its salient properties.

Consider the linear Gaussian state space model

$$x_{k+1} = A_k x_k + f_k u_k + w_k, \quad x_0 \sim \pi_0 \tag{3.29}$$

$$y_k = C_k x_k + g_k u_k + v_k, \tag{3.30}$$

where $x_k \in \mathcal{X} = \mathbb{R}^X$, $y_k \in \mathcal{Y} = \mathbb{R}^Y$, $w_k \sim N(0, Q_k)$, $v_k \sim N(0, R_k)$ and initial density $\pi_0 \sim N(\hat{x}_0, \Sigma_0)$. (Recall the notation (2.15) for the normal density.) Assume $\{w_k\}$, $\{v_k\}$ are i.i.d. processes and $\{w_k\}$, $\{v_k\}$ and initial condition x_0 are statistically independent. Finally, $\{u_k\}$ is a known exogenous sequence.

3.4.1 The Kalman filter equations

The equations of the celebrated Kalman filter are described below (proof given in §3.4.2).

THEOREM 3.4.1 (Kalman filter) *Consider the linear Gaussian state space model (3.29), (3.30) with known parameters* $(A_k, C_k, f_k, g_k, Q_k, R_k, \hat{x}_0, \Sigma_0)$ *and known prior distribution* $\pi_0 \sim N(\hat{x}_0, \Sigma_0)$. *Then, given a sequence of observations* $y_1, y_2, \ldots,$ *the filtered density recursion (3.15) for* $\pi_k(x) = p(x_k = x|y_{1:k})$ *has the following explicit solution:*

$$\pi_k \sim N(\hat{x}_k, \Sigma_k), \qquad \text{where } \hat{x}_k = \mathbb{E}\{x_k|y_{1:k}\},\ \Sigma_k = \mathbb{E}\{(x_k - \hat{x}_k)(x_k - \hat{x}_k)'\}.$$

Here, $\hat{x}_k, \Sigma_k$ *are computed recursively via the Kalman filter, Algorithm 1.*

Algorithm 1 Kalman filter

Given known parameters $(A_k, C_k, f_k, g_k, Q_k, R_k)$ and prior $\pi_0 = N(\hat{x}_0, \Sigma_0)$
For time $k = 1, 2, \ldots,$ given observation y_k, compute

$$\hat{x}_{k+1|k} = A_k\hat{x}_k + f_k u_k, \quad y_{k+1|k} = C_{k+1}\hat{x}_{k+1|k} + g_{k+1}u_{k+1} \tag{3.31}$$

$$\Sigma_{k+1|k} = A_k\Sigma_k A_k' + Q_k \tag{3.32}$$

$$S_{k+1} = C_{k+1}\Sigma_{k+1|k}C_{k+1}' + R_{k+1} \tag{3.33}$$

$$\hat{x}_{k+1} = \hat{x}_{k+1|k} + \Sigma_{k+1|k}C_{k+1}'S_{k+1}^{-1}(y_{k+1} - y_{k+1|k}) \tag{3.34}$$

$$\Sigma_{k+1} = \Sigma_{k+1|k} - \Sigma_{k+1|k}C_{k+1}'S_{k+1}^{-1}C_{k+1}\Sigma_{k+1|k} \tag{3.35}$$

Thus the Kalman filter is an exact solution to the integral equation (3.15) in the special case of a linear state space model with Gaussian state noise, Gaussian observation noise and Gaussian initial condition.

The Kalman filter equations in Algorithm 1 consist of the prediction step (3.31), (3.32) and the measurement update step (3.33), (3.34), (3.35). The prediction step yields the predicted density

$$p(x_{k+1}|y_{1:k}) = N(\hat{x}_{k+1|k}, \Sigma_{k+1|k}) \text{ where} \tag{3.36}$$
$$\hat{x}_{k+1|k} = \mathbb{E}\{x_{k+1}|y_{1:k}\}, \quad \Sigma_{k+1|k} = \mathbb{E}\{(\hat{x}_{k+1|k} - x_{k+1})(\hat{x}_{k+1|k} - x_{k+1})'\}$$

are the one step-ahead predicted mean and predicted covariance. We have already encountered these predictor equations in §2.3.2. Recall (3.32) is the *Lyapunov equation* (2.20) for the one step ahead predictor of the state covariance.

By substituting (3.33), (3.34) and (3.35), the Kalman filter equations (3.31) and (3.32) are often expressed as

$$\hat{x}_{k+1|k} = (A_k - K_kC_k)\hat{x}_{k|k-1} + K_k\,(y_k - g_ku_k) + f_ku_k, \tag{3.37}$$

$$K_k = A_k\Sigma_{k|k-1}C_k'(C_k\Sigma_{k|k-1}C_k' + R_k)^{-1} \tag{3.38}$$

$$\Sigma_{k+1|k} = A_k\left(\Sigma_{k|k-1} - \Sigma_{k|k-1}C_k'(C_k\Sigma_{k|k-1}C_k' + R_k)^{-1}C_k\Sigma_{k|k-1}\right)A_k' + Q_k \tag{3.39}$$

initialized by $\hat{x}_{0|-1} = \hat{x}_0$, $\Sigma_{0|-1} = \Sigma_0$. The matrix K_k defined in (3.38) is called the *gain* of the Kalman filter. The update equation (3.39) for the predictor covariance is called the discrete-time *Riccati equation*. Not surprisingly, if the observations yield no information about the state then the Riccati equation becomes the Lyapunov equation (3.32). (This is verified in the scalar case by setting the observation noise variance $R_k \to \infty$.)

A useful property of the Kalman filter is that the covariances Σ_k and $\Sigma_{k+1|k}$ are unconditional covariances – that is, these are not conditional on the observations $y_{1:k}$. That is, the Kalman filter provides automatically a performance metric on how good its state estimate is. In comparison, for the standard nonlinear filtering recursion (3.15), one can only obtain an expression for the *conditional* covariance of the state estimate (conditioned on the observation history) – evaluating the unconditional covariance of the filtered estimate requires integration over all possible observation trajectories.

Summary: The Kalman filter is a *finite dimensional filter* since the filtered density $\pi_k(x)$ is Gaussian and therefore completely characterized by the sufficient statistic of the mean and covariance, namely, $(\hat{x}_k, \Sigma_k)$. The Kalman filter is a *linear filter* since the state update (3.34) or (3.37) is linear in the observation y_k.

Sometimes the covariance matrix can become close to singular leading to numerical ill-conditioning. There are several alternative ways of implementing the Kalman filter equations with better numerical properties such as the square root algorithm (which guarantees a positive definite covariance matrix). We refer the reader to [17] for details.

3.4.2 Derivation of the Kalman filter

To keep the notation in this section manageable, we omit the exogenous input $\{u_k\}$ in the state space equations (3.29), (3.30) throughout this subsection.

Preliminaries

Recall the X-variate Gaussian density with mean μ and positive definite covariance P:

$$N(x; \mu, P) = (2\pi)^{-X/2} |P|^{-1/2} \exp\left(-\frac{1}{2}(x-\mu)' P^{-1} (x-\mu)\right),$$

where $|\cdot|$ denotes the determinant. We first introduce the following elementary but useful result involving products of Gaussian density functions. It can be regarded as the "Swiss Army knife" for dealing with Gaussian densities and their marginalization; see [32] for proof.

THEOREM 3.4.2 (Swiss Army Knife for Gaussians) *Consider two multivariate Gaussian densities $N(y; Cx, R)$ and $N(x; \mu, P)$ where matrix C and vector μ are compatible for matrix vector multiplication. Then*

$$N(y; Cx, R)\, N(x; \mu, P) = N(y; C\mu, CPC' + R)\; N(x; m + \bar{K}(y - C\mu), P - \bar{K}CP) \quad (3.40)$$

where in the right-hand side of the above equation

$$\bar{K} = PC'(CPC' + R)^{-1}$$
$$m = \mu + \bar{K}(y - C\mu).$$

As a result, the following hold:

$$\int_{\mathcal{X}} N(y; Cx, R)N(x; \mu, P)dx = N(y; C\mu, CPC' + R) \tag{3.41}$$

$$\frac{N(y; Cx, R)N(x; \mu, P)}{\int_{\mathcal{X}} N(y; Cx, R)N(x; \mu, P)dx} = N(x; m + \bar{K}(y - C\mu), P - \bar{K}CP). \tag{3.42}$$

The main point of the above theorem is that the first Gaussian density on the right-hand side of (3.40) only involves y and not x. Therefore, to marginalize the product of Gaussians by integrating over x, we can take the first term on the right-hand side of (3.40) outside the integral and the second term on the right-hand side integrates to 1 – this yields (3.41). Similarly, dividing the product of Gaussians by the marginalized product (which arises in Bayes' formula) results in the second term of the right-hand side of (3.40) – this yields (3.42).

Example. Evaluation of $p(y_k|y_{1:k-1})$: To illustrate the above theorem, let us evaluate $p(y_k|y_{1:k-1})$ for the linear Gaussian system (3.29), (3.30) using the Kalman filter estimates. Clearly,

$$p(y_k|y_{1:k-1}) = \int_{\mathcal{X}} p(y_k|x_k)\, p(x_k|y_{1:k-1})\, dx_k. \tag{3.43}$$

From (3.30) and (3.36), we know that

$$p(y_k|x_k) = N(y_k; Cx_k, R_k), \quad p(x_k|y_{1:k-1}) = N(x_k; \hat{x}_{k|k-1}, \Sigma_{k|k-1}).$$

So applying (3.41) of Theorem 3.4.2 yields

$$p(y_k|y_{1:k-1}) = \int_{\mathcal{X}} p(y_k|x_k)\, p(x_k|y_{1:k-1})\, dx_k = N(y_k; C\hat{x}_{k|k-1}, C\Sigma_{k|k-1}C' + R_k). \tag{3.44}$$

□

Several additional examples that use the above theorem will be presented in this book, including the derivation of the Kalman filter below and optimal importance sampling distributions in §3.9.3.

Derivation of the Kalman filter

Recall for the linear Gaussian state space model (2.18) that

$$p(x_{k+1}|x_k) = N(x_{k+1}; A_k x_k, Q_k), \quad p(y_{k+1}|x_{k+1}) = N(y_{k+1}; C_{k+1}x_{k+1}, R_{k+1}).$$

The Kalman filter equations of Algorithm 1 are derived by applying Theorem 3.4.2 to the general filtering recursion (3.18), (3.19). We prove this by mathematical induction. By assumption, the initial prior is $\pi_0 = N(x_0; \hat{x}_0, \Sigma_0)$. Next assume at time k that the filtered density is Gaussian: $p(x_k|y_{1:k}) = N(x_k; \hat{x}_k, \Sigma_k)$. Then applying (3.41) of Theorem 3.4.2 to the prediction step update (3.18) yields

$$\begin{aligned} p(x_{k+1}|y_{1:k}) &= \int_{\mathcal{X}} N(x_{k+1}; A_k x_k, Q_k)\, N(x_k; \hat{x}_k, \Sigma_k)dx_k \\ &= N(x_{k+1}, \hat{x}_{k+1|k}, \Sigma_{k+1|k}). \end{aligned}$$

Applying the measurement update step (3.19) together with (3.42) of Theorem 3.4.2 yields

$$
\begin{aligned}
p(x_{k+1}|y_{1:k+1}) &= \frac{N(y_{k+1}; C_{k+1}x_{k+1}, R_{k+1})N(x_{k+1}, \hat{x}_{k+1|k}, \Sigma_{k+1|k})}{\int_{\mathcal{X}} N(y_{k+1}; C_{k+1}x_{k+1}, R_{k+1})N(x_{k+1}, \hat{x}_{k+1|k}, \Sigma_{k+1|k})dx_{k+1}} \\
&= N(x_{k+1}; \hat{x}_{k+1|k} + \bar{K}_{k+1}(y_{k+1} - y_{k+1|k}), \Sigma_{k+1|k} - \bar{K}_{k+1}C_{k+1}\Sigma_{k+1|k})
\end{aligned}
$$

where $\bar{K}_{k+1} = \Sigma_{k+1|k}C'_{k+1}S_{k+1}^{-1}$ and S_{k+1} is defined in (3.33). Thus the filtered density at time $k+1$ is Gaussian and the mean and covariance are given by Algorithm 1. This completes the induction step.

3.4.3 Stationary Kalman filter

It is intuitively plausible that for a linear time invariant Gaussian state space model, i.e. when (A, C, Q, R) are constants in (3.29), (3.30), then the Kalman filter equations should simplify. This is indeed the case under suitable conditions and leads to the so-called stationary Kalman filter.

The stationary Kalman filter for a time invariant Gaussian state space model (A, C, Q, R) is as follows:

$$\hat{x}_{k+1} = \hat{x}_{k+1|k} + \Sigma_\infty C'(C\Sigma_\infty C' + R)^{-1}(y_{k+1} - C\hat{x}_{k+1|k}) \tag{3.45}$$

$$\bar{\Sigma}_\infty = \Sigma_\infty - \Sigma_\infty C'\left(C\Sigma_\infty C' + R\right)^{-1} C\Sigma_\infty \tag{3.46}$$

$$\hat{x}_{k+1|k} = (A - K_\infty C)\hat{x}_{k|k-1} + K_\infty y_k, \tag{3.47}$$

where $\bar{\Sigma}_\infty = \lim_{k\to\infty}\Sigma_k$ and $\Sigma_\infty = \lim_{k\to\infty}\Sigma_{k+1|k}$. The stationary Kalman gain K_∞ and predictor covariance Σ_∞ satisfy

$$K_\infty = A\Sigma_\infty C'\left(C\Sigma_\infty C' + R\right)^{-1} \tag{3.48}$$

$$\Sigma_\infty = A\left(\Sigma_\infty - \Sigma_\infty C'\left(C\Sigma_\infty C' + R\right)^{-1} C\Sigma_\infty\right)A' + Q. \tag{3.49}$$

Equation (3.49) is the fixed point of (3.39) (if such a fixed point exists), and is called the *algebraic Riccati equation*. There are many numerical methods available to solve (3.49) for Σ_∞; see, for example, [17].

From a computational point of view, implementing the stationary Kalman filter requires only pre-computing and storing the constant matrices K_∞ and Σ_∞. Only the state estimates (3.45) and (3.47) need to be computed in real time.

Under what conditions does the Kalman filter converge to the stationary Kalman filter? The following is a well-known result [17, Chapter 4].

THEOREM 3.4.3 *Consider the linear time invariant Gaussian state space model (3.29), (3.30) with parameters $(A, C, Q, R, \hat{x}_0, \Sigma_0)$. Then the covariance $\Sigma_{k+1|k} \to \Sigma_\infty$ and Kalman gain $K_k \to K_\infty$ if any one of the following conditions hold:*

1. *All eigenvalues of system matrix A are smaller than 1 in magnitude.*
2. *The matrix $\left[Q^{1/2}, AQ^{1/2}, \cdots, A^{X-1}Q^{1/2}\right]$ has rank X.*

In linear systems theory, Condition 1 means that the model is asymptotically stable. Condition 2 means that the pair $[A, Q^{1/2}]$ is reachable.

It is instructive to compare the above stationary Kalman filter with the stationary predictor of §2.3.3. The stationary covariance Σ_∞ now satisfies the algebraic Riccati equation instead of the algebraic Lyapunov equation. Intuitively, in the scalar case setting $R \to \infty$ in (3.49), i.e. the observations contain no information about the state, yields the algebraic Lyapunov equation.

3.4.4 The Kalman filter is the linear minimum mean square error estimator

Theorem 3.4.1 says that for Gaussian initial conditions, linear Gaussian dynamics and linear measurements with Gaussian noise, the Kalman filter is the optimal estimator for the state (in the minimum variance sense). What can one say if the state and/or measurement noise is non-Gaussian? The Kalman filter is the optimal *linear* minimum variance estimator – that is, amongst estimators that are linear in the measurements, the Kalman filter is the optimal estimator.

To show this result, we already know from §2.3.2, that even for zero mean i.i.d. non-Gaussian state noise, the mean and covariance are given by the Kalman predictor. Given observation y_k, assume a linear estimator of the form

$$\hat{x}_k = \hat{x}_{k|k-1} + G_k(y_k - C_k x_{k|k-1}). \tag{3.50}$$

Here G_k is a gain that we wish to determine to minimize the mean square error

$$\text{trace}[\Sigma_k] = \mathbb{E}\{(x_k - \hat{x}_k)'(x_k - \hat{x}_k)\}. \tag{3.51}$$

Choose the gain as $G_k = \bar{K}_k + \bar{G}_k$ where $\bar{K}_k = \Sigma_{k|k-1} C_k' S_k^{-1}$; see (3.33), (3.34). (Note the Kalman gain defined in (3.38) is $K_k = A_k \bar{K}_k$.) We now show that the optimal choice of the gain satisfies $\bar{G}_k = 0$. This implies that the optimal choice of the gain G_k is $\bar{K}_k$, which is exactly what is used in the Kalman filter (3.34). Equations (3.50) and (3.51) imply that

$$\text{trace}(\Sigma_k) = \text{trace}[\Sigma_{k|k-1} - \bar{K}_k C_k \Sigma_{k|k-1} + \bar{G}_k(C_k \Sigma_{k|k-1} C_k' + R_k)\bar{G}_k'].$$

Since $(C_k \Sigma_{k|k-1} C_k' + R_k)$ is positive definite, clearly, $\text{trace}[\bar{G}_k(C_k \Sigma_{k|k-1} C_k' + R_k)\bar{G}_k'] \geq 0$. Therefore the $\bar{G}_k = 0$ minimizes $\text{trace}(\Sigma_k)$.

3.4.5 Kalman predictor and smoothers

Kalman predictor

The aim is to compute the Δ-step conditional density $\pi_{k+\Delta|k}(x) = p(x_{k+\Delta} = x|y_{1:k})$. This is obtained as $\pi_{k+\Delta|k} = N(\hat{x}_{k+\Delta|k}, \Sigma_{k+\Delta|k})$ where for $i = 0, \ldots, \Delta - 1$, applying (3.31) and (3.32) yields

$$\hat{x}_{k+i+1|k} = A_{k+i}\hat{x}_{k+i}, \qquad \Sigma_{k+i+1|k} = A_{k+i}\Sigma_{k+i|k}A_{k+i}' + Q_{k+i}. \tag{3.52}$$

These are the Chapman–Kolmogorov equations obtained in (2.19), (2.20).

Fixed-interval Kalman smoother

As discussed in §3.3.5, the fixed-interval smoother is obtained by combining densities computed using forward and backward filters. The fixed-interval Kalman smoother is implemented as described in Algorithm 2, which is popularly called the Rauch–Tung–Striebel smoother.

Algorithm 2 Fixed-interval Kalman smoother

Given known parameters $(A_k, C_k, f_k, g_k, Q_k, R_k)$, $0 \leq k \leq N-1$ and prior $\pi_0 = N(\hat{x}_0, \Sigma_0)$
Given observations $y_{1:N} = (y_1, \ldots, y_N)$:

- Run Kalman filter Algorithm 1. Compute and store the estimates $\hat{x}_k$, Σ_k for $k = 0, \ldots, N$.
- For $k = N-1, \ldots, 0$, run the following backward recursion:

$$\hat{x}_{k|N} = \hat{x}_k + G_k(\hat{x}_{k+1|N} - \hat{x}_{k+1|k}), \tag{3.53}$$

$$\Sigma_{k|N} = \Sigma_k + G_k(\Sigma_{k+1|N} - \Sigma_{k+1|k})G'_k, \tag{3.54}$$

$$G_k = \Sigma_k A_k \Sigma^{-1}_{k+1|k}. \tag{3.55}$$

(Initial conditions $\hat{x}_{N|N} = \hat{x}_N$, $\Sigma_{N|N} = \Sigma_N$ are computed by the Kalman filter at time N.)

3.5 Hidden Markov model (HMM) filter

This section describes the HMM filter for estimating the underlying state of a Markov chain given noisy observations. The HMM filter is a key building block of a POMDP and will be used extensively in this book.

Consider a finite state Markov chain with state space $\mathcal{X} = \{1, \ldots, X\}$ with initial distribution π_0. Recall from §2.4 that an HMM is a noisily observed Markov chain specified by the parameters (P, B, C, π_0). Here P is the transition matrix of the underlying Markov chain and B are the observation probabilities defined as

$$P_{ij} = \mathbb{P}(x_{k+1} = j | x_k = i), \quad B_{xy} = p(y_k = y | x_k = x).$$

For example, for i.i.d. continuous-valued measurement noise, B_{xy} is given by (2.11). For discrete-valued measurement noise, B_{xy} are the symbol probabilities. The vector $C = \left[C(1), \ldots, C(X)\right]'$ denotes the physical state levels corresponding to states $1, \ldots, X$.

3.5.1 The HMM filter equations

Consider the filtering update (3.15). Since the state space of a HMM is finite, the posterior

$$\pi_k(i) = \mathbb{P}(x_k = i | y_{1:k}), \quad i = 1, \ldots, X$$

is an X-dimensional probability mass function. Moreover, the integration in (3.15) reduces to a summation over the finite state space $\mathcal{X}$, yielding the HMM filter:

$$\pi_{k+1}(j) = \frac{p(y_{k+1}|x_{k+1}=j)\sum_{i=1}^{X} P_{ij}\pi_k(i)}{\sum_{l=1}^{X} p(y_{k+1}|x_{k+1}=l)\sum_{i=1}^{X} P_{il}\pi_k(i)}, \quad j = 1,\dots,X, \tag{3.56}$$

initialized by the probability mass function π_0.

It is convenient to express the HMM filter (3.56) in matrix-vector notation. Given observation y_k, define the $X \times X$ diagonal matrix of likelihoods

$$B_{y_k} = \operatorname{diag}\big(p(y_k|x_k = 1),\dots,p(y_k|x_k = X)\big).$$

Define the X-dimensional posterior vector (this will be called the *belief state* later in the book)

$$\pi_k = \big[\pi_k(1) \quad \cdots \quad \pi_k(X)\big]'.$$

For observations y_k, $k = 1,2,\dots$, the HMM filter is summarized in Algorithm 3. If the observations start at time 0, then replace π_0 with $B_{y_0}\pi_0/\mathbf{1}'B_0\pi_0$.

Algorithm 3 Hidden Markov model filter

Given known parameters (P, B, C) and initial prior π_0

For time $k = 0,1,\dots$, given observation y_{k+1}, update the X-dimensional posterior

$$\pi_{k+1} = T(\pi_k, y_{k+1}) = \frac{B_{y_{k+1}}P'\pi_k}{\sigma(\pi_k, y_{k+1})}, \quad \sigma(\pi_k, y_{k+1}) = \mathbf{1}'B_{y_{k+1}}P'\pi_k. \tag{3.57}$$

Compute the conditional mean estimate of the state level at time $k+1$ as

$$\hat{C}(x_{k+1}) = \mathbb{E}\{C(x_{k+1}) \mid y_{1:k+1}\} = C'\pi_{k+1}. \tag{3.58}$$

Numerical implementation of the HMM filter (3.57) involves a matrix vector multiplication at each time instant and therefore requires $O(X^2)$ multiplications.

Un-normalized HMM filter and Forward algorithm

In analogy to (3.20) define the X-dimensional vector q_k with elements $q_k(i) = p(x_k = i, y_{1:k})$, $i = 1,2,\dots,X$. Then using (3.21), the HMM filter in the un-normalized form reads

$$q_{k+1} = B_{y_{k+1}}P'q_k \tag{3.59}$$

and the conditional mean estimate of the state level is computed as

$$\hat{C}(x_{k+1}) = \mathbb{E}\{C(x_{k+1})|y_{1:k+1}\} = \frac{C'q_{k+1}}{\mathbf{1}'q_{k+1}}.$$

In the signal processing literature, the un-normalized HMM filter update (3.59) is called the *forward* algorithm. In numerical implementation, it is necessary to scale q_{k+1} to prevent numerical underflow because (3.59) requires evaluating the product of the transition probabilities and observation likelihoods, which are smaller than one in magnitude. So the recursion (3.59) leads to all the components of q_k decaying to zero exponentially fast with k, eventually leading to an underflow error on a computer. The underflow problem is straightforwardly remedied by scaling all the elements of q_k by any arbitrary positive number. Since $\hat{x}_k$ involves the ratio of q_k with $\mathbf{1}'q_k$, this scaling factor cancels out in the computation of $\hat{x}_k$. Of course, if the scale factor is chosen as the normalization term $\sigma(\pi, y)$ at each time, then we obtain the normalized HMM filter (3.57).

3.5.2 HMM predictor and smoothers

Using the formulas in §3.3.4 and 3.3.5, we now discuss the optimal HMM predictor and smoother.

The Δ-step ahead HMM predictor satisfies the Chapman–Kolmogorov equation (2.32)

$$\pi_{k+\Delta|k} = P'^{\Delta} \pi_k. \tag{3.60}$$

Given the observation sequence $y_{1:N}$, to obtain the fixed-interval HMM smoother over a time interval N, define the backward variable $\beta_{k|N}$ as in (3.27), namely

$$\beta_{k|N}(x) = p(y_{k+1:N}|x_k = x), \quad \beta_{k|N} = \left[\beta_{k|N}(1), \dots, \beta_{k|N}(X)\right]'.$$

Then the X-dimensional vector $\beta_{k|N}$ is computed via the backward recursion (3.28) which can be expressed in matrix-vector notation as

$$\beta_{k|N} = PB_{y_{k+1}} \beta_{k+1|N}, \quad k = N-1, \dots, 1 \tag{3.61}$$

initialized by $\beta_{N|N} = \mathbf{1}$. The fixed-interval smoothed probabilities are computed by combining the HMM filter π_k with the backward variables $\beta_{k|N}$ as follows: for $k = 0, \dots, N$,

$$\pi_{k|N}(i) = \mathbb{P}(x_k = i|y_{1:N}) = \frac{\pi_k(i)\beta_{k|N}(i)}{\sum_{l=1}^{X} \pi_k(l)\beta_{k|N}(l)}, \quad i \in \{1, \dots, X\}. \tag{3.62}$$

In similar vein, one can compute the following smoothed estimate:

$$\pi_{k|N}(i,j) = \mathbb{P}(x_k = i, x_{k+1} = j|y_{1:N}) = \frac{\pi_k(i)P_{ij}B_{jy_{k+1}}\beta_{k|N}(j)}{\sum_{l=1}^{X} \pi_k(l)P_{lj}B_{jy_{k+1}}\beta_{k|N}(j)}. \tag{3.63}$$

To summarize, the HMM smoother requires computing the backward variables $\beta_{k|N}(i)$, $i \in \{1, \dots, X\}$, $k = 0, \dots, N$ and storing them in memory. Then the smoothed probability mass functions $\pi_{k|N}$, $k = 0, \dots, N$ are computed via (3.62), where the forward variables π_k are computed using the HMM filter (3.57).

3.5.3 MAP sequence estimation for HMMs

The optimal filter discussed in previous sections yields the state estimate that is optimal in the minimum mean square sense. In this subsection, we take a small diversion and define a different criterion for optimal state estimation. Suppose we are given a sequence of measurements $y_{1:N}$. Instead of minimizing the mean square error, another possibility is to compute the maximum a posterior (MAP) state sequence estimate:

$$x^*_{0:N} = \operatorname*{argmax}_{x_{0:N}} p(x_{0:N}, y_{1:N}).$$

Here $p(x_{0:N}, y_{1:N})$ denotes the joint probability of the state and observation sequence. Since $p(x_{0:N}, y_{1:N}) = p(x_{0:N}|y_{1:N}) \times p(y_{1:N})$ and $p(y_{1:N})$ is functionally independent of $x_{0:N}$, therefore

$$\underset{x_{0:N}}{\operatorname{argmax}}\, p(x_{0:N}, y_{1:N}) = \underset{x_{0:N}}{\operatorname{argmax}}\, p(x_{0:N}|y_{1:N}).$$

Due to the right-hand side, $x^*_{0:N}$ is called the MAP sequence estimate. Since

$$p(y_{1:N}, x_{0:N}) = \prod_{k=1}^{N} p(y_k|x_k)p(x_k|x_{k-1})\pi_0$$

one can compute $\operatorname{argmax}_{x_{0:N}} p(y_{1:N}, x_{0:N})$ by forward dynamic programming. For an HMM, this is called the Viterbi algorithm and is listed in Algorithm 4.

Algorithm 4 Viterbi algorithm for HMM

Given known parameters (P, B, C) and initial prior π_0

- Initialize $\delta_0(i) = \pi_0(i)$. For $k = 0, 1, \ldots, N-1$ compute the Viterbi score:

$$\delta_{k+1}(j) = \max_{i\in\{1,\ldots,X\}} \big(\delta_k(i)P_{ij}\big)\, p(y_{k+1}|x_k = j)$$

$$u_{k+1}(j) = \underset{i\in\{1,\ldots,X\}}{\operatorname{argmax}} \big(\delta_k(i)P_{ij}\big)\, p(y_{k+1}|x_k = j).$$

- Backtracking: Set $x^*_N = \operatorname{argmax}_i \delta_N(i)$. Then compute MAP sequence by backtracking as $x^*_k = u_{k+1}(x^*_{k+1})$.

The Viterbi algorithm is an off-line (non-causal) algorithm that operates on a sequence of observations $y_{1:N}$. It generates "hard estimates" – that is, the state estimate x^*_k takes on a value in the finite state space $\{1, 2, \ldots, X\}$ at each time k. (In comparison the filtered estimate $\hat{x}_{k|N} = \mathbb{E}\{x_k|y_{1:N}\}$ is a "soft estimate" since in general it does not take on a value in the state space.)

In numerical implementations of the Viterbi algorithm, to avoid numerical underflow, one uses the logarithm of the recursion:

$$\bar{\delta}_{k+1}(j) = \max_i \big(\bar{\delta}_k(i) + \log P_{ij}\big) + \log p(y_{k+1}|x_{k+1} = j).$$

Remark. MAP sequence estimation for linear Gaussian models: For the linear Gaussian state space model, the MAP sequence estimate is given by a Kalman smoother. Indeed $\arg\max_{x_{0:N}} p(y_{1:N}, x_{0:N}) = (\hat{x}_{1|N}, \ldots, \hat{x}_{N|N})$ where the fixed-interval smoothed estimates $\hat{x}_{k|N}$ are computed via the Kalman smoother Algorithm 2. This holds because for a Gaussian, the mean and mode coincide.

3.6 Examples: Markov modulated time series, out of sequence measurements and reciprocal processes

Having discussed the HMM filter, here we consider some elementary extensions.

3.6.1 Example 1: Markov modulated auto-regressive time series

Suppose, given observations $y_1, \ldots, y_{k-1}$, the distribution of y_k depends not only on the state x_k of the Markov chain but also on $y_{k-1}, \ldots, y_{k-d}$ for some integer-valued delay

d. These models are called *autoregressive processes with Markov regime* or *switching autoregressions*. An important example are linear autoregressions with Markov regime:

$$y_k + a_1(x_k)\, y_{k-1} + \cdots + a_d(x_k)\, y_{k-d} = \Gamma(x_k) w_k,$$

where $\{w_k\}$ is an i.i.d. noise sequence. Autoregressions with Markov regime are used in econometric modeling [133] and modeling maneuvering targets [32].

Suppose the Markov chain $\{x_k\}$ has transition matrix P. Then given the sequence of observations $y_{1:k}$, the filter for estimating x_k is identical to the HMM filter of Algorithm 3, except that the observation density now is

$$B_{xy_k} \propto p_w\left(\Gamma^{-1}(x)(y_k + a_1(x)\, y_{k-1} + \cdots + a_d(x)\, y_{k-d})\right), \quad x \in \{1, \ldots, X\}.$$

3.6.2 Example 2: HMM filters for out of sequence measurements

Consider estimating a finite state Markov chain when the noisy measurements are obtained with random time delays leading to a possible reordering of the measurements. With the delays modeled as a finite state Markov chain, the problem can be reformulated as a standard HMM filtering problem.

The motivation for such models stem from distributed sensing and processing systems in sensor networks where observations are gathered and communicated via cluster nodes to a fusion center; see also [109, 153]. The schematic setup is illustrated in the following figure.

$$x_k \rightarrow \boxed{\text{sensor}} \rightarrow y_k \rightarrow \boxed{\begin{array}{c}\text{Network}\\ \text{Delay } \Delta_k\end{array}} \rightarrow z_k \rightarrow \boxed{\text{Filter}} \rightarrow \hat{x}_k$$

A sensor obtains noisy observations $y_k = Cx_k + v_k$ where $\{x_k\}$ is a Markov chain with transition matrix P and $\{v_k\}$ is an i.i.d. noise process.

The filter receives observation y_k after a random delay Δ_k. The delay process $\{\Delta_k\}$ is modeled as an L-state Markov chain with state space $\{1, 2, \ldots, L\}$ denoting delays $\Delta^{(1)} \ldots, \Delta^{(L)}$ and transition probability matrix $P^{(\Delta)}$ with elements $P^{(\Delta)}_{ij} = \mathbb{P}(\Delta_{k+1} = j | \Delta_k = i)$. Assume that the delay process $\{\Delta_k\}$ is independent of the state process $\{x_k\}$ and the sensor observation noise process $\{v_k\}$. At the filter, we use the index k to denote the k-th measurement to arrive. This measurement is denoted as z_k, $k = 1, 2, \ldots$.

Given the measurement sequence $\{z_k\}$, the aim is to compute the posterior $\mathbb{P}(x_k = i | z_{1:k})$, $i \in \{1, \ldots, X\}$ and therefore the conditional mean state estimate $\mathbb{E}\{x_k | z_{1:k}\}$.

The problem can be reformulated as an HMM as follows. The k-th measurement to arrive at the processor z_k is an element of the sensor measurement set $\{y_{k-\Delta_1}, \ldots, y_{k-\Delta_L}\}$. In particular,

$$\begin{aligned} z_k &= \begin{bmatrix} y_{k-\Delta^{(1)}} & \cdots & y_{k-\Delta^{(L)}} \end{bmatrix} e_{\Delta_k} \\ &= \begin{bmatrix} Cx_{k-\Delta^{(1)}} & \cdots & Cx_{k-\Delta^{(L)}} \end{bmatrix} \Delta_k + \begin{bmatrix} v_{k-\Delta^{(1)}} & \cdots & v_{k-\Delta^{(L)}} \end{bmatrix} e_{\Delta_k} \end{aligned} \tag{3.64}$$

where e_{Δ_k} is a unit indicator vector in $\{e_1, e_2, \ldots, e_L\}$. The problem reduces to a standard HMM over the augmented state space $\mathcal{X}^L \times \{e_1, e_2, \ldots, e_L\}$ with transition matrix

$P \otimes \cdots \otimes P \otimes P^{(\Delta)}$ where $\otimes$ denotes Kronecker product (see footnote 6 on page 18). As a result, the HMM filter of Algorithm 3 can be applied to the model (3.64) with a computational cost of $O(X^{2L} \times L^2)$ at each time instant. Since many of the transition probabilities of the transition matrix of the augmented process are zero, the computational cost can be reduced significantly.

3.6.3 Example 3: Filtering of hidden reciprocal Markov processes

A reciprocal process (RP) is a one-dimensional version of a Markov random field (MRF). In tracking targets on the timescale of minutes, most real world targets are "destination-aware" – they have a well-defined destination. From a modeling point of view, being destination-aware means that the initial and final target states (in terms of position) need to be chosen from a joint distribution before specifying the transition law of the target dynamics. RPs are useful for modeling such destination-aware targets[5] [112]. Here we show that filtering of finite state RPs observed in noise is similar to the HMM filter; see also [335].

Hidden reciprocal process

Consider a discrete-time process $\{x_k, k = 0, \dots, N\}$ with finite state space $\mathcal{X}$. The process is said to be *reciprocal* [159] if at each time $k = 1, \dots, N-1$

$$\mathbb{P}(x_k | x_n, n \neq k) = \mathbb{P}(x_k | x_{k-1}, x_{k+1}). \tag{3.65}$$

Thus, given its neighbors x_{k-1} and x_{k+1}, state x_k is conditionally independent of $x_0, \dots, x_{k-2}, x_{k+2}, \dots, x_N$. The RP model is specified by the three-point transition functions (3.65) together with a given joint distribution on the end points $\pi_0(x_0, x_N) \stackrel{\text{defn}}{=} \mathbb{P}(x_0, x_N)$. In this section, to simplify notation, we consider homogeneous three-point transitions, denoted by

$$Q_{ijl} = \mathbb{P}(x_k = j | x_{k-1} = i, x_{k+1} = l), \quad i, j, l \in \mathcal{X} \tag{3.66}$$

where Q is not an explicit function of time.

Suppose the reciprocal process is observed via the observation process $\{y_k, k = 1, \dots, N\}$. Assume the observations are conditionally dependent on the reciprocal process with the conditional independence property $p(y_1, \dots, y_N | x_{0:N}) = \prod_{k=1}^{N} p(y_k | x_k)$. The process $\{y_k, k = 1, \dots, N\}$ is called a *hidden reciprocal process* in analogy to HMMs. Let $B_{iy} = p(y_k = y | x_k = i)$ denote the observation probabilities. The observations may be either discrete or continuous-valued.

Aim: Given the observation sequence $y_{1:N}$ of hidden reciprocal process, obtain a filtering recursion for the posterior $\pi_k(i) = \mathbb{P}(x_k = i | y_{1:k})$.

[5] Why not use a Markov chain with absorbing states to model destination-aware targets? The main reason is that a reciprocal process guarantees that the destination state is reached at pre-specified deterministic time point. A Markov chain, on the other hand, enters an absorbing state at a geometrically distributed change time.

Filtering of reciprocal process and Markov bridge

In order to derive a filtering recursion for the hidden reciprocal process, it is convenient to compute the posterior as

$$\pi_k(i) = \mathbb{P}(x_k = i|y_{1:k}) = \sum_{x=1}^{X} \mathbb{P}(x_k = i|y_{1:k}, x_N = x)\,\mathbb{P}(x_N = x). \tag{3.67}$$

In other words, first clamp the reciprocal process at time N to a state $x \in \mathcal{X}$ and then evaluate the posterior of this clamped process. This leads to the following definition: A *Markov bridge* is a Markov process with the end point x_N clamped to a specified value $x \in \mathcal{X}$. With this definition, we can evaluate the filtering recursion for a hidden reciprocal process in three steps:

1. Use the three-point transitions Q_{ijl} to evaluate the transition probabilities $P^x_{ij}(k) = \mathbb{P}(x_{k+1} = j|x_k = i, x_N = x)$, $k = 1, 2, \dots, N$ of the Markov bridge.
2. Use the HMM filter for this Markov bridge to evaluate $\mathbb{P}(x_k = i|y_{1:k}, x_N = x)$. This involves applying the HMM filter with time varying transition probability matrix $P^x_{ij}(k)$, $k = 1, 2, \dots, N$.
3. Finally use formula (3.67) to evaluate $\pi_k(i)$ for the reciprocal process.

It only remains to elaborate on Step 1. The Markov bridge with final state x is initialized with probability distribution

$$\pi^x_{0,N}(i) = \mathbb{P}(x_0 = i|x_N = x) = \frac{\mathbb{P}(x_0 = i, x_N = x)}{\sum_{i=1}^{X} \mathbb{P}(x_0 = i, x_N = x)} \tag{3.68}$$

where $\mathbb{P}(x_0 = i, x_N = x)$ is the specified end point distribution of the RP.

The time varying transition probabilities $P^x_{ij}(k)$ for the Markov bridge in terms of Q are constructed as follows. Start with the following property of a reciprocal process ([159], property (a3), p. 80).

$$\mathbb{P}(x_k|x_{k-1}, x_N) = \frac{\mathbb{P}(x_k|x_{k-1}, x_{k+1})}{\mathbb{P}(x_{k+1}|x_k, x_N)}\,\mathbb{P}(x_{k+1}|x_{k-1}, x_N). \tag{3.69}$$

Then (3.69) yields a backwards recursion $k = N-2, N-3, \dots$ for

$$P^x_{ij}(k) = \mathbb{P}(x_{k+1} = j|x_k = i, x_N = x) = \frac{Q_{ijl}}{P^x_{jl}(k+1)}\left(\sum_{m=1}^{X} \frac{Q_{iml}}{P^x_{m,l}(k+1)}\right)^{-1} \tag{3.70}$$

initialized with $P^x_{ij}(N-1) = 1$ if $j = x$ and zero otherwise. Equation (3.70) is the reciprocal process equivalent of the Chapman–Kolmogorov equation.

Observe that the above Chapman–Kolmogorov equation (3.70) has an additional degree of freedom in terms of the index l. The left-hand side of (3.70) is independent of the index l. Since $P^x(k)$ is a stochastic matrix (with elements $P^x_{ij}(k)$), it follows that for each j, x and time k there is at least one index l for which $P^x_{jl}(k+1)$ is nonzero. Thus (3.70) is well defined.

3.7 Geometric ergodicity of HMM filter

Thus far, this chapter has dealt with devising filtering algorithms. In this section we analyze the dynamics of the HMM filter. Consider two HMMs (P, B, π_0) and $(P, B, \bar{\pi}_0)$. They have identical transition matrices P and observation distributions B but different initial conditions, namely π_0 and $\bar{\pi}_0$. Let π_k and $\bar{\pi}_k$ denote the filtered density at time k computed using these different initial conditions. The aim of this section is to give sufficient conditions on P, B so that the HMM filtered density forgets the initial condition geometrically fast. This is termed "geometric ergodicity" of the HMM filter. From a practical point of view, geometric ergodicity is important since the initial conditions are rarely known. If the filter forgets its initial condition geometrically fast, then for long segments of data, the choice of initial condition becomes irrelevant. From a theoretical point of view, such forgetting of the initial condition geometrically fast is crucial to establish laws of large numbers for statistical inference. The results presented below are inspired by the seminal book [293] on positive matrices.

3.7.1 Strong mixing assumptions and main result

To proceed, we start with the following assumptions on the HMM: assume that the transition matrix P and observation distribution B satisfy

$$\sigma^{-}\mu_j \leq P_{ij} \leq \sigma^{+}\mu_j \tag{3.71}$$

$$\text{and } 0 < \sum_{j=1}^{X} \mu_j B_{jy} < \infty \text{ for all } y \in \mathcal{Y}. \tag{3.72}$$

Here $\sigma^{-} \leq \sigma^{+}$ are positive real numbers and μ is a probability mass function (pmf) so that $\sum_{j=1}^{X} \mu_j = 1$, $\mu_j \geq 0$.

Remark: If the transition matrix P has strictly positive elements, then (3.71) holds trivially. Also (3.72) holds for most examples of observation distributions B_{xy} considered in engineering applications such as Gaussian, exponential, etc. For the finite observation case, (3.72) holds if each observation y has a nonzero probability for at least one state x.

The above conditions (3.71), (3.72) are the so-called "strong mixing conditions" introduced in [25]. In what follows, we will relate the conditions to the Doeblin conditions (2.49) introduced in §2.7.3 and then prove that the HMM filter forgets its initial condition geometrically fast.

THEOREM 3.7.1 *Consider two HMMs (P, B, π_0) and $(P, B, \bar{\pi}_0)$ with different initial conditions π_0 and $\bar{\pi}_0$. Consider the HMM filter (3.57). Let π_k and $\bar{\pi}_k$, respectively denote the HMM filtered pmf at time k corresponding to the initial conditions π_0, $\bar{\pi}_0$, respectively. Then under conditions (3.71), (3.72),*

$$\|\pi_k - \bar{\pi}_k\|_{TV} \leq \frac{\sigma^{+}}{\sigma^{-}} \left(1 - \frac{\sigma^{-}}{\sigma^{+}}\right)^k \|\pi_0 - \bar{\pi}_0\|_{TV} \tag{3.73}$$

where the variational distance between $\pi_0, \bar{\pi}_0 \in \Pi$ is defined as

$$\|\pi_0 - \bar{\pi}_0\|_{TV} = \frac{1}{2}\|\pi_0 - \bar{\pi}_0\|_1 = \frac{1}{2}\sum_{i\in\mathcal{X}} |\pi_0(i) - \bar{\pi}_0(i)|.$$

Proof The proof constructs an inhomogeneous Markov chain involving the smoothed posterior of the Markov chain state and then characterizing its Dobrushin coefficient of ergodicity along the lines discussed in §2.7.3. The reader should review §2.7.3 before proceeding.

Define the inhomogeneous Markov chain with time varying transition probability matrix $P(n|k)$ comprising of elements

$$P_{ij}(n|k) = \mathbb{P}(x_n = j|x_{n-1} = i, x_{0:n-2}, y_{1:k}), \quad i,j \in \{1,\dots,X\}. \tag{3.74}$$

Define the fixed-interval smoothed conditional probability vector

$$\pi_{n|k} = \begin{bmatrix}\mathbb{P}(x_n = 1|y_{1:k}) & \cdots & \mathbb{P}(x_n = X|y_{1:k})\end{bmatrix}' \tag{3.75}$$

where $n \leq k$. It is straightforwardly seen that for $n \leq k$, the following version of the Chapman–Kolomogorov equation holds for the fixed-interval smoother

$$\pi_{n|k} = P'(n|k)\pi_{n-1|k} = \Big(\prod_{l=1}^{n} P'(l|k)\Big)'\pi_{0|k}.$$

Define $\bar{\pi}_{n|k}$ as in (3.75) but with initial condition $\bar{\pi}_0$. Applying properties 4 and 5 of the Dobrushin coefficient of Theorem 2.7.2 on page 28 yields

$$\begin{aligned}\|\pi_{n|k} - \bar{\pi}_{n|k}\|_{\text{TV}} &= \|\Big(\prod_{l=1}^{n} P(l|k)\Big)'\pi_{0|k} - \Big(\prod_{l=1}^{n} P(l|k)\Big)'\bar{\pi}_{0|k}\|_{\text{TV}} \\ &\leq \|\pi_{0|k} - \bar{\pi}_{0|k}\|_{\text{TV}} \prod_{l=1}^{n} \rho(P(l|k)), \quad n = 1, 2\dots, k.\end{aligned} \tag{3.76}$$

Examining the right-hand side of (3.76), to complete the proof we need to:

Step 1: Show that the Dobrushin coefficients $\rho(P(l|k))$ appearing in (3.76) are strictly smaller than 1. This is identical to what we did in (2.48) and (2.50). Lemma 3.7.2 below establishes that $\rho(P(l|k)) < 1$ (proof in appendix).

LEMMA 3.7.2 *The Dobrushin coefficients $\rho(P(l|k))$ are strictly smaller than 1. They satisfy*

$$\rho(P(l|k)) \leq 1 - \frac{\sigma^-}{\sigma^+} \tag{3.77}$$

where the positive constants σ^- and σ^+ are defined in (3.71).

Step 2: Establish an upper bound for $\|\pi_{0|k} - \bar{\pi}_{0|k}\|_{\text{TV}}$ in (3.76). Since the variational distance between two probability measures is always smaller than 1, if we only wanted to demonstrate forgetting of the initial condition, then Step 1 suffices, i.e. it suffices to show that $\rho(P(l|k)) < 1$. However, it is desirable to give a sharper bound for $\|\pi_{0|k} - \bar{\pi}_{0|k}\|_{\text{TV}}$ in terms of the initial conditions $\pi_0, \bar{\pi}_0$. In Lemma 3.7.3 below, we establish such a bound for $\|\pi_{0|k} - \bar{\pi}_{0|k}\|_{\text{TV}}$ in terms of $\|\pi_0 - \bar{\pi}_0\|_{\text{TV}}$ (proof in appendix).

LEMMA 3.7.3 *The total variational distance $\|\pi_{0|k} - \bar{\pi}_{0|k}\|_{TV}$ in (3.76) satisfies*

$$\|\pi_{0|k} - \bar{\pi}_{0|k}\|_{TV} \leq \frac{\sigma^+}{\sigma^-} \|\pi_0 - \bar{\pi}_0\|_{TV}$$

where π_0 and $\bar{\pi}_0$ denote the two initial conditions of the HMM filter.

Using Lemma 3.7.2 and Lemma 3.7.3 in (3.76) at time $n = k$ concludes the proof. □

3.7.2 Key technical result theorem 3.7.5

Here we state two important result, namely, Lemma 3.7.4 and Theorem 3.7.5 which are used in the proof of Lemma 3.7.3 above.

LEMMA 3.7.4 *For any vector $f \in \mathbb{R}^X$, and probability vectors $\pi, \bar{\pi} \in \Pi(X)$,*

1. $|f'(\pi - \bar{\pi})| \leq \max_{i,j} |f_i - f_j| \|\pi - \bar{\pi}\|_{TV}$.
2. $\|f\|_1 = \max_{|g|_\infty = 1} |g' f|$ *for* $g \in \mathbb{R}^X$.

Statement 1 is proved in Lemma 2.7.3 on page 29. Recall that it is tighter than Holder's inequality which states $|f'(\pi - \bar{\pi})| \leq \|f\|_\infty \|\pi - \bar{\pi}\|_1 = 2 \max_i |f_i| \|\pi - \bar{\pi}\|_{TV}$. Statement 2 follows from simple linear algebra [148].

Below we use the notation $T(\pi; B_y)$ to denote Bayes' formula

$$T(\pi; B_y) = \frac{B_y \pi}{\mathbf{1}' B_y \pi}, \quad \text{where } B_y = \text{diag}(B_{1y}, \cdots, B_{Xy}). \tag{3.78}$$

In particular, π is the prior, B_y are the observation likelihoods and $T(\pi; B_y)$ is the posterior, and the transition probability matrix is identity.

THEOREM 3.7.5 *(i) For posteriors $\pi, \bar{\pi} \in \Pi$, the variational distance satisfies*

$$\|T(\pi; B_y) - T(\bar{\pi}; B_y)\|_{TV} \leq \frac{\max_i B_{i,y}}{\mathbf{1}' B_y \pi} \|\pi - \bar{\pi}\|_{TV}.$$

(ii) Let p and q be any two probability mass functions on the state space $\mathcal{X}$. Then under conditions (3.71), (3.72), the backward recursion smoother satisfies

$$\frac{\sigma^-}{\sigma^+} \leq \frac{\sum_{l=1}^X p(l) \beta_{n|k}(l)}{\sum_{l=1}^X q(l) \beta_{n|k}(l)} \leq \frac{\sigma^+}{\sigma^-}.$$

The proof is in the appendix. The theorem bounds the difference in the beliefs after one step of the standard Bayes' formula. It constitutes the core of the proof of geometric ergodicity. By choosing $\pi = P\pi_1$ and $\bar{\pi} = P\pi_2$ for any two beliefs π_1, π_2, it bounds the difference between the beliefs after one step of the HMM filter update.

3.8 Suboptimal filters

Recalling the discussion in §3.3.3, how does one implement the optimal filtering recursion (3.15) for a general state space model with possibly non-Gaussian state and observation noise? Since a finite dimensional statistic is not available for parameterizing the posterior $\pi_k(x)$, one needs to introduce some sort of approximation resulting in a suboptimal filtering algorithm.

Widely used suboptimal[6] filtering algorithms include:

- The extended/unscented/cubature Kalman filter.
- The point mass filter.
- The sequential Markov chain Monte Carlo (particle) filter.

Suboptimal filters use a finite basis function approximation to the filtering recursion (3.15). Two types of basis approximation are widely used: Gaussian densities and Dirac delta functions (finite grids).

The extended/unscented/cubature Kalman filters approximate the posterior at each stage by a Gaussian density. The extended Kalman filter does a linearization (first-order Taylor series) approximation to the state and observation equation nonlinearities and then runs a standard Kalman filter on the linearized model. The unscented [333] and cubature [154] Kalman filter are higher order Taylor series expansions that run a Kalman filter-type algorithm. The Interacting Multiple Model (IMM) filter algorithm approximates the posterior by a convex combination of Gaussians (Gaussian mixture).

The point mass filter and particle filter algorithms approximate the posterior by a finite number of grid points, i.e. they use a delta function basis approximation of the posterior. Below we give a brief description of the point mass filter and particle filter algorithm.

3.8.1 Point mass filter

The idea here is to discretize the state space $\mathcal{X}$ to a predefined finite grid so that the integrals in the filtering recursion become finite summations. Assuming that the discretized state process is Markovian, the filtering problem then reduces to a hidden Markov model filtering problem discussed in §3.5. The main problem with such a deterministic grid approximation is the "curse of dimensionality". For state dimension X, the approximation error when using N grid points is $O(N^{-1/X})$ so that the number of grid points required to maintain a prescribed level of accuracy increases exponentially with state dimension X.

Consider the nonlinear state space model with additive noise

$$x_{k+1} = A(x_k) + \Gamma(x_k)w_k, \quad x_0 \sim \pi_0 \tag{3.79}$$

$$y_k = C(x_k) + D(x_k)v_k. \tag{3.80}$$

Let $\left[x^{(1)}, \dots, x^{(N)}\right]$ denote the N grid points in the state space $\mathcal{X}$. By quantizing x_k to these grid points, the transition kernel (2.10) and observation distribution (2.11), which we have assumed time-invariant for notational convenience, are

[6] Suboptimal filtering algorithms are tangential to the focus of this book since they are purely computational methods. From Part II onwards, we deal with the deeper issue of exploiting the structure of the Bayesian recursion to control its dynamics, rather than suboptimal numerical methods. Once the structure of the optimal controller is determined, any numerical filtering approximation can be used. Chapter 10 develops reduced complexity HMM filters that provably lower and upper bound the sample path of the posterior. Also, Chapter 17 shows that the state of a slow HMM can be estimated via stochastic gradient algorithms with no knowledge of the underlying parameters.

$$
\begin{aligned}
P_{ij} &\stackrel{\text{defn}}{=} p(x_{k+1} = x^{(j)} | x_k = x^{(i)}) \propto p_w \left(\Gamma^{-1}(x^{(i)}) \left[x^{(j)} - A(x^{(i)}) \right] \right) \\
B_{iy} &\stackrel{\text{defn}}{=} p(y_k = y | x_k = x^{(i)}) \propto p_v \left(D^{-1}(x^{(i)}) \left[y - C(x^{(i)}) \right] \right).
\end{aligned}
\tag{3.81}
$$

The point mass filter uses (3.81) in the HMM filter of Algorithm 3 on page 51. To compute the conditional mean estimate use $\left[x^1, \ldots, x^N\right]$ for C in (3.58) of Algorithm 3. Instead of using point mass (Dirac delta functions), other basis functions such as Gaussian mixtures can be used to yield suboptimal filters.

3.9 Particle filter

Particle filters are suboptimal filtering algorithms based on sequential Markov chain Monte Carlo methods. They are tangential to the focus of this book (see footnote 6) and we will only present the basic ideas. Discussion on particle smoothers and convergence proofs are omitted, see the references at the end of this chapter.

The particle filter is based on three simple ideas: First, propagating a large number N of random samples (called particles) from a distribution is approximately equivalent to propagating the distribution itself.

The second idea is that each of the N particles at time k is actually a random trajectory from the probability density $p(x_{0:k}|y_{1:k})$. These particles are propagated over time k. This is in contrast to the filters we have derived so far where the marginalized posterior $p(x_k|y_{1:k})$ is propagated. Because no marginalization is done, one can get away with $O(N)$ computational cost instead of $O(N^2)$. (Marginalizing over a finite state space of dimension N is equivalent to a matrix vector multiplication requiring $O(N^2)$ multiplications.)

To implement this second idea, one needs to go from the joint distribution $p(x_{0:k+1}, y_{1:k+1})$ to the conditional distribution $p(x_{0:k+1}|y_{1:k+1})$. This is achieved using self-normalized importance sampling described in Appendix A. The third key idea of the particle filter is to implement this self-normalized importance sampling recursively over time k. This is done via a recursive update over time for the importance weights – the entire procedure is called *sequential importance sampling with resampling*.

Particle filters use the Dirac delta function basis approximation

$$
p(x_{0:k}|y_{1:k}) \approx \sum_{i=1}^{N} \tilde{\omega}_k^{(i)} \delta(x_{0:k}^{(i)}). \tag{3.82}
$$

The trajectories $x_{0:k}^{(i)}$ of each of the N particles are propagated randomly according to the system dynamics. The scalar valued importance weights $\tilde{\omega}_k^{(i)}$ of each trajectory are updated recursively via Bayes' formula.

Given the particle representation (3.82), any functional $\phi(x_{0:k})$ of the state is estimated as

$$
\mathbb{E}\{\phi(x_{0:k}|y_{1:k})\} = \sum_{i=1}^{N} \phi(x_{0:k}^{(i)}) \, \tilde{\omega}_k^{(i)}.
$$

For example, choosing $\phi(x_{0:k}) = x_k$, yields the particle filter conditional mean state estimate

$$\hat{x}_k = \mathbb{E}\{x_k|y_{1:k}\} = \sum_{i=1}^{N} x_k^{(i)} \tilde{\omega}_k^{(i)}.$$

Objective: Since the particle filter is a purely computational tool, we consider a more general model than the partially observed Markov model (3.1), (3.2). Motivated by more general dynamics with correlated noise, assume the state transition and observation distribution have the following more general form:

$$p(x_{k+1}|x_{0:k}), \quad p(y_k|y_{1:k-1}, x_{1:k}).$$

The aim is to compute the conditional expectation $\mathbb{E}\{\phi(x_{0:k})|y_{1:k}\}$ for any function ϕ. This is done by estimating the density $p(x_{0:k}|y_{1:k})$, $k = 1, 2, \ldots$.

3.9.1 Bayesian importance sampling

We start with Bayesian importance sampling. Let $\pi(x_{0:k}|y_{1:k})$ denote an arbitrary importance distribution from which it is easy to obtain samples, such that $p(x_{0:k}|y_{1:k}) > 0$ implies $\pi(x_{0:k}|y_{1:k}) > 0$. Clearly,

$$\mathbb{E}\{\phi(x_{0:k})|y_{1:k}\} = \int \phi(x_{0:k}) \frac{p(x_{0:k}|y_{1:k})}{\pi(x_{0:k}|y_{1:k})} \pi(x_{0:k}|y_{1:k}) dx_{0:k}.$$

Then $\mathbb{E}\{\phi(x_{0:k}|y_{1:k})\}$ can be estimated by standard Monte Carlo methods. Simulate N i.i.d. trajectories $x_{0:k}^{(i)} \sim \pi(x_{0:k}|y_{1:k})$, $i = 1, \ldots, N$ and compute

$$\hat{\phi}_N(x_{0:k}) \stackrel{\text{defn}}{=} \sum_{i=1}^{N} \phi(x_{0:k}^{(i)}) \frac{\omega_k(x_{0:k}^{(i)})}{\sum_j \omega_k(x_{0:k}^{(j)})} = \sum_{i=1}^{N} \phi(x_{0:k}^{(i)}) \tilde{\omega}_k^{(i)}. \tag{3.83}$$

This is simply the self-normalized importance sampling described in Appendix A. The importance weight ω_k and normalized important weight $\tilde{\omega}_k^{(i)}$ are

$$\omega_k(x_{0:k}^{(i)}) = \frac{p(x_{0:k}^{(i)}|y_{1:k})}{\pi(x_{0:k}^{(i)}|y_{1:k})}, \quad \tilde{\omega}_k^{(i)} = \frac{\omega_k(x_{0:k}^{(i)})}{\sum_j \omega_k(x_{0:k}^{(j)})}.$$

By the strong law of large numbers, as the number of trajectories $N \to \infty$, the Monte Carlo estimate $\hat{\phi}_N(x_{0:k}) \to \mathbb{E}\{\phi(x_{0:k})|y_{1:k}\}$ with probability one.

3.9.2 Sequential Bayesian importance sampling

Since we are interested in real-time filtering, how can the above Bayesian importance sampling be done sequentially over time k? Clearly

$$\pi(x_{0:k}|y_{1:k}) = \pi(x_0|y_{1:k}) \prod_{t=1}^{k} \pi(x_t|x_{0:t-1}, y_{1:k}). \tag{3.84}$$

The main requirements on the importance function to facilitate real-time filtering are: (i) to obtain at any time k an estimate of the distribution $p(x_{0:k}|y_{1:k})$ and (ii) to be able to propagate this estimate in time without modifying the past simulated trajectories

$x_{0:k}^{(i)}$, $i = 1, \ldots, N$. To satisfy these requirements, we choose the importance density $\pi(x_{0:k+1}|y_{1:k+1})$ to admit $\pi(x_{0:k}|y_{1:k})$ as a marginal distribution at time k. So we restrict importance functions to the following form:

$$\pi(x_{0:k}|y_{1:k}) = \pi(x_0) \prod_{t=1}^{k} \pi(x_t|x_{0:t-1}, y_{1:t}). \tag{3.85}$$

Such an importance function allows for a recursive evaluation over time k of the weights $\omega_k^{(i)}$ and thus the normalized weights $\tilde{\omega}_k^{(i)}$, $i = 1, \ldots, N$:

$$\begin{aligned}\omega_k(x_{0:k}^{(i)}) &= \frac{p(x_{0:k}^{(i)}|y_{1:k})}{\pi(x_{0:k}^{(i)}|y_{1:k})} \propto \frac{p(y_k|y_{1:k-1}, x_{0:k}^{(i)})p(x_k^{(i)}|x_{0:k-1}^{(i)})}{\pi(x_k^{(i)}|x_{0:k-1}^{(i)}, y_{1:k})} \frac{p(x_{0:k-1}^{(i)}, |y_{1:k-1})}{\pi(x_{0:k-1}^{(i)}, |y_{1:k-1})} \\ &\propto \frac{p(y_k|y_{1:k-1}, x_{0:k}^{(i)})p(x_k^{(i)}|x_{0:k-1}^{(i)})}{\pi(x_k^{(i)}|x_{0:k-1}^{(i)}, y_{1:k})} \omega_{k-1}(x_{0:k-1}^{(i)}).\end{aligned}$$

To summarize, the particle filter representation of filtered density is

$$p(x_{0:k}|y_{1:k}) \approx \sum_{i=1}^{N} \frac{\omega_k(x_{0:k}^{(i)})}{\sum_{j=1}^{N} \omega_k(x_{0:k}^{(j)})} \delta(x_{0:k}^{(i)}).$$

Algorithm 5 gives the entire particle filtering algorithm. The rest of this section discusses two key issues in Algorithm 5:

- How to choose a good importance function (§3.9.3).
- How to avoid degeneracy of particles using a selection step (§3.9.5).

Algorithm 5 Particle filter (sequential importance sampling with resampling)

At each time k

1. *Sample particle positions*: For $i = 1, \ldots, N$:
   - Sample particles $\tilde{x}_k^{(i)} \sim \pi(x_k|x_{0:k-1}^{(i)}, y_{1:k})$.
   - Set $\tilde{x}_{0:k}^{(i)} = (x_{0:k-1}^{(i)}, \tilde{x}_k^{(i)})$.
2. *Update particle weights*: Update importance weights ω_k and normalized importance weights $\tilde{\omega}_k^{(i)}$ of particles

$$\omega_k(\tilde{x}_{0:k}^{(i)}) \propto \frac{p(y_k|y_{1:k-1}, \tilde{x}_{0:k}^{(i)})p(\tilde{x}_k^{(i)}|\tilde{x}_{0:k-1}^{(i)})}{\pi(\tilde{x}_k^{(i)}|x_{0:k-1}^{(i)}, y_{1:k})} \omega_{k-1}(x_{0:k-1}^{(i)}), \quad \tilde{\omega}_k^{(i)} = \frac{\omega_k(\tilde{x}_{0:k}^{(i)})}{\sum_{j=1}^{N} \omega_k(\tilde{x}_{0:k}^{(j)})}.$$

3. *Estimate conditional mean*: Estimate $\mathbb{E}\{\phi(x_{0:k})|y_{1:k}\}$ as (see (3.83))

$$\hat{\phi}_N(x_{0:k}) = \sum_{i=1}^{N} \phi(x_{0:k}^{(i)})\tilde{\omega}_k^{(i)}.$$

4. *Selection/resampling step*: Compute the effective number of particles as $\hat{N} = \frac{1}{\sum_{i=1}^{N} \tilde{\omega}_k^{(i)}}$. If $\hat{N}$ is smaller than a prescribed threshold, then
   - Multiply/discard particles $\tilde{x}_{0:k}^{(i)}$, $i = 1, \ldots, N$ with high/low normalized importance weights $\tilde{\omega}_k^{(i)}$ to obtain N new particles $x_{0:k}^{(i)}$, $i = 1, \ldots, N$.

3.9.3 Choice of importance distribution

Naturally, there are infinitely many possible choices for the importance function $\pi(x_{0:t}|y_{1:t})$ as long as $p(x_{0:k}|y_{1:k}) > 0$ implies $\pi(x_{0:k}|y_{1:k}) > 0$.

1. Optimal importance function: A sensible criterion is to choose a distribution that minimizes the variance of the importance weights at time k, given $x_{0:k-1}$ and $y_{1:k}$. From Appendix A.2, choosing the importance function

$$\pi(x_k|x_{0:k-1}, y_{1:k}) = p(x_k|x_{0:k-1}, y_{1:k}) \tag{3.86}$$

minimizes the variance of the importance weights conditional upon $x_{0:k-1}, y_{1:k}$.

Let us illustrate this optimal choice for the special case when the state transition kernel and observation distribution are

$$p(x_{k+1}|x_{0:k}) = p(x_{k+1}|x_k), \quad p(y_k|y_{1:k-1}, x_{1:k}) = p(y_k|x_k). \tag{3.87}$$

Then the optimal importance function (3.86) becomes

$$\pi(x_k|x_{0:k-1}, y_{1:k}) = p(x_k|x_{0:k-1}, y_{1:k}) = \frac{p(y_k|x_k)p(x_k|x_{k-1})}{p(y_k|x_{k-1})} = p(x_k|x_{k-1}, y_k). \tag{3.88}$$

So, the recursive update of the importance weights in the particle filter algorithm becomes

$$\omega_k^{(i)} = \omega_{k-1}^{(i)}\, p(y_k|x_{k-1}^{(i)}). \tag{3.89}$$

This is appealing from a computational point of view because $\omega_k^{(i)}$ no longer depends on the state $x_k^{(i)}$. So $\omega_k^{(i)}$ can be computed and $x_k^{(i)}$ sampled in parallel.

To summarize, in order to use the optimal importance function, one needs to:

1. Sample particles from $p(x_k|x_{0:k-1}^{(i)}, y_{1:k})$. Recall, if the model satisfies the conditional independence properties (3.87), then $p(x_k|x_{0:k-1}^{(i)}, y_{1:k}) = p(x_k|x_{k-1}^{(i)}, y_k)$.
2. Compute $p(y_k|x_{k-1})$ in closed form so as to compute the importance weights according to (3.88).

Example: For a Gaussian state space model with nonlinear dynamics and linear observations, the above two requirements for using the optimal importance function are met. Indeed, consider the model

$$\begin{aligned} x_{k+1} &= A(x_k) + w_k, \qquad & w_k &\sim N(0, Q) \\ y_k &= Cx_k + v_k, \qquad & v_k &\sim N(0, R) \end{aligned}$$

which satisfies the conditional independence (3.87) by construction. Therefore the optimal importance function satisfies (3.88). It turns out that this optimal importance function can be evaluated via a Kalman filter as follows:

$$
\begin{aligned}
\pi(x_k|x_{0:k-1}, y_{1:k}) = p(x_k|x_{0:k-1}, y_{1:k}) &= \frac{p(y_k|x_k)p(x_k|x_{k-1})}{\int_{\mathcal{X}} p(y_k|x_k = z)p(x_k = z|x_{k-1})dz} \\
&= \frac{N(y_k; Cx_k; R)N(x_k; A(x_{k-1}), Q)}{\int_{\mathcal{X}} N(y_k; Cz; R)N(z; A(x_{k-1}), Q)\, dz}.
\end{aligned} \tag{3.90}
$$

Applying (3.42) of Theorem 3.4.2 then yields

$$
p(x_k|x_{k-1}, y_k) = N(x_k; A(x_{k-1}) + K(y_k - CA(x_{k-1})), Q - KCQ).
$$

Finally, the importance weights (3.89) for the particle filter are the denominator of (3.90). Applying (3.41) of Theorem 3.4.2 yields

$$
p(y_k|x_{k-1}) = N(y_k; CA(x_{k-1}), CQC' + R).
$$

2. Prior importance function – bootstrap particle filter: This approach was originally proposed in the 1960s [134]. The importance function is chosen independent of the observations. That is, $\pi(x_k|x_{0:k-1}, y_{1:k}) = p(x_k|x_{k-1})$. Then the weights are updated as $\omega_k^{(i)} = \omega_{k-1}^{(i)} p(y_k|x_k^{(i)})$. The resulting filter is called the bootstrap particle filter. Since the particles are sampled independent of the observation, the resulting implementation is sensitive to outliers.

3. Fixed importance function: An even simpler choice of importance function is to choose $\pi(x_k|x_{0:k-1}, y_{1:k}) = p(x_k)$. That is, the particles are sampled independent of the observations and prior trajectory.

4. Auxiliary particle filter: This uses an importance density involving the current observation y_k:

$$
\pi(x_k|x_{0:k-1}, y_{1:k}) = p(x_k|x_{k-1}^{(i)}) \frac{\omega_k^{(i)} \bar{p}(y_k|x_{k-1}^{(i)})}{\sum_{j=1}^{N} \omega_k^{(i)} \bar{p}(y_k|x_{k-1}^{(i)})}.
$$

Here $\bar{p}(y_k|x_{k-1}^{(i)})$ is chosen as an approximation to $p(y_k|x_{k-1})$. For example, if the state dynamics are $x_{k+1} = A(x_k) + w_k$, one can choose

$$
\bar{p}(y_k|x_{k-1}^{(i)}) = p\big(y_k \mid x_k = A(x_{k-1}^{(i)}) + w_{k-1}^{(i)}\big),
$$

where $w_k^{(i)}$ is simulated from pdf p_w.

3.9.4 Perspective: comparison with HMM filter

How does the particle filter compare with an HMM filter when estimating the state of a finite state Markov chain in noise? Using the above importance functions, it is instructive to compare the particle filter with the HMM filter.

Consider an HMM has state space $\mathcal{X} = \{e_1, \ldots, e_X\}$ (where e_i denotes the unit vector with 1 in the i-th position), transition matrix P, observation kernel B_{xy} and state levels C. Then the particle filter with optimal importance sampling function reads:
For particles $i = 1, 2, \ldots, N$:

1. Sample $\tilde{x}_k^{(i)}$ with $\mathbb{P}(\tilde{x}_k^{(i)} = e_x) \propto B_{xy_k} P_{\tilde{x}_{k-1}^{(i)}, x}$ for $x \in \{1, 2, \ldots, X\}$.
2. Update the importance weight as $\omega_k^{(i)} = \omega_{k-1}^{(i)} \sum_j B_{j,y_k} P_{\tilde{x}_{k-1}^{(i)}, j}$.

Finally, the estimate of conditional mean $\mathbb{E}\{x_k|y_{1:k}\}$ is computed as

$$\hat{x}_k = \frac{\sum_{i=1}^N C'\tilde{x}_k^{(i)}\omega_k^{(i)}}{\sum_{i=1}^N \omega_k^{(i)}}.$$

Compare this with the HMM filter (3.57) and conditional mean estimate (3.58).

3.9.5 Degeneracy and resampling step

For importance functions of the form (3.85), the unconditional variance (with the observations $y_{1:k}$ being interpreted as random variables) of the weights ω_k increases (stochastically) over time. It is thus impossible to avoid a degeneracy phenomenon. After a few iterations of Algorithm 5 (if the selection step is not used), all but one of the normalized importance weights are very close to zero and a large computational burden is devoted to updating trajectories whose contribution to the final estimate is negligible. It is necessary to introduce a selection step to discard the particles $x_{0:t}^{(i)}$ with low normalized importance weights $\tilde{\omega}_t^{(i)}$ and increase the number of particles with high importance weight. This facilitates avoiding degeneracy and allows the algorithm to explore interesting zones where the conditional density is larger. Each time a selection step is used the weights are reset to N^{-1}. Of course, it is essential that this selection step keeps the distribution of the particles unchanged.

A selection procedure associates to each particle (trajectory), say $\tilde{x}_{0:t}^{(i)}$ $(i = 1, \ldots, N)$, a number of "children" N_i, such that $\sum_{i=1}^N N_i = N$, to obtain N new particles $x_{0:t}^{(i)}$. If $N_i = 0$, then the particle $\tilde{x}_{0:t}^{(i)}$ is discarded, otherwise it has N_i "children" at time t. To summarize, if the selection scheme is used at time t, then before the selection scheme, we have the particle approximation of the filtered density as $\sum_{i=1}^N \tilde{\omega}_t^{(i)}\delta(\tilde{x}_{0:t}^{(i)})$ and, after the selection step, we have $N^{-1}\sum_{i=1}^N \delta(x_{0:t}^{(i)})$.

Here are selection schemes that can be implemented in $O(N)$ iterations.

- *Sampling Importance Resampling (SIR)/Multinomial Sampling procedure*. This procedure, introduced originally in [128] samples N times from $\sum_{i=1}^N \tilde{\omega}_t^{(i)}\delta(\tilde{x}_{0:t}^{(i)})$ to obtain N new particles.
- *Residual Resampling* [213]. Set $\widetilde{N}_i = \left\lfloor N\widetilde{\omega}_t^{(i)}\right\rfloor$ then perform a SIR procedure to select the remaining $\overline{N}_t = N - \sum_{i=1}^N \widetilde{N}_i$ samples with the new weights $\omega_t'^{(i)} = \left(\widetilde{\omega}_t^{(i)}N - \widetilde{N}_i\right)/\overline{N}_t$, finally add the results to the current $\widetilde{N}_i$.

Convergence: The particle filter algorithm uses a fixed number of particles (N Dirac delta functions) to approximate the multivariate density $p(x_{0:k}|y_{1:k})$ which grows in dimension with time k. At first sight, this is a recipe for disaster. For ergodic systems, however, where the previous states and observations have geometrically diminishing effect on the marginal $p(x_k|y_{1:k})$, one would expect that by choosing N sufficiently large, the particle filter would yield satisfactory estimates. The convergence analysis of the particle filter Algorithm 5 is nontrivial since the resampling step makes the particles statistically dependent. However, under mild regularity conditions [245, pp. 300–306] a central limit theorem is established; see also [76].

3.10 Example: Jump Markov linear systems (JMLS)

As described in §2.6, JMLS are used widely to model maneuvering targets in tracking applications. Here we discuss particle filters for estimating the state of the JMLS introduced in §2.5, namely:

$$z_{k+1} = A(r_{k+1})\, z_k + \Gamma(r_{k+1})\, w_{k+1} + f(r_{k+1})\, u_{k+1} \tag{3.91}$$

$$y_k = C(r_k)\, z_k + D(r_k)\, v_k + g(r_k) u_k. \tag{3.92}$$

Here the unobserved X-state discrete-time Markov chain r_k has an $X \times X$ transition probability matrix P, and z_k is an unobserved continuous-valued process. u_k is a known exogenous input, $w_k \sim N(0, I)$ and $v_k \sim N(0, I)$ are i.i.d. Gaussian noise processes, $z_0 \sim N(\hat{z}_0, \Sigma_0)$ where Σ_0 is positive definite. Assume z_0, v_k and w_k be mutually independent for all k.

Given the observations $y_{1:k} = (y_1, \ldots, y_k)$, $k = 1, 2, \ldots$, the aim is to compute for any function $\phi(\cdot)$, the conditional mean state estimates

$$\phi_k \overset{\text{defn}}{=} \mathbb{E}\{\phi(r_k, z_k) | y_{1:k}\}.$$

Computing the filtered estimates exactly is intractable since evaluating

$$p(z_k | y_{1:k}) = \sum_{l=1}^{X^k} p(z_k | y_{1:k}, r_{0:k}^{(l)})\, \mathbb{P}(r_{0:k}^{(l)} y_{1:k})$$

requires X^k computational cost. Therefore, it is imperative to use a suboptimal filtering algorithm.

The Interacting Multiple Model (IMM) [32] algorithm constructs an X-dimensional Gaussian mixture approximation to the filtered density of a JMLS at each time. It uses a bank of X Kalman filters at each time. Below we describe a particle filtering algorithm.

3.10.1 Rao–Blackwellized particle filter

We now describe one possible particle filter implementation for estimating the state of a JMLS. By choosing $x_k = (z_k, r_k)$, one can directly use the particle filter in Algorithm 5 on page 63 to estimate $p(r_{0:k}, z_{0:k} | y_{1:k})$. However, instead of naively estimating $p(r_{0:k}, z_{0:k} | y_{1:k})$, due to the structure of the JMLS, it suffices to estimate the smaller dimension posterior $p(r_{0:k} | y_{1:k})$. This is because given an approximation of $p(r_{0:k} | y_{1:k})$, one can compute

$$p(r_{0:k}, z_k | y_{1:k}) = p(z_k | y_{1:k}, r_{0:k})\, p(r_{0:k} | y_{1:k}) \tag{3.93}$$

where $p(z_k | y_{1:k}, r_{0:k})$ is a Gaussian distribution whose parameters can be evaluated exactly using a Kalman filter. From (3.93) the following more statistically efficient Bayesian importance sampling estimate of ϕ_k can be computed:

$$\hat{\phi}_k = \frac{\sum_{i=1}^{N} \mathbb{E}\{\phi(r_k^{(i)}, z_k)|y_{1:k}, r_{0:k}^{(i)}\}\, \omega(r_{0:k}^{(i)})}{\sum_{i=1}^{N} \omega(r_{0:k}^{(i)})},$$
$$\omega(r_{0:k}) = \frac{p(r_{0:k}|y_{1:k})}{\pi(r_{0:k}|y_{1:k})} \text{ and } \pi(r_{0:k}|y_{1:k}) = \int \pi(r_{0:k}, z_{0:k}|y_{1:k}) dz_{0:k}. \tag{3.94}$$

This is to be compared with the standard particle filter estimate denoted as

$$\bar{\phi}_k = \frac{\sum_{i=1}^{N} \phi(r_k^{(i)}, z_k^{(i)})\, \omega(r_{0:k}^{(i)}, z_{0:k}^{(i)})}{\sum_{i=1}^{N} \omega(r_{0:k}^{(i)}, z_{0:k}^{(i)})}. \tag{3.95}$$

The conditional expectation in (3.94) is the essence of Rao–Blackwellization: a conditional expectation always has a smaller variance than an unconditional one; see Appendix A.2.2 on page 436.

In light of the above discussion, the particle filter Algorithm 5 on page 63 can be used to estimate $p(r_{0:k}|y_{1:k})$ with weights

$$\omega(r_{0:k}) \propto \frac{p\left(y_k|\, y_{1:k-1}, r_{0:k}\right) p\left(r_k|\, r_{k-1}\right)}{\pi\left(r_k|\, y_{1:k}, r_{0:k-1}\right)} \omega(r_{0:k-1}). \tag{3.96}$$

From §3.9.3, the "optimal" importance function is $p(r_k|r_{0:k-1}, y_{1:k})$. For any state $j \in \{1, , \ldots, X\}$ of the Markov chain,

$$p\left(r_k = j|\, r_{0:k-1}, y_{1:k}\right) = \frac{p\left(y_k|\, y_{1:k-1}, r_{0:k-1}, r_k = j\right) p\left(r_k = j|\, r_{k-1}\right)}{p\left(y_k|\, y_{1:k-1}, r_{0:k-1}\right)}.$$

So from (3.96), the associated importance weight $\omega(r_{0:k})$ is proportional to

$$p\left(y_k|\, y_{1:k-1}, r_{0:k-1}\right) = \sum_{j=1}^{X} p\left(y_k|\, y_{1:k-1}, r_{0:k-1}, r_k = j\right) \mathbb{P}\left(r_k = j|\, r_{k-1}\right). \tag{3.97}$$

Given the Markov chain trajectory $r_{0:k}$, the JMLS (3.92) becomes a standard linear Gaussian state space model. Using the same notation as the Kalman filter of Algorithm 1 on page 45, define $Q(r_k) = \Gamma(r_k)\Gamma'(r_k)$ and $R(r_k) = D(r_k)D'(r_k)$. Recall for the jump Markov linear system, we use the notation z_k instead of x_k for the continuous state. Below we use the notation $\hat{z}_k$ and $\hat{z}_{k|k-1}$ to denote the filtered and one-step ahead predicted state estimates.

We have already shown how to evaluate $p(y_k|y_{1:k-1})$ for a linear Gaussian system in (3.44). Applying this with our updated notation, we have

$$p(y_k|y_{1:k-1}, r_{0:k-1}, r_k = j) = N(y_k; C(j)\hat{z}_{k|k-1}, C(j)\Sigma_{k|k-1}C'(j) + R(j)).$$

Therefore from (3.97), it follows that $p\left(y_k|\, y_{1:k-1}, r_{0:k-1}\right)$ is an X-component Gaussian mixture:

$$p\left(y_k|\, y_{1:k-1}, r_{0:k-1}\right) = \sum_{j=1}^{X} N(y_k; C(j)\hat{z}_{k|k-1}, C(j)\Sigma_{k|k-1}C'(j) + R(j))\, P_{r_{k-1}j}.$$

In summary, the optimal importance function is evaluated using a bank of N Kalman filters, one for each particle. In each Kalman filter, the one-step ahead predicted state $\hat{z}_{k|k-1}(r_{0:k-1})$ and covariance $\Sigma_{k|k-1}$ are evaluated using (3.31) and (3.32).

Numerical examples illustrating the performance of this particle filtering algorithm are given in [100].

3.11 Complements and sources

Statistical estimation theory has its roots in the work of Kolmogorov and Wiener in the 1940s [175, 176, 338]. Wiener [338] considered the problem of finding the linear minimum mean squared error estimate of a continuous-time signal based on noisy observations of the signal. He assumed that the processes were jointly stationary with given covariance structure and that the observation interval was semi-infinite. The result was the Wiener filter with impulse response specified as the solution of a Wiener–Hopf integral equation.

In the early 1960s Kalman and Bucy proposed a new framework for the filtering problem [164, 166]. Rather than considering stationary processes and specifying the covariance structure, the signal was modeled using a stochastic state space system. In particular, when the signal or state was a vector Gauss–Markov process (linear dynamics and white Gaussian noise) and the observation process was a linear function of the state perturbed by white Gaussian noise, then the minimum mean squared error state estimate was obtained as the solution of a linear stochastic differential (continuous-time) or difference (discrete-time) equation driven by the observations. The Wiener–Hopf equation was effectively replaced by a matrix Riccati equation. The new framework readily handled nonstationarity and finite time intervals, and produced recursive algorithms suitable for numerical implementation.

The HMM filter goes back to [38]. Jazwinski [160] gives filtering recursions for partially observed state space models. The use of measure change ideas has been pioneered by Elliott [104]; see also [340]. In terms of modeling, reciprocal processes are useful for destination-aware targets. Another powerful class of models are stochastic context free grammars (SCFGs). These are more general than HMMs and widely used in natural language processing [231] and syntactic pattern analysis. Bayesian algorithms for SCFGs are similar to HMMs [112]. Particle filters are developed for SCFG modulated linear systems in [111].

Sequential Markov chain Monte Carlo (aka particle) filters have revolutionized computational Bayesian filtering during the last 20 years. The oldest known case of a particle filter dates back to [134]. Highly influential papers and books in particle filters include [128, 99, 98, 279, 101]. We have omitted discussion of the important topic of particle-based smoothing algorithms and convergence[101].

We have not discussed performance analysis of filters; see [319] for posterior Cramér–Rao bounds for discrete-time filtering. Also [131] develops explicit formulas for the error probabilities of filters using information theoretic methods when the underlying noise is Gaussian noise.

Bayesian filtering assumes that the underlying model is fully specified. In a series of remarkable papers, Weissman [314] develops filtering (denoising) algorithms based on universal data compression methods.

Continuous-time filtering: Continuous-time HMM filters are discussed briefly in Appendix B. There has been much cross-fertilization between discrete-time and continuous-time filtering. In continuous time, when the state process is modeled by a nonlinear stochastic differential equation, and the observation process involves possible nonlinearities, the situation becomes much more difficult from a technical viewpoint [340, 46]. The aim is to develop equations for conditional distributions of the state and obtain finite dimensional solutions. The first such results were obtained by Stratonovich [311] and Kushner [196] in the framework of the Stratonovich calculus and the Itô calculus respectively. The result was a nonlinear stochastic partial differential equation for the conditional (filtered) density, the *Kushner–Stratonovich equation*. The continuous-time HMM filter was first obtained by Wonham [341].

An alternative approach based on the work of Girsanov on the use of measure transformations in stochastic differential equations [352] leads to an equation for the unnormalized conditional density which is a linear stochastic partial differential equation driven by the observations commonly called the *Zakai equation*. Two known cases where this stochastic partial differential equation reduces to a finite set of stochastic ordinary differential equations are the Kalman filter [164] and the Beneš filter [45]. Issues of finite dimensionality have been investigated using Lie algebraic ideas [233, 70, 232, 86].

Appendix 3.A Proof of Lemma 3.7.2

To determine the Dobrushin coefficients, let us examine the elements of the transition matrix $P(l|k)$ defined in (3.74) more carefully. From the HMM fixed-interval smoother equations (3.62) and Bayes' formula, it follows that

$$P_{ij}(l|k) = \mathbb{P}(x_l = j|x_{l-1} = i, x_{0:l-2}, y_{1:k}) = \frac{P_{ij} B_{jy_l} \beta_{l|k}(j)}{\sum_{x=1}^{X} P_{ix} B_{xy_l} \beta_{l|k}(x)}.$$

Here $\beta_{l|k}(j) = \mathbb{P}(y_{l+1:k}|x_l = j)$ is the backward variable whose computation is detailed in the HMM smoother given in §3.5.2.

Next, we invoke conditions (3.71), (3.72). Then

$$P_{ij}(l|k) = \frac{P_{ij} B_{jy_l} \beta_{l|k}(j)}{\sum_{x=1}^{X} P_{ix} B_{xy_l} \beta_{l|k}(x)} \geq \frac{\sigma^-}{\sigma^+} \frac{\mu_j B_{jy_l} \beta_{l|k}(j)}{\sum_{x=1}^{X} \mu_x B_{xy_l} \beta_{l|k}(x)}. \tag{3.98}$$

In analogy to §2.7.3, denote

$$\epsilon = \frac{\sigma^-}{\sigma^+}, \quad \kappa_j = \frac{\mu_j B_{jy_l} \beta_{l|k}(j)}{\sum_{x=1}^{X} \mu_x B_{xy_l} \beta_{l|k}(x)}.$$

Clearly κ_j is a probability mass function and $\epsilon \in (0, 1]$. So (3.98) is simply the Doeblin condition (2.49), namely $P_{ij}(l|k) \geq \epsilon \kappa_j$. From Theorem 2.7.4 on page 29, the Dobrushin coefficients satisfy $\rho(P(l|k)) \leq 1 - \epsilon$.

Appendix 3.B Proof of Lemma 3.7.3

$$\pi_{0|k}(i) = \frac{\pi_0(i)\beta_{0|k}(i)}{\sum_{l=1}^X \pi_0(l)\beta_{0|k}(l)}, \quad \bar{\pi}_{0|k}(i) = \frac{\bar{\pi}_0(i)\beta_{0|k}(i)}{\sum_{l=1}^X \bar{\pi}_0(l)\beta_{0|k}(l)}. \tag{3.99}$$

Then applying Theorem 3.7.5(i) (which is stated and proved below) yields

$$\|\pi_{0|k} - \bar{\pi}_{0|k}\|_{\text{TV}} \leq \frac{\max_i \beta_{0|k}(i)}{\sum_{l=1}^X \pi_0(l)\beta_{0|k}(l)} \|\pi_0 - \bar{\pi}_0\|_{\text{TV}}. \tag{3.100}$$

Consider the right-hand side. Denoting $i^* = \text{argmax}_i\, \beta_{0|k}(i)$, we can express

$$\frac{\max_i \beta_{0|k}(i)}{\sum_{l=1}^X \pi_0(l)\beta_{0|k}(l)} = \frac{\sum_{l=1}^X I(l - i^*)\, \beta_{0|k}(l)}{\sum_{l=1}^X \pi_0(l)\beta_{0|k}(l)} \tag{3.101}$$

where $I(\cdot)$ denotes the indicator function. Next applying Theorem 3.7.5(ii) yields

$$\frac{\sum_{l=1}^X I(l - i^*)\beta_{0|k}(l)}{\sum_{l=1}^X \pi_0(l)\beta_{0|k}(l)} \leq \frac{\sigma^+}{\sigma^-}. \tag{3.102}$$

Using (3.101), (3.102) in the right-hand side of (3.100) proves the result.

Appendix 3.C Proof of Theorem 3.7.5

Statement (i): For any $g \in \mathbb{R}^X$,

$$\begin{aligned} g'\left(T(\pi; B_y) - T(\bar{\pi}; B_y)\right) &= g'\left(T(\pi; B_y) - \frac{B_y\bar{\pi}}{\mathbf{1}'B_y\pi} + \frac{B_y\bar{\pi}}{\mathbf{1}'B_y\pi} - T(\bar{\pi}; B_y)\right) \\ &= \frac{1}{\mathbf{1}'B_y\pi}\, g'\left[I - T(\bar{\pi}; B_y)\mathbf{1}'\right] B_y\,(\pi - \bar{\pi}). \end{aligned} \tag{3.103}$$

Applying Statement 1 of Lemma 3.7.4 to the right-hand side of (3.103) yields

$$|g'\left(T(\pi; B_y) - T(\bar{\pi}; B_y)\right)| \leq \frac{1}{\mathbf{1}'B_y\pi} \max_{i,j} |f_i - f_j|\; \|\pi - \bar{\pi}\|_{\text{TV}}$$

where $f_i = g'\left[I - T(\bar{\pi}; B_y)\mathbf{1}'\right] B_y e_i$ and $f_j = g'\left[I - T(\bar{\pi}; B_y)\mathbf{1}'\right] B_y e_j$. So

$$|f_i - f_j| = |g_i B_{i,y} - g'T(\bar{\pi}; B_y)B_{i,y} - (g_j B_{j,y} - g'T(\bar{\pi}; B_y)B_{j,y})|.$$

Since $T(\bar{\pi}; B_y)$ is a probability vector, clearly $\max_i |g'T(\bar{\pi}; B_y)| \leq \max_i |g_i|$. This together with the fact that $B_{i,y}$ are nonnegative implies

$$\max_{i,j} |f_i - f_j| \leq 2 \max_i |g_i| \max_i B_{i,y}.$$

So denoting $\|g\|_\infty = \max_i |g_i|$, we have

$$|g'\left(T(\pi; B_y) - T(\bar{\pi}; B_y)\right)| \leq 2\, \frac{\|g\|_\infty \max_i B_{i,y}}{\mathbf{1}'B_y\pi} \|\pi - \bar{\pi}\|_{\text{TV}}.$$

Finally applying Statement 2 of Lemma 3.7.4 yields

$$\|T(\pi;B_y) - T(\bar{\pi};B_y)\|_1 = \max_{|g|_\infty=1} |g'\left(T(\pi;B_y) - T(\bar{\pi};B_y)\right)|$$
$$\leq \max_{|g|_\infty=1} 2\frac{\|g\|_\infty \max_i B_{i,y}}{\mathbf{1}'B_y\pi}\|\pi - \bar{\pi}\|_{\text{TV}}.$$

Since the variational norm is half the l_1 norm, the proof is completed.
Statement (ii): Recall that the backward recursion for the smoother satisfies

$$\beta_{n|k}(i) = \sum_{l=1}^{X} \beta_{n+1|k}(l)P_{il}B_{ly_{n+1}}.$$

Invoking conditions (3.71), (3.72), yields

$$\sigma^- \sum_{l=1}^{X} \beta_{n+1|k}(l)\mu_l B_{ly_{n+1}} \leq \sum_{l=1}^{X} \beta_{n+1|k}(l)P_{il}B_{ly_{n+1}} \leq \sigma^+ \sum_{l=1}^{X} \beta_{n+1|k}(l)\mu_l B_{ly_{n+1}}.$$

4 Algorithms for maximum likelihood parameter estimation

Contents

Chapter 3 discussed filtering algorithms for estimating the state of a stochastic system. This chapter discusses maximum likelihood parameter estimation algorithms for linear Gaussian state space models and hidden Markov models. Computing the maximum likelihood parameter estimate for these models involves numerical optimization algorithms. These algorithms make extensive use of the filters and smoothers developed in Chapter 3. Thus, algorithms for maximum likelihood parameter estimation serve as an important application for the filters and smoothers derived in Chapter 3.

4.1 Maximum likelihood estimation criterion

Suppose a block of data $y_{1:N} = (y_1, \ldots, y_N)$ is generated by a model parameter vector θ^o. Assume that the dimension p of the vector θ^o is known and that θ^o lies in a set Θ which is a compact subset of $\mathbb{R}^p$.

Given the data $y_{1:N}$, the aim is to estimate the model that best fits the data in the likelihood sense. That is, compute the parameter $\theta_N^* \in \Theta$ which maximizes[1] the *likelihood function* $L_N(\theta)$:

$$\theta_N^* = \operatorname*{argmax}_{\theta \in \Theta} L_N(\theta), \qquad \text{where } L_N(\theta) = p(y_{1:N}|\theta). \tag{4.1}$$

The likelihood function $L_N(\theta)$ is the joint probability density (mass) function of the observations $y_{1:N}$ given the model parametrized by θ. The parameter estimate θ_N^* is called the *maximum likelihood estimate (MLE)*. It is often more convenient to maximize the log likelihood

[1] In the examples considered in this chapter, the likelihood $L_N(\theta)$ is a continuous function of θ. So the maximum exists but may not be unique.

$$\mathcal{L}_N(\theta) = \log L_N(\theta). \tag{4.2}$$

Clearly the maximizers of $L_N(\theta)$ and $\mathcal{L}_N(\theta)$ coincide.

There are several reasons why the likelihood function is a useful criterion for parameter estimation. For models such as linear Gaussian state space models and HMMs, under reasonable conditions, the MLE is strongly consistent, i.e. as the length of the data block $N \to \infty$, the MLE parameter estimate $\theta_N^* \to \theta^o$ with probability one, where θ^o denotes the true parameter value. Moreover, under suitable conditions, the MLE is also asymptotically normal and efficient and satisfies the Cramér–Rao bound. This chapter focuses on computation of the MLE as an application of the filters/smoothers derived in Chapter 3 – we will not consider statistical properties of the MLE; see §4.6.2 for a short discussion.[2]

4.1.1 Example: MLE of transition probability for Markov chain

As a warmup example, given the observed trajectory $x_{0:N}$ of an X-state Markov chain, we compute the MLE of the transition matrix $\theta = P$. (So the observations are $y_k = x_k$.) Note the MLE has to be a valid stochastic matrix satisfying the constraints: $\sum_{j=1}^X P_{ij} = 1$ and $0 \le P_{ij} \le 1$. With $I(\cdot)$ denoting the indicator function, the log likelihood function is

$$\begin{aligned}
\mathcal{L}_N(\theta) = \log L_N(\theta) = \log p(x_{0:N}|\theta) &= \sum_{k=1}^N \log p(x_k|x_{k-1},\theta) + \log p(x_0|\theta) \\
&= \sum_{k=1}^N \sum_{i=1}^X \sum_{j=1}^X I(x_{k-1} = i, x_k = j) \log P_{ij} + \sum_{i=1}^X I(x_0 = i) \log \pi_0(i) \\
&= \sum_{i=1}^X \sum_{j=1}^X J_{ij}(N) \log P_{ij} + \sum_{i=1}^X I(x_0 = i) \log \pi_0(i)
\end{aligned} \tag{4.3}$$

where $J_{ij}(N) = \sum_{k=1}^N I(x_{k-1} = i, x_k = j)$ denotes the number of times the Markov chain jumps from state i to state j in the time duration 1 to N. Then maximizing $\log L_N(\theta)$ with respect to $\theta = P$ subject to the constraints $\sum_{j=1}^X P_{ij} = 1$ and $0 \le P_{ij} \le 1$ yields[3] the MLE for the transition matrix as

$$P_{ij}^* = \frac{J_{ij}(N)}{\sum_{j=1}^X J_{ij}(X)} = \frac{J_{ij}(N)}{D_i(N)}, \quad i,j = 1,\ldots,X \tag{4.4}$$

[2] As discussed in [325], the basic problem in probability theory is: given a probabilistic model $(\Omega, \mathcal{F}, \mathbb{P})$ and an event $A \in \mathcal{F}$, determine (or estimate) the distribution function $\mathbb{P}(y(A) \le z)$ for some random variable $y_{1:N}$ defined for event A. The basic problem of statistics is: given $(\Omega, \mathcal{F})$, and data $y_{1:N}$ obtained under probability measure $\mathbb{P}$, estimate this probability measure for all $A \in \mathcal{F}$. Filtering and smoothing of Chapter 3 qualify as special cases of the basic problem in probability, while determining the MLE is a special case of the basic problem in statistics.

[3] Consider optimizing (4.3) subject to the constraint $\sum_m P_{lm} = 1$. The Lagrangian is $\mathcal{L}_N(\theta) - \sum_l \lambda_l(\sum_m P_{lm} - 1)$. Setting the derivative with respect to P_{ij} to zero yields $P_{ij} = J_{ij}(N)/\lambda_i$. Summing this over j yields $\lambda_i = \sum_j J_{ij}(N)$. So the constrained optimizer is $P_{ij}^* = J_{ij}(N)/\sum_j J_{ij}(N)$, namely (4.4). This optimizer automatically satisfies the constraint $P_{ij}^* \in [0,1]$.

where $D_i(N) = \sum_{k=1}^{N} I(x_{k-1} = i)$ denotes the amount of time the Markov chain spends in state i in the time duration 1 to N. Thus the MLE P_{ij}^* is the ratio of the number of jumps from state i to j divided by the duration in state i.

It is shown in [57] that the MLE of the transition matrix is strongly consistent if the true model is regular (aperiodic and irreducible).

4.2 MLE of partially observed models

The rest of this chapter considers MLE for two types of partially observed models, namely linear Gaussian state space models and hidden Markov models. Unlike the above fully observed example, in general it is not possible to obtain a closed form expression for the MLE of partially observed models. The MLE needs to be computed by a numerical optimization algorithm. This section discusses how general purpose optimization algorithms can be used to compute the MLE. The next section, §4.3 discusses the Expectation Maximization algorithm for computing the MLE. These algorithms use the filters and smoothers developed in Chapter 3.

4.2.1 General purpose optimization algorithm

Consider the partially observed stochastic model (2.6), (2.7) of §2.1. Given a batch of data $y_{1:N} = (y_1, \dots, y_N)$, in order to compute the MLE θ_N^* via an optimization algorithm, it is first necessary to evaluate the likelihood $L_N(\theta)$ at any value $\theta \in \Theta$. Numerical evaluation of the likelihood for a partially observed model is intimately linked with optimal filtering. Recall from (3.20) that the un-normalized filtered density at time N is defined as

$$q_N^\theta(x) = p(x_N = x, y_{1:N}|\theta)$$

where we explicitly include the model θ in the notation. Therefore, the likelihood can be computed by integrating the un-normalized filtered density at final time N over the state space:

$$L_N(\theta) = p(y_{1:N}|\theta) = \int_{\mathcal{X}} q_N^\theta(x)dx. \tag{4.5}$$

For any partially observed model where a finite dimensional filter exists, the model likelihood $L_N(\theta)$ can be evaluated exactly using (4.5). Then one can use a general purpose numerical optimization algorithm such as `fmincon` in Matlab to compute the MLE. This uses the interior reflective Newton method [88] for large scale problems and sequential quadratic programming [50] for medium scale problems. In general $L_N(\theta)$ is non-concave in θ and the constraint set Θ is non-convex. So the algorithm at best will converge to a local stationary point of the likelihood function.

4.2.2 Evaluation of gradient and Hessian

General purpose optimizers such as `fmincon` in Matlab allow for the gradient $\nabla_\theta L_N(\theta)$ and Hessian $\nabla_\theta^2 L_N(\theta)$ of the likelihood to be inputted in the algorithm. We now discuss

how to evaluate the gradient and Hessian. Since the maximizers of $L_N(\theta)$ and log likelihood $\mathcal{L}_N(\theta)$ coincide, instead of the gradient $\nabla_\theta L_N(\theta)$, and Hessian $\nabla^2_\theta L_N(\theta)$, an optimization algorithm can also use the gradient $\nabla_\theta \mathcal{L}_N(\theta)$ and Hessian $\nabla^2_\theta \mathcal{L}_N(\theta)$. There are two equivalent methods for evaluating the gradient and Hessian of the likelihood:
1. Sensitivity equations: From (4.5), one method of computing the gradient is

$$\nabla_\theta L_N(\theta) = \int_{\mathcal{X}} \nabla_\theta q_N^\theta(x) dx. \tag{4.6}$$

By running the un-normalized filter (3.20) for $k = 1, 2, \dots, N$, one can evaluate

$$\begin{aligned}\nabla_\theta q_{k+1}^\theta(x_{k+1}) = \nabla_\theta p(y_{k+1}|x_{k+1}, \theta) &\int_{\mathcal{X}} p(x_{k+1}|x_k, \theta) q_k^\theta(x_k) dx_k \\ + p(y_{k+1}|x_{k+1}, \theta) &\int_{\mathcal{X}} \nabla_\theta p(x_{k+1}|x_k, \theta) q_k^\theta(x_k) dx_k \\ + p(y_{k+1}|x_{k+1}, \theta) &\int_{\mathcal{X}} p(x_{k+1}|x_k, \theta) \nabla_\theta q_k^\theta(x_k) dx_k. \end{aligned} \tag{4.7}$$

The above are called the *sensitivity equations* of the Bayesian filter with respect to the parameter θ. At time $k = N$, substituting $\nabla_\theta q_N^\theta(x)$ in (4.6) yields the gradient of the likelihood. The Hessian can then be evaluated by differentiating the above sensitivity equations.
2. Fisher and Louis identity: Alternatively, the gradient and Hessian of the log likelihood can be evaluated using Fisher's identity[4] and Louis' identity:

$$\text{Fisher:} \quad \nabla_\theta \log L_N(\theta) = \mathbb{E}\{\nabla_\theta \log p(y_{1:N}, x_{0:N}|\theta) |\, y_{1:N}, \theta\} \tag{4.8}$$

$$\begin{aligned}\text{Louis:} \quad \nabla^2_\theta \log L_N(\theta) = &\ \mathbb{E}\left\{\nabla^2_\theta \log p(y_{1:N}, x_{0:N}|\theta) \middle|\, y_{1:N}, \theta\right\} \\ &- \mathbb{E}\left\{\nabla^2_\theta \log p(x_{0:N}|y_{1:N}, \theta) \middle|\, y_{1:N}, \theta\right\}. \end{aligned} \tag{4.9}$$

Equation (4.8) can be written as follows:

$$\nabla_\theta \log L_N(\theta) = \nabla_\theta \mathcal{Q}(\bar\theta, \theta), \quad \text{then set } \bar\theta = \theta, \tag{4.10}$$

where the function $\mathcal{Q}(\bar\theta, \theta) = \mathbb{E}\left\{\log p(y_{1:N}, x_{0:N}|\theta) |\, y_{1:N}, \bar\theta\right\}$ is called the auxiliary log likelihood. It turns out that $\mathcal{Q}(\bar\theta, \theta)$ is evaluated explicitly in the EM algorithm discussed in §4.3 using an optimal smoother.

In summary, two methods have been outlined above for evaluating the gradient and Hessian of the likelihood. They can be used in a general purpose optimization algorithms such as `fmincon` in Matlab.

[4] In more transparent notation, Fisher's identity says that for random variables x, y:

$$\nabla_\theta \log p(y|\theta) = \int [\nabla_\theta \log p(y, x|\theta)]\, p(x|y, \theta) dx = \mathbb{E}\{\nabla_\theta \log p(y, x|\theta)|y, \theta\}.$$

4.2.3 Example 1: MLE of linear Gaussian state space model

Here we describe how a general purpose optimization algorithm can be used to compute the MLE of a linear Gaussian state space model. Consider the highly stylized setting of a scalar auto-regressive Gaussian processes with noisy measurements. With $a = (a_1, \dots, a_M)'$ denoting the auto-regressive coefficients, let $\{s_k\}$ denote an M-th order autoregressive Gaussian process

$$s_k = \sum_{i=1}^{M} a_i s_{k-i} + \bar{w}_k, \qquad \bar{w}_k \sim N(0, \sigma^2) \text{ i.i.d.} \tag{4.11}$$

Suppose $\{s_k\}$ is observed via the noisy scalar-valued measurement process

$$y_k = s_k + v_k, \qquad v_k \sim N(0, R) \text{ i.i.d.} \tag{4.12}$$

Assume that $\{\bar{w}_k\}$, $\{v_k\}$ are independent processes. Define the $M + 2$-dimension parameter vector

$$\theta = (\sigma^2, R, a). \tag{4.13}$$

Aim: Given the sequence of observations $y_{1:N}$ generated by the true model θ^o, compute the MLE θ^*_N. We assume that the dimension of θ^o is known.

Strong consistency of the MLE of linear Gaussian state space models is proved in Chapter 7 of [74] under the conditions that the parameters of the true model lie in a compact set (of Euclidean space) and that the roots of $z^M - a_1 z^{M-1} - \cdots - a_{M-1} z - a_M = 0$ lie inside the unit circle (i.e. the process $\{s_k\}$ is stationary); see [135] for extensive results in linear time series analysis.

Defining the state vector $x_k = [s_k, \dots, s_{k-M}]'$, and $w_k = [\bar{w}_k, 0_{1\times M}]'$, the model (4.11), (4.12) can be rewritten as a linear Gaussian state space model[5]

$$x_{k+1} = Ax_k + w_k, \quad w_k \sim N(0, Q) \tag{4.14}$$

$$y_k = Cx_k + v_k, \quad v_k \sim N(0, R) \tag{4.15}$$

$$A = \begin{bmatrix} a & 0 \\ I_{M\times M} & 0_{M\times 1} \end{bmatrix}, \quad Q = \begin{bmatrix} \sigma^2 & 0_{1\times M} \\ 0_{M\times 1} & 0_{M\times M} \end{bmatrix}, \quad C = [1 \ \ 0_{1\times M}].$$

With (4.14) and (4.15), the model likelihood can be evaluated using the Kalman filter of Algorithm 1 on page 45. The log likelihood $\mathcal{L}_N(\theta)$ is

$$\log p(y_1|\theta) + \sum_{k=2}^{N} \log p(y_k|y_{k-1}, \theta) = \sum_{k=1}^{N} \log N(y_k; C\hat{x}_{k|k-1}, C\Sigma_{k|k-1}C' + R)$$

$$= -\frac{N}{2}\log 2\pi - \frac{1}{2}\sum_{k=1}^{N} \log \left|C\Sigma_{k|k-1}C' + R\right|$$

$$- \frac{1}{2}\sum_{k=1}^{N} (y_k - \hat{y}_{k|k-1})'(C\Sigma_{k|k-1}C' + R)^{-1}(y_k - \hat{y}_{k|k-1}).$$

[5] We have deliberately chosen the state dimension as $M + 1$, when a state dimension of M would have sufficed for an auto-regressive-M process. This augmented state dimension will be exploited in the EM algorithm presented later in this chapter.

The second equality follows from (3.44) for evaluating $p(y_k|y_{k-1},\theta)$. Given the above expression for the likelihood one can use the general purpose Matlab optimizer `fmincon` to compute a local maximum of the MLE. Ensuring that the estimates of a are such that the roots of $z^M - a_1 z^{M-1} - \cdots - a_{M-1}z - a_M = 0$ lie inside the unit circle is nontrivial; see [217].

The gradient and Hessian of the likelihood can be evaluated using either the sensitivity equations or Fisher's identity described in §4.2.2; see [292] for the detailed equations. If Fisher's identity is used, the conditional expectations are computed using a fixed-interval Kalman smoother applied to the state space model of (4.14), (4.15) as will be detailed in M-step of the EM algorithm in §4.3.2.

4.2.4 Example 2: MLE of HMMs

Here we discuss computing the MLE of an HMM using a general purpose optimization algorithm. Let $\mathcal{X} = \{1, 2, \ldots, X\}$ denote the state space of the underlying Markov chain. Recall from §2.4 that an HMM is a noisily observed Markov chain specified by (P, B, C, π_0).

Assume that the observation probability distribution B is parametrized by the finite dimensional vector η. The HMM parameters we wish to estimate are

$$\theta = (P, \eta).$$

For example, if the Markov chain is in additive zero mean Gaussian i.i.d. noise

$$y_k = C(x_k) + v_k, \quad v_k \sim \mathbf{N}(0, R),$$

then $\eta = (C, R)$ and the HMM parameters are $\theta = (P, \eta) = (P, C, R)$.

Aim: Given the sequence of observations $y_{1:N}$ generated by the true model θ^o, compute the MLE θ^*_N. We assume that the dimension of θ^o is known.

Using the un-normalized HMM filter $q_N(i) = p(x_N = i, y_{1:N}|\theta)$ (see (3.59)), the likelihood function for the HMM parameter θ is

$$L_N(\theta) = \sum_{i=1}^{X} q_N(i) = \mathbf{1}' q_N. \tag{4.16}$$

By repeated substitution in the HMM filter (3.57), the normalized (conditional) posterior π_N can be expressed in terms of q_N as

$$\pi_N = \frac{B_{y_N} P' B_{y_{N-1}} P' \cdots B_{y_1} P' \pi_0}{\prod_{k=0}^{N-1} \sigma(\pi_k, y_{k+1})} = \frac{q_N}{\prod_{k=0}^{N-1} \sigma(\pi_k, y_{k+1})}.$$

So the log likelihood is evaluated by running the HMM filter for $k = 1, \ldots, N$:

$$\mathcal{L}_N(\theta) = \log \mathbf{1}' q_N = \log\big(\mathbf{1}' \pi_N \prod_{k=0}^{N-1} \sigma(\pi_k, y_{k+1})\big) = \sum_{k=1}^{N} \log \sigma(\pi_{k-1}, y_k) \tag{4.17}$$

where $\sigma(\pi, y)$ is computed as the denominator (normalization term) in the HMM filter Algorithm 3 on page 51. Using the log likelihood $\mathcal{L}_N(\theta)$ evaluated via (4.17), the Matlab optimization solver `fmincon` can be used with constraints[6] $P'\mathbf{1} = \mathbf{1}$ and $P_{ij} \geq 0$, $i,j \in \{1,\dots,X\}$ to compute the MLE for an HMM.

The gradient and Hessian of the likelihood can be evaluated either via the filter sensitivity equations (4.7) or using (4.8), (4.9). If using (4.8), then the gradient $\nabla_\theta \mathcal{L}_N(\theta)$ is computed using (4.10) where the $\mathcal{Q}$ function is evaluated in (4.26) as described in the EM algorithm of §4.3.3.

4.3 Expectation Maximization (EM) algorithm

This section presents the EM algorithm for computing the MLE of a partially observed stochastic system (2.6), (2.7) of §2.1. Given the sequence of observations $y_{1:N}$ generated by the true model θ^o, the aim is to compute the MLE θ^*_N.

Similar to the general purpose optimization algorithms, the EM algorithm is an iterative hill-climbing algorithm. However, instead of directly working on the log likelihood function, the EM algorithm works on an alternative function called the auxiliary or complete likelihood at each iteration. A useful property of the EM algorithm is that by optimizing this auxiliary likelihood at each iteration, the EM algorithm climbs up the surface of the log likelihood, i.e. each iteration yields a model with a better or equal likelihood compared to the previous iteration; see §4.3.4 for other properties of EM.

4.3.1 EM algorithm

Given an initial parameter estimate $\theta^{(0)}$, the EM algorithm generates a sequence of estimates $\theta^{(n)}$, $n = 1, 2, \dots$ as follows:

Algorithm 6 Expectation Maximization (EM) algorithm

- Initialize $\theta^{(0)} \in \Theta$.
- For iterations $n = 1, 2, \dots,$ perform the following E and M steps
  - *Expectation (E) step*: Evaluate auxiliary log likelihood

$$\mathcal{Q}(\theta^{(n-1)}, \theta) = \mathbb{E}\{\log p(x_{0:N}, y_{1:N}|\theta) \mid y_{1:N}, \theta^{(n-1)}\} \tag{4.18}$$

  - *Maximization (M) step*: Compute updated parameter estimate by maximizing the auxiliary likelihood:

$$\theta^{(n)} = \operatorname*{argmax}_{\theta \in \Theta} \mathcal{Q}(\theta^{(n-1)}, \theta).$$

Some terminology: the underlying state sequence $x_{0:N}$ are called the *latent variables*, the joint state and observation sequence $(x_{0:N}, y_{1:N})$ is called the *complete data*, $p(x_{0:N}, y_{1:N}; \theta)$ is the complete data likelihood. So $\mathcal{Q}(\theta^{(n-1)}, \theta)$ is the conditional expectation of the complete log likelihood with respect to the observations and model estimate

[6] These constraints can be automatically incorporated by using spherical coordinates; see §16.2.1. In the 2-state case, choose $P = \begin{bmatrix} \sin^2\theta_1 & \cos^2\theta_1 \\ \cos^2\theta_2 & \sin^2\theta_2 \end{bmatrix}$, where $\theta_1, \theta_2 \in \mathbb{R}$ are unconstrained.

$\theta^{(n-1)}$. The observation sequence $y_{1:N}$ is called *incomplete data*. Recall $\mathcal{L}_N(\theta)$ defined in (4.2) is the actual model log likelihood corresponding to the incomplete data.

How is the auxiliary log likelihood (4.18) evaluated? Clearly,

$$p(x_{0:N}, y_{1:N}|\theta) = \prod_{k=1}^{N} p(y_k|x_k, \theta)\, p(x_k|x_{k-1}, \theta)\; p(x_0).$$

Since $p(x_0)$ is not a function of θ, it is irrelevant to computing θ^*_N. We denote it as a constant. So the auxiliary log likelihood $Q(\theta^{(n-1)}, \theta)$ in (4.18) is

$$Q(\theta^{(n-1)}, \theta) = \sum_{k=1}^{N} \int \log p(y_k|x_k, \theta)\, p(x_k|y_{1:N}, \theta^{(n-1)}) dx_k \tag{4.19}$$

$$+ \sum_{k=1}^{N} \int\int \log p(x_k, |x_{k-1}, \theta)\, p(x_{k-1}, x_k|y_{1:N}, \theta^{(n-1)}) dx_{k-1} dx_k + \text{ constant.}$$

Computing expressions such as $p(x_k|y_{1:N}, \theta^{(n-1)})$ involves fixed-interval smoothing algorithms described in Chapter 3. In subsequent sections, we will show how to evaluate (4.19) for linear Gaussian models and HMMs.

Why does the EM algorithm work? Theorem 4.3.1 below is the main result.

THEOREM 4.3.1 *EM Algorithm 6 yields a sequence of models $\theta^{(n)}$, $n = 1, 2 \ldots$, with non-decreasing model likelihoods. That is,*

$$\theta^{(n)} = \arg\max_{\theta} Q(\theta^{(n-1)}, \theta) \implies p(y_{1:N}|\theta^{(n)}) \geq p(y_{1:N}|\theta^{(n-1)}).$$

The proof follows from the following lemma.

LEMMA 4.3.2 $Q(\theta^{(n-1)}, \theta) - Q(\theta^{(n-1)}, \theta^{(n-1)}) \leq \log p(y_{1:N}|\theta) - \log p(y_{1:N}|\theta^{(n-1)})$.

Lemma 4.3.2 says that changes in the auxiliary likelihood function (namely $Q(\theta^{(n-1)}, \theta) - Q(\theta^{(n-1)}, \theta^{(n-1)})$) with respect to any choice of θ are smaller than changes in the likelihood function (namely $\log p(y_{1:N}|\theta) - \log p(y_{1:N}|\theta^{(n-1)})$).

From Lemma 4.3.2, Theorem 4.3.1 follows straightforwardly. Suppose we choose $\theta = \theta^{(n)}$ in Lemma 4.3.2 such that

$$Q(\theta^{(n-1)}, \theta^{(n)}) \geq Q(\theta^{(n-1)}, \theta^{(n-1)}). \tag{4.20}$$

Then Lemma 4.3.2 implies that $p(y_{1:N}|\theta^{(n)}) \geq p(y_{1:N}|\theta^{(n-1)})$. That is, any choice of $\theta^{(n)}$ that satisfies (4.20) will result in the likelihood of model $\theta^{(n)}$ being greater or equal to the likelihood of model $\theta^{(n-1)}$. While any choice of $\theta^{(n)}$ such that (4.20) holds will suffice, clearly the best choice is $\theta^{(n)} = \arg\max_\theta Q(\theta^{(n-1)}, \theta)$. Thus Theorem 4.3.1 is proved.

It only remains to prove Lemma 4.3.2.

$$\begin{aligned} Q(\theta^{(n)}, \theta) - Q(\theta^{(n)}, \theta^{(n)}) &= \mathbb{E}\{\log \frac{p(y_{1:N}, x_{0:N}|\theta)}{p(y_{1:N}, x_{0:N}|\theta^{(n)})} | y_{1:N}, \theta^{(n)}\} \\ &\leq \log \mathbb{E}\{\frac{p(y_{1:N}, x_{0:N}|\theta)}{p(y_{1:N}, x_{0:N}|\theta^{(n)})} | y_{1:N}, \theta^{(n)}\} \quad \text{(Jensen's inequality)} \end{aligned}$$

$$
\begin{aligned}
&= \log \int \frac{p(y_{1:N}, x_{0:N}|\theta)}{p(y_{1:N}, x_{0:N}|\theta^{(n)})} p(x_{0:N}|y_{1:N}, \theta^{(n)}) dx_{0:N} \\
&= \log \int \frac{p(y_{1:N}, x_{0:N}|\theta)}{p(x_{0:N}|y_{1:N}, \theta^{(n)}) p(y_{1:N}|\theta^{(n)})} p(x_{0:N}|y_{1:N}, \theta^{(n)}) dx_{0:N} \\
&= \log \int \frac{p(y_{1:N}, x_{0:N}|\theta)}{p(y_{1:N}|\theta^{(n)})} dx_{0:N} = \log \frac{p(y_{1:N}|\theta)}{p(y_{1:N}|\theta^{(n)})}.
\end{aligned}
$$

For a concave function $f(x)$, Jensen's inequality states $\mathbb{E}\{f(x)\} \leq f(\mathbb{E}\{x\})$.

Remark: Just because likelihoods of the model estimates obtained from the EM algorithm are monotone (non-decreasing) does not imply that the sequence of estimates $\{\theta^{(n)}\}$ converges. Convergence of the model estimates require continuity of $\mathcal{Q}$ and compactness of Θ; see [342].

4.3.2 Example 1: EM algorithm for MLE of linear Gaussian models

Here we illustrate how the EM algorithm (Algorithm 6) can be used to compute the MLE of the linear Gaussian state space model considered in §4.2.3:

$$
\begin{aligned}
s_k &= \sum_{i=1}^{M} a_i s_{k-i} + \bar{w}_k, \qquad \bar{w}_k \sim \mathbf{N}(0, \sigma^2) \\
y_k &= s_k + v_k, \qquad v_k \sim \mathbf{N}(0, R).
\end{aligned}
$$

With $a = (a_1, \ldots, a_M)'$, the model is parametrized by $\theta = (\sigma^2, R, a)$.

Aim: Given the sequence of observations $y_{1:N}$ generated by the true model θ^o, compute the MLE θ^*_N. We assume that the dimension of θ^o is known.

Recall the system was written in state space form in (4.14), (4.15) with state

$$
x_k = [s_k, \ldots, s_{k-M}]'.
$$

We now detail the E and M steps in Algorithm 6 to compute the updated model estimate $\theta^{(n+1)}$. Given observations $y_{1:N}$ and current model estimate $\theta^{(n)}$, the Kalman smoother of Algorithm 2 is run on the state space model (4.14), (4.15) to compute the conditional mean and covariance $\hat{x}_{k|N}$, $\Sigma_{k|N}$, $k = 1, \ldots, N$.

E-Step: Evaluating the auxiliary likelihood (4.19) yields

$$
\begin{aligned}
\mathcal{Q}(\theta^{(n)}, \theta) = &-\frac{N}{2} \log \sigma^2 - \frac{1}{2\sigma^2} \sum_{k=1}^{N} \mathbb{E}\{(s_k - \sum_{i=1}^{M} a_i s_{k-i})^2 | y_{1:N}, \theta^{(n)}\} \\
&- \frac{N}{2} \log R - \frac{1}{2R} \sum_{k=1}^{N} \mathbb{E}\{(y_k - s_k)^2 | y_{1:N}, \theta^{(n)}\} + \text{const}(\theta^{(n)}).
\end{aligned} \tag{4.21}
$$

where $\text{const}(\theta^{(n)})$ does not involve θ.

M-step: Recall $\theta = (a, \sigma^2, R)$. Maximizing $\mathcal{Q}(\theta^{(n)}, \theta)$ with respect to a by solving $\partial \mathcal{Q}(\theta^{(n)}, \theta)/\partial a = 0$ yields the M-step update for the auto-regressive parameter a as

$$
a = H^{-1} h. \tag{4.22}
$$

The elements of the $p \times p$ matrix H and $p \times 1$ vector h in terms of the Kalman smoother estimates for $i,j \in \{1,\dots,M\}$ are

$$\begin{aligned} H_{ij} &= \sum_{k=1}^{N} \mathbb{E}\{s_{k-i}s_{k-j}|y_{1:N},\theta^{(n)}\} = \sum_{k=1}^{N} \Sigma_{k|N}(i,j) + \hat{x}_{k|N}(i)\hat{x}_{k|N}(j) \\ h_i &= \sum_{k=1}^{N} \mathbb{E}\{s_k s_{k-i}|y_{1:N},\theta^{(n)}\} = \sum_{k=1}^{N} \Sigma_{k|N}(1,i+1) + \hat{x}_{k|N}(i)\hat{x}_{k|N}(i+1). \end{aligned} \tag{4.23}$$

The above update for h_M involves the $(1, M+1)$ element of the Kalman filter covariance matrix – this is the reason why we formulated (4.14), (4.15) as an $M+1$-dimensional state space model (instead of an M-dimensional one).

Solving for $\partial Q(\theta^{(n)},\theta)/\partial\sigma^2 = 0$ yields the M-step update for the variance

$$\sigma^2 = \frac{1}{N}\sum_{k=1}^{N} \mathbb{E}\{(s_k - \sum_{i=1}^{M} a_i\, s_{k-i})^2|y_{1:N},\theta^{(n)}\}.$$

Writing $\bar{a} = [1, a']$, we can rewrite the above expression as

$$\sigma^2 = \frac{1}{N}\sum_{k=1}^{N} \bar{a}' \left(\Sigma_{k|N} + \hat{x}_{k|N}\hat{x}'_{k|N}\right)\bar{a}. \tag{4.24}$$

in terms of the Kalman smoother estimates. Finally, $\partial Q(\theta^{(n)},\theta)/\partial R = 0$ yields

$$R = \frac{1}{N}\sum_{k=1}^{N} \left((y_k - \hat{x}^2_{k|N}(1))^2 + \Sigma_{k|N}(1,1)\right). \tag{4.25}$$

Setting $\theta^{n+1} = (a, \sigma^2, R)$ computed via (4.22), (4.24), (4.25) completes the M-step.

Discussion

(i) The EM algorithm estimates of the variances σ^2 and R, namely, (4.24) and (4.25), are nonnegative by construction. This is unlike a general purpose optimization algorithm where explicit constraints are required to ensure nonnegativity.

(ii) Consider the special case of a fully observed auto-regressive system where $y_k = s_k$, i.e. the observation noise variance $R = 0$. Then the MLE for a and σ^2 are given by (4.22) and (4.24) with the conditional expectations removed. The intuition is that since for a partially observed model, the second-order moments $\mathbb{E}\{s_{k-i}s_{k-j}\}$ are not known, they are replaced by the conditional mean estimates $\mathbb{E}\{s_{k-i}s_{k-j}|y_{1:n}\}$ in the EM algorithm update.

(iii) Consider the fully observed case where $y_k = s_k$. For large N, by the law of large numbers and asymptotic stationarity of the auto-regressive process, the elements of H have the property $H_{ij} = H_{|i-j|}$ implying that H is a Toeplitz matrix. The resulting system of equations (4.22) are called the Yule–Walker equations in classical signal processing and can be solved in $O(M^2)$ multiplications using the Levinson–Durbin algorithm.

4.3.3 Example 2: EM algorithm for MLE of HMMs

Here we describe the EM algorithm (Algorithm 6) for computing the MLE of the parameters of an HMM considered in §4.2.4. The HMM is parametrized by $\theta = (P, \eta)$ where P denotes the transition matrix and η parametrizes the observation probability distribution B.

Aim: Given the sequence of observations $y_{1:N}$ generated by the true model θ^o, compute the MLE θ^*_N. We assume that the dimension of θ^o is known.

Given observations $y_{1:N}$ and current model estimate $\theta^{(n)}$, we now describe the $(n+1)$-th iteration of EM Algorithm 6 for an HMM:

E-Step: Evaluating the auxiliary log likelihood (4.19) yields

$$
\begin{aligned}
Q(\theta^{(n)}, \theta) &= \mathbb{E}\{\sum_{k=1}^{N}\sum_{i=1}^{X} I(x_k = i)\ \log B_{iy_k}(\eta) \\
&\quad + \sum_{k=1}^{N}\sum_{i=1}^{X}\sum_{j=1}^{X} I(x_{k-1} = i, x_k = j) \log P_{ij} | y_{1:N}, \theta^{(n)}\} + \text{const}(\theta^{(n)}) \\
&= \sum_{k=1}^{N}\sum_{i=1}^{X} \pi_{k|N}(i)\ \log B_{iy_k}(\eta) + \sum_{k=1}^{N}\sum_{i=1}^{X}\sum_{j=1}^{X} \pi_{k|N}(i,j) \log P_{ij} + \text{const.}
\end{aligned} \tag{4.26}
$$

Here, for $i, j \in \{1, 2, \ldots, X\}$,

$$
\pi_{k|N}(i) = \mathbb{P}(x_k = i | y_{1:N}, \theta^{(n)}) \text{ and } \pi_{k|N}(i,j) = \mathbb{P}(x_k = i, x_{k+1} = j | y_{1:N}, \theta^{(n)}).
$$

To evaluate (4.26) given $y_{1:N}$ and $\theta^{(n)}$, the HMM fixed-interval smoother (3.62), (3.63) is run for $k = 1, \ldots, N$ to compute $\pi_{k|N}(i)$ and $\pi_{k|N}(i,j)$.

M-Step: Maximizing $Q(\theta^{(n)}, \theta)$ with respect to P subject to the constraint that P is a stochastic matrix yields (see Footnote 3 on page 74)

$$
P_{ij} = \frac{\sum_{k=1}^{N} \pi_{k|N}(i,j)}{\sum_{k=1}^{N} \pi_{k|N}(i)} = \frac{\mathbb{E}\{J_{ij}(N) | y_{1:k}, \theta^{(n)}\}}{\mathbb{E}\{D_i(N) | y_{1:k}, \theta^{(n)}\}}, \quad i, j \in \{1, 2 \ldots, X\}. \tag{4.27}
$$

Here, $J_{ij}(N) = \sum_{k=1}^{N} I(x_{k-1} = i, x_k = j)$ denotes the number of jumps from state i to state j and $D_i(N) = \sum_{k=1}^{N} I(x_{k-1} = i)$ is the duration time in state i until time N.

The M-step update for η is obtained by maximizing $Q(\theta^{(n)}, \theta)$ with respect to η. We now present two examples of this.

Example (i). HMM with Gaussian noise: Suppose

$$
y_k = C(x_k) + v_k, \quad v_k \sim N(0, R) \text{ i.i.d.} \tag{4.28}
$$

Then $\eta = (C(1), \ldots, C(X), R)$. The first term on the right-hand side of (4.26) is

$$
\sum_{k=1}^{N}\sum_{i=1}^{X} \pi_{k|N}(i)\ \log B_{iy_k}(\eta) = -\frac{N}{2}\log R - \frac{1}{2R}\sum_{k=1}^{N}\sum_{i=1}^{X} \big(y_k - C(i)\big)^2 \pi_{k|N}(i).
$$

Then setting $\partial Q(\theta^{(n)},\theta)/\partial C(i) = 0$ and $\partial Q(\theta^{(n)},\theta)/\partial R = 0$ yields

$$C(i) = \frac{\sum_{k=1}^{N} \pi_{k|N}(i) y_k}{\sum_{k=1}^{N} \pi_{k|N}(i)}, \quad R = \frac{1}{N}\sum_{k=1}^{N}\sum_{i=1}^{X} \pi_{k|N}(i)(y_k - C(i))^2. \tag{4.29}$$

Setting $\theta^{(n+1)} = (P, C, R)$ according to (4.27) and (4.29) completes the M-step for an HMM with Gaussian noise.

Example (ii). HMM with finite observation set: If $y_k \in \{1, \dots, Y\}$, then $\eta = B$ where B is an $X \times Y$ stochastic matrix. Maximizing the first term on the right-hand side of (4.26) subject to the constraint $\sum_{y=1}^{Y} B_{iy} = 1$ yields

$$B_{iy} = \frac{\sum_{k=1}^{N} \pi_{k|N}(i)\, I(y_k = y)}{\sum_{k=1}^{N} \pi_{k|N}(i)}, \quad i \in \{1, 2 \dots, X\}, y \in \{1, 2, \dots, Y\}. \tag{4.30}$$

Setting $\theta^{(n+1)} = (P, B)$ according to (4.27) and (4.30) completes the M-step for an HMM with finite observation set.

The EM algorithm for HMMs is also called the Baum–Welch algorithm.

Discussion

(i) The EM update for the transition matrix P in (4.27) yields a valid transition matrix by construction. Also the variance R in (4.29) is nonnegative by construction. It is instructive to compare the M-step update (4.27) of an HMM with the maximum likelihood estimate (4.4) for a fully observed Markov chain. Intuitively, since for an HMM the number of jumps and duration time are not known exactly, the M-step replaces these with their conditional mean estimates.

(ii) The EM algorithm applies, of course, to other types of observation models. Consider a sinusoid disturbance added to the Markov chain in Gaussian noise:

$$y_k = Cx_k + v_k + A\sin(\omega k + \phi), \quad v_k \sim N(0, R) \text{ i.i.d.}$$

The sinusoid models narrowband noise, while v_k is wideband noise. The EM algorithm can be used to estimate the parameter $\theta = (P, C, R, A, \omega, \phi)$.

Another example is ML parameter estimation of a Markov modulated auto-regressive process considered in §3.6.1. The EM algorithm can be used to estimate the transition matrix and Markov modulated AR parameters.

4.3.4 Discussion: general purpose optimization algorithms vs EM

Both EM and general purpose optimization algorithms such as Newton-type methods are hill-climbing algorithms. At best they converge to a local stationary point of the likelihood function. So one needs to initialize and run these algorithms from several starting points.

A useful property of the EM algorithm is that successive iterations generate model estimates with non-decreasing likelihood values. In contrast, practical implementations of general purpose optimization algorithms such as Newton methods require careful choice of the step size.

As shown in the two examples above, the EM algorithm yields valid transition matrices and nonnegative variances by construction. In the general purpose optimization algorithms, additional constraints need to be introduced.

The advantage of a general purpose optimization algorithm such as the Newton methods compared to the EM algorithm is the superior convergence rate. Methods for improving the convergence rate of EM have been studied extensively in the literature; see §4.6. Another advantage of a general purpose optimization algorithm is that its implementation code has been tested carefully (in terms of numerical conditioning).

4.4 Forward-only filter-based EM algorithms

The E-step of the EM algorithm in §4.3 involves a forward-backward smoother. This section shows how the E-step can be implemented by using filters instead of smoothers. So the E-step can be implemented going forwards only.

From an abstract point of view, the E-step in §4.3 requires computing conditional mean estimates of additive *functionals* of the state of the form $S(x_{0:N}) = \sum_{n=0}^{N} \phi(x_n)$ where $\phi(\cdot)$ is some pre-specified function of the state. These estimates were computed in §4.3 by forward-backward fixed-interval smoothing:

$$\mathbb{E}\{\sum_{n=0}^{N} \phi_n(x_n)|y_{1:N}\} = \sum_{n=0}^{N} \mathbb{E}\{\phi(x_n)|y_{1:N}\} = \sum_{n=0}^{N} \int_{\mathcal{X}} \phi(x_n)p(x_n|y_{1:N})dx_n. \tag{4.31}$$

In comparison, this section derives *filtering* algorithms to compute (4.31).

4.4.1 Non-universal filtering

Given the observations sequence $y_{1:k}$, the aim is to compute recursively via a filtering algorithm

$$\mathbb{E}\{S(x_{0:k})|y_{1:k}\} = \mathbb{E}\{\sum_{n=0}^{k} \phi(x_n)|y_{1:k}\}, \quad k = 1, 2, \ldots. \tag{4.32}$$

At final time instant N this filtered estimate coincides with the smoother-based E-step of (4.31). To give some perspective on (4.32), note that the conditional mean of a *function* $\phi(x_k)$, namely $\mathbb{E}\{\phi(x_k)|y_{1:k}\}$, can be computed from the optimal filter as

$$\mathbb{E}\{\phi(x_k)|y_{1:k}\} = \int_{\mathcal{X}} \phi(x)p(x_k|y_{1:k})dx_k.$$

In comparison, our aim in (4.32) is to estimate at each time k the *functional* $\mathbb{E}\{S(x_{0:k})|y_{1:k}\}$ which contains a running sum from time 0 to k.

The methodology in this section is different to Chapter 3 where we first obtained the posterior distribution and then computed the filtered estimate by integration of this posterior distribution. Unlike Chapter 3, for functionals $S(x_{0:k}) = \sum_{n=0}^{k} \phi(x_n)$, in general it is not possible to compute the posterior distribution $p(S(x_{0:k})|y_{1:k})$, $k = 1, 2, \ldots$, in terms of a finite dimensional filter. Instead, for specific type of functions ϕ, we will obtain a finite dimensional recursion for the conditional mean $\mathbb{E}\{S(x_{0:k})|y_{1:k}\}$ directly. Such filters are called *non-universal filters* since they only hold for specific functions $\phi(\cdot)$.

Recall the un-normalized filtered density $q_k(x) = p(x_k = x, y_{1:k})$ defined in §3.3.2. The main result is as follows:

THEOREM 4.4.1 *Consider the filtering problem (4.32). Then*

$$\mathbb{E}\{S(x_{0:k})|y_{1:k}\} = \frac{\int_{\mathcal{X}} \mu_k(x)dx}{\int_{\mathcal{X}} q_k(x)dx}. \tag{4.33}$$

Here the measure-valued density μ_k and un-normalized filtered density q_k are updated via the recursions

$$\mu_k(x_k) = \phi(x_k)\, q_k(x_k) + p(y_k|x_k) \int_{\mathcal{X}} p(x_k|x_{k-1})\, \mu_{k-1}(x_{k-1})\, dx_{k-1} \tag{4.34}$$

$$q_k(x_k) = p(y_k|x_k) \int_{\mathcal{X}} p(x_k|x_{k-1})\, q_{k-1}(x_{k-1})\, dx_{k-1}. \tag{4.35}$$

Proof Let $\int_{\mathcal{X}^k}$ denote the integral $\int_{\mathcal{X}}$ repeated k times. By definition,

$$\begin{aligned}\mathbb{E}\{S(x_{0:k})|y_{1:k}\} &= \int_{\mathcal{X}^{k+1}} S(x_{0:k})\, p(x_{0:k}|y_{1:k})\, dx_{0:k} \\ &= \frac{\int_{\mathcal{X}} \int_{\mathcal{X}^k} S(x_{0:k-1}, x_k)\, p(x_{0:k-1}, x_k, y_{1:k})\, dx_{0:k-1} dx_k}{\int_{\mathcal{X}} p(x_k, y_{1:k}) dx_k}.\end{aligned} \tag{4.36}$$

Clearly the denominator is $\int_{\mathcal{X}} q_k(x)dx$ where q_k satisfies the recursion (4.35) as discussed in §3.3.2. So we focus on the numerator of (4.36). Define

$$\mu_k(x_k) \stackrel{\text{defn}}{=} \int_{\mathcal{X}^k} S(x_{0:k-1}, x_k)\, p(x_{0:k-1}, x_k, y_{1:k})\, dx_{0:k-1}. \tag{4.37}$$

Then (4.36) is equivalent to the right-hand side of (4.33) providing μ_k defined in (4.37) satisfies the recursion (4.34). So we only need to show that μ_k defined in (4.37) satisfies the recursion (4.34). Now

$$\mu_k(x_k) = \int_{\mathcal{X}^k} \big[\phi(x_k) + S(x_{0:k-1})\big] p(y_k|x_k) p(x_k|x_{k-1}) p(x_{0:k-1}, y_{1:k-1}) dx_{0:k-1} \tag{4.38}$$

$$\begin{aligned}&= \int_{\mathcal{X}} \int_{\mathcal{X}^{k-1}} \phi(x_k) p(y_k|x_k) p(x_k|x_{k-1}) p(x_{0:k-2}, x_{k-1}, y_{1:k-1}) dx_{0:k-2} dx_{k-1} \\ &+ \int_{\mathcal{X}} \int_{\mathcal{X}^{k-1}} S(x_{0:k-2}, x_{k-1}) p(y_k|x_k) p(x_k|x_{k-1}) p(x_{0:k-2}, x_{k-1}, y_{1:k-1}) dx_{0:k-2} dx_{k-1}.\end{aligned} \tag{4.39}$$

Clearly, (4.38) is equivalent to $\phi(x_k) q_k(x_k)$. Also, (4.39) is equivalent to the second term of the right-hand side of (4.34) since by definition

$$\mu_{k-1}(x_{k-1}) = \int_{\mathcal{X}^{k-1}} S(x_{0:k-2}, x_{k-1}) p(x_{0:k-2}, x_{k-1}, y_{1:k-1}) dx_{0:k-2}.$$

□

Apart from the above result, to implement a filtered E-step, we also need to compute filtered estimates of functionals of the form $S(x_{0:k}) = \sum_{n=0}^{k-1} \phi(x_n, x_{n+1})$. That is, we wish to compute

$$\mathbb{E}\{S(x_{0:k})|y_{1:k}\} = \mathbb{E}\{\sum_{n=0}^{k-1} \phi(x_n, x_{n+1})|y_{1:k}\}, \quad k = 1, 2, \ldots. \tag{4.40}$$

By virtually an identical proof to Theorem 4.4.1, the following result holds.

COROLLARY 4.4.2 *Consider the filtering problem (4.40). Then*

$$\mathbb{E}\{S(x_{0:k})|y_{1:k}\} = \frac{\int_{\mathcal{X}} \int_{\mathcal{X}} \mu_k(x_{k-1}, x_k) dx_{k-1} dx_k}{\int_{\mathcal{X}} q_k(x_k) dx_k} \quad \textit{where}$$

$$\mu_k(x_{k-1}, x_k) = \phi(x_{k-1}, x_k) q_k(x_{k-1}, x_k) + \int p(y_k|x_k) p(x_k|x_{k-1}) \mu_{k-1}(x_{k-1}, x_{k-2}) dx_{k-2}$$

$$q_k(x_{k-1}, x_k) = p(y_k|x_k) p(x_k|x_{k-1}) q_{k-1}(x_{k-1})$$

$$q_k(x_k) = p(y_k|x_k) \int_{\mathcal{X}} p(x_k|x_{k-1})\, q_{k-1}(x_{k-1})\, dx_{k-1}.$$

4.4.2 Forward-only Expectation-step for HMM

This section shows how the filters for functionals of the state in Theorem 4.4.1 and Corollary 4.4.2, can be used to implement a forward-only (filtered) E-step for an HMM with Gaussian observation noise (4.28). (In comparison, §4.3.3 gave forward-backward HMM fixed-interval smoothers for the E-step.)

Occupation time: The occupation time in any state m until time k is

$$D_m(k) = \sum_{n=0}^{k} I(x_n = m), \quad m \in \{1, 2, \ldots, X\}.$$

Applying Theorem 4.4.1 with $\phi(x_n) = I(x_n = m)$ yields the following filtered estimate for the occupation time in state $m \in \mathcal{X}$:

$$\mathbb{E}\{D_m(k)|y_{1:k}\} = \frac{\sum_i \mu_k(i)}{\sum_i q_k(i)} \quad \text{where}$$

$$\mu_k = \operatorname{diag}(e_m) q_k + B_{y_k} P' \mu_{k-1}$$

$$q_k = B_{y_k} P' q_{k-1}. \tag{4.41}$$

Recall the above recursion for q_k is the un-normalized HMM filter (3.59).

Weighted occupation time: In similar vein, filters for functionals

$$W_m(k) = \sum_{n=0}^{k} I(x_n = m) y_n \text{ and } V_m(k) = \sum_{n=0}^{k} I(x_n = m)(y_n - C_m)^2$$

are obtained. Indeed, they are identical to (4.41) with $\operatorname{diag}(e_m)$ replaced by $\operatorname{diag}(e_m)\, y_n$ and $\operatorname{diag}(e_m)\,(y_n - C_m)^2$, respectively.

Jumps: Recall from §4.1.1 that for any two states l and m, the number of jumps from state l to m until time k is

$$J_{lm}(k) = \sum_{n=0}^{k-1} I(x_n = l, x_{n+1} = m), \quad l, m \in \{1, 2, \dots, X\}.$$

Applying Corollary 4.4.2 with $\phi(x_n, x_{n+1}) = I(x_n = l, x_{n+1} = m)$ yields the filtered estimate for the number of jumps until time k from state m to n:

$$\mathbb{E}\{J_{lm}(k)|y_{1:k}\} = \frac{\sum_i \sum_j \mu_k(i,j)}{\sum_i q_k(i)} \quad \text{where}$$

$$\mu_k(i,j) = I(i = l, j = m) q_k(i,j) + B_{j,y_k} P_{ij} \sum_{h=1}^{X} \mu_{k-1}(h,i)$$

$$q_k(i,j) = B_{j,y_k} P_{ij} q_{k-1}(i), \quad q_k(i) = B_{j,y_k} \sum_{i=1}^{X} P_{ij} q_{k-1}(i).$$

Forward EM equations for HMM: With the above filters, we have all the quantities required for the EM algorithm to estimate the parameters of an HMM comprising of a Markov chain in Gaussian noise. Running the above filters from time $k = 1, \dots, N$, the M-step update (4.27), (4.29) of the EM algorithm reads: For $i, j \in \{1, 2, \dots, X\}$,

$$\begin{aligned} P_{ij} &= \frac{\mathbb{E}\{J_{ij}(N)|y_{1:N}, \theta^{(n)}\}}{\mathbb{E}\{D_i(N)|y_{1:N}, \theta^{(n)}\}}, \\ C(i) &= \frac{\mathbb{E}\{W_i(N)|y_{1:N}, \theta^{(n)}\}}{\mathbb{E}\{D_i(N)|y_{1:N}, \theta^{(n)}\}}, \quad R = \frac{1}{N} \sum_{i=1}^{X} \mathbb{E}\{V_i(N)|y_{1:N}, \theta^{(n)}\}. \end{aligned} \tag{4.42}$$

Notice that the above terms are evaluated via (forward) filters.

4.5 Method of moments estimator for HMMs

This chapter has focused on maximum likelihood estimation. Here, motivated by recent work in spectral methods for HMM estimation [151], we briefly summarize a method of moments estimator for estimating the transition matrix assuming that the HMM observation space $\mathcal{Y} = \{1, 2, \dots, Y\}$ is finite. An advantage is that if one wishes to estimate the transition probabilities P (assuming the observation probabilities B are known) then the problem can be formulated as convex optimization problem implying a unique solution.

Assume P is a regular stochastic matrix (see Definition 2.4.1 in §2.4.3) with unique stationary distribution π_∞. Define the $Y \times Y$ matrix S with elements $S_{y,y'} = \mathbb{P}(y_{k-1} = y, y_k = y')$, $y, y' \in \{1, 2, \dots, Y\}$. In terms of the HMM parameters, S can be expressed as

$$S = B' P' \text{diag}(\pi_\infty) B$$

where B is the $X \times Y$ matrix with elements $B_{xy} = \mathbb{P}(y_k = y | x_k = x)$ and π_∞ is the stationary distribution of P. Also S can be estimated straightforwardly from the HMM observations $y_{1:N}$ via the following normalized counter:

$$\hat{S}_{yy'} = \frac{1}{N-1} \sum_{k=1}^{N-1} I(y_k = y, y_{k+1} = y'), \quad y, y' \in \{1, 2, \ldots, Y\}.$$

Denoting $A = P'\text{diag}(\pi_\infty)$, one can formulate estimation of A as equivalent to solving the convex optimization problem:

$$\begin{aligned} &\text{Compute } \hat{A} = \underset{A}{\text{argmin}} \, \|\hat{S} - B'AB\|_F \\ &\text{subject to } A \geq 0 \text{ (nonnegative matrix)}, \quad \mathbf{1}'A\mathbf{1} = 1 \end{aligned} \tag{4.43}$$

where $\|\cdot\|_F$ denotes the Frobenius norm. Once $\hat{A}$ is computed, then the $\hat{\pi}_\infty = \hat{A}'\mathbf{1}$ and the transition matrix is estimated as $\hat{P} = \hat{A}\,\text{diag}^{-1}(\hat{\pi}_\infty)$. Notice also that evaluating the optimum of (4.43) does not require optimal filtering.

4.6 Complements and sources

Most of the algorithms given in this chapter are considered to be classical today. [274] is a highly influential tutorial paper on HMMs. The Fisher and Louis identities are discussed in [94, p.29] and [222], respectively. The seminal paper [94] formulated the general theme of EM algorithms; see also [236]. The Baum–Welch algorithm [38, 39] is simply the EM algorithm applied to an HMM. The convergence of EM algorithms is studied in [342], which also corrects an error in the convergence proof of [94]. [236] is devoted to EM algorithms. [315, 212] presents stochastic EM algorithms and data augmentation algorithms. Forward-only EM algorithms for HMMs where the E-step is implemented using filters (instead of smoothers) were pioneered by Elliott [104]. These were generalized to forward-only EM algorithms for Gaussian state space models in [105, 106]. Forward-backward (smoother-based) EM algorithms for Gaussian state space models go back to [299, 124]. The EM abstraction serves as a useful interpretation of turbo-coding in communication systems [254]. The EM algorithm is used in natural language processing for estimating the probabilities of production rules associated with stochastic context-free grammars [231].

This chapter assumes that the model dimension is known. When the model dimension is not known, penalized likelihood methods can be used; see [123] for the HMM case. A big omission in this chapter is the discussion of Cramér–Rao bounds and asymptotic efficiency of the MLE; for the HMM case see [76].

4.6.1 Variants of EM

There are numerous variants of the EM algorithm such as the space-alternating generalized Expectation-Maximization (SAGE) algorithm [115] and the Expectation Conditional Maximization Either (ECME) algorithm [211, 239]. ECME updates some parameters using the EM algorithm and others by direct maximization of the likelihood function via the Newton–Raphson algorithm.

Markov chain Monte Carlo EM (MCEM) algorithms constitute another class of EM-based algorithms. Recall that the E-step of the EM algorithm evaluates

$$Q(\theta^{(n)},\theta) = \int \log p(y_{1:N}, x_{0:N}|\theta) p(x_{0:N}|y_{1:N},\theta^{(n)}) dx_{0:N}.$$

When the E-step is difficult to implement exactly, $Q(\theta^{(n)},\theta)$ can be estimated via Monte Carlo simulation. This is the basis of the Markov chain EM (MCEM) algorithm. In the simplest setting, MCEM simulates i.i.d. samples $x_{0:N}^{(m,n)} \sim p(x_{0:N}|y_{1:N},\theta^{(n)})$, for $m = 1,2,\dots,M$. Then $Q(\theta^{(n)},\theta)$ is estimated as

$$\hat{Q}(\theta^{(n)},\theta) = \frac{1}{M}\sum_{m=1}^{M} \log p(y_{1:N}, x_{0:N}^{(m,n)}|\theta).$$

More sophisticated implementations of the MCEM algorithm obtain the samples $\{x_{0:N}^{(m,n)}, m = 1,\dots,M\}$ from a Markov chain Monte Carlo method such as Gibbs sampling or the Metropolis–Hastings algorithm with stationary distribution $p(x_{0:N}|y_{1:N},\theta^{(n)})$; see [62] and [206].

4.6.2 Consistency of the MLE

This chapter has described how to compute the MLE θ_N^* given a batch of data $y_{1:N} = (y_1,\dots,y_N)$ generated by the true model θ^o. Here we briefly discuss consistency of the MLE. The MLE is said to be strongly consistent if $\theta_N^* \to \theta^o$ with probability one as $N \to \infty$.

Consistency of MLE for HMM

We say that two HMMs θ and $\bar{\theta}$ are equivalent if their likelihoods satisfy $L_N(\theta) = L_N(\bar{\theta})$ for any N and observation sequence $y_{1:N}$. It is important to keep in mind that, for an HMM, any permutation of the state indices results in an equivalent model. For example, the following two HMMs are permutation equivalent:

$$\left(\begin{bmatrix} P_{11} & P_{12} \\ P_{21} & P_{22} \end{bmatrix}, \begin{bmatrix} B_{11} & B_{12} \\ B_{21} & B_{22} \end{bmatrix}\right) \equiv \left(\begin{bmatrix} P_{22} & P_{21} \\ P_{12} & P_{11} \end{bmatrix}, \begin{bmatrix} B_{22} & B_{21} \\ B_{12} & B_{11} \end{bmatrix}\right).$$

So consistency of the MLE for an HMM means that the MLE converges with probability one to any of the permutations of the true model.

Consider an HMM with parameter vector $\theta \in \Theta$ where Θ is a compact subset of Euclidean space. Assume the true model $\theta^o \in \Theta$ and the dimension of θ^o is known. For a practitioner, the following conditions are sufficient for strong consistency of the MLE.

(L1) The true transition matrix $P(\theta^o)$ is aperiodic and irreducible.

(L2) The HMM is identifiable. For any two probability vectors π and $\bar{\pi}$,

$$\sum_{i=1}^{X} \pi_i B_{iy}(\theta) = \sum_{i=1}^{X} \bar{\pi}_i B_{iy}(\bar{\theta}) \implies (\pi,\theta) \text{ and } (\bar{\pi},\bar{\theta}) \text{ are permutation equivalent.}$$

This identifiability property holds for observation probability distributions such as Gaussians, exponentials and Poisson [205].

(L3) The true observation probabilities satisfy $\sum_i B_{iy}(\theta^o) > 0$ for each y.
(L4) The parameterized transition matrix $P(\theta)$ is continuous in θ.

THEOREM 4.6.1 *If L1–L4 hold for an HMM, then the MLE is strongly consistent if either one of the following hold:*

(L5a) The observation space $\mathcal{Y}$ is finite. This is proved in [38].
(L5b) For each i, the observation density $B_{iy}(\theta)$ is positive, continuous and bounded and has finite second moment for each $\theta \in \Theta$. This is proved in [97] for more general HMMs and Markov modulated time series.

The above conditions hold for HMMs in Gaussian noise where $B_{iy} \sim N(\mu_i, \sigma_i^2)$ and σ_i^2 is finite and positive.

Outline of consistency proof

We now give some insight into how consistency of the MLE is proved. The proof dates back to Wald [332]. Define the normalized log likelihood as

$$l_N(\theta) = \frac{1}{N} \log p(y_{1:N}|\theta). \tag{4.44}$$

Proving the consistency of a MLE involves the following four steps:
Step 1. Show that the following *uniform* law of large numbers result holds:

$$\sup_{\theta \in \Theta} l_N(\theta) \to \sup_{\theta \in \Theta} l(\theta, \theta^o) \text{ with probability one as } N \to \infty. \tag{4.45}$$

Here $l(\theta, \theta^o)$ is a deterministic function of θ and the true model θ^o. Recall that θ^o generates the data $y_{1:N}$.

To get some insight into Step 1, if Θ is a finite set, then it suffices for the law of large numbers to hold for $l_N(\theta)$ for each $\theta \in \Theta$. When Θ is a compact subset of Euclidean space, then we can approximate Θ by a finite grid and impose that $l_N(\theta)$ is well behaved between the grid points. The key concept ensuring this well-behaved property is *stochastic equicontinuity* of the sequence of random functions $\{l_N(\theta)\}$ for $\theta \in \Theta$ – this implies that the limit $l(\theta)$ is continuous, thereby establishing uniform, almost sure convergence; see [270].

To carry out Step 1 for an HMM, the normalized log likelihood can be expressed as the following arithmetic mean in terms of the HMM prediction filter:

$$l_N(\theta) = \frac{1}{N} \sum_{k=1}^{N} \log p(y_k|y_{1:k-1}, \theta) = \frac{1}{N} \sum_{k=1}^{N} \log \left[\mathbf{1}' B_{y_k}(\theta) \pi_{k|k-1}^{\theta} \right], \quad \pi_{0|-1} = \pi_0. \tag{4.46}$$

From §3.5, $\pi_{k|k-1}^{\theta} = (\mathbb{P}_\theta(x_k = i|y_{1:k-1}), i = 1, \ldots, X)$ is computed as

$$\pi_{k+1|k}^{\theta} = \frac{P'(\theta) B_{y_k}(\theta) \pi_{k|k-1}^{\theta}}{\mathbf{1}' B_{y_k}(\theta) \pi_{k|k-1}^{\theta}}.$$

Define the Markov process $z_k = \{x_k, y_k, \pi_{k|k-1}^{\theta}\}$. Conditions (L1) and (L2) imply that (3.71), (3.72) hold and so z_k is geometrically ergodic as in §3.7.

Next assume that z_0 starts at its invariant distribution. This is done by the following trick. Assume $\{x_k, y_k\}$ runs from time $-\infty$ so that when it reaches time 0, z_k it is stationary. Due to the geometric ergodicity, it follows that as $N \to \infty$, the difference between evaluating (4.46) starting with initial condition π_0 vs initial condition $\mathbb{P}(x_0|y_{-m:-1})$ vanishes for any m. Therefore as $N \to \infty$,

$$l_N(\theta) = \frac{1}{N}\sum_{k=1}^{N} \log p(y_k|y_{-m:k-1}, x_{-m}, \theta), \quad m > 0.$$

By considering the infinite past, i.e. $m \to \infty$, $p(y_k|y_{-m:k-1}, x_{-m}, \theta)$ is a stationary sequence. Thus the strong law of large numbers applies and so, as $N \to \infty$,

$$l_N(\theta) \to l(\theta, \theta^o) = \lim_{m\to\infty} \mathbb{E}^o_\theta\{\log p(y_k|y_{-m:k-1}, x_{-m}, \theta)\} \text{ w.p.1.} \tag{4.47}$$

Next, (L2) and (L5b) are sufficient for uniform convergence of (4.47), i.e. for (4.45) to hold. So Step 1 of the consistency proof holds. (See [204, 97] for details; [205] used the sub-additive ergodic theorem to show consistency.)

Step 2. Show that $l(\theta, \theta^o)$ is maximized at $\theta = \theta^o$. This is shown as follows:

$$\begin{aligned} l(\theta, \theta^o) - l(\theta^o, \theta^o) &= \lim_{m\to\infty} \mathbb{E}_{\theta^o}\{\log \frac{p(y_k|y_{-m:k-1}, x_{-m}, \theta)}{p(y_k|y_{-m:k-1}, x_{-m}, \theta^o)}\} \\ &= \lim_{m\to\infty} \mathbb{E}_{\theta^o}\left\{\mathbb{E}_{\theta^o}\{\log \frac{p(y_k|y_{-m:k-1}, x_{-m}, \theta)}{p(y_k|y_{-m:k-1}, x_{-m}, \theta^o)}|y_{-m:k-1}, x_{-m}, \theta^o\}\right\}. \end{aligned} \tag{4.48}$$

The inner conditional expectation is the Kullback–Leibler (KL) divergence and is maximized[7] at $\theta = \theta^o$. Thus the overall expression is maximized at $\theta = \theta^o$.

Step 3. Show that θ^o is the unique maximizer of $l(\theta, \theta^o)$. Establishing such unique identifiability is often challenging since it can be difficult to characterize the structure of the function l explicitly.

For the HMM case, (L2) is sufficient for unique identifiability.

Step 4. It follows from Step 1 that

$$\operatorname*{argmax}_{\theta\in\Theta} l_N(\theta) \to \operatorname*{argmax}_{\theta\in\Theta} l(\theta, \theta^o) \text{ with probability one as } N \to \infty.$$

The LHS is the MLE θ^* while the RHS from Steps 2 and 3 is θ^o. Hence $\theta^*_N \to \theta^o$ with probability one thereby establishing strong consistency of the MLE.

[7] This follows since, in simplified notation, $\mathbb{E}_{\theta^o}\{\log \frac{p(y|\theta)}{p(y|\theta^o)}\} \leq \log \mathbb{E}_{\theta^o}\{\frac{p(y|\theta)}{p(y|\theta^o)}\} = 0$. Here the inequality follows from Jensen's inequality, namely $\mathbb{E}\{\phi(x)\} \leq \phi(\mathbb{E}\{x\})$ for concave function ϕ.

5 Multi-agent sensing: social learning and data incest

Contents

This chapter discusses multi-agent filtering over a graph (social network). We consider online reputation and polling systems where individuals make recommendations based on their private observations and recommendations of friends. Such interaction of individuals and their social influence is modeled as Bayesian social learning on a directed acyclic graph. Data incest (misinformation propagation) occurs due to unintentional reuse of identical actions in the formation of public belief in social learning; the information gathered by each agent is mistakenly considered to be independent. This results in overconfidence and bias in estimates of the state. Necessary and sufficient conditions are given on the structure of information exchange graph to mitigate data incest. Incest removal algorithms are presented. The incest removal algorithms are illustrated in an expectation polling system where participants in a poll respond with a summary of their friends' beliefs.

This chapter differs from previous chapters in that we focus on the structure of the Bayesian update and information structures rather than numerical computation.

5.1 Social sensing

We use the term *social sensor* or *human-based sensor* to denote an agent that provides information about its environment (state of nature) on a social network after interaction with other agents. Examples of such social sensors include Twitter posts,

Facebook status updates and ratings on online reputation systems like Yelp and TripAdvisor. Online reputation systems (Yelp, TripAdvisor) can be viewed as sensors of social opinion – they go beyond physical sensors since user opinions/ratings (such as human evaluation of a restaurant or movie) are impossible to measure via physical sensors.

Devising a fair online reputation system involves constructing a data fusion system that combines estimates of individuals to generate an unbiased estimate. This presents unique challenges from a statistical signal processing and data fusion point of view. First, humans interact with and influence other humans since ratings posted on online reputation systems strongly influence the behavior of individuals. (For example, a one-star increase in the Yelp rating maps to five- to nine-percent revenue increase [228].) This can result in correlations introduced by the structure of the underlying social network. Second, due to privacy concerns, humans rarely reveal raw observations of the underlying state of nature. Instead, they reveal their decisions (ratings, recommendations, votes) which can be viewed as a low resolution (quantized) function of their belief.

In this chapter, *social learning* [29, 56, 84] serves as a useful mathematical abstraction for modeling the interaction of social sensors. Social learning in multi-agent systems deals with the following question: how do decisions made by agents affect decisions made by subsequent agents? Each agent chooses its action by optimizing its local utility function. Subsequent agents then use their private observations together with the actions of previous agents to estimate (learn) an underlying state. The setup is fundamentally different from classical signal processing in which sensors use noisy observations to compute estimates. In social learning, agents use noisy observations together with decisions made by previous agents, to estimate the underlying state of nature. A key result in social learning of a random variable is that agents herd [56]; that is, they eventually end up choosing the same action irrespective of their private observations.

Social learning has been used widely in economics, marketing, political science and sociology to model the behavior of financial markets, crowds, social groups and social networks; see [29, 56, 3, 84, 219] and numerous references therein. Related models have been studied in the context of sequential decision making in information theory [89, 142] and statistical signal processing [85, 188]. Seminal books in social networks include [156, 329].

Motivation: Consider the following example of a multi-agent system where agents seek to estimate an underlying state of nature. An agent visits a restaurant and obtains a noisy private measurement of the state (quality of food). She then rates the restaurant as excellent on an online reputation website. Another agent is influenced by this rating, visits the restaurant, and also gives a good rating on the online reputation website. The first agent visits the reputation site and notices that another agent has also given the restaurant a good rating – this double confirms her rating and she enters another good rating. In a fair reputation system, such "double counting" or "data incest" should have been prevented by making the first agent aware that the rating of the second agent was influenced by her own rating. The information exchange between the agents is represented by the directed graph of Figure 5.1. The fact that there are two distinct paths (denoted in red) between Agent 1 at time 1 and Agent 1 at time 3 implies that the information

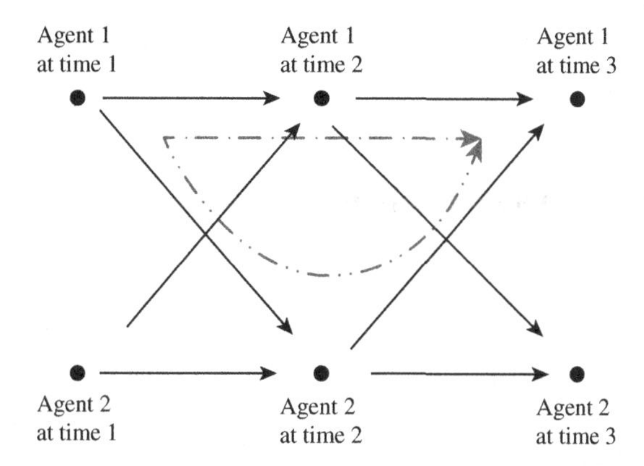

Figure 5.1 Example of the information flow in a social network with two agents and over three event epochs. The arrows represent exchange of information.

of Agent 1 at time 1 is double counted thereby leading to a data incest event. Such data incest results in a bias in the estimate of the underlying state.

5.1.1 Main results and organization

This chapter consists of three parts. The first part, namely §5.2 and §5.3, deals with multi-agent social learning. The social learning Bayesian filter and insight into herding and formation of information cascades is provided.

The second part (§5.4 and §5.5) deals with the design of a fair online reputation system. We formulate the data incest problem in a multi-agent system where individual agents perform social learning and exchange information on a sequence of directed acyclic graphs. The aim is to develop a distributed data fusion protocol which incorporates social influence constraints and provides an unbiased estimate of the state of nature at each node.

The third part of the chapter (§5.6) describes how the data incest problem formulation and incest removal algorithms can be applied to an expectation polling system. Polls seek to estimate the fraction of a population that support a political party, executive decision, etc. In *intent* polling, individuals are sampled and asked who they intend to vote for. In *expectation* polling [285] individuals are sampled and asked who they believe will win the election. It is intuitive that expectation polling is more accurate than intent polling; since in expectation polling an individual considers its own intent together with the intents of its friends.[1] However, the variance and bias of the estimate depend on the social network structure, and data incest can strongly bias the estimate. We illustrate

[1] [285] analyzes all US presidential electoral college results from 1952–2008 where both intention and expectation polling were conducted and shows a remarkable result: In 77 cases where expectation and intent polling pointed to different winners, expectation polling was accurate 78 percent of the time. The dataset from the American National Election Studies comprised of voter responses to two questions:

Intention: Who will you vote for in the election for President?
Expectation: Who do you think will be elected President in November?

how the posterior distribution of the leading candidate in the poll can be estimated based on incestuous estimates.

5.2 Multi-agent social learning

5.2.1 Motivation: what is social learning?

In social learning, agents estimate the underlying state of nature not only from their local measurements, but also from the actions of previous agents.

Consider a countable number of agents performing social learning to estimate the state of an underlying finite state Markov chain x. Let $\mathcal{X} = \{1, 2, \dots, X\}$ denote a finite state space, P the transition matrix and π_0 the initial distribution of the Markov chain.

Each agent acts once in a predetermined sequential order indexed by $k = 1, 2, \dots$ The index k can also be viewed as the discrete-time instant when agent k acts. A multi-agent system seeks to estimate x_0. Assume at the beginning of iteration k, all agents have access to the public belief π_{k-1} defined in Step (iv) below. The social learning protocol proceeds as follows:

(i) *Private Observation*: At time k, agent k records a private observation $y_k \in \mathbf{Y}$ from the observation distribution $B_{iy} = \mathbb{P}(y|x = i)$, $i \in \mathcal{X}$. Throughout this section we assume that $\mathbf{Y} = \{1, 2, \dots, Y\}$ is finite.

(ii) *Private Belief*: Using the public belief π_{k-1} available at time $k - 1$ (defined in Step (iv) below), agent k updates its private posterior belief $\eta_k(i) = \mathbb{P}(x_k = i|a_1, \dots, a_{k-1}, y_k)$ via the HMM filter:

$$\eta_k = \frac{B_{y_k} P' \pi_{k-1}}{\mathbf{1}' B_y P' \pi_{k-1}}, \quad B_{y_k} = \operatorname{diag}\big(\mathbb{P}(y_k|x = 1), \cdots, \mathbb{P}(y_k|x = X)\big). \tag{5.1}$$

Here $\mathbf{1}$ denotes the X-dimensional vector of ones, η_k is an X-dimensional probability mass function (pmf) and P' denotes transpose of the matrix P.

(iii) *Myopic Action*: Agent k takes action $a_k \in \mathbb{A} = \{1, 2, \dots, A\}$ to minimize its expected cost

$$a_k = \arg\min_{a \in \mathbb{A}} \mathbb{E}\{c(x, a)|a_1, \dots, a_{k-1}, y_k\} = \arg\min_{a \in \mathbb{A}}\{c_a' \eta_k\}. \tag{5.2}$$

Here $c_a = [c(1, a), \dots, c(X, a)]$ where $c(i, a)$ denotes the cost incurred when the underlying state is i and the agent chooses action a.

Agent k then broadcasts its action a_k to subsequent agents.

(iv) *Social Learning Filter*: Define the public belief at time k as the vector

$$\pi_k = [\pi_k(1), \dots, \pi_k(X)]', \quad \text{where } \pi_k(i) = \mathbb{P}(x_k = i|a_1, \dots a_k).$$

Given the action a_k of agent k, and the public belief π_{k-1}, each subsequent agent $k' > k$ performs social learning to compute the public belief π_k according to the following "social learning filter" initialized at time $k = 0$ with π_0:

$$\pi_k = T(\pi_{k-1}, a_k), \text{ where } T(\pi, a) = \frac{R_a^\pi P' \pi}{\sigma(\pi, a)}, \quad \sigma(\pi, a) = \mathbf{1}_X' R_a^\pi P' \pi, \tag{5.3}$$

where $R_a^\pi = \text{diag}\big(\mathbb{P}(a|x=1,\pi),\ldots,\mathbb{P}(a|x=X,\pi)\big)$

$$\mathbb{P}(a_k = a|x_k = i, \pi_{k-1} = \pi) = \sum_{y\in\mathbf{Y}} \mathbb{P}(a|y,\pi)\mathbb{P}(y|x_k = i) \tag{5.4}$$

$$\mathbb{P}(a_k = a|y,\pi) = \begin{cases} 1 \text{ if } c_a' B_y P'\pi \le c_{\tilde{u}}' B_y P'\pi, \ \tilde{u} \in \mathbb{A} \\ 0 \text{ otherwise.} \end{cases}$$

The derivation of the social learning filter (5.3) is given in the discussion below.

5.2.2 Discussion

Information exchange structure: Figure 5.2 illustrates the above social learning protocol in which the information exchange is sequential. Agents send their hard decisions (actions) to subsequent agents. In the social learning protocol we have assumed that each agent acts once. Another way of viewing the social learning protocol is that there are finitely many agents that act repeatedly in some pre-defined order. If each agent chooses its local decision using the current public belief, then the setting is identical to the social learning setup. See [3] for results in social learning over several types of network adjacency matrices.

Filtering with hard decisions: Social learning can be viewed as agents making *hard* decision estimates at each time and sending these estimates to subsequent agents. In conventional Bayesian state estimation, a *soft* decision is made, namely, the posterior distribution (or equivalently, observation) is sent to subsequent agents. For example, if $\mathbb{A} = \mathcal{X}$, and the costs are chosen as $c_a = -e_a$ where e_a denotes the unit indicator with 1 in the a-th position, then $\text{argmin}_a\, c_a'\pi = \text{argmax}_a\, \pi(a)$, i.e. the maximum aposteriori probability (MAP) state estimate. For this example, social learning is equivalent to agents sending the hard MAP estimates to subsequent agents.

Instead of social learning, if each agent chooses its action as $a_k = y_k$ (that is, agents send their private observations), then the right-hand side of (5.4) is $\sum_{y\in\mathbf{Y}} I(y = y_k)\,\mathbb{P}(y|x_k = i) = \mathbb{P}(y_k|x_k = i)$ where $I(\cdot)$ denotes the indicator function. Then the social learning filter (5.3) becomes the HMM filter (3.57).

Dependence of observation likelihood on prior: In standard state estimation via a Bayesian filter, the observation likelihood is functionally independent of the current prior distribution. An unusual feature of the social learning filter is that the likelihood of the action given the state (denoted by R_a^π) is an explicit function of the prior π! Not only does the action likelihood depend on the prior, but it is also a discontinuous function, due to the presence of the argmin in (5.2).

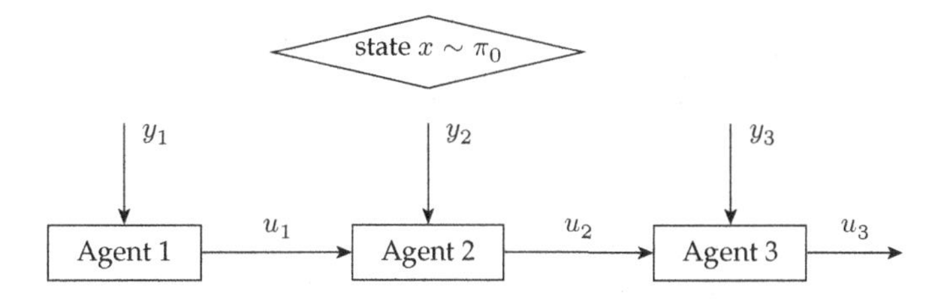

Figure 5.2 Information exchange structure in social learning.

Derivation of social learning filter: The derivation of the social learning filter (5.3) is as follows: Define the posterior as $\pi_k(j) = \mathbb{P}(x_k = j|a_1, \ldots, a_k)$. Then

$$\begin{aligned}\pi_k(j) &= \frac{1}{\sigma(\pi_{k-1}, a_k)} \mathbb{P}(a_k|x_k = j, a_1, \ldots, a_{k-1}) \sum_i P_{ij} \mathbb{P}(x_{k-1} = i|a_1, \ldots, a_{k-1}) \\ &= \frac{1}{\sigma(\pi_{k-1}, a_k)} \sum_y \mathbb{P}(a_k|y_k = y, a_1, \ldots, a_{k-1}) \mathbb{P}(y_k = y|x_k = j) \sum_i P_{ij} \pi_{k-1}(i) \\ &= \frac{1}{\sigma(\pi_{k-1}, a_k)} \sum_y \mathbb{P}(a_k|y_k = y, \pi_{k-1}) \mathbb{P}(y_k = y|x_k = j) \sum_i P_{ij} \pi_{k-1}(i),\end{aligned}$$

where the normalization term is

$$\sigma(\pi_{k-1}, a_k) = \sum_j \sum_y \mathbb{P}(a_k|y_k = y, \pi_{k-1}) \mathbb{P}(y_k = y|x_k = j) \sum_i P_{ij} \pi_{k-1}(i).$$

Risk-averse social learning filter: Instead of minimizing the expected cost (5.2), agents can minimize a risk averse measure such as the conditional value at risk (CVaR):

$$\begin{aligned}a_k &= \operatorname*{argmin}_{a \in \mathbb{A}} \mathrm{CVaR}\big(c(x, a)\big) \\ &= \operatorname*{argmin}_{a \in \mathbb{A}} \inf_{z \in \mathbb{R}} \Big\{ z + \frac{1}{\alpha} \sum_{i=1}^{X} \max\{c(i, a) - z, 0\}\, \eta_k(i) \Big\}, \quad \alpha \in (0, 1].\end{aligned}$$

CVaR is used widely in financial modeling [281]; see §8.6.2. The smaller α is, the more risk averse the agent is. The risk averse social learning filter is then given by (5.4) with

$$\mathbb{P}(a_k = a|y, \pi) = \begin{cases} 1 \text{ if } a_k = \operatorname*{argmin}_{a \in \mathbb{A}} \mathrm{CVaR}\big(c(x, a)\big) \\ 0 \text{ otherwise.} \end{cases}$$

5.3 Information cascades and constrained social learning

The social learning protocol of §5.2 and social learning filter (5.3) result in remarkable behavior amongst the agents. Agents form herds and information cascades and blindly follow previous agents. This is discussed below. Also we discuss how social learning can be optimized.

5.3.1 Rational agents form information cascades

The first consequence of the unusual nature of the social learning filter (5.3) is that social learning can result in multiple rational agents taking the same action independently of their observations. To illustrate this behavior, throughout this subsection, we assume that x is a finite state random variable (instead of a Markov chain) with prior distribution π_0.

We start with the following definitions:

- An individual agent k *herds* on the public belief π_{k-1} if it chooses its action $a_k = a(\pi_{k-1}, y_k)$ in (5.2) independently of its observation y_k.
- A *herd of agents* takes place at time $\bar{k}$, if the actions of all agents after time $\bar{k}$ are identical, i.e. $a_k = a_{\bar{k}}$ for all time $k > \bar{k}$.

- An *information cascade* occurs at time $\bar{k}$, if the public beliefs of all agents after time $\bar{k}$ are identical, i.e. $\pi_k = \pi_{\bar{k}}$ for all $k > \bar{k}$.

When an information cascade occurs, the public belief freezes and so social learning ceases. From the above definitions it is clear that an information cascade implies a herd of agents, but the reverse is not true.

The following result which is well known in the economics literature [56, 84] states that if agents follow the above social learning protocol, then after some finite time $\bar{k}$, an *information cascade* occurs.[2]

THEOREM 5.3.1 ([56]) *The social learning protocol described in §5.2.1 leads to an information cascade in finite time with probability 1. That is there exists a finite time $\bar{k}$ after which social learning ceases, i.e. public belief $\pi_{k+1} = \pi_k$, $k \geq \bar{k}$, and all agents choose the same action, i.e. $a_{k+1} = a_k$, $k \geq \bar{k}$.* □

The proof is in the appendix follows via an elementary application of the martingale convergence theorem. Let us give some insight as to why Theorem 5.3.1 holds. The martingale convergence theorem implies that at some finite time $k = k^*$, the agent's probability $\mathbb{P}(a_k|y_k, \pi_{k-1})$ becomes independent of the private observation y_k. Then clearly from (5.4), $\mathbb{P}(a_k = a|x_k = i, \pi_{k-1}) = \mathbb{P}(a_k = a|\pi)$. Substituting this into the social learning filter (5.3), we see that $\pi_k = \pi_{k-1}$. Thus after some finite time k^*, the social learning filter hits a fixed point and social learning stops. As a result, all subsequent agents $k > k^*$ completely disregard their private observations and take the same action a_{k^*}, thereby forming an information cascade (and therefore a herd).

5.3.2 Numerical example

To fix ideas, consider the social learning model with $X = 2$ (state space), $Y = 2$ (observation space), $\mathbb{A} = \{1, 2\}$ (local actions), $P = I$,

$$B = \begin{bmatrix} 0.8 & 0.2 \\ 0.2 & 0.8 \end{bmatrix}, \quad c = \begin{bmatrix} 2 & 3 \\ 0 & 4 \end{bmatrix}.$$

Since $\pi = (1 - \pi(2), \pi(2))'$, it suffices to consider the scalar $\pi(2) \in [0, 1]$ to represent the space of beliefs. This space of beliefs $[0, 1]$ can be decomposed into three disjoint intervals denoted $\mathcal{P}_1, \mathcal{P}_2, \mathcal{P}_3$ such that on each of these intervals, R^π is a constant. This is illustrated in Fig. 5.3. The three possible decision likelihood matrices R^π with elements $R^\pi_{ia} = \mathbb{P}(a|x = i, \pi)$ in (5.4), for $\pi(2) \in \mathcal{P}_1, \mathcal{P}_2, \mathcal{P}_3$, are

$$R^1 = \begin{bmatrix} 0 & 1 \\ 0 & 1 \end{bmatrix}, \quad R^2 = \begin{bmatrix} B_{11} & B_{12} \\ B_{21} & B_{22} \end{bmatrix}, \quad R^3 = \begin{bmatrix} 1 & 0 \\ 1 & 0 \end{bmatrix}. \tag{5.5}$$

[2] A nice analogy is provided in [84]. If I see someone walking down the street with an umbrella, I assume (based on rationality) that he has checked the weather forecast and is carrying an umbrella since it might rain. Therefore, I also take an umbrella. So now there are two people walking down the street carrying umbrellas. A third person sees two people with umbrellas and based on the same inference logic, also takes an umbrella. Even though each individual is rational, such herding behavior might be irrational since the first person who took the umbrella, may not have checked the weather forecast. Another example is that of patrons who decide to choose a restaurant. Despite their menu preferences, each patron chooses the restaurant with the most customers. So eventually all patrons herd to one restaurant.

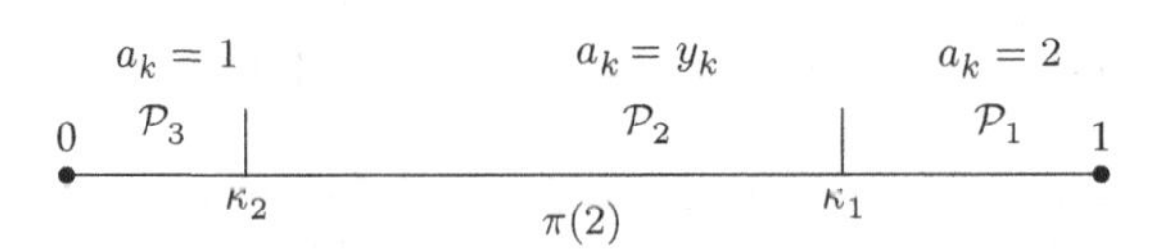

Figure 5.3 In social learning, the action likelihoods R^π are a piecewise constant in the belief π. For the model parameters specified in §5.3.2, there are three distinct likelihoods; see (5.5). The boundary points of the regions are $\kappa_1 = 0.667$ and $\kappa_2 = 0.112$. Information cascades occur in regions $\mathcal{P}_1$ and $\mathcal{P}_3$.

That is, in region $\mathcal{P}_1$ the agent always chooses $a_k = 1$ irrespective of its private observation, similarly in $\mathcal{P}_3$ the agent always choose $a_k = 2$. These regions form information cascades. In the region $\mathcal{P}_2$, the agent choses $a_k = y_k$. In Theorem 5.5.1 below we will show that a_k is increasing with public belief $\pi_{k-1}(2)$. Therefore there are three distinct regions for R^π as shown in Figure 5.3. As an exercise, the reader should evaluate the corresponding regions for the risk averse social learning filter.

5.3.3 Constrained social sensing: individual privacy vs group reputation

The above social learning protocol can be interpreted as follows. Agents seek to estimate an underlying state of nature but reveal their actions by maximizing their privacy according to the optimization (5.2). This leads to an information cascade and social learning stops. In other words, agents are interested in optimizing their own costs (such as maximizing privacy) and ignore the information benefits their action provides to others.

Partially observed Markov decision process formulation

We now describe an optimized social learning procedure that delays herding. This approach is motivated by the following question: how can agents assist social learning by choosing their actions to trade off individual privacy (local costs) with optimizing the reputation of the entire social group?

Suppose agents seek to maximize the reputation of their social group by minimizing the following social welfare cost involving all agents in the social group (compared to the myopic objective (5.2) used in standard social learning):

$$J_\mu(\pi) = \mathbb{E}_\mu\left\{\sum_{k=1}^{\infty} \rho^{k-1} c'_{a(\pi_{k-1}, y_k, \mu(\pi_{k-1}))} \eta_k\right\}. \tag{5.6}$$

In (5.6), $a(\pi, y, \mu(\pi)))$ denotes the decision rule that agents use to choose their actions as will be explained below. Also $\rho \in [0, 1)$ is an economic discount factor and π_0 denotes the initial probability (prior) of the state x. $\mathbb{E}_\mu$ denotes the expectation of the evolution of the observations and underlying state which are strategy dependent.

The key attribute of (5.6) is that each agent k chooses its action according to the privacy constrained rule

$$a_k = a(\pi_{k-1}, y_k, \mu(\pi_{k-1})). \tag{5.7}$$

Here, the policy $\mu : \pi_{k-1} \to \{1, 2 \ldots, L\}$ maps the available public belief to the set of L privacy values. The higher the privacy value, the less the agent reveals through its action. This is in contrast to standard social learning (5.2) in which the action is chosen myopically by each agent.

The above formulation can be interpreted as follows: individual agents seek to maximize their privacy according to social learning (5.7) but also seek to maximize the reputation of their entire social group (5.6). Determining the policy μ^* that minimizes (5.6), and thereby maximizes the social group reputation, is equivalent to solving a stochastic control problem that is called a partially observed Markov decision process (POMDP) problem. We will study POMDPs starting from Chapter 7.

Structure of privacy constrained sensing policy

In general, POMDPs are computationally intractable to solve and therefore one cannot say anything useful about the structure of the optimal policy μ^*. However, useful insight can be obtained by considering the following extreme case of the above problem. Suppose there are two privacy values and each agent k chooses action

$$a_k = \begin{cases} y_k & \text{if } \mu(\pi_k) = 1 \text{ (no privacy)} \\ \arg\min_a c_a' \pi_{k-1} & \text{if } \mu(\pi_k) = 2 \text{ (full privacy).} \end{cases}$$

That is, an agent either reveals its raw observation (no privacy) or chooses its action by completely neglecting its observation (full privacy). Once an agent chooses the full privacy option, then all subsequent agents choose exactly the same option and therefore herd: this follows since each agent's action reveals nothing about the underlying state of nature. Therefore, for this extreme example, determining the optimal policy $\mu^*(\pi)$ is equivalent to solving a stopping time problem: determine the earliest time for agents to herd (maintain full privacy) subject to maximizing the social group reputation.

For such a quickest herding stopping time problem, §13.4 shows that under reasonable conditions the optimal policy μ^* has the following structure: choose increased privacy when belief is close to the target state.

5.4 Data incest in online reputation systems

This section generalizes the previous section by considering social learning in a social network. How can multiple social sensors interacting over a social network estimate an underlying state of nature? The state could be the position coordinates of an event [289] or the quality of a social parameter such as quality of a restaurant or political party. In comparison to the previous section, we now consider social learning on a family of time dependent directed acyclic graphs. In such cases, apart from herding, the phenomenon of data incest arises.

5.4.1 Information exchange graph in social network

Consider an online reputation system comprised of social sensors $\{1, 2, \dots, S\}$ that aim to estimate an underlying state of nature (a random variable). Let $x \in \mathcal{X} = \{1, 2, \dots, X\}$ represent the state of nature (such as the quality of a hotel) with known prior distribution

π_0. Let $k = 1, 2, 3, \ldots$ depict epochs at which events occur. These events involve taking observations, evaluating beliefs and choosing actions as described below. The index k marks the historical order of events and not necessarily absolute time. For simplicity, we refer to k as "time".

To model the information exchange in the social network, we will use a family of directed acyclic graphs. It is convenient also to reduce the coordinates of time k and agent s to a single integer index n which denotes agent s at time k:

$$n = s + S(k-1), \quad s \in \{1, \ldots, S\}, \; k = 1, 2, 3, \ldots . \tag{5.8}$$

We refer to n as a "node" of a time dependent information flow graph G_n which we now define. Let

$$G_n = (V_n, E_n), \quad n = 1, 2, \ldots \tag{5.9}$$

denote a sequence of time-dependent *directed acyclic graphs (DAGs)*[3] of information flow in the social network until and including time k. Each vertex in V_n represents an agent s' at time k' and each edge (n', n'') in $E_n \subseteq V_n \times V_n$ shows that the information (action) of node n' (agent s' at time k') reaches node n'' (agent s'' at time k''). It is clear that G_n is a sub-graph of G_{n+1}.

The adjacency matrix A_n of G_n is an $n \times n$ matrix with elements

$$A_n(i,j) = \begin{cases} 1 & \text{if } (v_j, v_i) \in E \\ 0 & \text{otherwise} \end{cases}, \quad A_n(i,i) = 0. \tag{5.10}$$

The transitive closure matrix T_n is the $n \times n$ matrix

$$T_n = \operatorname{sgn}((\mathbf{I}_n - A_n)^{-1}) \tag{5.11}$$

where for matrix M, the matrix $\operatorname{sgn}(M)$ has elements

$$\operatorname{sgn}(M)(i,j) = \begin{cases} 0 & \text{if } M(i,j) = 0\,, \\ 1 & \text{if } M(i,j) \neq 0. \end{cases}$$

Note that $A_n(i,j) = 1$ if there is a single hop path between nodes i and j, In comparison, $T_n(i,j) = 1$ if there exists a path (possible multi-hop) between i and j.

The information reaching node n depends on the information flow graph G_n. The following two sets will be used to specify the incest removal algorithms below:

$$\mathcal{H}_n = \{m : A_n(m,n) = 1\} \tag{5.12}$$
$$\mathcal{F}_n = \{m : T_n(m,n) = 1\}. \tag{5.13}$$

Thus $\mathcal{H}_n$ denotes the set of previous nodes m that communicate with node n in a single-hop. In comparison, $\mathcal{F}_n$ denotes the set of previous nodes m whose information eventually arrives at node n. Thus $\mathcal{F}_n$ contains all possible multi-hop connections by which information from node m eventually reaches node n.

[3] A DAG is a directed graph with no directed cycles. The ordering of nodes $n = 1, 2, \ldots,$ proposed here is a special case of the well known result that the nodes of a DAG are partially orderable; see §5.6.

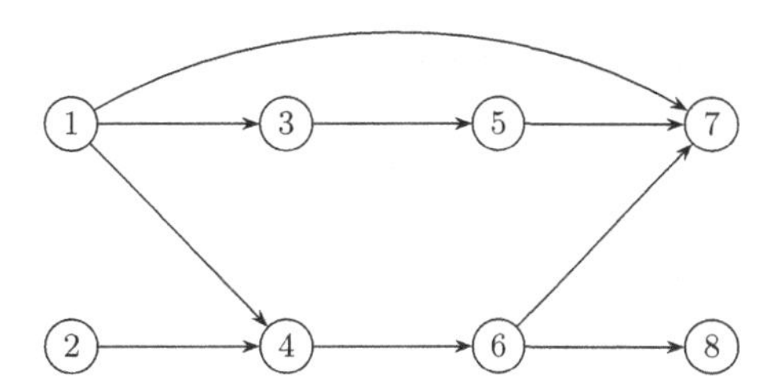

Figure 5.4 Example of an information flow network with $S = 2$ two agents, namely $s \in \{1, 2\}$ and time points $k = 1, 2, 3, 4$. Circles represent the nodes indexed by $n = s + S(k - 1)$ in the social network and each edge depicts a communication link between two nodes.

Properties of A_n and T_n

Due to causality with respect to the time index k (information sent by an agent can only arrive at another agent at a later time instant), the following obvious properties hold (proof omitted):

LEMMA 5.4.1 *Consider the sequence of DAGs G_n, $n = 1, 2, \ldots$*
(i) The adjacency matrices A_n are upper triangular.
A_n is the upper left $n \times n$ submatrix of A_{n+1}.
(ii) The transitive closure matrices T_n are upper triangular with ones on the diagonal. Hence, T_n is invertible.
(iii) Classical social learning of §5.2.1, is a trivial example with adjacency matrix $A_n(i,j) = 1$ for $j = i + 1$ and $A_n(i,j) = 0$ elsewhere. □

5.4.2 Illustrative example

To illustrate the above notation consider a social network consisting of $S = 2$ agents with the information flow graph of Figure 5.4 for time points $k = 1, 2, 3, 4$ characterized by the family of DAGs $\{G_1, \ldots, G_7\}$. The adjacency matrices $A_1, \ldots, A_7$ are constructed recursively as follows: A_n is the upper left $n \times n$ submatrix of A_{n+1}, $n = 1, \ldots, 6$ initialized by

$$A_7 = \begin{bmatrix} 0 & 0 & 1 & 1 & 0 & 0 & 1 \\ 0 & 0 & 0 & 1 & 0 & 0 & 0 \\ 0 & 0 & 0 & 0 & 1 & 0 & 0 \\ 0 & 0 & 0 & 0 & 0 & 1 & 0 \\ 0 & 0 & 0 & 0 & 0 & 0 & 1 \\ 0 & 0 & 0 & 0 & 0 & 0 & 1 \\ 0 & 0 & 0 & 0 & 0 & 0 & 0 \end{bmatrix}.$$

Let us explain these matrices. Since nodes 1 and 2 do not communicate, clearly A_1 and A_2 are zero matrices. Nodes 1 and 3 communicate, hence A_3 has a single one, etc. Note that if nodes 1, 3, 4 and 7 are assumed to be the same individual, then at node 7, the individual remembers what happened at node 5 and node 1, but not node 3. This models the case where the individual has selective memory and remembers certain highlights of past events. From (5.12) and (5.13),

$$\mathcal{H}_7 = \{1, 5, 6\}, \quad \mathcal{F}_7 = \{1, 2, 3, 4, 5, 6\}$$

where $\mathcal{H}_7$ denotes all one hop links to node 7 while $\mathcal{F}_7$ denotes all multihop links to node 7.

Using (5.11), the transitive closure matrices $T_1, \dots, T_7$ are given by: T_n is the upper left $n \times n$ submatrix of T_{n+1} and

$$T_7 = \begin{bmatrix} 1 & 0 & 1 & 1 & 1 & 1 & 1 \\ 0 & 1 & 0 & 1 & 0 & 1 & 1 \\ 0 & 0 & 1 & 0 & 1 & 0 & 1 \\ 0 & 0 & 0 & 1 & 0 & 1 & 1 \\ 0 & 0 & 0 & 0 & 1 & 0 & 1 \\ 0 & 0 & 0 & 0 & 0 & 1 & 1 \\ 0 & 0 & 0 & 0 & 0 & 0 & 1 \end{bmatrix}.$$

Note that $T_n(i,j)$ is nonzero only for $i \geq j$ due to causality since information sent by an agent can only arrive at another social group at a later time instant.

5.5 Fair online reputation system

5.5.1 Data incest model and social influence constraint

Each node n receives recommendations from its immediate friends (one hop neighbors) according to the information flow graph defined above. That is, node n receives actions $\{a_m, m \in \mathcal{H}_n\}$ from nodes $m \in \mathcal{H}_n$ and then seeks to compute the associated public beliefs $\pi_m, m \in \mathcal{H}_n$. If node n naively (incorrectly) assumes that the public beliefs $\pi_m, m \in \mathcal{H}_n$ are independent, then it would fuse these as

$$\pi_{n-} = \frac{\prod_{m \in \mathcal{H}_n} \pi_m}{\mathbf{1}_X' \prod_{m \in \mathcal{H}_n} \pi_m}. \tag{5.14}$$

This naive data fusion would result in data incest.

Aim: The aim is to provide each node n the true posterior distribution

$$\pi_{n-}^0(i) = \mathbb{P}(x = i | \{a_m, m \in \mathcal{F}_n\}) \tag{5.15}$$

subject to the following *social influence constraint*: there exists a fusion algorithm $\mathcal{A}$ such that

$$\pi_{n-}^0 = \mathcal{A}(\pi_m, m \in \mathcal{H}_n). \tag{5.16}$$

Discussion: Fair rating and social influence

We briefly pause to discuss (5.15) and (5.16).

(i) We call π_{n-}^0 in (5.15) the *true* or *fair online rating* available to node n since $\mathcal{F}_n$ defined in (5.13) denotes all information (multi-hop links) available to node n. By definition π_{n-}^0 is incest free and is the desired conditional probability that agent n needs.

Indeed, if node n combines π_{n-}^0 together with its own private observation via social learning, then clearly

$$\eta_n(i) = \mathbb{P}(x = i | \{a_m, m \in \mathcal{F}_n\}, y_n), \quad i \in \mathcal{X},$$
$$\pi_n(i) = \mathbb{P}(x = i | \{a_m, m \in \mathcal{F}_n\}, a_n), \quad i \in \mathcal{X},$$

are, respectively, the correct (incest free) private belief for node n and the correct after-action public belief. If agent n does not use π^0_{n-}, then incest can propagate; for example if agent n naively uses (5.14).

Why should an individual n agree to use π^0_{n-} to combine with its private message? It is here that the social influence constraint (5.16) is important. $\mathcal{H}_n$ can be viewed as the "*social message*", i.e. personal friends of node n since they directly communicate to node n, while the associated beliefs can be viewed as the "*informational message*". As described in [61], the social message from personal friends exerts a large social influence – it provides significant incentive (peer pressure) for individual n to comply with the protocol of combining its estimate with π^0_{n-} and thereby prevent incest. [61] shows that receiving messages from known friends has significantly more influence on an individual than the information in the messages. This study includes a comparison of information messages and social messages on Facebook and their direct effect on voting behavior. To quote [61], "The effect of social transmission on real-world voting was greater than the direct effect of the messages themselves..."

5.5.2 Protocol for fair online reputation system

Protocol 7 evaluates the fair online rating by eliminating data incest in a social network. The aim is to achieve (5.15) subject to (5.16).

At this stage, the public rating π_{n-} computed in (5.17) of Protocol 7 is not necessarily the fair online rating π^0_{n-} of (5.15). Without careful design of algorithm $\mathcal{A}$ in (5.17), due to interdependencies of actions on previous actions, π_{n-} can be substantially different from π^0_{n-}. Then η_n computed via (5.19) will not be the correct private belief and incest will propagate in the network. In other words, η_n, π_{n-} and π_n are defined purely in terms of their computational expressions in Protocol 7; they are not necessarily the desired conditional probabilities, unless algorithm $\mathcal{A}$ is properly designed to remove incest. Note also that algorithm $\mathcal{A}$ needs to satisfy the social influence constraint (5.16).

5.5.3 Ordinal decision making in Protocol 7

Protocol 7 assumes that each agent is a Bayesian utility optimizer. The following discussion shows that under reasonable conditions, such a Bayesian model is a useful idealization.

Humans typically make *monotone* decisions: the more favorable the private observation, the higher the recommendation. Humans make *ordinal* decisions[4] since humans tend to think in symbolic ordinal terms. Under what conditions is the recommendation a_n made by node n *monotone increasing* in its observation y_n and *ordinal*? Recall from Steps (iii) and (iv) of Protocol 7 that the recommendation of node n is

$$a_n(\pi^0_{n-}, y_n) = \underset{a}{\operatorname{argmin}}\, c'_a B_{y_n} \pi^0_{n-}$$

[4] Humans typically convert numerical attributes to ordinal scales before making decisions. For example, it does not matter if the cost of a meal at a restaurant is \$200 or \$205; an individual would classify this cost as "high". Also credit rating agencies use ordinal symbols such as AAA, AA, A.

Protocol 7 Incest removal in online reputation system

(i) *Information from social network*:

1. *Social message from friends*: Node n receives social message $\mathcal{H}_n$ comprising the names or photos of friends that have made recommendations.
2. *Informational message from friends*: The reputation system fuses recommendations $\{a_m, m \in \mathcal{H}_n\}$ into the single informational message π_{n-} and presents this to node n. The reputation system computes π_{n-} as follows:
   (a) $\{a_m, m \in \mathcal{H}_n\}$ are used to compute public beliefs $\{\pi_m, m \in \mathcal{H}_n\}$ using Step (v) below.
   (b) $\{\pi_m, m \in \mathcal{H}_n\}$ are fused into π_{n-} as

$$\pi_{n-} = \mathcal{A}(\pi_m, m \in \mathcal{H}_n). \tag{5.17}$$

   In §5.5.5, fusion algorithm $\mathcal{A}$ is designed as

$$\underline{\pi}_{n-}(i) = \sum_{m \in \mathcal{H}_n} w_n(m)\, \underline{\pi}_m(i), \quad i \in \mathcal{X}. \tag{5.18}$$

   Here $\underline{\pi}_m(i) = \log \pi_m(i)$ and $w_n(m)$ are weights.

(ii) *Observation*: Node n records private observation y_n from distribution $B_{iy} = \mathbb{P}(y|x = i)$, $i \in \mathcal{X}$.

(iii) *Private belief*: Node n uses y_n and informational message π_{n-} to update its private belief via Bayes rule:

$$\eta_n = \frac{B_{y_n}\pi_{n-}}{\mathbf{1}'_X B_y \pi_{n-}}. \tag{5.19}$$

(iv) *Recommendation*: Node n makes recommendation $a_n = \arg\min_a c'_a \eta_n$ and records this on the reputation system.

(v) *Public belief update by network administrator*: Based on recommendation a_n, the reputation system computes the public belief π_n using the social learning filter (5.3).

So an equivalent question is: under what conditions is the argmin increasing in observation y_n? Note that an increasing argmin is an *ordinal* property – that is, $\text{argmin}_a\, c'_a B_{y_n} \pi^0_{n-}$ increasing in y implies $\text{argmin}_a\, \phi(c'_a B_{y_n} \pi^0_{n-})$ is also increasing in y for any monotone function $\phi(\cdot)$.

The following result gives sufficient conditions for each agent to give a recommendation that is monotone and ordinal in its private observation:

THEOREM 5.5.1 *Suppose the observation probabilities and costs satisfy the following conditions:*

(A1) B_{iy} *are TP2 (totally positive of order 2); that is,* $B_{i+1,y}B_{i,y+1} \leq B_{i,y}B_{i+1,y+1}$.
(A2) $c(x,a)$ *is submodular. That is,* $c(x,a+1) - c(x,a) \leq c(x+1,a+1) - c(x+1,a)$.

Then

1. *Under (A1) and (A2), the recommendation* $a_n(\pi^0_{n-}, y_n)$ *made by agent n is increasing and hence ordinal in observation* y_n, *for any* π^0_{n-}.
2. *Under (A2),* $a_n(\pi^0_{n-}, y_n)$ *is increasing in belief* π^0_{n-} *with respect to the monotone likelihood ratio (MLR) stochastic order*[5] *for any observation* y_n. □

The proof is in the appendix. Theorem 5.5.1 can be interpreted as follows. If agents makes recommendations that are monotone and ordinal in the observations and monotone in the belief, then they mimic Bayesian social learning. Even if the agent does not exactly follow a Bayesian social learning model, its monotone ordinal behavior implies that such a model is a useful idealization.

Condition (A1) is widely studied in monotone decision making and will be discussed in §10.5 where it is called Assumption (F1).

Condition (A2) is a submodularity condition that will be discussed in Chapter 12. Actually (A2) is a stronger version of the more general single-crossing condition (see appendix and also §9.1.3)

$$(c_{a+1} - c_a)' B_{y+1} \pi^0 \geq 0 \implies (c_{a+1} - c_a)' B_y \pi^0 \geq 0.$$

This single crossing condition is ordinal, since for any monotone function ϕ, it is equivalent to

$$\phi((c_{a+1} - c_a)' B_{y+1} \pi^0) \geq 0 \implies \phi((c_{a+1} - c_a)' B_y \pi^0) \geq 0.$$

(A2) also makes sense in a reputation system for the costs to be well posed. Suppose the recommendations in action set $\mathbb{A}$ are arranged in increasing order and also the states in $\mathcal{X}$ for the underlying state are arranged in ascending order. Then (A2) says: if recommendation $a + 1$ is more accurate than recommendation a for state x; then recommendation $a + 1$ is also more accurate than recommendation a for state $x + 1$ (which is a higher quality state than x).

For the example in §5.3.2, it is easily verified that conditions (A1) and (A2) in Theorem 5.5.1 hold. Hence, the local action a_k is increasing with public belief π_{k-1}. So there are 3 distinct regions for R^π as shown in Figure 5.3 on page 100.

5.5.4 Discussion of Protocol 7

Individuals have selective memory: Protocol 7 allows for cases where each node can remember some (or all) of its past actions or none. This models cases where people forget most of the past except for specific highlights. For example, in the information flow graph of the illustrative example in the appendix (Figure 5.4), if nodes 1,3,4 and 7 are assumed to be the same individual, then at node 7, the individual remembers what happened at node 5 and node 1, but not node 3.

Automated Recommender System: Steps (i) and (v) of Protocol 7 can be combined into an automated recommender system that maps previous actions of agents in the social

[5] Given probability mass functions $\{p_i\}$ and $\{q_i\}$, $i = 1, \ldots, X$ then p MLR dominates q if $\log p_i - \log p_{i+1} \leq \log q_i - \log q_{i+1}$. In §10.1, MLR dominance is discussed in much more detail.

group to a single recommendation (rating) π_{n-} of (5.17). This recommender system can operate completely opaquely to the actual user (node n). Node n simply uses the automated rating π_{n-} as the current best available rating from the reputation system. Actually Algorithm $\mathcal{A}$ presented below fuses the beliefs in a linear fashion. A human node n receiving an informational message comprised of a linear combination of recommendation of friends, along with the social message has incentive to follow the protocol.

Agent reputation: The cost function minimization in Step (iv) can be interpreted in terms of the reputation of agents in online reputation systems. If an agent continues to write bad reviews for high quality restaurants on Yelp, her reputation becomes lower among the users. Consequently, other people ignore reviews of that (low-reputation) agent in evaluating their opinions about the social unit under study (restaurant). Therefore, agents minimize the penalty of writing inaccurate reviews.

Utility maximizers: Given data sets of the social influence and response of an agent (or aggregate behavior of groups of agents), one can determine is the agent is a utility maximizer (or cost minimizer) using the principle of revealed preferences. This is discussed in §6.7.3.

5.5.5 Incest removal algorithm

It only remains to describe algorithm $\mathcal{A}$ in Step 2b of Protocol 7 so that

$$\pi_{n-}(i) = \pi^0_{n-}(i), \text{ where } \pi^0_{n-}(i) = \mathbb{P}(x = i | \{a_m, m \in \mathcal{F}_n\}), \quad i \in \mathcal{X}. \tag{5.20}$$

Recall π^0_{n-} in (5.15) is the *fair online rating* available to node n. It is convenient to work with the logarithm of the un-normalized belief. Define

$$\underline{\pi}_n(i) \propto \log \pi_n(i), \quad \underline{\pi}_{n-}(i) \propto \log \pi_{n-}(i), \quad i \in \mathcal{X}.$$

Define the $n-1$ dimensional weight vector:

$$w_n = T^{-1}_{n-1} t_n. \tag{5.21}$$

Recall that t_n denotes the first $n-1$ elements of the nth column of the transitive closure matrix T_n. Thus the weights are purely a function of the adjacency matrix of the graph and do not depend on the observed data.

We present algorithm $\mathcal{A}$ in two steps: first, the actual computation is given in Theorem 5.5.2, second, necessary and sufficient conditions on the information flow graph for the existence of such an algorithm to achieve the social influence constraint (5.16).

THEOREM 5.5.2 (Fair Rating Algorithm) *Consider the reputation system with Protocol 7. Suppose the network administrator runs the following algorithm in (5.17):*

$$\underline{\pi}_{n-}(i) = \sum_{m=1}^{n-1} w_n(m) \underline{\pi}_m(i) \tag{5.22}$$

where the weights w_n are chosen according to (5.21).

Then $\underline{\pi}_{n-}(i) \propto \log \pi^0_{n-}(i)$. *That is, the fair rating* $\log \pi^0_{n-}(i)$ *defined in (5.15) is obtained via (5.22).* □

Theorem 5.5.2 says that the fair rating π^0_{n-} can be expressed as a linear function of the action log-likelihoods in terms of the transitive closure matrix T_n of graph G_n.

Achievability of fair rating by Protocol 7

1. Algorithm $\mathcal{A}$ at node n specified by (5.17) needs to satisfy the social influence constraint (5.16) – that is, it needs to operate on beliefs $\underline{\pi}_m, m \in \mathcal{H}_n$.
2. On the other hand, to provide incest free estimates, algorithm $\mathcal{A}$ specified in (5.22) requires all previous beliefs $l_{1:n-1}(i)$ that are specified by the nonzero elements of the vector w_n.

The only way to reconcile points 1 and 2 is to ensure that $A_n(j,n) = 0$ implies $w_n(j) = 0$ for $j = 1, \ldots, n-1$. This condition means that the single hop past estimates $\underline{\pi}_m, m \in \mathcal{H}_n$ available at node n according to (5.17) in Protocol 7 provide all the information required to compute $w'_n \underline{\pi}_{1:n-1}$ in (5.22). We formalize this condition in the following theorem.

THEOREM 5.5.3 (Achievability of fair rating) *Consider the fair rating algorithm specified by (5.22). For Protocol 7 using the social influence constraint information* $(\pi_m, m \in \mathcal{H}_n)$ *to achieve the estimates* $\underline{\pi}_{n-}$ *of algorithm (5.22), a necessary and sufficient condition on the information flow graph* G_n *is*

$$A_n(j,n) = 0 \implies w_n(j) = 0. \tag{5.23}$$

Therefore for Protocol 7 to generate incest free estimates for nodes $n = 1, 2, \ldots$, *condition (5.23) needs to hold for each* n. *(Recall* w_n *is specified in (5.22).)* □

Summary: Algorithm (5.22) together with the condition (5.23) ensure that incest free estimates are generated by Protocol 7.

Illustrative example (continued)

Let us continue with the example of Figure 5.4 where we already specified the adjacency matrices and transitive closer matrices. The weight vectors are obtained from (5.22) as

$$\begin{aligned} &w_2 = \begin{bmatrix} 0 \end{bmatrix}, \quad w_3 = \begin{bmatrix} 1 & 0 \end{bmatrix}', \quad w_4 = \begin{bmatrix} 1 & 1 & 0 \end{bmatrix}', \\ &w_5 = \begin{bmatrix} 0 & 0 & 1 & 0 \end{bmatrix}', \quad w_6 = \begin{bmatrix} 0 & 0 & 0 & 1 & 0 \end{bmatrix}', \\ &w_7 = \begin{bmatrix} -1 & 0 & 0 & 0 & 1 & 1 \end{bmatrix}'. \end{aligned}$$

w_2 means that node 2 does not use the estimate from node 1. This formula is consistent with the constraints on information flow because the estimate from node 1 is not available to node 2; see Figure 5.4. w_3 means that node 3 uses estimates from nodes 1; w_4 means that node 4 uses estimates only from node 1 and node 2. As shown in Figure 5.4, the misinformation propagation occurs at node 7 since there are multiple paths from node 1 to node 7. The vector w_7 says that node 7 adds estimates from nodes

5 and 6 and removes estimates from node 1 to avoid triple counting of these estimates already integrated into estimates from nodes 3 and 4. Using the algorithm (5.22), incest is completely prevented in this example.

Here is an example in which exact incest removal is impossible. Consider the information flow graph of Figure 5.4 but with the edge between node 1 and node 7 deleted. Then $A_7(1,7) = 0$ while $w_7(1) \neq 0$, and therefore the condition (5.23) does not hold. Hence, exact incest removal is not possible for this case. In §5.6 we compute the Bayesian estimate of the underlying state when incest cannot be removed. Additional numerical examples are in [186].

5.6 Belief-based expectation polling

Recall from §5.1.1 that in expectation polling, individuals are sampled and asked who they believe will win the election; as opposed to intent polling where individuals are sampled and asked who they intend to vote for. In this section, we consider data incest in expectation polling. Unlike previous sections, the agents no longer are assumed to eliminate incest at each step. So incest propagates in the network. Given the incestuous beliefs, the aim is to compute the posterior of the state. We illustrate how the results of §5.2 can be used to eliminate data incest (and therefore bias) in expectation polling systems. The bias of the estimate from expectation polling depends strongly on the social network structure. The approach below is Bayesian: we compute the posterior and therefore the conditional mean estimate.

We consider two formulations below:

1. An expectation polling system where in addition to specific polled voters, the minimal number of additional voters are recruited. The pollster then is able to use the incestuous beliefs to compute the posterior conditioned on the observations and thereby eliminate incest (§5.6.1 below).
2. An expectation polling system when it is not possible to recruit additional voters. The pollster then can only compute the posterior conditioned on the incestuous beliefs. (§5.6.2 below).

Suppose X candidates contest an election. Let $x \in \mathcal{X} = \{1, 2, \ldots, X\}$ denote the candidate that is leading amongst the voters, i.e. x is the true state of nature. There are N voters. These voters communicate their expectations via a social network according to the steps listed in Protocol 8. We index the voters as nodes $n \in \{1, \ldots, N\}$ as follows: $m < n$ if there exists a directed path from node m to node n in the directed acyclic graph (DAG). It is well known that such a labeling of nodes in a DAG constitute a partially ordered set. The remaining nodes can be indexed arbitrarily, providing the above partial ordering holds.

Protocol 8 is similar to Protocol 7, except that voters exchange their expectations (beliefs) of who the leading candidate is, instead of recommendations. (So unlike the previous section, agents do not perform social learning.)

Protocol 8 Belief-based expectation polling protocol

1. *Intention from friends*: Node n receives the beliefs $\{\pi_m, m \in \mathcal{H}_n\}$ of who is leading from its immediate friends, namely nodes $\mathcal{H}_n$.
2. *Naive data fusion*: Node n fuses the estimates of its friends naively (and therefore with incest) as
$$\pi_{n-} = \frac{\prod_{m \in \mathcal{H}_n} \pi_m}{\mathbf{1}'_X \prod_{m \in \mathcal{H}_n} \pi_m}.$$
3. *Observation*: Node n records its private observation y_n from the distribution $B_{iy} = \mathbb{P}(y|x = i)$, $i \in \mathcal{X}$ of who the leading candidate is.
4. *Belief update*: Node n uses y_n to update its belief via Bayes formula:
$$\pi_n = \frac{B_{y_n} \pi_{n-}}{\mathbf{1}'_X B_{y_n} \pi_{n-}}. \tag{5.24}$$
5. Node n sends its belief π_n to subsequent nodes as specified by the social network graph.

Remark: Similar to Theorem 5.5.1, the Bayesian update (5.24) in Protocol 8 can be viewed as an idealized model; under assumption (A1) the belief π_n increases with respect to the observation y_n (in terms of the monotone likelihood ratio order). So even if agents are not actually Bayesian, if they choose their belief to be monotone in the observation, they mimic the Bayesian update.

5.6.1 Exact incest removal in expectation polling

Assuming that the voters follow Protocol 8, the pollster seeks to estimate the leading candidate (state of nature x) by sampling a subset of the N voters.

Let $\mathcal{R} = \{\mathcal{R}_1, \mathcal{R}_2, \ldots, \mathcal{R}_L\}$ denote the $L - 1$ sampled *recruited* voters together with node $\mathcal{R}_L$ that denotes the pollster. For example the $L - 1$ voters could be volunteers or paid recruits that have already signed up for the polling agency. Since, by Protocol 8, these voters have naively combined the intentions of their friends (by ignoring dependencies), the pollster needs to poll additional voters to eliminate incest. Let $\mathcal{E}$ denote this set of *extra* (additional) voters to poll. Clearly the choice of $\mathcal{E}$ will depend on $\mathcal{R}$ and the structure of the social network.

What is the smallest set $\mathcal{E}$ of additional voters to poll in order to compute the posterior $\mathbb{P}(x|y_1, \ldots, y_{\mathcal{R}_L})$ of state x (and therefore eliminate data incest)?

Assume that all the recruited $L - 1$ nodes in $\mathcal{R}$ report to a central polling node $\mathcal{R}_L$. We assume that the pollster has complete knowledge of the network.

Using the formulation of §5.4, the pollster constructs the directed acyclic graph $G_n = (V_n, E_n)$, $n \in \{1, \ldots, \mathcal{R}_L\}$. The methodology of §5.5.5 is straightforwardly used to determine the minimal set of extra (additional) voters $\mathcal{E}$. The procedure is as follows:

1. For each $n = 1, \ldots, \mathcal{R}_L$, compute the weight vector $w_n = T_{n-1}^{-1} t_n$; see (5.21).
2. The indices of the nonzero elements in w_n that are not in V_n, constitute the minimal additional voters (nodes) that need to be polled. This is due to the necessity and sufficiency of (5.23) in Theorem 5.5.3.

3. Once the additional voters in $\mathcal{E}$ are polled for their beliefs, the polling agency corrects belief π_n reported by node $n \in \mathcal{R}$ to remove incest as:

$$l_n^0(i) = \underline{\pi}_n(i) + \sum_{m \in \mathcal{E}, m < n} w_n(m) \underline{\pi}_m(i). \tag{5.25}$$

5.6.2 Bayesian expectation polling using incestuous beliefs

Consider Protocol 8. However, unlike the previous subsection, assume that the pollster *cannot* sample additional voters $\mathcal{E}$ to remove incest; e.g. the pollster is unable to reach marginalized sections of the population.

Given incest containing beliefs $\pi_{\mathcal{R}_1}, \ldots, \pi_{\mathcal{R}_L}$ of the L recruits and pollster, how can the pollster compute the posterior $\mathbb{P}(x|\pi_{\mathcal{R}_1}, \ldots, \pi_{\mathcal{R}_L})$ and hence the unbiased conditional mean estimate $\mathbb{E}\{x|\pi_{\mathcal{R}_1}, \ldots, \pi_{\mathcal{R}_L}\}$ where x is the state of nature?

Optimal estimation using incestuous beliefs: Define the following notation

$$\begin{aligned} &\pi_{\mathcal{R}} = \begin{bmatrix} \pi_{\mathcal{R}_1} & \cdots & \pi_{\mathcal{R}_L} \end{bmatrix}', \; \underline{\pi}_{\mathcal{R}} = \begin{bmatrix} \underline{\pi}_{\mathcal{R}_1} & \cdots & \underline{\pi}_{\mathcal{R}_L} \end{bmatrix}', \\ &Y_{\mathcal{R}} = \begin{bmatrix} y_1 & \cdots & y_{\mathcal{R}_L} \end{bmatrix}', \; o(x) = \begin{bmatrix} \log b_{xy_1} \ldots, \log b_{xy_n} \end{bmatrix}'. \end{aligned} \tag{5.26}$$

THEOREM 5.6.1 *Consider the beliefs $\pi_{\mathcal{R}_1}, \ldots, \pi_{\mathcal{R}_L}$ of the L recruits and pollster in an expectation poll operating according to Protocol 8. Then for each $x \in \mathcal{X}$, the posterior is evaluated as*

$$\mathbb{P}(x|\pi_{\mathcal{R}}) \propto \sum_{Y_{\mathcal{R}} \in \mathcal{Y}} \prod_{m=1}^{\mathcal{R}_L} B_{xy_m} \pi_0(x). \tag{5.27}$$

Here $\mathcal{Y}$ denotes the set of sequences $\{Y_{\mathcal{R}}\}$ that satisfy

$$\begin{aligned} &\mathcal{O}\, o(x) = \underline{\pi}_{\mathcal{R}}(x) - \mathcal{O} e_1 \underline{\pi}_0(x) \\ &\text{where, } \mathcal{O} = \begin{bmatrix} e_{\mathcal{R}_1} & e_{\mathcal{R}_2} & \cdots & e_{\mathcal{R}_L} \end{bmatrix}' (I - A')^{-1}. \end{aligned} \tag{5.28}$$

Recall, $o(x)$ is defined in (5.26), A is the adjacency matrix and e_m denotes the unit $\mathcal{R}_L$ dimension vector with 1 in the m-th position. □

The above theorem asserts that given the incest containing beliefs of the L recruits and pollster, the posterior distribution of the candidates can be computed via (5.27). The conditional mean estimate can then be computed.

5.6.3 An extreme example of incest in expectation polling

The following example is a Bayesian version of polling in a social network described in [91]. We show that due to data incest, expectation polling can be significantly biased. Then the methods of §5.6.1 and 5.6.2 for eliminating incest are illustrated.

Consider the social network Figure 5.5 with represents an expectation polling system. The $L-1$ recruited nodes are denoted as $\mathcal{R} = \{2, 3, \ldots, L\}$. These sampled nodes report their beliefs to the polling node $L+1$. Since the poll only considers sampled voters, for

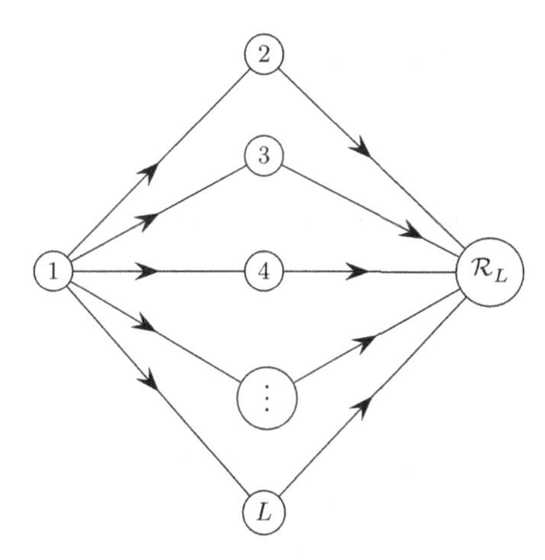

Figure 5.5 Expectation polling where social network structure can result in significant bias. Node 1 has undue influence on the beliefs of other nodes. Node $\mathcal{R}_L = L + 1$ represents the pollster. Only sampled voters are shown.

notational convenience, we ignore labeling the remaining voters; therefore Figure 5.5 only shows $L + 1$ nodes.

An important feature of the graph of Figure 5.5 is that all recruited nodes are influenced by node 1. This unduly affects the estimates reported by every other node. For large L, even though node 1 constitutes a negligible fraction of the total number of voters, it significantly biases the estimate of x due to incest.[6]

To illustrate an extreme case of bias due to data incest, suppose that $\mathcal{X} = \{1, 2\}$ (so there are $X = 2$ candidates). Consider Figure 5.5 with a total of $L = 6$ recruited nodes; so $\mathcal{R} = \{2, \ldots, 8\}$. Assume the private observations recorded at the nodes $1, 2, \ldots, 8$ are, respectively, $[2, 1, 1, 1, 1, 1, 1, 2]$. Suppose the true state of nature is $x = 1$, that is, candidate 1 is leading the poll. The nodes exchange and compute their beliefs according to Protocol 8. For prior $\pi_0 = [0.5, 0.5]'$ and observation matrix $B = \begin{bmatrix} 0.8 & 0.2 \\ 0.2 & 0.8 \end{bmatrix}$, it is easily verified that $\pi_8(1) = 0.2$, $\pi_8(2) = 0.8$. That is, even though all six samples recorded candidate 1 as winning (based on private observations) and the true state is 1, the belief π_8 of the pollster is significantly biased toward candidate 2.

Next, we examine the two methods of incest removal proposed in §5.6.1 and §5.6.2, respectively. Due to the structure of adjacency matrix A of the network in Figure 5.5, condition (5.23) does not hold and therefore exact incest removal is not possible unless node 1 is also sampled. Accordingly, suppose node 1 is sampled in addition to nodes $\{2, \ldots, 7\}$ and data incest is removed via algorithm (5.25). Then the incest free estimate is $\mathbb{P}(x = 1 | y_1, \ldots, y_8) = 0.9961$.

[6] [91] achieves unbiased expectation polling by weighing the estimate of each node by the reciprocal of its degree distribution, or alternatively sample nodes with probability inversely proportional to their degree distribution. Then highly influential nodes such as node 1 in Figure 5.5 cannot bias the estimate. This chapter is motivated by Bayesian considerations where we are interested in estimating the conditional mean estimate which by definition is unbiased.

Finally, consider the case where node 1 cannot be sampled. Then using (5.27), the posterior is computed as $\mathbb{P}(x = 1|\pi_8) = 0.5544$.

Comparing the three estimates for candidate 1, namely, 0.2 (naive implementation of Protocol 8 with incest), 0.9961 (with optimal incest removal), and 0.5544 (incest removal based on π_8), one can see that naive expectation polling is significantly biased. Recall the ground truth is that candidate 1 is leading.

5.7 Complements and sources

Throughout this chapter, numerous references have been given to social learning in economics, marketing, political science and sociology. The book [84] is a remarkable exposition of social learning in economics models. Many of the social learning models considered in subsequent chapters are motivated by [84]. For results on data incest and herding based on an actual psychology experiment conducted by our colleagues at the Department of Psychology of University of British Columbia in 2013, please see [132] for details. Parts of this chapter are based on [186] where several additional numerical examples are presented.

The expectation polling in §5.6) is a form of *social sampling* [91] where participants in a poll respond with a summary of their friends responses. [91] analyzes the effect of the social network structure on the bias and variance of expectation polls. Social sampling has interesting parallels with the so-called Keynesian beauty contest; see, for example, `https://en.wikipedia.org/wiki/Keynesian_beauty_contest` for a discussion.

Wisdom of crowds

Surowiecki's book [312] is an influential popular piece that explains the wisdom-of-crowds hypothesis. The wisdom-of-crowds hypothesis predicts that the independent judgments of a crowd of individuals (as measured by any form of central tendency) will be relatively accurate, even when most of the individuals in the crowd are ignorant and error prone. The book also studies situations (such as rational bubbles) in which crowds are not wiser than individuals. Collect enough people on a street corner staring at the sky, and everyone who walks past will look up. Such herding behavior is typical in social learning. In [343] examples are given that show that if just 10 percent of the population holds an unshakable belief, their belief will be adopted by the majority of society.

In which order should agents act?

In the social learning protocol, we assumed that the agents act sequentially in a predefined order. However, in many social networking applications, it is important to optimize the order in which agents act. For example, consider an online review site where individual reviewers with different reputations make their reviews publicly available. If a reviewer with high reputation publishes her review first, this review will unduly affect the decision of a reviewer with lower reputation. In other words, if the most senior agent "speaks" first it would unduly affect the decisions of more junior agents. This could lead to an increase in bias of the underlying state estimate. On the other hand, if the most junior agent is polled first, then since its variance is large, several agents would need to be polled in order to reduce the variance. We refer the reader to [255] for an interesting description of who should speak first in a public debate.[7] It turns out that for two agents, the seniority rule is always optimal for any prior – that is, the senior agent speaks first followed by the junior agent; see [255] for the proof. However, for more than two agents, the optimal order depends on the prior, and the observations in general.

Respondent-driven sampling

An important sampling methodology for social networks is *respondent-driven sampling* (RDS). RDS can be viewed as a Markov chain Monte Carlo sampling strategy and was introduced by Heckathorn [140, 141, 203] sampling from hidden populations in social networks. As mentioned in [126], the US Centers for Disease Control and Prevention (CDC) recently selected RDS for a 25-city study of injection drug users that is part of the National HIV Behavioral Surveillance System [202]. RDS is a variant of the well known method of snowball sampling where current sample members recruit future sample members. The RDS procedure is as follows: a small number of people in the target population serve as seeds. After participating in the study, the seeds recruit other people they know through the social network in the target population. The sampling continues according to this procedure with current sample members recruiting the next wave of sample members until the desired sampling size is reached.

Appendix 5.A Proofs

5.A.1 Proof of Theorem 5.3.1

It is convenient to deal with the ratio of elements of $\pi_k \in \Pi$. Accordingly, define $\lambda_k(i,j) = \log(\pi_k(i)/\pi_k(j))$, $i,j \in \mathcal{X}$. Note from (5.3),

$$\lambda_{k+1}(i,j) = \lambda_k(i,j) + \gamma_k(i,j), \text{ where } \gamma_{k+1}(i,j) = \log \frac{P(a_{k+1}|x=i)}{P(a_{k+1}|x=j)}.$$

Using (13.27), define the set $\bar{\mathbf{Y}}_k \subseteq \mathbf{Y}$ as the set of observation symbols for which $\prod_{\tilde{u}\in\mathbb{A}-\{a\}} I(c_a' B_y \pi_{k-1} > c_{\tilde{u}}' B_y \pi_{k-1}) = 0$. Then we can write

$$\gamma_k(i,j) = \log \frac{1 - \sum_{y\in\bar{\mathbf{Y}}_k} P(y|x=i)}{1 - \sum_{y\in\bar{\mathbf{Y}}_k} P(y|x=j)}.$$

If a cascade forms, then $\bar{\mathbf{Y}}_k$ is the empty set, so $\gamma_k(,i,j) = 0$ and no social learning takes place. If $\bar{\mathbf{Y}}_k$ is nonempty, then $|\gamma_k(i,j)| > K$ for some positive constant K (which depends on the model parameters $P(y|x)$, $x \in X$) and a cascade cannot form.

[7] As described in [255], seniority is considered in the rules of debate and voting in the US Supreme Court. "In the past, a vote was taken after the newest justice to the Court spoke, with the justices voting in order of ascending seniority largely, it was said, to avoid the pressure from long-term members of the Court on their junior colleagues."

With the above setup, we are ready to prove the result. Define the filtration sigma-algebra $\mathcal{A}_k = \sigma(a_1, \ldots, a_k)$. Clearly for any $i \in \mathcal{X}$, $\pi_k(i) = P(x = i|a_1, \ldots, a_k) = \mathbb{E}\{I(x = i)|\mathcal{A}_k\}$ is an $\mathcal{A}_k$ martingale, since $\mathbb{E}\{\pi_{k+1}(i)|\mathcal{A}_k\} = \mathbb{E}\{\mathbb{E}\{I(x = i|\mathcal{A}_{k+1}\}|\mathcal{A}_k\} = \mathbb{E}\{I(x = i|\mathcal{A}_k\}$ (via the smoothing property of conditional expectations). So, by the martingale convergence theorem, there exists a random variable π_∞, such that $\pi_k \to \pi_\infty$ with probability 1 (w.p.1). Therefore $\lambda_k(i,j) \to \lambda_\infty(i,j)$ w.p.1. Now since $\lambda_k(i,j) \to \lambda_\infty(i,j)$ w.p.1, there exists some $\bar{k}$ such for all $k \geq \bar{k}$, that $|\lambda_k(i,j) - \lambda_\infty(i,j)| < K/3$.

$$\text{Therefore, } |\lambda_{k+1}(i,j) - \lambda_k(i,j)| < 2K/3. \tag{5.29}$$

With the above results, we prove that an information cascade occurs by contradiction. Suppose a cascade does not form. Then $P(a|x = i, \pi)$ would be different from $P(a|x = j, \pi)$ for at least one pair $i, j \in X$, $i \neq j$. From (13.27) this implies that the set $\bar{\mathbf{Y}}_k$ is nonempty and so

$$|\gamma_{k+1}(i,j)| = |\lambda_{k+1}(i,j) - \lambda_k(i,j)| \geq K. \tag{5.30}$$

So (5.29), (5.30) constitute a contradiction. Therefore $\bar{\mathbf{Y}}_k$, $k \geq \bar{k}$ is the empty set or equivalently, $\gamma_k(i,j) = 1$, i.e. $R^\pi_{a_k} = I$, which is equivalent to a cascade.

5.A.2 Proof of Theorem 5.5.1

The proof uses MLR stochastic dominance (defined in footnote 5) and the following single crossing condition:

DEFINITION 5.A.1 (Single crossing [15]) $g : \mathbf{Y} \times \mathbb{A} \to \mathbb{R}$ *satisfies a single crossing condition in* (y, a) *if* $g(y, a) - g(y, \bar{a}) \geq 0$ *implies* $g(\bar{y}, a) - g(\bar{y}, \bar{a}) \geq 0$ *for* $\bar{a} > a$ *and* $\bar{y} > y$. *Then* $a^*(y) = \operatorname{argmin}_a g(y, a)$ *is increasing in* y. □

By (A1) it is verified that the Bayesian update satisfies $\frac{B_y\pi}{\mathbf{1}'B_y\pi} \leq_r \frac{B_{y+1}\pi}{\mathbf{1}'B_{y+1}\pi}$ where $\leq_r$ is the MLR stochastic order. (The MLR order is closed under conditional expectation and is discussed in detail in Chapter 10.) By submodular assumption (A2), $c_{a+1} - c_a$ is a vector with decreasing elements. Therefore

$$(c_{a+1} - c_a)' \frac{B_y\pi}{\mathbf{1}'B_y\pi} \geq (c_{a+1} - c_a)' \frac{B_{y+1}\pi}{\mathbf{1}'B_{y+1}\pi}.$$

Since the denominator is nonnegative, it follows that $(c_{a+1} - c_a)'B_{y+1}\pi \geq 0 \implies (c_{a+1} - c_a)'B_y\pi \geq 0$. This implies that $c_a'B_y\pi$ satisfies a single crossing condition in (y, a). Therefore $a_n(\pi, y) = \operatorname{argmin}_a c_a'B_y\pi$ is increasing in y.

5.A.3 Proof of Theorem 5.5.2

The local estimate at node n is given by (5.22), namely,

$$\underline{\pi}_{n-}(i) = w_n'\underline{\pi}_{1:n-1}(i). \tag{5.31}$$

Define $\bar{R}^{\pi}_{ia} = \log \mathbb{P}(a|x = i, \pi)$ and the $n-1$ dimensional vector $\bar{R}_{1:n-1}(i) = [\bar{R}^{\pi_1}_{i,a_1}, \ldots, \bar{R}^{\pi_{n-1}}_{i,a_{n-1}}]$. From the structure of transitive closure matrix T_n,

$$\underline{\pi}_{1:n-1}(i) = T'_{n-1}\,\bar{R}_{1:n-1}(i),\ \underline{\pi}_{n-}(i) = t'_n\bar{R}_{1:n-1}(i). \tag{5.32}$$

Substituting the first equation in (5.32) into (5.31) yields $\underline{\pi}_{n-}(i) = w'_n T'_{n-1}\,\bar{R}_{1:n-1}(i)$. Equating this with the second equation in (5.32) yields $w_n = T^{-1}_{n-1} t_n$. (Recall that by Lemma 5.4.1 on page 103, T_{n-1} is invertible.)

5.A.4 Proof of Theorem 5.6.1

Given n nodes, from Bayes formula and the structure of adjacency matrix A

$$\underline{\pi}_{1:n}(i) = o_{1:n} + e_1\underline{\pi}_0(i) + A'\underline{\pi}_{1:n}(i).$$

Since $I - A$ is invertible by construction,

$$\underline{\pi}_{1:n}(i) = (I - A')^{-1} o_{1:n} + (I - A')^{-1} e_1 \underline{\pi}_0(i).$$

Then $\underline{\pi}_{\mathcal{R}}(i) = \begin{bmatrix} e_{\mathcal{R}_1} & e_{\mathcal{R}_2} & \cdots & e_{\mathcal{R}_L}\end{bmatrix}' \underline{\pi}_{1:n}(i)$ satisfies (5.28). Finally $\mathbb{P}(x|\pi_{\mathcal{R}}) \propto \sum_{Y_{1:n}\in\mathcal{Y}} \mathbb{P}(Y_{1:n}|x)\,\pi_0(x)$. Here $\mathcal{Y}$ is the set of n dimensional vectors satisfying (5.28).

Part II

Partially observed Markov decision processes: models and applications

Part II of the book describes the Markov decision process (MDP) problem and the partially observed Markov decision process (POMDP) problem.

Chapter 6 gives a concise description of finite state MDPs, Bellman's stochastic dynamic programming equation for solving MDPs, and solution methods for infinite horizon Markov decision processes. Also constrained Markov decision processes are discussed briefly.

Chapter 7 describes the POMDP problem, Bellman's dynamic programming equation and associated algorithms, and discusses optimal search theory (which is a useful POMDP example).

Chapter 8 discusses several examples of POMDPs in controlled sensing including radar scheduling, social learning and risk-averse formulations.

To give some perspective, from a control theory point of view, one can consider three frameworks:

(i) Centralized control comprised of a single controller;
(ii) Team theory where there are several controllers but a single objective;
(iii) Game theory where multiple selfish decision-makers have their own individual utilities.

In this book we focus only on centralized control, apart from social learning examples where the interaction of multiple agents is discussed.

6 Fully observed Markov decision processes

Contents

A Markov decision process (MDP) is a Markov process with feedback control. That is, as illustrated in Figure 6.1, a decision-maker (controller) uses the state x_k of the Markov process at each time k to choose an action u_k. This action is fed back to the Markov process and controls the transition matrix $P(u_k)$. This in turn determines the probability that the Markov process jumps to a particular state x_{k+1} at time $k+1$ and so on. The aim of the decision-maker is to choose a sequence of actions over a time horizon to minimize a cumulative cost function associated with the expected value of the trajectory of the Markov process.

MDPs arise in stochastic optimization models in telecommunication networks, discrete event systems, inventory control, finance, investment and health planning. Also POMDPs can be viewed as continuous state MDPs.

This chapter gives a brief description of MDPs which provides a starting point for POMDPs. The main result is that optimal choice of actions by the controller in Figure 6.1 is obtained by solving a backward stochastic dynamic programming problem.

6.1 Finite state finite horizon MDP

Let $k = 0, 1, \dots, N$ denote discrete time. N is called the time horizon or planning horizon. In this section we consider MDPs where the horizon N is finite.

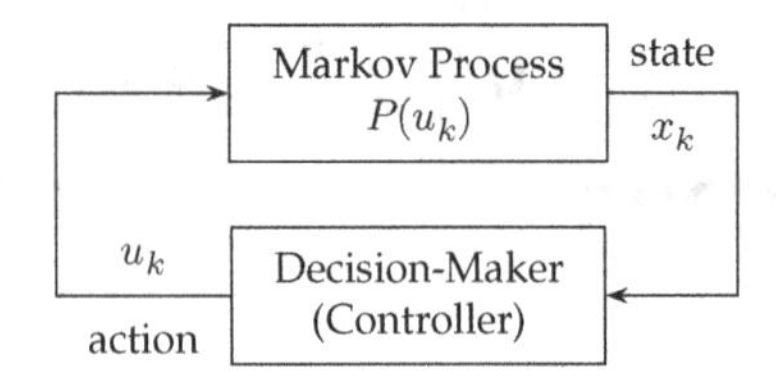

Figure 6.1 Markov decision process schematic illustrating a controlled Markov process.

The finite state MDP model consists of the following ingredients:

1. $\mathcal{X} = \{1, 2, \ldots, X\}$ denotes the state space and $x_k \in \mathcal{X}$ denotes the state of the controlled Markov chain at time $k = 0, 1, \ldots, N$.
2. $\mathcal{U} = \{1, 2, \ldots, U\}$ denotes the action space.[1] The elements $u \in \mathcal{U}$ are called actions. In particular, $u_k \in \mathcal{U}$ denotes the action chosen at time k.
3. For each action $u \in \mathcal{U}$ and time $k \in \{0, \ldots, N-1\}$, $P(u, k)$ denotes an $X \times X$ transition probability matrix with elements

$$P_{ij}(u, k) = \mathbb{P}(x_{k+1} = j | x_k = i, u_k = u), \quad i, j \in \mathcal{X}. \tag{6.1}$$

4. For each state $i \in \mathcal{X}$, action $u \in \mathcal{U}$ and time $k \in \{0, 1, \ldots, N-1\}$, the scalar $c(i, u, k)$ denotes the one-stage cost incurred by the decision-maker (controller).
5. At time N, for each state $i \in \mathcal{X}$, $c_N(i)$ denotes the terminal cost.

Thus an MDP model is the 5-tuple

$$(\mathcal{X}, \mathcal{U}, P_{ij}(u, k), c(i, u, k), c_N(i)), \qquad i, j \in \mathcal{X}, u \in \mathcal{U} \tag{6.2}$$

with $k = 0, 1, \ldots, N-1$. Throughout this chapter we assume that the model (6.2) is known.

Note that the MDP considered in this chapter is a "fully observed" model meaning that each time k the decision-maker observes the state x_k exactly (without noise or delay). Using the MDP model (6.2), Algorithm 9 describes the dynamics of an MDP. It involves at each time k choosing an action u_k, accruing an instantaneous cost $c(x_k, u_k, k)$ and evolution of the state x_k.

As described in (6.3), at each time k, the decision maker uses all the information available until time k (namely, $\mathcal{I}_k$) to choose action $u_k = \mu_k(\mathcal{I}_k)$ using policy μ_k. A *policy* (also called decision rule or strategy) μ_k is a mapping from the information set $\mathcal{I}_k$ to the action space. It is notationally convenient to denote the sequence of policies that the decision-maker uses from time 0 to $N-1$ as

$$\boldsymbol{\mu} = (\mu_0, \mu_1, \ldots, \mu_{N-1}). \tag{6.5}$$

[1] More generally, the action space $\mathcal{U}$ can be state dependent, i.e. of the form $\mathcal{U}(x)$, $x \in \mathcal{X}$. This facilitates modeling local constraints on the set of allowable actions. To simplify notation we do not consider state dependent action spaces; see point 4 in §6.2.1.

Algorithm 9 Dynamics of finite state MDP

At time $k = 0$, the state x_0 is simulated from initial distribution π_0.
For time $k = 0, 1, \ldots, N-1$:

1. The decision-maker chooses action

$$u_k = \mu_k(\mathcal{I}_k) \in \mathcal{U}, \quad k = 0, 1, \ldots, N-1. \tag{6.3}$$

   based on available information

$$\mathcal{I}_0 = \{x_0\}, \ \mathcal{I}_k = \{x_0, u_0, \ldots, x_{k-1}, u_{k-1}, x_k\}. \tag{6.4}$$

   Here μ_k denotes a policy (strategy) that the decision-maker uses at time k.
2. The decision-maker incurs an instantaneous cost $c(x_k, u_k, k)$ for choosing action u_k at time k.
3. The state evolves to x_{k+1} with transition probability matrix $P(u_k, k)$ in (6.1).
4. The decision-maker updates the available information as

$$\mathcal{I}_{k+1} = \mathcal{I}_k \cup \{u_k, x_{k+1}\}.$$

   If $k < N$, then set k to $k+1$ and go back to Step 1.
   If $k = N$, then the decision-maker pays terminal cost $c_N(x_N)$ and the process terminates.

6.1.1 Objective

To completely specify an MDP, in addition to the model (6.2), dynamics in Algorithm 9 and policy sequence (6.5), we need to specify a performance criterion or objective function. This section considers the finite horizon objective

$$J_\mu(x) = \mathbb{E}_\mu \left\{ \sum_{k=0}^{N-1} c(x_k, \mu_k(\mathcal{I}_k), k) + c_N(x_N) \mid x_0 = x \right\}. \tag{6.6}$$

which is the expected cumulative cost incurred by using policy $\boldsymbol{\mu}$ up to time N. Here, $\mathbb{E}_\mu$ denotes expectation with respect to the probability distribution induced by $\mathcal{I}_N = \{x_0, u_0, x_1, u_1, \ldots, x_{N-1}, u_{N-1}, x_N\}$.

Given the model (6.2), and subject to the dynamics specified in Algorithm 9, the goal of the decision-maker is to determine the optimal policy

$$\boldsymbol{\mu}^* = \operatorname*{argmin}_{\boldsymbol{\mu}} J_\mu(x). \tag{6.7}$$

In words: find the policy sequence $\boldsymbol{\mu}$ that minimizes the expected cumulative cost (6.6) for *every* initial state x. If $\mathcal{X}$ and $\mathcal{U}$ are finite, the minimum always exists. An optimal policy $\boldsymbol{\mu}^*$ attains the smallest expected cumulative cost amongst all policies for every initial state. Obviously, the optimal policy sequence $\boldsymbol{\mu}^*$ may not be unique. (For example, if the costs $c(x, u)$ are identical for all x and u, then all policies are optimal.)

The above MDP is an instance of a stochastic control problem. Such problems differ from static optimization problems in three ways. A decision made at time k affects:

(i) the costs incurred in the future; (ii) the choice of actions in the future; and (iii) the values of the state in the future. Note that the action at time k is determined dynamically by using all information $\mathcal{I}_k$ available until time k. This is the essence of *closed loop* control. In comparison, in static optimization (open loop), the actions $u_0, \ldots, u_{N-1}$ are determined at time 0.

6.1.2 Classes of policies

To solve the MDP (6.7) for an optimal policy $\boldsymbol{\mu}^*$, we need to examine the space of policies more carefully. We distinguish three types of policies:

1. *General policies*: The most general class of policies $\boldsymbol{\mu} = (\mu_0, \mu_1, \ldots, \mu_{N-1})$ are randomized history dependent. That is, at each time k, action u_k is chosen according to probability distribution $\mu_k(\mathcal{I}_k)$ where $\mathcal{I}_k$ is defined in (6.4). So u_k is a probabilistic function of $\mathcal{I}_k$.
2. *Randomized Markovian policies*: Here u_k is chosen according to probability distribution $\mu_k(x_k)$. That is, u_k is a probabilistic function of state x_k only.
3. *Deterministic Markovian policies*: Here indexpolicies! deterministic Markovian $u_k = \mu_k(x_k)$ where μ_k denotes a deterministic mapping from the state space $\mathcal{X}$ to action space $\mathcal{U}$.

Clearly, deterministic Markovian policies are a subset of randomized Markovian policies when the probabilities are degenerate (either 0 or 1). Also, randomized Markov policies are a subset of general policies.

An important feature of the MDP problem (6.7) is that it suffices to consider deterministic Markovian policies to achieve the minimum (shown below). This means that the expectation $\mathbb{E}_{\boldsymbol{\mu}}$ in (6.6) is computed only with respect to $\{x_0, x_1, \ldots, x_N\}$ since $u_k = \mu_k(\mathcal{I}_k)$ is a deterministic function of x_k.

6.2 Bellman's stochastic dynamic programming algorithm

Consider the MDP problem (6.6) with objective (6.7). The optimal policy $\boldsymbol{\mu}^*$ is obtained via Bellman's stochastic dynamic programming algorithm.[2]

THEOREM 6.2.1 (Bellman's dynamic programming algorithm) *The optimal policy $\boldsymbol{\mu}^* = (\mu_0, \mu_1, \ldots, \mu_{N-1})$ for the finite horizon MDP (6.2) can be obtained as the solution of the following backward recursion:*
Initialize $J_N(i) = c(i, N)$. Then for $k = N-1, \ldots, 0$ evaluate

$$J_k(i) = \min_{u \in \mathcal{U}} \{c(i,u,k) + \sum_j P_{ij}(u,k) J_{k+1}(j)\}$$

$$\mu_k^*(i) = \operatorname*{argmin}_{u \in \mathcal{U}} \{c(i,u,k) + \sum_j P_{ij}(u,k) J_{k+1}(j)\}. \tag{6.8}$$

[2] The study of dynamic programming dates back to Richard Bellman, who wrote the first book on this subject in 1957 [40] and gave it its name.

For any initial state $i \in \{1,\ldots,X\}$, the expected cumulative cost of the optimal policy $\boldsymbol{\mu}^*$, namely $J_{\mu^*}(i)$ in (6.7) is obtained as $J_0(i)$ from (6.8).

The proof of Theorem 6.2.1 relies on the principle of optimality:

From any point on an optimal trajectory, the remaining trajectory is optimal for the corresponding problem initiated at that point.

THEOREM 6.2.2 (Principle of optimality) *Consider the finite state Markov decision process (6.6) with objective (6.7). Then*

1. *The deterministic Markovian policy $\boldsymbol{\mu}^*$ obtained via the stochastic dynamic programming recursion (6.8) is optimal in the sense of (6.7).*
2. *Conversely, if a deterministic Markovian policy $\boldsymbol{\mu}^* = (\mu_0^*, \mu_1^*, \ldots, \mu_{N-1}^*)$ is optimal in the sense of (6.7) then it satisfies (6.8).*

The proof is presented in the appendix of this chapter and uses the notation of continuous-valued MDPs introduced in §6.3.

6.2.1 Remarks on Bellman's dynamic programming algorithm (6.8)

1. Bellman's algorithm (6.8) gives an explicit construction of the optimal policy μ^* and also the expected cumulative cost attained by the optimal policy, namely $J_{\mu^*}(i)$. The optimal policy is *deterministic (nonrandomized) Markovian* by construction.
2. The dynamic programming algorithm (6.8) runs offline (it does not depend on the sample paths; it only depends on the MDP model parameters) and backwards in time to compute the optimal policies μ_k^*, $k = N-1,\ldots,0$. Implementing (6.8) involves $O(X^2U)$ multiplications at each iteration. The optimal policy $\boldsymbol{\mu}^*$ can be stored in an $X \times U \times N$ lookup table.

Once the optimal policy is computed (offline) via (6.8), then the real-time controller is implemented as illustrated in Figure 6.1. The decision-maker (real-time controller) runs Algorithm 9. The optimal actions u_k, $k = 0, 1, \ldots, N-1$ are chosen in real-time k as

$$u_k = \mu_k^*(x_k).$$

That is, in Step 1 of Algorithm 9, the decision-maker (controller) views the current state x_k, then uses the lookup table (policy μ_k^*) to choose action u_k.
3. The main intuition is that the dynamic programming recursion reduces an optimization over policy space (6.7) to a sequence of minimizations carried out over the set of actions (6.8). The function $J_k(x)$ is called the *value function* at iteration k. It is the optimal expected cumulative cost for the $N-k$ horizon problem that starts in state x at time k and terminates at time N. That is to say, for $k = 0, 1, \ldots, N$,

$$J_k(x) = \min_{\mu_k,\ldots,\mu_{N-1}} \mathbb{E}_{\mu^*}\left\{\sum_{l=k}^{N-1} c(x_l, u_l, l) + c(x_N, N) \mid x_k = x\right\}. \tag{6.9}$$

For this reason, J_k is also called the *optimal cost-to-go function* at time k. Here $\mathbb{E}_{\mu^*}$ denotes expectation with respect to $x_k, \ldots, x_N$ using the optimal policy $\boldsymbol{\mu}^*$.

4. Continuing with footnote 1 on page 122, if the action space is state dependent of the form $\mathcal{U}(x)$, then Theorem 6.2.1 continues to hold with $\mathcal{U}$ in min and argmin of (6.8) replaced by $\mathcal{U}(x)$.
5. More generally, if the costs are of the form $c(\zeta_k, u_k, k)$, where ζ_k is possibly some noisy function of the state, the dynamic programming recursion reads

$$J_k(x_k) = \min_{u \in \mathcal{U}} \{\mathbb{E}\{c(\zeta, u, k) + J_{k+1}(x_{k+1}) \mid \mathcal{I}_k\}\}.$$

where $\mathcal{I}_k = (x_0, u_0, x_1, u_1, \dots, x_{k-1}, u_{k-1}, x_k)$ denotes all the information available to the decision-maker up to time k.

For example, suppose the cost is of the form $c(x_k, x_{k+1}, u_k, k)$. That is, it depends on the current and future state. Then the expected cost $\mathbb{E}\{c(x_k, x_{k+1}, u_k, k) \mid \mathcal{I}_k\}$ is straightforwardly obtained from the optimal predictor as

$$c(x, u_k, k) \overset{\text{defn}}{=} \mathbb{E}\{c(x_k, x_{k+1}, u_k, k) \mid \mathcal{I}_k\} = \sum_{j=1}^{X} c(x_k = i, x_{k+1} = j, u_k, k) P_{ij}(u_k). \quad (6.10)$$

6.2.2 Forward recursion

Bellman's equation (6.8) is a backward dynamic programming recursion initialized with J_N and proceeding backwards for $k = N-1, \dots, 0$. It is sometimes convenient to re-express Bellman's equation in terms of a forward iteration[3] of indices. Define the value function

$$V_n(x) = J_{N-n}(x), \quad 0 \leq n \leq N, \ x \in \mathcal{X}. \quad (6.11)$$

It follows from (6.8) that $V_n(x)$ satisfies the dynamic programming equation

$$V_n(i) = \min_u \left\{ c(i, u, N-n) + \sum_{j \in \mathcal{X}} P_{ij}(u, N-n)\, V_{n-1}(j) \right\} \quad (6.12)$$

initialized by $V_0(x) = c_N(x)$ which is the terminal cost.

6.2.3 Example: Machine replacement

Machine replacement is often used as a benchmark problem in MDPs. The state space $\mathcal{X} = \{1, 2, \dots, X\}$ corresponds to the machine quality. State 1 denotes the worst machine performance while X corresponds to the best state.

The action space is $\mathcal{U} \in \{1 \text{ (replace machine)}, 2 \text{ (use existing machine)}\}$. When action $u = 1$ is chosen, the resulting brand new machine starts at state X. When action $u = 2$ is chosen, the machine state will either remain the same or deteriorate by one unit. Let θ denote the probability that the machine state will deteriorate by one unit.

[3] Please note that this is still a backward dynamic programming formulation; we have simply specified a set of new variables that iterate with a forward index. In particular, this is **not** forward dynamic programming, which was used in the Viterbi Algorithm 4 on page 53 in Chapter 3.

Suppose the number of states $X = 3$. Then the transition probabilities $P(u)$, $u \in \mathcal{U}$, of the MDP are (we denote $P(u,k)$ by $P(u)$ since they are time invariant):

$$P(1) = \begin{bmatrix} 0 & 0 & 1 \\ 0 & 0 & 1 \\ 0 & 0 & 1 \end{bmatrix}, \quad P(2) = \begin{bmatrix} 1 & 0 & 0 \\ \theta & 1-\theta & 0 \\ 0 & \theta & 1-\theta \end{bmatrix}. \tag{6.13}$$

Operating a machine in state x incurs a cost $c(x, u = 2)$ due to possible loss of productivity and poor quality of the machine output. On the other hand, replacing the machine at any state x costs R. So $c(x, u = 1) = R$. The aim is to minimize the expected cumulative cost $\mathbb{E}_\mu\{\sum_{k=0}^{N-1} c(x_k, u_k)|x_0\}$ for some specified horizon N. Bellman's equation (6.8) reads: initialize $J_N(i) = 0$. For $k = N-1, \dots, 0$:

$$J_k(i) = \min\{R + J_{k+1}(1), c(i,2) + \sum_{j=1}^{X} P_{ij}(2) J_{k+1}(j)\}, \quad i \in \{1, \dots, X\}$$

The optimal policy μ_k^*, which is the argmin of the right-hand side, is stored in a lookup table. Then the real-time controller operates as in Algorithm 9.

6.3 Continuous-state MDP

In this section we discuss *continuous-state* MDPs and the corresponding Bellman's stochastic dynamic programming equation. We will see in Chapter 7 that a POMDP can be formulated as a continuous-state MDP; so the dynamic programming equation described in this section is of relevance.

In §2.1 we described stochastic state space models. Here we consider a controlled version of this model that evolves over time $k = 0, 1, \dots, N-1$ as

$$x_{k+1} = \phi_k(x_k, u_k, w_k), \tag{6.14}$$

initialized by state x_0. Here $\{w_k\}$ is an i.i.d. random process with probability density (or mass) p_w that is statistically independent of initial state x_0. Also, $u_k \in \mathcal{U}$ denotes the action (control) taken by a decision-maker (controller) at each time k. $\mathcal{U}$ denotes a set from which the action is chosen.[4] If action u_k is chosen at time k, an instantaneous cost $c(x_k, u_k, k)$ is incurred at time k. Finally a terminal cost $c_N(x_N)$ is incurred at time N.

As in Chapter 2, we can write the state space model (6.14) as a Markov process with action dependent transition probabilities

$$\mathbb{P}(x_{k+1} \in S | x_k = x, u_k = u) = \int I(\phi_k(x,u,w) \in S)\, p_w(w) dw$$

where S denotes any measurable subset of $\mathbb{R}^X$ and $I(\cdot)$ denotes the indicator function. Alternatively, with δ denoting the Dirac-delta function, in transition density form this reads

[4] Continuous-state MDPs with continuous valued actions is a highly technical area where measurability issues arise [52] and is outside the scope of this book. We consider a finite action space and assume that the state space is a bounded set. For example in POMDPs, the action space is finite and the state space is the unit simplex.

$$p(x_{k+1}|x_k = x, u_k = u) = \int \delta(x_{k+1} - \phi_k(x, u, w))p_w(w)dw.$$

In summary, the continuous-state MDP is a 5-tuple

$$(\mathcal{X}, \mathcal{U}, p(x_{k+1}|x_k, u_k), c(x_k, u_k, k), c_N(x_N)). \tag{6.15}$$

The objective is to determine the policy $\boldsymbol{\mu}^*$ which minimizes the expected cumulative cost (6.6). This optimal policy satisfies the following Bellman's stochastic dynamic programming equation: Initialize $J_N(x) = c_N(x)$ (terminal cost). Then for $k = N-1, \ldots, 0$, the optimal policy is evaluated as $\boldsymbol{\mu}^* = (\mu_0^*, \mu_1^*, \ldots, \mu_{N-1}^*)$ where

$$\boxed{\begin{aligned} J_k(x) &= \min_{u \in \mathcal{U}} \left\{ c(x, u, k) + \int J_{k+1}\left(\phi_k(x, u, w)\right) p_w(w)\, dw \right\}, \\ \mu_k^*(x) &= \operatorname*{argmin}_{u \in \mathcal{U}} \left\{ c(x, u, k) + \int J_{k+1}\left(\phi_k(x, u, w)\right) p_w(w)\, dw \right\}. \end{aligned}} \tag{6.16}$$

The expected cumulative cost associated with optimal policy $\boldsymbol{\mu}^*$, namely $J_{\boldsymbol{\mu}^*}(x)$ in (6.7) is obtained from (6.16) as $J_0(x)$ for any initial state x. The proof that (6.16 achieves the minimum is given the appendix (proof of Theorem 6.2.2). Of course, since x is continuous valued, one cannot in general solve (6.16) in closed form. The state either needs to be discretized over a grid or the structure of the dynamics needs to be exploited to determine (or approximate) a solution to (6.16).

6.4 Infinite horizon discounted cost

In this section we consider the case where the horizon length $N = \infty$. Also, the transition probabilities and costs are assumed not to be explicit functions of time, and there is no terminal cost. The infinite horizon discounted MDP model considered here is the 5-tuple

$$(\mathcal{X}, \mathcal{U}, P_{ij}(u), c(i, u), \rho), \qquad i, j \in \mathcal{X}, u \in \mathcal{U}.$$

where $\rho \in [0, 1)$ is an economic discount factor. The discount factor ρ weights the costs in the following manner: the cost incurred by the decision-maker at time k is $\rho^k c(x_k, u_k)$. Therefore, the first few decisions are much more important than subsequent decisions.

6.4.1 Objective and dynamic programming equation

The aim is to determine the optimal policy $\boldsymbol{\mu}^* = \operatorname{argmin}_{\boldsymbol{\mu}} J_{\boldsymbol{\mu}}(i)$ where $J_{\boldsymbol{\mu}}(i)$ denotes the infinite horizon discounted cumulative cost

$$J_{\boldsymbol{\mu}}(i) = \mathbb{E}_{\boldsymbol{\mu}} \left\{ \sum_{k=0}^{\infty} \rho^k\, c(x_k, u_k) \mid x_0 = i \right\}. \tag{6.17}$$

Here $\boldsymbol{\mu} = (\mu_0, \mu_1, \ldots)$ is a sequence of policies where μ_k at time k maps $\mathcal{I}_k = \{x_0, u_0, \ldots, x_{k-1}, u_{k-1}, x_k\}$ to action u_k. The discount factor ρ ensures that $J_{\boldsymbol{\mu}}(i)$ is always bounded since clearly,

$$J_{\boldsymbol{\mu}}(i) \leq \max_{i,u} \frac{|c(i,u)|}{1-\rho}. \tag{6.18}$$

It is intuitive that the first few decisions are insensitive to horizon N if N is large. So as $N \to \infty$, one expects that the optimal policy at time 0, namely μ_0^*, loses its dependence on N. As a result, it is intuitive that the optimal policy μ_0^* is also optimal at subsequent times $k = 1, 2, \ldots$. Such a time invariant policy is called a stationary policy.

Define a *stationary* policy sequence as $\boldsymbol{\mu} = (\mu, \mu, \mu, \cdots)$. Here each individual policy μ at time k is no longer an explicit function of k. As in §6.1.2, define a stationary *Markovian* policy as one where the decision-maker chooses the action $u_k = \mu(x_k)$ instead of $u_k = \mu(\mathcal{I}_k)$. For notational convenience when denoting stationary policies we will use μ instead of $\boldsymbol{\mu}$. The cumulative cost (6.17) associated with stationary policy μ will be denoted as $J_\mu(i)$.

To get some intuition on what the dynamic programming equation for an infinite horizon discounted MDP might look like, consider the case when the horizon N is finite. From the dynamic programming algorithm of Theorem 6.2.2, it follows that

$$J_k(i) = \min_{u \in \mathcal{U}} \{\rho^k \, c(i,u) + \sum_{j \in \mathcal{X}} P_{ij}(u) J_{k+1}(j)\}$$

initialized by $J_N(i) = 0$. For discounted cost problems, it is more convenient to work with the following value function $V_n(i)$ defined as

$$V_n(i) = \rho^{n-N} J_{N-n}(i), \quad 0 \leq n \leq N, \; i \in \mathcal{X}.$$

Then $V_n(i)$ satisfies the dynamic programming equation

$$V_n(i) = \min_u \{c(i,u) + \rho \sum_{j \in \mathcal{X}} P_{ij}(u) V_{n-1}(j)\}, \quad V_0(i) = 0. \tag{6.19}$$

It is intuitive that the dynamic programming equation for an infinite horizon discounted MDP is obtained as the limit $n \to \infty$ in (6.19). The main result for discounted cost MDPs is stated in the following three assertions:

THEOREM 6.4.1 *Consider an infinite horizon discounted cost Markov decision process with discount factor $\rho \in [0, 1)$. Then*

1. *For any initial state i, the optimal cumulative cost $J_{\mu^*}(i)$ is attained by the value function $V(i)$ which satisfies Bellman's equation (6.20).*
2. *For any initial state i, the optimal cumulative cost $J_{\mu^*}(i)$ achieved by the stationary deterministic Markovian policy μ^* which satisfies Bellman's equation (6.20).*
3. *The value function V is the unique solution to Bellman's equation (6.20). (Of course, the optimal policy may not be unique.)*

$$V(i) = \min_{u \in \mathcal{U}} Q(i,u), \quad \mu^*(i) = \operatorname*{argmin}_{u \in \mathcal{U}} Q(i,u),$$
$$Q(i,u) = c(i,u) + \rho \sum_j P_{ij}(u) V(j). \tag{6.20}$$

The proof of the first statement is very similar to the proof of statement 2 of Theorem 6.2.2 in the appendix and hence omitted. To prove the second assertion note that (6.20) can be expressed as

$$
\begin{aligned}
V(i) &= \mathbb{E}\{c(x_0, \mu^*(x_0))\} + \rho \sum_{k=1}^{\infty} \mathbb{E}\{V(x_1)|x_0 = i\} \\
&= \mathbb{E}\{\sum_{k=0}^{N-1} \rho^k \, c(x_k, \mu^*(x_k)\} + \rho^N \sum_{k=N}^{\infty} \mathbb{E}\{V(x_N)|x_0 = i\}.
\end{aligned}
$$

Since $V(x_N)$ is bounded from (6.18) and $\rho^N \to 0$ as $N \to \infty$ it follows that $V(i) = \mathbb{E}\{\sum_{k=0}^{\infty} \rho^k \, c(x_k, \mu^*(x_k)\}$. The third assertion (uniqueness of value function) follows from contraction mapping theorem for the value iteration algorithm used to prove Theorem 6.4.2.

6.4.2 Numerical methods

Here we outline three classical methods for solving infinite horizon discounted cost Markov decision processes.

Value iteration algorithm

The value iteration algorithm is a successive approximation algorithm to compute the value function V of Bellman's equation (6.20). It proceeds as follows. Choose the number of iterations N (typically large). Initialize $V_0(i) = 0$. Then for iterations $k = 1, 2, \ldots, N$, compute

$$
\begin{aligned}
V_k(i) &= \min_{u \in \mathcal{U}} Q_k(i, u), \quad \mu_k^*(i) = \operatorname{argmin}_{u \in \mathcal{U}} Q_k(i, u), \\
Q_k(i, u) &= c(i, u) + \rho \sum_j P_{ij}(u) V_{k-1}(j).
\end{aligned}
\tag{6.21}
$$

Then use the stationary policy μ_N^* at each time instant in the real-time controller of Algorithm 9. Note that the value iteration algorithm is identical to the finite horizon dynamic programming recursion described in (6.19), except that in the controller implementation the stationary policy μ_N^* is used at each time.

THEOREM 6.4.2 *The value iteration algorithm converges geometrically fast to the optimal value function V (6.20) of Bellman's equation. In particular,*

$$
|V(i) - V_k(i)| \leq \frac{\rho^{k+1}}{1-\rho} \max_{i,u} |c(i, u)|.
$$

The proof in the context of POMDPs appears in Theorem 7.6.2, §7.6.

Policy iteration algorithm

This is an iterative algorithm that computes an improved policy at each iteration compared to that of the previous iteration. Since for an U-action, X-state Markov decision process, there are a finite number X^U possible stationary policies, the policy iteration algorithm converges to the optimal policy in a finite number of iterations.

The policy iteration algorithm proceeds as follows.
Assume a stationary policy μ_{n-1} and its associated cumulative cost $J_{\mu_{n-1}}$ from iteration $n-1$ are given. At iteration n:

1. *Policy improvement*: Compute stationary policy μ_n as

$$\mu_n(i) = \underset{u}{\operatorname{argmin}} \left[c(i,u) + \rho \sum_j P_{ij}(u) J_{\mu_{n-1}}(j) \right], \quad i \in \mathcal{X}.$$

2. *Policy evaluation*: Given policy μ_n, compute the discounted cumulative cost associated with this policy as

$$J_{\mu_n}(i) = c(i, \mu_n(i)) + \rho \sum_j P_{ij}(\mu_n(i)) J_{\mu_n}(j), \quad i \in \mathcal{X}. \tag{6.22}$$

   This is a linear system of equations and can be solved for J_{μ_n}.
   If $J_{\mu_n}(i) < J_{\mu_{n-1}}(i)$ for all i, then set $n = n+1$ and continue.
   Else stop.

Linear programming

Here we formulate Bellman's equation as a linear programming optimization problem. Define $\underline{V}(i)$ such that

$$\underline{V}(i) \leq \min_u c(i,u) + \rho \sum_{j \in \mathcal{X}} P_{ij}(u) \underline{V}(j)$$

It is easily shown by mathematical induction on the value iteration algorithm that $\underline{V}(i) \leq V(i)$ where V is the value function from Bellman's equation. Therefore the value function V is the largest $\underline{V}$ that satisfies the above inequality. So V is the solution of the optimization problem

$$\max_{\underline{V}} \sum_i \alpha_i \underline{V}(i) \quad \text{subject to } \underline{V}(i) \leq \min_u c(i,u) + \rho \sum_{j \in \mathcal{X}} P_{ij}(u) \underline{V}(j), \quad i \in \mathcal{X}.$$

where α_i, $i \in X$ are any arbitrary nonnegative scalars. Equivalently, V is the solution of the following linear programming problem:

$$\begin{aligned} &\max_{\underline{V}} \sum_i \alpha_i \underline{V}(i) \\ &\text{subject to } \underline{V}(i) \leq c(i,u) + \rho \sum_{j \in \mathcal{X}} P_{ij}(u) \underline{V}(j), \quad i \in \mathcal{X}, u \in \mathcal{U}. \end{aligned} \tag{6.23}$$

The dual problem to (6.23) is the linear program

$$\begin{aligned} &\text{Minimize } \sum_{i \in \mathcal{X}} \sum_{u \in \mathcal{U}} c(i,u)\, \pi(i,u) \text{ with respect to } \pi \\ &\text{subject to } \pi(i,u) \geq 0, \quad i \in \mathcal{X}, u \in \mathcal{U} \end{aligned}$$

$$\sum_u \pi(j,u) = \rho \sum_i \sum_u P_{ij}(u)\pi(i,u) + \alpha_j, \; j \in \mathcal{X}. \tag{6.24}$$

If the nonnegative parameters α_j are chosen as $\alpha_j = \mathbb{P}(x_0 = j)$, then $\pi(i,u)$ can be interpreted as $\pi(i,u) = \sum_{k=0}^{\infty} \rho^n \, \mathbb{P}(x_k = i, u_k = u)$ which is the expected discounted time of being in state i and taking action a. The optimal policy is

$$\mu^*(i) = u \text{ if } \pi(i,u) > 0, \quad u \in \mathcal{U}.$$

Since the optimal policy is nonrandomized (dynamic programming yields a nonrandomized Markovian policy by construction), for each i, only one $u \in \mathcal{U}$ will have a nonzero $\pi(i,u)$.

6.5 Infinite horizon average cost

The discounted cost criterion discussed in §6.4 weighs the costs involved with the first few decisions much more heavily than those in the long run. In comparison, the average cost criterion is the expected cost per unit time over the long run. So the decisions from the first few steps do not influence the overall infinite horizon cumulative cost.

6.5.1 Objective and dynamic programming equation

As in the discounted cost case, by a stationary policy we mean the sequence of policies $\boldsymbol{\mu} = (\mu, \mu, \mu, \ldots)$ and for notational convenience we denote $\boldsymbol{\mu}$ by μ. For any stationary policy μ, let $\mathbb{E}_\mu$ denote the corresponding expectation and define the infinite horizon average cost as

$$J_\mu(x_0) = \lim_{N \to \infty} \frac{1}{N+1} \mathbb{E}_\mu \left[\sum_{k=0}^{N} c(x_k, u_k) \mid x_0 \right]. \tag{6.25}$$

We say that a stationary policy μ^* is optimal if $J_\mu^*(x_0) \leq J_\mu(x_0)$ for all states $x_0 \in \mathcal{X}$.

Unlike discounted infinite horizon problems, average cost problems are plagued by technicalities. Depending on the structure of the transition matrices $P(u)$, an optimal stationary policy need not even exist. Such existence problems are highly technical – particularly for infinite state spaces (denumerable or uncountable). We refer the reader to [21, 294] for general sufficient conditions for the existence of optimal stationary policies for average cost problems.

Our description here is much more modest. Since we are considering finite state Markov chains the following condition is sufficient for our needs

DEFINITION 6.5.1 (Unichain Condition) *An MDP is said to be unichain if every deterministic stationary policy yields a Markov chain with a single recurrent class (that is, one communicating class of states) and a set of transient states which may be empty.*

It turns out that if an average cost MDP is unichain, then an optimal stationary policy μ^* always exists. Checking the unichain condition can be computationally intractable though since there are X^U possible stationary policies; for each such policy one needs to check that the resulting transition matrix $(P_{ij}(\mu(i)),\ i,j \in \mathcal{X})$ forms a recurrent class. If all the transition probabilities $P_{ij}(u)$ are strictly positive for each μ, then the unichain condition trivially holds.

The following analog of Bellman's dynamic programming equation holds:

THEOREM 6.5.2 *Consider a finite-state finite-action unichain average cost Markov decision process. Then the optimal policy μ^* satisfies*

$$g + V(i) = \min_u \{c(i,u) + \sum_j P_{ij}(u)V(j)\} \tag{6.26}$$

$$\mu^*(i) = \operatorname*{argmin}_u \{c(i,u) + \sum_j P_{ij}(u)V(j)\}. \tag{6.27}$$

Here μ^* is a stationary nonrandomized policy. Also $g = J_{\mu^*}(x_0)$ denotes the average cost achieved by the optimal policy μ^* and is independent of the initial state x_0. Furthermore g is unique while if $V(i), i \in \mathcal{X}$ satisfies (6.26) then so does $V(i) + K$ for any constant K.

(6.26) is called the *average cost optimal equation* (ACOE). The proof of the above theorem proceeds in two steps.

THEOREM 6.5.3 *If a bounded function $V(i)$ and a constant g exist that satisfy ACOE (6.26), then the optimal cost for the average cost MDP is achieved by the stationary nonrandomized policy μ^* defined in (6.27).*

THEOREM 6.5.4 *For a unichain finite state MDP, the conditions of Theorem 6.5.3 hold.*

The proofs of both theorems are in the appendix to this chapter. Theorem 6.5.3 can be proved as an existence result for the solution of *Poisson's equation*.[5] For a Markov chain with transition probability matrix P, Poisson's equation reads:

$$\phi + h(i) = c(i) + \sum_j P_{ij}h(i), \quad i \in \mathcal{X} \tag{6.28}$$

where c is an input. Comparing Poisson's equation (6.28) and ACOE (6.26), the two equations are equivalent for the optimal policy μ^* with variables

$$\phi \to g,\ h \to V,\ P_{ij}(\mu^*(i)) \to P_{ij} \text{ and } c(i,\mu^*(i)) \to c(i).$$

[5] Suppose P is a regular transition matrix of a Markov chain $\{x_n\}$ with stationary distribution π. Then there exists a process $\{\nu_n\}$ satisfying Poisson's equation, namely,

$$x_{n+1} - \pi = (I - P)'\nu_{n+1} \text{ or equivalently } \nu_{n+1} = \sum_{i=0}^{\infty}(P^i - 1\pi')'x_{n+1}$$

where $\nu_{n+1} - P'\nu_n$ is a martingale difference. That is, $\mathbb{E}\{\nu_{n+1}|x_0,\ldots,x_n\} - P'\nu_n = 0$. The proof follows by substituting the martingale decomposition (2.28) for x_{n+1} into Poisson's equation. So Poisson's equation can be viewed as a martingale decomposition for a Markov process.

Poisson's equation leads naturally to the average cost criterion: it will be shown in the appendix that if the solution (ϕ, h) to Poisson's equation is such that h is bounded, then ϕ satisfies

$$\phi = \lim_{N\to\infty} \frac{1}{N+1}\mathbb{E}\left[\sum_{k=0}^{N} c(x_k) \mid x_0\right].$$

The proof of Theorem 6.5.4 uses the so-called "vanishing discount factor" idea which relates an average cost MDP to a discounted cost MDP as follows: Let V^ρ denote the value function of a discounted cost MDP with discount factor ρ (this was defined in (6.20)). Then (g, V) of the ACOE (6.26) are obtained as

$$\begin{aligned} g &= \lim_{\rho\to 1}(1-\rho)V^\rho(i) \\ V(i) &= \lim_{\rho\to 1}\left(V^\rho(i) - V^\rho(j)\right) \quad \text{for any state } j \end{aligned} \tag{6.29}$$

The proof of Theorem 6.5.4 shows that the above limits are well defined.

Remark: There are two deep relationships between discounted cost problems and average cost problems that we will not explore in detail; see [272]. The first is the Tauberian theorem in infinite series. Translated to our setting it reads: if $\lim_{N\to\infty}\frac{1}{N+1}\sum_{n=0}^{N} c_n$ exists, then $\lim_{\rho\uparrow 1}(1-\rho)\sum_{n=0}^{\infty}\rho^n c_n$ exists and

$$\lim_{\rho\to 1}(1-\rho)\sum_{n=0}^{\infty}\rho^n c_n = \frac{1}{N+1}\sum_{n=0}^{N} c_n.$$

The second relation deals with Blackwell optimal policies. Let μ^*_ρ denote the optimal policy of a discounted MDP with discount factor ρ. Let μ^* denote the optimal policy of an average cost MDP. Under the conditions of Theorem 6.5.4, it follows that given a sequence $\rho_n \to 1$, then $\mu^*_{\rho_n} \to \mu^*$. Since there are only a finite number of possible policies, the convergence implies that: there exists a discount factor $\bar{\rho} < 1$, such that for all $\rho \in [\bar{\rho}, 1)$, the optimal policy for a discounted cost MDP coincides with that of an average cost MDP. Therefore an average cost MDP behaves like a discounted cost MDP when the discount factor ρ is close to 1 (but determining $\bar{\rho}$ may not be not straightforward). Such policies μ^*_ρ that coincide with μ^* are called Blackwell optimal policies [49].

6.5.2 Numerical methods

Relative value iteration algorithm

Since by Theorem 6.5.2 any constant added to the value function also satisfies (6.26), it is is convenient to rewrite (6.26) in the following form that is anchored at state 1. (One could choose any state instead of 1). Define $\tilde{V}(i) = V(i) - V(1)$. So obviously $\tilde{V}(1) = 0$. Then ACOE (6.26) can be rewritten relative to state 1 as

$$g + \tilde{V}(i) = \min_u\{c(i,u) + \sum_{j>1} P_{ij}(u)\tilde{V}(j)\}, \quad i > 1,$$

$$g + \cancel{V(1)} = \min_u \{c(1,u) + \sum_j P_{1j}(u)\tilde{V}(j)\} + \cancel{V(1)}. \tag{6.30}$$

In analogy to the value iteration algorithm described earlier for the discounted cost case, (6.30) yields the following *relative value iteration algorithm* that operates for $k = 1, 2, \ldots$

$$\begin{aligned} \tilde{V}_k(i) &= \min_u \{c(i,u) + \sum_{j>1} P_{ij}(u)\tilde{V}_{k-1}(j)\} - g_k, \quad i > 1, \\ g_k &= \min_u \{c(1,u) + \sum_j P_{1j}(u)\tilde{V}_{k-1}(j)\}. \end{aligned} \tag{6.31}$$

Linear programming

An average cost MDP can be formulated as a linear programming problem and solved using interior point methods [68] or simplex algorithms. From Theorem 6.5.2, solving Bellman's equation is equivalent to the following linear program

$$\begin{aligned} &\text{Maximize } g \\ &\text{subject to } g + V(i) \leq c(i,u) + \sum_{j \in \mathcal{X}} P_{ij}(u)V(j). \end{aligned} \tag{6.32}$$

The dual of the above problem can be formulated as a linear programming problem which we will now describe. Let $\pi(x,u) = \mathbb{P}(x_k = x, u_k = u)$, $x \in \mathcal{X}$, $u \in \mathcal{U}$, denote the joint action state probabilities.

THEOREM 6.5.5 *Consider a finite-state finite-action unichain average cost Markov decision process. Then the optimal policy μ^* is*

$$\mu^*(x) = u \text{ with probability } \theta_{x,u} = \frac{\pi^*(x,u)}{\sum_{i \in \mathcal{X}} \pi^*(i,u)}, \quad x \in \mathcal{X}. \tag{6.33}$$

The $X \times U$ elements of π^* are the solution of the linear programming problem:

$$\textit{Minimize} \sum_{i \in \mathcal{X}} \sum_{u \in \mathcal{U}} c(i,u)\pi(i,u) \ \textit{with respect to } \pi \tag{6.34}$$

$$\textit{subject to } \pi(i,u) \geq 0, \quad i \in \mathcal{X}, u \in \mathcal{U}$$

$$\sum_i \sum_u \pi(i,u) = 1, \tag{6.35}$$

$$\sum_u \pi(j,u) = \sum_i \sum_u \pi(i,u)P_{ij}(u), \quad j \in \{1, \ldots, X\}. \tag{6.36}$$

Although Theorem 6.5.5 specifies μ^* as a randomized policy, in fact μ^* is deterministic with $\pi(x,u) > 0$ if $u = \mu^*(i)$ and zero otherwise. This follows since the primal (6.32) is equivalent to dynamic programming which has a deterministic Markovian optimal policy by construction. Compared to the linear program for the discounted model (6.24), the above linear program includes the additional constraint (6.35) and sets the scalars $\alpha_j = 0$. Also one of the equations in (6.36) is redundant since $P(u)$ has an eigenvalue at 1.

Proof Define the randomized optimal policy for the above constrained MDP in terms of the following action probabilities θ:

$$\mathbb{P}[u_k = u | x_k = i] = \theta_{i,u}, \text{ where } \theta_{i,u} \geq 0, \quad \sum_{u \in \mathcal{U}} \theta_{i,u} = 1.$$

Consider the augmented (homogeneous) Markov chain $z_k \stackrel{\text{defn}}{=} (x_k, u_k)$ with state space $\mathcal{Z} = S \times \mathcal{U}$ and transition probabilities

$$P_{i,u,j,\bar{u}}(\theta) = \mathbb{P}(x_{n+1} = j, u_{n+1} = \bar{u} \mid x_k = i, u_k = u) = \theta_{j,\bar{u}} P_{ij}(u). \tag{6.37}$$

If the chain $\{z_k\}$ is ergodic, it possesses a unique invariant probability measure $\pi_\theta(i, a); i \in \mathcal{X}, u \in \mathcal{U}$. Let $\mathbb{E}_{\pi(\alpha)}$ denote expectation with regard to measure π_θ parameterized by θ. From (6.25) we have

$$J_\mu(x_0) = \sum_{i \in \mathcal{X}} \sum_{u \in \mathcal{U}} \pi_\theta(i, u) c(i, u).$$

Also since π is the stationary distribution with respect to transition probabilities $P_{i,u,j,\bar{u}}(\theta) = \theta_{j,\bar{u}} P_{ij}(u)$, it satisfies $\sum_i \sum_u \pi(i, u) = 1$ and

$$\sum_i \sum_u \pi(i, u) P_{ij}(u) \theta_{j,\bar{u}} = \pi(j, \bar{u})$$

Summing this over $\bar{u}$ yields the last equation of (6.36). □

6.6 Average cost constrained Markov decision process

In this section we consider constrained Markov decision Processes (CMDP). Consider the unichain average cost formulation above. For any stationary policy μ, define the infinite horizon average cost

$$J_\mu(x_0) = \lim_{N \to \infty} \frac{1}{N+1} \mathbb{E}_\mu \left[\sum_{k=0}^{N} c(x_k, u_k) \mid x_0 \right]. \tag{6.38}$$

Motivated by wireless network resource allocation problems such as admission control [351, 253], consider the cost (6.38), subject to L constraints

$$\lim_{N \to \infty} \frac{1}{N+1} \mathbb{E}_\mu \left[\sum_{k=0}^{N} \beta_l(x_k, u_k) \right] \leq \gamma_l, \quad l = 1, \ldots, L. \tag{6.39}$$

These constraints are used in telecommunication networks to depict quality of service (QoS) constraints and long-term fairness constraints. The aim is to compute an optimal policy μ^* for the CMDP that satisfies

$$J_{\mu^*}(x_0) = \inf_\mu J_\mu(x_0) \quad \forall x_0 \in \mathcal{X}, \tag{6.40}$$

subject to the L constraints (6.39).

The constraints (6.39) are called *global constraints* since they involve the entire sample path of the process. In comparison, a *local constraint* is a constraint on the action at each time k such as, for example, a state dependent action space. Local constraints are straightforwardly dealt with in dynamic programming by taking the minimum at each stage subject to the constraint as described in §6.2.1. However, dynamic programming cannot deal with global constraints. The reason is that global constraints can imply that the optimal policy is *randomized* – whereas dynamic programming by construction computes a deterministic policy. Below we discuss two approaches to solving constrained MDPs. The first is to formulate a CMDP as a linear programming problem. The second approach is to formulate a CMDP as a Lagrangian dynamic programming problem.

6.6.1 Linear programming

It can be shown [11] that if the number of constraints $L > 0$ then the optimal policy μ^* is *randomized* for at most L of the states.

THEOREM 6.6.1 *Consider an average cost unichain CMDP with L constraints. Then*
(i) The optimal policy μ^* is randomized for at most L states.
(ii) For state $x \in \mathcal{X}$, the optimal policy is

$$\mu^*(x) = u \text{ with probability } \theta_{xu} = \frac{\pi^*(x,u)}{\sum_{x\in\mathcal{X}} \pi^*(x,u)}. \tag{6.41}$$

Here, the $X \times U$ elements of π^* are the solution of the linear programming problem:

$$\begin{aligned}
&\text{Minimize} \sum_{i\in\mathcal{X}} \sum_{u\in\mathcal{U}} c(i,u)\,\pi(i,u) \ \text{ with respect to } \pi \\
&\text{subject to } \pi(i,u) \geq 0, \quad i \in \mathcal{X}, u \in \mathcal{U} \\
&\qquad \sum_i \sum_u \pi(i,u) = 1, \\
&\qquad \sum_u \pi(j,u) = \sum_i \sum_u \pi(i,u) P_{ij}(u), \quad j \in \{1,\dots,X\}, \\
&\qquad \sum_i \sum_u \pi(i,u)\,\beta_l(i,u) \leq \gamma_l, \quad l = 1,2,\dots,L.
\end{aligned} \tag{6.42}$$

The proof of the second assertion follows directly from that of Theorem 6.5.5 which dealt with the unconstrained average cost case. The L constraints (6.39) appear as the last equation in (6.42). Thus the optimization problem (6.40) with constraints (6.39) can be written as the linear program in the above theorem.

6.6.2 Lagrangian dynamic programming formulation

Here we outline the Lagrangian dynamic programming formulation of a CMDP. The main idea is that the optimal cost and (randomized) policy of the CMDP can be expressed in terms of an unconstrained MDP called the Lagrangian MDP – the Lagrange multipliers of the Lagrangian MDP determine the randomized policy of the CMDP. The formulation is important since it will allow us to use structural results developed for unconstrained MDPs for constrained MDPs. Lagrangian dynamic programming is described extensively in the book [11].

Consider a CMDP with a single global constraint, i.e. $L = 1$ in (6.39). Compute

$$\mu^* = \operatorname*{argmin}_{\mu} J_\mu(x_0) \text{ where} \tag{6.43}$$

$$J_\mu(x_0) = \lim_{N\to\infty} \frac{1}{N+1} \mathbb{E}_\mu \left[\sum_{k=0}^{N} c(x_k, u_k) \mid x_0 \right],$$

$$\text{subject to } B_{x_0}(\mu) \stackrel{\text{defn}}{=} \lim_{N\to\infty} \frac{1}{N+1} \mathbb{E}_\mu \left[\sum_{k=0}^{N} \beta(x_k, u_k) \mid x_0 \right] \leq \gamma. \tag{6.44}$$

Define the Lagrangian average cost criterion as

$$J(\mu, \lambda) = \lim_{N\to\infty} \frac{1}{N+1} \mathbb{E}_\mu \left[\sum_{k=0}^{N} c(x_k, u_k; \lambda) \right] \tag{6.45}$$

where $c(x, u; \lambda)$ is the Lagrangian instantaneous cost defined as

$$c(x, u; \lambda) = c(x, u) + \lambda\, \beta(x, u) \tag{6.46}$$

for a scalar valued Lagrangian multiplier $\lambda \geq 0$. We will call the infinite horizon MDP with objective (6.45) and instantaneous costs (6.46) the *Lagrangian MDP*. For any fixed value of λ, let

$$\mu^*_\lambda = \operatorname*{argmax}_{\mu} J(\mu, \lambda)$$

denote the optimal policy of the Lagrangian MDP. Since the Lagrangian MDP (6.45) has no constraints, clearly μ^*_λ is a deterministic policy.

The following theorem (Theorem 12.7 from [11]) characterizes the optimal (possibly randomized) policy of a CMDP in terms of the deterministic policies of the Lagrangian MDP. Below $\boldsymbol{\mu}$ and $\boldsymbol{\mu}_D$ denote the set of randomized and deterministic Markovian policies, respectively.

THEOREM 6.6.2 *Consider a unichain CMDP with a single global constraint. Then*

1. *The optimal cost function of the CMDP satisfies*

$$J_{\mu^*}(x_0) = \min_{\mu \in \boldsymbol{\mu}} \sup_{\lambda \geq 0} J(\mu, \lambda) - \lambda\gamma = \sup_{\lambda \geq 0} \min_{\mu \in \boldsymbol{\mu}_D} J(\mu, \lambda) - \lambda\gamma \tag{6.47}$$

   where $J(\mu, \lambda)$ is the average cost of the Lagrangian MDP (6.45).

2. *The optimal policy μ^* of the CMDP is a randomized mixture of two pure policies of the Lagrangian MDP. That is,*

$$\mu^* = \alpha \mu^*_{\lambda_1} + (1 - \alpha)\mu^*_{\lambda_2}, \tag{6.48}$$

   where $\alpha \in [0, 1]$ denotes the randomized mixture components and $\mu^*_{\lambda_1}, \mu^*_{\lambda_2}$ are the optimal deterministic policies of the Lagrangian MDP with multipliers λ_1 and λ_2, respectively.

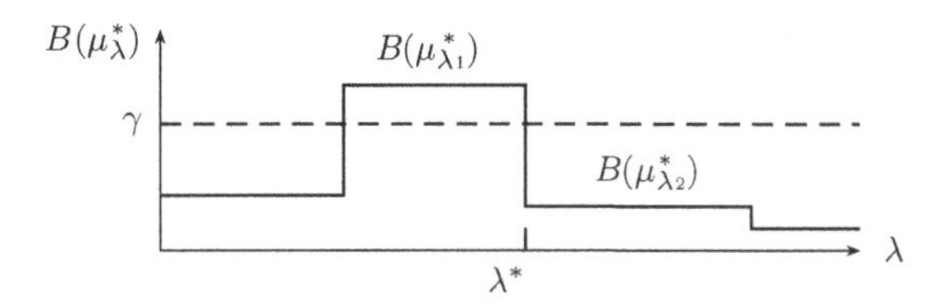

Figure 6.2 Illustration of mixture components for optimal randomized policy μ^* of a constrained Markov decision process. For the constraint to be active $\alpha B_{x_0}(\mu^*_{\lambda_1}) + (1-\alpha)B_{x_0}(\mu^*_{\lambda_2}) = \gamma$ where $\alpha \in [0, 1]$ is the mixture component of μ^*.

Note that the inner minimization in the rightmost expression in (6.47) is over the set of pure policies. Theorem 6.6.2 establishes that the CMDP problem can be solved in two steps.

- Step 1: Solve the Lagrangian MDP problem using relative value or policy iteration algorithms available for unconstrained MDPs.
- Step 2: Compute the optimal randomized policy for the constrained MDP as a probabilistic mixture of two pure policies.

Typically, for the CMDP to be well posed, the constraint (6.44) is active at the optimal policy, i.e. $B_{x_0}(\mu^*) = \gamma$. This means that the optimal policy $\mu^*_{\lambda^*}$ for the Lagrangian MDP satisfies $B_{x_0}(\mu^*_{\lambda^*}) = \gamma$ where λ^* denotes the optimal Lagrange multiplier. We then have the following explicit expression for the mixture component α of the randomized optimal policy (6.48):

$$\alpha = \frac{\gamma - B_{x_0}(\mu^*_{\lambda_2})}{B_{x_0}(\mu^*_{\lambda_1}) - B_{x_0}(\mu^*_{\lambda_2})}. \tag{6.49}$$

To obtain (6.49), note that $B_{x_0}(\mu^*_\lambda)$ versus λ is a piecewise constant function. This is because there are only a finite number of possible policies and so as λ varies μ^*_λ assumes a finite number of values. So if the constraint is active, $\alpha B_{x_0}(\mu^*_{\lambda_1}) + (1-\alpha)B_{x_0}(\mu^*_{\lambda_2}) = \gamma$, implying that the randomized mixture component satisfies (6.49). Figure 6.2 illustrates this interpretation.

6.7 Inverse optimal control and revealed preferences

In an MDP, the aim is to compute the optimal policy given the cost and transition probabilities. This section discusses three related inverse problems: the inverse optimal control problem of determining the cost of an MDP given the optimal policy and transition probabilities, the inverse linear programming problem of determining the cost of an MDP closest to the original cost when additional information becomes available, and the classical problem of revealed preferences of determining if the response of an agent (in a static system) is consistent with cost minimization (utility maximization). All three problems boil down to solving linear programs.

6.7.1 Inverse optimal control

Suppose an agent chooses its actions according to the optimal policy of an MDP. By observing the actions and states of the agent over a period of time, one can estimate the optimal policy of the agent. Given the optimal policy and transition matrices (or estimates), the inverse optimal control problem for an MDP seeks to compute the costs. The problem can be viewed as learning the behavior of an agent by observing its optimized actions. Its applications include apprenticeship and imitation learning, modeling animal and human behavior [28] and model predictive control [235].

We discuss inverse optimal control for a discounted cost MDP with discount factor ρ. Consider the special case where the cost is only state dependent so that $c(x,u) = c(x)$. Then (6.20) implies that

$$\cancel{c(i)} + \rho \sum_j P_{ij}(u) V_j \leq \cancel{c(i)} + \rho \sum_j P_{ij}(\mu^*(i)) V_j, \quad u \in \{1,2,\ldots,U\}.$$

Also with $P(\mu^*)$ denoting the matrix with i-th row $[P_{i1}(\mu^*(i),\ldots,P_{iX}(\mu^*(i)]$, $i = 1,2,\ldots,X$, it follows from (6.22) that

$$V = (I - \rho P(\mu^*))^{-1} c, \quad c = [c(1),\ldots,c(x)]'.$$

Therefore the following U linear inequalities

$$(P(\mu^*) - P(u))(I - \rho P(\mu^*))^{-1} c \leq 0, \quad u \in \{1,2,\ldots,U\} \tag{6.50}$$

characterize the set of feasible cost vectors c for which μ^* is optimal. In [28] several criteria are given to choose a specific element from the set of feasible cost vectors. Inverse optimal control dates back to 1964 [165].

6.7.2 Inverse linear programming for refining cost estimate

Consider a discounted cost MDP expressed as the linear program (6.24). Suppose the cost vector c, transition matrices $P(u)$ and so the optimal policy μ^* (equivalently π^*) is known. These model the aggregate parameters of a social group. Once an individual agent joins the social group, additional information about the preferences of the individual become known. The aim is to refine (re-compute) the cost vector to incorporate this additional information.

For (6.24), define the set $\mathcal{C}$ of feasible cost vectors that have π^* as the optimum:

$$\mathcal{C} = \left\{ d : \boxed{\begin{array}{l} \operatorname{argmin}_\pi \sum_{i\in\mathcal{X}} \sum_{u\in\mathcal{U}} d(i,u)\,\pi(i,u) \\ \text{subject to } \pi \geq 0, \\ \sum_u \pi(j,u) = \rho \sum_i \sum_u P_{ij}(u)\,\pi(i,u) + \alpha_j,\ j \in \mathcal{X} \end{array}} = \pi^* \right\}.$$

The aim is to compute the cost vector $\hat{c} \in \mathcal{C}$ of the individual agent that is closest to the aggregate cost c of the social group. Such problems are called *inverse linear programming problems* [7].

The refined cost estimate $\hat{c}$ is the solution of the following linear program:

$$\text{Compute } \min \|\hat{c} - c\|_1 \text{ subject to}$$
$$\begin{cases} V(i) = \hat{c}(i,u) + \rho \sum_{j\in\mathcal{X}} P_{ij}(u)V(j), & i \in \mathcal{X}, u = \mu^*(i) \\ V(i) \le \hat{c}(i,u) + \rho \sum_{j\in\mathcal{X}} P_{ij}(u)V(j), & i \in \mathcal{X}, u \neq \mu^*(i) \\ L\hat{c} \ge l \end{cases} \tag{6.51}$$

The linear programming variables are $\hat{c}$ and V. The constraint $L\hat{c} \geq l$ models the additional information available regarding the individual preferences of the agent. For example, if it is known after an agent joins the social group that its costs are increasing in the state, then the constraint is $\hat{c}(i+1,u) \geq \hat{c}(i,u)$.

Equation (6.51) follows from the linear complementarity conditions of linear programming.[6] The inverse linear programming problem (6.51) finds the closest cost vector $\hat{c} \in \mathcal{C}$ to the aggregate cost c.

6.7.3 Revealed preferences and Afriat's theorem

A key assumption in most of the models in this book is that agents are utility maximizers (equivalently, cost minimizers). Given a dataset of decisions made by an agent, is it possible to determine if the agent is a utility maximizer? In microeconomics, the principle of *revealed preferences* provides a constructive test to determine if an agent is an utility maximizer subject to budget constraints based on observing its choices over time.

Given a time series of data $\mathcal{D} = \{(\pi_k, a_k), k \in \{1, 2, \ldots, N\}\}$ where $\pi_k \in \mathbb{R}^X$ denotes the input and $a_k \in \mathbb{R}^X$ denotes the response of agent, is it possible to detect if the agent is a *utility maximizer*? An agent is a *utility maximizer* if for every input π_k, the chosen response a_k satisfies

$$a_k(\pi_k) \in \underset{\{\pi_k' a \le I_k\}}{\arg\max}\, V(a) \tag{6.52}$$

with $V(a)$ a *non-satiated* utility function. Non-satiated means that an increase in any element of response a results in the utility function increasing.[7]

In (6.52) the budget constraint $\pi_k' a_k \leq I_k$ models the total amount of resources available to the agent for choosing response a_k to input π_t.

The celebrated Afriat's theorem [4, 5] provides a necessary and sufficient condition for a finite dataset $\mathcal{D}$ to have originated from an utility maximizer.

THEOREM 6.7.1 (Afriat's theorem) *Given a dataset* $\mathcal{D} = \{(\pi_k, a_k) : k \in \{1, 2, \ldots, N\}\}$, *the following statements are equivalent:*

[6] Given a linear program $\min l'z$ subject to $Az = b, z \geq 0$, then the dual is $\max b'\lambda$ subject to $A'\lambda \leq l$. Let A_j denote the j-th column of A, $J = \{j : z(j) > 0\}$, $\bar{J} = \{j : z(j) = 0\}$. Then the linear complementary slackness conditions state that π^* is an optimal solution if both the following conditions hold: (i) $A_j'\lambda \leq l_j$ for $j \in \bar{J}$ and (ii) $A_j'\lambda = l_j$ for $j \in J$.

[7] More formally, a utility function $V(\cdot)$ is said to be non-satiated if for every $a \in \mathbb{R}^X$ there exists $\bar{a}$ in $\mathbb{R}^X$ such that $\|a - \bar{a}\| \leq \epsilon$ implies that $V(\bar{a}) > V(a)$.
The non-satiated assumption rules out trivial cases such as a constant utility function which can be optimized by any response.

1. *The agent is a utility maximizer and there exists a non-satiated and concave utility function that satisfies (6.52).*
2. *For scalars u_t and $\lambda_t > 0$ the following set of inequalities has a feasible solution:*

$$V_\tau - V_k - \lambda_k \pi_k'(a_\tau - a_k) \leq 0 \text{ for } k, \tau \in \{1, 2, \ldots, N\}. \tag{6.53}$$

3. *A non-satiated and concave utility function that satisfies (6.52) is given by:*

$$V(a) = \min_{k \in \{1,\ldots,N\}} \{u_k + \lambda_k \pi_k'(a - a_k)\}. \tag{6.54}$$

4. *The dataset $\mathcal{D}$ satisfies the Generalized Axiom of Revealed Preference (GARP), namely for any $k \leq N$, $\pi_t' a_t \geq \pi_t' a_{t+1}$ for all $t \leq k-1$ implies that $\pi_k' a_k \leq \pi_k' a_1$.*

□

As pointed out in [326], a remarkable feature of Afriat's theorem is that if the dataset can be rationalized by a nontrivial utility function, then it can be rationalized by a continuous, concave, monotonic utility function. That is, violations of continuity, concavity, or monotonicity cannot be detected with a finite number of observations.

Verifying GARP (statement 4 of Theorem 6.7.1) on a dataset $\mathcal{D}$ comprising T points can be done using Warshall's algorithm with $O(N^3)$ [326, 118] computations. Alternatively, determining if Afriat's inequalities (6.53) are feasible can be done via a linear programming feasibility test (using for example interior point methods [68]). Note that the utility (6.54) is not unique and is ordinal by construction. Ordinal means that any monotone increasing transformation of the utility function will also satisfy Afriat's theorem. Therefore the utility mimics the ordinal behavior of humans; see also §5.5.3. The estimated utility (6.54) is the lower envelop of a finite number of hyperplanes that is consistent with the dataset $\mathcal{D}$.

In 5.5.4, we mentioned that revealed preferences can be used for modeling social sensors as utility maximizers; experimental studies the demonstrate these tests in social media networks appear in [186].

6.8 Complements and sources

Excellent textbooks on MDPs include [149, 150, 193, 283, 272, 49]. For a comprehensive treatment of constrained MDPs and Lagrangian dynamic programming, see [11]. MDPs with general state and action spaces are studied in [52, 143] but require a lot of technical sophistication. We refer the reader to [21, 294] for general sufficient conditions for the existence of optimal stationary policies for average cost problems. [114] provides a detailed description of several topics in MDPs including applications in communication and queuing systems.

Inverse optimal control goes back to Kalman [165]. Varian has written several highly influential papers that deal with revealed preferences and Afriat's theorem; see [327, 328].

The formulation in this chapter assumes that the state is observed without delay, the action takes effect immediately, and the cost incurred by the action is accrued immediately. [171] considers random delays.

Appendix 6.A Proof of Theorem 6.2.2

We prove Theorem 6.2.2 for the continuous-valued MDP case using the notation of §6.3 (so that the proof can be reused in Chapter 7). Consider an arbitrary randomized policy $\boldsymbol{\mu} = (\mu_0, \ldots, \mu_{N-1})$ where μ_k maps $\mathcal{I}_k$ to action u_k for $k = 0, \ldots, N-1$. Denote this as $u_k \sim p_k(u|\mathcal{I}_k)$ where p_k denotes the probability mass function that characterizes the randomized policy μ_k. Define the cost-to-go of an arbitrary randomized policy $\boldsymbol{\mu}$ as

$$J_{\mu,k}(x) = \mathbb{E}_{\mu}\left\{\sum_{l=k}^{N} c(x_l, u_l, l) + c(x_N, N) \mid x_k = x\right\}. \tag{6.55}$$

Obviously $J_{\mu,k}(x)$ can be expressed recursively as

$$J_{\mu,k}(x) = \sum_{u\in\mathcal{U}}\left[c(x,u,k) + \int J_{\mu,k+1}(\phi_k(x,u,w))p_w(w)\,dw\right]p_k(u|\mathcal{I}_k, x_k = x). \tag{6.56}$$

Note that $J_{\mu^*,k}(x)$ is the optimal cost-to-go function $J_k(x)$ defined in (6.9).

Statement 1: To establish statement 1, we show by backward induction for $k = N, \ldots, 0$ that $J_{\mu,k}(x) \geq J_k(x)$ with equality achieved at $\boldsymbol{\mu} = \boldsymbol{\mu}^*$ where $\boldsymbol{\mu}^*$ is obtained via (6.16).

Clearly at $k = N$, $J_{\mu,N}(x) = J_N(x) = c(x_N, N)$. Assume next that $J_{\mu,k+1}(x) \geq J_{k+1}(x)$ for some $k+1$ (inductive hypothesis). Then

$$\begin{aligned}
J_{\mu,k}(x) &= \sum_{u\in\mathcal{U}}\left[c(x,u,k) + \int J_{\mu,k+1}(\phi_k(x,u,w))p_w(w)\,dw\right]p_k(u|\mathcal{I}_k, x_k = x) \\
&\geq \sum_{u\in\mathcal{U}}\left[c(x,u,k) + \int J_{\mu^*,k+1}(\phi_k(x,u,w))p_w(w)\,dw\right]p_k(u|\mathcal{I}_k, x_k = x) \\
&\geq \min_{u\in\mathcal{U}}\left[c(x,u,k) + \int J_{\mu^*,k+1}(\phi_k(x,u,w))p_w(w)\,dw\right] \\
&= J_{\mu^*,k}(x).
\end{aligned}$$

The last inequality follows since a convex combination is always larger than the minimum. Also choosing $\boldsymbol{\mu} = \boldsymbol{\mu}^*$ with $p_k(u|\mathcal{I}_k, x_k = x) = 1$ if $u = \mu^*(x)$, all the inequalities above become equalities.

Statement 2: Let $\tilde{J}_k(x)$ denote the optimal cumulative cost incurred from time k to N, i.e. $\tilde{J}_k(x) = \min_{\mu} J_{\mu,k}(x)$ where $J_{\mu,k}(x)$ is defined in (6.55). We need to show that $\tilde{J}_k(x)$ satisfies Bellman's equation.

Let $\boldsymbol{\mu} = (\mu_k, \mu^*_{k+1} \ldots, \mu^*_{N-1})$ where the policy μ_k is arbitrary and $\mu^*_{k+1} \ldots, \mu^*_{N-1}$ is optimal for the cost from time $k+1$ to N. Then from (6.56),

$$J_{\mu,k}(x) = \sum_{u \in \mathcal{U}} \left[c(x,u,k) + \int \tilde{J}_{k+1}(\phi_k(x,u,w)) p_w(w)\, dw \right] p_k(u|\mathcal{I}_k, x_k = x)$$
$$\geq \min_u \left[c(x,u,k) + \int \tilde{J}_{k+1}(\phi_k(x,u,w)) p_w(w)\, dw \right].$$

Therefore optimizing with respect to μ_k yields

$$\tilde{J}_k(x) \geq \min_u \left[c(x,u,k) + \int \tilde{J}_{k+1}(\phi_k(x,u,w)) p_w(w)\, dw \right].$$

One can obtain the reverse inequality as follows: By definition (6.55)

$$\tilde{J}_k(x) \leq \sum_{u \in \mathcal{U}} \left[c(x,u,k) + \int \tilde{J}_{k+1}(\phi_k(x,u,w)) p_w(w)\, dw \right] p_k(u|\mathcal{I}_k, x_k = x).$$

Since $p_k(u|\mathcal{I}_k, x_k = x)$ specifies an arbitrary randomized policy, choosing it to be a Dirac measure concentrated on $\min_u \left[c(x,u,k) + \int \tilde{J}_{k+1}(\phi_k(x,u,w)) p_w(w)\, dw \right]$ implies that

$$\tilde{J}_k(x) \leq \min_{u \in \mathcal{U}} \left[c(x,u,k) + \int \tilde{J}_{k+1}(\phi_k(x,u,w)) p_w(w)\, dw \right].$$

Therefore the optimal cost to go $\tilde{J}_k(x)$ satisfies Bellman's equation implying that the associated optimal policy satisfies Bellman's equation.

Appendix 6.B Proof of Theorems 6.5.3 and 6.5.4

Proof of Theorem 6.5.3

To simplify notation, for policy μ, denote $P_{ij}(\mu(i))$ as P_{ij} and $c(i, \mu(i))$ as $c(i)$.

From the Markov property of $\{x_k\}$, Poisson's equation can be rewritten as

$$h(x_k) + \phi = c(x_k) + \mathbb{E}\{h(x_{k+1})|x_0, \dots, x_k\}.$$

Note also that using ACOE,

$$h(x_k) + g \leq c(x_k) + \mathbb{E}\{h(x_{k+1})|x_0, \dots, x_k\}. \tag{6.57}$$

with equality holding for the optimal policy μ^*. Define the random variables $\{M_k, k = 0, 1, \dots\}$ with $M_0 = h(x_0)$ and

$$M_{k+1} = h(x_{k+1}) + \sum_{n=0}^{k} c(x_n) - (k+1)g.$$

Then

$$\mathbb{E}\{M_{k+1}|x_0, \dots, x_k\} = \mathbb{E}\{h(x_{k+1})|x_0, \dots, x_k\} + \sum_{n=0}^{k} c(x_n) - (k+1)g.$$

Substituting (6.57) for $\mathbb{E}\{h(x_{k+1})|x_0, \dots, x_k\}$ into the above equation yields

$$\mathbb{E}\{M_{k+1}|x_0, \dots, x_k\} \geq M_k.$$

Therefore, if $h(\cdot)$ is bounded then M_k is a submartingale sequence.[8] Since a submartingale has increasing mean, $\mathbb{E}\{M_0|x_0\} \leq \mathbb{E}\{M_{k+1}|x_0\}$ implying that

$$\mathbb{E}\{h(x_0)|x_0\} \leq \mathbb{E}\{h(x_{k+1})|x_0\} + \mathbb{E}\{\sum_{n=0}^{k} c(x_n)|x_0\} - (k+1)g.$$

Rearranging this yields the following inequality (with equality holding for μ^*):

$$\mathbb{E}\{\frac{1}{k+1}\sum_{n=0}^{k} c(x_n)|x_0\} \geq g - \frac{1}{k+1}\left[\mathbb{E}\{h(x_{k+1})|x_0\} - \mathbb{E}\{h(x_0)|x_0\}\right].$$

Taking the limit $k \to \infty$, since $h(\cdot)$ is bounded, yields

$$g \leq \lim_{k\to\infty} \frac{1}{k+1}\mathbb{E}\left[\sum_{n=0}^{k} c(x_n) \mid x_0\right]$$

with equality holding for the optimal policy μ^*.

Proof of Theorem 6.5.4

The proof is from [283, Chapter V]. For a discounted cost MDP with discount factor ρ, denote the value function as V^ρ. Define $\tilde{V}^\rho(i) = V^\rho(i) - V^\rho(1)$. The proof is in two steps:

Step 1: If $|\tilde{V}^\rho(i)| < M$ for some constant M for each i and $\rho \in (0,1)$, then there exist $V(i)$ and g that satisfy (6.26).

Proof: Since $|\tilde{V}^\rho(i)| < M$, there exists a sequence $\{\rho_n\} \to 1$ such that $\lim_{n\to\infty} \tilde{V}^{\rho_n}(i)$ exists for each i. Also because the rewards are bounded, $(1-\rho_n)V^{\rho_n}(1)$ is bounded. So there is a subsequence $\{\rho_{\bar{n}}\}$ of $\{\rho_n\}$ such that $\rho_{\bar{n}} \to 1$ and $g \stackrel{\text{defn}}{=} \lim_{\bar{n}\to\infty}(1-\rho_{\bar{n}})V^{\rho_{\bar{n}}}(1)$ exists. Next from Bellman's equation (6.20) for discounted cost MDPs

$$(1-\rho_{\bar{n}})V^{\rho_{\bar{n}}}(1) + \tilde{V}^{\rho_{\bar{n}}}(i) = \min_u\{c(i,u) + \rho_{\bar{n}}\sum_{j\in\mathcal{X}} P_{ij}(u)\tilde{V}^{\rho_{\bar{n}}}(j)\}.$$

Finally, take the limit as $\bar{n} \to \infty$. Since $\tilde{V}^{\rho_{\bar{n}}}(j)$ is bounded, the limit can be moved inside the summation to obtain (6.26).

Step 2: A unichain MDP satisfies $|\tilde{V}^\rho(i)| < M$ for some constant M.

Proof: Let $x_0 = 1$ and denote $T = \min\{n > 0 : x_n = 1\}$. For a finite state unichain MDP T is finite. Assume all costs are nonnegative without loss of generality. Then the value function associated with the optimal policy μ^* is

[8] A sequence of random variables $\{M_n\}$ is a martingale with respect to a sequence $\{X_n\}$ if $\mathbb{E}\{|M_n|\} < \infty$ and $\mathbb{E}\{M_{n+1}|X_1,\ldots,X_n\} = M_n$. Also, $\{M_n\}$ is a submartingale if $\mathbb{E}\{M_{n+1}|X_1,\ldots,X_n\} \geq M_n$. From the smoothing property of conditional expectations it follows that for a martingale $\mathbb{E}\{M_n\}$ is a constant; while for a submartingale $\mathbb{E}\{M_n\}$ is increasing with n.

$$V^\rho(i)=\mathbb{E}_{\mu^*}\{\sum_{n-0}^{T-1}\rho^n c(x,u)|x_0=i)\}+\mathbb{E}_{\mu^*}\{\sum_{n=T}^{\infty}\rho^n c(x,u)|x_0=i)\}$$

$$=\mathbb{E}_{\mu^*}\{\sum_{n=0}^{T-1}\rho^n c(x,u)|x_0=i)\}+V^\rho(1)\,\mathbb{E}_{\mu^*}\{\rho^T\} \tag{6.58}$$

$$\leq c_{\max}\,\mathbb{E}_{\mu^*}\{T\}+V^\rho(1)\,\mathbb{E}_{\mu^*}\{\rho^T\}$$

$$\leq \text{positive constant }+V^\rho(1). \tag{6.59}$$

where (6.58) holds since $x_T=1$ by definition of T.

The reverse inequality can be obtained as follows. From (6.58), it follows that

$$V^\rho(i)\geq V^\rho(1)\mathbb{E}_{\mu^*}\{\rho^T\}$$

since the first term on the right-hand side of (6.58) is nonnegative. Equivalently,

$$V^\rho(1)\leq V^\rho(i)+(1-\mathbb{E}_{\mu^*}\{\rho^T\})V^\rho(1)\stackrel{b}{\leq}V^\rho(i)+(1-\rho^{\mathbb{E}_{\mu^*}\{T\}})V^\rho(1)$$

where (b) follows from Jensen's inequality. So

$$V^\rho(1)\leq V^\rho(i)+\text{positive constant}. \tag{6.60}$$

The result then follows from (6.59) and (6.60).

7 Partially observed Markov decision processes (POMDPs)

Contents

A POMDP is a *controlled* HMM. Recall from §2.4 that an HMM consists of an X-state Markov chain $\{x_k\}$ observed via a noisy observation process $\{y_k\}$. Figure 7.1 displays the schematic setup of a POMDP where the action u_k affects the state and/or observation (sensing) process of the HMM. The HMM filter (discussed extensively in Chapter 3) computes the posterior distribution π_k of the state. The posterior π_k is called the *belief state*. In a POMDP, the stochastic controller depicted in Figure 7.1 uses the belief state to choose the next action.

This chapter is organized as follows. §7.1 describes the POMDP model. Then §7.2 gives the belief state formulation and the Bellman's dynamic programming equation for the optimal policy of a POMDP. It is shown that a POMDP is equivalent to a *continuous-state* MDP where the states are belief states (posteriors). Bellman's equation for continuous-state MDP was discussed in §6.3. §7.3 gives a toy example of a POMDP. Despite being a continuous-state MDP, §7.4 shows that for finite horizon POMDPs, Bellman's equation has a finite dimensional characterization. §7.5 discusses several algorithms that exploit this finite dimensional characterization to compute the optimal policy. §7.6 considers discounted cost infinite horizon POMDPs. As an example of a POMDP, optimal search of a moving target is discussed in §7.7.

7.1 Finite horizon POMDP

A POMDP model with finite horizon N is a 7-tuple

$$(\mathcal{X}, \mathcal{U}, \mathcal{Y}, P(u), B(u), c(u), c_N). \tag{7.1}$$

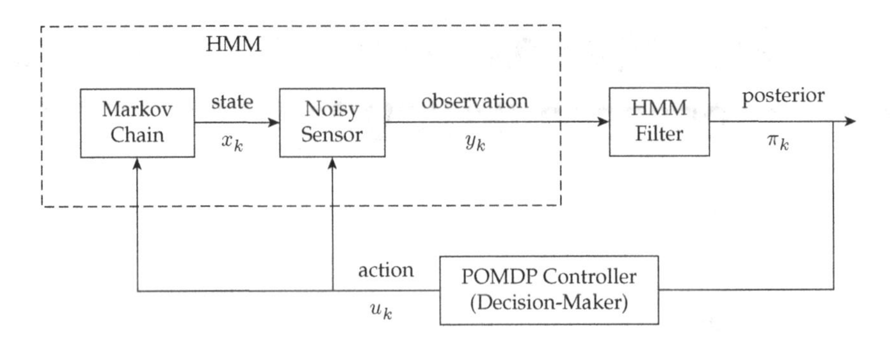

Figure 7.1 Partially observed Markov decision process (POMDP) schematic setup. The Markov system together with noisy sensor constitute a hidden Markov model (HMM). The HMM filter computes the posterior (belief state) π_k of the state of the Markov chain. The controller (decision-maker) then chooses the action u_k at time k based on π_k.

1. $\mathcal{X} = \{1, 2, \ldots, X\}$ denotes the state space and $x_k \in \mathcal{X}$ denotes the state of a controlled Markov chain at time $k = 0, 1, \ldots, N$.
2. $\mathcal{U} = \{1, 2, \ldots, U\}$ denotes the action space with $u_k \in \mathcal{U}$ denoting the action chosen at time k by the controller.
3. $\mathcal{Y}$ denotes the observation space which can either be finite or a subset of $\mathbb{R}$. $y_k \in \mathcal{Y}$ denotes the observation recorded at each time $k \in \{1, 2, \ldots, N\}$.
4. For each action $u \in \mathcal{U}$, $P(u)$ denotes a $X \times X$ transition probability matrix with elements

$$P_{ij}(u) = \mathbb{P}(x_{k+1} = j | x_k = i, u_k = u), \quad i, j \in \mathcal{X}. \tag{7.2}$$

5. For each action $u \in \mathcal{U}$, $B(u)$ denotes the observation distribution with

$$B_{iy}(u) = \mathbb{P}(y_{k+1} = y | x_{k+1} = i, u_k = u), \quad i \in \mathcal{X}, y \in \mathcal{Y}. \tag{7.3}$$

6. For state x_k and action u_k, the decision-maker incurs a cost $c(x_k, u_k)$.
7. Finally, at terminal time N, a terminal cost $c_N(x_N)$ is incurred.

The POMDP model (7.1) is a *partially observed* model since the decision-maker does not observe the state x_k. It only observes noisy observations y_k that depend on the action and the state specified by the probabilities in (7.3). Recall that an HMM is characterized by $(\mathcal{X}, \mathcal{Y}, P, B)$; so a POMDP is a controlled HMM with the additional ingredients of action space $\mathcal{U}$, action-dependent transition probabilities, action-dependent observation probabilities and costs. In general, the transition matrix, observation distribution and cost can be explicit functions of time; however, to simplify notation, we have omitted this time dependency.

Given the model (7.1), the dynamics of a POMDP proceed according to Algorithm 10. This involves at each time k choosing an action u_k, accruing an instantaneous cost $c(x_k, u_k)$, evolution of the state from x_k to x_{k+1} and observing x_{k+1} in noise as y_{k+1}.

As depicted in (7.5), at each time k, the decision-maker uses all the information available until time k (namely, $\mathcal{I}_k$) to choose action $u_k = \mu_k(\mathcal{I}_k)$ using policy μ_k.

Algorithm 10 Dynamics of partially observed Markov decision process

At time $k = 0$, the state x_0 is simulated from initial distribution π_0.
For time $k = 0, 1, \ldots, N-1$:

1. Based on available information
$$\mathcal{I}_0 = \{\pi_0\}, \ \mathcal{I}_k = \{\pi_0, u_0, y_1, \ldots, u_{k-1}, y_k\}, \tag{7.4}$$
the decision-maker chooses action
$$u_k = \mu_k(\mathcal{I}_k) \in \mathcal{U}, \quad k = 0, 1, \ldots, N-1. \tag{7.5}$$
Here, μ_k denotes a policy that the decision-maker uses at time k.
2. The decision-maker incurs a cost $c(x_k, u_k)$ for choosing action u_k.
3. The state evolves randomly with transition probability $P_{x_k x_{k+1}}(u_k)$ to the next state x_{k+1} at time $k+1$. Here
$$P_{ij}(u) = \mathbb{P}(x_{k+1} = j | x_k = i, u_k = u).$$
4. The decision-maker records a noisy observation $y_{k+1} \in \mathcal{Y}$ of the state x_{k+1} according to
$$\mathbb{P}(y_{k+1} = y | x_{k+1} = i, u_k = u) = B_{iy}(u).$$
5. The decision-maker updates its available information as
$$\mathcal{I}_{k+1} = \mathcal{I}_k \cup \{u_k, y_{k+1}\}.$$
If $k < N$, then set k to $k+1$ and go back to Step 1.
If $k = N$, then the decision-maker pays a terminal cost $c_N(x_N)$ and the process terminates.

With the dynamics specified by Algorithm 10, denote the sequence of policies that the decision-maker uses from time 0 to $N-1$ as $\boldsymbol{\mu} = (\mu_0, \mu_1, \ldots, \mu_{N-1})$.

Objective

To specify a POMDP completely, in addition to the model (7.1), dynamics in Algorithm 10 and policy sequence[1] $\boldsymbol{\mu}$, we need to specify a performance criterion or objective function. This section considers the finite horizon objective
$$J_{\mu}(\pi_0) = \mathbb{E}_{\mu}\left\{\sum_{k=0}^{N-1} c(x_k, u_k) + c_N(x_N) \mid \pi_0\right\}. \tag{7.6}$$
which is the expected cumulative cost incurred by the decision-maker when using policy $\boldsymbol{\mu}$ up to time N given the initial distribution π_0 of the Markov chain. Here, $\mathbb{E}_{\mu}$ denotes expectation with respect to the joint probability distribution of

[1] §7.2 shows that a POMDP is equivalent to a continuous-state MDP. So from §6.3, it suffices to consider nonrandomized policies to achieve the minimum in (7.7).

$(x_0, y_0, x_1, y_1, \dots, x_{N-1}, y_{N-1}, x_N, y_N)$. The goal of the decision-maker is to determine the optimal policy sequence

$$\boldsymbol{\mu}^* = \underset{\boldsymbol{\mu}}{\operatorname{argmin}}\, J_{\boldsymbol{\mu}}(\pi_0), \qquad \text{for any initial prior } \pi_0 \tag{7.7}$$

that minimizes the expected cumulative cost. Of course, the optimal policy sequence $\boldsymbol{\mu}^*$ may not be unique.

Remarks

1. The decision-maker does not observe the state x_k. It only observes noisy observations y_k that depend on the action and the state via Step 4. Also, the decision-maker knows the cost matrix $c(x, u)$ for all possible states and actions in $\mathcal{X}$, $\mathcal{U}$. But since the decision-maker does not know the state x_k at time k, it does not know the cost accrued at time k in Step 2 or terminal cost in Step 5. Of course, the decision-maker can estimate the cost by using the noisy observations of the state.
2. The term POMDP is usually reserved for the case when the observation space $\mathcal{Y}$ is finite. However, we consider both finite and continuous valued observations.
3. The action u_k affects the evolution of the state (Step 3) and observation distribution (Step 4). In controlled sensing applications such as radars and sensor networks, the action only affects the observation distribution and not the evolution of the target.
4. More generally, the cost can be of the form $\bar{c}(x_k = i, x_{k+1} = j, y_k = y, y_{k+1} = \bar{y}, u_k = u)$. This is equivalent to the cost (see (7.13) below)

$$c(i, u) = \sum_{y \in \mathcal{Y}} \sum_{\bar{y} \in \mathcal{Y}} \sum_{j \in \mathcal{X}} \bar{c}(i, j, , y, \bar{y}, u) P_{ij}(u) B_{j\bar{y}}(u) B_{iy}(u). \tag{7.8}$$

7.2 Belief state formulation and dynamic programming

This section details a crucial step in the formulation and solution of a POMDP, namely, the belief state formulation. In this formulation, a POMDP is equivalent to a continuous-state MDP with states being the belief states. These belief states are simply the posterior state distributions computed via the HMM filter described in Part 1 of the book. We then formulate the optimal policy as the solution to Bellman's dynamic programming recursion written in terms of the belief state. The main outcome of this section is the formulation of the POMDP dynamics in terms of the belief state in Algorithm 11.

7.2.1 Belief state formulation of POMDP

Recall from §6.2 that for a fully observed MDP, the optimal policy is Markovian and the optimal action $u_k = \mu_k^*(x_k)$. In comparison, for a POMDP the optimal action chosen by the decision-maker is in general

$$u_k = \mu_k^*(\mathcal{I}_k), \qquad \text{where} \qquad \mathcal{I}_k = (\pi_0, u_0, y_1, \dots, u_{k-1}, y_k). \tag{7.9}$$

Since $\mathcal{I}_k$ is increasing in dimension with k, to implement a controller, it is useful to obtain a sufficient statistic that does not grow in dimension. The posterior distribution π_k

computed via the HMM filter of §3.5 is a sufficient statistic for $\mathcal{I}_k$. Define the posterior distribution of the Markov chain given $\mathcal{I}_k$ as

$$\pi_k(i) = \mathbb{P}(x_k = i|\mathcal{I}_k), \; i \in \mathcal{X} \quad \text{where } \mathcal{I}_k = \{\pi_0, u_0, y_1, \ldots, u_{k-1}, y_k\}. \tag{7.10}$$

We will call the X-dimensional probability vector $\pi_k = [\pi_k(1), \ldots, \pi_k(X)]'$ as the *belief state* or *information state* at time k. It is computed via the HMM filter (Algorithm 3), namely $\pi_k = T(\pi_{k-1}, y_k, u_{k-1})$ where

$$T(\pi, y, u) = \frac{B_y(u)P'(u)\pi}{\sigma(\pi, y, u)}, \quad \text{where } \sigma(\pi, y, u) = \mathbf{1}'_X B_y(u)P'(u)\pi, \tag{7.11}$$
$$B_y(u) = \operatorname{diag}\big(B_{1y}(u), \cdots, B_{Xy}(u)\big).$$

The main point established below in Theorem 7.2.1 is that (7.9) is equivalent to

$$u_k = \mu_k^*(\pi_k). \tag{7.12}$$

In other words, the optimal controller operates on the belief state π_k (HMM filter posterior) to determine the action u_k.

In light of (7.12), let us first define the space where π_k lives in. The beliefs $\pi_k, k = 0, 1, \ldots$ defined in (7.10) are X-dimensional probability vectors. Therefore they lie in the $X - 1$ dimensional unit simplex denoted as

$$\Pi(X) \stackrel{\text{defn}}{=} \left\{\pi \in \mathbb{R}^X : \mathbf{1}'\pi = 1, \quad 0 \leq \pi(i) \leq 1 \text{ for all } i \in \mathcal{X} = \{1, 2, \ldots, X\}\right\}.$$

$\Pi(X)$ is called the *belief space*. $\Pi(2)$ is a one-dimensional simplex (unit line segment), As shown in Figure 7.2, $\Pi(3)$ is a two-dimensional simplex (equilateral triangle); $\Pi(4)$ is a tetrahedron, etc. Note that the unit vector states $e_1, e_2, \ldots, e_X$ of the underlying Markov chain x are the vertices of $\Pi(X)$.

We now formulate the POMDP objective (7.6) in terms of the belief state. Consider the objective (7.6). Then

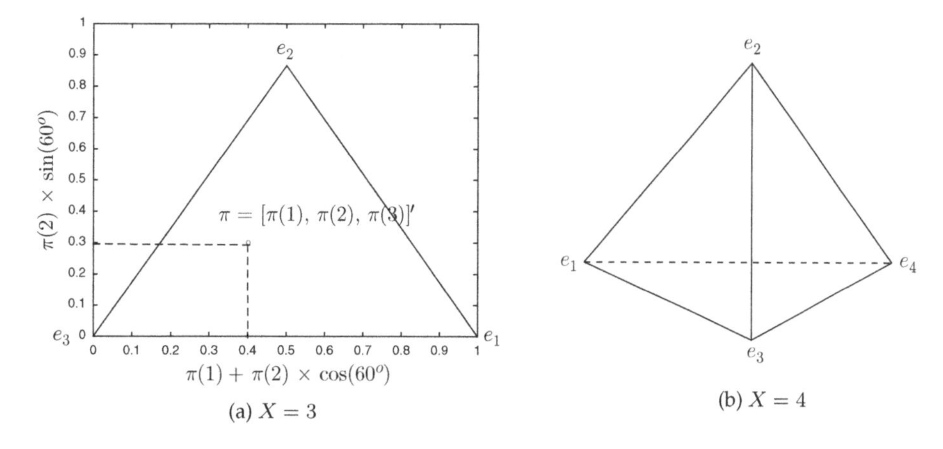

(a) $X = 3$ (b) $X = 4$

Figure 7.2 For an X-state Markov chain, the belief space $\Pi(X)$ in a $X - 1$-dimensional unit simplex $\Pi(X)$. The figure shows $\Pi(X)$ for $X = 3, 4$.

$$
\begin{aligned}
J_\mu(\pi_0) &= \mathbb{E}_\mu\Big\{\sum_{k=0}^{N-1} c(x_k, u_k) + c_N(x_N) \mid \pi_0\Big\} \\
&\overset{(a)}{=} \mathbb{E}_\mu\Big\{\sum_{k=0}^{N-1} \mathbb{E}\{c(x_k, u_k) \mid \mathcal{I}_k\} + \mathbb{E}\{c_N(x_N) \mid \mathcal{I}_N\} \mid \pi_0\Big\} \\
&= \mathbb{E}_\mu\Big\{\sum_{k=0}^{N-1}\sum_{i=1}^{X} c(i, u_k)\,\pi_k(i) + \sum_{i=1}^{X} c_N(i)\,\pi_N(i) \mid \pi_0\Big\} \\
&= \mathbb{E}_\mu\Big\{\sum_{k=0}^{N-1} c'_{u_k}\pi_k + c'_N\pi_N \mid \pi_0\Big\}
\end{aligned}
\tag{7.13}
$$

where (a) uses the smoothing property of conditional expectations (3.11). In (7.13), the X-dimensional cost vectors $c_u(k)$ and terminal cost vector c_N are defined as

$$
c_u = \begin{bmatrix} c(1,u) & \cdots & c(X,u)\end{bmatrix}', \quad c_N = \begin{bmatrix} c_N(1) & \cdots & c_N(X)\end{bmatrix}'. \tag{7.14}
$$

Summary: The POMDP has been expressed as a continuous-state (fully observed) MDP with dynamics (7.11) given by the HMM filter and objective function (7.13). This continuous-state MDP has belief state π_k which lies in unit simplex belief space $\Pi(X)$. Thus we have the following useful decomposition illustrated in Figure 7.1:

- An HMM filter uses the noisy observations y_k to compute the belief state π_k
- The POMDP controller then maps the belief state π_k to the action u_k.

Determining the optimal policy for a POMDP is equivalent to partitioning $\Pi(X)$ into regions where a particular action $u \in \{1, 2, \ldots, U\}$ is optimal.

As a result, the real time POMDP controller in Algorithm 10 can be re-expressed in terms of the belief state as described in Algorithm 11.

Algorithm 11 Real Time POMDP Controller

At time $k = 0$, the state x_0 is simulated from initial distribution π_0.

For time $k = 0, 1, \ldots, N-1$:

- Step 1: Based on belief state π_k, choose action $u_k = \mu_k(\pi_k) \in \mathcal{U}$.
- Step 2: The decision maker incurs a cost $c'_{u_k}\pi_k$.
- Steps 3 and 4: identical to Algorithm 10
- Step 5: Update the belief state $\pi_{k+1} = T(\pi_k, y_{k+1}, u_k)$ using HMM filter (7.11).

7.2.2 Stochastic dynamic programming for POMDP

Since a POMDP is a continuous-state MDP with state space being the unit simplex, we can straightforwardly write down the dynamic programming equation for the optimal policy as we did in §6.3 for continuous-state MDPs.

THEOREM 7.2.1 *For a finite horizon POMDP with model (7.1) and dynamics given by Algorithm 10 or equivalently Algorithm 11:*

1. The minimum expected cumulative cost $J_{\mu^*}(\pi)$ is achieved by deterministic policies

$$
\mu^* = (\mu_0^*, \mu_1^*, \ldots, \mu_{N-1}^*), \quad \text{where } u_k = \mu_k^*(\pi_k).
$$

2. *The optimal policy* $\boldsymbol{\mu}^* = (\mu_0, \mu_1, \ldots, \mu_{N-1})$ *for a POMDP is the solution of the following Bellman's dynamic programming backward recursion: Initialize* $J_N(\pi) = c_N' \pi$ *and then for* $k = N-1, \ldots, 0$

$$J_k(\pi) = \min_{u \in \mathcal{U}} \{c_u' \pi + \sum_{y \in \mathcal{Y}} J_{k+1}\left(T(\pi, y, u)\right) \sigma(\pi, y, u)\}$$

$$\mu_k^*(\pi) = \operatorname*{argmin}_{u \in \mathcal{U}} \{c_u' \pi + \sum_{y \in \mathcal{Y}} J_{k+1}\left(T(\pi, y, u)\right) \sigma(\pi, y, u)\}. \tag{7.15}$$

The expected cumulative cost $J_{\boldsymbol{\mu}^*}(\pi)$ *(7.13) of the optimal policy* $\boldsymbol{\mu}^*$ *is given by the value function* $J_0(\pi)$ *for any initial belief* $\pi \in \Pi(X)$.

The proof of the above theorem follows from Theorem 6.2.2 in Chapter 6. Since the belief space $\Pi(X)$ is uncountable, the above dynamic programming recursion does not translate into practical solution methodologies. $J_k(\pi)$ needs to be evaluated at each $\pi \in \Pi(X)$, an uncountable set.[2]

7.3 Machine replacement POMDP: toy example

To illustrate the POMDP model and dynamic programming recursion described above, consider a toy example involving the machine replacement problem. The fully observed version of this problem was discussed in §6.2.3. Here we describe the two-state version of the problem with noisy observations.

The state space is $\mathcal{X} = \{1, 2\}$ where state 1 corresponds to a poorly performing machine while state 2 corresponds to a brand new machine. The action space is $\mathcal{U} \in \{1, 2\}$ where action 2 denotes keep using the machine, while action 1 denotes replace the machine with a brand new one which starts in state 2. The transition probabilities of the machine state are

$$P(1) = \begin{bmatrix} 0 & 1 \\ 0 & 1 \end{bmatrix}, \quad P(2) = \begin{bmatrix} 1 & 0 \\ \theta & 1-\theta \end{bmatrix}.$$

where $\theta \in [0, 1]$ denotes the probability that machine deteriorates.

Assume that the state of the machine x_k is indirectly observed via the quality of the product $y_k \in \mathcal{Y} = \{1, 2\}$ generated by the machine. Let p denote the probability that the machine operating in the good state produces a high quality product, and q denote the probability that a deteriorated machine produces a poor quality product. Then the observation probability matrix is

$$B = \begin{bmatrix} p & 1-p \\ 1-q & q \end{bmatrix}.$$

Operating the machine in state x incurs an operating cost $c(x, u = 2)$. On the other hand, replacing the machine at any state x, costs R, that is, $c(x, u = 1) = R$. The aim is to minimize the cumulative expected cumulative cost $\mathbb{E}_{\boldsymbol{\mu}}\{\sum_{k=0}^{N-1} c(x_k, u_k) | \pi_0\}$ for some specified horizon N. Here π_0 denotes the initial distribution of the state of the machine at time 0.

[2] For the reader familiar with linear quadratic Gaussian (LQG) control, it turns out in that case $J_k(\pi)$ is quadratic in the conditional mean estimate of the state $\hat{x}_k$ and the optimal control law is $u_k = -L_k \hat{x}_k$ where $\hat{x}_k$ is computed using the Kalman filter. LQG state and measurement control is described in §8.3.

Bellman's equation (7.15) reads: initialize $J_N(\pi) = 0$ (since there is no terminal cost) and for $k = N-1, \ldots, 0$:

$$J_k(\pi) = \min\{c_1'\pi + J_{k+1}(e_1),\; c_2'\pi + \sum_{y \in \{1,2\}} J_{k+1}(T(\pi, y, 2))\sigma(\pi, y, 2)\}$$

$$\text{where } T(\pi, y, 2) = \frac{B_y P'(2)\pi}{\sigma(\pi, y, 2)}, \quad \sigma(\pi, y, 2) = \mathbf{1}' B_y P'(2)\pi, \; y \in \{1, 2\},$$

$$B_1 = \begin{bmatrix} p & 0 \\ 0 & 1-q \end{bmatrix}, \quad B_2 = \begin{bmatrix} 1-p & 0 \\ 0 & q \end{bmatrix}.$$

Since the number of states is $X = 2$, the belief space $\Pi(X)$ is a one-dimensional simplex, namely the interval $[0, 1]$. So $J_k(\pi)$ can be expressed in terms of $\pi_2 \in [0, 1]$, because $\pi_1 = 1 - \pi_2$. Denote this as $J_k(\pi_2)$.

One can then implement the dynamic programming recursion numerically by discretizing π_2 in the interval $[0, 1]$ over a finite grid and running the Bellman's equation over this finite grid. Although this numerical implementation is somewhat naive, the reader should do this to visualize the value function and optimal policy. The reader would notice that the value function $J_k(\pi_2)$ is piecewise linear and concave in π_2. The main result in the next section is that for a finite horizon POMDP, the value function is always piecewise linear and concave, and the value function and optimal policy can be determined exactly (therefore a grid approximation is not required).

7.4 Finite dimensional controller for finite horizon POMDP

Despite the belief space $\Pi(X)$ being continuum, the following remarkable result due to Sondik [306, 302] shows that Bellman's equation (7.15) for a finite horizon POMDP has a finite dimensional characterization when the observation space $\mathcal{Y}$ is finite.

THEOREM 7.4.1 *Consider the POMDP model (7.1) with finite action space $\mathcal{U} = \{1, 2, \ldots, U\}$ and finite observation space $\mathcal{Y} = \{1, 2, \ldots, Y\}$. At each time k the value function $J_k(\pi)$ of Bellman's equation (7.15) and associated optimal policy $\mu_k^*(\pi)$ have the following finite dimensional characterization:*

1. *$J_k(\pi)$ is piecewise linear and concave with respect to $\pi \in \Pi(X)$. That is,*

$$J_k(\pi) = \min_{\gamma \in \Gamma_k} \gamma'\pi. \tag{7.16}$$

   Here, Γ_k at iteration k is a finite set of X-dimensional vectors.
   Note $J_N(\pi) = c_N'\pi$ and $\Gamma_N = \{c_N\}$ where c_N denotes the terminal cost vector.
2. *The optimal policy $\mu_k^*(\pi)$ has the following finite dimensional characterization: The belief space $\Pi(X)$ can be partitioned into at most $|\Gamma_k|$ convex polytopes. In each such polytope $\mathcal{R}_l = \{\pi : J_k(\pi) = \gamma_l'\pi\}$, the optimal policy $\mu_k^*(\pi)$ is a constant corresponding to a single action. That is, for belief $\pi \in \mathcal{R}_l$ the optimal policy is*

$$\mu_k^*(\pi) = u(\operatorname*{argmin}_{\gamma_l \in \Gamma_k} \gamma_l'\pi)$$

   where the right-hand side is the action associated with polytope $\mathcal{R}_l$.

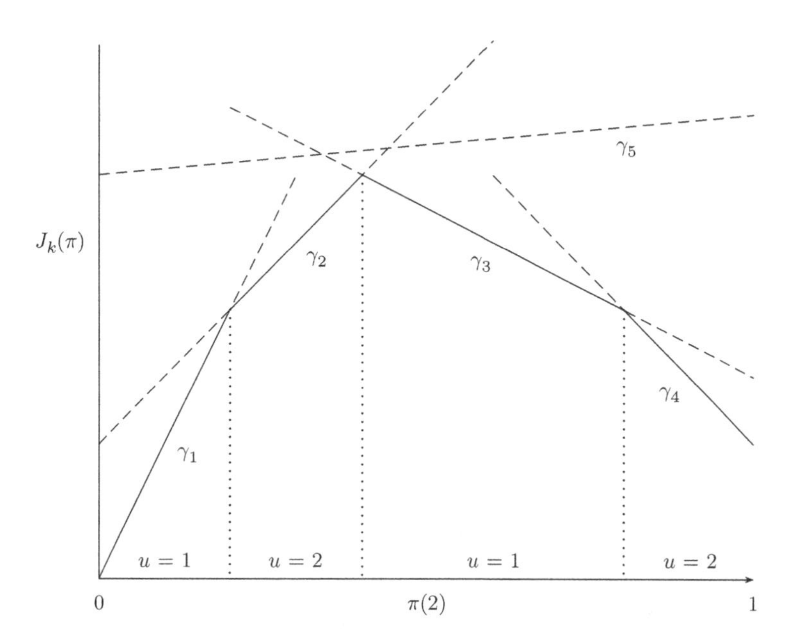

Figure 7.3 Example of piecewise linear concave value function $J_k(\pi)$ of a POMDP with a 2-state underlying Markov chain ($X = 2$). Here $J_k(\pi) = \min\{\gamma_1'\pi, \gamma_2'\pi, \gamma_3'\pi, \gamma_4'\pi\}$ is depicted by solid lines. The figure also shows that the belief space can be partitioned into four regions. Each region where line segment $\gamma_l'\pi$ is active (i.e. is equal to the solid line) corresponds to a single action, $u = 1$ or $u = 2$. Note that γ_5 is never active.

The above theorem says that for a POMDP, the explicit solution to Bellman's equation is a piecewise linear concave function of the belief state π. At iteration k, this function $J_k(\pi)$ has at most $|\Gamma_k|$ linear segments. Each linear segment corresponds to a convex polytope within the belief space $\Pi(X)$ where a particular single action is optimal. In § 7.5 we will give algorithms that compute the piecewise linear concave function $J_k(\pi)$ and optimal policy $\mu_k^*(\pi)$. The optimal real time POMDP controller then proceeds according to Algorithm 11.

Figure 7.3 illustrates the piecewise linear concave structure of the value function $J_k(\pi)$ for the case of a two-state Markov chain ($X = 2$). In this case, the belief state $\pi = \begin{bmatrix} 1 - \pi(2) \\ \pi(2) \end{bmatrix}$ is parametrized by the scalar $\pi(2) \in [0, 1]$ and the belief space is the one-dimensional simplex $\Pi(2) = [0, 1]$. Figure 7.3 also illustrates the finite dimensional structure of the optimal policy $\mu_k^*(\pi)$ asserted by the above theorem. In each region of belief space where $\gamma_l'\pi$ is active, the optimal policy takes on a single action.

7.4.1 Proof of Theorem 7.4.1

The proof of Theorem 7.4.1 is important since it gives an explicit construction of the value function. Exact algorithms for solving POMDPs that will be described in §7.5 are based on this construction.

Theorem 7.4.1 is proved by backward induction for $k = N, \dots, 0$. Obviously, $J_N(\pi) = c_N'\pi$ is linear in π. Next assume $J_{k+1}(\pi)$ is piecewise linear and concave in π: so $J_{k+1}(\pi) = \min_{\bar{\gamma} \in \Gamma_{k+1}} \bar{\gamma}'\pi$. Substituting this in (7.15) yields

$$
\begin{aligned}
J_k(\pi) &= \min_{u\in\mathcal{U}}\{c_u'\pi + \sum_{y\in\mathcal{Y}} \min_{\bar{\gamma}\in\Gamma_{k+1}} \frac{\bar{\gamma}' B_y(u) P'(u)\pi}{\cancel{\sigma(\pi,u,y)}} \cancel{\sigma(\pi,u,y)}\} \\
&= \min_{u\in\mathcal{U}}\Big\{\sum_{y\in\mathcal{Y}} \min_{\bar{\gamma}\in\Gamma_{k+1}} \Big\{\Big[\frac{c_u}{Y} + P(u)B_y(u)\bar{\gamma}\Big]'\pi\Big\}\Big\}.
\end{aligned}
\tag{7.17}
$$

The right-hand side is the minimum (over u) of the sum (over y) of piecewise linear concave functions. Both these operations preserve the piecewise linear concave property. This implies $J_k(\pi)$ is piecewise linear and concave of the form

$$
J_k(\pi) = \min_{\gamma\in\Gamma_k} \gamma'\pi, \quad \text{where } \Gamma_k = \cup_{u\in\mathcal{U}} \oplus_{y\in\mathcal{Y}} \Big\{\frac{c_u}{Y} + P(u)B_y(u)\bar{\gamma} \mid \bar{\gamma}\in\Gamma_{k+1}\Big\}.
\tag{7.18}
$$

Here $\oplus$ denotes the cross-sum operator: given two sets of vectors A and B, $A \oplus B$ consists of all pairwise additions of vectors from these two sets. Recall $\mathcal{U} = \{1, 2, \ldots, U\}$ and $\mathcal{Y} = \{1, 2, \ldots, Y\}$ are finite sets. A more detailed explanation of going from (7.17) to (7.18) is given in (7.21), (7.22).

7.4.2 POMDPs with continuous-value observations

So far we have considered POMDPs with finite-valued observation spaces $\mathcal{Y}$. Suppose the observation space is a continuum, and the observations are continuous-valued generated from a probability density function $B_{x,y}(u) = p(y_{k+1} = y | x_{k+1} = x, u_k = u)$. For example, this could be a normal density implying that $\mathcal{Y} = \mathbb{R}$ and the Markov chain is observed in Gaussian noise. Bellman's dynamic programming recursion then reads

$$
J_k(\pi) = \min_{u\in\mathcal{U}}\{c_u'\pi + \int_{y\in\mathcal{Y}} J_{k+1}\left(T(\pi, y, u)\right)\sigma(\pi, y, u)dy\}, \quad J_N(\pi) = c_N'\pi. \tag{7.19}
$$

THEOREM 7.4.2 *Consider a POMDP model (7.1) with continuous-valued observations. Then the value function $J_k(\pi)$ is concave with respect to $\pi \in \Pi(X)$ for $k = N, \ldots, 0$.*

Sadly, even though the value function is concave, the finite dimensional piecewise linear representation of Theorem 7.4.1 no longer holds for continuous valued observations. So the dynamic programming equation (7.19) does not translate into a useful algorithm.[3]

To prove Theorem 7.4.2 we start with the following lemma which applies to both continuous and discrete-valued observations.

LEMMA 7.4.3 *The value function of a POMDP is positive homogeneous. That is, for any positive constant α, it follows that $J_n(\alpha\pi) = \alpha J_n(\pi)$. As a result, Bellman's equation (7.19) becomes*

$$
J_k(\pi) = \min_{u\in\mathcal{U}}\{c_u'\pi + \int_{y\in\mathcal{Y}} J_{k+1}\left(B_y(u)P'(u)\pi\right)dy\}. \tag{7.20}
$$

[3] §14.7 shows that the concavity of the value function allows us to develop useful structural results for the optimal policy in terms of Blackwell dominance.

Proof From the dynamic programming recursion (7.15), for $\alpha > 0$,

$$J_k(\alpha\pi) = \min_{u \in \mathcal{U}} \left\{ C(\alpha\pi, u) + \int_{\mathcal{Y}} J_{k+1}\left(\frac{B_y(u)P'(u)\cancel{\alpha}\pi}{\mathbf{1}'B_y(u)P'(u)\cancel{\alpha}\pi} \right) \mathbf{1}'B_y(u)P(u)'\alpha\pi\, dy \right\} = \alpha\, J_k(\pi).$$

Then choosing $\alpha = \sigma(\pi, y, u)$, we can use the above positive homogeneity to write

$$\begin{aligned} J_{k+1}\big(T(\pi, y, u)\big)\,\sigma(\pi, y, u) &= J_{k+1}\big(T(\pi, y, u)\sigma(\pi, y, u)\big) \\ &= J_{k+1}\left(\frac{B_y(u)P'(u)\pi}{\cancel{\sigma(\pi, y, u)}} \cancel{\sigma(\pi, y, u)} \right). \end{aligned}$$

So the dynamic programming recursion (7.19) can be written as (7.20). □

The proof of Theorem 7.4.2 then follows by induction. Consider (7.20). Assume $J_{k+1}(\pi)$ is concave. Since the composition of concave function and increasing linear function is concave, therefore $J_{k+1}\left(B_y(u)P'(u)\pi\right)$ is concave. Since concavity is preserved by integration [68, p.79], $\int_{y \in \mathcal{Y}} J_{k+1}\left(B_y(u)P'(u)\pi\right) dy$ is concave. Therefore $c_u'\pi + \int_{\mathcal{Y}} J_{k+1}\left(B_y(u)P'(u)\pi\right) dy$ is concave. Finally, the pointwise minimum of concave functions is concave implying that the right-hand side of (7.20) is concave. Therefore $J_k(\pi)$ is concave, thereby completing the inductive proof.

7.5 Algorithms for finite horizon POMDPs with finite observation space

This section discusses algorithms for solving[4] a finite horizon POMDP when the observation set $\mathcal{Y}$ is finite. These algorithms exploit the finite dimensional characterization of the value function and optimal policy given in Theorem 7.4.1.

Consider the POMDP model (7.1) with finite action space $\mathcal{U} = \{1, 2, \ldots, U\}$ and finite observation set $\mathcal{Y} = \{1, 2, \ldots, Y\}$. Given the finite dimensional characterization in Theorem 7.4.1, the next step is to compute the set of vectors Γ_k that determine the piecewise linear segments of the value function $J_k(\pi)$ at each time k. Unfortunately, the number of piecewise linear segments can increase exponentially with the action space dimension U and double exponentially with time k. This is seen from the fact that given the set of vectors Γ_{k+1} that characterizes the value function at time $k+1$, a single step of the dynamic programming recursion yields that the set of all vectors at time k are $U|\Gamma_{k+1}|^Y$. (Of these it is possible that many vectors are never active such as γ_5 in Figure 7.3.) Therefore, exact computation of the optimal policy is only computationally tractable for small state dimension X, small action space dimension U and small observation space dimension Y. Computational complexity theory gives worst case bounds for solving a problem. It is shown in [256] that solving a POMDP is a PSPACE complete problem. [209] gives examples of POMDPs that exhibit this worst case behavior.

[4] By "solving" we mean solving Bellman's dynamic programming equation (7.15) for the optimal policy $\mu_k^*(\pi)$, $k = 0, \ldots, N-1$. Once the optimal policy is obtained, then the real-time controller is implemented according to Algorithm 11.

7.5.1 Exact algorithms: incremental pruning, Monahan and Witness

Exact[5] algorithms for solving finite horizon POMDPs are based on the finite dimensional characterization of the value function provided by Theorem 7.4.1. The first exact algorithm for solving finite horizon POMDPs was proposed by Sondik [306]; see [227, 78, 79, 210] for several algorithms. Bellman's dynamic programming recursion (7.15) can be expressed as the following three steps:

$$\begin{aligned} Q_k(\pi, u, y) &= \frac{c_u'\pi}{Y} + J_{k+1}\left(T(\pi, y, u)\right)\sigma(\pi, y, u) \\ Q_k(\pi, u) &= \sum_{y \in \mathcal{Y}} Q_k(\pi, u, y) \\ J_k(\pi) &= \min_u Q_k(\pi, u). \end{aligned} \tag{7.21}$$

Based on the above three steps, the set of vectors Γ_k that form the piecewise linear value function in Theorem 7.4.1, can be constructed as

$$\begin{aligned} \Gamma_k(u, y) &= \left\{ \frac{c_u}{Y} + P(u)B_y(u)\,\gamma \mid \gamma \in \Gamma^{(k+1)} \right\} \\ \Gamma_k(u) &= \oplus_y \Gamma_k(u, y) \\ \Gamma_k &= \cup_{u \in \mathcal{U}} \Gamma_k(u). \end{aligned} \tag{7.22}$$

Here $\oplus$ denotes the cross-sum operator: given two sets of vectors A and B, $A \oplus B$ consists of all pairwise additions of vectors from these two sets.

In general, the set Γ_k constructed according to (7.22) may contain superfluous vectors (we call them "inactive vectors" below) that never arise in the value function $J_k(\pi) = \min_{\gamma_l \in \Gamma_k} \gamma_l'\pi$. The algorithms listed below seek to eliminate such useless vectors by pruning Γ_k to maintain a parsimonious set of vectors.

Incremental pruning algorithm: We start with the incremental pruning algorithm described in Algorithm 12. The code is freely downloadable from [77].

Algorithm 12 Incremental pruning algorithm for solving POMDP

Given set Γ_{k+1} generate Γ_k as follows:
Initialize $\Gamma_k(u, y), \Gamma_k(u), \Gamma_k$ as empty sets
For each $u \in \mathcal{U}$
  For each $y \in \mathcal{Y}$

$$\Gamma_k(u, y) \leftarrow \text{prune}\left(\left\{ \frac{c_u}{Y} + P(u)B_y(u)\gamma \mid \gamma \in \Gamma^{(k+1)} \right\}\right)$$

$$\Gamma_k(u) \leftarrow \text{prune}\left(\Gamma_k(u) \oplus \Gamma_k(u, y)\right)$$

  $\Gamma_k \leftarrow \text{prune}\left(\Gamma_k \cup \Gamma_k(u)\right)$

Let us explain the "prune" function in Algorithm 12. Recall the piecewise linear concave characterization of the value function $J_k(\pi) = \min_{\gamma \in \Gamma_k} \gamma'\pi$ with set of vectors Γ_k. Suppose there is a vector $\gamma \in \Gamma_k$ such that for all $\pi \in \Pi(X)$, it holds that

[5] Exact here means that there is no approximation involved in the dynamic programming algorithm. However, the algorithm is still subject to numerical round-off and finite precision effects.

$\gamma'\pi \geq \bar{\gamma}'\pi$ for all vectors $\bar{\gamma} \in \Gamma_k - \{\gamma\}$. Then γ dominates every other vector in Γ_k and is never active. For example, in Figure 7.3, γ_5 is never active. The prune function in Algorithm 12 eliminates such inactive vectors γ and so reduces the computational cost of the algorithm.

Given a set of vectors Γ, how can an inactive vector be identified and therefore pruned (eliminated)? The following linear programming dominance test can be used to identify inactive vectors:

$$\begin{aligned} &\min\ x \\ \text{subject to: } &(\gamma - \bar{\gamma})'\pi \geq x, \quad \forall \bar{\gamma} \in \Gamma - \{\gamma\} \\ &\pi(i) \geq 0, i \in \mathcal{X}, \quad \mathbf{1}'\pi = 1, \quad \text{i.e. } \pi \in \Pi(X). \end{aligned} \tag{7.23}$$

Clearly, if the above linear PROGRAM yields a solution $x \geq 0$, then γ dominates all other vectors in $\Gamma - \{\gamma\}$. Then vector γ is inactive and can be eliminated from Γ. In the worst case, it is possible that all vectors are active and none can be pruned.

Monahan's algorithm: Monahan [244] proposed an algorithm that is identical to Algorithm 12 except that the prune steps in computing $\Gamma_k(u, y)$ and $\Gamma_k(u)$ are omitted. So $\Gamma_k(u)$ comprises of $U|\Gamma_{k+1}|^Y$ vectors and these are then pruned according to the last step of Algorithm 12.

Witness algorithm: The Witness algorithm [80], constructs $\Gamma_k(u)$ associated with $Q_k(\pi, u)$ (7.21) in polynomial time with respect to X, U, Y and $|\Gamma_{k+1}|$. [81] shows that the incremental pruning Algorithm 12 has the same computational cost as the Witness algorithm and can outperform it by a constant factor.

7.5.2 Lovejoy's suboptimal algorithm

Computing the value function and therefore optimal policy of a POMDP via the exact algorithms given above is intractable apart from small toy examples. Lovejoy [226] proposed an ingenious suboptimal algorithm that computes upper and lower bounds to the value function of a POMDP. The intuition behind this algorithm is depicted in Figure 7.4 and is as follows: Let $\bar{J}_k$ and $\underline{J}_k$, respectively, denote upper and lower bounds to J_k. It is obvious that by considering only a subset of the piecewise linear segments in Γ_k and discarding the other segments, one gets an upper bound $\bar{J}_k$. That is, for any $\bar{\Gamma}_k \subset \Gamma_k$,

$$\bar{J}_k(\pi) = \min_{\gamma_l \in \bar{\Gamma}_k} \gamma_l'\pi \geq \min_{\gamma_l \in \Gamma_k} \gamma_l'\pi = J_k(\pi).$$

In Figure 7.4, J_k is characterized by line segments in $\Gamma_k = \{\gamma_1, \gamma_2, \gamma_3, \gamma_4\}$ and the upper bound $\bar{J}_k$ is constructed from line segments in $\bar{\Gamma}_k = \{\gamma_2, \gamma_3\}$, i.e. discarding segments γ_1 and γ_4. This upper bound is displayed in dashed lines.

By choosing $\bar{\Gamma}_k$ with small cardinality at each iteration k, one can reduce the computational cost of computing $\bar{J}_k$. This is the basis of Lovejoy's [226] lower bound approximation. Lovejoy's algorithm [226] operates as follows:

Initialize: $\bar{\Gamma}_N = \Gamma_N = \{c_N\}$. Recall c_N is the terminal cost vector.

Step 1: Given a set of vectors Γ_k, construct the set $\bar{\Gamma}_k$ by pruning Γ_k as follows: Pick any R belief states $\pi_1, \pi_2, \ldots, \pi_R$ in the belief simplex $\Pi(X)$. (Typically, one often picks the

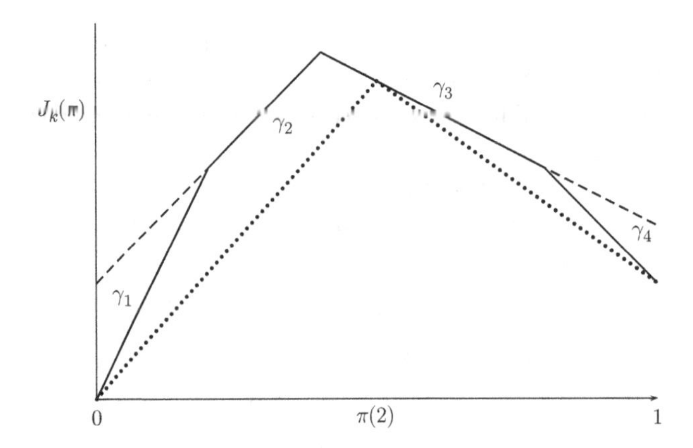

Figure 7.4 Intuition behind Lovejoy's suboptimal algorithm for solving a POMDP for $X = 2$. The piecewise linear concave value function J_k is denoted by unbroken lines. Interpolation (dotted lines) yields a lower bound to the value function. Omitting any of the piecewise linear segments leads to an upper bound (dashed lines). The main point is that the dotted line (lower bound) and dashed line (upper bound) sandwich the value function (unbroken line).

R points based on a uniform Freudenthal triangulization of $\Pi(X)$; see [226] for details). Then set

$$\bar{\Gamma}_k = \{\arg\min_{\gamma \in \Gamma_k} \gamma' \pi_r, \quad r = 1, 2, \ldots, R\}.$$

Step 2: Given $\bar{\Gamma}_k$, compute the set of vectors Γ_{k-1} using a standard POMDP algorithm.
Step 3: $k \to k - 1$ and go to Step 1.

Notice that $\bar{J}_k(\pi) = \min_{\gamma \in \bar{\Gamma}_k} \gamma' \pi$ is represented completely by R piecewise linear segments. Lovejoy [226] shows that for all k, $\bar{J}_k$ is an upper bound to the optimal value function J_k, Thus Lovejoy's algorithm gives a suboptimal policy that yields an upper bound to the value function at a computational cost of no more than R evaluations per iteration k.

So far we have discussed Lovejoy's upper bound. Lovejoy [226] also provides a constructive procedure for computing a lower bound to the optimal value function. The intuition behind the lower bound is displayed in Figure 7.4 and is as follows: a linear interpolation to a concave function lies below the concave function. Choose any R belief states $\pi_1, \pi_2, \ldots, \pi_R$. Then construct $\underline{J}_k(\pi)$ depicted in dotted lines in Figure 7.4 as the linear interpolation between the points $(\pi_i, J_k(\pi_i))$, $i = 1, 2, \ldots, R$. Clearly due to concavity of J_k, it follows that $\underline{J}_k(\pi) \leq J_k(\pi)$ for all $\pi \in \Pi(X)$.

7.5.3 Point-based value iteration methods

Point-based value iteration methods seek to compute an approximation of the value function at special points in the belief space. The main idea is to compute solutions only for those belief states that have been visited by running the POMDP. This motivates the development of approximate solution techniques that use a sampled set of belief states on which the POMDP is solved [137, 296, 194].

As mentioned above, Lovejoy [226] uses Freudenthal triangulation to form a grid on the belief space and then computes the approximate policy at these belief states. Another possibility is to take all extreme points of the belief simplex or to use a random grid. Yet another option is to include belief states that are encountered when simulating the POMDP. Trajectories can be generated in the belief simplex by sampling random actions and observations at each time [262, 308]. More sophisticated schemes for belief sampling have been proposed in [296]. The SARSOP approach of [194] performs successive approximations of the reachable belief space from following the optimal policy.

7.5.4 Belief compression POMDPs

An interesting class of suboptimal POMDP algorithms [286] involves reducing the dimension of the belief space $\Pi(X)$. The X-dimension belief states π are projected to $\bar{X}$-dimension belief states $\bar{\pi}$ that live in the reduced dimension simplex $\Pi(\bar{X})$ where $\bar{X} \ll X$. The principal component analysis (PCA) algorithm is used to achieve this belief compression as follows: suppose a sequence of beliefs $\pi_{1:N} = (\pi_1, \ldots, \pi_N)$ is generated; this is a $X \times N$ matrix. Then perform a singular value decomposition $\pi_{1:N} = UDV'$ and choose the largest $\bar{X}$ singular values. Then the original belief π and compressed belief $\bar{\pi}$ are related by $\pi = U_{\bar{X}}\bar{\pi}$ or $\bar{\pi} = U'_{\bar{X}}\pi$ where $U_{\bar{X}}$ and $D_{\bar{X}}$ denote the truncated matrices corresponding to the largest $\bar{X}$ singular values. (PCA is suited to dimensionality reduction when the data lies near a linear manifold. However, POMDP belief manifolds are rarely linear and so [286] proposes an exponential family PCA.) The next step is to quantize the low-dimensional belief state space $\Pi(\bar{X})$ into a finite state space $\bar{\mathcal{S}} = \{\bar{q}_1, \ldots, \bar{q}_L\}$. The corresponding full-dimensional beliefs are $\mathcal{S} = \{q_1, \ldots, q_L\} = \{U_{\bar{X}}\bar{q}_1, \ldots, U_{\bar{X}}\bar{q}_L\}$. The reduced-dimension dynamic programming recursion then is identical to that of a finite-state MDP. It reads

$$V_{k+1}(\bar{q}_i) = \min_{u \in \mathcal{U}} \{ \tilde{c}(\bar{q}_i, u) + \sum_{j=1}^{L} \tilde{T}(\bar{q}_i, \bar{q}_j, u) V_k(\bar{q}_j) \}$$

$$\text{where } \tilde{c}(\bar{q}_i, u) = c'_u q_i, \quad \tilde{T}(\bar{q}_i, \bar{q}_j, u) = \sum_{y} I\big(W(U'_{\bar{X}} T(q_i, y, u)) = \bar{q}_j\big)\sigma(q_i, y, u).$$

The operator W quantizes the low dimensional belief to the nearest belief in $\bar{\mathcal{S}}$. Also T and σ are the full-dimensional HMM filter and normalization term defined in (7.11). The above equation holds since $\tilde{T}(\bar{q}_i, \bar{q}_j, u) = \mathbb{P}(\bar{q}_j | \bar{q}_i, u) = \mathbb{P}(\{y : W(U'_{\bar{X}} T(q_i, y, u)) = \bar{q}_j\})$.

Chapter 10 presents algorithms for approximating the belief with provable bounds. Also §17.3 presents stochastic gradient algorithms for estimating the underlying state of an HMM directly. Either of these algorithms can be used in the above reduced-dimensional dynamic programming algorithm instead of PCA-type compression.

7.5.5 Open loop feedback control (OLFC) algorithm

Open loop feedback control (OLFC) is a suboptimal algorithm that can be used for both finite and continuous-valued observation spaces $\mathcal{Y}$. OLFC proceeds as follows: given the belief π_n at time $n = 0, 1, 2, \ldots, N-1$, compute the *open loop* action sequence $u_n^*, \ldots, u_{N-1}^*$ that minimizes

$$C_n(u_n, \dots, u_{N-1}) = \mathbb{E}\{\sum_{k=n}^{N-1} c(x_k, u_k) + c_N(x_N)|\mathcal{I}_n\}, \tag{7.24}$$
$$\text{where } \mathcal{I}_n = (\pi_0, u_0, y_1, \dots, u_{n-1}, y_n).$$

Use the action u_n^* to obtain observation y_{n+1} according to the POMDP simulation Algorithm 10. Then compute the belief state $\pi_{n+1} = T(\pi_n, y_{n+1}, u_n^*)$ at time $n+1$ using HMM filter (7.11). The procedure is then repeated with belief π_{n+1} at time $n+1$ and so on. Since π_{n+1} depends on the action u_n^* and in turn affects the choice of action u_{n+1}^* there is *feedback control* in the algorithm. Hence the name "open loop feedback control".

The open loop feedback control algorithm is detailed in Algorithm 13.

Algorithm 13 Open loop feedback control (OLFC) for POMDP with finite horizon N

For $n = 0, \dots, N-1$

1. Given belief π_n, evaluate expected cumulative cost
$$C_n(u_n, \dots, u_{N-1}) = c'_{u_n}\pi_n + c'_{u_{n+1}}P'_{u_n}\pi_n + c'_{u_{n+2}}P'_{u_{n+1}}P'_{u_n}\pi_n + \cdots + c'_N P'_{u_{N-1}} \cdots P'_{u_{n+2}}P'_{u_{n+1}}\pi_n$$
for each of the $|U|^{N-n}$ possible sequences $u_n, \dots, u_{N-1}$.
2. Evaluate minimizing sequence $(u_n^*, \dots, u_{N-1}^*) = \operatorname{argmin} C_n(u_n, \dots, u_{N-1})$
3. Use action u_n^* to obtain observation y_{n+1} according to POMDP simulation Algorithm 10.
4. Evaluate updated belief $\pi_{n+1} = T(\pi_n, y_{n+1}, u_n^*)$ using HMM filter (7.11).

End for

Note that *open loop control* is a special case of OLFC. In open loop control, only the first iteration $n = 0$ of Algorithm 13 is performed and the resulting action sequence $(u_0^*, \dots, u_{N-1}^*)$ is then used for the entire horizon. In comparison, OLFC recomputes the optimal action sequence $(u_n^*, \dots, u_{N-1}^*)$ at each iteration $n = 0, \dots, N-1$ based on the updated belief π_n.

OLFC can be applied as a suboptimal control method for nonlinear partially observed state space models and also jump Markov linear systems. Evaluating the expected cost in step 1 of Algorithm 13 simply involves propagating the Chapman–Kolmogorov equation described in §2.2 – this can be computed by stochastic simulation as explained in §A.1.3. The belief update $T(\pi, y, u)$ in Step 4 of Algorithm 13 can be evaluated via the particle filtering algorithm of Chapter 3.

A useful property of OLFC is that its performance is provably at least as good as open loop control.

THEOREM 7.5.1 *OLFC results in an expected cumulative cost that is at least as small as open loop control.*

The result is interesting since it means that however poor the feedback control might be in OLFC (after all OLFC is suboptimal), it is always superior to open loop control. To prove Theorem 7.5.1, introduce the following notation:

- Let $\bar{J}_0(\pi_0)$ denote the expected cumulative cost incurred with OLFC and $(\bar{\mu}_0, \ldots, \bar{\mu}_{N-1})$ denote the corresponding policy. Here $\bar{\mu}_n$ is defined implicitly in terms of Step 2 of Algorithm 13 as $u_n^* = \bar{\mu}_n(\pi_n)$. Clearly $\bar{J}_0(\pi_0)$ can be obtained by the backward recursion:

$$\bar{J}_n(\pi_n) = c'_{\bar{\mu}_n(\pi_n)}\pi_n + \bar{J}_{n+1}(T(\pi_n, y_{n+1}, \bar{\mu}_n(\pi_n))), \quad n = N-1, \ldots, 0$$

  initialized by terminal cost $\bar{J}_N(\pi_N) = c'_N \pi_N$.
- Let $J_n^{\mathrm{h}}(\pi_0)$ denote the expected cumulative cost incurred by the following hybrid strategy:
  - Apply OLFC from time 0 to $n-1$ and compute π_n.
  - Then apply loop control from time n to $N-1$.

  So clearly, the open loop control cumulative cost is $J_0^{\mathrm{h}}(\pi_0)$.

Then proving Theorem 7.5.1 is equivalent to showing that $\bar{J}_0(\pi_0) \leq J_0^{\mathrm{h}}(\pi_0)$ for all $\pi_0 \in \Pi(X)$. We will prove $\bar{J}_n(\pi_0) \leq J_n^{\mathrm{h}}(\pi_0)$ by backward induction for $n = N, \ldots 0$. By definition $\bar{J}_N(\pi_0) = J_N^{\mathrm{h}}(\pi_0)$. Next, assume the induction hypothesis $\bar{J}_{n+1}(\pi) \leq J_{n+1}^{\mathrm{h}}(\pi)$ for all $\pi \in \Pi(X)$. Then

$$\begin{aligned}
\bar{J}_n(\pi_n) &= c'_{\bar{\mu}_n(\pi_n)}\pi_n + \bar{J}_{n+1}(T(\pi_n, y_{n+1}, \bar{\mu}_n(\pi_n))) \\
&\leq c'_{\bar{\mu}_n(\pi_n)}\pi_n + J_{n+1}^{\mathrm{h}}(T(\pi_n, y_{n+1}, \bar{\mu}_n(\pi_n))) \quad \text{(by induction hypothesis)} \\
&= c'_{\bar{\mu}_n(\pi_n)}\pi_n + \mathbb{E}\left\{ \min_{u_{n+1},\ldots,u_{N-1}} C_{n+1}(u_{n+1}, \ldots, u_{N-1})|\mathcal{I}_n \right\} \quad \text{(see (7.24))} \\
&= c'_{\bar{\mu}_n(\pi_n)}\pi_n + \mathbb{E}\left\{ \min_{u_{n+1},\ldots,u_{N-1}} \mathbb{E}\{ \sum_{k=n+1}^{N-1} c(x_k, u_k) + c_N(x_N)|\mathcal{I}_{n+1}\} \mid \mathcal{I}_n \right\} \quad \text{(by (7.24))} \\
&\leq c'_{\bar{\mu}_n(\pi_n)}\pi_n + \min_{u_{n+1},\ldots,u_{N-1}} \mathbb{E}\left\{ \sum_{k=n+1}^{N-1} c(x_k, u_k) + c_N(x_N)|\mathcal{I}_n \right\} \\
&= J_n^{\mathrm{h}}(\pi_n)
\end{aligned}$$

The last inequality above follows from the smoothing property of conditional expectations[6] and the fact that $\mathbb{E}\{\min[\cdot]\} \leq \min[\mathbb{E}\{\cdot\}]$. The last equality follows since $\bar{\mu}_n$ is the open loop policy applied at time n which coincides with OLFC policy at time n.

7.6 Discounted infinite horizon POMDPs

So far we have considered finite horizon POMDPs. This section consider infinite horizon discounted cost POMDPs. The discounted POMDP model is a 7-tuple

[6] The smoothing property of conditional expectations says: if $\mathcal{G}_1 \subset \mathcal{G}_2$, then $\mathbb{E}\{\mathbb{E}\{X|\mathcal{G}_2\}|\mathcal{G}_1\} = \mathbb{E}\{X|\mathcal{G}_1\}$; see (3.8) of Chapter 3.

$(\mathcal{X}, \mathcal{U}, \mathcal{Y}, P(u), B(u), c(u), \rho)$ where $P(u)$, $B(u)$ and c are no longer explicit functions of time and $\rho \in [0, 1)$ in an economic discount factor. Also, compared to (7.1). there is no terminal cost c_N.

As in §6.4, define a stationary policy sequence as $\boldsymbol{\mu} = (\mu, \mu, \mu, \cdots)$ where μ is not an explicit function of time k. We will use μ instead of $\boldsymbol{\mu}$ to simply notation. For stationary policy $\mu : \Pi(X) \to \mathcal{U}$, initial belief $\pi_0 \in \Pi(X)$, discount factor $\rho \in [0, 1)$, define the objective function as the discounted expected cost:

$$J_\mu(\pi_0) = \mathbb{E}_\mu \left\{ \sum_{k=0}^{\infty} \rho^k c(x_k, u_k) \right\}, \quad \text{where } u_k = \mu(\pi_k).$$

As in §7.2 we can re-express this objective in terms of the belief state as

$$J_\mu(\pi_0) = \mathbb{E}_\mu \left\{ \sum_{k=0}^{\infty} \rho^k c'_{\mu(\pi_k)} \pi_k \right\}, \tag{7.25}$$

where $c_u = [c(1, u), \ldots, c(X, u)]'$, $u \in \mathcal{U}$ is the cost vector for each action, and the belief state evolves according to the HMM filter $\pi_k = T(\pi_{k-1}, y_k, u_{k-1})$ given in (7.11).

The aim is to compute the optimal stationary policy $\mu^* : \Pi(X) \to \mathcal{U}$ such that $J_{\mu^*}(\pi_0) \leq J_\mu(\pi_0)$ for all $\pi_0 \in \Pi(X)$ where $\Pi(X)$ denotes the belief space defined in § 7.2. From the dynamic programming recursion, we have for any finite horizon N that

$$J_k(\pi) = \min_{u \in \mathcal{U}} \{ \rho^k c'_u \pi + \sum_{y \in \mathcal{Y}} J_{k+1} \left(T(\pi, y, u) \right) \sigma(\pi, y, u) \}$$

initialized by $J_N(\pi) = 0$. For discounted cost problems, it is more convenient to work with a forward iteration of indices. Accordingly, define the following value function $V_n(\pi)$:

$$V_n(\pi) = \rho^{n-N} J_{N-n}(\pi), \quad 0 \leq n \leq N, \ \pi \in \Pi(X).$$

Then it is easily seen that $V_n(\pi)$ satisfies the dynamic programming equation

$$V_n(\pi) = c'_u \pi + \rho \sum_{y \in Y} V_{n-1} \left(T(\pi, y, u) \right) \sigma(\pi, y, u), \quad V_0(\pi) = 0. \tag{7.26}$$

7.6.1 Bellman's equation for discounted infinite horizon POMDPs

The main result for infinite horizon discounted cost POMDPs is as follows:

THEOREM 7.6.1 *Consider an infinite horizon discounted cost POMDP with discount factor $\rho \in [0, 1)$. Then with $\pi \in \Pi(X)$ denoting the belief state,*

1. *The optimal expected cumulative cost is achieved by a stationary deterministic Markovian policy μ^*.*
2. *The optimal policy $\mu^*(\pi)$ and value function $V(\pi)$ satisfy Bellman's dynamic programming equation*

$$\mu^*(\pi) = \underset{u \in \mathcal{U}}{\operatorname{argmin}}\, Q(\pi, u), \quad J_{\mu^*}(\pi_0) = V(\pi_0) \tag{7.27}$$
$$V(\pi) = \min_{u \in \mathcal{U}} Q(\pi, u), \quad Q(\pi, u) = c_u' \pi + \rho \sum_{y \in Y} V\left(T(\pi, y, u)\right) \sigma(\pi, y, u),$$

where $T(\pi, y, u)$ and $\sigma(\pi, y, u)$ are the HMM filter and normalization (7.11). The expected cumulative cost incurred by the optimal policy is $J_{\mu^*}(\pi) = V(\pi)$.

3. *The value function $V(\pi)$ is continuous and concave in $\pi \in \Pi(X)$.*

Statements 1 and 2 follow from the proof of Theorem 6.2.2 in Chapter 6. Statement 3 follows since $V_N(\pi)$ is continuous and concave in π (Theorem 7.4.1) and the value iteration algorithm generates a sequence of functions $\{V_N(\pi), N = 1, 2, \ldots\}$ that converges uniformly over $\Pi(X)$ to V (shown in Theorem 7.6.3 below). Actually for concavity of the limit, pointwise convergence suffices.

We now present a classical result regarding the existence and uniqueness of a solution V to Bellman's equation (7.27) for the infinite horizon discounted cost POMDP. This result shows that Bellman's equation is a contraction mapping. So by the fixed point theorem a unique solution exists. This forms the basis of the value iteration algorithm presented in §7.6.2.

For any bounded function ϕ on $\Pi(X)$ denote the dynamic programming operator $L : \phi \to \mathbb{R}$ as

$$L\phi(\pi) = \min_u \left\{ c_u' \pi + \rho \sum_{y \in Y} \phi\left(T(\pi, y, u)\right) \sigma(\pi, y, u) \right\}. \tag{7.28}$$

Let $\mathcal{B}(X)$ denote the set of bounded real-valued functions on $\Pi(X)$. Then for any ϕ and $\psi \in \mathcal{B}(X)$, define the sup-norm metric

$$\|\phi - \psi\|_\infty = \sup_{\pi \in \Pi(X)} |\phi(\pi) - \psi(\pi)|.$$

Then the function space $\mathcal{B}(X)$ is a Banach space [229], i.e. a complete metric space.

THEOREM 7.6.2 *For discount factor $\rho \in [0, 1)$, the dynamic programming operator L is a contraction mapping. That is, for any $\phi, \psi \in \mathcal{B}(X)$,*

$$\|L\phi - L\psi\|_\infty \le \rho \|\phi - \psi\|_\infty.$$

As a result, it follows from Banach's fixed point theorem that there exists a unique solution V satisfying Bellman's equation $V = LV$.

Proof Suppose ϕ and ψ are such that for a fixed π, $L\psi(\pi) \ge L\phi(\pi)$. Let u^* denote the minimizer for $L\phi(\pi)$, that is,

$$u^* = \underset{u}{\operatorname{argmin}} \left\{ c_u' \pi + \rho \sum_{y \in Y} \phi\left(T(\pi, y, u)\right) \sigma(\pi, y, u) \right\}.$$

Then clearly, $L\psi(\pi) \leq c'_{u^*}\pi + \rho \sum_{y \in Y} \psi\left(T\left(\pi, y, u^*\right)\right) \sigma\left(\pi, y, u^*\right)$ since u^* is not necessarily the minimizer for $L\psi(\pi)$. So

$$
\begin{aligned}
0 \leq L\psi(\pi) - L\phi(\pi) &\leq \rho \sum_{y \in Y} \left[\psi\left(T\left(\pi, y, u^*\right)\right) - \phi\left(T\left(\pi, y, u^*\right)\right)\right] \sigma\left(\pi, y, u^*\right) \\
&\leq \rho \|\psi - \phi\|_\infty \sum_{y \in Y} \sigma\left(\pi, y, u^*\right) \\
&= \rho \|\psi - \phi\|_\infty .
\end{aligned}
$$

A similar argument holds for the set of beliefs π for which $L\psi(\pi) \leq L\phi(\pi)$. Therefore, for all $\pi \in \Pi(X)$,

$$
|L\psi(\pi) - L\phi(\pi)| \leq \rho \|\psi - \phi\|_\infty .
$$

Taking the supremum over $\pi \in \Pi(X)$ in the above expression proves that L is a contraction mapping.

Having shown that L is a contraction mapping, the Banach fixed point theorem [229] asserts the existence and uniqueness of a solution. □

7.6.2 Value iteration algorithm for discounted cost POMDPs

Let $n = 1, 2, \ldots, N$ denote iteration number. The value iteration algorithm for a discounted cost POMDP is a successive approximation algorithm for computing the value function $V(\pi)$ of Bellman's equation (7.27) and proceeds as follows: initialize $V_0(\pi) = 0$. For iterations $n = 1, 2, \ldots, N$, evaluate

$$
\begin{aligned}
V_n(\pi) &= \min_{u \in \mathcal{U}} Q_n(\pi, u), \quad \mu_n^*(\pi) = \operatorname*{argmin}_{u \in \mathcal{U}} Q_n(\pi, u), \\
Q_n(\pi, u) &= c'_u \pi + \rho \sum_{y \in Y} V_{n-1}\left(T\left(\pi, y, u\right)\right) \sigma\left(\pi, y, u\right).
\end{aligned}
\tag{7.29}
$$

Finally, the stationary policy μ_N^* is used at each time instant k in the real-time controller of Algorithm 11. The obvious advantage of the stationary policy is that only the policy $\mu_N^*(\pi)$ needs to be stored for real-time implementation of the controller in Algorithm 11.

Summary: The POMDP value iteration algorithm (7.29) is identical to the finite horizon dynamic programming recursion (7.15). So at each iteration n, $V_n(\pi)$ is piecewise linear and concave in π (by Theorem 7.2.1) and can be computed using any of the POMDP algorithms discussed in §7.5. The number of piecewise linear segments that characterize $V_n(\pi)$ can grow exponentially with iteration n. Therefore, except for small state, action and observation spaces, suboptimal algorithms (such as those discussed in §7.5) need to be used.[7]

How are the number of iterations N chosen in the value iteration algorithm (7.29)? The value iteration algorithm (7.29) generates a sequence of value functions $\{V_n\}$ that will

[7] In Part III of the book, we will show how the value iteration algorithm can be used not as a computational algorithm, but as a mathematical tool to establish the monotone structure of the optimal policy; then stochastic gradient algorithms that exploit this monotone structure will be used to estimate the optimal policy.

converge uniformly (sup-norm metric) as $N \to \infty$ to $V(\pi)$, the optimal value function of Bellman's equation. The number of iterations N in (7.29) can be chosen as follows: let $\epsilon > 0$ denote a specified tolerance.

THEOREM 7.6.3 *Consider the value iteration algorithm with discount factor ρ and N iterations. Then:*

1. $\sup_\pi |V_N(\pi) - V_{N-1}(\pi)| \leq \epsilon$ *implies that* $\sup_\pi |V_N(\pi) - V(\pi)| \leq \frac{\epsilon\rho}{1-\rho}$.
2. $|V_N(\pi) - V(\pi)| \leq \frac{\rho^{N+1}}{1-\rho} \max_{x,u} |c(x,u)|$.

Proof 1. Recall that V is the fixed point of Bellman's equation, i.e. $V = LV$ where Bellman's operator L is defined in (7.28). Therefore

$$\begin{aligned}\|V - V_N\| &= \|LV - LV_N + LV_N - V_N\| \\ &\leq \|LV - LV_N\| + \|LV_N - V_N\| \\ &= \|LV - LV_N\| + \|LV_N - LV_{N-1}\| \\ &\leq \rho\|V - V_N\| + \rho\|V_N - V_{N-1}\|\end{aligned}$$

where the last step follows from Theorem 7.6.2. Therefore,

$$\|V - V_N\| \leq \frac{\rho\|V_N - V_{N-1}\|}{1-\rho}.$$

2. To show the second assertion, start with

$$V(\pi) = V_N(\pi) + \mathbb{E}\{\sum_{k=N+1}^{\infty} \rho^k c'_{\mu^*(\pi_k)}\pi_k \mid \pi_0 = \pi\}.$$

Then noting that $c(x,u) \geq \min_{x,u} c(x,u) \geq -\max_{x,u} |c(x,u)|$, it follows that

$$V(\pi) \geq V_N(\pi) + \mathbb{E}\{\sum_{k=N+1}^{\infty} \rho^k \min_{x,u} c(x,u)\} \geq V_N(\pi) - \mathbb{E}\{\sum_{k=N+1}^{\infty} \rho^k \max_{x,u} |c(x,u)|\}$$

$$V(\pi) \leq V_N(\pi) + \mathbb{E}\{\sum_{k=N+1}^{\infty} \rho^k \max_{x,u} c(x,u)\} \leq V_N(\pi) + \mathbb{E}\{\sum_{k=N+1}^{\infty} \rho^k \max_{x,u} |c(x,u)|\}.$$

□

Actually, similar to the value iteration algorithm, one can evaluate the expected discounted cumulative cost of an arbitrary stationary policy (not necessarily the optimal policy) as follows:

COROLLARY 7.6.4 (Policy evaluation) *For any stationary policy μ, the associated expected discounted cumulative cost $J_\mu(\pi)$ defined in (7.25) for a POMDP satisfies*

$$J_\mu(\pi) = c'_{\mu(\pi)}\pi + \rho \sum_{y \in \mathcal{Y}} J_\mu\big(T(\pi, y, \mu(\pi))\big)\, \sigma(\pi, y, \mu(\pi)). \tag{7.30}$$

Similar to the value iteration algorithm (7.29), $J_\mu(\pi)$ can be obtained as $J_\mu(\pi) = \lim_{n\to\infty} V_{\mu,n}(\pi)$. Here $V_{\mu,n}(\pi)$, $n = 1,2,\dots$ satisfies the recursion

$$V_{\mu,n}(\pi) = c'_{\mu(\pi)}\pi + \rho \sum_{y\in\mathcal{Y}} V_{\mu,n-1}\big(T(\pi,y,\mu(\pi))\big)\,\sigma(\pi,y,\mu(\pi)), \quad V_{\mu,0}(\pi) = 0.$$

Also Theorem 7.6.3 holds for $J_\mu(\pi)$. □

The proof is omitted since it is similar to the corresponding theorems for the optimal policy given above. Note that (7.30) can be written as

$$J_\mu(\pi) = \gamma'_{\mu(\pi)}\pi, \quad \text{where } \gamma_{\mu(\pi)} = c_u + \rho \sum_{y\in\mathcal{Y}} \gamma_{\mu(T(\pi,y,u))} P(u) B_y(u), \tag{7.31}$$

where $u = \mu(\pi)$ on the right-hand side. We will use this representation below.

7.6.3 Finitely transient policies

Consider an infinite horizon discounted cost POMDP. As discussed previously, the number of piecewise linear segments that characterize the value function $V_n(\pi)$ can grow exponentially with iteration n. From Corollary 7.6.4, a similar result holds for the n-stage cost $V_{\mu,n}(\pi)$ for an arbitrary stationary policy μ.

Are there stationary policies for POMDPs where after some finite number of iterations n^*, the number of piecewise linear segments that characterize $V_{\mu,n}$, $n \geq n^*$ remain constant? The answer is a qualified "yes". This property holds for stationary policies that are *finitely transient*. For a finitely transient policy μ, the infinite horizon cumulative cost $J_\mu(\pi) = \lim_{n\to\infty} V_{\mu,n}(\pi)$ is piecewise linear and concave with a finite number of segments on the belief space $\Pi(X)$ and can be computed exactly as the solution of a linear system of equations. Since $J_\mu(\pi)$ can be evaluated exactly for a finitely transient policy, one can then devise a policy improvement algorithm to compute better policies similar to the policy iteration algorithm of §6.4.2. This is the basis of the policy improvement algorithm in [307]. Before proceeding, we emphasize that finitely transient policies are an interesting theoretical construct with limited practical value in solving large-scale POMDPs since n^* could be arbitrarily large.

We start by first defining a Markov partition of the belief space $\Pi(X)$.

DEFINITION 7.6.5 (Markov partition) *Given a stationary policy μ, a partition $\mathcal{R} = \{\mathcal{R}_1, \mathcal{R}_2, \dots, \mathcal{R}_L\}$ of $\Pi(X)$ is said to be Markov if for any $l \in \{1,2\dots,L\}$:*

1. *All beliefs in $\mathcal{R}_l$ are assigned to the same action by policy μ. That is, $\pi_1,\pi_2 \in \mathcal{R}_l$ implies that $\mu(\pi_1) = \mu(\pi_2)$.*
2. *Under the belief update $T(\pi,y,\mu(\pi))$, all beliefs in $\mathcal{R}_l$ map into the same set. That is, $\pi_1,\pi_2 \in \mathcal{R}_l$ implies that $T(\pi_1,y,\mu(\pi_1))$ and $T(\pi_2,y,\mu(\pi_2))$ lie in the same set $\mathcal{R}_{m(y)}$ where $m(y) \in \{1,2,\dots,L\}$.*

If a POMDP satisfies the Markov partition property it behaves like an MDP with L states. Indeed, if a stationary policy μ induces a Markov partition, then from (7.31) and $\pi_0 \in \mathcal{R}_l$, the cumulative discounted cost is obtained exactly as

$$J_\mu(\pi_0) = \gamma'_l\pi_0, \quad \text{if } \pi_0 \in \mathcal{R}_l,$$

where $\gamma_l \in \mathbb{R}^X$ satisfies the linear system of L equations

$$\gamma_l = c_{\mu(l)} + \rho \sum_{y=1}^{Y} P(l)\, B_y(l)\, \gamma_{m(y)}, \qquad l, m(y) \in \{1, 2, \dots, L\}. \tag{7.32}$$

In (7.32), we have used $\mu(l)$ to denote $\mu(\pi)$, $\pi \in \mathcal{R}_l$.

We are now in a position to define a finitely transient policy. Let D_μ denote the smallest closed set of discontinuities[8] of stationary policy $\mu(\pi)$. That is

$$D_\mu = \text{closure}\{\pi \in \Pi(X) : \mu(\pi) \text{ is discontinuous at } \pi\}.$$

Define the sequence of sets D^n as follows: $D^0 = D_\mu$, and for $n \geq 1$,

$$D^n = \{\pi \in \Pi(X) : T(\pi, y, \mu(\pi)) \in D^{n-1} \text{ for some } y \in \mathcal{Y}\}.$$

THEOREM 7.6.6 (Sondik [307]) *Consider an infinite horizon discounted cost POMDP.*

1. *A stationary policy μ is finitely transient if and only if there exists an integer n such that $D^n = \emptyset$. The smallest such integer n^* is called the index of the policy.*
2. *A finitely transient policy μ induces a Markov partition of $\Pi(X)$ and therefore the associated expected cumulative cost $J_\mu(\pi_0)$ can be evaluated using (7.32).*

Sondik [307] gives an explicit characterization of the Markov partition (asserted in statement 2 of Theorem 7.6.6) as follows. Define for $k = 0.1, \dots,$

$$\mathcal{P}^k = \text{ partition of } \Pi(X) \text{ with partition specified by } \cup_{n=0}^{k} D^k.$$

Then Sondik [307] shows that $\mathcal{P}^{n^*}$ is a Markov partition for the finitely transient policy μ.

In general, it is difficult to construct finitely transient policies for a POMDP. In Part III of the book, we will use a different methodology (lattice programming) to directly characterize the structure of the optimal policy.

7.7 Example: Optimal search for a Markovian moving target

In this section we discuss optimal search of a moving target. This serves as a useful illustrative example of a POMDP. From an abstract point of view, many resource allocation problems involving controlled sensing and communication with noisy information can be formulated as an optimal search problem; see for example [162] where opportunistic transmission over a fading channel is formulated as an optimal search problem.

A target moves among X cells according to a Markov chain with transition matrix P. At time instants $k \in \{0, 1, 2, \dots\}$, the searcher must choose an action from the action space $\mathcal{U}$. The set $\mathcal{U}$ contains actions that search a particular cell or a group of cells simultaneously. Assuming action u is selected by the searcher at time k, it is executed with probability $1 - q(u)$. If the action cannot be executed, the searcher is said to be *blocked* for time k. This blocking event with probability $q(u)$ models the scenario when the search sensors are a shared resource and not enough resources are available to carry

[8] This is a fancy way of saying that D_μ is the set of beliefs π at which the policy $\mu(\pi)$ switches from one action to another action.

out the search at time k. If the searcher is not blocked and action u searches the cell that the target is in, the target is detected with probability $1-\beta(u)$; failure to detect the target when it is in the cell searched is called an *overlook*. So the overlook probability in cell u is $\beta(u)$.

If the decision-maker knows P, β, q, in which order should it search the cells to find the moving target with minimum expected effort?

7.7.1 Formulation of finite horizon search problem

It is assumed here that the searcher has a total of N attempts to find the moving target. Given a target that moves between X cells, let $\bar{\mathcal{X}} = \{1, 2, \ldots, X, T\}$ denote the augmented state space. Here T corresponds to a fictitious *terminal state* that is added as a means of terminating search if the target is detected prior to exhausting the N search horizon.

Denote the observation space as $\mathcal{Y} = \{F, \bar{F}, b\}$. Here F denotes "target found", $\bar{F}$ denotes "target not found" and b denotes "search blocked".

An optimal search problem consists of the following ingredients:

1. Markov state dynamics: The location of the target is modeled as a finite state Markov chain. The target moves amongst the X cells according to transition probability matrix P. Let $x_k \in \bar{\mathcal{X}}$ denote the state (location) of the target at the start of search epoch k where $k = 0, 1, 2, \ldots, N-1$. To model termination of the search process after the target is found, we model the target dynamics by the *observation* dependent transition probability matrices P^y, $y \in \mathcal{Y} = \{F, \bar{F}, b\}$:

$$P^F = \begin{bmatrix} 0 & 0 & \cdots & 1 \\ 0 & 0 & \cdots & 1 \\ \vdots & \vdots & \ddots & \vdots \\ 0 & 0 & \cdots & 1 \end{bmatrix}, \quad P^{\bar{F}} = P^b = \begin{bmatrix} P & \mathbf{0} \\ \mathbf{0}' & 1 \end{bmatrix}. \tag{7.33}$$

That is, $\mathbb{P}(x_{k+1} = j | x_k = i, y_k = y) = P^y_{ij}$. As can be seen from the transition matrices above, the terminal state T is designed to be absorbing. A transition to T occurs only when the target is detected. The initial state of the target is sampled from $\pi_0(i)$, $i \in \{1, \ldots, X\}$.

2. Action: At each time k, the decision-maker chooses action u_k from the finite set of search actions $\mathcal{U}$. In addition to searching the target in one of the X cells, $\mathcal{U}$ may contain actions that specify the simultaneous search in a number of cells.

3. Observation: Let $y_k \in \mathcal{Y} = \{F, \bar{F}, b\}$ denote the observation received at time k upon choosing action u_k. Here

$$y_k = \begin{cases} F & \text{target is found,} \\ \bar{F} & \text{target is not found,} \\ b & \text{search action is blocked due to insufficient available resources.} \end{cases}$$

Define the *blocking* probabilities $q(u)$ and *overlook* probabilities $\beta(u)$, $u \in \mathcal{U}$ as:

$$\begin{aligned} q(u) &= \mathbb{P}(\text{insufficient resources to perform action } u \text{ at epoch } k), \\ \beta(u) &= \mathbb{P}(\text{target not found} | \text{target is in the cell } u). \end{aligned} \tag{7.34}$$

Then, the observation y_k received is characterized probabilistically as follows. For all $u \in \mathcal{U}$ and $j = 1, \ldots, X$,

$$
\begin{aligned}
\mathbb{P}(y_k = F | x_k = j, u_k = u) &= \begin{cases} (1 - q(u))(1 - \beta(u)) & \text{if action } u \text{ searches cell } j, \\ 0 & \text{otherwise,} \end{cases} \\
\mathbb{P}(y_k = \bar{F} | x_k = j, u_k = u) &= \begin{cases} 1 - q(u) & \text{if action } u \text{ does not search cell } j, \\ \beta(u)(1 - q(u)) & \text{otherwise,} \end{cases} \\
\mathbb{P}(y_k = b | x_k = j, u_k = u) &= q(u).
\end{aligned} \tag{7.35}
$$

Finally, for the fictitious terminal state T, the observation F is always received regardless of the action taken, so that

$$\mathbb{P}(y_k = F | x_k = T, u_k = u) = 1.$$

4. Cost: Let $c(x_k, u_k)$ denote the instantaneous cost for choosing action u_k when the target's state is x_k. Three types of instantaneous costs are of interest.

1. Maximize probability of detection [265, 103]. The instantaneous reward is the probability of detecting the target (obtaining observation F) for the current state and action. This constitutes a negative cost. So

$$
\begin{aligned}
c(x_k = j, u_k = u) &= -\mathbb{P}(y_k = F | x_k = j, u_k = u) \quad \text{for } j = 1, \ldots, X, \\
c(x_k = T, u_k = u) &= 0.
\end{aligned} \tag{7.36}
$$

2. Minimize search delay [265]. An instantaneous cost of 1 unit is accrued for every action taken until the target is found, i.e. until the target reaches the terminal state T:

$$
\begin{aligned}
c(x_k = j, u_k = u) &= 1 \quad \text{for } j = 1, \ldots, X, \\
c(x_k = T, u_k = u) &= 0.
\end{aligned} \tag{7.37}
$$

3. Minimize search cost. The instantaneous cost depends only on the action taken. Let $c(u)$ denote the positive cost incurred for action u, then

$$
\begin{aligned}
c(x_k = j, u_k = u) &= c(u) \quad \text{for } j = 1, \ldots, X, \\
c(x_k = T, u_k = u) &= 0.
\end{aligned} \tag{7.38}
$$

5. Performance criterion: Let $\mathcal{I}_k$ denote the information (history) available at the start of search epoch k:

$$\mathcal{I}_0 = \{\pi_0\}, \quad \mathcal{I}_k = \{\pi_0, u_0, y_0, \ldots, u_{k-1}, y_{k-1}\} \quad \text{for } k = 1, \ldots, N. \tag{7.39}$$

$\mathcal{I}_k$ contains the initial probability distribution π_0, the actions taken and observations received prior to search time k. A *search policy* $\boldsymbol{\mu}$ is a sequence of *decision rules* $\boldsymbol{\mu} = \{\mu_0, \ldots, \mu_{N-1}\}$ where each decision rule $\mu_k : \mathcal{I}_k \to \mathcal{U}$. The performance criterion considered is

$$J_{\boldsymbol{\mu}}(\pi_0) = \mathbb{E}_{\boldsymbol{\mu}} \left\{ \sum_{k=0}^{N-1} c(x_k, \mu_k(\mathcal{I}_k)) \big| \pi_0 \right\}. \tag{7.40}$$

This is the expected cost accrued after N time points using search policy μ when the initial distribution of the target is π_0. The *optimal search problem* is to find the policy μ^* that minimizes (7.40) for all initial distributions, i.e.

$$\mu^* = \underset{\mu \in \mathcal{U}}{\operatorname{argmin}}\, J_\mu(\pi_0), \qquad \forall\, \pi_0 \in \Pi(X). \tag{7.41}$$

Similar to (7.13), we can express the objective in terms of the belief state as

$$\begin{aligned} J_\mu(\pi_0) &= \mathbb{E}_\mu\left\{\sum_{k=0}^{N} c(x_k, \mu_k(\mathcal{I}_k))\big|\pi_0\right\}, \\ &= \mathbb{E}_\mu\left\{\sum_{k=0}^{N} \mathbb{E}\{c(x_k, \mu_k(\mathcal{I}_k))|\mathcal{I}_k\}\big|\pi_0\right\} = \sum_{k=0}^{N} \mathbb{E}\{c'_{u_k}\pi_k\} \end{aligned} \tag{7.42}$$

where belief state $\pi_k = [\pi_k(1), \dots, \pi_k(X+1)]'$ is defined as $\pi_k(i) = \mathbb{P}(x_k = i|\mathcal{I}_k)$. The belief state is updated by the HMM predictor[9] as follows:

$$\begin{aligned} \pi_{k+1} &= T(\pi_k, y_k, u_k) = \frac{P^{y_k\prime}\tilde{B}_{y_k}(u_k)\pi_k}{\sigma(\pi_k, y_k, u_k)}, \qquad \sigma(\pi, y, u) = \mathbf{1}'\tilde{B}_y(u)\pi \\ \tilde{B}_y(u) &= \operatorname{diag}(\mathbb{P}(y_k = y|x_k = 1, u_k = u), \dots, \mathbb{P}(y_k = y|x_k = X, u_k = u), \\ &\qquad\qquad \mathbb{P}(y_k = y|x_k = T, u_k = u)). \end{aligned} \tag{7.43}$$

Recall the observation dependent transition probabilities P^y are defined in (7.33).

The optimal policy μ^* can be computed via the dynamic programming recursion (7.15) where T and σ are defined in (7.43).

7.7.2 Formulation of optimal search as a POMDP

In order to use POMDP software to solve the search problem (7.41) via dynamic programming, it is necessary to express the search problem as a POMDP. The search problem in §7.7.1 differs from a standard POMDP in two ways:

Timing of the events: In a POMDP, the observation y_{k+1} received upon adopting action u_k is with regards to the new state of the system, x_{k+1}. In the search problem, the observation y_k received for action u_k conveys information about the current state of the target (prior to transition), x_k.

Transition to the new state: In a POMDP the probability distribution that characterizes the system's new state, x_{k+1}, is a function of its current state x_k and the action u_k adopted. However, in the search problem, the distribution of the new state, x_{k+1}, is a function of x_k and observation y_k.

The search problem of §7.7.1 can be reformulated as a POMDP (7.1) as follows: define the augmented state process $s_{k+1} = (y_k, x_{k+1})$.

Then consider the following POMDP with $2X + 1$ underlying states.

[9] The reader should note the difference between the information pattern of a standard POMDP (7.9), namely, $\mathcal{I}_k = (\pi_0, u_0, y_1, \dots, u_{k-1}, y_k)$ and the information pattern $\mathcal{I}_k$ for the search problem in (7.39). In the search problem, $\mathcal{I}_k$ has observations until time $k-1$, thereby requiring the HMM predictor (7.43) to evaluate the inner conditional expectation of the cost in (7.42).

State space: $S = \{(\bar{F},1),(\bar{F},2),\ldots,(\bar{F},X),(b,1),(b,2),\ldots,(b,X),(F,T)\}$.
Action space: $\mathcal{U}$ (same as search problem in §7.7.1).
Observation space: $\mathcal{Y} = \{F,\bar{F},b\}$ (same as search problem in §7.7.1).
Transition probabilities: For each action $u \in \mathcal{U}$, define

$$P(u) = \begin{bmatrix} B_{\bar{F}}(u)P & B_b(u)P & B_F(u)\mathbf{1} \\ B_{\bar{F}}(u)P & B_b(u)P & B_F(u)\mathbf{1} \\ \mathbf{0}' & \mathbf{0}' & 1 \end{bmatrix} \tag{7.44}$$

where P is the transition matrix of the moving target and for $y \in \{F,\bar{F},b\}$

$$B_y(u) = \operatorname{diag}\big(\mathbb{P}(y_k = y|x_k = 1, u_k = u),\ldots,\mathbb{P}(y_k = y|x_k = X, u_k = u)\big). \tag{7.45}$$

Recall these are computed in terms of the blocking and overlook probabilities using (7.35).
Observation probabilities: For each action $u \in \mathcal{U}$, define the observation probabilities $R_{sy}(u) = \mathbb{P}(y_k = y|s_{k+1} = s, u_k = u)$ where for $s = (\bar{y},x)$,

$$R_{\bar{y}x,y}(u) = \begin{cases} 1 & \bar{y} = y, \\ 0 & \text{otherwise} \end{cases}, \quad y,\bar{y} \in \mathcal{Y},\ x \in \{1,2,\ldots,X\}.$$

Costs: For each action $u \in \mathcal{U}$, define the POMDP instantaneous costs as

$$g(i,u) = \begin{cases} c(i,u) & \text{for } i = 1,\ldots,X, \\ c(i-X,u) & \text{for } i = X+1,\ldots,2X, \\ 0 & \text{for } i = 2X+1. \end{cases} \tag{7.46}$$

To summarize, optimal search over a finite horizon is equivalent to the finite horizon POMDP $(S,\mathcal{U},\mathcal{Y},P(u),R(u),g(u))$.

7.7.3 Optimal search over an infinite horizon

So far we have considered optimal search over a finite horizon. This section discusses optimal search over an infinite horizon. However, instead of having a discounted cost, we consider the following more general undiscounted cost:

$$J_{\mu}(\pi_0) = \lim_{N\to\infty} \mathbb{E}_{\mu}\left\{\sum_{k=0}^{N-1} c'_{\mu_k(\pi_k)}\pi_k|\pi_0\right\}. \tag{7.47}$$

The policy is $\boldsymbol{\mu} = \{\mu_0,\mu_1,\ldots\}$. Assuming that the limit in (7.47) is well defined, the aim is to compute the optimal policy

$$\boldsymbol{\mu}^* = \operatorname*{argmin}_{\boldsymbol{\mu}} J_{\boldsymbol{\mu}}(\pi), \quad \forall \pi \in \Pi(X). \tag{7.48}$$

Several reasons motivate such undiscounted problems. First, they allow formulating search problems that find the moving target with minimum delay (7.37) or at minimum cost (7.38). These criteria cannot be captured by the finite horizon reward (7.40) since then the additional constraint that the search should terminate in N steps is imposed.

Also, these criteria do not have a discount factor and so, introducing a discount factor $\rho \in [0, 1)$ is unnatural. Second, a moving target may never be found[10] and so the cumulative cost (7.47) may be unbounded unless additional assumptions are made on the search model. These additional assumptions give insight into the dynamics of the search process. In particular, under the assumptions discussed below, the optimal search policy is stationary even though $\rho = 1$ (no discount factor).

For a discounted infinite horizon POMDP, the value iteration algorithm over a finite horizon *uniformly* approximates the value function (Theorem 7.6.3). The aim below is to show that when $\rho = 1$ (i.e. no discounting), the same solution methodology and result applies for an infinite horizon search problem. This is so because there is inherent discounting in the search problem in a sense to be made precise below. Such infinite horizon search problems belong to the class of *partially observed stochastic shortest path problems*; see [52, 49] for a comprehensive treatment.

Existence and characterization of optimal stationary policy

(A1) There exists a positive integer l and $\alpha \in (0, 1)$ such that $\mathbb{P}(\pi_l = e_T|\pi_0) \geq \alpha$ for all π_0. (Recall T is the artificial terminal state and e_T is the unit vector.)

(A2) In (7.35), for all $u \in \mathcal{U}$, $\mathbb{P}(y_k = F|x_k = i, u_k = u) \neq 0$ for some $i \in \{1, \ldots, X\}$.

The following is the main result; see Proposition 4.1, [52, Chap. 4] for proof.

PROPOSITION 7.7.1 *Consider the undiscounted optimal search problem (7.47). Under assumptions (A1) and (A2)*
(i) There exists a stationary policy that is optimal.
(ii) Let μ_N denote the policy obtained from the value iteration algorithm ((7.29) with $\rho = 1$) after N iterations. For any $\epsilon > 0$, there exists a positive integer $\bar{N}$ such that for all choice of $N > \bar{N}$, the stationary policy μ_N satisfies $\sup_\pi |V_{\mu_N}(\pi) - J_{\mu^*}(\pi)|_\infty < \epsilon$.

(A1) says that after l time points, the probability of termination (finding the target) is nonzero. Under (A1), the search problem has an inherent discount factor of the form $1 - \alpha$. Indeed,

$$\begin{aligned}\mathbb{P}(\pi_{kl} \neq e_T|\pi_0) &= \mathbb{P}(\pi_{kl} \neq e_T|\pi_{(k-1)l} \neq e_T)\mathbb{P}(\pi_{(k-1)l} \neq e_T|\pi_0)\\ &\leq (1-\alpha)\mathbb{P}(\pi_{(k-1)l} \neq e_T|\pi_0) \leq (1-\alpha)^k.\end{aligned}$$

As a result, since the terminal state T incurs zero cost, $J_\mu(\pi_0)$ is finite:

$$\begin{aligned}J_\mu(\pi_0) &= \lim_{N\to\infty} \mathbb{E}_\mu\left\{\sum_{k=0}^{N-1} c'_{\mu_k(u_k)}\pi_k|\pi_0\right\}\\ &\leq \min_{x,u} |c(x,u)| \lim_{N\to\infty} \sum_{k=0}^{N-1} \mathbb{P}(\pi_k \neq e_T).\end{aligned}$$

[10] If the target jumps periodically between two cells as $1, 2, 1, 2, \ldots$ and the searcher examines the cells periodically as $2, 1, 2, 1, \ldots$, then the searcher will never find the target.

For assumption (A2) to be satisfied, every search action should be "useful". Search action u is said to be useful if there exists a cell i such that adopting u while the target is in cell i yields a nonzero probability of discovery.

The following theorem specifies useful sufficient conditions for (A1) to hold.

(C1) The transition matrix P of the moving target is primitive. That is, for some integer $l > 0$, $P^l_{ij} > 0, \forall\, i,j \in \{1,\ldots,X\}$ where $P^l_{ij} = \mathbb{P}(x_l = j|x_0 = i)$ (see §2.4.3).

(C2) In (7.34), for all $u \in \mathcal{U}$, the blocking probabilities satisfy $q(u) \in (0,1)$.

(C3) In (7.34), for all $u \in \mathcal{U}$, the overlook probabilities satisfy $\beta(u) \in (0,1)$.

THEOREM 7.7.2 *The following are sufficient conditions for (A1):*
(i) (C1) holds for $l = 1$.
(ii) (C1) holds for $l > 1$ and (C2), (C3) hold.

Proof We only prove the first statement; see [301] for the more general case.

$$\begin{aligned}\mathbb{P}(x_2 = T|\pi_0) &= \mathbb{P}(x_0 = T|\pi_0) + \mathbb{P}(x_1 = T|x_0 \neq T, \pi_0) + \mathbb{P}(x_2 = T|x_1 \neq T, \pi_0)\\ &= \pi_0(T) + \mathbf{1}'B_F(u_0)\tilde{\pi}_0 + \mathbf{1}'B_F(u_1)P'B_{\bar{F}}(u_0)\tilde{\pi}_0 + \mathbf{1}'B_F(u_1)\,P'B_b(u_0)\tilde{\pi}_0\\ &\geq \pi_0(T) + \underline{B}_F \min_{ij} P_{ij}\mathbf{1}'(B_F(u_0) + B_{\bar{F}}(u_0) + B_b(u_0))\tilde{\pi}_0\\ &\geq \underline{B}_F \min_{ij} P_{ij} > 0.\end{aligned}$$

where $\tilde{\pi}_0 = [\pi_0(1),\ldots,\pi_0(X)]'$, $B_y(u)$ is defined in (7.45) and $\underline{B}_F = \min_{i,u} B_{iF}(u)$. Note $\min_{ij} P_{ij} > 0$ by assumption (C1) with $l = 1$ and $\underline{B}_F > 0$ by (A2). □

Summary: For optimal search over an infinite horizon (without discount factor), under (A1), (A2), the optimal cost is achieved by a stationary policy. This can be computed via the value iteration algorithm (7.29) with $\rho = 1$.

7.7.4 Classical optimal search for non-moving target

As shown above, optimal search for a Markovian moving target is a POMDP and has all the complexity associated with solving a POMDP. We now consider the "classical" search problem where the target does not move so that the transition matrix P is the identity matrix. For this special case of a non-moving target, the problem simplifies enormously.[11]

Assume $\mathcal{X} = \mathcal{U} = \{1,2,\ldots,X\}$. Suppose the search cost is as in (7.38) so that searching location u costs $c(u)$. Assume that the overlook probabilities are $\beta(u)$, $u \in \mathcal{U}$, the blocking probabilities $q(u) = 0$ in (7.34) and observations $y \in \{F, \bar{F}\}$.

Since $P = I$ (identity matrix), using (7.43), the belief state update given the observation $y_k = \bar{F}$ is

[11] Optimal search for a non-moving target can be formulated as a multi-armed bandit problem [125]. §17.4 presents random search stochastic optimization algorithms for estimating a slowly moving target.

$$\pi_{k+1} = T(\pi_k, y_k = \bar{F}, u_k) = \frac{B_{\bar{F}}(u_k)\pi_k}{\sigma(\pi_k, \bar{F}, u_k)} = \begin{cases} \frac{\pi_k(i)}{\sigma(\pi_k,\bar{F},u_k)}, & i \neq u_k \\ \frac{\pi_k(i)\beta(u_k)}{\sigma(\pi_k,\bar{F},u_k)}, & i = u_k \end{cases}$$

$$\sigma(\pi, \bar{F}, u) = \mathbf{1}' B_{\bar{F}}(u)\pi = 1 - \pi(u)(1 - \beta(u)) \tag{7.49}$$

initialized by prior belief π_0. Of course if $y_k = F$, then the target is found and the problem terminates (the state jumps to terminal state T).

Then given the prior belief π_0 (prior distribution of the location of the target), the optimal search strategy can be characterized explicitly as follows:

THEOREM 7.7.3 *Given belief π_k at time k, the optimal strategy is to search location*

$$u_k = \mu^*(\pi_k) = \operatorname*{argmax}_{i \in \mathcal{U}} \frac{\pi_k(i)(1 - \beta(i))}{c(i)}$$

where the belief π_k is updated according to Bayes rule (7.49).

As a result, optimal search for a non-moving target is much simpler than that for a moving target which involves solving a POMDP.

The proof of Theorem 7.7.3 uses the "interchange argument". Let $J_{u_1,u_2,\boldsymbol{\mu}}$ denote the expected cumulative cost of a search policy that first searches cell i, then cell j and then searches according to $\boldsymbol{\mu}$. Next let $J_{u_2,u_1,\boldsymbol{\mu}}$ denote the corresponding cost of a search policy that first searches cell j, then cell i and then searches according to $\boldsymbol{\mu}$.

LEMMA 7.7.4

$$J_{u_1,u_2,\boldsymbol{\mu}} \leq J_{u_2,u_1,\boldsymbol{\mu}} \iff \frac{\pi(u_1)(1 - \beta(u_1))}{c(u_1)} \geq \frac{\pi(u_2)(1 - \beta(u_2))}{c(u_2)}$$

Proof $J_{u_1,u_2,\boldsymbol{\mu}} = c'_{u_1}\pi + \sum_{y_1} c'_{u_2} T(\pi, y_1, u_1)\sigma(\pi, y_1, u_1)$

$$+ \sum_{y_1}\sum_{y_2} J_{\mu}(T(T(\pi, y_1, u_1), y_2, u_2))\,\sigma(\pi, y_1, u_1)\,\sigma(T(\pi, y_1, u_1), y_2, u_2)$$

$$= c(u_1) + c(u_2)\,\sigma(\pi, \bar{F}, u_1) + K \tag{7.50}$$

where K denotes the double summation term. The last equality follows since $c_{u_1} = c(u_1)\mathbf{1}$, $c_{u_2} = c(u_2)\mathbf{1}$ if $y_1 = \bar{F}$ and zero if $y_1 = F$ by definition. Note the following important property of the belief update $T(\pi, y, u)$ in (7.49) when the transition matrix $P = I$: For $y_1 = y_2 = \bar{F}$,

$$T(T(\pi, y_1, u_1), y_2, u_2)) = T(T(\pi, y_2, u_2), y_1, u_1))$$
$$\sigma(\pi, y_1, u_1)\,\sigma(T(\pi, y_1, u_1), y_2, u_2) = \sigma(\pi, y_2, u_2)\,\sigma(T(\pi, y_2, u_2), y_1, u_1).$$

since the numerator in the belief update comprises of the diagonal matrix $B^u(\bar{F})$ and product of diagonal matrices commute. (Equivalently, since $P = I$, the observations are i.i.d. and so the filtered updates after permuting the order of the observations remain invariant.) So

$$J_{u_1,u_2,\boldsymbol{\mu}} \leq J_{u_2,u_1,\boldsymbol{\mu}} \iff c(u_1) + c(u_2)\sigma(\pi, \bar{F}, u_1) \leq c(u_2) + c(u_1)\sigma(\pi, \bar{F}, u_2).$$

Substituting $\sigma(\pi, \bar{F}, u) = 1 - \pi(u)(1 - \beta(u))$ implies that

$$J_{u_1,u_2,\mu} \leq J_{u_2,u_1,\mu} \iff \frac{\pi(u_1)(1-\beta(u_1))}{c(u_1)} \geq \frac{\pi(u_2)(1-\beta(u_2))}{c(u_2)}.$$

□

With Lemma 7.7.4, the proof of Theorem 7.7.3 proceeds as follows. Suppose

$$\frac{\pi(1)(1-\beta(1))}{c(1)} = \max_{u \in \mathcal{U}} \frac{\pi(u)(1-\beta(u))}{c(u)}.$$

From Lemma 7.7.4 it follows that a policy which does not immediately search cell 1 has a larger cumulative cost than the policy that does search cell 1.

7.8 Complements and sources

The following websites are repositories of papers and software for solving POMDPs:

www.pomdp.org
bigbird.comp.nus.edu.sg/pmwiki/farm/appl/

The belief state formulation in partially observed stochastic control goes back to the early 1960s; see the seminal works of Stratonovich [311], Astrom [24] and Dynkin [102]. Sondik [306] first showed that Bellman's equation for a finite horizon POMDP has a finite dimensional piecewise linear concave solution. This led to the influential papers [302, 307]; see [244, 227, 79, 297] for surveys. POMDPs have been applied in dynamic spectrum management for cognitive radio [138, 353], adaptive radars [246, 184]. Due to the finite action space and finite state space of the underlying Markov chain we have avoided deeper measurability issues that arise in stochastic control. [52, 143] deal with such issues.

[51] contains a survey of open loop feedback control and other suboptimal methods including limited lookahead policies and model predictive control (which is widely used by the process control community).

Optimal search theory is a well studied problem [310, 155]. Early papers in the area include [103, 265]. The paper [230] shows that in many cases optimal search for a target moving between two cells has a threshold type optimal policy – this verifies a conjecture made by Ross in [283]. The proof that the search problem for a moving target is a stochastic shortest path problem is given in [260, 301]. When the target does not move, optimal search is equivalent to a multi-armed bandit [125]. §17.4 discusses random search algorithms.

Average cost POMDPs

Average cost POMDPs are riddled with technicalities and not considered in this book. In analogy to the MDP case considered in Theorem 6.5.2 on page 133, Bellman's dynamic programming equation for an average cost POMDP reads

$$g + V(\pi) = \min_u \{c_u' \pi + \sum_y V(T(\pi, y, u)) \sigma(\pi, y, u)\}. \tag{7.51}$$

If a bounded solution (g^*, V^*) exists to (7.51), then any stationary policy that attains the minimum on the right-hand side is average cost optimal. Analogous to the fully observed case (6.29), the vanishing discount approach yields

$$g = \lim_{\rho \to 1} (1 - \rho) V^{\rho}(\pi), \quad V(\pi) = \lim_{\rho \to 1} (V^{\rho}(\pi) - V^{\rho}(\bar{\pi}))$$

where $\rho \in (0, 1)$ denotes the discount factor and $\bar{\pi}$ is an arbitrary belief state. Necessary and sufficient conditions for the existence of a solution to (7.51) based on the vanishing discount approach are studied in [282, 264, 152]. The results in [282] translated to average cost POMDPs, assert that equicontinuity of the discounted value functions is sufficient for (7.51) to have a bounded solution. For finite-state space $\mathcal{X}$, observation space $\mathcal{Y}$, and action space $\mathcal{U}$, [264] shows that a necessary and sufficient condition for a bounded solution to (7.51) is the uniform boundedness of $|V^{\rho}(\pi) - V^{\rho}(\bar{\pi})|$ for any arbitrary fixed belief $\bar{\pi}$. [152] establish similar necessary and sufficient conditions for POMDPs when $\mathcal{U}$ is compact. In terms of the transition matrix, [264] gives a reachability and detectability condition.

8 POMDPs in controlled sensing and sensor scheduling

Contents

8.1 Introduction

Statistical signal processing deals with extracting signals from noisy measurements. Motivated by physical, communication and social constraints, this chapter addresses the deeper issue of how to dynamically control and optimize signal processing resources to extract signals from noisy measurements. In such controlled sensing problems, several types of sensors (or sensing modes) are available for measuring a given process. Associated with each sensor is a per unit-of-time measurement cost, reflecting the fact that measurements that are more costly to make typically contain more reliable information. *Which sensor (or sensing mode) should the decision-maker choose at each time instant to provide the next measurement*? Such problems are motivated by technological advances in the design of flexible sensors such as sophisticated multi-function radars which can be configured to operate in one of many modes for each measurement.

The controlled sensing problem considered in this chapter is also called the *sensor scheduling problem*, *measurement control problem* or *active sensing problem*. In the context of signal processing, the phrase *sensor-adaptive* signal processing is apt since the sensors adapt (reconfigure) their sensing modes in real time. Controlled sensing arises in numerous applications including adaptive radar (how much resources to devote to each target ([246]), cognitive radio (how to sense the radio spectrum for available channels [353]), and social networks (how do social sensors affect marketing and advertising strategies [185]).

This chapter discusses the formulation of controlled sensing problems with three examples. The first example considers state and measurement control of a linear Gaussian state space model. As an application, radar scheduling is discussed. The second example deals with POMDPs in controlled sensing. Unlike the formulation in Chapter 7, POMDPs arising in controlled sensing have instantaneous costs which are explicit functions of the belief. As a result, the cost in terms of the belief state is *nonlinear*. The third example discusses POMDPs in social learning. In social learning the observation probabilities are an explicit function of the belief. POMDPs with social learning involve the interaction of local and global decision-makers. The chapter concludes with a discussion of risk-averse MDPs and POMDPs which are of recent interest in mathematical finance.

8.2 State and sensor control for state space models

Consider state and measurement control for the state space model discussed in Chapter 2. This general formulation serves as a useful abstraction for specific examples considered subsequently. With $\{w_k\}$ and $\{v_k\}$ denoting i.i.d. state and observation noise processes, consider the controlled state space model[1]

$$x_{k+1} = \phi_k(x_k, u_k^x, w_k), \quad x_0 \sim \pi_0 \tag{8.1}$$

$$y_k = \psi_k(x_k, u_{k-1}^y, v_k). \tag{8.2}$$

Here u_k^x denotes state control applied to the dynamics of the process and u_k^y denotes the measurement control applied to the sensor. The formulation below assumes that u^x is a real valued vector (unconstrained) while u^y takes on values from a finite set. The schematic setup is illustrated in Figure 8.1.

Define the action vector u_k, information history $\mathcal{I}_k$ and belief state π_k as

$$\begin{aligned} &u_k = \begin{bmatrix} u_k^x & u_k^y \end{bmatrix}', \quad k = 0, \ldots, N-1, \\ &\mathcal{I}_0 = \{\pi_0\}, \quad \mathcal{I}_k = \{\pi_0, u_0, y_1, \ldots, u_{k-1}, y_k\} \quad \text{for } k = 1, \ldots, N-1, \\ &\pi_k = p(x_k|\mathcal{I}_k). \end{aligned} \tag{8.3}$$

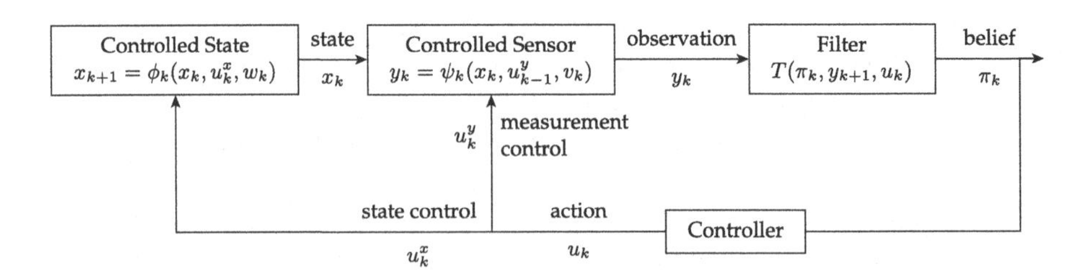

Figure 8.1 State and Sensor Control Schematic Setup of §8.2.

[1] The timing used here is consistent with Chapter 7 where action u_{k-1} determines state x_k and observation y_k.

As discussed extensively in Chapter 3, the belief state evolves according to the optimal filtering density update (3.15) for $k = 0, \ldots, N-1$:

$$\pi_{k+1} = T(\pi_k, y_{k+1}, u_k) = \frac{p(y_{k+1}|x_{k+1}, u_k^y) \int_{\mathcal{X}} p(x_{k+1}|x_k, u_k^x)\, \pi_k(x_k)\, dx_k}{\int_{\mathcal{X}} p(y_{k+1}|x_{k+1}, u_k^y) \int_{\mathcal{X}} p(x_{k+1}|x_k, u_k^x)\, \pi_k(x_k)\, dx_k dx_{k+1}},$$

where the denominator is denoted as $\sigma(\pi_k, y_{k+1}, u_k)$.

Consider the following finite horizon performance criterion for the controlled sensing problem which involves the state and measurement actions:

$$J_{\boldsymbol{\mu}}(\pi_0) = \mathbb{E}_{\boldsymbol{\mu}}\Big\{\sum_{k=0}^{N-1} c(x_k, u_k^x, u_k^y) + c_N(x_N) \mid \pi_0\Big\}. \tag{8.4}$$

As in the POMDP case (discussed in Chapter 7), the cost function $c(x, u^x, u^y)$ is known ahead of time, but the instantaneous costs accrued at each time k are not known since the state x_k is not observed. In (8.4), $\boldsymbol{\mu} = (\mu_0, \ldots, \mu_{N-1})$ denotes the policy sequence and μ_k maps $\mathcal{I}_k$ to u_k; equivalently, as shown in Theorem 7.2.1, μ_k maps belief π_k to action u_k. As in the POMDP case detailed in (7.13), the performance criterion (8.4) can be expressed in terms of the belief state as

$$J_{\boldsymbol{\mu}}(\pi_0) = \mathbb{E}_{\boldsymbol{\mu}}\Big\{\sum_{k=0}^{N-1} \langle c_{u_k}, \pi_k\rangle + \langle c_N, \pi_N\rangle \mid \pi_0\Big\}$$

$$\langle c_{u_k}, \pi_k\rangle = \int_{\mathcal{X}} c(x, u_k^x, u_k^y)\pi_k(x)dx, \quad \langle c_N, \pi_N\rangle = \int_{\mathcal{X}} c_N(x)\pi_N(x)dx.$$

The aim is to determine the optimal measurement and state control policy sequence $\boldsymbol{\mu}^*$ so that $J_{\boldsymbol{\mu}^*}(\pi_0) \leq J_{\boldsymbol{\mu}}(\pi_0)$ for any initial belief π_0.

As in Theorem 7.2.1, the optimal policy $\boldsymbol{\mu}^*$ satisfies Bellman's backward dynamic programming recursion: For $k = N-1, \ldots, 0$,

$$\begin{aligned} \mu_k^* &= \operatorname*{argmin}_u Q_k(\pi, u), \quad J_k(\pi) = \min_u Q_k(\pi, u), \\ J_k(\pi) &= \min_u \left\{ \langle c_u, \pi\rangle + \int_y J_{k+1}(T(\pi, y, u))\, \sigma(\pi, y, u)\, dy \right\}. \end{aligned} \tag{8.5}$$

In general computing the belief state π cannot be implemented via a finite dimensional filter (§3.3.3) and solving dynamic programming over the continuum π is intractable. The POMDP described in Chapter 7 is a special case where the belief is finite dimensional and computed via the HMM filter, and the value function has the explicit piecewise linear concave representation. Another special case where a finite-dimensional characterization holds for (8.5) is the linear Gaussian state and measurement control problem discussed in §8.3 below.

Open loop feedback control

Open loop feedback control (Algorithm 13 in §7.5.5) can be used as a suboptimal method for solving (8.5). The expected cumulative cost in Step 1 of Algorithm 13 can be evaluated by propagating the Chapman–Kolmogorov equation §2.2 via stochastic simulation as explained in §A.1.3. The belief update $T(\pi, y, u)$ in Step 4 of Algorithm 13 can be approximated by a particle filter described in Chapter 3.

8.3 Example 1: Linear Gaussian control and controlled radars

Consider (8.1), (8.2) when the controlled dynamics and observation equations are linear in the state:[2]

$$\begin{aligned} x_{k+1} &= A x_k + f u_k^x + w_k \\ y_k &= H(u_{k-1}^y) x_k + v_k. \end{aligned} \tag{8.6}$$

Assume that the state and observation noise processes and initial state are Gaussian; so $w_k \sim N(0, Q)$, $v_k \sim N(0, R)$, $x_0 \sim \pi_0 = N(\hat{x}_0, \Sigma_0)$.

Define the finite horizon objective as the following special case of (8.4):

$$J_\mu(\pi_0) = \mathbb{E}_\mu \left\{ \sum_{k=0}^{N-1} \left[x_k' M_k x_k + u_k^{x\prime} N_k u_k^x + c(u_k^y) \right] + x_N' M_N x_N \right\} \tag{8.7}$$

where M_k and N_k are symmetric positive definite matrices. Recall that u_k^x is unconstrained while u_k^y takes on values in a finite set. The aim is to compute the optimal policy $\boldsymbol{\mu}^* = (\mu_0^*, \ldots, \mu_{N-1}^*)$ for (8.6), (8.7).

If the measurement control cost $c(u_k^y)$ is omitted, then (8.6), (8.7) becomes the classical linear quadratic Gaussian (LQG) control problem [18]. The quadratic terms in (8.7) model the control and state energy. The objective of LQG control is to compute the optimal policy μ^* to steer the state process $\{x_k\}$ to minimize the total energy consumed. In comparison to classical LQG, the objective (8.6) considers controlling both the state and measurement equations.

8.3.1 Optimal controller

To introduce the main result, first consider the following equations. For a specified measurement control sequence $\{u_k^y\}$, the covariance of the state estimate for the linear Gaussian system (8.6) is obtained from the Kalman filter covariance equation (3.35) which is repeated below with the measurement control explicitly indicated: For $k = 0, \ldots, N-1$,

$$\begin{aligned} \Sigma_{k+1|k} &= A \Sigma_k A' + Q \\ \Sigma_{k+1} &= \Sigma_{k+1|k} - \Sigma_{k+1|k} H'(u_k^y) \left(H(u_k^y) \Sigma_{k+1|k} H'(u_k^y) + R \right)^{-1} H(u_k^y) \Sigma_{k+1|k}. \end{aligned} \tag{8.8}$$

Next define the backward Riccati equation for $k = N-1, \ldots, 0$ as

$$\begin{aligned} \Gamma_k &= A' \Gamma_{k+1} A + M_k - \Gamma_{k+1}^*, \quad \Gamma_N = M_N, \\ \text{where } \Gamma_{k+1}^* &= A' \Gamma_{k+1} f (f' \Gamma_{k+1} f + N_k)^{-1} f' \Gamma_{k+1} A. \end{aligned} \tag{8.9}$$

[2] To avoid a notational clash with costs denoted as c, we use H for the observation matrix.

The main result is as follows (the proof is in the appendix).

THEOREM 8.3.1 *Consider the linear Gaussian system (8.6) with state and measurement control. Then the optimal policy* $\boldsymbol{\mu}^* = (\boldsymbol{\mu}^{x*}, \boldsymbol{\mu}^{y*})$ *which minimizes the objective (8.7) is obtained as follows:*

1. *The optimal measurement policy* $\boldsymbol{\mu}^{y*}$ *is the solution of the following deterministic optimal control problem: with* Σ_k *evolving according to the nonlinear deterministic dynamics (8.8),*

$$\boldsymbol{\mu}^{y*} = \underset{\boldsymbol{\mu}^y}{\operatorname{argmin}}\, \bar{J}_{\boldsymbol{\mu}^y} \quad where \quad \bar{J}_{\boldsymbol{\mu}^y} = \sum_{k=1}^{N} \operatorname{trace}[\Gamma_k^* \Sigma_k] + c(u_{k-1}^y). \tag{8.10}$$

2. *The optimal state control policy* $\boldsymbol{\mu}^{x*}$ *is given explicitly by*

$$u_k^x = \mu_k^*(\pi_0, u_0, y_1, \ldots, u_{k-1}, y_k) = -L_k \hat{x}_k. \tag{8.11}$$

Here, $\hat{x}_k$ *is the state estimate obtained in (3.34) via the Kalman filter Algorithm 1 of §3.4 together with covariance matrix* Σ_k *at each time k. The feedback gain matrix* L_k *in (8.11) is evaluated as*

$$L_k = (f'\Gamma_{k+1} f + N_k)^{-1} f'\Gamma_{k+1} A. \tag{8.12}$$

8.3.2 Discussion of Theorem 8.3.1

The theorem says that the optimal state control policy $\boldsymbol{\mu}^{x*}$ which specifies the actions $u_0^x, \ldots u_{N-1}^x$, can be determined separately from the optimal measurement control policy $\boldsymbol{\mu}^{y*}$ which specifies the measurement actions $u_0^y, \ldots, u_{N-1}^y$. Statement 1 of the theorem says that $\boldsymbol{\mu}^{y*}$ is obtained by solving a nonlinear deterministic optimal control problem.

Statement 2 characterizes the optimal state control policy for the classical linear quadratic Gaussian control problem. The optimal policy is linear in the estimated state obtained from a Kalman filter. The feedback gain L_k is evaluated via (8.12) with Γ_{k+1} obtained from the backward Riccati equation (8.9). It can be shown that the state control problem can be solved by assuming that the state is observed exactly (without any noise), designing the controller for this special case, and then simply replacing the fully observed state by its estimate from the Kalman filter. This property is called the *separation principle* of linear quadratic control design since the control and estimator can be designed separately. The state and measurement control solution in Theorem 8.3.1 can be viewed as a *double separation principle.*

In the special case when there is no state control u^x, then the measurement control problem is: determine the optimal policy $\boldsymbol{\mu}^{y*}$ to minimize the objective (8.7) where the term $u_k^{x\prime} N_k u_k^x$ is omitted and $f u_k^x$ is omitted in the Gaussian model (8.6). Then the objective $\bar{J}_{\boldsymbol{\mu}^y}$ in (8.10) becomes

$$\bar{J}_{\boldsymbol{\mu}^y} = \sum_{k=1}^{N} \operatorname{trace}[\Sigma_k] + c(u_{k-1}^y). \tag{8.13}$$

Here the covariance matrix Σ_k acts as the state and evolves according to the dynamics (8.8). This again is a deterministic optimal control problem and the optimal measurement control sequence $\{u_k^y\}$ is independent of the sample path realization of the measurements $\{y_k\}$. This is unlike a POMDP, where the observation sample path determines the sequence of actions.[3]

8.3.3 Example: Radar scheduling for tracking maneuvering target

In light of (8.13), an intuitive suboptimal myopic policy for measurement control is to choose at each time k the sensor

$$u_k^y = \underset{u \in \mathcal{U}}{\operatorname{argmin}} \big\{ \operatorname{trace}(\Sigma_{k+1}(u^y)) + c(u^y) \big\}. \tag{8.14}$$

In this section we illustrate this myopic policy[4] for multi-sensor tracking of a single maneuvering target modeled as the jump Markov linear system (2.40), (2.41) in Chapter 2. For the target dynamics $z_k \in \mathbb{R}^4$, $\Delta = 1$ second (sampling rate) and $Q = \operatorname{diag}(0.1, 0.1, 0.01.0.01)$.

The action space is $\mathcal{U} = \{1, 2, 3\}$. This corresponds to three 2-D (range-azimuth) radar sensors that are used to track the target. In terms of the observation equation (2.41), for each sensor $u \in \{1, 2, 3\}$, the range and bearings measurements (polar coordinates) translate to cartesian coordinates [32] with

$$H(u) = \begin{bmatrix} 1 & 0 & 0 & 0 \\ 0 & 0 & 1 & 0 \end{bmatrix},$$

$$R(u) = \begin{bmatrix} d_{k,u}^2 \sigma_{\theta,u}^2 \sin^2 \theta_{k,u} + \sigma_d^2 \cos^2 \theta_{k,u} & (\sigma_d^2 - d_{k,u}^2 \sigma_{\theta,u}^2) \sin \theta_{k,u} \cos \theta_{k,u} \\ (\sigma_d^2 - d_{k,u}^2 \sigma_{\theta,u}^2) \sin \theta_{k,u} \cos \theta_{k,u} & d_{k,u}^2 \sigma_{\theta,u}^2 \cos^2 \theta_{k,u} + \sigma_d^2 \sin^2 \theta_{k,u} \end{bmatrix}.$$

Here $d_{k,u}$, $\theta_{k,u}$ denote respectively, the range and bearing of the target with respect to sensor u. Sensors 1 and 3 have azimuth accuracy $\sigma_{\theta,u}^2$ of 1 degree squared while sensor 2 has $\sigma_{\theta,u}^2$ of 0.2 degree squared. All sensors have range measurement accuracy variance σ_d^2 of 20m squared.

We simulated the target from time $k = 1, \ldots, 60$ seconds. The target starts at (x, y) coordinates (100km, 200km) with (x, y) speed[5] components 0 m/s and −2000 m/s. After 25 seconds the target executes a sharp maneuver for 3 seconds after which its (x, y) speed components are 500 m/s and −1600 m/s. During this maneuver period, the state noise covariance Q is scaled by 10 to reflect additional uncertainty during the maneuver.

[3] The optimal *policy* is always independent of the sample path of the observations since dynamic programming is a deterministic recursion. What we are saying here is that the optimal measurement *action* u_k^y at each time k is predetermined and independent of the sample path of the observation process $\{y_k\}$. This is not surprising since the Kalman filter covariance Σ_k is independent of the observation sequence as discussed in Chapter 3.

[4] Since the model is now a jump Markov linear system, Σ_{k+1} is no longer deterministic. It can be estimated using the IMM algorithm or a particle filter. Also, instead of the myopic policy (8.14), one can use open loop feedback control discussed in §7.5.5 and §8.2.

[5] These speeds are typical for medium range surface-to-air missiles.

The three radar sensors communicate with a centralized tracker which knows the location of the radars and their measurement models. The tracker needs to decide at each time which sensor to choose to obtain the next measurement.[6]

The optimal policy for a jump Markov linear system satisfies Bellman's equation (8.5) where $x_k = (z_k, r_k)$. Since solving (8.5) is intractable, we consider here the following heuristic scheduling policy:

Step 1: Compute the conditional covariance of the state z_{k+1} given information $\mathcal{I}_k$ (defined in (8.3)) as

$$\hat{\Sigma}_{k+1}(u^y) = \sum_{i=1} \sum_{j=1} \Sigma_{k+1}(j, u^y) P_{ij} \, \mathbb{P}(r_k = i | \mathcal{I}_k), \quad u^y \in \{1, 2, \ldots, U\}.$$

Here P is the transition matrix of the underlying Markov chain r_k (which models the maneuvers), $\Sigma_{k+1}(j, u^y)$ is the state covariance associated with sensor u^y given the target maneuver mode is $r_{k+1} = j$, and $\mathbb{P}(r_k = i|\mathcal{I}_k)$ can be estimated via a particle filter of §3.10 or an interacting multiple model (IMM) algorithm.

Step 2: Choose the sensor u_k^y using (8.14) to provide the next observation y_{k+1}.

We simulated three scheduling algorithms: the best single sensor, a round-robin scheduling policy, and the above adaptive scheduling policy that uses (8.14) with an IMM algorithm (bank of Kalman filters) for estimating $\hat{\Sigma}_{k+1}(u^y)$. The cost $c(u^y)$ was chosen as a constant independent of u^y. To compare the performance of the three scheduling policies, we illustrate their performance as the location of Sensor 3 is randomly changed. The x- and y-coordinates of the location of Sensor 3 are randomly varied as:

$$(x, y) = (150, 100) + (n_x, n_y) \tag{8.15}$$

where n_x, n_y are two independent uniformly distributed random variables with distribution $U[-a, a]$ and a is a parameter that is varied below. The following table gives average variance estimates of the x-coordinate of the tracking error for the three scheduling policies, for each value of a, the variance is averaged from time 45 to 60 and then averaged over 200 independent realizations.

Even with the simplistic myopic policy (8.14), Table 8.1 shows that the performance gain is significant. In §12.7 we will revisit the problem of radar control and develop structural results for the optimal policy.

8.4 Example 2: POMDPs in controlled sensing

The controlled sensing problem considered in this section is an instance of a finite horizon POMDP covered in Chapter 7. However, unlike Chapter 7 where the instantaneous cost $c'_{u_k}\pi_k$ was linear in the belief state π_k, in controlled sensing problems a nonlinear

[6] A related example is scheduling a multi-mode radar where the mode $u = 1$ measures the position of a target accurately but the velocity poorly, and vice versa for mode $u = 2$. The aim is to choose the radar mode at each time to minimize the tracking error. Other examples include controlling radar transmit waveforms and beam direction.

Table 8.1 Comparison of performance of three scheduling policies for randomly located Sensor 3. The location of the sensor is specified by (8.15).

Sensor 3 location parameter a (km)	Average Variance (m^2)		
	Single	Round Robin	Adaptive Sched
0	108×10^{-4}	48×10^{-4}	0.527×10^{-4}
1.5	123×10^{-4}	49×10^{-4}	0.532×10^{-4}
15	568×10^{-4}	113×10^{-4}	15×10^{-4}
150	12.47	1.02	0.45

cost in π_k is more appropriate so as to penalize the uncertainty in the state estimate at each time k.

Suppose a set of sensors (or sensing modes) $\mathcal{U} = \{1, 2, \ldots, U\}$ is available for measuring the state of a finite state Markov chain in noise. The aim is to devise a controlled sensing strategy that dynamically selects which sensor to use at each time instant to optimize a performance index and provide an estimate of the state of the Markov chain. The performance index is formulated so that more accurate sensors are more expensive to use. Thus we seek to achieve an optimal trade-off between the sensor performance and sensor usage cost.

The setup is as specified in the POMDP protocol of Algorithm 10. In controlled sensing, typically the action only affects the observation probabilities and not the state transition probabilities (Step 3 of Algorithm 10). (For example, choosing a specific radar operating mode does not affect the dynamics of the aircraft being tracked.) However, this is not always the case. In controlled sampling, the aim is to determine when to make a noisy observation of the Markov chain and the actions affect the transition matrices as described in §13.5.

8.4.1 Nonlinear cost POMDP formulation

For notational convenience, we assume throughout this section that the Markov chain's states are unit vectors. So the state space is

$$\mathcal{X} = \{e_1, e_2, \ldots, e_X\}.$$

To incorporate uncertainty of the state estimate, consider the following instantaneous cost at each time k in Step 2 of Algorithm 11 on page 152:

$$c(x_k, u_k) + d(x_k, \pi_k, u_k), \quad u_k \in \mathcal{U} = \{1, 2, \ldots, U\}.$$

This cost comprises of two terms (recall the notation in §7.2 for belief π_k and belief space $\Pi(X)$):

(i) *Sensor usage cost*: $c(x_k, u_k)$ denotes the instantaneous cost of using sensor u_k at time k when the Markov chain is in state x_k.

(ii) *Sensor performance loss*: $d(x_k, \pi_k, u_k)$ models the performance loss when using sensor u_k. This loss is modeled as an explicit function of the belief state π_k to capture the uncertainty in the state estimate.

Typically there is trade-off between the sensor usage cost and performance loss. Accurate sensors have high usage cost but small performance loss.

Denote the information history available at time k to the controller as

$$\mathcal{I}_k = \{\pi_0, u_0, y_1, \dots, u_{k-1}, y_k\}.$$

Then in terms of the belief state π_k, the instantaneous cost can be expressed as

$$\begin{aligned} C(\pi_k, u_k) &= \mathbb{E}\{c(x_k, u_k) + d(x_k, \pi_k, u_k) | \mathcal{I}_k\} \\ &= c'_{u_k}\pi_k + D(\pi_k, u_k), \\ \text{where } c_u &= (c(e_1, u), \dots, c(e_X, u))', \\ D(\pi_k, u_k) &\overset{\text{defn}}{=} \mathbb{E}\{d(x_k, \pi_k, u_k)|\mathcal{I}_k\} = \sum_{i=1}^{X} d(e_i, \pi_k, u_k)\, \pi_k(i). \end{aligned} \tag{8.16}$$

Similarly, define the terminal cost at time N as $c_N(x_N) + d_N(x_N, \pi_N)$. In terms of the belief π, the terminal cost is

$$\begin{aligned} D_N(\pi_N) &= \mathbb{E}\{d_N(x, \pi_N)|\mathcal{I}_N\} = \sum_{i=1}^{X} d_N(e_i, \pi_N)\pi_N(i), \\ C_N(\pi_N) &= c'_N\pi_N + D_N(\pi_N). \end{aligned} \tag{8.17}$$

Examining (8.16) and (8.17), it is clear that $D(\pi, u)$ and $D_N(\pi)$ are in general, *nonlinear* in the belief π.

In complete analogy to Chapter 7, the aim in controlled sensing is to determine the optimal sensing policy μ^* to minimize the finite horizon[7] cumulative cost J_μ over the set of admissible policies. That is, compute

$$\begin{aligned} J_{\mu^*} &= \inf_{\mu} J_{\mu} \text{ where } \boldsymbol{\mu} = (\mu_0, \dots \mu_{N-1}), \\ J_\mu &= \mathbb{E}_\mu \left\{ \sum_{k=0}^{N-1} \Big[c\big(x_k, u_k\big) + d\big(x_k, \pi_k, u_k\big)\Big] + c_N(x_N) + d_N(x_N, u_N)|\pi_0 \right\} \\ &= \mathbb{E}_\mu \left\{ \sum_{k=0}^{N-1} \Big[c'_{u_k}\pi_k + D(\pi_k, u_k) \Big] + C_N(\pi)|\pi_0 \right\}, \quad u_k = \mu_k(\pi_k). \end{aligned} \tag{8.18}$$

8.4.2 Examples of nonlinear cost POMDP

The nonstandard feature of the objective (8.18) is the nonlinear performance loss terms $D(\pi, u)$ and $D_N(\pi)$. These costs[8] should be chosen so that they are zero at the vertices e_i of the belief space $\Pi(X)$ defined in §7.2 (reflecting perfect state estimation) and largest at the centroid of the belief space (most uncertain estimate). We now discuss examples of $d(x, \pi, u)$ and its conditional expectation $D(\pi, u)$.

[7] Alternatively, one can pose the problem as an infinite horizon discounted cost problem.

[8] A linear function $c'_u\pi$ cannot attain its maximum at the centroid of a simplex since a linear function achieves it maximum at a boundary point. Alternatively, a risk averse objective function can be used; see §8.6 and §13.3.

(i) Piecewise linear cost: Here we choose the performance loss as

$$d(x,\pi,u) = \begin{cases} 0 & \text{if } \|x-\pi\|_\infty \leq \epsilon \\ \epsilon & \text{if } \epsilon \leq \|x-\pi\|_\infty \leq 1-\epsilon\,, \\ 1 & \text{if } \|x-\pi\|_\infty \geq 1-\epsilon \end{cases} \quad \epsilon \in [0, 0.5]. \tag{8.19}$$

Then $D(\pi,u)$ is piecewise linear and concave. This cost is useful for subjective decision making; e.g. the distance of a target to a radar is quantized into three regions: close, medium and far.

(ii) Mean square, l_1 and l_∞ performance loss: Suppose in (8.18) we choose

$$d(x,\pi,u) = \alpha(u)(x-\pi)'M(x-\pi) + \beta(u), \; x \in \{e_1,\ldots,e_X\}, \pi \in \Pi. \tag{8.20}$$

Here M is a user-defined positive semi-definite symmetric matrix, $\alpha(u)$ and $\beta(u)$, $u \in \mathcal{U}$ are user-defined positive scalar weights that allow different sensors (sensing modes) to be weighed differently. So (8.20) is the squared error of the Bayesian estimator (weighted by M, scaled by $\alpha(u)$ and translated by $\beta(u)$). In terms of the belief state, the mean square performance loss (8.20) is

$$D(\pi_k,u_k) = \mathbb{E}\{d(x_k,\pi_k,u_k)|\mathcal{I}_k\} = \alpha(u_k)\Big(\sum_{i=1}^{X} M_{ii}\pi_k(i) - \pi_k' M \pi_k\Big) + \beta(u_k) \tag{8.21}$$

because $\mathbb{E}\{(x_k-\pi_k)'M(x_k-\pi_k)|\mathcal{I}_k\} = \sum_{i=1}^{X}(e_i-\pi)'M(e_i-\pi)\pi(i)$. The cost (8.21) is quadratic and concave in the belief.

Alternatively, if $d(x,\pi,u) = \|x-\pi\|_1$ then $D(\pi,u) = 2(1-\pi'\pi)$ is also quadratic in the belief. Also, choosing $d(x,\pi,u) = \|x-\pi\|_\infty$ yields $D(\pi,u) = (1-\pi'\pi)$.

(iii) Entropy-based performance loss: Here in (8.18) we choose

$$D(\pi,u) = -\alpha(u)\sum_{i=1}^{S} \pi(i)\log_2 \pi(i) + \beta(u), \qquad \pi \in \Pi. \tag{8.22}$$

The intuition is that an inaccurate sensor with cheap usage cost yields a Bayesian estimate π with a higher entropy compared to an accurate sensor.

(iv) Predictive sensor performance loss: Since the action (sensing mode) u_k affects the choice of observation y_{k+1} and therefore affects π_{k+1}, one can also consider a sensor performance cost of the form $d(x_{k+1},\pi_{k+1},u_k)$ instead of $d(x_k,\pi_k,u_k)$. Given information history $\mathcal{I}_k = \{\pi_0,u_0,y_1,\ldots,u_{k-1},y_k\}$, the expected cost given $\mathcal{I}_k$, namely $\mathbb{E}\{d(x_{k+1},\pi_{k+1},u_k)|\mathcal{I}_k\}$, can be viewed as a *predictive performance cost* since it predicts the performance loss at time $k+1$ given information until time k. In terms of the belief, the cost can be expressed as

$$\begin{aligned} D(\pi_k,u_k) &= \mathbb{E}\{d(x_{k+1},\pi_{k+1},u_k)|\mathcal{I}_k\} \\ &= \sum_{i=1}^{X}\sum_{j=1}^{X}\sum_{y} d(j, T(\pi_k,y,u_k))\,\sigma(\pi_k,y,u_k)P_{ij}\pi_k(i) \end{aligned} \tag{8.23}$$

where $T(\pi,y,u)$ is the HMM filter and $\sigma(\pi,y,u)$ is its normalizing constant defined in (7.11).

Summary: The above discussion gave examples of nonlinear sensor *performance loss* $D(\pi, u)$ in (8.18) where π denotes the belief state. Recall that in all these examples, the overall instantaneous cost is $C(\pi, u) = c_u'\pi + D(\pi, u)$ defined in (8.16) where the linear term $c_u'\pi$ is the sensor *usage cost*. The papers [183, 339, 33] use such nonlinear costs in the belief state for sensor scheduling. Informally, one can view the non-negative scaling constant α in the above examples as a Lagrange multiplier for a nonlinear performance cost constraint when the objective is to minimize the sensor usage cost.

8.4.3 Concavity of value function

In Chapter 7 it was shown that the value function of a standard POMDP (with linear costs) was piecewise linear and concave in the belief state. We now consider a POMDP with possibly nonlinear instantaneous costs $C(\pi, u)$ and terminal cost $C_N(\pi)$ defined in (8.16) and (8.17). Given the action space $\mathcal{U} = \{1, \ldots, U\}$, Bellman's backward dynamic programming equation (7.15) reads:

$$J_k(\pi) = \min_{u \in \mathcal{U}} \{C(\pi, u) + \sum_{y \in \mathcal{Y}} J_{k+1}(T(\pi, y, u))\, \sigma(\pi, y, u)\}, \quad J_N(\pi) = C_N(\pi). \tag{8.24}$$

The aim is to show that the value function $J_k(\pi)$, $k = N, \ldots 0$ is concave providing that the costs $C(\pi, u)$ and $C_N(\pi)$ are concave.

THEOREM 8.4.1 *Consider a POMDP with possibly continuous-valued observations. Assume that for each action u, the instantaneous cost $C(\pi, u)$ and terminal cost $C_N(\pi, u)$ defined in (8.16) and (8.17) are concave and continuous with respect to $\pi \in \Pi(X)$. Then the value function $J_k(\pi)$ is concave in π.*

The proof is in the appendix. Even though $J_k(\pi)$ is concave, it has no finite dimensional characterization in general, unlike the linear cost POMDP considered in §7.2 where $J_k(\pi)$ was piecewise linear and concave. Nevertheless, in §14.7 it will be shown that the concavity of the value function can be exploited to develop useful structural results for the optimal policy in terms of Blackwell dominance.

Returning to the sensor performance costs described above, since the piecewise linear, mean square error (8.21) and entropy costs (8.22) are concave, Theorem 8.4.1 implies that the corresponding POMDP value function is concave.

8.4.4 Algorithms for controlled sensing POMDPs

The POMDP algorithms of §7.5 can be applied to nonlinear cost POMDPs. For the case of piecewise linear performance loss (8.19), using an identical proof to Theorem 7.4.1, it can be shown that the value function of the resulting POMDP is piecewise linear and concave. So the exact algorithms and suboptimal algorithms of §7.5 can be used to solve such POMDPs and obtain the optimal policy. For the other examples of nonlinear cost POMDPs proposed in this section, the nonlinear cost $C(\pi, u)$ can be approximated by piecewise linear costs in π and then the algorithms of §7.5 used.

Finally, if $C(\pi, u)$ is concave, then the value function is concave by Theorem 8.4.1 and Lovejoy's suboptimal algorithm of §7.5.2 can be used to construct upper and lower bounds to the value function and a suboptimal policy.

8.4.5 Example: Scheduling between active, coarse and passive sensors

Here we illustrate POMDPs with nonlinear costs in a controlled sensing application. Consider an aircraft flying toward a base station. Using various sensors available at the base station, the task is to determine if the aircraft is a threat. The choice of deciding between various sensors arises because the better sensors make the location of the base station visible to the aircraft, while the more stealthy sensors tend to be more inaccurate. The sensors give information about the aircraft's type and distance. The example below is set up as a POMDP with six states ($X = 6$), six observations ($Y = 6$) and three sensors ($U = 3$).

Setup: The state space has two components: aircraft type – either the aircraft is a `friend` or `hostile`; and the current aircraft distance from the base station discretized into three distinct distances $d_1 = 10, d_2 = 5, d_3 = 1$. These six states are represented by unit vectors $e_1, \ldots, e_6$, where e_i, $1 \leq i \leq 3$ corresponds to the aircraft being at distance d_i and friendly; while e_{i+3}, $1 \leq i \leq 3$ corresponds to the aircraft being at a distance d_i and hostile.

The distance of the aircraft evolves according to the transition probability matrix P_1 below. The aircraft type never changes with time (i.e. a friendly aircraft cannot become hostile and vice versa) – hence the overall transition probability matrix is block diagonal:

$$P = \begin{bmatrix} P_1 & 0 \\ 0 & P_1 \end{bmatrix}, \qquad P_1 = \begin{bmatrix} 0.8 & 0.2 & 0 \\ 0.1 & 0.8 & 0.1 \\ 0 & 0.2 & 0.8 \end{bmatrix}. \tag{8.25}$$

Assume that three sensors are available, i.e. $\mathcal{U} = 3$. These are:

- `active`: This active sensor (e.g. radar) yields accurate measurements of the distance of the aircraft but renders the base more visible to the aircraft. The active sensor also yields less accurate information on the aircraft type (`friend` or `hostile`). The measurements feed to an HMM filter.
- `passive`: This is a passive imaging sensor. It yields accurate information on the aircraft type but less accurate measurements of the aircraft distance. The passive sensor is "low probability of intercept" – it does not make the base station too visible to the incoming aircraft.
- `predict`: Employ no sensor – predict the state of the aircraft using an HMM state predictor.

The observation y_k at each time k consists of two independent components: the aircraft type (`friend` or `hostile`) and its distance to the base station d_i. For distance measurements, assume the following observation probabilities

$$B(\texttt{active}) = \begin{bmatrix} 0.95 & 0.05 & 0 \\ 0.025 & 0.95 & 0.025 \\ 0 & 0.05 & 0.95 \end{bmatrix}, \; B(\texttt{passive}) = \begin{bmatrix} 0.9-q & 0.1+q & 0 \\ \frac{0.1+q}{2} & 0.9-q & \frac{0.1+q}{2} \\ 0 & 0.1+q & 0.9-q \end{bmatrix} \tag{8.26}$$

and $B_{iy}(\texttt{predict}) = 1/3$ for all i, y. We will vary the parameter q in $B(\texttt{passive})$.

The sensors' detection of the aircraft type (`friend` or `hostile`) is independent of the distances reported by the sensors. The `active` sensor detects the correct type with probability $0.5 + q$ and the `passive` sensor with probability 0.9.

The sensor usage costs are as follows: For $i \in \{1, 2, 3\}$

$$c(x_k = e_i, \texttt{active}) = 8, \quad c(x_k = e_{i+3}, \texttt{active}) = \frac{2}{d_i} + 8,$$

$$c(x_k = e_i, \texttt{passive}) = 6, \quad c(x_k = e_{i+3}, \texttt{passive}) = \frac{5}{d_i} + 6,$$

$$c(x_k = e_i, \texttt{predict}) = 1, \quad c(x_k = e_{i+3}, \texttt{predict}) = \frac{1}{d_i} + 1.$$

So for a friendly aircraft, the cost of using a sensor is independent of the distance of the aircraft. For a hostile aircraft, the cost incurred is inversely proportional to the distance of the aircraft.

For the sensor performance loss, we chose the mean square error cost $d(x, \pi, u)$ in (8.20) with $M = I$, $\alpha = 200$. In terms of the belief, we constructed the following piecewise linear approximation of $D(\pi, u)$ comprising of seven segments:

$$\|x_k - \pi_k\| = \begin{cases} g_i' \pi_k & \text{if } \|e_i - \pi_k\|_\infty < 0.3, \quad i = 1, \dots, 6 \\ 84 & \text{otherwise.} \end{cases} \tag{8.27}$$

Here g_i is six-dimensional vector with i-th element 1 and all other elements 282.333.

Results: With the above setup, the POMDP program available from the website [77] was used to solve the HMM sensor scheduling problem. The POMDP program solves Bellman's equation (8.24) at each time instant by outputting the solution vector set Γ_k. However, the POMDP program is designed only for a linear cost. To deal with the piecewise linear cost function, we wrote a preprocessing program which at each time k, takes Γ_k from the POMDP program and adds our piecewise linear cost function. The resulting augmented set of vectors is inputted to the POMDP program at the next time iteration $k - 1$ (and so on).

The POMDP program allows the user to choose from several available algorithms. We used the "Incremental Pruning" algorithm developed by Cassandra, et al. in 1997 [81]. The POMDP program was run for the above parameters over a horizon of $N = 7$ for different values of q (probability of detection). In all cases, no apriori assumption was made on the initial distance (state) of the target – so $\pi_0(i) = 1/6$.

For a horizon length of $N = 7$ and $d_N = 0$, Figure 8.2 compares the performance of the optimal sensor scheduling algorithm versus using the `predict`, `passive` or `active` sensor alone. The performance is plotted versus the parameter q in (8.26). It is seen that the optimal sensor schedule (which dynamically selects between `active` `passive` and `predict`) incurs the smallest cost.

8.5 Example 3: Multi-agent controlled sensing with social learning

Previous sections considered controlled sensing using POMDPs where the instantaneous cost was an explicit function of the belief. So the expected cost was a nonlinear function

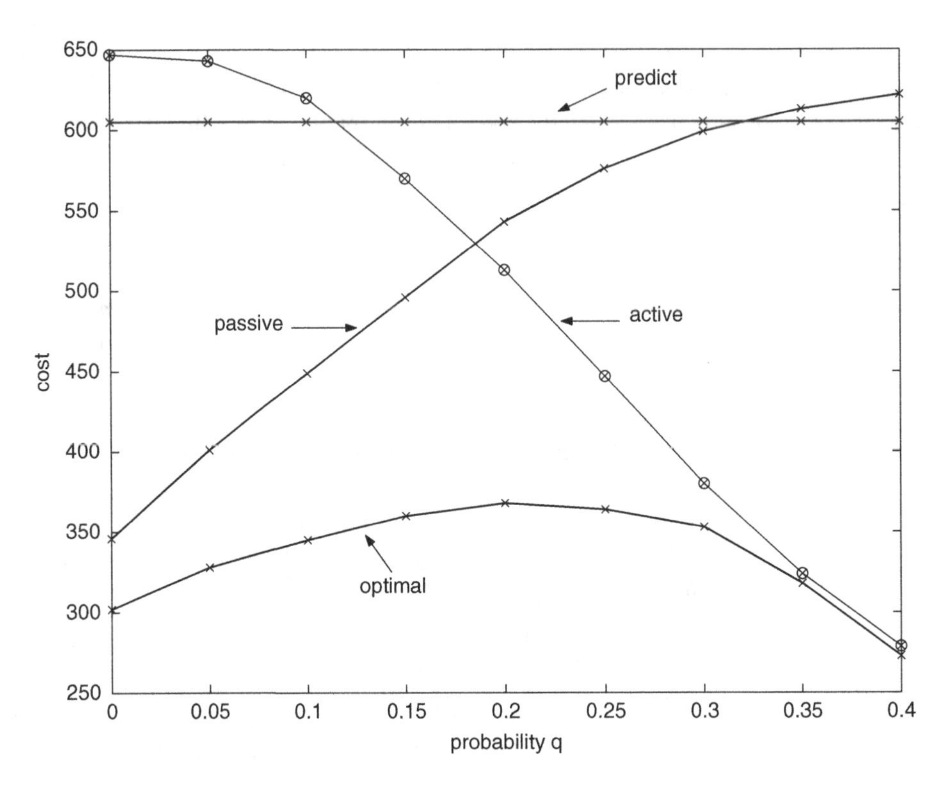

Figure 8.2 Finite horizon HMM sensor scheduling with three sensors: `active`, `predict` and `passive`. Dynamically switching between these sensors according to the optimal policy results in a lower cost than using any of the sensors individually. The parameter q is defined in (8.26).

of the belief. This section considers POMDPs where the belief is updated via social learning. Recall from Chapter 5 that in social learning the observation probability $B(u)$ is an explicit function of the belief. (It is this property that leads to unusual behavior such as herding of agents.) It is shown below that in social learning POMDPs, because the observation probabilities are an explicit function of the belief, the value function is not necessarily concave. (§13.4 discusses the effect of this unusual property on the optimal policy.)

Figure 8.3 displays the schematic setup. Suppose that a multi-agent system performs social learning as described in Chapter 5 to estimate the underlying state x_k of a Markov chain. Agent k records noisy observation y_k of the Markov chain. Recall that in social learning, instead of revealing its private belief π_k, each agent k reveals its local decision a_k to subsequent agents. The agent chooses its local decision by myopically optimizing a local utility function (which depends on the public belief of the state and its local observation). Subsequent agents update their public belief based on these local decisions (in a Bayesian setting), and the sequential procedure continues. Given these local decisions, how can a global decision-maker choose its decisions u_k to perform stochastic control with respect to a global objective?

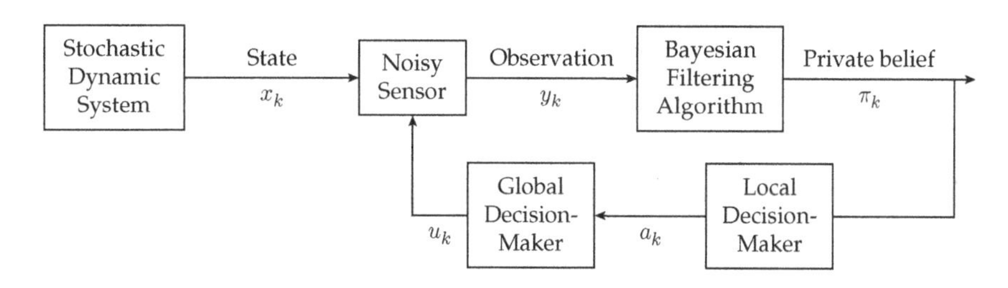

Figure 8.3 Controlled sensing with social learning.

As a related example, consider a multi-sensor system where each adaptive sensor is equipped with a local sensor manager (controller). Based on the existing belief of the underlying state, the local sensor-manager chooses the sensor mode, e.g. low resolution or high resolution. The sensor then views the world based on this mode. Given the belief states and local sensor-manager decisions, how can such a multi-agent system achieve a global objective such as quickest time change detection? The information flow pattern is similar to social learning. In social learning, the sequence of events is

$$\text{prior} \rightarrow \text{observation} \rightarrow \text{local decision} \rightarrow \text{posterior}.$$

In the multi-sensor system with sensor management, the sequence of events is

$$\text{prior} \rightarrow \text{local decision} \rightarrow \text{observation} \rightarrow \text{posterior}.$$

Such problems arise in automated tracking and surveillance systems [8, 16].

8.5.1 POMDP formulation for multi-agent social learning

Consider the global decision-maker's action space $\mathcal{U} = \{1, 2, \ldots, U\}$ and local decision-maker's action space $\mathcal{A} = \{1, 2. \ldots, A\}$. Then Bellman's equation for the optimal policy of the global decision-maker is

$$\begin{aligned} V(\pi) &= \min_{u \in \mathcal{U}} Q(\pi, u), \quad \mu^*(\pi) = \operatorname*{argmin}_{u \in \mathcal{U}} Q(\pi, u), \pi \in \Pi(X), \\ Q(\pi, u) &= C(\pi, u) + \rho \sum_{a \in \mathcal{A}} V\left(T(\pi, a, u)\right) \sigma(\pi, a, u). \end{aligned} \tag{8.28}$$

Here $C(\pi, u) = c_u' \pi$ are the costs incurred by the global decision-maker for taking action u when the public belief is π.

The public belief π is computed by the local decision-makers using the social learning filter (see (5.3) in §5.2.1):

$$\begin{aligned} \pi_{k+1} &= T(\pi_k, a_{k+1}, u_k), \text{ where } T(\pi, a, u) = \frac{R_a^\pi(u) P' \pi}{\sigma(\pi, a, u)}, \ \sigma(\pi, a, u) = \mathbf{1}' R_a^\pi(u) P' \pi, \\ R_a^\pi &= \operatorname{diag}(R_{i,a}^\pi, \ i \in X), \ R_{i,a}^\pi(u) = P(a_{k+1} = a | x_{k+1} = i, \pi_k = \pi, u), \\ R^\pi &= B M^\pi \text{ where } M_{y,a}^\pi \stackrel{\text{defn}}{=} P(a | y, \pi) = \prod_{\tilde{a} \in \mathcal{A} - \{a\}} I(l_a' B_y P' \pi < l_{\tilde{a}}' B_y P' \pi). \end{aligned} \tag{8.29}$$

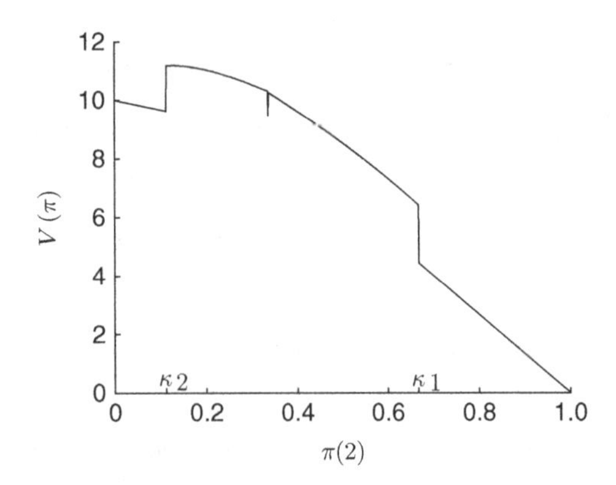

Figure 8.4 Value function $V(\pi)$ in (8.28) for the POMDP model with social learning is not concave. The parameters are specified in §8.5.2. $\kappa_1 = 0.112$ and $\kappa_2 = 0.667$.

Here R^π is a $Y \times A$ matrix. For each local decision $a \in \mathcal{A}$, l_a is a cost vector with elements $l(i,a), i \in \{1, \ldots, X\}$ that denote the costs incurred when a local decision-maker chooses local decision a in state i. (We have denoted the local decision-maker's costs as l to distinguish from the global decision-maker's cost c_u.)

8.5.2 Non-concave value function

Unlike standard POMDPs, for POMDPs with social learning, the value function $V(\pi)$ in Bellman's equation (8.28) is not necessarily concave. We illustrate this in a numerical example below. An important consequence is that global decision-making based on local decisions is in general nontrivial; this is discussed further in §13.4.

Consider the social learning POMDP model with two states ($X = 2$), two observations ($Y = 2$), two global actions ($U = 2$), local decision space $\mathcal{A} = \{1, 2\}$, discount factor $\rho = 0.7$, transition matrices $P(1) = P(2) = I$,

$$B(1) = B(2) = \begin{bmatrix} 0.8 & 0.2 \\ 0.2 & 0.8 \end{bmatrix}, \quad c = l = \begin{bmatrix} 2 & 3 \\ 0 & 4 \end{bmatrix}.$$

In this simplistic numerical example, the global action u only affects the costs and does not affect the observation or transition probabilities. Please review the social learning part of this model considered in §5.3.2.

Figure 8.4 displays the value function $V(\pi)$ in (8.28) computed using the value iteration algorithm (with the public belief $\pi(2)$ quantized to 1000 values in the interval $[0, 1]$). Clearly the value function is discontinuous and non-concave. The points $\kappa_1 = 0.112$ and $\kappa_2 = 0.667$ are defined in Figure 5.3 in §5.3.2.

Let us show that for this example $V(\pi)$ is not necessarily concave. Suppose at the k-th iteration of the value iteration algorithm $V_k(\pi) = \gamma'\pi$. Then

$$V_{k+1}(\pi) = \min_u c_u'\pi + \rho\, \gamma'\pi\, I(\pi \in \mathcal{P}_1) + \rho\, \gamma' B_y \pi\, I(\pi \in \mathcal{P}_2) + \rho\, \gamma'\pi\, I(\pi \in \mathcal{P}_3).$$

The linear segment in region $\mathcal{P}_2$ has different slope and is discontinuous from other two regions, implying that $V_{k+1}(\pi)$ is not concave. To summarize, the structure of the solution to Bellman's equation for social learning POMDPs is very different to that of standard POMDPs.

8.6 Risk-averse MDPs and POMDPs

The theories of *coherent risk* and *dynamic risk measures* have been developed extensively over the last decade [161, 276, 288] with applications in mathematical finance. The aim of this section is to describe the key ideas in risk-averse MDPs and POMDPs. In simple terms, we will replace the expectation (which is an additive operator) in the cumulative cost with a more general risk operator which is sub-additive.

We start with a fully observed finite state MDP with state space $\mathcal{X} = \{1, 2, \dots, X\}$ and action space $\mathcal{U} = \{1, 2, \dots, U\}$. Recall from §6.2 that a finite horizon MDP (with horizon length N) assumes a cumulative cost

$$J_\mu(x_0) = \mathbb{E}_\mu\Big\{\sum_{k=0}^{N} c(x_k, \mu_k(x_k)) + c_N(x_N) \mid x_0\Big\}, \tag{8.30}$$

and the optimal policy $\boldsymbol{\mu}^* = (\mu_0^*, \mu_1^*, \dots, \mu_{N-1}^*)$ satisfies Bellman's equation

$$J_k(i) = \min_{u\in\mathcal{U}}\{c(i,u) + \mathbb{E}\{J_{k+1}(x_{k+1})|x_k, u_k\}, \quad \mu_k^*(i) = \operatorname*{argmin}_{u\in\mathcal{U}}\{\cdot\}. \tag{8.31}$$

The idea behind risk-averse control is to replace the conditional expectations $\mathbb{E}_\mu\{\cdot|x_0\}$ and $\mathbb{E}\{\cdot|x_k, u_k\}$ in (8.30) and (8.31) by more general risk measures. Accordingly, instead of (8.30), consider an MDP with the nested objective function:

$$J_\mu(x_0) = \mathbf{R}_\mu^0\Big\{c(x_0, u_0)+\mathbf{R}_\mu^1\{c(x_1, u_1)+\cdots+\mathbf{R}_\mu^{N-1}\{c(x_{N-1}, u_{N-1})+c_N(x_N)\}\}\Big\}. \tag{8.32}$$

Here the one step risk measures $\mathbf{R}_\mu^0, \mathbf{R}_\mu^1, \dots$ replace $\mathbb{E}_\mu$ and will be discussed below. In general, the optimal policy $\boldsymbol{\mu}^*$ that minimizes (8.32) is non-Markovian. We call an MDP with nested objective (10.14) as a *risk-averse MDP*. A standard MDP with objective (8.30) is called a *risk neutral MDP*.

Several applications motivate risk-averse MDPs. Risk-neutral MDPs are not appropriate in portfolio allocation problems where agents are interested in increasing the long term profit but also reducing the variance in profit. In organ transplant decisions, risk neutrality is not a realistic assumption [83].

8.6.1 Example: Machine replacement with insurance

We now motivate risk-averse MDPs by considering the following variation of the classical machine replacement problem discussed in §6.2.3. Suppose that at each time instant, the machine owner has the possibility of purchasing insurance against machine failure at the next time step. Such insurance decreases the replacement cost R by some amount $C > 0$. If the current machine state is i, then the price of insurance is $w(i)$. The action space is $\mathcal{U} = \{1, 2, 3, 4\}$ which denote respectively, buy insurance and replace, don't buy insurance and replace, buy insurance and continue, don't buy insurance and continue.

Recall from §6.2.3 that the state space is $\{1, 2 \ldots, X\}$ where state 1 denotes machine failure without insurance. Introduce the additional state 0 corresponding to a failed machine with insurance. The insurance affects the transition probabilities as follows: For action 3, the system moves to state 0 instead of state 1 with probability P_{i1}. For action 4, the system moves to state 1 with P_{i1}. Here P_{ij}, $i, j \in \{1, \ldots, X\}$ denote the transition probabilities for action 2 in §6.2.3.

The operating cost of the machine for actions 3 and 4 are identical to the operating cost $c(i, 2)$ in §6.2.3. (The operating cost is independent of insurance.) So the costs of the $X + 1$ state MDP are given by

$$\begin{aligned} \bar{c}(i, u) &= c(i, 2), \quad u \in \{2, 4\}, i \geq 2, \\ \bar{c}(i, u) &= c(i, 2) + w(i), \quad u \in \{1, 3\}, i \geq 2, \\ \bar{c}(0, u) &= R - C, \quad \bar{c}(1, u) = R, \quad u \in \{1, 2\}. \end{aligned}$$

Unlike §6.2.3, assume that the machine is always replaced when it fails, i.e. the action space is state dependent with $\mathcal{U}(i) = \{1, 2\}$ for $i = 0, 1$ denoting the set of possible actions for state i. Then Bellman's equation is

$$\begin{aligned} J_k(i) = \min \Big\{ & R + w(i) + J_{k+1}(X), R + J_{k+1}(X), \\ & c(i, 2) + w(i) + \sum_{\substack{l=0 \\ l \neq 1}}^{X} p_{il} J_{k+1}(l), c(i, 2) + \sum_{\substack{l=0 \\ l \neq 0}}^{X} p_{il} J_{k+1}(l) \Big\}, i \geq 2 \end{aligned} \tag{8.33}$$

$$J_k(1) = R + J_{k+1}(X), \quad J_k(0) = R - C + J_{k+1}(X). \tag{8.34}$$

If the insurance company is in the profit making business, then $w(i) > P_{i1} C$. In such a case, it is clear from Bellman's equation (8.33) that one would never buy insurance since actions 1 and 3 incur higher cost-to-go than 2 and 4. (Use (8.34) to substitute $J_{k+1}(1)$ and $J_{k+1}(0)$ expressed in terms of J_{k+2} into (8.33).) However, in real life people do buy insurance, meaning that the risk-neutral MDP formulation (8.33) is not realistic when dealing with risk-averse behavior. Therefore a more useful risk measure is required.

8.6.2 Markov risk transition mapping

With the above motivating example, now return to the risk-averse MDP with nested objective (8.32). For the formulation to be practically useful, we must impose conditions on the risk measures so that a Markovian policy achieves the minimum of the objective (8.32). The key assumption is this: for any function $\phi(x_k, u_k, x_{k+1})$, assume that

$$\mathbf{R}^k_\mu \{\phi(x_k, u_k, x_{k+1})\} = \mathcal{R}\{\phi(x_k, u_k, \cdot), x_k, P_{x_k}(u_k)\}, \quad \text{where } u_k = \mu_k(x_k). \tag{8.35}$$

Here $\mathcal{R}$ is called a *Markov risk transition mapping* and as explained below is a generalization of the conditional expectation $\mathbb{E}\{\cdot | x_k, u_k\}$ in (8.31).

To explain the notation in (8.35), note that the risk transition mapping depends on the function ϕ, the state x and the probability vector $P_x(u)$; so it is of the form $\mathcal{R}(\phi, x, P_x(u))$. The dot in $\phi(x_k, u_k, \cdot)$ on the right-hand side of (8.35) represents a dummy variable that is integrated out with respect to the x_k-th row of the transition matrix $P(u_k)$. The simplest case is when the risk transition mapping $\mathcal{R}$ is chosen as the conditional expectation $\mathbb{E}\{\cdot|x_k, u_k\}$. Then

$$\begin{aligned}\mathcal{R}\big\{\phi(x_k, u_k, \cdot), x_k, P_{x_k}(u_k)\big\} &= \mathbb{E}\{\phi(x_k, u_k, x_{k+1})|x_k, u_k\} \\ &= \sum_{j=1}^{X} \phi(x_k, u_k, j) P_{x_k j}(u_k). \end{aligned} \tag{8.36}$$

Below, we will consider three more general examples of Markov risk transition mappings $\mathcal{R}$. But first, we need to specify conditions on $\mathcal{R}$ for a Markov policy to be optimal. We assume that the risk transition mapping $\mathcal{R}\{\phi, x, P_x(u)\}$ in (8.35) is chosen to satisfy the following properties (we omit the arguments x and P_x below to simplify notation):

DEFINITION 8.6.1 (Coherent risk measure) *For any function $\phi(x, u, \bar{x})$, the Markov risk transition mapping $\mathcal{R}\{\phi, x, P_x(u)\}$ is said to be a coherent risk measure if*

1. *Convexity: $\mathcal{R}\{\alpha\phi + (1-\alpha)\bar{\phi}\} \leq \alpha\mathcal{R}\{\phi\} + (1-\alpha)\mathcal{R}\{\bar{\phi}\}$ for functions ϕ and $\bar{\phi}$, and constant $\alpha \in [0, 1]$.*
2. *Monotonicity: $\phi \leq \bar{\phi}$ implies that $\mathcal{R}\{\phi\} \leq \mathcal{R}\{\bar{\phi}\}$.*
3. *$\mathcal{R}\{\alpha + \phi\} = \alpha + \mathcal{R}\{\phi\}$ for all constant $\alpha \in \mathbb{R}$.*
4. *Positive homogeneity: $\mathcal{R}\{\alpha\phi\} = \alpha\mathcal{R}\{\phi\}$ for $\alpha \geq 0$.*

Obviously, the conditional expectation $\mathbb{E}\{\cdot|x_k, u_k\}$ satisfies the properties of Definition 8.6.1 as can be verified from (8.36). More general examples of coherent risk measures include:

Example 1: Mean semi-deviation risk: For $\alpha \in [0, 1]$, define the transition mapping

$$\mathcal{R}\{\phi, x, P_x(u)\} = \sum_{j=1}^{X} \phi(j) P_{xj}(u) + \alpha \sum_{j=1}^{X} \max\big\{\phi(j) - \sum_{l=1}^{X} \phi(l) P_{xl}(u), 0\big\} P_{xj}(u) \tag{8.37}$$

in (8.35) and consider the resulting MDP with objective (8.32).

Example 2: Conditional value at risk (CVaR): For $\alpha \in (0, 1]$, define[9]

$$\mathcal{R}\{\phi, x, P_x(u)\} = \inf_{z \in \mathbb{R}} \big\{z + \frac{1}{\alpha} \sum_j \max\{\phi(j) - z, 0\} P_{xj}(u)\big\} \tag{8.38}$$

in (8.35) and consider the resulting MDP with objective (8.32).

Example 3: Exponential risk: Define

$$\mathcal{R}\{\phi, x, P_x(u)\} = \log\big(\sum_j \exp(\phi(j)) P_{xj}(u)\big). \tag{8.39}$$

Using (8.35), by repeated substitution of (8.39) into (8.32), it follows that for the exponential risk case, the objective can be expressed as

[9] For a random variable x with cdf F_x, $\text{CVaR}_\alpha(x) = \min_z\{z + \frac{1}{\alpha}\mathbb{E}\{\max\{x - z, 0\}\}$ where $\alpha \in [0, 1]$. The value at risk (VaR) is the percentile loss namely, $\text{VaR}_\alpha(x) = \min\{z : F_x(z) \geq \alpha\}$. While CVaR is a coherent risk measure, VaR is not convex and so not coherent. CVaR has other remarkable properties [281]: it is continuous in α and jointly convex in (x, α). For continuous cdf F_x, $\text{cVaR}_\alpha(x) = \mathbb{E}\{X|X > \text{VaR}_\alpha(x)\}$. Note that the variance is not a coherent risk measure.

$$J_{\boldsymbol{\mu}}(x_0) = \log \mathbb{E}_{\boldsymbol{\mu}}\big\{ \exp\big(\sum_{k=1}^{N-1} c(x_k, u_k) + c_N(x_N)\big)\big\}.$$

8.6.3 Bellman's equation for risk-averse MDPs

THEOREM 8.6.2 (Risk-averse MDP) *Consider a risk-averse MDP with nested objective (8.32). Assume that the one-step risk measures $\mathbf{R}^k$ satisfy (8.35) and that the Markov risk transition mapping $\mathcal{R}$ is a coherent risk measure (Definition 8.6.1). Then the optimal policy $\boldsymbol{\mu}^* = (\mu_0^*, \mu_1^*, \ldots, \mu_{N-1}^*)$ is evaluated via Bellman's equation*

$$\begin{aligned} J_k(i) &= \min_{u \in \mathcal{U}} \big\{c(i,u) + \mathcal{R}\big\{J_{k+1}, i, P_i(u)\big\}\big\}, \quad J_N(i) = c_N(i), \\ \mu_k^*(i) &= \operatorname*{argmin}_{u \in \mathcal{U}} \big\{c(i,u) + \mathcal{R}\big\{J_{k+1}, i, P_i(u)\big\}\big\} \end{aligned} \tag{8.40}$$

where $P_i(u)$ denotes the i-th row of transition matrix $P(u)$.

Notice that compared to the standard Bellman's equation (8.31), in (8.40) $\mathcal{R}$ replaces $\mathbb{E}\{\cdot|x_k, u_k\}$. The proof of Theorem 8.6.2 presented in [288] is similar to that of Theorem 6.2.2, and uses the coherent risk measure properties of Definition 8.6.1. The setup can be extended to the infinite horizon discounted cost case in which case (8.40) serves as the value iteration algorithm.

Examples: Consider a risk averse MDP with either the mean semi-deviation risk (8.37), or average value at risk (8.38), or exponential risk (8.39). Then Bellman's equation (8.40) holds for each of these examples with the second term $\mathcal{R}\big\{J_{k+1}, i, P_i(u)\big\}$ given by (8.37), (8.38), (8.39), respectively, with $\phi = J_{k+1}$.

8.6.4 Bellman's equation for risk-averse POMDPs

A POMDP is a controlled Markov process with state $(\pi_k, y_k) \in \Pi(X) \times \mathcal{Y}$ where π_k is the belief and y_k is the observation. The controlled transition kernel is

$$\begin{aligned} p(\pi_{k+1}, y_{k+1}|\pi_k, y_k, u_k) &= p(\pi_{k+1}|y_{k+1}, \pi_k, u_k)\, p(y_{k+1}|\pi_k, u_k) \\ &= \delta\big(\pi_{k+1} - T(\pi_k, y_{k+1}, u_k)\big)\, \sigma(\pi_k, y_{k+1}, u_k) \end{aligned} \tag{8.41}$$

where T is the Bayesian filter and σ is the normalization term defined in (7.11). For this controlled process with instantaneous costs $C(\pi_k, u_k)$ (which can be nonlinear in π_k), one can define the nested objective involving risks as in (8.32). A coherent risk measure can be defined involving (π_k, y_k) just as in Definition 8.6.1. For example, Bellman's equation for a POMDP with CVaR risk reads

$$\mu_k^*(\pi) = \operatorname*{argmin}_u Q_k(\pi, u), \quad J_k(\pi) = \min_u Q_k(\pi, u), \quad \text{where} \tag{8.42}$$

$$Q_k(\pi, u) = C(\pi, u) + \inf_{z \in \mathbb{R}} \big\{z + \frac{1}{\alpha} \sum_y \max\{J_{k+1}(T(\pi, y, u)) - z, 0\} \sigma(\pi, y, u)\big\}.$$

To summarize, this section shows that a modified version of Bellman's equation holds for MDPs and POMDPs formulated in terms of dynamic coherent risk measures. §13.3 discusses stopping time POMDPs with exponential risk (8.39).

8.7 Notes and sources

This chapter has described highly stylized results in sensor scheduling in the context of POMDPs (scheduling of Bayesian filters). Scheduling theory itself is a mature area with an enormous body of literature; see, for example, [263]. The formulation in §8.3 dates back to [237]. The radar numerical example is from [110]; see [246, 184] for adaptive radars. We will revisit radar control in §12.7.

Regarding POMDPs in controlled sensing discussed in §8.4, nonlinear costs (in terms of the belief state) are used in [33], which deals with sensor scheduling in continuous time, and more recently in [178, 183, 339, 179]. Piecewise linear costs also arise in social learning; see [84] and the references therein.

Social learning was described in Chapter 5. The result of §8.5 (POMDPs with social learning) is discussed in [84, Chapter 4.5] in the context of pricing of information externalities in economic models. We will return to this in §13.4.

Seminal references for coherent risk measures include [22, 23]. The literature in risk-averse MDPs can be categorized into three approaches: (i) utility functions, (ii) mean variance models [116] (dating back to Markowitz [234]) and (iii) exponential models [37, 158]. Conditional value at risk (CVaR) was introduced in [281]. The description in §8.6 and [288, 83] generalizes the utility function and exponential models to consider dynamic risk measures. In §13.3, we will discuss structural results for stopping time POMDPs with exponential risk.

Appendix 8.A Proof of theorems

8.A.1 Proof of Theorem 8.3.1

Recall that for a linear Gaussian system the belief $\pi_k = N(\hat{x}_k, \Sigma_k)$ where $\hat{x}_k$ and Σ_k are given by the Kalman filter. Using the formula $\mathbb{E}\{x_k' M_k x_k | \mathcal{I}_k\} = \hat{x}_k' M_k \hat{x}_k + \operatorname{trace}(\Sigma_k M_k)$, the instantaneous and terminal costs in terms of the belief state are

$$\begin{aligned} C(\pi, u_k) &= \hat{x}_k' M_k \hat{x}_k + \operatorname{trace}(\Sigma_k M_k) + u_k^x N_k u_k^x + c(u_k^y), \\ C_N(\pi) &= \hat{x}_N' M_N \hat{x}_N + \operatorname{trace}(\Sigma_N M_N). \end{aligned} \tag{8.43}$$

Suppose (with considerable hindsight) we guess that the value function at time $k+1$ is the following quadratic function of the state estimate:

$$J_{k+1}(\pi) = \hat{x}_{k+1}' M_{k+1} \hat{x}_{k+1} + \operatorname{trace}(\Sigma_{k+1} \Gamma_{k+1}) + J_{k+1}^y(\Sigma_{k+1}) + \kappa_{k+1}, \tag{8.44}$$

where κ_{k+1} is independent of u_k^x and u_k^y. Note that

$$\begin{aligned} \mathbb{E}\{\hat{x}_{k+1}' M_{k+1} \hat{x}_{k+1} | x_k, \Sigma_k\} = (A\hat{x}_k + f u_k^x)' \Gamma_{k+1} (A\hat{x}_k + f u_k^x) + \operatorname{trace}(\Gamma_{k+1} \bar{K}_{k+1} \\ \big(R_{k+1} + H(u_k) \Sigma_{k+1|k} H'(u_k)\big) \bar{K}_{k+1} \end{aligned} \tag{8.45}$$

where $\bar{K}_{k+1} = \Sigma_{k+1|k} H'(u_k^y) \big(H(u_k^y) \Sigma_{k+1|k} H'(u_k^y) + R_{k+1}\big)^{-1}$ see (3.34). (Note the Kalman gain defined in (3.38) is $K_k = A_k \bar{K}_k$.) Using (8.9), the last line of (8.45) can be expressed as $\operatorname{trace}((\Gamma^*_{k+1} + \Gamma_k - M_k)\Sigma_k + \Gamma_{k+1}(Q_k - \Sigma_{k+1}))$. Then substituting (8.44), (8.45) into the dynamic programming recursion yields

$$\begin{aligned}
J_k(\pi_k) &= \min_{u_k} C(\pi_k, u_k) + \mathbb{E}_{y_{k+1}}\{J_{k+1}(\pi_{k+1})|\pi_k = N(\hat{x}_k, \Sigma_k)\} \\
&= \min_{u_k} C(\pi_k, u_k) + (A\hat{x}_k + fu^x_k)'\Gamma_{k+1}(A\hat{x}_k + fu^x_k) + \operatorname{trace}(\Sigma_{k+1}\Gamma^*_{k+1}) \\
&\quad + J^y_{k+1}(\Sigma_{k+1}) + \kappa_{k+1} \\
&= \min_{u^x_k}\{\hat{x}'_k M_k \hat{x}_k + u^x_k N_k u^x_k + (A\hat{x}_k + fu^x_k)'\Gamma_{k+1}(A\hat{x}_k + fu^x_k)\} \\
&\quad + \min_{u^y_k}\{c(u^y_k) + \operatorname{trace}(\Sigma_{k+1}\Gamma^*_{k+1}) + J^y_{k+1}(\Sigma_{k+1})\} + \operatorname{trace}(\Sigma_k M_k) + \kappa_{k+1}.
\end{aligned}$$

Note Σ_k is independent of u^y_k whereas from (8.8), Σ_{k+1} depends on u^y_k. The minimization over u^x_k can be performed by elementary calculus to yield (8.11). Therefore, the dynamic programming recursion for u^y_k is

$$J^y_k(\Sigma_k) = \min_{u^x_k}\{c(u^y_k) + \operatorname{trace}(\Sigma_{k+1}\Gamma^*_{k+1}) + J^y_{k+1}(\Sigma_{k+1})\}, \quad J^y_N = \operatorname{trace}(\Sigma_N \Gamma^*_N).$$

8.A.2 Proof of Theorem 8.4.1

The proof is by induction. By assumption $J_N(\pi) = C_N(\pi)$ is concave. Next assume $J_{k+1}(\pi)$ is concave. Therefore, $J_{k+1}(\pi)$ can be upper bounded by piecewise linear concave functions. Indeed, consider L arbitrary but distinct belief states $\pi^1, \ldots, \pi^L \in \Pi(X)$. Let $\bar{\gamma}_l$ denote the gradient vector of $J_{k+1}(\pi)$ at $\pi = \pi^l$, $l = 1, \ldots, L$. Now construct a piecewise linear concave cost out of these gradient vectors as $J^{(L)}_{k+1}(\pi) = \min_{l\in\{1,\ldots L\}} \bar{\gamma}'_l \pi$. Since $J_{k+1}(\pi)$ is concave by assumption, it follows that $J^{(L)}_{k+1}(\pi) \geq J_{k+1}(\pi)$ for all $\pi \in \Pi(X)$. (The piecewise linear function composed of tangents always upper bounds a concave function.) Substituting this expression into (8.24) yields

$$J^{(L)}_k(\pi) = \min_{u\in\mathcal{U}} \left\{ C(\pi, u) + \sum_{y\in\mathcal{Y}} \min_l \bar{\gamma}'_l B_y(u) P'(u)\pi \right\}. \tag{8.46}$$

Clearly $\min_l \bar{\gamma}'_l B_y(u) P'(u)\pi$ is piecewise linear and concave. Since concavity is preserved by summation (and also integration for continuous valued observations) [68, p.79], $\sum_{y\in\mathcal{Y}} \min_l \bar{\gamma}'_l B_y(u)P'(u)\pi$ is concave. So $C(\pi, u) + \sum_{y\in\mathcal{Y}} \min_l \bar{\gamma}'_l B_y(u)P'(u)\pi$ is concave since $C(\pi, u)$ is concave by assumption. Finally, the pointwise minimum of concave functions is concave implying that $J^{(L)}_k(\pi)$ is concave.

Finally, as $L \to \infty$, the piecewise linear approximation $J^{(L)}_{k+1}(\pi)$ converges uniformly[10] to $J_{k+1}(\pi)$. So from (8.46), $J^{(L)}_k(\pi)$ converges uniformly to $J_k(\pi)$ as $L \to \infty$. Hence $J_k(\pi)$ is concave, thereby completing the inductive step.

[10] $J^{(L)}_{k+1}(\pi)$ is decreasing with L and $J^{(L)}_{k+1}(\pi) \geq J_{k+1}(\pi)$; $\Pi(X)$ is compact. Therefore the convergence is uniform on $\Pi(X)$ [287, Theorem 7.13]. Actually, only pointwise convergence is required for the limit to be concave.

Part III

Partially observed Markov decision processes: structural results

Part III develops structural results for the optimal policy of a POMDP. That is, without brute-force computation, the aim is to characterize the structure of the solution of Bellman's dynamic programming equation by making assumptions on the POMDP model.

- Chapter 9 starts with the special case of fully observed finite state MDPs.
- Chapter 10 develops important structural results for the belief update using the HMM filter. These results form a crucial step in formulating the structural results for POMDPs. We will illustrate these results by developing reduced complexity filtering algorithms that provably lower and upper bound the true HMM filter posterior distribution.
- Once structural results are established for the HMM filter, the next step is to give conditions under which the value function, obtained as the solution of Bellman's equation, has a monotone structure in the belief state. This is the subject of Chapter 11. To illustrate this structural result, Chapter 11 gives two examples. The first example gives conditions under which the optimal policy of a two-state POMDP is monotone. The second example shows how POMDP multi-armed bandit problems can be solved efficiently.
- For stopping time POMDPs, Chapter 12 gives conditions under which the stopping set is convex and the optimal policy is monotone with respect to the monotone likelihood ratio stochastic order. Chapter 13 covers several examples of stopping time POMDPs such as quickest change detection and social learning that exhibit these structural results for their optimal policy.
- Finally, Chapter 14 gives conditions for general POMDPs so that the optimal policy is provably lower and upper bounded by myopic policies. In regions of the belief space where the upper and lower bounds overlap, they coincide with the optimal policy. It is shown that the volume of the belief space where these bounds overlap can be maximized via a linear programming problem.

9 Structural results for Markov decision processes

Contents

For finite state MDPs with large dimensional state spaces, computing the optimal policy by solving Bellman's dynamic programming recursion or the associated linear programming problem can be prohibitively expensive. Structural results give sufficient conditions on the MDP model to ensure that the optimal policy $\mu^*(x)$ is increasing (or decreasing) in the state x. Such policies will be called *monotone policies*. To see why monotone policies are important, consider an MDP with two actions $\mathcal{U} = \{1, 2\}$ and a large state space $\mathcal{X} = \{1, 2, \dots, X\}$. If the optimal policy $\mu^*(x)$ is increasing[1] in x, then it has to be a step function of the form

$$\mu^*(x) = \begin{cases} 1 & x < x^* \\ 2 & x \geq x^*. \end{cases} \tag{9.1}$$

Here $x^* \in \mathcal{X}$ is some fixed state at which the optimal policy switches from action 1 to action 2. A policy of the form (9.1) will be called a *threshold policy* and x^* will be called the threshold state. Figure 9.1 illustrates a threshold policy.

Note that x^* completely characterizes the threshold policy (9.1). Therefore, if one can prove that the optimal policy $\mu^*(x)$ is increasing in x, then one only needs to compute the threshold state x^*. Computing (estimating) the single point x^* is often more efficient than solving Bellman's equation. Also real-time implementation of a controller

[1] Throughout this book, increasing is used in the weak sense to mean increasing.

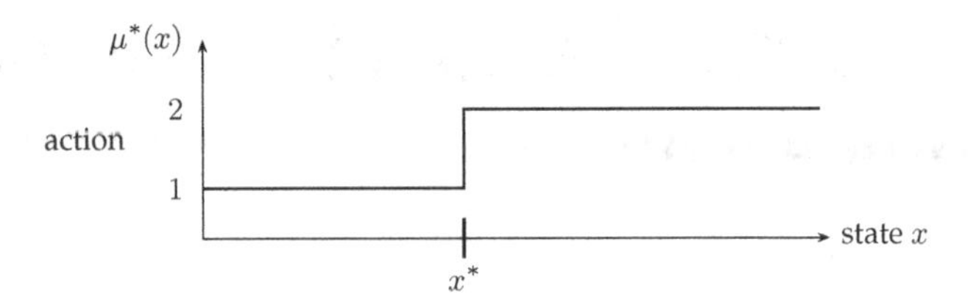

Figure 9.1 Monotone increasing threshold policy $\mu^*(x)$. Here, x^* is the threshold state at which the policy switches from 1 to 2.

with monotone policy (9.1) is simple; only x^* needs to be stored in a lookup table in Algorithm 9 on page 123.

Recall (Chapter 6.2) that for a finite horizon MDP, Bellman's equation for the optimal policy $\mu_k^*(x)$ reads:

$$Q_k(x,u) \stackrel{\text{defn}}{=} c(x,u,k) + J'_{k+1} P_x(u) \tag{9.2}$$
$$J_k(x) = \min_{u \in \mathcal{U}} Q_k(x,u), \quad \mu_k^*(x) = \operatorname*{argmin}_{u \in \mathcal{U}} Q_k(x,u)$$

where $J_{k+1} = \left[J_{k+1}(1), \ldots, J_{k+1}(X)\right]'$ denotes the value function. The key point is that the optimal policy is $\mu_k^*(x) = \operatorname{argmin}_u Q_k(x,u)$. What are sufficient conditions on the MDP model to ensure that the optimal policy $\mu_k^*(x)$ is increasing in x (as shown in Figure 9.1)? The answer to this question lies in the area of *monotone comparative statics* – which studies how the argmin or argmax of a function behaves as one of the variables changes. The main result of this chapter is to show that $Q_k(x,u)$ in (9.2) being *submodular* in (x,u) is a sufficient condition for $\mu^*(x)$ to increase in x. Since $Q_k(x,u)$ is the conditional expectation of the cost to go given the current state (see (6.9)), giving conditions on the MDP model to ensure that $Q_k(x,u)$ is submodular requires characterizing how expectations vary as the state varies. For this we will use *stochastic dominance*.

In the next two sections we introduce these two important tools, namely, *submodularity/supermodularity* and *stochastic dominance*. They will be used to give conditions under which an MDP has monotone optimal policies.

9.1 Submodularity and supermodularity

Throughout this chapter we assume that the state space $\mathcal{X} = \{1, 2, \ldots, X\}$ and action space $\mathcal{U} = \{1, 2, \ldots, U\}$ are finite.

9.1.1 Definition and examples

A real-valued function $\phi(x,u)$ is *submodular* in (x,u) if

$$\phi(x, u+1) - \phi(x,u) \geq \phi(x+1, u+1) - \phi(x+1, u). \tag{9.3}$$

In other words, $\phi(x, u+1) - \phi(x, u)$ has decreasing differences with respect to x. A function $\phi(x, u)$ is *supermodular* if $-\phi(x, u)$ is submodular.[2]

Note that submodularity and supermodularity treat x and u symmetrically. That is, an equivalent definition is $\phi(x, u)$ is submodular if $\phi(x+1, u) - \phi(x, u)$ has decreasing differences with respect to u.

Examples: The following are submodular in (x, u)

(i) $\phi(x, u) = -xu$. (ii) $\phi(x, u) = \max(x, u)$.

(iii) Any function of one variable such as $\phi(x)$ or $\phi(u)$ is trivially submodular.

(iv) The sum of submodular functions is submodular.

(v) Although we consider finite state and action spaces, the following intuition is useful. If $(x, u) \in \mathbb{R}^2$, then $\phi(x, u)$ is supermodular if the mixed derivative $\partial^2\phi/\partial x \partial u \geq 0$. This is obtained from (9.3) by taking limits of the finite differences.

Continuing with the example of $(x, u) \in \mathbb{R}^2$, it is instructive to compare supermodularity with convexity. Convexity of a function $\phi(x, u)$ for $(x, u) \in \mathbb{R}^2$ requires the Hessian matrix $\begin{bmatrix} \partial^2\phi/\partial x^2 & \partial^2\phi/\partial x \partial u \\ \partial^2\phi/\partial x \partial u & \partial^2\phi/\partial u^2 \end{bmatrix}$ to be positive semidefinite; whereas supermodularity only requires the mixed derivative terms to be nonnegative.

(vi) In economics, supermodularity is equivalent to commodities being *complementary*, while submodularity is equivalent to commodities being *substitutes*. For example, let $\phi(x, u)$ be the utility function associated with the sale of x music players and u earphones. Then supermodularity (increasing differences) implies

$$\phi(x+1, u+1) - \phi(x+1, u) \geq \phi(x, u+1) - \phi(x, u). \tag{9.4}$$

The left-hand side is the marginal value of an earphone when $x+1$ music players have been sold. The right-hand side is the marginal value of an earphone when x music players have been sold. Thus (9.4) says that the marginal value of an earphone increases with the sales of music players. So music players and earphones are complementary products. On the other hand, iPhones and Blackberry smartphones are substitute products; increased sales of iPhones results in decreased marginal value of Blackberry smartphones. So the utility function would be submodular.

9.1.2 Topkis' monotonicity theorem

Let $u^*(x)$ denote the set of possible minimizers of $\phi(x, u)$ with respect to u:

$$u^*(x) = \{\operatorname*{argmin}_{u \in \mathcal{U}} \phi(x, u)\}, \quad x \in \mathcal{X}.$$

In general there might not be a unique minimizer and then $u^*(x)$ has multiple elements. Let $\bar{u}^*(x)$ and $\underline{u}^*(x)$ denote the maximum and minimum elements of this set. We call these, respectively, the *maximum and minimum selection* of $u^*(x)$.

The key result is the following Topkis' monotonicity theorem.[3]

[2] Chapter 12 deals with POMDPs where the state space comprises of probability vectors. Appendix 12.A on page 279 gives a more general definition of submodularity on a lattice.

[3] Supermodularity for structural results in MDPs and game theory was introduced by Topkis in the seminal paper [321]. Chapter 12 gives a more general statement in terms of lattices.

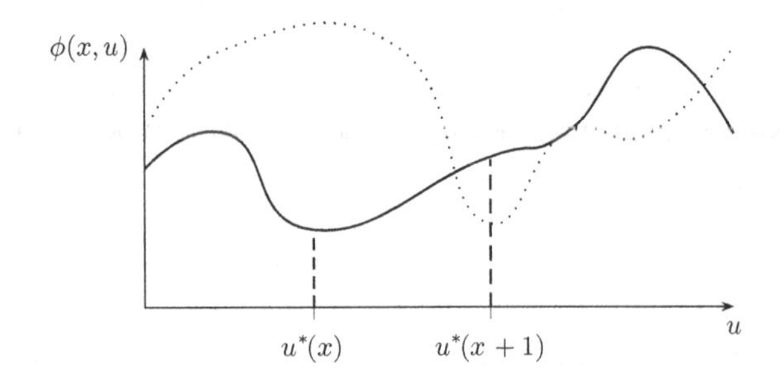

Figure 9.2 Visual illustration of main idea of proof of Theorem 9.1.1. The solid (respectively, dotted) curve represents $\phi(x,u)$ (respectively $\phi(x+1,u)$) plotted versus u. Also, $u^*(x+1) = \operatorname{argmin}_u \phi(x+1,u)$. If for all values of u, $\phi(x,u)$ to the right of $u^*(x+1)$ is larger than $\phi(x,u)$ to left of $u^*(x+1)$, then clearly the $\operatorname{argmin}_u \phi(x,u)$ must lie to the left of $u^*(x+1)$. That is, $u^*(x) \leq u^*(x+1)$.

THEOREM 9.1.1 *Consider a function $\phi : \mathcal{X} \times \mathcal{U} \to \mathbb{R}$.*

1. *If $\phi(x,u)$ is submodular, then the maximum and minimal selections $\bar{u}^*(x)$ and $\underline{u}^*(x)$ are increasing*[4] *in x.*
2. *If $\phi(x,u)$ is supermodular, then the maximum and minimal selections $\bar{u}^*(x)$ and $\underline{u}^*(x)$ are decreasing in x.*

Proof We prove statement 1. The proof is illustrated visually in Figure 9.2. Fix x and consider $\phi(x,u)$ as a function of u. Clearly, if $\phi(x,u)$ is larger than $\phi(x,\bar{u}^*(x+1))$ for u to the "right" of $\bar{u}^*(x+1)$, then $\operatorname{argmin}_u \phi(x,u)$ must lie to the "left" of $\bar{u}^*(x+1)$ (see Figure 9.2). Therefore,

$$\phi(x,\bar{u}^*(x+1)) \leq \phi(x,u) \text{ for } u \geq \bar{u}^*(x+1) \implies u^*(x) \leq \bar{u}^*(x+1).$$

Let us write this sufficient condition as

$$\phi(x,u) - \phi(x,\bar{u}^*(x+1)) \geq 0 \text{ for } u \geq \bar{u}^*(x+1).$$

Also by definition $\bar{u}^*(x+1) \in \operatorname{argmin}_u \phi(x+1,u)$. So clearly

$$\phi(x+1,u) - \phi(x+1,\bar{u}^*(x+1)) \geq 0 \text{ for all } u \in \mathcal{U}.$$

From the above two inequalities, it is sufficient that

$$\phi(x,u) - \phi(x,\bar{u}^*(x+1)) \geq \phi(x+1,u) - \phi(x+1,\bar{u}^*(x+1)) \text{ for } u \geq \bar{u}^*(x+1).$$

A sufficient condition for this is that for any $\bar{u} \in \mathcal{U}$,

$$\phi(x,u) - \phi(x,\bar{u}) \geq \phi(x+1,u) - \phi(x+1,\bar{u}) \text{ for } u \geq \bar{u}.$$

□

[4] We stress again that increasing is used in the weak sense to mean increasing.

9.1.3 Single crossing condition

The "single crossing condition" is a more general condition than submodularity for the argmin to increase. The Milgrom–Shannon theorem [242] states: $u^*(x) = \operatorname{argmin}_u \phi(x, u)$ is increasing in x if the following single crossing condition holds

$$\phi(x+1, u+1) - \phi(x+1, u) \geq 0 \implies \phi(x, u+1) - \phi(x, u) \geq 0. \qquad (9.5)$$

Obviously, submodularity (9.3) implies this single crossing condition. (9.5) is called the single crossing condition because it ensures that $f(x) = \phi(x, u+1) - \phi(x, u)$ versus x crosses 0 only once. Unlike submodularity, the single crossing property in (x, u) does not imply a single crossing property in (u, x).

The single crossing property is used widely in game theory to characterize Nash equilibria. It is an ordinal condition meaning that $\phi(x, u)$ is single crossing implies that for an increasing function g, the composition $g \circ \phi$ is also single crossing. (This does not hold in general for submodular functions.) The single crossing condition, however, is not useful in stochastic control since the property is not closed under summation. Recall that $Q(x, u)$ in dynamic programming involves the sum of the current cost and the expected cost to go. For such problems submodularity is more useful since it is closed under summation.

9.2 First-order stochastic dominance

Stochastic dominance is the next tool that will be used to develop MDP structural results.

DEFINITION 9.2.1 (First-order stochastic dominance) *Let π_1, π_2 denote two pmfs or pdfs*[5] *with distribution functions F_1, F_2, respectively. Then π_1 is said to first-order stochastically dominate π_2 (written as $\pi_1 \geq_s \pi_2$ or $\pi_2 \leq_s \pi_1$) if*

$$1 - F_1(x) \geq 1 - F_2(x), \quad \text{for all } x \in \mathbb{R}.$$

Equivalently, $\pi_1 \geq_s \pi_2$ if

$$F_1(x) \leq F_2(x), \quad \text{for all } x \in \mathbb{R}.$$

In this chapter, we consider pmfs. For pmfs with support on $\mathcal{X} = \{1, 2, \ldots, X\}$, Definition 9.2.1 is equivalent to the following property of the tail sums:

$$\pi_1 \geq_s \pi_2 \text{ if } \sum_{i=j}^{X} \pi_1(i) \geq \sum_{i=j}^{X} \pi_2(i), \quad j \in \mathcal{X}.$$

For state space dimension $X = 2$, first-order stochastic dominance is a complete order since $\pi = [1 - \pi(2),\ \pi(2)]'$ and so $\pi_1 \geq_s \pi_2$ if $\pi_1(2) \geq \pi_2(2)$. Therefore, for $X = 2$, any two pmfs are first-order stochastic orderable.

[5] Recall the acronyms pmf (probability mass function) and pdf (probability density function).

For $X \geq 3$, first-order stochastic dominance is a *partial order* since it is not always possible to order any two belief pmfs π_1 and π_2. For example,

If $\pi_1 = [0.3,\ 0.2,\ 0.5]'$ and $\pi_2 = [0.05,\ 0.15, 0.8]'$, then $\pi_1 \leq_s \pi_2$.
If $\pi_1 = [0.3,\ 0.2,\ 0.5]'$ and $\pi_2 = [0.2,\ 0.4,\ 0.4]'$, then π_1 and π_2 are not orderable.

The following is an equivalent characterization of first-order dominance:

THEOREM 9.2.2 ([249]) *Let $\mathcal{V}$ denote the set of all X dimensional vectors v with increasing components, i.e. $v_1 \leq v_2 \leq \cdots v_X$. Then $\pi_1 \geq_s \pi_2$ if and only if, for all $v \in \mathcal{V}$, $v'\pi_1 \geq v'\pi_2$. Similarly, pdf $\pi_1 \geq_s \pi_2$ if and only if $\int \phi(x)\pi_1(x)dx \geq \int \phi(x)\pi_2(x)dx$ for any increasing function $\phi(\cdot)$.*

In other words, $\pi_1 \geq_s \pi_2$ if and only if, for any increasing function $\phi(\cdot)$, $\mathbb{E}_{\pi_1}\{\phi(x)\} \geq \mathbb{E}_{\pi_2}\{\phi(x)\}$, where $\mathbb{E}_\pi$ denotes expectation with respect to the pmf (pdf) π. As a trivial consequence, choosing $\phi(x) = x$, it follows that $\pi_1 \geq_s \pi_2$ implies that the mean of pmf π_1 is larger than that of pmf π_2.

Finally, we need the following concept that combines supermodularity with stochastic dominance. We say that a transition probabilities $P_{ij}(u)$ are *tail-sum supermodular* in (i, u) if $\sum_{j \geq l} P_{ij}(u)$ is supermodular in (i, u), i.e.

$$\sum_{j=l}^{X} \big(P_{ij}(u+1) - P_{ij}(u)\big) \text{ is increasing in } i, \quad i \in \mathcal{X}, u \in \mathcal{U}. \tag{9.6}$$

In terms of first-order stochastic dominance, (9.6) can be rewritten as

$$\frac{1}{2}\big(P_{i+1}(u+1) + P_i(u)\big) \geq_s \frac{1}{2}\big(P_i(u+1) + P_{i+1}(u)\big), \quad i \in \mathcal{X}, u \in \mathcal{U}, \tag{9.7}$$

where $P_i(u)$ denotes the i-th row of the matrix $P(u)$. Due to the term $1/2$ both sides are valid probability mass functions. Thus we have the following result.

THEOREM 9.2.3 *Let $\bar{\mathcal{V}}$ denote the set of X dimensional vectors v with decreasing components, i.e. $v_1 \geq v_2 \geq \cdots \geq v_X$. Then $P_{ij}(u)$ is tail-sum supermodular in (i, u) iff for all $v \in \mathcal{V}$, $v'P_i(u)$ is submodular in (i, u); that is:*

$$v'\big(P_{i+1}(u+1) - P_{i+1}(u)\big) \leq v'\big(P_i(u+1) - P_i(u)\big), \quad i \in \mathcal{X}, u \in \mathcal{U}.$$

The proof follows immediately from Theorem 9.2.2 and (9.7).

9.3 Monotone optimal policies for MDPs

With the above two tools, we now give sufficient conditions for an MDP to have a monotone optimal policy.

For finite horizon MDPs, recall from (6.2) that the model is the 5-tuple

$$(\mathcal{X}, \mathcal{U}, P_{ij}(u,k), c(i,u,k), c_N(i)), \qquad i,j \in \mathcal{X}, u \in \mathcal{U}. \tag{9.8}$$

Assume the MDP model satisfies the following four conditions:

(A1) Costs $c(x, u, k)$ are decreasing in x.
The terminal cost $c_N(x)$ is decreasing in x.

(A2) $P_i(u, k) \leq_s P_{i+1}(u, k)$ for each i and u. Here $P_i(u, k)$ denotes the i-th row of the transition matrix for action u at time k.

(A3) $c(x, u, k)$ is submodular in (x, u) at each time k. That is:
$c(x, u + 1, k) - c(x, u, k)$ is decreasing in x.

(A4) $P_{ij}(u, k)$ is tail-sum supermodular in (i, u) in the sense of (9.6). That is,
$\sum_{j \geq l} \big(P_{ij}(u + 1, k) - P_{ij}(u, k)\big)$ is increasing in i.

For infinite horizon discounted cost and average cost MDPs, identical conditions will be used except that the instantaneous costs $c(x, u)$ and transition matrix $P(u)$ are time invariant and there is no terminal cost.

Note that (A1) and (A2) deal with different states for a fixed action u, while (A3) and (A4) involve different actions and states.

The following is the main structural result for an MDP.

THEOREM 9.3.1 *1. Assume that a finite horizon MDP satisfies conditions (A1), (A2), (A3) and (A4). Then at each time $k = 0, 1, \ldots, N - 1$, there exists an optimal policy $\mu_k^*(x)$ that is increasing in state $x \in \mathcal{X}$.*

2. Consider a finite horizon risk averse MDP with CVaR or exponential cost with Bellman's equation given by (8.40). Then Statement 1 holds.

3. Assume that a discounted infinite horizon cost problem or unichain average cost problem satisfies (A1), (A2), (A3) and (A4). Then there exists an optimal stationary policy $\mu^*(x)$ that is increasing in state $x \in \mathcal{X}$.

4. Consider the constrained MDP with a single global constraint (§6.6.2). Suppose the Lagrangian costs $c(x, u; \lambda)$ defined in (6.46) satisfy (A1) and (A3). Assume also that (A2), (A4) hold. Then there exists an optimal policy $\mu^*(x)$ that is a randomized mixture of two increasing policies. That is

$$\mu^*(x) = \alpha \mu_{\lambda_1}^*(x) + (1 - \alpha)\mu_{\lambda_2}^*(x), \quad \alpha \in [0, 1],$$

where $\mu_{\lambda_1}^*(x)$ and $\mu_{\lambda_2}^*(x)$ are monotone increasing in state $x \in \mathcal{X}$.

Proof **Statement 1**: To prove statement 1, write Bellman's equation (6.8) as

$$Q_k(i, u) \stackrel{\text{defn}}{=} c(i, u, k) + J_{k+1}' P_i(u, k) \tag{9.9}$$

$$J_k(i) = \min_{u \in \mathcal{U}} Q_k(i, u), \quad \mu_k^*(i) = \operatorname*{argmin}_{u \in \mathcal{U}} Q_k(i, u)$$

where $J_{k+1} = \big[J_{k+1}(1), \ldots, J_{k+1}(X)\big]'$ denotes the value function. The proof proceeds in two steps.

Step 1: Monotone value function. Assuming (A1) and (A2), we show via mathematical induction that $Q_k(i, u)$ is decreasing in i for each $u \in \mathcal{U}$. So the value function $J_k(i)$ is decreasing in i for $k = N, N - 1, \ldots, 0$.

Clearly $Q_N(i, u) = c_N(i)$ is decreasing in i by (A1). Now for the induction step. Suppose $Q_{k+1}(i, u)$ is decreasing in $i \in \mathcal{X}$ for each u. Then $J_{k+1}(i) = \min_u Q_{k+1}(i, u)$

is decreasing in i, since the minimum of decreasing functions is decreasing. So the X-dimensional vector J_{k+1} has decreasing elements.

Next $P_i(u,k) \leq_s P_{i+1}(u,k)$ by (A2). Hence from Theorem 9.2.2, $J'_{k+1} P_i(u,k) \geq J'_{k+1} P_{i+1}(u,k)$. Finally since $c(i,u,k)$ is decreasing in i by (A1), it follows that

$$c(i,u,k) + J'_{k+1} P_i(u,k) \geq c(i+1,u,k) + J'_{k+1} P_{i+1}(u,k).$$

Therefore, $Q_k(i,u) \geq Q_k(i+1,u)$ implying that $Q_k(i,u)$ is decreasing in i for each $u \in \mathcal{U}$. (This in turn implies that $J_k(i) = \min_u Q_k(i,u)$ is decreasing in i.) Hence the induction step is complete.

Step 2: Monotone policy. Assuming (A3) and (A4) and using the fact that $J_k(i)$ is decreasing in i (Step 1), we show that $Q_k(i,u)$ is submodular in (i,u).

By (A3), $c(i,u,k)$ is submodular in (i,u). By assumption (A4), since J_{k+1} is a vector with decreasing elements (by Step 1), it follows from Theorem 9.2.3 that $J'_{k+1} P_i(u,k)$ is submodular in (i,u). Since the sum of submodular functions is submodular, it follows that $Q_k(i,u) = c(i,u,k) + J'_{k+1} P_i(u,k)$ is submodular in (i,u).

Since $Q_k(i,u)$ is submodular, it then follows from Theorem 9.1.1 that $\mu^*_k(i) = \operatorname{argmin}_{u \in \mathcal{U}} Q_k(i,u)$ is increasing in i. (More precisely, if μ^* is not unique, then there exists a version of $\mu^*_k(i)$ that is increasing in i.)

Statement 2: The proof is similar to that of Statement 1. For the CVaR risk averse MDP with risk specified in (8.38), note that since $Q_k(i,u,z)$ is submodular in (i,u) for each z, then $\inf_{z \in \mathbb{R}} Q_k(i,u,z)$ is submodular in (i,u).

Statement 3 of Theorem 9.3.1 follows by applying the proof of Statement 1 to the value iteration algorithm (6.21) or relative value iteration algorithm (6.31). Recall that for discounted and average cost problems, we used the value function notation V_k with forward indices $k = 0, 1, \ldots$. From Theorem 6.4.2, we know that $\lim_{k\to\infty} Q_k(i,u) = Q(i,u)$ and $\lim_{k\to\infty} V_k = V$. Therefore $Q(i,u)$ and V inherit the properties of $Q_k(i,u)$ and V_k respectively. As a consequence, Step 1 of statement 1 implies that $Q(i,u)$ and $V(i)$ are decreasing in i, and Step 2 implies that $Q(i,u)$ is submodular. (More generally, the pointwise limit of submodular functions is submodular [322, Lemma 2.6.1].)

Statement 4 of Theorem 9.3.1 follows similarly to Statement 2 by applying Lagrangian dynamic programming of Theorem 6.6.2. □

State-dependent action space: If the action space is state dependent (see Footnote 1 on page 122) of the form $\mathcal{U}(x)$, $x \in \mathcal{X} = \{1,2,\ldots,X\}$, then the following additional conditions are required for Theorem 9.3.1 to hold:

- *Ascending*: $u \in \mathcal{U}(x)$ and $\bar{u} \in \mathcal{U}(x+1)$ implies $\min\{u,\bar{u}\} \in \mathcal{U}(x)$, $\max\{u,\bar{u}\} \in \mathcal{U}(x+1)$.
- *Contracting*: $\mathcal{U}(x+1) \subseteq \mathcal{U}(x)$.

9.4 How does the optimal cost depend on the transition matrix?

How does the optimal expected cumulative cost J_{μ^*} of an MDP vary with transition matrix? Can the transition matrices be ordered so that the larger they are (with respect to some ordering), the larger the optimal cost? Such a result would allow us to compare the optimal performance of different MDPs, even though computing these via dynamic programming can be numerically expensive.

Consider two distinct MDP models with transition matrices $P(u)$ and $\bar{P}(u)$, $u \in \mathcal{U}$, respectively. Let $\mu^*(P)$ and $\mu^*(\bar{P})$ denote the optimal policies for these two different MDP models. Let $J_{\mu^*(P)}(x; P)$ and $J_{\mu^*(\bar{P})}(x; \bar{P})$ denote the optimal value functions corresponding to applying the respective optimal policies.

Introduce the following ordering on the transition matrices of the two MDPs.

(A5) Each row of $P(u)$ first-order stochastic dominates the corresponding row of $\bar{P}(u)$ for $u \in \mathcal{U}$. That is, $P_i(u) \geq_s \bar{P}_i(u)$ for $i \in \mathcal{X}$, $u \in \mathcal{U}$.

THEOREM 9.4.1 *Consider two distinct MDPs with transition matrices $P(u)$ and $\bar{P}(u)$, $u \in \mathcal{U}$. If (A1), (A2), (A5) hold, then the expected cumulative costs incurred by the optimal policies satisfy $J_{\mu^*(P)}(x; P) \leq J_{\mu^*(\bar{P})}(x; \bar{P})$.*

The theorem says that controlling an MDP with transition matrices $P(u)$, $u \in \mathcal{U}$ is always cheaper than an MDP with transition matrices $\bar{P}(u)$, $u \in \mathcal{U}$ if (A1), (A2) and (A5) hold. Note that Theorem 9.4.1 does need require numerical evaluation of the optimal policies or value functions.

Proof Suppose

$$Q_k(i, u) = c(i, u, k) + J'_{k+1} P_i(u), \quad \bar{Q}_k(i, u) = c(i, u, k) + \bar{J}'_{k+1} \bar{P}_i(u).$$

The proof is by induction. Clearly $J_N(i) = \bar{J}_N(i) = c_N(i)$ for all $i \in \mathcal{X}$. Now for the inductive step. Suppose $J_{k+1}(i) \leq \bar{J}_{k+1}(i)$ for all $i \in \mathcal{X}$. Therefore $J'_{k+1} P_i(u) \leq \bar{J}'_{k+1} P_i(u)$. By (A1), (A2), $\bar{J}_{k+1}(i)$ is decreasing in i. By (A5), $P_i \geq_s \bar{P}_i$. Therefore $\bar{J}'_{k+1} P_i \leq \bar{J}'_{k+1} \bar{P}_i$. So $c(i, u, k) + J'_{k+1} P_i(u) \leq c(i, u, k) + \bar{J}'_{k+1} \bar{P}_i(u)$ or equivalently, $Q_k(i, u) \leq \bar{Q}_k(i, u)$. Thus $\min_u Q_k(i, u) \leq \min_u \bar{Q}_k(i, u)$, or equivalently, $J_k(i) \leq \bar{J}_k(i)$ thereby completing the induction step. □

9.5 Algorithms for monotone policies – exploiting sparsity

Consider an average cost MDP. Assume that the costs and transition matrices satisfy (A1)–(A4). Then by Theorem 9.3.1 the optimal stationary policy $\mu^*(x)$ is increasing in x. How can this monotonicity property be exploited to compute (estimate) the optimal policy? This section discusses several approaches. (These approaches also apply to discounted cost MDPs.)

9.5.1 Policy search and Q-learning with submodular constraints

Suppose, for example, $\mathcal{U} = \{1, 2\}$ so that the monotone optimal stationary policy is a step function of the form (9.1) (see Figure 9.1) and is completely defined by the threshold state x^*.

If the MDP parameters are known, then the policy iteration algorithm of §6.4.2 can be used. If the policy μ_{n-1} at iteration $n - 1$ is monotone then under the assumptions of (A1), (A2) of Theorem 9.3.1, the policy evaluation step yields $J_{\mu_{n-1}}$ as a decreasing vector. Then under (A1)–(A4), an identical proof to Theorem 9.3.1 implies that the policy improvement step yields μ_n that is monotone. So the policy iteration algorithm will automatically be confined to monotone policies if initialized by a monotone policy.

In many cases, the optimal policy is monotone since the model structure satisfies (A1)-(A4) even though the MDP parameter values are not known. Then we need an algorithm to search for x^* over the finite state space $\mathcal{X}$. Chapter 17.4 presents discrete-valued stochastic optimization algorithms for implementing this. Another possibility is to solve a continuous-valued relaxation as follows: define the parametrized policy $\mu_\psi(x)$ where $\psi \in \mathbb{R}_+^2$ denotes the parameter vector. Also define the sample path cumulative cost estimate $\hat{C}_N(\psi)$ over some fixed time horizon N as

$$\mu_\psi(x) = \begin{cases} 1 & \frac{1}{1+\exp(-\psi_1(x-\psi_2))} < 0.5 \\ 2 & \text{otherwise} \end{cases}, \quad \hat{C}_N(\psi) = \frac{1}{N+1} \sum_{k=0}^{N} c(x_k, \mu_\psi(x_k)).$$

Note μ_ψ is a sigmoidal approximation to the step function (9.1). Consider the stochastic optimization problem: compute $\psi^* = \operatorname{argmin}_\psi \mathbb{E}\{\hat{C}_N(\psi)\}$. This can be solved readily via simulation-based gradient algorithms detailed in Chapter 15 such as the SPSA Algorithm 17 on page 350.

Alternatively, instead of exploiting the monotone structure in policy space, the submodular structure of the value function can be exploited. From Theorem 9.3.1, the Q-function $Q(x, u)$ in (9.9) is submodular. This submodularity can be exploited in Q-learning algorithms as will be described in Chapter 16.1.3.

The above methods operate without requiring explicit knowledge of transition matrices. Such methods are useful in transmission scheduling in wireless communication where often by modeling assumptions (A1)–(A4) hold, but the actual values of the transition matrices are not known.

9.5.2 Sparsity exploiting linear programming

Here we describe how the linear programming formulation of Theorem 6.5.5 for the optimal policy can exploit the monotone structure of the optimal policy. Suppose the number of actions U is small but the number of states X is large. Then the monotone optimal policy $\mu^*(x)$ is *sparse* in the sense that it is a piecewise constant function of the state x that jumps upwards at most at $U-1$ values (where by assumption U is small). In other words, $\mu^*(x+1) - \mu^*(x)$ is nonzero for at most $U-1$ values of x. In comparison, an unstructured policy can jump between states at arbitrary values and is therefore not sparse.

How can this sparsity property of a monotone optimal policy be exploited to compute the optimal policy? A convenient way of parametrizing sparsity in a monotone policy is in terms of the *conditional* probabilities $\theta_{x,u}$ defined in (6.33). Indeed $\theta_{x,u} - \theta_{x-1,u}$ as a function of x for fixed u is nonzero for up to only two values of $x \in \mathcal{X}$. A natural idea, arising from sparse estimation and compressed sensing [136], is to add a Lagrangian sum-of-norms term[6]

$$\lambda \sum_{x \geq 2} \|\theta_x - \theta_{x-1}\|_2, \quad \lambda \geq 0, \tag{9.10}$$

[6] Each term in the summation is a l_2 norm, the overall expression is the sum of norms. This is similar to the l_1 norm which is the sum of absolute values.

to a cost function whose minimum yields the optimal policy. Here $\theta_x = [\theta_{x,1}, \ldots, \theta_{x,U}]'$. The term (9.10) is a variant of the *fused lasso*[7] or *total variation* penalty, and can be interpreted as a convex relaxation of a penalty on the number of changes of conditional probability θ (as a function of state x).

Unfortunately, including (9.10) directly in the linear program (6.34) yields a non-convex problem since the linear program is formulated in terms of the stationary probabilities $\pi(x,u)$ instead of the conditional probability θ. Instead, consider the following equivalent formulation of the MDP which deals with the marginalized stationary probabilities $p(x) = \sum_{u \in \mathcal{U}} \pi(x,u)$:

$$\begin{aligned}
&\underset{p,\theta}{\text{Minimize}} \sum_{x,u} p(x)\theta_{x,u}\, c(x,u) + \lambda \sum_{x \geq 2} \|\theta_x - \theta_{x-1}\|_2 \\
&\text{subject to } p(x) \geq 0,\ \theta_{x,u} \geq 0, \quad x \in \mathcal{X}, u \in \mathcal{U} \\
&\qquad \sum_x p(x) = 1, \quad \sum_u \theta_{\bar{x},u} = 1, \qquad \bar{x} \in \mathcal{X} \\
&\qquad \pi(x) = \sum_{\bar{x},u} p(\bar{x})\theta_{\bar{x},u} P_{\bar{x}x}(u), \quad x \in \mathcal{X}.
\end{aligned} \tag{9.11}$$

To combine the advantages of the linear programming formulation (6.34) with the sparsity enhancing formulation (9.11), one can combine iterative minimization schemes for both formulations. Specifically, the linear program (6.34) can be solved iteratively via the so-called *alternating direction method of multipliers* (ADMM) [67], which is well suited for large-scale convex optimization problems. The combined procedure is as follows:

Step 1: Run a pre-specified number of iterations of the ADMM algorithm. This yields the estimate $\pi(x,u)$. Set $p(x) = \sum_u \pi(x,u)$.

Step 2: Run one iteration of a subgradient algorithm for optimizing (9.11) with respect to θ with p fixed. Set the updated estimate $\pi(x,u) = \theta_{x,u}\, p(x)$.

Go back to Step 1 and repeat.

In this way, minimization of (6.34) ensures global optimality, while the iterative steps to minimize (9.11) enforce sparsity. Numerical studies in [190] show that the resulting sparsity-enhanced ADMM has improved convergence; see also [331] for the use of ADMM in total variation estimation problems.

9.6 Example: Transmission scheduling over wireless channel

The conditions given in §9.3 are *sufficient* for the optimal policy to have a monotone structure. We conclude this chapter by describing an example where the sufficient conditions in §9.3 do not hold. However, a somewhat more sophisticated proof shows that the optimal policy is monotone. The formulation below generalizes the classical result of [96, 283] to the case of Markovian dynamics.

Consider the transmission of time-sensitive video (multimedia) packets in a wireless communication system with the use of an ARQ protocol for retransmission. Suppose L

[7] The lasso (least absolute shrinkage and selection operator) estimator was originally proposed in [318]. This is one of the most influential papers in statistics since the 1990s. It seeks to determine $\theta^* = \operatorname{argmax}_\theta \|y - A\theta\|_2^2 + \lambda\|\theta\|_1$ given an observation vector y, input matrix A and scalar $\lambda > 0$.

such packets stored in a buffer need to be transmitted over $N \geq L$ time slots. At each time slot, assuming the channel state is known, the transmission controller decides whether to attempt a transmission. The quality of the wireless channel (which evolves due to fading) is represented abstractly by a finite state Markov chain. The channel quality affects the error probability of successfully transmitting a packet. If a transmission is attempted, the result (an ACK or NACK of whether successful transmission was achieved) is received. If a packet is transmitted but not successfully received, it remains in the buffer and may be retransmitted. At the end of all N time slots, no more transmission is allowed and a penalty cost is incurred for packets that remain in the buffer.

How should a transmission controller decide at which time slots to transmit the packets? If the current channel quality is poor, should the controller wait for the channel quality to improve, keeping in mind the deadline of N time slots? It is shown below that the optimal transmission scheduling policy is a monotone (threshold) function of time and buffer size. The framework is applicable to any delay-sensitive real-time packet transmission system.

9.6.1 MDP model for transmission control

We formulate the above transmission scheduling problem as a finite horizon MDP with a penalty terminal cost. The wireless fading channel is modeled as a finite state Markov chain. Let $s_k \in S = \{\gamma_1, \ldots, \gamma_K\}$ denote the channel state at time slot k. Assume s_k evolves as a Markov chain according to transition probability matrix $P = (P_{ss'} : s, s' = 1, 2, \ldots, K)$, where $P_{ss'} = \mathbb{P}(s_{k+1} = \gamma_{s'} | s_k = \gamma_s)$. Here the higher the state s, the better the quality of the channel.

Let $\mathcal{U} = \{u_0 = 0 \text{ (do not transmit)}, u_1 = 1 \text{ (transmit)}\}$ denote the action space. In a time slot, if action u is selected, a cost $c(u)$ is accrued, where $c(\cdot)$ is an increasing function. The probability that a transmission is successful is an increasing function of the action u and channel state s:

$$\gamma(u, s) = \begin{cases} 0 & \text{If } u = 0 \\ 1 - P_e(s) & \text{If } u = 1. \end{cases} \tag{9.12}$$

Here $P_e(s)$ denotes the error probability for channel state s and is a decreasing function of s.

Let n denotes the residual transmission time: $n = N, N-1, \ldots, 0$. At the end of all N time slots, i.e. when $n = 0$, a terminal penalty cost $c_N(i)$ is paid if i untransmitted packets remain in the buffer. It is assumed that $c_N(i)$ is increasing in i and $c_N(0) = 0$.

The optimal scheduling policy $\mu_n^*(i, s)$ is the solution of Bellman's equation:[8]

$$V_n(i, s) = \min_{u \in U} Q_n(i, s, u), \quad \mu_n^*(i, s) = \arg\min_{u \in U} Q_n(i, s, u), \tag{9.13}$$

$$Q_n(i, s, u) = \left\{ c(u) + \sum_{s' \in S} P_{ss'} \left[\gamma(u, s) V_{n-1}(i-1, s') + (1 - \gamma(u, s)) \, V_{n-1}(i, s') \right] \right\}$$

[8] It is notationally convenient here to use Bellman's equation with forward indices. So we use $V_n = J_{N-n}$ for the value function. This notation was used previously for MDPs in (6.11).

initialized with $V_n(0,s) = 0, V_0(i,s) = c_N(i)$. A larger terminal cost $c_N(i)$ emphasizes delay sensitivity while a larger action cost $c(u)$ emphasizes energy consumption. If $P_{ss} = 1$ then the problem reduces to that considered in [96, 283].

9.6.2 Monotone structure of optimal transmission policy

THEOREM 9.6.1 *The optimal transmission policy $\mu_n^*(i,s)$ in (9.13) has the following monotone structure:*

1. *If the terminal cost $c_N(i)$ is increasing in the buffer state i, then $\mu_n^*(i,s)$ is decreasing in the number of transmission time slots remaining n.*
2. *If $c_N(i)$ is increasing in the buffer state i and is integer convex, i.e.*

$$c_N(i+2) - c_N(i+1) \geq c_N(i+1) - c_N(i) \; \forall i \geq 0, \tag{9.14}$$

then $\mu_n^*(i,s)$ is a threshold policy of the form:

$$\mu_n^*(i,s) = \begin{cases} 0 & i < i_{n,s}^* \\ 1 & i \geq i_{n,s}^* \end{cases}.$$

Here the threshold buffer state $i_{n,s}^*$ depends on n (time remaining) and s (channel state). Furthermore, the threshold $i_{n,s}^*$ is increasing in n.

The theorem says that the optimal transmission policy is *aggressive* since it is optimal to transmit more often when the residual transmission time is less or the buffer occupancy is larger. The threshold structure of the optimal transmission scheduling policy can be used to reduce the computational cost in solving the dynamic programming problem or the memory required to store the solutions. For example, the total number of transmission policies given L packets, N time slots and K channel states is 2^{NLK}. In comparison, the number of transmission policies that are monotone in the number of transmission time slots remaining and the buffer state is NL^K, which can be substantially smaller. [144] gives several algorithms (e.g. (modified) value iteration, policy iteration) that exploit monotone results to efficiently compute the optimal policies.

To prove Theorem 9.6.1, the following results are established in the appendix.

LEMMA 9.6.2 *The value function $V_n(i,s)$ defined by (9.13) is increasing in the number of remaining packets i and decreasing in the number of remaining time slots n.*

LEMMA 9.6.3 *If $c_N(\cdot)$ is an increasing function (of the terminal buffer state) then the value function $V_n(i,s)$ satisfies the following submodularity condition:*

$$V_n(i+1,s) - V_n(i,s) \geq V_{n+1}(i+1,s) - V_{n+1}(i,s), \tag{9.15}$$

for all $i \geq 0$, $s \in S$. Hence, $Q_n(i,s,u)$ in (9.13) is supermodular in (n,u). Furthermore, if the penalty cost $c_N(\cdot)$ is an increasing function and satisfies (9.14) then $V_n(i,s)$ has increasing differences in the number of remaining packets:

$$V_n(i+2,s) - V_n(i+1,s) \geq V_n(i+1,s) - V_n(i,s), \tag{9.16}$$

for all $i \geq 0$, $s \in S$. Hence $Q_n(i,s,u)$ is submodular in (i,u).

With the above two lemmas, the proof of Theorem 9.6.1 is as follows:

First statement: If $c_N(i)$ is increasing in i then $V_n(i,s)$ satisfies (9.15) and $Q_n(i,s,u)$ given by (9.13) is supermodular in (u,n) (provided that $\gamma(u,s)$ defined in (9.12) is an increasing function of the action u). Therefore, $\mu_n^*(i,s)$ is decreasing in n.

Second statement: Due to Lemma 9.6.3, if $c_N(i)$ is increasing in i and satisfies (9.14) then $V_n(i,h)$ satisfies (9.16) and $Q_n(i,h,u)$ is submodular in (u,i) (provided that $\gamma(u,h)$ increases in u). The submodularity of $Q_n(i,h,u)$ in (u,i) implies that $\mu_n^*(i,h)$ is increasing in i. The result that $i^*_{\bar{n},\bar{h}}$ is increasing in $\bar{n}$ follows since $\mu_n^*(i,h)$ is decreasing in n (first statement).

9.7 Complements and sources

The use of supermodularity for structural results in MDPs and game theory was pioneered by Topkis in the seminal paper [321] culminating in the book [322]. We refer the reader to [15] for a tutorial description of supermodularity with applications in economics. [144, Chapter 8] has an insightful treatment of submodularity in MDPs. Excellent books in stochastic dominance include [249, 295]. [303] covers several cases of monotone MDPs. The paper [242] is highly influential in the area of monotone comparative statics (determining how the argmax or argmin behaves as a parameter varies) and discusses the single crossing condition. [273] has some recent results on conditions where the single crossing property is closed under addition. A more general version of Theorem 9.4.1 is proved in [248]. The example in §9.6 is expanded in [252] with detailed numerical examples. Also [253] considers the average cost version of the transmission scheduling problem on a countable state space (to model an infinite buffer). [345] studies supermodularity and monotone policies in discrete event systems. [12] covers deeper results in multimodularity, supermodularity and extensions of convexity to discrete spaces for discrete event systems.

Appendix 9.A Proofs of theorems

9.A.1 Proof of Lemma 9.6.2

The monotonicity of $V_n(i,s)$ in i follows from the monotonicity of $c_N(\cdot)$ and the definition of $V_n(i,s)$. It is clear that $V_1(i,s) \leq V_0(i,s)$ since $V_1(i,s) \leq Q_1(i,s,0) = \sum_{s' \in S} P_{ss'} V_0(i,s') = c_N(i) = V_0(i,s)$. The monotonicity of $V_n(i,s)$ in n then follows straightforwardly from the definition given by (9.13) of $V_n(i,s)$.

9.A.2 Proof of Lemma 9.6.3

Rewrite $Q_n(i,s,u)$ in (9.13) as

$$Q_n(i,s,u) = c(u) + \sum_{s' \in S} P_{ss'} \gamma(u,s) \Big(V_{n-1}(i-1,s') - V_{n-1}(i,s') \Big) + \sum_{s' \in S} P_{ss'} V_{n-1}(i,s'). \tag{9.17}$$

From (9.17), the reader should verify that by evaluating $Q_{n+1}(i,s,u) - Q_n(i,s,u)$, it follows that $Q_n(i,s,u)$ is supermodular in (n,u) if $V_n(i,s)$ is submodular in (n,i) given that $\gamma(u,s)$ (defined by (9.12)) increases in u. Similarly, $Q_n(i,s,u)$ is submodular in (i,u) if $V_n(i,s)$ has increasing differences in i.

The proof is divided into two parts as below.

Part 1: $V_n(i,s)$ is submodular in (i,n) and hence $Q_n(i,s,u))$ is supermodular in (n,u): The proof is by mathematical induction. (9.15) holds for $n+i=0$ since $V_n(i,s)$ is decreasing in n. Assume that (9.15) holds for $n+i=k$. We will prove that it holds for $n+i=k+1$. Let $V_{n+1}(i+1,s) = Q_{n+1}(i+1,s,u_{11})$, $V_{n+1}(i,s) = Q_{n+1}(i,s,u_{10})$, $V_n(i+1,s) = Q_n(i+1,s,u_{01})$, $V_n(i,s) = Q_n(i,s,u_{00})$ for some $u_{00}, u_{01}, u_{10}, u_{11}$. We have to prove that

$$
\begin{aligned}
& Q_{n+1}(i+1,s,u_{11}) - Q_{n+1}(i,s,u_{10}) - Q_n(i+1,s,u_{01}) + Q_n(i,s,u_{00}) \le 0 \\
\Leftrightarrow & Q_{n+1}(i+1,s,u_{11}) - Q_n(i+1,s,u_{01}) - Q_{n+1}(i,s,u_{10}) + Q_n(i,s,u_{00}) \le 0 \\
\Leftrightarrow & \underbrace{Q_{n+1}(i+1,s,u_{11}) - Q_{n+1}(i+1,s,u_{01})}_{\le 0 \text{ (By optimality)}} + \underbrace{Q_{n+1}(i+1,s,u_{01}) - Q_n(i+1,s,u_{01})}_{A} \\
& - \underbrace{(Q_{n+1}(i,s,u_{10}) - Q_n(i,s,u_{10}))}_{B} + \underbrace{(-Q_n(i,s,u_{10}) + Q_n(i,s,u_{00}))}_{\le 0 \text{ (By optimality)}} \le 0.
\end{aligned}
$$

By induction hypothesis we have

$$
\begin{aligned}
A = & \sum_{s' \in S} P_{ss'} \Big(\gamma(u_{01},s) \left[V_n(i,s') - V_{n-1}(i,s') \right] \\
& \qquad + (1-\gamma(u_{01},s)) \left[V_n(i+1,s') - V_{n-1}(i+1,s') \right] \Big) \\
\le & \sum_{s' \in S} P_{ss'} \left[V_n(i,s') - V_{n-1}(i,s') \right].
\end{aligned}
$$

Similarly, $B \ge \sum_{s' \in S} P_{ss'} \left[V_n(i,s') - V_{n-1}(i,s') \right]$. Hence, $B \ge A$.

Therefore, $V_n(i,s)$ satisfies (9.15), which implies that $Q_n(i,s,u)$ given by (9.17) is supermodular in (n,u). (See the remarks below (9.17).)

Part 2: $V_n(i,s)$ has increasing differences in i hence $Q_n(i,s)$ given by (9.17) is submodular in (u,i) The proof is by mathematical induction on n. First, (9.16) holds for $n=0$ due to (9.14). Assume (9.16) holds for $n=k$. We will prove that it holds for $n=k+1$. Let $V_{k+1}(i+2,s) = Q_{k+1}(i+2,s,u_2)$, $V_{k+1}(i+1,s) = Q_{k+1}(i+1,s,u_1)$, $V_{k+1}(i,s) = Q_{k+1}(i,s,u_0)$ for some u_0, u_1, u_2. We then have to prove that

$$
\begin{aligned}
& Q_{k+1}(i+2,s,u_2) - Q_{k+1}(i+1,s,u_1) - Q_{k+1}(i+1,s,u_1) + Q_{k+1}(i,s,u_0) \ge 0 \\
\Leftrightarrow & \underbrace{Q_{k+1}(i+2,s,u_2) - Q_{k+1}(i+1,s,u_2)}_{A} + \underbrace{Q_{k+1}(i+1,s,u_2) - Q_{k+1}(i+1,s,u_1)}_{\ge 0 \text{ (By optimality)}} \\
& \underbrace{-Q_{k+1}(i+1,s,u_1) + Q_{k+1}(i+1,s,u_0)}_{\ge 0 \text{ (By optimality)}} - \underbrace{(Q_{k+1}(i+1,s,u_0) - Q_{k+1}(i,s,u_0))}_{B} \ge 0.
\end{aligned}
$$

In addition, it follows from the induction hypothesis that

$$A = \sum_{s' \in S} P_{ss'} \Big[\gamma(u, s)(V_k(i + 1, s') - V_k(i, s'))$$

$$+ (1 - \gamma(u, s)) (V_k(i + 2, s') - V_k(i + 1, s')) \Big]$$

$$\geq \sum_{s' \in S} P_{ss'}(V_k(i + 1, s') - V_k(i, s')).$$

Similarly, $B \leq \sum_{s' \in S} P_{ss'}(V_k(i + 1, s') - V_k(i, s'))$. Hence, $A - B \geq 0$.

Therefore, $V_n(i, s)$ satisfies (9.16), which implies that $Q_n(i, s, u)$ given by (9.17) is submodular in (i, u). (See the remarks below (9.17).)

10 Structural results for optimal filters

Contents

This chapter and the following four chapters develop structural results for the optimal policy of a POMDP. In Chapter 9, we used first-order stochastic dominance to characterize the structure of optimal policies for finite state MDPs. However, first-order stochastic dominance is not preserved under Bayes rule for the belief state update. So a stronger stochastic order is required to obtain structural results for POMDPs. We will use the *monotone likelihood ratio (MLR) stochastic order* to order belief states. This chapter develops important structural results for the HMM filter using the MLR order. These results form a crucial step in formulating the structural results for POMDPs.

The plan of this and the next three chapters is displayed in Figure 10.1 Please also see the short description at the beginning of Part III.

Outline of this chapter

Recall from Chapter 7 that for a POMDP with controlled transition matrix $P(u)$ and observation probabilities $B_{xy}(u) = p(y|x, u)$, given the observation y_{k+1}, the HMM filter recursion (7.11) for the belief state π_{k+1} in terms of π_k reads

$$\pi_{k+1} = T(\pi_k, y_{k+1}, u_k) = \frac{B_{y_{k+1}}(u_k)P'(u_k)\pi_k}{\sigma(\pi_k, y_{k+1}, u_k)}, \; \sigma(\pi, y, u) = \mathbf{1}'B_y(u)P'(u)\pi,$$

$$B_y(u) = \operatorname{diag}(B_{1y}(u), \ldots, B_{Xy}(u)), \quad u \in \mathcal{U} = \{1, 2, \ldots, U\}, y \in \mathcal{Y}. \tag{10.1}$$

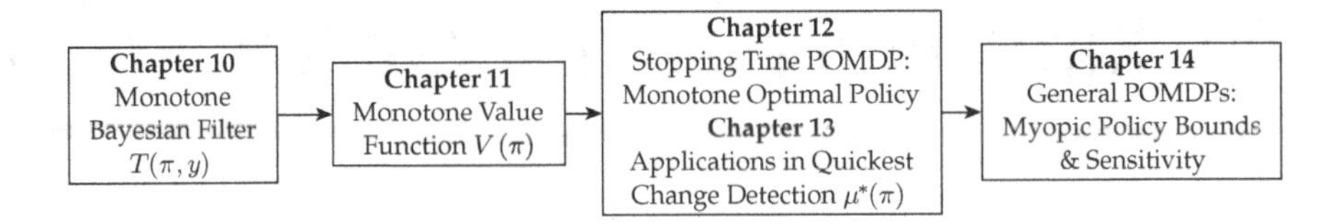

Figure 10.1 Organization of Chapters 10 to 14 on POMDP structural results.

The two main questions addressed in this chapter are:

1. How can beliefs (posterior distributions) π computed by the HMM filter be ordered within the belief space (unit simplex) $\Pi(X)$?
2. Under what conditions does the HMM filter $T(\pi, y, u)$ increase with belief π, observation y and action u?

Answering the first question is crucial to define what it means for a POMDP to have an optimal policy $\mu^*(\pi)$ increasing with π. Recall from Chapter 9 that for the fully observed MDP case, we gave conditions under which the optimal policy is increasing with scalar state x – ordering the states was trivial since they were scalars. However, for a POMDP, we need to order the belief states π which are probability mass functions (vectors) in the unit simplex $\Pi(X)$.

The first question above will be answered by using the monotone likelihood ratio (MLR) stochastic order to order belief states. The MLR is a *partial order* that is ideally suited for Bayesian estimation (optimal filtering) since it is preserved under conditional expectations. §10.1 discusses the MLR order.

The answer to the second question is essential for giving conditions on a POMDP so that the optimal policy $\mu^*(\pi)$ increases in π. Recall that Bellman's equation (7.27) for a POMDP involves the term $\sum_y V(T(\pi, y, u))\sigma(\pi, y, u)$. So to show that the optimal policy is monotone, it is necessary to characterize the behavior of the HMM filter $T(\pi, y, u)$ and normalization term $\sigma(\pi, y, u)$. §10.2 to §10.5 discuss structural properties of the HMM filter.

Besides their importance in establishing structural results for POMDPs, the structural results for HMM filters developed in this chapter are also useful for constructing reduced complexity HMM filtering algorithms that provably upper and lower bound the optimal posterior (with respect to the MLR order). We shall describe the construction of such reduced complexity HMM filters in §10.6. Sparse rank transition matrices for the low complexity filters will be constructed via nuclear norm minimization (convex optimization).

10.1 Monotone likelihood ratio (MLR) stochastic order

For dealing with POMDPs, the MLR stochastic order is the main concept that will be used to order belief states in the unit $X-1$-dimensional unit simplex

$$\Pi(X) = \left\{\pi \in \mathbb{R}^X : \mathbf{1}'\pi = 1, \quad 0 \leq \pi(i) \leq 1,\ i \in \mathcal{X} = \{1, 2, \ldots, X\}\right\}.$$

10.1.1 Definition

DEFINITION 10.1.1 (Monotone likelihood ratio (MLR) ordering) *Let $\pi_1, \pi_2 \in \Pi(X)$ denote two belief state vectors. Then π_1 dominates π_2 with respect to the MLR order, denoted as $\pi_1 \geq_r \pi_2$, if*

$$\pi_1(i)\pi_2(j) \leq \pi_2(i)\pi_1(j), \quad i < j, i, j \in \{1, \ldots, X\}. \tag{10.2}$$

Similarly $\pi_1 \leq_r \pi_2$ if $\leq$ in (10.2) is replaced by a $\geq$.

Equivalently, $\pi_1 \geq_r \pi_2$ if the likelihood ratio $\pi_1(i)/\pi_2(i)$ is increasing[1] monotonically with i (providing the ratio is well defined). Similarly, for the case of pdfs, define $\pi_1 \geq_r \pi_2$ if their ratio $\pi_1(x)/\pi_2(x)$ is increasing in $x \in \mathbb{R}$.

DEFINITION 10.1.2 *A function $\phi : \Pi(X) \to \mathbb{R}$ is said to be MLR increasing if $\pi_1 \geq_r \pi_2$ implies $\phi(\pi_1) \geq \phi(\pi_2)$. Also ϕ is MLR decreasing if $-\phi$ is MLR increasing.*

Recall the definition of first-order stochastic dominance (Definition 9.2.1 in Chapter 9). MLR dominance is a stronger condition than first-order dominance.

THEOREM 10.1.3 *For pmfs or pdfs π_1 and π_2, then $\pi_1 \geq_r \pi_2$ implies $\pi_1 \geq_s \pi_2$.*

Proof $\pi_1 \geq_r \pi_2$ implies $\pi_1(x)/\pi_2(x)$ is increasing in x. Denote the corresponding cdfs as F_1, F_2. Define $t = \{\sup x : \pi_1(x) \leq \pi_2(x)\}$. Then $\pi_1 \geq_r \pi_2$ implies that for $x \leq t$, $\pi_1(x) \leq \pi_2(x)$ and for $x \geq t$, $\pi_1(x) \geq \pi_2(x)$. So for $x \leq t$, $F_1(x) \leq F_2(x)$. Also for $x > t$, $\pi_1(x) \geq \pi_2(x)$ implies $1 - \int_x^\infty \pi_1(x)dx \leq 1 - \int_x^\infty \pi_2(x)dx$ or equivalently, $F_1(x) \leq F_2(x)$. Therefore $\pi_1 \geq_s \pi_2$. □

For state space dimension $X = 2$, MLR is a *complete* order and coincides with first-order stochastic dominance. The reason is that for $X = 2$, since $\pi(1) + \pi(2) = 1$, it suffices to choose the second component $\pi(2)$ to order π. Indeed, for

$$X = 2, \quad \pi_1 \geq_r \pi_2 \iff \pi_1 \geq_s \pi_2 \iff \pi_1(2) \geq \pi_2(2).$$

For state space dimension $X \geq 3$, MLR and first-order dominance are *partial orders* on the belief space $\Pi(X)$. Indeed, $[\Pi(X), \geq_r]$ is a partially ordered set (poset) since it is not always possible to order any two belief states in $\Pi(X)$.
Example (i): $[0.2, 0.3, 0.5]' \geq_r [0.4, 0.5, 0.1]'$
Example (ii): $[0.3, 0.2, 0.5]'$ and $[0.4.0.5.0.1]'$ are not MLR comparable.
Figure 10.2 gives a geometric interpretation of first-order and MLR dominance for $X = 3$.

10.1.2 Why MLR ordering?

The MLR stochastic order is useful in filtering and POMDPs since it is preserved under conditional expectations (or more naively, application of Bayes' rule).

[1] Throughout Part III of the book, increasing is used in the weak sense to denote non-decreasing.

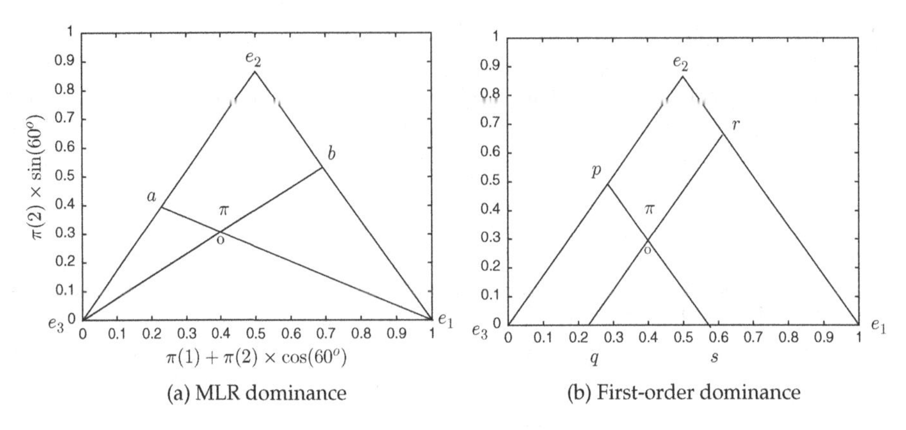

(a) MLR dominance

(b) First-order dominance

Figure 10.2 Geometric interpretation of MLR and first-order dominance partial orders in the belief space $\Pi(X)$ for $X = 3$. **Figure (a)**: Given belief state π, all beliefs within the triangular region with vertices (e_3, a, π) MLR dominate π. Here a is the point at which the line passing through e_1 and π intersects the line between e_2, e_3. Also π MLR dominates all beliefs within the triangular region with vertices (e_1, b, π). Here b is the point at which the line passing through e_3 and π intersects the line between e_1, e_2. All other beliefs in the unit simplex are not MLR orderable with respect to π. **Figure (b)**: All beliefs within the polytope (π, p, e_3, q) first order dominate π. Here the line (p, s) passing through π is constructed parallel to the line (e_2, e_1). Also π_1 first order dominates all beliefs in the polytope (π, r, e_1, s). The line (r, q) passing through π is constructed parallel to the line (e_2, e_3). All other beliefs are not first-order comparable with respect to π.

THEOREM 10.1.4 *MLR dominance is preserved under Bayes' rule: for continuous or discrete-valued observation y with observation likelihoods* $B_y = \text{diag}(B_{1y}, \ldots, B_{Xy})$, $B_{xy} = p(y|x)$, *given two beliefs* $\pi_1, \pi_2 \in \Pi(X)$, *then*

$$\pi_1 \geq_r \pi_2 \iff \frac{B_y \pi_1}{\mathbf{1}' B_y \pi_1} \geq_r \frac{B_y \pi_2}{\mathbf{1}' B_y \pi_2}$$

providing $\mathbf{1}' B_y \pi_1$ *and* $\mathbf{1}' B_y \pi_2$ *are nonzero.*[2]

Proof By definition of MLR dominance, the right-hand side is

$$\cancel{B_{iy} B_{i+1,y}} \pi_1(i) \pi_2(i+1) \leq \cancel{B_{iy} B_{i+1,y}} \pi_1(i+1) \pi_2(i)$$

which is equivalent to $\pi_1 \geq_r \pi_2$. □

Notice that in the fully observed MDP of Chapter 9, we used first-order stochastic dominance. However, first-order stochastic dominance is not preserved under conditional expectations and so is not useful for POMDPs.

[2] A notationally elegant way of saying this is: given two random variables X and Y, then $X \leq_r Y$ iff $X|X \in A \leq_r Y|Y \in A$ for all events A providing $P(X \in A) > 0$ and $P(Y \in A) > 0$. Requiring $\mathbf{1}' B_y \pi > 0$ avoids pathological cases such as $\pi = [1, 0]'$ and $B_y = \text{diag}(0, 1)$, i.e. prior says state 1 with certainty, while observation says state 2 with certainty.

10.1.3 Examples

1. *First-order stochastic dominance is not closed under Bayes' rule*: Consider beliefs $\pi_1 = (\frac{1}{3}, \frac{1}{3}, \frac{1}{3})'$, $\pi_2 = (0, \frac{2}{3}, \frac{1}{3})'$. Then clearly $\pi_1 \leq_s \pi_2$. Suppose $P = I$ and the observation likelihoods B_{xy} have values: $\mathbb{P}(y|x = 1) = 0$, $\mathbb{P}(y|x = 2) = 0.5$, $\mathbb{P}(y|x = 3) = 0.5$. Then the filtered updates are $T(\pi_1, y, u) = (0, \frac{1}{2}, \frac{1}{2})'$ and $T(\pi_2, y, u) = (0, \frac{2}{3}, \frac{1}{3})'$. Thus $T(\pi_1, y, u) \geq_s T(\pi_2, y.u)$ showing that first-order stochastic dominance is not preserved after a Bayesian update.

The MLR order is stronger than first-order dominance. For example, choosing $\pi_3 = (0, 1/3, 2/3)'$, then $\pi_1 \leq_r \pi_3$. The filtered updates are $T(\pi_1, y, u) = (0, \frac{1}{2}, \frac{1}{2})'$ and $T(\pi_3, y, u) = (0, \frac{1}{3}, \frac{2}{3})'$ and it is seen that $T(\pi_1, y, u) \leq_r T(\pi_3, y, u)$.

2. Examples of pmfs that satisfy MLR dominance are ([249] has a detailed list):

$$\text{Poisson:} \quad \frac{\lambda_1^k}{k!} \exp(\lambda_1) \leq_r \frac{\lambda_2^k}{k!} \exp(\lambda_2), \quad \lambda_1 \leq \lambda_2.$$

$$\text{Binomial:} \quad \binom{n_1}{k} p_1^k (1 - p_1)^{n_1 - k} \leq_r \binom{n_2}{k} p_1^k (1 - p_2)^{n_2 - k}, \quad n_1 \leq n_2, p_1 \leq p_2.$$

$$\text{Geometric:} \quad (1 - p_1) p_1^k \leq_r (1 - p_2) p_2^k, \quad p_1 \leq p_2.$$

3. The MLR order is also defined for probability density functions (pdfs): $p \geq_r q$ if $p(x)/q(x)$ is increasing in x. If the pdfs are differentiable, this is equivalent to saying $\frac{d}{dx} \frac{p(x)}{q(x)} \geq 0$. Examples include:

$$\text{Normal:} \quad N(x; \mu_1, \sigma^2) \leq_r N(x; \mu_2, \sigma^2), \quad \mu_1 \leq \mu_2.$$

$$\text{Exponential:} \quad \lambda_1 \exp(\lambda_1(x - a_1)) \leq_r \lambda_2 \exp(\lambda_2(x - a_2)), \quad a_1 \leq a_2, \lambda_1 \geq \lambda_2.$$

Uniform pdfs $U[a, b] = I(x \in [a, b])/(b - a)$ are not MLR comparable with respect to a or b.

10.2 Total positivity and copositivity

This section defines the concepts of total positivity and copositivity. These are crucial concepts in obtaining monotone properties of the optimal filter with respect to the MLR order. These properties will be used in later chapters to obtain structural results for POMDPs.

DEFINITION 10.2.1 (Totally positive of order 2 (TP2)) *A stochastic matrix M is TP2 if all its second-order minors are nonnegative. That is, determinants*

$$\begin{vmatrix} M_{i_1 j_1} & M_{i_1 j_2} \\ M_{i_2 j_1} & M_{i_2 j_2} \end{vmatrix} \geq 0 \text{ for } i_2 \geq i_1, j_2 \geq j_1. \tag{10.3}$$

Equivalently, a transition or observation kernel[3] *denoted M is TP2 if the $i+1$-th row MLR dominates the i-th row: that is, $M_{i,:} \geq_r M_{j,:}$ for every $i > j$.*

[3] We use the term "kernel" to allow for continuous and discrete valued observation spaces $\mathcal{Y}$. If $\mathcal{Y}$ is discrete, then $B(u)$ is a $X \times Y$ stochastic matrix. If $\mathcal{Y} \subseteq \mathbb{R}$, then $B_{xy}(u) = p(y|x, u)$ is a probability density function.

Next, we define a copositive ordering. Recall from §3.5 that an HMM is parameterized by the transition and observation probabilities (P, B). Start with the following notation. Given an HMM $(P(u), B(u))$ and another HMM[4] $(P(u+1), B(u+1))$, define the sequence of $X \times X$ dimensional symmetric matrices $\Gamma^{j,u,y}, j = 1, \ldots, X-1, y \in \mathcal{Y}$ as

$$\Gamma^{j,u,y} = \frac{1}{2}\left[\gamma_{mn}^{j,u,y} + \gamma_{nm}^{j,u,y}\right]_{X\times X}, \quad \text{where} \tag{10.4}$$

$$\begin{aligned}\gamma_{mn}^{j,u,y} &= B_{j,y}(u)B_{j+1,y}(u+1)P_{m,j}(u)P_{n,j+1}(u+1) \\ &\quad - B_{j+1,y}(u)B_{j,y}(u+1)P_{m,j+1}(u)P_{n,j}(u+1).\end{aligned}$$

DEFINITION 10.2.2 (Copositive ordering $\preceq$ of transition and observations probabilities) *Given $(P(u), B(u))$ and $(P(u+1), B(u+1))$, we say that*

$$(P(u), B(u)) \preceq (P(u+1), B(u+1))$$

if the sequence of $X \times X$ matrices $\Gamma^{j,u,y}, j = 1, \ldots, X-1, y \in \mathcal{Y}$ are copositive.[5] *That is,*

$$\pi' \Gamma^{j,u,y} \pi \geq 0, \forall \pi \in \Pi(X), \textit{ for each } j, y. \tag{10.5}$$

The above notation $(P(u), B(u)) \preceq (P(u+1), B(u+1))$ is intuitive since it will be shown below that the copositive condition (10.5) is a necessary and sufficient condition for the HMM filter update to satisfy $T(\pi, y, u) \leq_r T(\pi, y, u+1)$ for any posterior $\pi \in \Pi(X)$ and observation $y \in \mathcal{Y}$. This will be denoted as Assumption (F3) below.

We are also interested in the special case of the optimal HMM predictors instead of optimal filters. Recall that optimal prediction is a special case of filtering obtained by choosing non-informative observation probabilities, i.e. all elements of $B_{x,y}$ versus x are identical. In analogy to Definition 10.2.2 we make the following definition.

DEFINITION 10.2.3 (Copositive ordering of transition matrices) *Given $P(u)$ and $P(u+1)$, we say that*

$$P(u) \preceq P(u+1)$$

if the sequence of $X \times X$ matrices $\Gamma^{j,u}, j = 1 \ldots, X-1$ are copositive, i.e.

$$\pi' \Gamma^{j,u} \pi \geq 0, \quad \forall \pi \in \Pi(X), \quad \textit{for each } j, \textit{ where} \tag{10.6}$$

$$\Gamma^{j,u} = \frac{1}{2}\left[\gamma_{mn}^{j,u} + \gamma_{nm}^{j,u}\right]_{X\times X}, \; \gamma_{mn}^{j,u} = P_{m,j}(u)P_{n,j+1}(u+1) - P_{m,j+1}(u)P_{n,j}(u+1).$$

10.3 Monotone properties of optimal filter

With the above definitions, we can now give sufficient conditions for the optimal filtering recursion $T(\pi, y, u)$ to be monotone with respect to the MLR order. The following are the main assumptions.

[4] The notation u and $u+1$ is used to distinguish between the two HMMs. Recall that in POMDPs, u denotes the actions taken by the controller.

[5] A symmetric matrix M is positive semidefinite if $x'Mx \geq 0$ for any vector x. In comparison, M is copositive if $\pi' M \pi \geq 0$ for any probability vector π. Clearly if a symmetric matrix is positive definite then it is copositive. Thus copositivity is a weaker condition than positive definiteness.

(F1) $B(u)$ with elements $B_{x,y}(u)$ is TP2 for each $u \in \mathcal{U}$ (see Definition 10.2.1).
(F2) $P(u)$ is TP2 for each action $u \in \mathcal{U}$.
(F3) $(P(u), B(u)) \preceq (P(u+1), B(u+1))$ (copositivity condition in Definition 10.2.2).
(F3′) All elements of the matrices $\Gamma^{j,u,y}$, are nonnegative. (This is sufficient[6] for (F3).)
($\underline{\text{F3}}$) $P(u) \preceq P(u+1)$ (copositivity condition in Definition 10.2.3).
($\underline{\text{F3}}'$) All the elements of $\Gamma^{j,u}$ are nonnegative. (This is sufficient for ($\underline{\text{F3}}$)).
(F4) $\sum_{y \leq \bar{y}} \sum_{j \in \mathcal{X}} \left[P_{i,j}(u) B_{j,y}(u) - P_{i,j}(u+1) B_{j,y}(u+1)\right] \leq 0$ for all $i \in \mathcal{X}$ and $\bar{y} \in \mathcal{Y}$.

Assumptions (F1), (F2) deal with the transition and observation probabilities for a fixed action u. In comparison, (F3), (F3′), ($\underline{\text{F3}}$), ($\underline{\text{F3}}'$), (F4) are conditions on the transition and observation probabilities of two different HMMs corresponding to the actions u and $u+1$.

Main result: The following theorem is the main result of this chapter. The theorem characterizes how the HMM filter $T(\pi, y, u)$ and normalization measure $\sigma(\pi, y, u)$ behave with increasing π, y and u. The theorem forms the basis of all the structural results for POMDPs presented in subsequent chapters.

THEOREM 10.3.1 (Structural result for filtering) *Consider the HMM filter $T(\pi, y, u)$ and normalization measure $\sigma(\pi, y, u)$ defined as*

$$\begin{aligned} T(\pi, y, u) &= \frac{B_y(u) P'(u) \pi}{\sigma(\pi, y, u)}, \quad \sigma(\pi, y, u) = \mathbf{1}' B_y(u) P'(u) \pi, \quad \textit{where} \\ B_y(u) &= \operatorname{diag}(B_{1y}(u), \ldots, B_{Xy}(u)), \quad u \in \mathcal{U} = \{1, 2, \ldots, U\}, y \in \mathcal{Y}. \end{aligned} \tag{10.7}$$

Suppose $\pi_1, \pi_2 \in \Pi(X)$ are arbitrary belief states. Let $\geq_r$ and $\geq_s$ denote MLR dominance and first order dominance, respectively. Then

1. *(a) For $\pi_1 \geq_r \pi_2$, the HMM predictor satisfies $P'(u)\pi_1 \geq_r P'(u)\pi_2$ iff (F2) holds.*
   (b) Therefore, for $\pi_1 \geq_r \pi_2$, the HMM filter satisfies $T(\pi_1, y, u) \geq_r T(\pi_2, y, u)$ for any observation y iff (F2) holds (since MLR dominance is preserved by Bayes' rule, Theorem 10.1.4).
2. *Under (F1), (F2), $\pi_1 \geq_r \pi_2$ implies $\sigma(\pi_1, u) \geq_s \sigma(\pi_2, u)$ where*
$$\sigma(\pi, u) \equiv [\sigma(\pi, 1, u), \cdots, \sigma(\pi, Y, u)].$$
3. *For $y, \bar{y} \in \mathcal{Y}$, $y > \bar{y}$ implies $T(\pi_1, y, u) \geq_r T(\pi_1, \bar{y}, u)$ iff (F1) holds.*
4. *Consider two HMMs $(P(u), B(u))$ and $(P(u+1), B(u+1))$. Then*
   (a) $T(\pi, y, u+1) \geq_r T(\pi, y, u)$ iff (F3) holds.
   (b) Under (F3′), $T(\pi, y, u+1) \geq_r T(\pi, y, u)$.
5. *Consider two HMMs $(P(u), B)$ and $(P(u+1), B)$. Then*
   (a) ($\underline{\text{F3}}$) is necessary and sufficient for $P'(u+1)\pi \geq_r P'(u)\pi$.
   (b) ($\underline{\text{F3}}'$) is sufficient for $P'(u+1)\pi \geq_r P'(u)\pi$.
   (c) Either ($\underline{\text{F3}}$) or ($\underline{\text{F3}}'$) are sufficient for the optimal filter to satisfy $T(\pi, y, u+1) \geq_r T(\pi, y, u)$ for any $y \in \mathcal{Y}$.

[6] Any square matrix M with nonnegative elements is copositive since $\pi' M \pi$ is always non-negative for any belief π. So a sufficient condition for the copositivity (10.5) is that the individual elements $\frac{1}{2}\left(\gamma_{mn}^{j,u,y} + \gamma_{nm}^{j,u,y}\right) \geq 0$.

6. *Under (F4),* $\sigma(\pi, u+1) \geq_s \sigma(\pi, u)$.
7. *Statement 10.3.1 holds for discrete valued observation space* $\mathcal{Y}$. *All the other statements hold for discrete and continuous-valued* $\mathcal{Y}$.

The proof is the appendix §10.A.1.

Statement 10.3.1 asserts that a TP2 transition matrix is sufficient for a one-step ahead HMM predictor to preserve MLR stochastic dominance with respect to π. As a consequence Statement 10.3.1 holds since applying Bayes' rule to $P'(u)\pi_1$ and $P'(u)\pi_2$, respectively, yields the filtered updates, and Bayes' rule preserves MLR dominance (recall Theorem 10.1.4).

Statement 10.3.1 asserts that the normalization measure of the HMM filter $\sigma(\pi, y, u)$ is monotone increasing in π (with respect to first-order dominance) if the observation kernel is TP2 (F1) and transition matrix is TP2 (F2).

Statement 10.3.1 asserts that the HMM filter $T(\pi, y, u)$ is monotone increasing in the observation y iff (F1) holds. That is, a larger observation yields a larger belief if and only if the observation kernel is TP2.

Finally, Statements 10.3.1, 10.3.1 and 10.3.1 compare the filter update and normalization measures for two different HMMs index by actions u and $u+1$. Statements 10.3.1 and 10.3.1 say that if $(P(u), B(u))$ and $(P(u+1), B(u+1))$ satisfy the specified conditions, then the belief update and normalization term with parameters $(P(u+1), B(u+1))$ dominate those with parameters $(P(u), B(u))$. Statement 10.3.1 gives a similar result for predictors and HMMs with identical observation probabilities.

10.4 Illustrative example

This section gives simple examples to illustrate Theorem 10.3.1. Suppose

$$P(1) = \begin{bmatrix} 0.6 & 0.3 & 0.1 \\ 0.2 & 0.5 & 0.3 \\ 0.1 & 0.3 & 0.6 \end{bmatrix}, P(2) = \begin{bmatrix} 0 & 1 & 0 \\ 1 & 0 & 0 \\ 0 & 0 & 1 \end{bmatrix}, \pi_1 = \begin{bmatrix} 0.2 \\ 0.2 \\ 0.6 \end{bmatrix}, \pi_2 = \begin{bmatrix} 0.3 \\ 0.2 \\ 0.5 \end{bmatrix}.$$

It can be checked that the transition matrix $P(1)$ is TP2 (and so (F2) holds). $P(2)$ is not TP2 since the second-order minor comprised of the (1,1), (1,2), (2,1) and (2,2) elements is -1 (and so strictly negative). Finally, $\pi_1 \geq_r \pi_2$ since the ratio of their elements $[2/3, 1, 6/5]$ is increasing.

Example (i): Statement 10.3.1 says that $P'(1)\pi_1 \geq_r P'(1)\pi_2$ which can be verified since the ratio of elements $[0.8148, 1, 1.1282]'$ is increasing. On the other hand, since $P(2)$ is not TP2, $P'(2)\pi_1$ is not MLR smaller than $P'(2)\pi_2$. The ratio of elements of $P'(2)\pi_1$ with $P'(2)\pi_2$ is $[1, 0.6667, 1.2]$ implying that they are not MLR orderable (since the ratio is neither increasing or decreasing).

Statement 10.3.1 (MLR order is preserved by Bayes' rule) was illustrated numerically in §10.1.3.

Example (ii): To illustrate Statement 10.3.1, suppose $B(1) = P(1)$ so that (F1), (F2) hold. Then

$$\sigma(\pi_1, 1) = [0.2440, 0.3680, 0.3880]', \quad \sigma(\pi_2, 1) = [0.2690, 0.3680, 0.3630]'.$$

Clearly $\sigma(\pi_1, 1) \geq_s \sigma(\pi_2, 1)$.
Example (iii): Regarding Statement 10.3.1, if $B = P(1)$ then $B_1 = \text{diag}(0.6, 0.2.0.1)$, $B_2 = \text{diag}(0.3, 0.5.0.3)$. Then writing $T(\pi, y, u)$ as $T(\pi, y)$,

$$T(\pi_1, y = 1) = [0.5410, 0.2787, 0.1803]', \quad T(\pi_1, y = 2) = [0.1793, 0.4620, 0.3587]'$$

implying that $T(\pi_1, y = 1) \leq_r T(\pi_1, y = 2)$.
Example (iv): Consider a cost vector

$$c = [c(x = 1), c(x = 2), c(x = 3)]' = \begin{bmatrix} 3 & 2 & 1 \end{bmatrix}'.$$

Suppose a random variable x has a prior π and is observed via noisy observations y with observation matrix $B(1)$ above. Then the expected cost after observing y is $c'T(\pi, y)$. It is intuitive that a larger observation y corresponds to a larger state and therefore a smaller expected cost (since the cost vector is decreasing in the state). From Theorem 10.3.1(10.3.1) this indeed is the case since $T(\pi, y) \geq_r T(\pi, \bar{y})$ for $y > \bar{y}$ which implies that $T(\pi, y) \geq_s T(\pi, \bar{y})$ and therefore $c'T(\pi, y)$ is decreasing in y.

10.5 Discussion and examples of Assumptions (F1)–(F4)

Since Assumptions (F1)–(F4) will be used a lot in subsequent chapters, we now discuss their motivation with examples.

Assumption (F1)

(F1) is required for preserving the MLR ordering with respect to observation y of the Bayesian filter update. (F1) is satisfied by numerous continuous and discrete distributions; see any classical detection theory book such as [269]. Since (F1) is equivalent to each row of B being MLR dominated by subsequent rows, any of the examples in §10.1.3 yield TP2 observation kernels. For example, if the i-th row of B is $N(y - x; \mu_i, \sigma^2)$ with $\mu_i < \mu_{i+1}$ then B is TP2. The same logic applies to Exponential, Binomial, Poisson, Geometric, etc. For a discrete distribution example, suppose each sensor obtains measurements y of the state x in quantized Gaussian noise. Define

$$\mathbb{P}(y|x = i) = \frac{\bar{b}_{iy}}{\sum_{y=1}^{Y} \bar{b}_{iy}} \quad \text{where } \bar{b}_{iy} = \frac{1}{\sqrt{2\pi\,\Sigma}} \exp\left(-\frac{1}{2}\frac{(y - g_i)^2}{\Sigma}\right). \tag{10.8}$$

Assume the state levels g_i are increasing in i. Also, $\Sigma \geq 0$ denotes the noise variance and reflects the quality of the measurements. It is easily verified that (A2) holds. As another example, consider equal dimensional observation and state spaces ($X = Y$) and suppose $P(y = i|x = i) = p_i$, $P(y = i - 1|x = i) = P(y = i + 1|x = i) = (1 - p_i)/2$. Then for $1/(\sqrt{2} + 1) \leq p_i \leq 1$, (A2) holds.

Assumption (F2)

(F2) is essential for the Bayesian update $T(\pi, y, u)$ preserving monotonicity with respect to π. TP2 stochastic orders and kernels have been studied in great detail in [169]. (F2) is satisfied by several classes of transition matrices; see [174, 172].

The left-to-right Bakis HMM used in speech recognition [274] has an upper triangular transition matrix which has a TP2 structure under mild conditions, e.g. if the upper triangular elements in row i are $(1 - P_{ii})/(X - i)$ then P is TP2 if $P_{ii} \leq 1/(X - i)$.

As another example, consider a tridiagonal transition probability matrix P with $P_{ij} = 0$ for $j \geq i + 2$ and $j \leq i - 2$. As shown in [121, pp. 99–100], a necessary and sufficient condition for tridiagonal P to be TP2 is that $P_{ii}P_{i+1,i+1} \geq P_{i,i+1}P_{i+1,i}$.

Karlin's classic book [170, p. 154] shows that the matrix exponential of any tridiagonal generator matrix is TP2. That is, $P = \exp(Qt)$ is TP2 if Q is a tridiagonal generator matrix (nonnegative off-diagonal entries and each row adds to 0) and $t > 0$.

The following lemmas give useful properties of TP2 transition matrices.

LEMMA 10.5.1 *If P is TP2, i.e. (F2) holds, then $P_{11} \geq P_{21} \geq P_{31} \geq \cdots \geq P_{X1}$.*

Proof We prove the contrapositive, that is, $P_{i1} < P_{i+1,1}$ implies P is not TP2. Recall from (A3-Ex1), TP2 means that $P_{i1}P_{i+1,j} \geq P_{i+1,1}P_{ij}$ for all j. So assuming $P_{i1} < P_{i+1,1}$, to show that P is not TP2, we need to show that there is at least one j such that $P_{i+1,j} < P_{ij}$. But $P_{i1} < P_{i+1,1}$ implies $\sum_{k \neq 1} P_{i+1,k} < \sum_{k \neq 1} P_{ik}$, which in turn implies that at least for one j, $P_{i+1,j} < P_{ij}$. □

LEMMA 10.5.2 *The product of two TP2 matrices is TP2.*

Proof Let A and B be TP2 matrices and $C = AB$. Then the second-order minor

$$C_{ij}C_{i+1,j+1} - C_{i+1,j}C_{i,j+1} = \sum_k \sum_l A_{ik}A_{i+1,l}B_{kj}B_{l,j+1} - A_{i+1,k}A_{il}B_{kj}B_{l,j+1}$$

$$= \sum_{k<l} \left(A_{ik}A_{i+1,l} - A_{i+1,k}A_{il}\right) B_{kj}B_{l,j+1} + \sum_{k>l} \left(A_{ik}A_{i+1,l} - A_{i+1,k}A_{il}\right) B_{kj}B_{l,j+1}$$

$$= \sum_{k \leq l} \left(A_{ik}A_{i+1,l} - A_{i+1,k}A_{il}\right) \left(B_{kj}B_{l,j+1} - B_{k,j+1}B_{lj}\right) > 0$$

since A and B are TP2. Similarly other second-order minors can be shown to be nonnegative. □

Lemma 10.5.2 lets us construct TP2 matrices by multiplying other TP2 matrices. The lemma is also used in the proof of the main Theorem 10.3.1.

Assumption (F3), (F3′) and optimal prediction

Assumption (F3′) is sufficient condition for the belief due to action $u + 1$ to MLR dominate the belief due to action u, i.e. in the terminology of [241], $u + 1$ yields a more "favorable outcome" than u. In general, the problem of verifying copositivity of a matrix is NP-complete [73]. Assumption (F3′) is a simpler but more restrictive sufficient condition than (F3) to ensure that $\Gamma^{j,u,y}$ in (10.5) is copositive. Here is an example of $(P(1), B(1)) \preceq (P(2), B(2))$ which satisfies (F3′):

$$P(1) = \begin{bmatrix} 0.8000 & 0.1000 & 0.1000 \\ 0.2823 & 0.1804 & 0.5373 \\ 0.1256 & 0.1968 & 0.6776 \end{bmatrix}, B(1) = \begin{bmatrix} 0.8000 & 0.1000 & 0.1000 \\ 0.0341 & 0.3665 & 0.5994 \\ 0.0101 & 0.2841 & 0.7058 \end{bmatrix},$$

$$P(2) = \begin{bmatrix} 0.0188 & 0.1981 & 0.7831 \\ 0.0051 & 0.1102 & 0.8847 \\ 0.0016 & 0.0626 & 0.9358 \end{bmatrix}, B(2) = \begin{bmatrix} 0.0041 & 0.1777 & 0.8182 \\ 0.0025 & 0.1750 & 0.8225 \\ 0.0008 & 0.1290 & 0.8701 \end{bmatrix}.$$

(F3) is necessary and sufficient for the optimal predictor with transition matrix $P(u+1)$ to MLR dominate the optimal predictor with transition matrix $P(u)$. (F3′) is a sufficient condition for (F3′) since it requires all the elements of the matrix to be nonnegative which trivially implies copositivity. We require MLR dominance of the predictor since then by Theorem 10.1.4 MLR dominance of the filter is assured for any observation distribution. (First-order dominance is not closed under Bayesian updates.)

A straightforward sufficient condition for (10.6) to hold is if all rows of $P(u+1)$ MLR dominate the last row of $P(u)$.

Remark: Here are examples of transition matrices $P(u), P(u+1)$ that satisfy (F2) and (F3′), i.e. $P(u) \preceq P(u+1)$. Several larger dimensional examples are in §14.6.

Example 1: $P(u+1) = \begin{bmatrix} 1 & 0 \\ 1 - P_{22}(u+1) & P_{22}(u+1) \end{bmatrix}, P(u) = \begin{bmatrix} 1 & 0 \\ 1 - P_{22}(u) & P_{22}(u) \end{bmatrix}$

where $P_{22}(u+1) \geq P_{22}(u)$.

Example 2: $P(u+1) = \begin{bmatrix} 1 & 0 & 0 \\ 0.5 & 0.3 & 0.2 \\ 0.3 & 0.4 & 0.3 \end{bmatrix}, P(u) = \begin{bmatrix} 1 & 0 & 0 \\ 0.9 & 0.1 & 0 \\ 0.8 & 0.2 & 0 \end{bmatrix}.$

Example 3: $P(u+1) = \begin{bmatrix} 0.2 & 0.8 \\ 0.1 & 0.9 \end{bmatrix}, \quad P(u) = \begin{bmatrix} 0.8 & 0.2 \\ 0.7 & 0.3 \end{bmatrix}.$

Assumption (F4)

This ensures that the normalized measure $\sigma(\pi, u+1)$ first-order stochastically dominates $\sigma(\pi, u)$.

Assumptions (F3′) and (F4) are relaxed versions of Assumptions (c), (e), (f) of [225, Proposition 2] and Assumption (i) of [277, Theorem 5.6] in the stochastic control literature. The assumptions (c), (e), (f) of [225] require that $P(u+1) \underset{\text{TP2}}{\geq} P(u)$ and $B(u+1) \underset{\text{TP2}}{\geq} B(u)$, where $\underset{\text{TP2}}{\geq}$ (TP2 stochastic ordering) is defined in Definition 10.6.2 on page 234, which is impossible for stochastic matrices, unless $P(u) = P(u+1)$, $B(u) = B(u+1)$ or the matrices $P(u), B(u)$ are rank 1 for all u meaning that the observations are non-informative.

10.6 Example: Reduced complexity HMM filtering with stochastic dominance bounds

The main result Theorem 10.3.1 can be exploited to design reduced complexity HMM filtering algorithms with provable sample path bounds. In this section we derive such reduced-complexity algorithms by using Assumptions (F2) and (F3′) with statement 10.3.1 of Theorem 10.3.1 for the transition matrix.

10.6.1 Upper and lower sample path bounds for optimal filter

Consider an HMM with $X \times X$ transition matrix P and observation matrix B with elements $B_{xy} = p(y_k = y|x_k = x)$. The observation space $\mathcal{Y}$ can be discrete or continuous-valued; so that B is either a pmf or a pdf. The HMM filter computes the posterior

$$\pi_{k+1} = T(\pi_k, y_{k+1}; P), \text{ where } T(\pi, y; P) = \frac{B_y P' \pi}{\mathbf{1}' B_y P' \pi}, \; B_y = \text{diag}(B_{1y}, \dots, B_{Xy}). \tag{10.9}$$

The above notation explicitly shows the dependence on the transition matrix P. Due to the matrix-vector multiplication $P'\pi$, the HMM filter involves $O(X^2)$ multiplications and can be excessive for large X.

The main idea of this section is to construct low rank transition matrices $\underline{P}$ and $\bar{P}$ such that the above filtering recursion using these matrices form lower and upper bounds to π_k in the MLR stochastic dominance sense. Since $\underline{P}$ and $\bar{P}$ are low rank (say r), the cost involved in computing these lower and upper bounds to π_k at each time k will be $O(Xr)$ where $r \ll X$.

Since that plan is to compute filtered estimates using $\underline{P}$ and $\bar{P}$ instead of the original transition matrix P, we need additional notation to distinguish between the posteriors and estimates computed using P, $\underline{P}$ and $\bar{P}$. Let

$$\underbrace{\pi_{k+1} = T(\pi_k, y_{k+1}; P)}_{\text{optimal}}, \quad \underbrace{\bar{\pi}_{k+1} = T(\bar{\pi}_k, y_{k+1}; \bar{P})}_{\text{upper bound}}, \quad \underbrace{\underline{\pi}_{k+1} = T(\underline{\pi}_k, y_{k+1}; \underline{P})}_{\text{lower bound}}$$

denote the posterior updated using the optimal filter (10.9) with transition matrices P, $\bar{P}$ and $\underline{P}$, respectively. Assuming that the state levels of the Markov chain are $g = (1, 2, \dots, X)'$, the conditional mean estimates of the underlying state computed using P, $\underline{P}$ and $\bar{P}$, respectively, will be denoted as

$$\hat{x}_k = \mathbb{E}\{x_k|y_{0:k}; P\} = g'\pi_k, \quad \underline{x}_k \stackrel{\text{defn}}{=} \mathbb{E}\{x_k|y_{0:k}; \underline{P}\} = g'\underline{\pi}_k,$$
$$\bar{x}_k \stackrel{\text{defn}}{=} \mathbb{E}\{x_k|y_{0:k}; \bar{P}\} = g'\bar{\pi}_k. \tag{10.10}$$

Also denote the maximum aposteriori (MAP) state estimates computed using $\underline{P}$ and $\bar{P}$ as

$$\hat{x}^{\text{MAP}} \stackrel{\text{defn}}{=} \underset{i}{\text{argmax}}\, \pi_k(i), \quad \underline{x}_k^{\text{MAP}} \stackrel{\text{defn}}{=} \underset{i}{\text{argmax}}\, \underline{\pi}_k(i), \quad \bar{x}_k^{\text{MAP}} \stackrel{\text{defn}}{=} \underset{i}{\text{argmax}}\, \bar{\pi}_k(i). \tag{10.11}$$

The following is the main result of this section (proof in appendix).

THEOREM 10.6.1 (Stochastic dominance sample-path bounds) *Consider the HMM filtering updates $T(\pi, y; P)$, $T(\pi, y; \bar{P})$ and $T(\pi, y; \underline{P})$ where $T(\cdot)$ is defined in (10.9) and P denotes the transition matrix of the HMM. (Recall $\geq_r$ denotes MLR dominance.)*

1. *For any transition matrix P, there exist transition matrices $\underline{P}$ and $\bar{P}$ such that $\underline{P} \preceq P \preceq \bar{P}$ (recall $\preceq$ is the copositive ordering defined in Definition 10.2.3).*

2. *Suppose transition matrices $\underline{P}$ and $\bar{P}$ are constructed such that $\underline{P} \preceq P \preceq \bar{P}$. Then for any observation y and belief $\pi \in \Pi(X)$, the filtering updates satisfy the sandwich result*

$$T(\pi, y; \underline{P}) \leq_r T(\pi, y; P) \leq_r T(\pi, y; \bar{P}).$$

3. *Suppose P is TP2 (Assumption (F2)). Assume the filters $T(\pi, y; P)$, $T(\pi, y; \bar{P})$ and $T(\pi, y; \underline{P})$ are initialized with common prior π_0. Then the posteriors satisfy*

$$\underline{\pi}_k \leq_r \pi_k \leq_r \bar{\pi}_k, \quad \textit{for all time } k = 1, 2, \ldots$$

As a consequence for all time $k = 1, 2, \ldots$,
(a) The conditional mean state estimates defined in (10.10) satisfy $\underline{x}_k \leq \hat{x}_k \leq \bar{x}_k$.
(b) The MAP state estimates defined in (10.11) satisfy $\underline{x}_k^{MAP} \leq \hat{x}_k^{MAP} \leq \bar{x}_k^{MAP}$.

□

Statement 1 says that for any transition matrix P, there always exist transition matrices $\underline{P}$ and $\bar{P}$ such that $\underline{P} \preceq P \preceq \bar{P}$ (copositivity dominance). Actually if P is TP2, then one can trivially construct the tightest rank 1 bounds $\underline{P}$ and $\bar{P}$ as shown below.

Given existence of $\underline{P}$ and $\bar{P}$, the next step is to optimize the choice of $\underline{P}$ and $\bar{P}$. This is discussed in §10.6.2 where nuclear norm minimization is used to construct sparse eigenvalue matrices $\underline{P}$ and $\bar{P}$.

Statement 2 says that for any prior π and observation y, the one-step filtering updates using $\underline{P}$ and $\bar{P}$ constitute lower and upper bounds to the original filtering problem. This is simply a consequence of (F3′) and Statement 10.3.1 of Theorem 10.3.1.

Statement 3 globalizes Statement 2 and asserts that with the additional assumption that the transition matrix P of the original filtering problem is TP2, then the upper and lower bounds hold for all time. Since MLR dominance implies first-order stochastic dominance (see Theorem 10.1.3), the conditional mean estimates satisfy $\underline{x}_k \leq \hat{x}_k \leq \bar{x}_k$.

10.6.2 Convex optimization to compute low rank transition matrices

It only remains to give algorithms for constructing low-rank transition matrices $\underline{P}$ and $\bar{P}$ that yield the lower and upper bounds $\underline{\pi}_k$ and $\bar{\pi}_k$ for the optimal filter posterior π_k. These involve convex optimization [113] for minimizing the nuclear norm. *The computation of $\underline{P}$ and $\bar{P}$ is independent of the observation sample path and so the associated computational cost is irrelevant to the real-time filtering.* Recall that the motivation is as follows: if $\underline{P}$ and $\bar{P}$ have rank r, then the computational cost of the filtering recursion is $O(rX)$ instead of $O(X^2)$ at each time k.

Construction of $\underline{P}$, $\bar{P}$ without rank constraint

Given a TP2 matrix P, the transition matrices $\underline{P}$ and $\bar{P}$ such that $\underline{P} \preceq P \preceq \bar{P}$ can be constructed straightforwardly via an LP solver. With $\underline{P}_1, \underline{P}_2, \ldots, \underline{P}_X$ denoting the rows of $\underline{P}$, a sufficient condition for $\underline{P} \preceq P$ is that $\underline{P}_i \leq_r P_1$ for any row i. Hence, the rows $\underline{P}_i$ satisfy linear constraints with respect to P_1 and can be straightforwardly constructed via an LP solver. A similar construction holds for the upper bound $\bar{P}$, where it is sufficient to construct $\bar{P}_i \geq_r P_X$.

Rank 1 bounds: If P is TP2, an obvious construction is to construct $\underline{P}$ and $\bar{P}$ as follows: choose rows $\underline{P}_i = P_1$ and $\bar{P}_i = P_X$ for $i = 1, 2, \ldots, X$. These yield rank 1 matrices $\underline{P}$ and $\bar{P}$. It is clear from Theorem 10.6.1 that $\underline{P}$ and $\bar{P}$ constructed in this manner are the tightest rank 1 lower and upper bounds.

Nuclear norm minimization algorithms to compute low rank transition matrices $\underline{P}$, $\bar{P}$

This subsection constructs $\underline{P}$ and $\bar{P}$ as low-rank transition matrices subject to the condition $\underline{P} \preceq P \preceq \bar{P}$. To save space we consider the lower bound transition matrix $\underline{P}$; construction of $\bar{P}$ is similar. Consider the following optimization problem for $\underline{P}$:

$$\text{Minimize rank of } X \times X \text{ matrix } \underline{P} \tag{10.12}$$

subject to the constraints $\mathbf{Cons}(\Pi(X), \underline{P}, m)$ for $m = 1, 2, \ldots, X-1$, where for $\epsilon > 0$,

$$\mathbf{Cons}(\Pi(X), \underline{P}, m) \equiv \begin{cases} \Gamma^{(m)} \text{ is copositive on } \Pi(X) & (10.13\text{a}) \\ \|P'\pi - \underline{P}'\pi\|_1 \leq \epsilon \text{ for all } \pi \in \Pi(X) & (10.13\text{b}) \\ \underline{P} \geq 0, \quad \underline{P}\mathbf{1} = \mathbf{1}. & (10.13\text{c}) \end{cases}$$

Recall Γ is defined in (10.6) and (10.13a) is equivalent to $\underline{P} \preceq P$. The constraints $\mathbf{Cons}(\Pi(X), \underline{P}, m)$ are convex in matrix $\underline{P}$, since (10.13a) and (10.13c) are linear in the elements of $\underline{P}$, and (10.13b) is convex (because norms are convex). The constraints (10.13a), (10.13c) are exactly the conditions of Theorem 10.6.1, namely that $\underline{P}$ is a stochastic matrix satisfying $\underline{P} \preceq P$.

The convex constraint (10.13b) is equivalent to $\|\underline{P} - P\|_1 \leq \epsilon$, where $\|\cdot\|_1$ denotes the induced 1-norm for matrices.[7]

To solve the above problem, we proceed in two steps:

1. The objective (10.12) is replaced with the reweighted nuclear norm (see §10.6.2 below).
2. Optimization over the copositive cone (10.13a) is achieved via a sequence of simplicial decompositions (see remark at end of §10.6.2).

Reweighted nuclear norm

Since the rank is a non-convex function of a matrix, direct minimization of the rank (10.12) is computationally intractable. Instead, we follow the approach developed by Boyd and coworkers [113] to minimize the iteratively reweighted nuclear norm. Inspired by Candès and Tao [75], there has been much recent interest in minimizing nuclear norms for constructing matrices with sparse eigenvalue sets or equivalently low rank. Here we compute $\underline{P}, \bar{P}$ by minimizing their nuclear norms subject to copositivity conditions that ensure $\underline{P} \preceq P \preceq \bar{P}$.

Let $\|\cdot\|_*$ denote the nuclear norm, which corresponds to the sum of the singular values of a matrix, The re-weighted nuclear norm minimization proceeds as a *sequence*

[7] The three statements $\|P'\pi - \underline{P}'\pi\|_1 \leq \epsilon$, $\|\underline{P} - P\|_1 \leq \epsilon$ and $\sum_{i=1}^{X} \|(P' - \underline{P}')_{:,i}\|_1 \pi(i) \leq \epsilon$ are all equivalent since $\|\pi\|_1 = 1$ because π is a probability vector (pmf).

of convex optimization problems indexed by $n = 0, 1, \ldots$. Initialize $\underline{P}^{(0)} = I$. For $n = 0, 1, \ldots$, compute $X \times X$ matrix

$$\underline{P}^{(n+1)} = \underset{\underline{P}}{\text{argmin}} \, \|\underline{W}_1^{(n)} \underline{P} \, \underline{W}_2^{(n)}\|_* \tag{10.14}$$

$$\text{subject to: constraints } \mathbf{Cons}(\Pi(X), \underline{P}, m),\ m = 1, \ldots, X-1$$
$$\text{namely, (10.13a), (10.13b), (10.13c).}$$

Notice that at iteration $n + 1$, the previous estimate, $\underline{P}^{(n)}$ appears in the cost function of (10.14) in terms of weighting matrices $\underline{W}_1^{(n)}$, $\underline{W}_2^{(n)}$. These weighting matrices are evaluated iteratively as

$$\underline{W}_1^{(n+1)} = ([\underline{W}_1^{(n)}]^{-1} U \Sigma U^T [\underline{W}_1^{(n)}]^{-1} + \delta I)^{-1/2},$$
$$\underline{W}_2^{(n+1)} = ([\underline{W}_2^{(n)}]^{-1} V \Sigma V^T [\underline{W}_2^{(n)}]^{-1} + \delta I)^{-1/2}. \tag{10.15}$$

Here $\underline{W}_1^{(n)} \underline{P}^{(n)} \underline{W}_2^{(n)} = U \Sigma V^T$ is a reduced singular value decomposition, starting with $\underline{W}_1^{(0)} = \underline{W}_2^{(0)} = I$ and $\underline{P}^0 = P$. Also δ is a small positive constant in the regularization term δI. In numerical examples of §10.6.4, we used YALMIP with MOSEK and CVX to solve the above convex optimization problem.

The intuition behind the reweighting iterations is that as the estimates $\underline{P}^{(n)}$ converge to the limit $\underline{P}^{(\infty)}$, the cost function becomes approximately equal to the rank of $\underline{P}^{(\infty)}$.

Remark: Problem (10.14) is a convex optimization problem in $\underline{P}$. However, one additional issue needs to be resolved: the constraints (10.13a) involve a copositive cone and cannot be solved directly by standard interior point methods. To deal with the copositive constraints (10.13a), one can use the state-of-the-art simplicial decomposition method detailed in [73]; see [189] for details.

10.6.3 Stochastic dominance bounds for multivariate HMMs

This section shows how the bounds in Theorem 10.6.1 can be generalized to multivariate HMMs – the main idea is that MLR dominance is replaced by the multivariate TP2 (totally positive of order 2) stochastic dominance [249, 336, 169]. We consider a highly stylized example which will serve as a reproducible way of constructing large-scale HMMs in numerical examples of §10.6.4.

Consider L independent Markov chains, $x_k^{(l)}$, $l = 1, 2 \ldots, L$ with transition matrices $A^{(l)}$. Define the joint process $x_k = (x_k^{(1)}, \ldots, x_k^{(L)})$. Let $\mathbf{i} = (i_1, \ldots, i_L)$ and $\mathbf{j} = (j_1, \ldots, j_L)$ denote the vector indices where each index $i_l, j_l \in \{1, \ldots, X\}$, $l = 1, \ldots, L$.

Suppose the observation process recorded at a sensor has the conditional probabilities $B_{\mathbf{i},y} = \mathbb{P}(y_k = y | x_k = \mathbf{i})$. Even though the individual Markov chains are independent of each other, since the observation process involves all L Markov chains, computing the filtered estimate of x_k, requires computing and propagating the joint posterior $P(x_k | y_{1:k})$. This is equivalent to HMM filtering of the process x_k with transition matrix $P = A^{(1)} \otimes \cdots \otimes A^{(L)}$ where $\otimes$ denotes Kronecker product. If each process $x^{(l)}$ has S states, then P is an $S^L \times S^L$ matrix and the computational cost of the HMM filter at each time is $O(S^{2L})$ which is excessive for large L.

A naive application of the results of the previous sections will not work, since the MLR ordering does not apply to the multivariate case (in general, mapping the vector index into a scalar index does not yield univariate distributions that are MLR orderable). We use the totally positive (TP2) stochastic order, which is a natural multivariate generalization of the MLR order [249, 336, 169]. Denote the element-wise minimum and maximum

$$\mathbf{i} \wedge \mathbf{j} = [\min(i_1, j_1), \ldots, \min(i_L, j_L)], \; \mathbf{i} \vee \mathbf{j} = [\max(i_1, j_1), \ldots, \max(i_L, j_L)].$$

Denote the L-variate posterior at time k as

$$\pi_k(\mathbf{i}) = \mathbb{P}(x_k^{(1)} = i_1, x_k^{(2)} = i_2, \ldots, x_k^{(L)} = i_L \mid y_{1:k})$$

and let $\Pi(X)$ denote the space of all such L-variate posteriors.

DEFINITION 10.6.2 (TP2 ordering[8]) *Let π_1 and π_2 denote any two L-variate probability mass functions. Then $\pi_1 \underset{TP2}{\geq} \pi_2$ if $\pi_1(\mathbf{i})\pi_2(\mathbf{j}) \leq \pi_1(\mathbf{i} \vee \mathbf{j})\pi_2(\mathbf{i} \wedge \mathbf{j})$.*

If π_1 and π_2 are univariate, then this definition is equivalent to the MLR ordering $\pi_1 \geq_r \pi_2$. Indeed, just like the MLR order, the TP2 order is closed under conditional expectations [169]. Next define $P'\pi$ such that its $\mathbf{j}$-th component is $\sum_{\mathbf{i}} P_{\mathbf{ij}}\pi(\mathbf{i})$. In analogy to Definition 10.2.3, given two transition matrices $\underline{P}$ and P, we say that

$$P \succeq_M \underline{P}, \text{ if } P'\pi \underset{\text{TP2}}{\geq} \underline{P}'\pi \text{ for all } \pi \in \Pi(X). \tag{10.16}$$

The main result regarding filtering of multivariate HMMs is as follows:

THEOREM 10.6.3 *Consider an L-variate HMM where each transition matrix satisfies $A^{(l)} \succeq \underline{A}^{(l)}$ for $l = 1, \ldots, L$ (where $\succeq$ is interpreted as in Definition 10.2.3). Then*

(i) $A^{(1)} \otimes \cdots \otimes A^{(L)} \succeq_M \underline{A}^{(1)} \otimes \cdots \otimes \underline{A}^{(L)}$.
(ii) Theorem 10.6.1 holds for the posterior and state estimates with $\geq_r$ replaced by $\underset{TP2}{\geq}$.

□

We need to qualify statement (ii) of Theorem 10.6.3 since for multivariate HMMs, the conditional mean $\hat{x}_k$ and MAP estimate $\hat{x}_k^{\text{MAP}}$ are L-dimensional vectors. The inequality $\underline{x}_k \leq \hat{x}_k$ of statement (ii) is interpreted as the component-wise partial order on $\mathbb{R}^L$, namely, $\underline{x}_k(l) \leq \hat{x}_k(l)$ for all $l = 1, \ldots, L$. (A similar result applies for the upper bounds.)

10.6.4 Numerical examples

Below we present numerical examples to illustrate the behavior of the reduced complexity HMM filtering algorithms proposed above. To give the reader an easily reproducible

[8] In general the TP2 order is not reflexive. A multivariate distribution π is said to be multivariate TP2 (MTP2) if $\pi \underset{\text{TP2}}{\geq} \pi$ holds, i.e. $\pi(\mathbf{i})\pi(\mathbf{j}) \leq \pi(\mathbf{i} \vee \mathbf{j})\pi(\mathbf{i} \wedge \mathbf{j})$. This definition of reflexivity also applies to stochastic matrices. That is, if $\mathbf{i}, \mathbf{j} \in \{1, \ldots, X\} \times \{1, \ldots, X\}$ are two-dimensional indices then MTP2 and TP2 (Definition 10.2.1) are identical for a stochastic matrix. To show this, note that $P \underset{\text{TP2}}{\geq} P$ means $P_{i_1,i_2}P_{j_1,j_2} \leq P_{\max(i_1,j_1),\max(i_2,j_2)}P_{\min(i_1,j_1),\min(i_2,j_2)}$. This is equivalent to P having all second-order minors nonnegative (Definition 10.2.1).

numerical example of large dimension, we construct a 3125 state Markov chain according to the multivariate HMM construction detailed in §10.6.3. Consider $L = 5$ independent Markov chains $x_k^{(l)}$, $l = 1, \ldots, 5$, each with five states. The observation process is

$$y_k = \sum_{l=1}^{5} x_k^{(l)} + v_k$$

where the observation noise v_k is zero mean i.i.d. Gaussian with variance σ_v^2. Since the observation process involves all five Markov chains, computing the filtered estimate requires propagating the joint posterior. This is equivalent to defining a $5^5 = 3125$-state Markov chain with transition matrix $P = \underbrace{A \otimes \cdots \otimes A}_{5 \text{ times}}$ where $\otimes$ denotes Kronecker product. The optimal HMM filter incurs $5^{10} \approx 10$ million computations at each time step k.

Generating TP2 transition matrix

To illustrate the reduced complexity global sample path bounds developed in Theorem 10.6.1, we consider the case where P is TP2. We used the following approach to generate P: First construct $A = \exp(Qt)$, where Q is a tridiagonal generator matrix (nonnegative off-diagonal entries and each row adds to 0) and $t > 0$. Karlin's book [170, p.154] shows that A is then TP2. Second, as shown in [169], the Kronecker products of A preserve the TP2 property implying that P is TP2.

Using the above procedure, we constructed a 3125×3125 TP2 transition matrix P as follows:

$$Q = \begin{bmatrix} -0.8147 & 0.8147 & 0 & 0 & 0 \\ 0.4529 & -0.5164 & 0.06350 & 0 & \\ 0 & 0.4567 & -0.7729 & 0.3162 & 0 \\ 0 & 0 & 0.0488 & -0.1880 & 0.1392 \\ 0 & 0 & 0 & 0.5469 & -0.5469 \end{bmatrix},$$
$$A = \exp(2Q), \quad P = \underbrace{A \otimes \cdots \otimes A}_{5 \text{ times}}. \tag{10.17}$$

Off-line optimization of lower bound via convex optimization

We used the semidefinite optimization `solvesdp` solver from `MOSEK` with `YALMIP` and `CVX` to solve[9] the convex optimization problem (10.14) for computing the upper and lower bound transition matrices $\underline{P}$ and $\bar{P}$. To estimate the rank of the resulting transition matrices, we consider the costs (10.14), which correspond approximately to the number of singular values larger than δ (defined in (10.15)). The reweighted nuclear norm algorithm is run for five iterations, and the simplicial algorithm is stopped as soon as the cost decreased by less than 0.01.

[9] See web.cvxr.com/cvx/doc/ for a complete documentation of CVX.

Table 10.1 Ranks of lower bound transition matrices $\underline{P}$ each of dimension 3125×3125 obtained as solutions of the nuclear norm minimization problem (10.14) for six different choices of ϵ appearing in constraint (10.13b). Note $\epsilon = 0$ corresponds to $\underline{P} = P$ and $\epsilon = 2$ corresponds to the i.i.d. case.

ϵ	0	0.4	0.8	1.2	1.6	2
r (rank of $\underline{P}$)	3125 ($\underline{P} = P$)	800	232	165	40	1 (i.i.d.)

To save space we present results only for the lower bounds. We computed[10] five different lower bound transition matrices $\underline{P}$ by solving the nuclear norm minimization problem (10.14) for five different choices of $\epsilon \in \{0.4, 0.8, 1.2, 1.6, 2\}$ defined in constraint (10.13b).

Table 10.1 displays the ranks of these five transition matrices $\underline{P}$, and also the rank of P which corresponds to the case $\epsilon = 0$. The low rank property of $\underline{P}$ can be visualized by displaying the singular values. Figure 10.3 displays the singular values of $\underline{P}$ and P. When $\epsilon = 2$, the rank of $\underline{P}$ is 1 and models an i.i.d. chain; $\underline{P}$ then simply comprises of repetitions of the first row of P. As ϵ is made smaller the number of singular values increases. For $\epsilon = 0$, $\underline{P}$ coincides with P.

Performance of lower complexity filters

At each time k, the reduced complexity filter $\underline{\pi}_k = T(\underline{\pi}_{k-1}, y_k; \underline{P})$ incurs computational cost of $O(Xr)$ where $X = 3125$ and r is specified in Table 10.1. For each matrix $\underline{P}$ and noise variances σ_v^2 in the range $(0, 2.25]$ we ran the reduced complexity HMM filter $T(\pi, y; \underline{P})$ for a million iterations and computed the average mean square error of the state estimate. These average mean square error values are displayed in Figure 10.4. As might be intuitively expected, Figure 10.4 shows that the reduced complexity filters yield a mean square error that lies between the i.i.d. approximation ($\epsilon = 2$) and the optimal filter ($\epsilon = 0$). In all cases, as mentioned in Theorem 10.6.1, the estimate $\underline{\pi}_k$ provably lower bounds the true posterior π_k as $\underline{\pi}_k \leq_r \pi_k$ for all time k. Therefore the conditional mean estimates satisfy $\underline{x}_k \leq \hat{x}_k$ for all k.

10.6.5 Discussion: Reduced complexity predictors

If one were interested in constructing reduced complexity HMM predictors (instead of filters), the results in this section are straightforwardly relaxed using first-order dominance $\leq_s$ instead of MLR dominance $\leq_r$ as follows: construct $\underline{P}$ by nuclear norm minimization as in (10.14), where (10.13a) is replaced by the linear constraints $\underline{P}_i \leq_s P_i$,

[10] In practice the following preprocessing is required: $\underline{P}$ computed via the semidefinite optimization solver has several singular values close to zero – this is the consequence of nuclear norm minimization. These small singular values need to be truncated exactly to zero thereby resulting in the computational savings associated with the low rank. How to choose the truncation threshold? We truncated those singular values of $\underline{P}$ to zero so that the resulting matrix $\hat{\underline{P}}$ (after rescaling to a stochastic matrix) satisfies the normalized error bound $\frac{\|\underline{P}'\pi - \hat{\underline{P}}\pi\|_2}{\|\underline{P}\|_2\|\pi\|_2} \leq \frac{\|\hat{\underline{P}} - \underline{P}\|_2}{\|\underline{P}\|_2} \leq 0.01$, i.e., negligible error. The rescaling to a stochastic matrix involved subtracting the minimum element of the matrix (so every element is nonnegative) and then normalizing the rows. This transformation does not affect the rank of the matrix thereby maintaining the low rank. For notational convenience, we continue to use $\underline{P}$ instead of $\hat{\underline{P}}$.

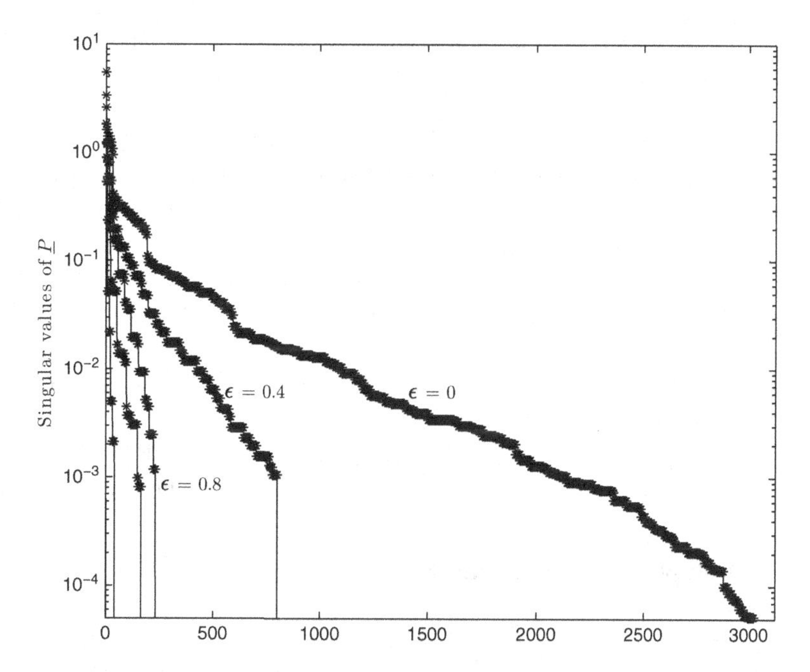

Figure 10.3 Plot of 3125 singular values of P and singular values of five different transition matrices $\underline{P}$ parametrized by ϵ in Table 10.1. The transition matrix P (corresponding to $\epsilon = 0$) of dimension 3125×3125 is specified in (10.17).

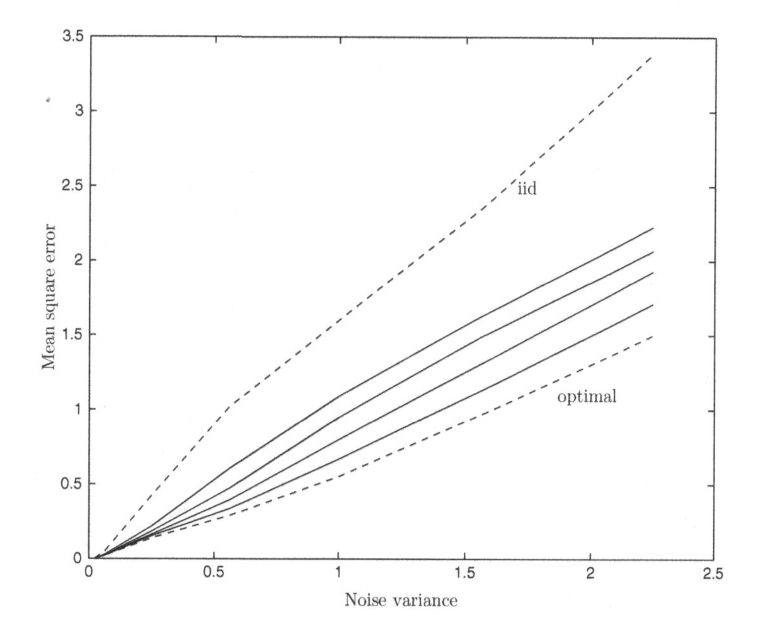

Figure 10.4 Mean square error of lower bound reduced complexity filters computed using five different transition matrices $\underline{P}$ summarized in Table 10.1. The transition matrix P of dimension 3125×3125 is specified in (10.17). The four solid lines (lowest to highest curve) are for $\epsilon = 0.4, 0.8, 1.2, 1.6$. The optimal filter corresponds to $\epsilon = 0$, while the i.i.d. approximation corresponds to $\epsilon = 2$.

on the rows $i = 1,\dots,X$, and (10.13b), (10.13c) hold. Thus the construction of $\underline{P}$ is a standard convex optimization problem and the bound $\underline{P}'\pi \leq_s P'\pi$ holds for the optimal predictor for all $\pi \in \Pi(X)$.

Further, if $\underline{P}$ is chosen so that its rows satisfy the linear constraints $\underline{P}_i \leq_s \underline{P}_{i+1}$, $i = 1,\dots,X-1$, then the following global bound holds for the optimal predictor: $(\underline{P}')^k\pi \leq_s (P')^k\pi$ for all time k and $\pi \in \Pi(X)$. A similar result holds for the upper bounds in terms of $\bar{P}$.

It is instructive to compare this with the filtering case, where we imposed a TP2 condition on P for the global bounds of Theorem 10.6.1(3) to hold wrt $\leq_r$. We could have equivalently imposed a TP2 constraint on $\underline{P}$ and allow P to be arbitrary for the global filtering bounds to hold, however the TP2 constraint is non-convex and so it is difficult to then optimize $\underline{P}$.

Finally, note that the predictor bounds in terms of $\leq_s$ do not hold if a filtering update is performed since $\leq_s$ is not closed with regard to conditional expectations.

10.7 Complements and sources

The books [249, 295] give comprehensive accounts of stochastic dominance. [336] discusses the TP2 stochastic order which is a multivariate generalization of the MLR order. Karlin's book [168] is a classic on totally positive matrices. The classic paper [169] studies multivariate TP2 orders; see also [277].

The material in §10.6 is based on [189]; where additional numerical results are given. Also it is shown in [189] how the reduced complexity bounds on the posterior can be exploited by using a Monte Carlo importance sampling filter. The approach in §10.6 of optimizing the nuclear norm as a surrogate for rank has been studied as a convex optimization problem in several papers [215]. Inspired by the seminal work of Candès and Tao [75], there has been much recent interest in minimizing nuclear norms in the context of sparse matrix completion problems. Algorithms for testing for copositive matrices and copositive programming have been studied recently in [72, 73].

There has been extensive work in signal processing on posterior Cramér–Rao bounds for nonlinear filtering [319]; see also [279] for a textbook treatment. These yield lower bounds to the achievable variance of the conditional mean estimate of the optimal filter. However, such posterior Cramér–Rao bounds do not give constructive algorithms for computing upper and lower bounds for the sample path of the filtered distribution. The sample path bounds proposed in this chapter have the attractive feature that they are guaranteed to yield lower and upper bounds to both hard and soft estimates of the optimal filter.

Appendix 10.A Proofs

10.A.1 Proof of Theorem 10.3.1

First recall from Definition 10.2.1 that $\pi_1 \geq_r \pi_2$ (MLR dominance) is equivalent to saying that the matrix $\begin{bmatrix} \pi_2' \\ \pi_1' \end{bmatrix}$ is TP2. This TP2 notation is more convenient for proofs.

Statement 1a: *If (F2) holds, then $\pi_1 \geq_r \pi_2$ implies $P'\pi_1 \geq_r P'\pi_2$*: Showing that $P'\pi_1 \geq_r P'\pi_2$ is equivalent to showing that $\begin{bmatrix} \pi_2' P \\ \pi_1' P \end{bmatrix}$ is TP2. But $\begin{bmatrix} \pi_2' P \\ \pi_1' P \end{bmatrix} = \begin{bmatrix} \pi_2' \\ \pi_1' \end{bmatrix} P$. Also since $\pi_1 \geq_r \pi_2$, the matrix $\begin{bmatrix} \pi_2' \\ \pi_1' \end{bmatrix}$ is TP2. By (F2), P is TP2. Since the product of TP2 matrices is TP2 (see Lemma 10.5.2), the result holds.

If $\pi_1 \geq_r \pi_2$ implies $P'\pi_1 \geq_r P'\pi_2$ then (F2) holds: Choose $\pi_1 = e_j$ and $\pi_2 = e_i$ where $j > i$, and as usual e_i is the unit vector with 1 in the i-th position. Clearly then $\pi_1 \geq_r \pi_2$. Also $P'e_i$ is the i-th row of P. So $P'e_j \geq_r P'e_i$ implies the j-th row is MLR larger than the i-th row of P. This implies that P is TP2 by definition 10.2.1.

Statement 1b: follows by applying Theorem 10.1.4 to 10.3.1.

Statement 2: Since MLR dominance implies first-order dominance, by (F1), $\sum_{y \geq \bar{y}} B_{x,y}(u)$ is increasing in x. By (F2), $(P_{i,1}, \dots P_{i,X}) \leq_s (P_{j,1}, \dots, P_{j,X})$ for $i \leq j$. Therefore $\sum_j P_{ij}(u) \sum_{y \geq \bar{y}} B_{j,y}(u)$ is increasing in $i \in \mathcal{X}$. Therefore $\pi_1 \geq_r \pi_2$ implies $\sigma(\pi_1, u) \geq_s \sigma(\pi_2, u)$.

Statement 3: Denote $P'(u)\pi_1 = \bar{\pi}$. Then $T(\pi_1, y, u) \geq_r T(\pi_1, \bar{y}, u)$ is equivalent to

$$\left(B_{i,y} B_{i+1,\bar{y}} - B_{i+1,y} B_{i,\bar{y}}\right) \cancel{\underline{\bar{\pi}(i)\bar{\pi}(i+1)}} \leq 0, \quad y > \bar{y}.$$

This is equivalent to B being TP2, namely condition (F1).

Statement 4a: By defintion of MLR dominance, $T(\pi, y, u) \leq_r T(\pi, y, u+1)$ is equivalent to

$$\begin{aligned} &\sum_m \sum_n B_{jy}(u+1) B_{j+1,y}(u) P_{nj}(u+1) \pi_n \pi_m \\ &\quad \leq \sum_m \sum_n B_{jy}(u) B_{j+1,y}(u+1) P_{mj}(u) P_{n,j+1}(u+1) \pi_m \pi_n \end{aligned}$$

and also

$$\begin{aligned} &\sum_m \sum_n B_{jy}(u+1) B_{j+1,y}(u) P_{nj}(u+1) \pi_n \pi_m \\ &\quad \leq \sum_m \sum_n B_{jy}(u) B_{j+1,y}(u+1) P_{nj}(u) P_{m,j+1}(u+1) \pi_m \pi_n \end{aligned}$$

This is equivalent to (F3$'$).

Statement 4b: follows since (F3$'$) is sufficient for (F3) .

The proofs of Statement 10.3.1 and 10.3.1 are very similar to Statement 10.3.1 and omitted.

10.A.2 Proof of Theorem 10.6.1

1. Choose $\underline{P} = \left[e_1, \dots, e_1\right]'$ and $\bar{P} = \left[e_X, \dots, e_X\right]'$ where e_i is the unit X-dimensional vector with 1 in the i-th position. Then clearly, $\underline{P} \succeq P \succeq \bar{P}$. These correspond to extreme points on the space of matrices with respect to copositive dominance.
2. Statement 2 is simply Statement 10.3.1 of Theorem 10.3.1.
3. Suppose $\underline{\pi}_k \leq_r \pi_k$. Then by Statement 2, $T(\underline{\pi}_k, y_{k+1}; \underline{P}) \leq_r T(\underline{\pi}_k, y_{k+1}; P)$. Next since P is TP2, it follows that $\underline{\pi}_k \leq_r \pi_k$ implies $T(\underline{\pi}_k, y_{k+1}; P) \leq_r T(\pi_k, y_{k+1}; P)$.

Combining the two inequalities yields $T(\underline{\pi}_k, y_{k+1}; \underline{P}) \leq_r T(\pi_k, y_{k+1}; P)$, or equivalently $\underline{\pi}_{k+1} \leq_r \pi_{k+1}$. Finally, MLR dominance implies first-order dominance which by Result 9.2.2 implies dominance of means thereby proving 3(a).

To prove 3(b) we need to show that $\underline{\pi} \leq_r \pi$ implies $\arg\max_i \underline{\pi}(i) \leq \arg\max_i \pi(i)$. This is shown by contradiction: Let $i^* = \text{argmax}_i \pi_i$ and $j^* = \text{argmax}_j \underline{\pi}_j$. Suppose $i^* \leq j^*$. Then $\pi \geq_r \underline{\pi}$ implies $\pi(i^*) \leq \frac{\underline{\pi}(i^*)}{\underline{\pi}(j^*)} \pi(j^*)$. Since $\frac{\underline{\pi}(i^*)}{\underline{\pi}(j^*)} \leq 1$, we have $\pi(i^*) \leq \pi(j^*)$ which is a contradiction since i^* is the argmax for $\pi(i)$.

10.A.3 Proof of Theorem 10.6.3

It suffices to show that $\underline{A} \preceq A \implies \underline{A} \otimes \underline{A} \preceq_M A \otimes A$. (The proof for repeated Kronecker products follows by induction.) Consider the TP2 ordering in Definition 10.6.2. The indices $\mathbf{i} = (j, n)$ and $\mathbf{j} = (f, g)$ are each two-dimensional. There are four cases: $(j < f, n < g)$, $(j < f, n > g)$, $(j > f, n < g)$, $(j > f, n > g)$. TP2 dominance for the first and last cases are trivial to establish. We now show TP2 dominance for the third case (the second case follows similarly): choosing $\mathbf{i} = (j, g-1)$ and $\mathbf{j} = (j-1, g)$, it follows that $\underline{A} \otimes \underline{A} \preceq A \otimes A$ is equivalent to

$$\sum_m \sum_l A_{m,g-1} \underline{A}_{l,g} \sum_i \sum_k (A_{ij} \underline{A}_{k,j+1} - A_{i,j+1} \underline{A}_{k,j}) \pi_{im} \pi_{kl} \leq 0$$

So a sufficient condition is that for any nonnegative numbers π_{im} and π_{kl}, $\sum_i \sum_k (A_{ij} \underline{A}_{k,j+1} - A_{i,j+1} \underline{A}_{k,j}) \pi_{im} \pi_{kl} \leq 0$ which is equivalent to $\underline{A} \preceq A$ by Definition 10.2.3.

11 Monotonicity of value function for POMDPs

Contents

This chapter gives sufficient conditions on the POMDP model so that the value function in Bellman's dynamic programming equation is decreasing with respect to the monotone likelihood ratio (MLR) stochastic order. That is, $\pi_1 \geq_r \pi_2$ (in terms of MLR dominance) implies $V(\pi_1) \leq V(\pi_2)$. To prove this result, we will use the structural properties of the optimal filter established in Chapter 10.

Giving conditions for a POMDP to have a monotone value function is useful for several reasons: it serves as an essential step in establishing sufficient conditions for a stopping time POMDPs to have a monotone optimal policy – this is discussed in Chapter 12. For more general POMDPs (discussed in Chapter 14), it allows us to upper and lower bound the optimal policy by judiciously constructed myopic policies. Please see Figure 10.1 for the sequence of chapters on POMDP structural results.

After giving sufficient conditions for a monotone value function, this chapter also gives two examples of POMDPs to illustrate the usefulness of this result:

- *Example 1: Monotone optimal policy for two-state POMDP*: §11.3 gives sufficient conditions for a two-state POMDP to have a monotone optimal policy. The optimal policy is characterized by at most $U - 1$ threshold belief states (where U denotes the number of possible actions). One only needs to compute (estimate) these $U - 1$ threshold belief states in order to determine the optimal policy. This is considerably easier than solving Bellman's equation. Also real-time implementation of a controller with a monotone policy is simple; only the threshold belief states need to be stored in a lookup table. Figure 11.1 illustrates a monotone policy for a two-state POMDP with $U = 3$.

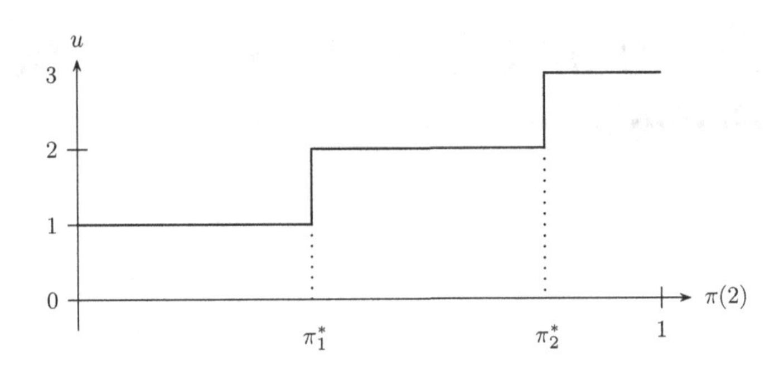

Figure 11.1 §11.3 gives sufficient conditions for a two-state POMDP to have a monotone optimal policy. The figure illustrates such a monotone policy for $\mathcal{U} = 3$. The optimal policy is completely determined by the threshold belief states π_1^* and π_2^*.

- *Example 2: POMDP Multi-armed bandits and opportunistic scheduling*: §11.4 discusses how monotone value functions can be used to solve POMDP multi-armed bandit problems efficiently. It is shown that for such problems, the optimal strategy is "opportunistic": choose the bandit with the largest belief state in terms of MLR order.

11.1 Model and assumptions

Consider a discrete-time, infinite horizon discounted cost[1] POMDP which was formulated in §7.6. The state space for the underlying Markov chain is $\mathcal{X} = \{1, 2, \dots, X\}$, the action space is $\mathcal{U} = \{1, 2, \dots, U\}$ and the belief space is the unit $X - 1$ dimensional unit simplex

$$\Pi(X) = \left\{\pi \in \mathbb{R}^X : \mathbf{1}'\pi = 1, \quad 0 \le \pi(i) \le 1, \; i \in \mathcal{X} = \{1, 2, \dots, X\}\right\}.$$

For stationary policy $\mu : \Pi(X) \to \mathcal{U}$, initial belief $\pi_0 \in \Pi(X)$, discount factor $\rho \in [0, 1)$, the discounted cost expressed in terms of the belief is (see (7.25))

$$J_\mu(\pi_0) = \mathbb{E}_\mu \left\{ \sum_{k=0}^{\infty} \rho^k C(\pi_k, \mu(\pi_k)) \right\}. \tag{11.1}$$

Here $C(\pi, u)$ is the cost accrued at each stage and is not necessarily linear in π. Recall that POMDPs with nonlinear costs $C(\pi, u)$ were discussed in §8.4 in the context of controlled sensing.

The belief evolves according to the HMM filter $\pi_{k+1} = T(\pi_k, y_{k+1}, u_k)$ where

$$T(\pi, y, u) = \frac{B_y(u)P'(u)\pi}{\sigma(\pi, y, u)}, \quad \sigma(\pi, y, u) = \mathbf{1}'_X B_y(u)P'(u)\pi,$$
$$B_y(u) = \text{diag}(B_{1y}(u), \cdots, B_{Xy}(u)), \quad \text{where } B_{xy}(u) = p(y|x, u). \tag{11.2}$$

[1] This chapter considers discounted cost POMDPs for notational convenience to avoid denoting the time dependencies of parameters and policies. The main result of this chapter, namely Theorem 11.2.1 also holds for finite horizon POMDPs providing conditions (C), (F1) and (F2) hold at each time instant for the time dependent cost, observation matrix and transition matrix.

Throughout this chapter $y \in \mathcal{Y}$ can be discrete-valued in which case B_{xy} in (11.2) is a pmf or continuous-valued in which case B_{xy} is a pdf.

The optimal stationary policy $\mu^* : \Pi(X) \rightarrow \mathcal{U}$ such that $J_{\mu^*}(\pi_0) \leq J_\mu(\pi_0)$ for all $\pi_0 \in \Pi(X)$ satisfies Bellman's dynamic programming equation (7.27)

$$\mu^*(\pi) = \operatorname*{argmin}_{u \in \mathcal{U}} Q(\pi, u), \quad J_{\mu^*}(\pi_0) = V(\pi_0) \tag{11.3}$$
$$V(\pi) = \min_{u \in \mathcal{U}} Q(\pi, u), \quad Q(\pi, u) = C(\pi, u) + \rho \sum_{y \in Y} V\left(T\left(\pi, y, u\right)\right) \sigma\left(\pi, y, u\right).$$

Assumptions

(C) The cost $C(\pi, u)$ is first-order stochastically decreasing with respect to π for each action $u \in \{1, 2, \ldots, U\}$. That is $\pi_1 \geq_s \pi_2$ implies $C(\pi_1, u) \leq C(\pi_2, u)$.

For linear costs $C(\pi, u) = c_u' \pi$, (C) is equivalent to the condition:

The instantaneous cost $c(x, u)$ is decreasing in x for each u.

(F1) The observation probability kernel $B(u)$ is TP2 for each action $u \in \{1, 2, \ldots, U\}$.

(F2) The transition matrix $P(u)$ is TP2 for each action $u \in \{1, 2, \ldots, U\}$.

Recall that assumptions (F1) and (F2) were discussed in Chapter 10; see (F1), (F2) on page 225. (F1) and (F2) are required for the Bayesian filter $T(\pi, y.u)$ to be monotone increasing with observation y and π with respect to the MLR order. This is a key step in showing $V(\pi)$ is MLR decreasing in π.

Sufficient conditions for (C)

We pause briefly to discuss Assumption (C), particularly in the context of nonlinear costs that arise in controlled sensing (discussed in §8.4).

For linear costs $C(\pi, u) = c_u' \pi$, obviously the elements of c_u decreasing is necessary and sufficient for $C(\pi, u)$ to be decreasing with respect to $\geq_s$.

For nonlinear costs, we can give the following sufficient condition for $C(\pi, u)$ to be decreasing in π with respect to first-order stochastic dominance. Consider the subset of $\mathbb{R}^X_+$ defined as $\Delta = \{\delta : 1 = \delta(1) \geq \delta(2) \cdots \geq \delta(X)\}$. Define the $X \times X$ matrix

$$\Psi = \begin{bmatrix} 1 & -1 & 0 & \cdots & 0 \\ 0 & 1 & -1 & \cdots & 0 \\ \vdots & \vdots & \vdots & \ddots & \vdots \\ 0 & 0 & 0 & \cdots & 1 \end{bmatrix}. \tag{11.4}$$

Clearly every $\pi \in \Pi(X)$ can be expressed as $\pi = \Psi \delta$ where $\delta \in \Delta$. Consider two beliefs $\pi_1 = \Psi \delta_1$ and $\pi_2 = \Psi \delta_2$ such that $\pi_1 \geq_s \pi_2$. The equivalent partial order induced on δ_1 and δ_2 is: $\delta_1 \succeq \delta_1$ where $\succeq$ is the componentwise partial order on $\mathbb{R}^X$.

LEMMA 11.1.1 *Consider a nonlinear cost $C(\pi_2, u)$ that is differentiable in π.*

1. *For $\pi_1 \geq_s \pi_2$, a sufficient condition for $C(\pi_1, u) \leq C(\pi_2, u)$ is $\frac{d}{d\delta} C(\Psi \delta) \leq 0$ element wise.*

2. *Consider the special case of a quadratic cost $C(\pi, u) = \phi_u'\pi - \alpha(h'\pi)^2$ where α is a nonnegative constant, $\phi_u, h \in \mathbb{R}^X$ with elements ϕ_{iu}, h_i, $i = 1, \ldots, X$. Assume h a vector of nonnegative elements that are either monotone increasing or decreasing. Then a sufficient condition for $C(\pi, u)$ to be first order decreasing in π is*

$$\phi_i - \phi_{i+1} \geq 2\alpha h_1(h_i - h_{i+1}). \tag{11.5}$$

Recall from §8.4 that POMDPs with quadratic costs arise in controlled sensing. Lemma 11.1.1 gives sufficient conditions for such costs to be decreasing with respect to first-order stochastic dominance.[2]

Proof It is sufficient to show under (C) that $C(\pi, u)$ is $\geq_s$ decreasing in $\pi \in \Pi(X), \forall u \in \mathcal{U}$. Since $C(\pi, u) \geq_s$ decreasing on $\pi \in \Pi(X)$ is equivalent to $C(\delta, u)$ $\succeq$ decreasing on $\delta \in \Delta$, a sufficient condition is $\dfrac{\partial C(\pi, u)}{\partial \delta(i)} \leq 0, i = 2, \cdots, X$. Evaluating this yields

$$\phi_{iu} - \phi_{i+1,u} \geq 2\alpha h'(h_i - h_{i+1}) \tag{11.6}$$

If h_i is either monotone increasing or decreasing in i, then a sufficient condition for (11.6) is $\phi_{iu} - \phi_{i+1,u} \geq 2\alpha h_1(h_i - h_{i+1})$.

To summarize, this section has described three important assumptions on the POMDP model that will be used in the main result to follow. □

11.2 Main result: monotone value function

The following is the main result of this chapter.

THEOREM 11.2.1 *Consider the infinite horizon discounted cost POMDP (11.1), (11.2) with continuous or discrete-valued observations. Then under (C), (F1), (F2), $Q(\pi, u)$ is MLR decreasing in π. As a result, the value function $V(\pi)$ in Bellman's equation (11.3) is MLR decreasing in π. That is, $\pi_1 \geq_r \pi_2$ implies that $V(\pi_1) \leq V(\pi_2)$.*

Proof The proof is by mathematical induction on the value iteration algorithm and makes extensive use of the structural properties of the HMM filter developed in Theorem 10.3.1. Recall from (7.29) that the value iteration algorithm proceeds as follows: Initialize $V_0(\pi) = 0$ and for iterations $n = 1, 2, \ldots,$

$$V_n(\pi) = \min_{u \in \mathcal{U}} Q_n(\pi, u), \quad Q_n(\pi, u) = C(\pi, u) + \rho \sum_{y \in Y} V_{n-1}\left(T(\pi, y, u)\right) \sigma(\pi, y, u).$$

Assume that $V_{n-1}(\pi)$ is MLR decreasing in π by the induction hypothesis. Under (F1), Theorem 10.3.1(10.3.1) says that $T(\pi, y, u)$ is MLR increasing in y. As a result, $V_{n-1}\left(T(\pi, y, u)\right)$ is decreasing in y. Under (F1), (F2), Theorem 10.3.1(10.3.1) says

$$\pi \geq_r \bar{\pi} \implies \sigma(\pi, u) \geq_s \sigma(\bar{\pi}, u), \tag{11.7}$$

[2] Note that $C(\pi, u)$ first order increasing in π implies that $C(\pi, u)$ is MLR increasing in π, since MLR dominance implies first-order dominance.

where $\geq_s$ denotes first order dominance. Next, $V_{n-1}(T(\pi, y, u))$ decreasing in y and the first-order dominance (11.7) implies using Theorem 9.2.2 that

$$\pi \geq_r \bar{\pi} \implies \sum_y V_{n-1}(T(\pi, y, u))\,\sigma(\pi, y, u) \leq \sum_y V_{n-1}(T(\pi, y, u))\,\sigma(\bar{\pi}, y, u). \tag{11.8}$$

From Theorem 10.3.1(10.3.1), it follows that under (F2),

$$\pi \geq_r \bar{\pi} \implies T(\pi, y, u) \geq_r T(\bar{\pi}, y, u).$$

Using the induction hypothesis that $V_{n-1}(\pi)$ is MLR decreasing in π implies

$$\pi \geq_r \bar{\pi} \implies V_{n-1}(T(\pi, y, u)) \leq V_{n-1}(T(\bar{\pi}, y, u)).$$

which in turn implies

$$\pi \geq_r \bar{\pi} \implies \sum_y V_{n-1}(T(\pi, y, u))\,\sigma(\bar{\pi}, y, u) \leq \sum_y V_{n-1}(T(\bar{\pi}, y, u))\,\sigma(\bar{\pi}, y, u). \tag{11.9}$$

Combining (11.8), (11.9), it follows that

$$\pi \geq_r \bar{\pi} \implies \sum_y V_{n-1}(T(\pi, y, u))\,\sigma(\pi, y, u) \leq \sum_y V_{n-1}(T(\bar{\pi}, y, u))\,\sigma(\bar{\pi}, y, u). \tag{11.10}$$

Finally, under (C), $C(\pi, u)$ is MLR decreasing (see Footnote 2)

$$\pi \geq_r \bar{\pi} \implies C(\pi, u) \leq C(\bar{\pi}, u). \tag{11.11}$$

Since the sum of decreasing functions is decreasing, it follows that

$$\begin{aligned} \pi \geq_r \bar{\pi} \implies & C(\pi, u) + \sum_y V_{n-1}(T(\pi, y, u))\,\sigma(\pi, y, u) \\ & \leq C(\bar{\pi}, u) + \sum_y V_{n-1}(T(\bar{\pi}, y, u))\,\sigma(\bar{\pi}, y, u) \end{aligned}$$

which is equivalent to $Q_n(\pi, u) \leq Q_n(\bar{\pi}, u)$. Therefore $Q_n(\pi, u)$ is MLR decreasing in π. Since the minimum of decreasing functions is decreasing, $V_n(\pi) = \min_u Q_n(\pi, u)$ is MLR decreasing in π. Finally, since V_n converges uniformly to V, it follows that $V(\pi)$ is also MLR decreasing. □

To summarize, although value iteration is not useful from a computational point of view for POMDPs, we have exploited its structure of prove the monotonicity of the value function. In the next two chapters, several examples will be given that exploit the monotone structure of the value function of a POMDP.

11.3 Example 1: Monotone policies for 2-state POMDPs

This section gives sufficient conditions for the optimal policy $\mu^*(\pi)$ to be monotone increasing in π when the underlying Markov chain has $X = 2$ states (see Figure 11.1). For $X = 2$, since π is a two-dimensional probability mass function with $\pi(1)+\pi(2) = 1$, it suffices to order the beliefs in terms of the second component $\pi(2)$ which lies in the interval $[0, 1]$.

Consider a discounted cost POMDP $(\mathcal{X}, \mathcal{U}, \mathcal{Y}, P(u), B(u), c(u), \rho)$ where state space $\mathcal{X} = \{1, 2\}$, action space $\mathcal{U} = \{1, 2, \ldots, U\}$, observation space $\mathcal{Y}$ can be continuous or discrete, and $\rho \in [0, 1)$. The main assumptions are as follows:

(C) $c(x, u)$ is decreasing in $x \in \{1, 2\}$ for each $u \in \mathcal{U}$.
(F1) B is totally positive of order 2 (TP2).
(F2) $P(u)$ is totally positive of order 2 (TP2).
($\underline{\text{F3}}'$) $P_{12}(u+1) - P_{12}(u) \leq P_{22}(u+1) - P_{22}(u)$ (tail-sum supermodularity).
(S) The costs are submodular: $c(1, u+1) - c(1, u) \geq c(2, u+1) - c(2, u)$.

Recall (C) on page 243 and (F1), (F2) on page 225. The main additional assumption above is the submodularity assumption (S). Apart from (F1), the above conditions are identical to the fully observed MDP case considered in Theorem 9.3.1 on page 209. Indeed (A2) and (A4) in Theorem 9.3.1 are equivalent to (F2) and ($\underline{\text{F3}}'$), respectively, for $X = 2$.

THEOREM 11.3.1 *Consider a POMDP with an underlying $X = 2$ state Markov chain. Under (C), (F1), (F2), ($\underline{\text{F3}}'$), (S), the optimal policy $\mu^*(\pi)$ is increasing in π. Thus $\mu^*(\pi(2))$ has the following finite dimensional characterization: there exist $U + 1$ thresholds (real numbers) $0 = \pi_0^* \leq \pi_1^* \leq \cdots \leq \pi_U^* \leq 1$ such that*

$$\mu^*(\pi) = \sum_{u \in \mathcal{U}} u \, I\left(\pi(2) \in (\pi_{u-1}^*, \pi_u^*]\right).$$

The theorem also applies to finite horizon problems. Then the optimal policy μ_k^* at each time k has the above structure.

The proof is in the appendix. It exploits the fact that the value function $V(\pi)$ is decreasing in π (Theorem 11.2.1 on page 244) and is concave to show that $Q(\pi, u)$ is submodular. That is

$$Q(\pi, u) - Q(\pi, \bar{u}) - Q(\bar{\pi}, u) + Q(\bar{\pi}, \bar{u}) \leq 0, \quad u > \bar{u}, \; \pi \geq_r \bar{\pi}. \tag{11.12}$$

where $Q(\pi, u)$ is defined in Bellman's equation (11.3). Recall that for $X = 2$, $\geq_s$ (first-order dominance) and $\geq_r$ (MLR dominance) coincide, implying that $\pi \geq_r \bar{\pi} \iff \pi \geq_s \bar{\pi} \iff \pi(2) \geq \bar{\pi}(2)$. As a result, for $X = 2$, submodularity of $Q(\pi, u)$ needs to be established with respect to $(\pi(2), u)$ where $\pi(2)$ is a scalar in the interval $[0, 1]$. Hence the same simplified definition of submodularity used for a fully observed MDP (Theorem 9.1.1 on page 206) can be used.

Summary: We have given sufficient conditions for a 2 state POMDP to have a monotone (threshold) optimal policy as illustrated in Figure 11.1. The threshold values can then be estimated via simulation-based policy gradient algorithm such as the SPSA Algorithm 17 on page 350; see also §12.4.2. Theorem 11.3.1 only holds for $X = 2$ states and does not generalize to $X \geq 3$. For $X \geq 3$, determining sufficient conditions for submodularity (11.12) to hold is an open problem. In Chapter 14, we will instead construct judicious myopic bounds for $X \geq 3$.

11.4 Example 2: POMDP multi-armed bandits structural results

In this section, we show how the monotone value function result of Theorem 11.2.1 facilitates solving POMDP multi-armed bandit problems efficiently.

The multi-armed bandit problem is a dynamic stochastic scheduling problem for optimizing in a sequential manner the allocation effort between a number of competing projects. Numerous applications of finite state Markov chain multi-armed bandit problems appear in the operations research and stochastic control literature: see [125], [337] for examples in job scheduling and resource allocation for manufacturing systems. The reason why multi-armed bandit problems are interesting is because their structure implies that the optimal policy can be found by a so-called Gittins index rule [125, 283]: At each time instant, the optimal action is to choose the process with the highest Gittins index, where the Gittins index of each process is a function of the state of that process. So the problem decouples into solving individual control problems for each process.

This section considers multi-armed bandit problems where the finite state Markov chain is not directly observed – instead the observations noisy measurements of the unobserved Markov chain. Such POMDP multi-armed bandits are a useful model in stochastic scheduling.

11.4.1 POMDP multi-armed bandit model

The POMDP multi-armed bandit has the following model: consider L independent projects $l = 1, \ldots, L$. Assume for convenience each project l has the same finite state space $\mathcal{X} = \{1, 2, \ldots, X\}$. Let $x_k^{(l)}$ denote the state of project l at discrete time $k = 0, 1 \ldots,$. At each time instant k only one of these projects can be worked on. The setup is as follows:

- If project l is worked on at time k:
  1. An instantaneous nonnegative reward $\rho^k\, r(x_k^{(l)})$ is accrued where $0 \leq \rho < 1$ denotes the discount factor.
  2. The state $x_k^{(l)}$ evolves according to an X-state homogeneous Markov chain with transition probability matrix P.
  3. The state of the active project l is observed via noisy measurements $y_{k+1}^{(l)} \in \mathcal{Y} = \{1, 2, \ldots, Y\}$ of the active project state $x_{k+1}^{(l)}$ with observation probability $B_{xy} = \mathbb{P}(y^{(l)} = y | x^{(l)} = x)$.
- The states of all the other $(L - 1)$ idle projects are unaffected, i.e. $x_{k+1}^{(l)} = x_k^{(l)}$, if project l is idle at time k. No observations are obtained for idle projects.

For notational convenience we assume all the projects have the same reward functions, transition and observation probabilities and state spaces. So the reward $r(x^{(l)}, l)$ is denoted as $r(x^{(l)})$, etc. All projects are initialized with $x_0^{(l)} \sim \pi_0^{(l)}$ where $\pi_0^{(l)}$ are specified initial distributions for $l = 1, \ldots, L$. Denote $\pi_0 = (\pi_0^{(1)}, \ldots, \pi_0^{(L)})$.

Let $u_k \in \{1, \ldots, L\}$ denote which project is worked on at time k. So $x_{k+1}^{(u_k)}$ is the state of the active project at time $k + 1$. Denote the history at time k as

$$\mathcal{I}_0 = \pi_0, \quad \mathcal{I}_k = \{\pi_0, y_1^{(u_0)}, \ldots, y_k^{(u_{k-1})}, u_0, \ldots, u_{k-1}\}.$$

Then the project at time k is chosen according to $u_k = \mu(\mathcal{I}_k)$, where the policy denoted as μ belongs to the class of stationary policies. The cumulative expected discounted reward over an infinite time horizon is given by

$$J_\mu(\pi) = \mathbb{E}_\mu\Big\{\sum_{k=0}^{\infty} \rho^k r\left(x_k^{(u_k)}\right) | \pi_0 = \pi\Big\}, \quad u_k = \mu(\mathcal{I}_k). \tag{11.13}$$

The aim is to determine the optimal stationary policy $\mu^*(\pi) = \arg\max_\mu J_\mu(\pi)$ *which yields the maximum reward in (11.13).*

Note that we have formulated the problem in terms of rewards rather than costs since typically the formulation involves maximizing rewards of active projects. Of course the formulation is equivalent to minimizing a cost.

At first sight, the POMDP (11.13) seems intractable since the equivalent state space dimension is X^L. The multi-armed bandit structure yields a remarkable simplification – the problem can be solved by considering L individual POMDPs each of dimension X. Actually with the structural result below, one only needs to evaluate the belief state for the L individual HMMs and choose the largest belief at each time (with respect to the MLR order).

11.4.2 Belief state formulation

We now formulate the POMDP multi-armed bandit in terms of the belief state. For each project l, denote by $\pi_k^{(l)}$ the belief at time k where

$$\pi_k^{(l)}(i) = \mathbb{P}(x_k^{(l)} = i|\mathcal{I}_k)$$

The POMDP multi-armed bandit problem can be viewed as the following scheduling problem: consider P parallel HMM state estimation filters, one for each project. The project l is active, an observation $y_{k+1}^{(l)}$ is obtained and the belief $\pi_{k+1}^{(l)}$ is computed recursively by the HMM state filter

$$\pi_{k+1}^{(l)} = T(\pi_k^{(l)}, y_{k+1}^{(l)}) \qquad \text{if project } l \text{ is worked on at time } k \tag{11.14}$$

$$\text{where } T(\pi^{(l)}, y^{(l)}) = \frac{B_{y^{(l)}} P' \pi^{(l)}}{\sigma(\pi^{(l)}, y^{(l)})}, \quad \sigma(x^{(l)}, y^{(l)}) = \mathbf{1}' B_{y^{(l)}} P' \pi^{(l)}.$$

In (11.14) $B_{y^{(l)}} = \operatorname{diag}\big(\mathbb{P}(y^{(l)}|x^{(l)} = 1), \ldots, \mathbb{P}(y^{(l)}|x^{(l)} = X)\big)$.

The beliefs of the other $L-1$ projects remain unaffected, i.e.

$$\pi_{k+1}^{(q)} = \pi_k^{(q)} \qquad \text{if project } q \text{ is not worked on,} \quad q \in \{1, \ldots, L\},\ q \neq l \tag{11.15}$$

Note that each belief $\pi^{(l)}$ lives in the unit simplex $\Pi(X)$.

Let r denote the X dimensional reward vector $[r(x_k^{(l)} = 1), \ldots, r(x_k^{(l)} = X)]'$. In terms of the belief state, the reward functional (11.13) can be re-written as

$$J_\mu(\pi) = \mathbb{E}\Big\{\sum_{k=0}^{\infty} \rho^k r' \pi_k^{(u_k)} \mid (\pi_0^{(1)}, \ldots, \pi_0^{(L)}) = \pi\Big\}, \quad u_k = \mu(\pi_k^{(1)}, \ldots, \pi_k^{(L)}). \tag{11.16}$$

The aim is to compute the optimal policy $\mu^*(\pi) = \arg\max_\mu J_\mu(\pi)$.

11.4.3 Gittins index rule

Define $\bar{M} \stackrel{\text{defn}}{=} \max_i r(i)/(1-\rho)$ and let M denote a scalar in the interval $[0, \bar{M}]$.

It is known that the optimal policy of a multi-armed bandit has an *indexable rule* [337]. Translated to the POMDP multi-armed bandit the result reads:

THEOREM 11.4.1 (Gittins index) *Consider the POMDP multi-armed bandit problem comprising L projects. For each project l there is a function* $\gamma(\pi_k^{(l)})$ *called the* Gittins index, *which is only a function of the parameters of project l and the information state* $\pi_k^{(l)}$, *whereby the optimal scheduling policy at time k is to work on the project with the largest Gittins index:*

$$\mu^*(\pi_k^{(1)}, \pi_k^{(2)}, \ldots, \pi_k^{(L)}) = \max_{l \in \{1,\ldots,L\}} \left\{ \gamma(\pi_k^{(l)}) \right\}. \tag{11.17}$$

The Gittins index of project l with belief $\pi^{(l)}$ *is*

$$\gamma(\pi^{(l)}) = \min\{M : V(\pi^{(l)}, M) = M\} \tag{11.18}$$

where $V(\pi^{(l)}, M)$ *satisfies Bellman's equation*

$$V(\pi^{(l)}, M) = \max\left\{ r'\pi^{(l)} + \rho \sum_{y=1}^{Y} V\left(T(\pi^{(l)}, y), M\right) \sigma(\pi^{(l)}, y),\ M \right\}. \tag{11.19}$$

Theorem 11.4.1 says that the optimal policy is "greedy": at each time choose the project with the largest Gittins index; and the Gittins index for each project can be obtained by solving a dynamic programming equation for that project. Theorem 11.4.1 is well known in the multi-armed bandit literature [125] and will not be proved here.

Bellman's equation (11.19) can be approximated over any finite horizon N via the value iteration algorithm. Since as described in Chapter 7, the value function of a POMDP at each iteration has a finite dimensional characterization, the Gittins index can be computed explicitly for any finite horizon using any of the exact POMDP algorithms in Chapter 7. Moreover, the error bounds for value iteration for horizon N (compared to infinite horizon) of Theorem 7.6.3 directly translate to error bounds in determining the Gittins index. However, for large dimensions, solving each individual POMDP to compute the Gittins index is computationally intractable.

11.4.4 Structural result: characterization of monotone Gittins index

Our focus below is to show how the monotone value function result of Theorem 11.2.1 facilitates solving (11.17), (11.18) efficiently. We show that under reasonable conditions on the rewards, transition matrix and observation probabilities, the Gittins index is monotone increasing in the belief (with respect to the MLR order). This means that if the information states of the L processes at a given time instant are MLR comparable, the optimal policy is to pick the process with the largest belief. This is straightforward to implement and makes the solution practically useful.

Since we are dealing rewards rather than costs, we say that assumption (C) holds if $r(i)$ is increasing in $i \in \mathcal{X}$. (This corresponds to the cost decreasing in i.)

THEOREM 11.4.2 *Consider the POMDP multi-armed bandit where all the L projects have identical transition and observation matrices and reward vectors. Suppose assumptions (C), (F1) and (F2) on page 225 hold for each project. Then the Gittins index $\gamma(\pi)$ is MLR increasing in π. Therefore, if the beliefs $\pi_k^{(l)}$ of the L projects are MLR comparable, then the optimal policy μ^* defined in (11.17) is opportunistic:*

$$u_k = \mu^*(\pi_k^{(1)}, \ldots, \pi_k^{(L)}) = \underset{l \in \{1,\ldots,L\}}{\operatorname{argmax}} \, \pi_k^{(l)}. \tag{11.20}$$

Proof First using exactly the same proof as Theorem 11.2.1, it follows that $V(\pi, M)$ is MLR increasing in π.

Given that the value function $V(\pi, M)$ is MLR increasing in π, we can now characterize the Gittins index. Recall from (11.18) that $\gamma(\pi) = \min\{M : V(\pi, M) - M = 0\}$. Suppose $\pi^{(1)} \geq_r \pi^{(2)}$. This implies $V(\pi^{(1)}, M) \geq V(\pi^{(2)}, M)$ for all M. So $V(\pi^{(1)}, \gamma(\pi^{(2)})) - \gamma(\pi^{(2)}) \geq V(\pi^{(2)}, \gamma(\pi^{(2)})) - \gamma(\pi^{(2)}) = 0$. Since $V(\pi, M) - M$ is decreasing in M (this is seen by subtracting M from both sides of (11.19)), it follows from the previous inequality that the point $\min\{M : V(\pi^{(1)}, M) - M = 0\} > \min\{M : V(\pi^{(2)}, M) - M = 0\}$. So $\gamma(\pi^{(1)}) \geq \gamma(\pi^{(2)})$. □

Discussion

It is instructive to compare (11.17) with (11.20). Theorem 11.4.2 says that instead of choosing the project with the largest Gittins index, it suffices to choose the project with the largest MLR belief (providing the beliefs of the L projects are MLR comparable). In other words, the optimal policy is *opportunistic* (other terms used are "greedy" or "myopic") with respect to the beliefs ranked by MLR order. The resulting optimal policy is trivial to implement and makes the solution practically useful. There is no need to compute the Gittins index.

The following examples yield trajectories of belief states which are MLR comparable across the L projects. As a result, under (C), (F1), (F2), the optimal policy is opportunistic and completely specified by Theorem 11.4.2.

Example 1: If $X = 2$, then all beliefs are MLR comparable.

Example 2: Suppose B is a bi-diagonal matrix. Then if $\pi_0^{(l)}$ is a unit indicator vector, then all subsequent beliefs are MLR comparable (since all beliefs comprise of two consecutive nonzero elements and the rest are zero elements).

Example 3: Suppose $B_{iy} = 1/Y$ for all i, y. Suppose all processes have same initial belief π_0 and pick P such that either $P'\pi_0 \geq_r \pi_0$ or $P'\pi_0 \leq_r \pi_0$. Then from Theorem 10.3.110.3.1, if P is TP2, all beliefs are MLR comparable.

When the trajectories of beliefs for the individual bandit processes are not MLR comparable, they can be projected to MLR comparable beliefs, and a suboptimal policy implemented as follows:

Assume at time instant k, the beliefs of all L processes are MLR comparable. Let $\sigma(1), \ldots, \sigma(L)$ denote the permutation of $(1, \ldots, L)$ so that

$$\pi_k^{\sigma(1)} \geq_r \pi_k^{\sigma(2)} \geq_r \cdots \geq_r \pi_k^{\sigma(L)}.$$

From Theorem 11.4.2, the optimal action is $u_k = \sigma(1)$. But the updated belief $\pi_{k+1}^{\sigma(1)}$ may not be MLR comparable with the other $L-1$ information states. So we project $\pi_{k+1}^{\sigma(1)}$ to the nearest belief denoted $\bar{\pi}$ in the simplex $\Pi(X)$ that is MLR comparable with the other $L-1$ information states. That is, at time $k+1$ solve the following L optimization problems: compute the projection distances

$$\mathcal{P}(\bar{\pi}^{(1)}) = \min_{\bar{\pi}\in\Pi(X)} \|\bar{\pi} - \pi_{k+1}^{\sigma(1)}\| \text{ subject to } \bar{\pi} \geq_r \pi_k^{\sigma(2)}$$

$$\mathcal{P}(\bar{\pi}^{(l)}) = \min_{\bar{\pi}\in\Pi(X)} \|\bar{\pi} - \pi_{k+1}^{\sigma(1)}\| \text{ subject to } \pi_k^{\sigma(l)} \geq_r \bar{\pi} \geq_r \pi_k^{\sigma(p+1)}, \; p = 2,\dots,L-1$$

$$\mathcal{P}(\bar{\pi}^{(L)}) = \min_{\bar{\pi}\in\Pi(X)} \|\bar{\pi} - \pi_{k+1}^{\sigma(1)}\| \text{ subject to } \pi_k^{\sigma(L)} \geq_r \bar{\pi}.$$

Here $\|\cdot\|$ denotes some norm, and $\mathcal{P}$, $\bar{\pi}^p$ denote, respectively, the minimizing value and minimizing solution of each of the problems. Finally set $\pi_{k+1}^{\sigma(1)} = \operatorname{argmin}_{\bar{\pi}_p} \mathcal{P}(\bar{\pi}_p)$. The above L problems are convex optimization problems and can be solved efficiently in real time. Thus all the beliefs at time $k+1$ are MLR comparable, the action u_{k+1} is chosen as the index of the largest belief.

Summary: For POMDP multi-armed bandits that satisfy the conditions of Theorem 11.4.2, the optimal policy is opportunistic: choose the project with the largest belief (in terms of the MLR order).

11.5 Complements and sources

The proof that under suitable conditions the POMDP value function is monotone with respect to the MLR order goes back to Lovejoy [225]. The MLR order is the natural setting for POMDPs due to the Bayesian nature of the problem. More generally, a similar proof holds for multivariate observation distributions – in this case the TP2 stochastic order (which is a multivariate version of the MLR order) is used – to establish sufficient conditions for monotone value function for a multivariate POMDP; see [277].

The result in §11.3 that establishes a threshold policy for two-state POMDPs is from [10]. However, the proof does not work for action-dependent (controlled) observation probabilities. In §14.7 we will use Blackwell ordering to deal with action dependent observation probabilities. For optimal search problems with two states, [230] proves the optimality of threshold policies under certain conditions. Other types of structural result for two-state POMDPs are in [130, 13].

The POMDP multi-armed bandit structural result in §11.4 is from [191] where several numerical examples are presented. [337] and [125] are classic works in Bayesian multi-armed bandits. More generally, [214] establishes the optimality of indexable policies for restless bandit POMDPs with two states. (In a restless bandit, the state of idle projects also evolves.) §17.6.2 gives a short discussion on non-Bayesian bandits.

Appendix 11.A Proof of Theorem 11.3.1

Start with the following lemma.

LEMMA 11.A.1 *Under (F1), (F3′), $\sigma(\pi,\cdot,u)$ is supermodular for $X=2$. That is*

$$\sum_{y\geq\bar{y}}\left[\sigma(\pi,y,u+1)-\sigma(\pi,y,u)\right]\geq\sum_{y\geq\bar{y}}\left[\sigma(\bar{\pi},y,u+1)-\sigma(\bar{\pi},y,u)\right],\quad \pi\geq_s\bar{\pi}.$$

Equivalently, with $\sigma(\pi,.,u)=(\sigma(\pi,y,u),y\in\mathcal{Y}))$ and $\geq_s$ denoting first-order dominance,

$$\frac{1}{2}\left(\sigma(\pi,\cdot,u+1)+\sigma(\bar{\pi},\cdot,u)\right)\geq_s\frac{1}{2}\left(\sigma(\bar{\pi},\cdot,u+1)+\sigma(\pi,.,u)\right). \tag{11.21}$$

Note that both sides of (11.21) are valid probability mass (density) functions.

Proof From (F3′) it follows that $\sum_{j\geq\bar{j}}P_{ij}(2)-P_{ij}(1)$ is increasing in i. Therefore

$$\sum_{j\geq\bar{j}}\sum_i(P_{ij}(2)-P_{ij}(1))\pi(i)\geq\sum_{j\geq\bar{j}}\sum_i(P_{ij}(2)-P_{ij}(1))\bar{\pi}(i),\quad \bar{j}\in\mathcal{X}.$$

This can be expressed as $\sum_{j\geq\bar{j}}p_j\geq\sum_{j\geq\bar{j}}q_j$, or equivalently, $p\geq_s q$ where

$$p_j=\frac{1}{2}\sum_i\left(P_{ij}(2)\pi(i)+P_{ij}(1)\bar{\pi}(i)\right),\quad q_j=\frac{1}{2}\left(\sum_i P_{ij}(2)\bar{\pi}(i)+P_{ij}(1)\pi(i)\right).$$

Note p and q are valid probability mass functions. Next (F1) implies that $\phi(j)\stackrel{\text{defn}}{=}\sum_{y\geq\bar{y}}B_{jy}$ is increasing in j. Therefore since $p\geq_s q$ it follows that $\sum_j\sum_{y\geq\bar{y}}B_{jy}p_j\geq\sum_j\sum_{y\geq\bar{y}}B_{jy}q_j$ which yields (11.21). □

Proof of Theorem 11.3.1

It suffices to prove that $Q(\pi,u)$ is submodular, that is,

$$Q(\pi,u)-Q(\pi,\bar{u})-Q(\bar{\pi},u)+Q(\bar{\pi},\bar{u})\leq 0,\quad \text{for } u>\bar{u},\ \pi(2)>\bar{\pi}(2), \tag{11.22}$$

where $Q(\pi,u)$ is defined in Bellman's equation (11.3). For notational convenience denote $u=2$, $\bar{u}=1$.

From Bellman's equation (11.3), the left-hand side of (11.22) is

$$\begin{aligned}
&C(\pi,2)-C(\pi,1)-C(\bar{\pi},2)+C(\bar{\pi},1) && \text{(a)}\\
&+\rho\sum_y V(T(\pi,y,2))\left[\sigma(\pi,y,2)-\sigma(\pi,y,1)-\sigma(\bar{\pi},y,2)+\sigma(\bar{\pi},y,1)\right] && \text{(b)}\\
&+\rho\sum_y\left[V(T(\bar{\pi},y,1))-V(T(\pi,y,2))\right]\sigma(\bar{\pi},y,1) && \text{(c)}\\
&+\rho\sum_y\left[V(T(\pi,y,2))-V(T(\bar{\pi},y,2))\right]\sigma(\bar{\pi},y,2) && \text{(d)}\\
&+\rho\sum_y\left[V(T(\pi,y,2))-V(T(\pi,y,1))\right]\sigma(\pi,y,1). && (11.23)
\end{aligned}$$

Since the cost is submodular by (S), the first line (a) of (11.23) is negative.

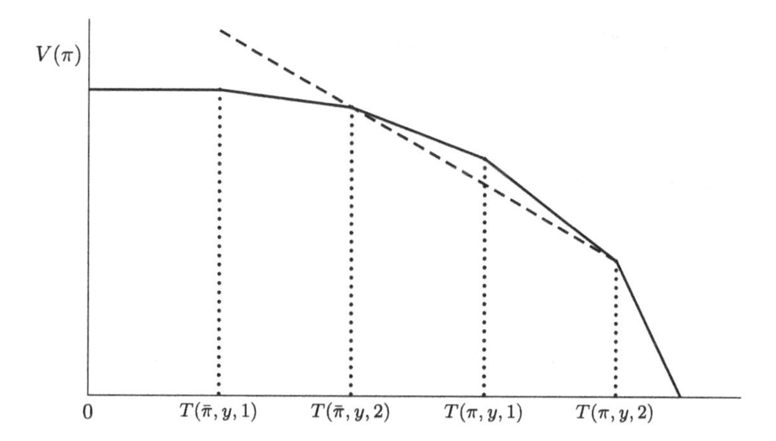

Figure 11.2 The ordering of belief states is as specified in (11.24). The dashed line connects $V(\bar{\pi}, y, 2)$ and $V(\pi, y, 2)$. The vertical coordinate of this line at $T(\bar{\pi}, y, 1)$ is the right-hand side of (11.25). Since $V(\pi)$ is concave and decreasing, this vertical coordinate at $T(\bar{\pi}, y, 1)$ is larger than $V(T(\bar{\pi}, y, 1))$. Similarly, the vertical coordinate of the line at $T(\pi, y, 1)$ is smaller than $V(T(\pi, y, 1))$ due to concavity – this yields (11.27).

Consider the second line (b) of (11.23). $V(\pi)$ is MLR decreasing from Theorem 11.2.1 on page 244 (under (C), (F1), (F2)). Also $T(\pi, y, u)$ is MLR increasing in y from Theorem 10.3.110.3.1 (under (F1)). It therefore follows that $V(T(\pi, y, u))$ is MLR increasing in y. Also under (F1), (F3′), $\sigma(\pi, \cdot, u)$ is supermodular using Lemma 11.A.1. Therefore, using (11.21), the second line of (11.23) is negative.

So to show that $Q(\pi, u)$ is submodular, it only remains to prove that the sum of (c), (d) and the last line of (11.23) is negative. From statement 10.3.1 of Theorem 10.3.1, it follows under (F3′) that $T(\pi, y, 2) \geq_r T(\pi, y, 1)$ and $T(\bar{\pi}, y, 2) \geq_r T(\bar{\pi}, y, 1)$ where $\geq_r$ denotes MLR dominance. From statement 10.3.1 of Theorem 10.3.1, under (F2), $\pi \geq_r \bar{\pi}$ implies $T(\pi, y, u) \geq_r T(\bar{\pi}, y, u)$. Putting these inequalities together we have

$$T(\pi, y, 2) \geq_r T(\pi, y, 1) \geq_r T(\bar{\pi}, y, 1), \quad T(\pi, y, 2) \geq_r T(\bar{\pi}, y, 2) \geq_r T(\bar{\pi}, y, 1).$$

Assume that $T(\pi, y, 1) \geq_r T(\bar{\pi}, y, 2)$ (the reverse ordering yields an identical result). Then

$$T(\pi, y, 2) \geq_r T(\pi, y, 1) \geq_r T(\bar{\pi}, y, 2) \geq_r T(\bar{\pi}, y, 1). \tag{11.24}$$

Now we use the fact that for $X = 2$, $\Pi(X)$ is a one dimensional simplex that can be represented by $\pi(2) \in [0, 1]$. So below we represent π, $T(\pi, y, u)$, etc. by their second elements.

We can express line (c) in (11.23) as follows:

$$V(T(\bar{\pi}, y, 1)) - V(T(\pi, y, 2)) = V(T(\bar{\pi}, y, 1)) - V(T(\bar{\pi}, y, 2)) + V(T(\bar{\pi}, y, 2)) - V(T(\pi, y, 2)).$$

Since $V(\pi)$ is concave and decreasing in $\pi(2)$ it follows that

$$V(T(\bar{\pi}, y, 1)) \leq \frac{T(\bar{\pi}, y, 1) - T(\bar{\pi}, y, 2)}{T(\bar{\pi}, y, 2) - T(\pi, y, 2)} \left(V(T(\bar{\pi}, y, 2)) - V(T(\pi, y, 2))\right) + V(T(\bar{\pi}, y, 2)). \tag{11.25}$$

Therefore, line (c) of (11.23) satisfies

$$V(T(\bar{\pi}, y, 1)) - V(T(\pi, y, 2)) \leq \left[1 + \frac{T(\bar{\pi}, y, 1) - T(\bar{\pi}, y, 2)}{T(\bar{\pi}, y, 2) - T(\pi, y, 2)}\right] \times \left[V(T(\bar{\pi}, y, 2)) - V(T(\pi, y, 2))\right]. \tag{11.26}$$

Using the fact that $V(\pi)$ is concave and decreasing in $\pi(2)$, the final line of (11.23) satisfies

$$V(T(\pi, y, 2)) - V(T(\pi, y, 1)) \leq \frac{T(\pi, y, 1) - T(\pi, y, 2)}{T(\bar{\pi}, y, 2) - T(\pi, y, 2)} \left[V(T(\pi, y, 2)) - V(T(\bar{\pi}, y, 2))\right]. \tag{11.27}$$

Using (11.26), (11.27), the summation of (c), (d) and the last line of (11.23) is upper bounded by

$$\sum_y \left[V(T(\pi, y, 2)) - V(T(\bar{\pi}, y, 2))\right] \left[\sigma(\bar{\pi}, y, 2) + \frac{T(\pi, y, 1) - T(\pi, y, 2)}{T(\bar{\pi}, y, 2) - T(\pi, y, 2)} \sigma(\pi, y, 1) + \left(\frac{T(\bar{\pi}, y, 2) - T(\bar{\pi}, y, 1)}{T(\pi, y, 2) - T(\bar{\pi}, y, 2)} - 1\right) \sigma(\bar{\pi}, y, 1)\right]. \tag{11.28}$$

Since $V(\pi)$ is MLR decreasing (Theorem 11.2.1) and $T(\pi, y, 2) \geq_r T(\bar{\pi}, y, 2)$ (11.24), clearly $V(T(\pi, y, 2)) - V(T(\bar{\pi}, y, 2)) \leq 0$. The term in square brackets in (11.28) can be expressed after some tedious manipulations as

$$B_{1y} B_{2y} \left[\pi(2) - \bar{\pi}(2)\right] \frac{\left[P_{22}(2) - P_{12}(2) - (P_{22}(1) - P_{12}(1))\right]}{\sigma(\pi, y, 2)\left[T(\pi, y, 2) - T(\bar{\pi}, y, 2)\right]} \geq 0.$$

Hence (11.28) is negative, implying that the sum of the (c), (d) and the last line of (11.23) is negative.

12 Structural results for stopping time POMDPs

Contents

12.1 Introduction

The previous chapter established conditions under which the value function of a POMDP is monotone with respect to the MLR order. Also conditions were given for the optimal policy for a two-state POMDP to be monotone (threshold). This and the next chapter develop structural results for the optimal policy of multi-state POMDPs. To establish the structural results, we will use submodularity, and stochastic dominance on the lattice[1] of belief states to analyze Bellman's dynamic programming equation – such analysis falls under the area of "Lattice Programming" [144]. Lattice programming and "monotone comparative statics" pioneered by Topkis [322] (see also [15, 26]) provide a general set of sufficient conditions for the existence of monotone strategies. Once a POMDP is shown to have a monotone policy, then gradient-based algorithms that exploit this structure can be designed to estimate this policy. This and the next two

[1] A lattice is a partially ordered set (in our case belief space $\Pi(X)$) in which every two elements have a supremum and infimum (in our case with respect to the monotone likelihood ratio ordering). The appendix of this chapter gives definitions of supermodularity on lattices.

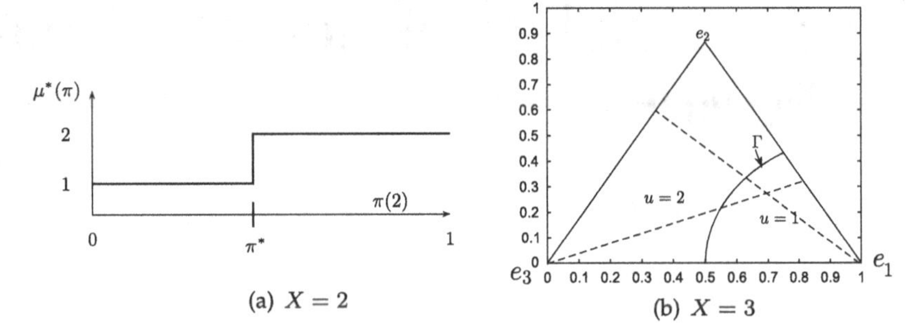

Figure 12.1 Illustration of the two main structural results for stopping time POMDP established in this chapter. Theorem 12.2.1 shows that the stopping set (where action $u = 1$ is optimal) is convex. Theorem 12.3.4 shows that the optimal policy is increasing on any line from e_1 to (e_2, e_3) and decreasing on any line e_3 to (e_1, e_2). Therefore, the stopping set includes state 1 (e_1). Also the boundary of the stopping set Γ intersects any line from e_1 to (e_2, e_3) at most once; similarly for any line from e_3 to (e_1, e_2). Thus the set of beliefs where $u = 2$ is optimal is a connected set. Figure 12.4 shows several types of stopping sets that are excluded by Theorem 12.3.4.

chapters rely heavily on the structural results for filtering (Chapter 10) and monotone value function (Chapter 11). Please see Figure 10.1 on page 220 for the context of this chapter.

12.1.1 Main results

This chapter deals with structural results for the optimal policy of *stopping time POMDPs*. Stopping time POMDPs have action space $\mathcal{U} = \{1 \text{ (stop)}, 2 \text{ (continue) }\}$. They arise in sequential detection such as quickest change detection and machine replacement. Establishing structural results for stopping time POMDPs are easier than that for general POMDPs (which is considered in the next chapter). The main structural results in this chapter regarding stopping time POMDPs are:

1. *Convexity of stopping region*: §12.2 shows that the set of beliefs where it is optimal to apply action 1 (stop) is a convex subset of the belief space. This result unifies several well known results about the convexity of the stopping set for sequential detection problems.
2. *Monotonicity of the optimal policy*: §12.3 gives conditions under which the optimal policy of a stopping time POMDP is monotone with respect to the monotone likelihood ratio (MLR) order. The MLR order is naturally suited for POMDPs since it is preserved under conditional expectations.

Figure 12.1 displays these structural results. For $X = 2$, we will show that stopping set is the interval $[\pi^*, 1]$ and the optimal policy $\mu^*(\pi)$ is a step function; see Figure 12.1(a)). So it is only necessary to compute the threshold state π^*.

Most of this chapter is devoted to characterizing the optimal policy for stopping time POMDPs when $X \geq 3$. The main result shown is that under suitable conditions, the optimal policy for a stopping time POMDP is MLR increasing and therefore has a threshold switching "curve" (denoted by Γ in Figure 12.1(b)). So one only needs to estimate this curve Γ, rather than solve a dynamic programming equation. We will show that the threshold curve Γ has useful properties: it can intersect any line from e_1 to the edge (e_2, e_3) only once. Similarly, it can intersect any line from e_3 to the edge (e_1, e_2) only once (these are the dashed lines in Figure 12.1(b)). It will be shown that the optimal MLR linear threshold policy, which approximates the curve Γ, can then be estimated via simulation-based stochastic approximation algorithms. Such a linear threshold policy is straightforward to implement in a real-time POMDP controller.

§12.6 discusses structural results for POMDPs with multivariate observations. The multivariate TP2 stochastic order is used.

This chapter also presents in §12.7, a stopping time problem involving Kalman filters with applications in radar scheduling. The result is conceptually similar to stopping time POMDPs; the optimal policy is shown to be monotone (with respect to positive definite covariance matrices) and the parametrized policy can be computed via simulation-based optimization.

The structural results presented in this chapter provide a unifying theme and insight into what might otherwise simply be a collection of techniques and results in sequential detection. In Chapter 13 we will present several examples of stopping time POMDPs in sequential quickest change detection, multi-agent social learning and controlled measurement sampling.

12.2 Stopping time POMDP and convexity of stopping set

A stopping time POMDP has action space $\mathcal{U} = \{1 \text{ (stop)}, 2 \text{ (continue)}\}$.

For continue action $u = 2$, the state $x \in \mathcal{X} = \{1, 2, \ldots, X\}$ evolves with transition matrix P and is observed via observations y with observation probabilities $B_{xy} = \mathbb{P}(y_k = y|x_k = x)$. An instantaneous cost $c(x, u = 2)$ is incurred.Thus for $u = 2$, the belief state evolves according to the HMM filter $\pi_k = T(\pi_{k-1}, y_k)$ defined in (7.11). Since action 1 is a stop action and has no dynamics, to simplify notation, we write $T(\pi, y, 2)$ as $T(\pi, y)$ and $\sigma(\pi, y, 2)$ as $\sigma(\pi, y)$ in this chapter.

The action 1 incurs a terminal cost of $c(x, u = 1)$ and the problem terminates.

We consider the class of stationary policies

$$u_k = \mu(\pi_k) \in \mathcal{U} = \{1 \text{ (stop)}, 2 \text{ (continue) }\}. \tag{12.1}$$

Let τ denote a stopping time adapted to $\{\pi_k\}$, $k \geq 0$. That is, with u_k determined by decision policy (12.1),

$$\tau = \{\inf k : u_k = 1\}. \tag{12.2}$$

Let $\Pi(X) = \left\{\pi \in \mathbb{R}^X : \mathbf{1}'\pi = 1, \quad 0 \leq \pi(i) \leq 1 \text{ for all } i \in \mathcal{X}\right\}$ denote the belief space. For stationary policy $\mu : \Pi(X) \to \mathcal{U}$, initial belief $\pi_0 \in \Pi(X)$, discount factor[2] $\rho \in [0, 1]$, the discounted cost objective is

[2] In stopping time POMDPs we allow for $\rho = 1$ as well.

$$J_\mu(\pi_0) = \mathbb{E}_\mu\left\{\sum_{k=0}^{\tau-1} \rho^k c(x_k, 2) + \rho^\tau c(x_\tau, 1)\right\} = \mathbb{E}_\mu\left\{\sum_{k=0}^{\tau-1} \rho^k c_2'\pi_k + \rho^\tau c_1'\pi_\tau\right\}, \quad (12.3)$$

where $c_u = [c(1,u), \dots, c(X,u)]'$. The aim is to determine the optimal stationary policy $\mu^* : \Pi(X) \to \mathcal{U}$ such that $J_{\mu^*}(\pi_0) \leq J_\mu(\pi_0)$ for all $\pi_0 \in \Pi(X)$.

For the above stopping time POMDP, μ^* is the solution of Bellman's equation which is of the form[3] (where $V(\pi)$ below denotes the value function):

$$\mu^*(\pi) = \operatorname*{argmin}_{u\in\mathcal{U}} Q(\pi, u), \quad V(\pi) = \min_{u\in\mathcal{U}} Q(\pi, u), \quad (12.4)$$
$$Q(\pi, 1) = c_1'\pi, \quad Q(\pi, 2) = c_2'\pi + \rho \sum_{y\in Y} V(T(\pi, y))\, \sigma(\pi, y)$$

where $T(\pi, y)$ and $\sigma(\pi, y)$ are the HMM filter and normalization (11.2).

12.2.1 Convexity of stopping region

We now present the first structural result for stopping time POMDPs: the stopping region for the optimal policy is convex. Define the stopping set $\mathcal{R}_1$ as the set of belief states for which stopping ($u = 1$) is the optimal action. Define $\mathcal{R}_2$ as the set of belief states for which continuing ($u = 2$) is the optimal action. That is

$$\mathcal{R}_1 = \{\pi : \mu^*(\pi) = 1 \text{ (stop)}\}, \quad \mathcal{R}_2 = \{\pi : \mu^*(\pi) = 2\} = \Pi(X) - \mathcal{R}_1. \quad (12.5)$$

The theorem below shows that the stopping set $\mathcal{R}_1$ is convex (and therefore a connected set). Recall that the value function $V(\pi)$ is concave on $\Pi(X)$. (This essential property of POMDPs was proved in Theorem 7.4.1.)

THEOREM 12.2.1 ([223]) *Consider the stopping-time POMDP with value function given by (12.4). Then the stopping set $\mathcal{R}_1$ is a convex subset of the belief space $\Pi(X)$.*

Proof Pick any two belief states $\pi_1, \pi_2 \in \mathcal{R}_1$. To demonstrate convexity of $\mathcal{R}_1$, we need to show for any $\lambda \in [0, 1]$, $\lambda\pi_1 + (1-\lambda)\pi_2 \in \mathcal{R}_1$. Since $V(\pi)$ is concave,

$$\begin{aligned} V(\lambda\pi_1 + (1-\lambda)\pi_2) &\geq \lambda V(\pi_1) + (1-\lambda)V(\pi_2) \\ &= \lambda Q(\pi_1, 1) + (1-\lambda)Q(\pi_2, 1) \text{ (since } \pi_1, \pi_2 \in \mathcal{R}_1) \\ &= Q(\lambda\pi_1 + (1-\lambda)\pi_2, 1) \text{ (since } Q_1(\pi, 1) \text{ is linear in } \pi) \\ &\geq V(\lambda\pi_1 + (1-\lambda)\pi_2) \text{ (since } V(\pi) \text{ is the optimal value function).} \end{aligned}$$

Thus all the inequalities above are equalities, and $\lambda\pi_1 + (1-\lambda)\pi_2 \in \mathcal{R}_1$. □

Note that the theorem says nothing about the "continue" region $\mathcal{R}_2$. In Theorem 12.3.4 below we will characterize both $\mathcal{R}_1$ and $\mathcal{R}_2$. Figure 12.1 illustrates the assertion of Theorem 12.2.1 for $X = 2$ and $X = 3$.

[3] The stopping time POMDP can be expressed as an infinite horizon POMDP. Augment $\Pi(X)$ to include the fictitious stopping state e_{X+1} which is cost free, i.e. $c(e_{X+1}, u) = 0$ for all $u \in \mathcal{U}$. When decision $u_k = 1$ is chosen, the belief state π_{k+1} transitions to e_{X+1} and remains there indefinitely. Then (12.3) is equivalent to $J_\mu(\pi) = \mathbb{E}_\mu\{\sum_{k=0}^{\tau-1} \rho^k c_2'\pi_k + \rho^\tau c_1'\pi + \sum_{k=\tau+1}^{\infty} \rho^k c(e_{X+1}, u_k\}$, where the last summation is zero.

12.2.2 Example 1: Classical quickest change detection

Quickest detection is a useful example of a stopping time POMDP that has applications in biomedical signal processing, machine monitoring and finance [268, 36]. The classical Bayesian quickest detection problem is as follows: An underlying discrete-time state process x jump changes at a geometrically distributed random time τ^0. Consider a sequence of random measurements $\{y_k, k \geq 1\}$, such that conditioned on the event $\{\tau^0 = t\}$, y_k, $\{k \leq t\}$ are i.i.d. random variables with distribution B_{1y} and $\{y_k, k > t\}$ are i.i.d. random variables with distribution B_{2y}. The quickest detection problem involves detecting the change time τ^0 with minimal cost. That is, at each time $k = 1, 2, \ldots$, a decision $u_k \in \{\text{continue}, \text{stop and announce change}\}$ needs to be made to optimize a tradeoff between false alarm frequency and linear delay penalty.[4]

A geometrically distributed change time τ^0 is realized by a two state ($X = 2$) Markov chain with absorbing transition matrix P and prior π_0 as follows:

$$P = \begin{bmatrix} 1 & 0 \\ 1 - P_{22} & P_{22} \end{bmatrix}, \quad \pi_0 = \begin{bmatrix} 0 \\ 1 \end{bmatrix}, \quad \tau^0 = \inf\{k : x_k = 1\}. \tag{12.6}$$

The system starts in state 2 and then jumps to the absorbing state 1 at time τ^0. Clearly τ^0 is geometrically distributed with mean $1/(1 - P_{22})$.

The cost criterion in classical quickest detection is the *Kolmogorov–Shiryayev criterion* for detection of disorder [298]

$$J_\mu(\pi) = d\,\mathbb{E}_\mu\{(\tau - \tau^0)^+\} + \mathbb{P}_\mu(\tau < \tau^0), \quad \pi_0 = \pi. \tag{12.7}$$

where μ denotes the decision policy. The first term is the delay penalty in making a decision at time $\tau > \tau^0$ and d is a positive real number. The second term is the false alarm penalty incurred in announcing a change at time $\tau < \tau^0$.

Stopping time POMDP: The quickest detection problem with penalty (12.7) is a stopping time POMDP with $\mathcal{U} = \{1 \text{ (announce change and stop)}, 2 \text{ (continue)}\}$, $\mathcal{X} = \{1, 2\}$, transition matrix in (12.6), arbitrary observation probabilities B_{xy}, cost vectors $c_1 = [0,\ 1]'$, $c_2 = [d,\ 0]'$ and discount factor $\rho = 1$.

Theorem 12.2.1 then implies the following structural result.

COROLLARY 12.2.2 *The optimal policy μ^* for classical quickest detection has a* threshold *structure: There exists a threshold point $\pi^* \in [0, 1]$ such that*

$$u_k = \mu^*(\pi_k) = \begin{cases} 2 \text{ (continue)} & \text{if } \pi_k(2) \in [\pi^*, 1] \\ 1 \text{ (stop and announce change)} & \text{if } \pi_k(2) \in [0, \pi^*). \end{cases} \tag{12.8}$$

[4] There are two general formulations for quickest time detection. In the first formulation, the change point τ^0 is an unknown deterministic time, and the goal is to determine a stopping rule such that a worst case delay penalty is minimized subject to a constraint on the false alarm frequency (see, e.g., [247, 267, 344, 268]). The second formulation, which is the formulation considered in this book (this chapter and also Chapter 13), is the Bayesian approach where the change time τ^0 is specified by a prior distribution.

Proof Since $X = 2$, $\Pi(X)$ is the interval $[0, 1]$, and $\pi(2) \in [0, 1]$ is the belief state. Theorem 12.2.1 implies that the stopping set $\mathcal{R}_1$ is convex. In one dimension this implies that $\mathcal{R}_1$ is an interval of the form $[a^*, \pi^*)$ for $0 \leq a < \pi^* \leq 1$. Since state 1 is absorbing, Bellman's equation (12.4) with $\rho = 1$ applied at $\pi = e_1$ implies

$$\mu^*(e_1) = \underset{u}{\operatorname{argmin}}\{\underbrace{c(1, u = 1),}_{0} \ d(1 - \pi(2)) + V(e_1)\} = 1.$$

So e_1 or equivalently $\pi(2) = 0$ belongs to $\mathcal{R}_1$. Therefore, $\mathcal{R}_1$ is an interval of the form $[0, \pi^*)$. Hence the optimal policy is of the form (12.8). □

Theorem 12.2.1 says that for quickest detection of a multi-state Markov chain, the stopping set $\mathcal{R}_1$ is convex. (Recall $\mathcal{R}_1$ is the set of beliefs where $u = 1 =$ stop is optimal.) What about the continuing set $\mathcal{R}_2 = \Pi(X) - \mathcal{R}_1$ where action $u = 2$ is optimal? For $X = 2$, using Corollary 12.2.2, $\mathcal{R}_2 = [\pi^*, 1]$ and is therefore convex. However, for $X > 2$, Theorem 12.2.1 does not say anything about the structure of $\mathcal{R}_2$; indeed, $\mathcal{R}_2$ could be a disconnected set. In §12.3, we will use more powerful POMDP structural results to give sufficient conditions for both $\mathcal{R}_1$ and $\mathcal{R}_2$ to be connected sets.

12.2.3 Example 2: Instruction problem

The book [49] discusses the following instruction problem. A student is instructed and examined repeatedly until some stopping time τ when the instruction is stopped. The action space is $\mathcal{U} = \{1 \text{ (stop)}, 2 \text{ (instruct)}\}$. The state space and observation spaces are

$$\mathcal{X} = \{1 \text{ (learnt)}, 2 \text{ (not learnt)}\}, \quad \mathcal{Y} = \{1 \text{ (correct answer)}, 2 \text{ (wrong answer)}\}.$$

Here, $x_k \in \mathcal{X}$ denotes the status of the student at time k. The observation $y_k \in \mathcal{Y}$ is the outcome of an exam at each time k. If action $u = 2$ (instruct) is chosen, then the transition probabilities, observation probabilities and cost are

$$P = \begin{bmatrix} 1 & 0 \\ 1-p & p \end{bmatrix}, \quad B = \begin{bmatrix} 1 & 0 \\ 1-q & q \end{bmatrix}, \quad c_2 = \begin{bmatrix} 1 \\ 1 \end{bmatrix}.$$

So $1-p$ is the probability that the student learns when instructed; $1-q$ is the probability that the student gives the correct answer in an exam when he has not yet learnt; and 1 is the cost of instruction. If instruction is terminated then the stopping cost vector is $c_1 = [0, f]'$ where $f > 0$. So $c(x = 2, u = 1) = f$ is the cost that the instruction was stopped but the student has not yet learnt.

This two-state stopping time POMDP is almost identical to quickest detection apart from the cost vectors being different. By Theorem 12.2.1 the stopping region is an interval. Since state 1 is absorbing, a similar proof to Corollary 12.2.2 implies that the optimal policy is of the form (12.8).

12.3 Monotone optimal policy for stopping time POMDP

We now consider the next major structural result: sufficient conditions to ensure that a stopping time POMDP has a monotone optimal policy.[5]

Consider a stopping time POMDP with state space and action space

$$\mathcal{X} = \{1, \ldots, X\}, \quad \mathcal{U} = \{1 \text{ (stop)}, 2 \text{ (continue)}\}.$$

Action 2 implies continue with transition matrix P, observation distribution B and cost $C(\pi, 2)$, while action 1 denotes stop with stopping cost $C(\pi, 1)$. So the model is almost identical to the previous section except that the costs $C(\pi, u)$ are in general nonlinear functions of the belief. Recall such nonlinear costs were motivated by controlled sensing applications in Chapter 8.4.

In terms of the belief state π, Bellman's equation reads

$$Q(\pi, u=1) = C(\pi, 1), \; Q(\pi, u=2) = C(\pi, 2) + \rho \sum_y V(T(\pi, y, 2))\sigma(\pi, y, 2),$$

$$V(\pi) = \min_{u \in \{1,2\}} Q(\pi, u), \quad \mu^*(\pi) = \operatorname{argmin}_{u \in \{1,2\}} Q(\pi, u). \tag{12.9}$$

12.3.1 Objective

One possible objective would be to give sufficient conditions on a stopping time POMDP so that the optimal policy $\mu^*(\pi)$ is MLR increasing on $\Pi(X)$. That is, with $\geq_r$ denoting MLR dominance,

$$\pi_1, \pi_2 \in \Pi(X), \quad \pi_1 \geq_r \pi_2 \implies \mu^*(\pi_1) \geq \mu^*(\pi_2). \tag{12.10}$$

However, because $\Pi(X)$ is only partially orderable with respect to the MLR order, it is difficult to exploit (12.10) for devising useful algorithms. Instead, in this section, our aim is to give (less restrictive) conditions that lead to

$$\pi_1, \pi_2 \in \mathcal{L}(e_i, \bar{\pi}), \quad \pi_1 \geq_r \pi_2 \implies \mu^*(\pi_1) \geq \mu^*(\pi_2), \quad \cdot i \in \{1, X\}. \tag{12.11}$$

Here $\mathcal{L}(e_i, \bar{\pi})$ denotes any line segment in $\Pi(X)$ which starts at e_1 and ends at any belief $\bar{\pi}$ in the subsimplex $\{e_2, , \ldots, e_X\}$; or any line segment which starts at e_X and ends at any belief $\bar{\pi}$ in the subsimplex $\{e_1, \ldots, e_{X-1}\}$. (These line segments are the dashed lines in Figure 12.1(b).) So instead of proving $\mu^*(\pi)$ is MLR increasing for any two beliefs in the belief space, we will prove that $\mu^*(\pi)$ is MLR increasing for any two beliefs on these special line segments $\mathcal{L}(e_i, \bar{\pi})$. The main reason is that the MLR order is a total order on such lines (not just a partial order) meaning that any two beliefs on $\mathcal{L}(e_i, \bar{\pi})$ are MLR orderable. Proving (12.11) yields two very useful results:

1. The optimal policy $\mu^*(\pi)$ of a stopping time POMDP is characterized by switching curve Γ; see Theorem 12.3.4. This is illustrated in Figure 12.1(b).

[5] The reader should review MLR dominance §10.1 and the structural results on monotone value functions in Chapter 11. To access the final result, the reader can jump directly to statement 2 of Theorem 12.3.4 and then proceed to §12.3.5.

2. The optimal linear approximation to switching curve Γ that preserves (12.11) can be estimated via a simulation-based stochastic approximation algorithm thereby facilitating a simple controller; see §12.4.

12.3.2 MLR dominance on lines

Since our plan is to prove (12.11) on line segments in the belief space, we formally define these line segments. Define the sub-simplices, $\mathcal{H}_1$ and $\mathcal{H}_X$:

$$\mathcal{H}_1 = \{\pi \in \Pi(X) : \pi(1) = 0\}, \ \mathcal{H}_X = \{\pi \in \Pi(X) : \pi(X) = 0\}. \tag{12.12}$$

Denote a generic belief state that lies in either $\mathcal{H}_1$ or $\mathcal{H}_X$ by $\bar{\pi}$. For each such $\bar{\pi} \in \mathcal{H}_i$, $i \in \{1, X\}$, construct the line segment $\mathcal{L}(e_i, \bar{\pi})$ that connects $\bar{\pi}$ to e_i. Thus each line segment $\mathcal{L}(e_i, \bar{\pi})$ comprises of belief states π of the form:

$$\mathcal{L}(e_i, \bar{\pi}) = \{\pi \in \Pi(X) : \pi = (1-\epsilon)\bar{\pi} + \epsilon e_i, \ 0 \le \epsilon \le 1\}, \bar{\pi} \in \mathcal{H}_i. \tag{12.13}$$

To visualize (12.12) and (12.13), Figure 12.2 illustrates the setup for $X = 3$. The sub-simplex $\mathcal{H}_1$ is simply the line segment (e_2, e_3); and $\mathcal{H}_3$ is the line segment $\{e_1, e_2\}$. Also shown are examples of line segments $\mathcal{L}(e_1, \bar{\pi})$ and $\mathcal{L}(e_3, \bar{\pi})$ for arbitrary points $\bar{\pi}_1$ and $\bar{\pi}_2$ in $\mathcal{H}_1$ and $\mathcal{H}_3$.

We now define the MLR order on such lines segments. Recall the definition of MLR order $\geq_r$ on the belief space $\Pi(X)$ in Definition 10.1.1 on page 221.

DEFINITION 12.3.1 (MLR ordering $\geq_{L_i}$ on lines) *π_1 is greater than π_2 with respect to the MLR ordering on the line $\mathcal{L}(e_i, \bar{\pi})$, $i \in \{1, X\}$ – denoted as $\pi_1 \geq_{L_i} \pi_2$, if $\pi_1, \pi_2 \in \mathcal{L}(e_i, \bar{\pi})$ for some $\bar{\pi} \in \mathcal{H}_i$, and $\pi_1 \geq_r \pi_2$.*

Appendix 12.B shows that the partially ordered sets $[\mathcal{L}(e_1, \bar{\pi}), \geq_{L_X}]$ and $[\mathcal{L}(e_X, \bar{\pi}), \geq_{L_1}]$ are chains, i.e. totally ordered sets. All elements $\pi_1, \pi_2 \in \mathcal{L}(e_X, \bar{\pi})$ are comparable, i.e. either $\pi_1 \geq_{L_X} \pi_2$ or $\pi_2 \geq_{L_X} \pi_1$ (and similarly for $\mathcal{L}(e_1, \bar{\pi})$). The largest element (supremum) of $[\mathcal{L}(e_1, \bar{\pi}), \geq_{L_X}]$ is $\bar{\pi}$ and the smallest element (infimum) is e_1.

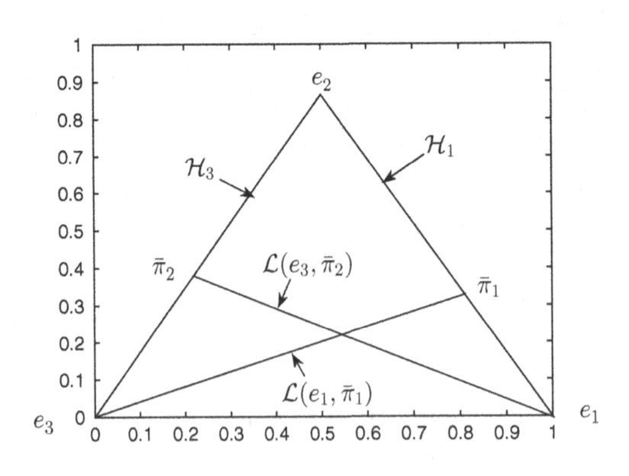

Figure 12.2 Examples of sub-simplices $\mathcal{H}_1$ and $\mathcal{H}_3$ and points $\bar{\pi}_1 \in \mathcal{H}_1$, $\bar{\pi}_2 \in \mathcal{H}_3$. Also shown are the lines $\mathcal{L}(e_1, \bar{\pi}_1)$ and $\mathcal{L}(e_3, \bar{\pi}_2)$ that connect these point to the vertices e_1 and e_3.

12.3.3 Submodularity with MLR order

To prove the structural result (12.11), we will show that $Q(\pi, u)$ in (12.9) is a submodular function on the chains $[\mathcal{L}(e_1, \bar{\pi}), \geq_{L_1}]$ and $[\mathcal{L}(e_1, \bar{\pi}), \geq_{L_X}]$. This requires less restrictive conditions than submodularity on the entire simplex $\Pi(X)$.

DEFINITION 12.3.2 (Submodular function) *Suppose $i = 1$ or X. Then $f : \mathcal{L}(e_i, \bar{\pi}) \times \mathcal{U} \to \mathbb{R}$ is submodular if $f(\pi, u) - f(\pi, \bar{u}) \leq f(\tilde{\pi}, u) - f(\tilde{\pi}, \bar{u})$, for $\bar{u} \leq u$, $\pi \geq_{L_i} \tilde{\pi}$.*

A more general definition of submodularity on a lattice is given in Appendix 12.A on page 279. Also Appendix 12.B contains additional properties that will be used in proving the main theorem below.

The following key result says that for a submodular function $Q(\pi, u)$, there exists a version of the optimal policy $\mu^*(\pi) = \operatorname{argmin}_u Q(\pi, u)$ that is MLR increasing on lines.

THEOREM 12.3.3 (Topkis theorem) *Suppose $i = 1$ or X. If $f : \mathcal{L}(e_i, \bar{\pi}) \times \mathcal{U} \to \mathbb{R}$ is submodular, then there exists a $\mu^*(\pi) = \operatorname{argmin}_{u \in \mathcal{U}} f(\pi, u)$, that is increasing on $[\mathcal{L}(e_i, \bar{\pi}), \geq_{L_i}]$, i.e. $\pi^0 \geq_{L_i} \pi \implies \mu^*(\pi) \leq \mu^*(\pi^0)$.*

12.3.4 Assumptions and main result

For convenience, we repeat (C) on page 243 and (F1), (F2) on page 225. The main additional assumption below is the submodularity assumption (S).

(C) $\pi_1 \geq_s \pi_2$ implies $C(\pi_1, u) \leq C(\pi_2, u)$ for each u.
For linear costs, the condition is: $c(x, u)$ is decreasing in x for each u.
(F1) B is totally positive of order 2 (TP2).
(F2) P is totally positive of order 2 (TP2).
(S) $C(\pi, u)$ is submodular on $[\mathcal{L}(e_X, \bar{\pi}), \geq_{L_X}]$ and $[\mathcal{L}(e_1, \bar{\pi}), \geq_{L_1}]$.
For linear costs the condition is $c(x, 2) - c(x, 1) \geq c(X, 2) - c(X, 1)$ and $c(1, 2) - c(1, 1) \geq c(x, 2) - c(x, 1)$.

THEOREM 12.3.4 (Switching curve optimal policy) *Assume (C), (F1), (F2) and (S) hold for a stopping time POMDP. Then:*

1. *There exists an optimal policy $\mu^*(\pi)$ that is $\geq_{L_X}$ increasing on lines $\mathcal{L}(e_X, \bar{\pi})$ and $\geq_{L_1}$ increasing on lines $\mathcal{L}(e_1, \bar{\pi})$.*
2. *Hence there exists a threshold switching curve Γ that partitions belief space $\Pi(X)$ into two individually connected[6] regions $\mathcal{R}_1$, $\mathcal{R}_2$, such that the optimal policy is*

$$\mu^*(\pi) = \begin{cases} \text{continue} = 2 & \text{if } \pi \in \mathcal{R}_2 \\ \text{stop} = 1 & \text{if } \pi \in \mathcal{R}_1 \end{cases} \qquad (12.14)$$

The threshold curve Γ intersects each line $\mathcal{L}(e_X, \bar{\pi})$ and $\mathcal{L}(e_1, \bar{\pi})$ at most once.

3. *There exists $i^* \in \{0, \ldots, X\}$, such that $e_1, \ldots, e_{i^*} \in \mathcal{R}_1$ and $e_{i^*+1}, \ldots, e_X \in \mathcal{R}_2$.*
4. *For the case $X = 2$, there exists a unique threshold point $\pi^*(2)$.*

□

[6] A set is connected if it cannot be expressed as the union of disjoint nonempty closed sets [287].

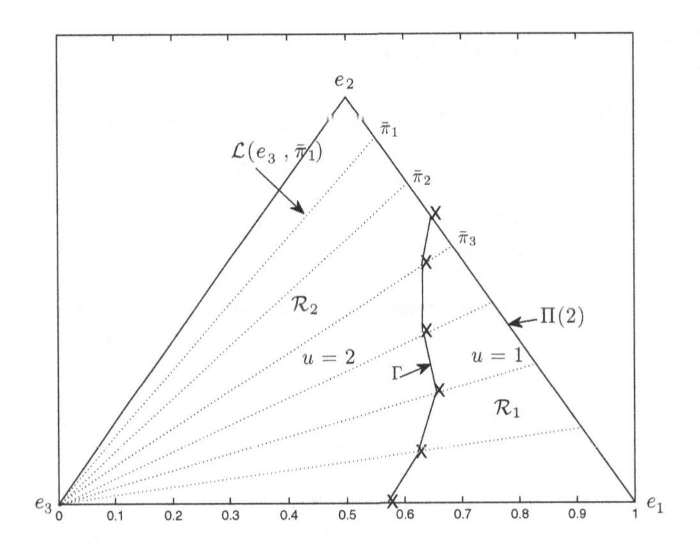

Figure 12.3 Illustration of switching decision curve Γ for optimal policy of a stopping time POMDP. Here $X = 3$ and hence $\Pi(X)$ is an equilateral triangle. Theorem 12.3.4 shows that for a nonlinear cost POMDP, the stopping region $\mathcal{R}_1$ is a connected set and $e_1 \in \mathcal{R}_1$. (Recall that for linear stopping cost, $\mathcal{R}_1$ is convex from Theorem 12.2.1.) Also $\mathcal{R}_2$ is connected. The lines segments $\mathcal{L}(e_X, \bar{\pi}_1)$ connecting the sub-simplex $\Pi(2)$ to e_3 are defined in (12.13). Theorem 12.3.4 says that the threshold curve Γ can intersect each line $\mathcal{L}(e_X, \bar{\pi})$ only once. Similarly, Γ can intersect each line $\mathcal{L}(e_1, \bar{\pi})$ only once (not shown).

Let us explain the intuition behind the proof of the theorem. As shown in Theorem 11.2.1 on page 244, (C), (F1) and (F2) are sufficient conditions for the value function $V(\pi)$ to be MLR decreasing in π.

(S) is sufficient for the costs $c_u'\pi$ to be submodular on lines $\mathcal{L}(e_X, \bar{\pi})$ and $\mathcal{L}(e_1, \bar{\pi})$. Finally (C), (F1) and (S) are sufficient for $Q(\pi, u)$ to be submodular on lines $\mathcal{L}(e_X, \bar{\pi})$ and $\mathcal{L}(e_1, \bar{\pi})$.

As a result, Topkis Theorem 12.3.3 implies that the optimal policy is monotone on each chain $[\mathcal{L}(e_X, \bar{\pi}), \geq_{L_X}]$. So there exists a threshold belief state on each line $\mathcal{L}(e_X, \bar{\pi})$ where the optimal policy switches from 1 to 2. (A similar argument holds for lines $[\mathcal{L}(e_1, \bar{\pi}), \geq_{L_1}]$).

The entire simplex $\Pi(X)$ can be covered by the union of lines $\mathcal{L}(e_X, \bar{\pi})$. The union of the resulting threshold belief states yields the switching curve $\Gamma(\pi)$. This is illustrated in Figure 12.3.

12.3.5 Some intuition

Recall for $\mathcal{X} = \{1, 2, 3\}$, the belief state space $\Pi(3)$ is an equilateral triangle. So on $\Pi(3)$, more insight can be given to visualize what the above theorem says. In Figure 12.4, six examples are given of decision regions that violate the theorem. To make these examples nontrivial, we have included $e_1 \in \mathcal{R}_1$ in all cases.

The decision regions in Figure 12.4(a) violate the condition that $\mu^*(\pi)$ is increasing on lines toward e_3. Even though $\mathcal{R}_1$ and $\mathcal{R}_2$ are individually connected regions, the

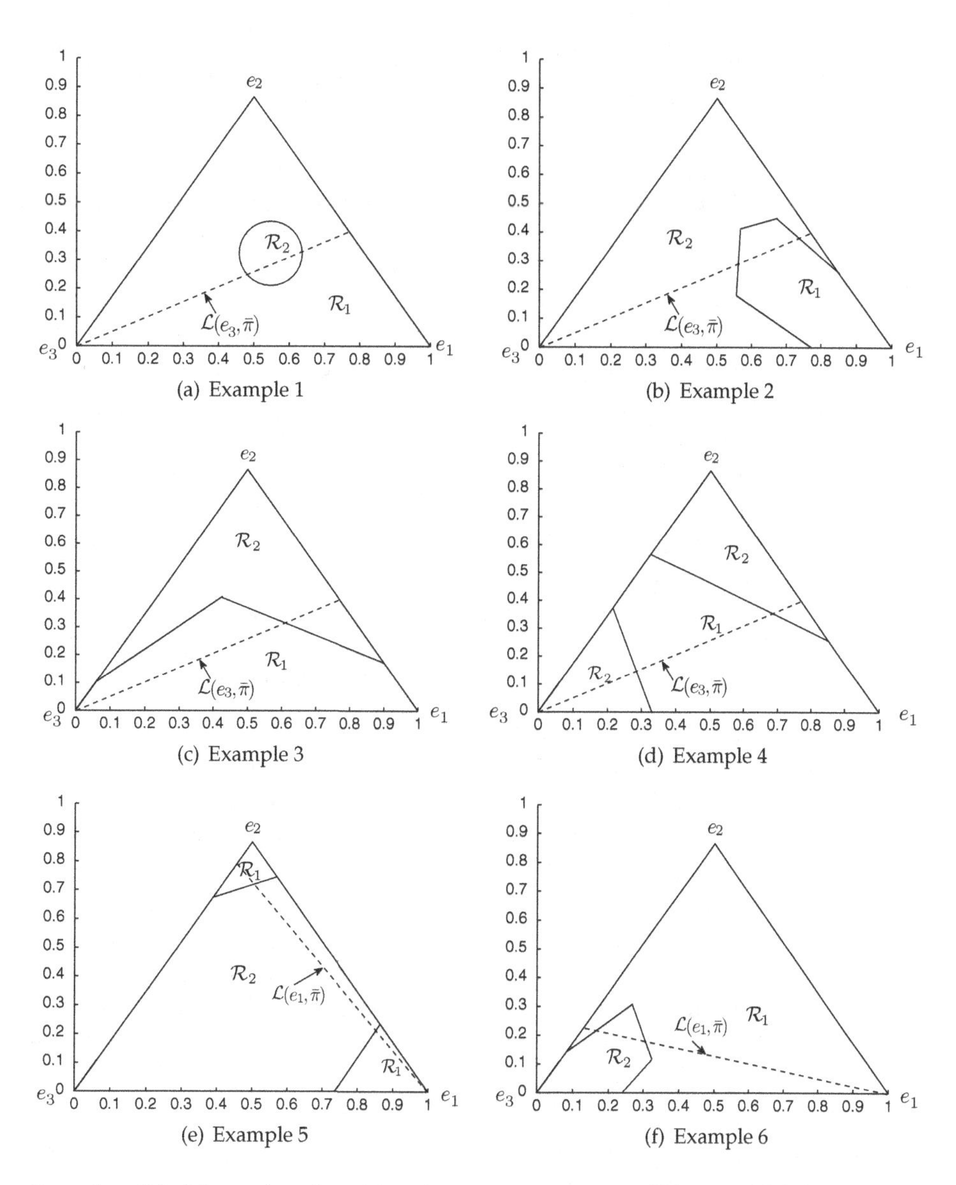

Figure 12.4 Examples of decision regions that violate the monotone property of Theorem 12.3.4 on belief space $\Pi(3)$. In each case the policy is not increasing on lines from e_3 to (e_1, e_2) or on lines from e_1 to (e_2, e_3). Such lines are denoted by broken lines.

depicted line $\mathcal{L}(e_3, \bar{\pi})$ intersects the boundary of $\mathcal{R}_2$ more than once (and so violates Theorem 12.3.4).

The decision regions in Figure 12.4(b) are assumed to satisfy Theorem 12.2.1 and the stopping set $\mathcal{R}_1$ is convex. As mentioned above, Theorem 12.3.4 gives more structure to $\mathcal{R}_1$ and $\mathcal{R}_2$. Indeed, the decision regions in Figure 12.4(b) violate Theorem 12.3.4. They violate the statement that the policy is increasing on lines toward e_3 since the boundary of $\mathcal{R}_1$ (i.e. switching curve Γ) cannot intersect a line from e_3 more than once. Therefore, Theorem 12.3.4 says a lot more about the structure of the boundary than convexity does. In particular, for linear costs, Theorems 12.3.4 and 12.2.1 together say that the switching curve Γ is convex and cannot intersect a line $\mathcal{L}(e_3, \bar{\pi})$ or a line $\mathcal{L}(e_1, \bar{\pi})$ more than once.

Figure 12.4(c) also assumes that Theorem 12.2.1 holds and so $\mathcal{R}_1$ is convex. But the decision regions in Figure 12.4(c) violate Statement (ii) of Theorem 12.3.4. In particular, if e_1 and e_3 lie in $\mathcal{R}_1$, then e_2 should also lie in $\mathcal{R}_1$. Again this reveals that Theorem 12.3.4 says a lot more about the structure of the stopping region even for the case of linear costs.

Figure 12.4(d) also satisfies Theorem 12.2.1 since $\mathcal{R}_1$ is convex; but does not satisfy Theorem 12.3.4 since $\mathcal{R}_2$ is not a connected set. Indeed when $\mathcal{R}_2$ is not connected as shown in the proof of Theorem 12.3.4, the policy $\mu^*(\pi)$ is not monotone on the line $\mathcal{L}(e_3, \bar{\pi})$ since it goes from 2 to 1 to 2.

Figure 12.4(e) and (f) violate Theorem 12.3.4 since the optimal policy $\mu^*(\pi)$ is not monotone on line $\mathcal{L}(e_1, \bar{\pi})$; it goes from 1 to 2 to 1. For the case $\alpha = 0$, Figure 12.4(e) and (f) violate Theorem 12.2.1 since the stopping region $\mathcal{R}_1$ is non-convex.

Since the conditions of Theorem 12.3.4 are sufficient conditions, what happens when they do not hold? It is straightforward to construct other examples where both $\mathcal{R}_1$ and $\mathcal{R}_2$ are disconnected regions when the assumptions of Theorem 12.3.4 are violated.

It is instructive to compare Theorem 12.3.4 with Theorem 12.2.1.

1. Theorem 12.2.1 shows that stopping $\mathcal{R}_1$ is convex for linear costs but says nothing about $\mathcal{R}_2$. Theorem 12.3.4 shows that both $\mathcal{R}_1$ and $\mathcal{R}_2$ are individually connected sets (even for nonlinear costs under suitable assumptions).
2. Theorem 12.3.4 shows that state 1 (namely belief state e_1) belongs to $\mathcal{R}_1$ (assuming $\mathcal{R}_1$ is nonempty) since the optimal policy is MLR increasing.
3. Considering Figure 12.3 for $X = 3$, the boundary of the switching curve Γ in Theorem 12.3.4 sweeps an angle from zero to sixty degrees in the equilateral triangle. By Lebesgue theorem, any such curve (being a function of a monotonically increasing angle from zero to sixty degrees) is differentiable almost everywhere. Of course from Theorem 12.2.1 for linear costs, since $\mathcal{R}_1$ is convex its boundary is also differentiable almost everywhere.

12.3.6 Example. Explicit stopping set for one-step-ahead property

We conclude this section by considering a special (restrictive) case where the stopping set $\mathcal{R}_1$ can be determined explicitly.

Consider the set of belief states

$$\mathcal{S}^o = \{\pi : c_1'\pi \leq c_2'\pi + \rho c_1' P'\pi\}. \tag{12.15}$$

Note that $c_2'\pi + \rho c_1' P'\pi$ is the cost incurred by proceeding one step ahead and then stopping, while $c_1'\pi$ is the cost of stopping immediately.

THEOREM 12.3.5 *Suppose the parameters (c_1, c_2, P, B, ρ) of a stopping time POMDP satisfy the following one-step-ahead property:*

$$\pi \in \mathcal{S}^o \implies T(\pi, y) \in \mathcal{S}^o, \quad \forall y \in \mathcal{Y}. \tag{12.16}$$

Then the optimal stopping set $\mathcal{R}_1 = \mathcal{S}^o$.

Proof Step 1: $\pi \in \mathcal{S}^o$ implies that $V(\pi) = c_1'\pi$ and so it is optimal to stop. This follows straightforwardly from induction via the value iteration algorithm.
Step 2: $\pi \notin \mathcal{S}^o$ implies that $V(\pi) < c_1'\pi$ and so it is optimal not to stop. This follows since if $\pi \notin \mathcal{S}^o$ then proceeding one step ahead and stopping (cost $c_2'\pi + \rho c_1' P'\pi$) is cheaper than stopping immediately (cost $c_1'\pi$). □

It is clear from Bellman's equation (12.4) that in general, $\mathcal{S}^o \subseteq \mathcal{R}_1$. Theorem 12.3.5 says that under condition (12.16), the two sets are actually equivalent.

At first sight (12.16) is not a useful condition since it needs to be checked for a continuum of beliefs and all possible observations. However, under assumptions (C), (F1), (F2), (S), it turns out that (12.16) has a finite dimensional characterization. Recall that under assumptions (F1), (F2), $T(\pi, y)$ is increasing in π and y. When (C) and (S) also hold then the optimal policy is monotone (Theorem 12.3.4). Then for finite observation space $\mathcal{Y} = \{1, 2, \ldots, Y\}$, condition (12.16) is equivalent to the following finite set of conditions:

$$T(\pi_i^*, Y) \in \mathcal{S}^o, \; i = 2, \ldots, X, \text{ where } \pi_i^* = \{\pi \in \mathcal{S}^0 : \pi(j) = 0, j \neq \{1, i\}\} \qquad (12.17)$$

Note that π_i^* are the $X - 1$ corner points at which the hyperplane $c_1'\pi = c_2'\pi + \rho c_1' P'\pi$ intersect the faces of the simplex $\Pi(X)$.

As an example, consider the quickest detection problem with geometric change time ($X = 2$) discussed in §12.2.2. If the observation space is finite then Theorem 12.3.5 says that the threshold state is $\pi^*(2) = d/(d + 1 - \rho P_{22})$ providing $e_2' T(\pi^*, Y) \leq \pi^*(2)$ (namely, condition (12.17)) holds.

12.4 Characterization of optimal linear decision threshold for stopping time POMDP

In this section, we assume (C), (F1), (F2) and (S) hold. Therefore Theorem 12.3.4 applies and the optimal policy $\mu^*(\pi)$ is characterized by a switching curve Γ as illustrated in Figure 12.3.

How can the switching curve Γ be estimated (computed)? In general, any user-defined basis function approximation can be used to parametrize this curve. However, such parametrized policy needs to capture the essential feature of Theorem 12.3.4: it needs to be MLR increasing on lines.

12.4.1 Linear threshold policies

We derive the optimal *linear* approximation to the switching curve Γ on simplex $\Pi(X)$. Such a linear decision threshold has two attractive properties: (i) Estimating it is computationally efficient. (ii) We give conditions on the coefficients of the linear threshold that are necessary and sufficient for the resulting policy to be MLR increasing on lines. Due to the necessity and sufficiency of the condition, optimizing over the space of linear thresholds on $\Pi(X)$ yields the "optimal" linear approximation to switching curve Γ.

Since $\Pi(X)$ is a subset of $\mathbb{R}^{X-1}$, a linear hyperplane on $\Pi(X)$ is parametrized by $X-1$ coefficients. So, on $\Pi(X)$, define the linear threshold policy $\mu_\theta(\pi)$ as

$$\mu_\theta(\pi) = \begin{cases} \text{stop} = 1 & \text{if } \begin{bmatrix} 0 & 1 & \theta' \end{bmatrix}' \begin{bmatrix} \pi \\ -1 \end{bmatrix} < 0 \\ \text{continue} = 2 & \text{otherwise} \end{cases} \quad \pi \in \Pi(X). \tag{12.18}$$

Here $\theta = (\theta(1), \ldots, \theta(X-1))' \in \mathbb{R}^{X-1}$ denotes the parameter vector of the linear threshold policy.

Theorem 12.4.1 below characterizes the optimal linear decision threshold approximation to the threshold switching curve on $\Pi(X)$. Assume conditions (C), (F1) and (S) hold so that Theorem 12.3.4 holds. Also assuming that the stopping region $\mathcal{R}_1$ is nonempty, then e_1 lies in the stopping set. This implies that the $(X-1)$-th component of θ satisfies $\theta(X-1) > 0$.

THEOREM 12.4.1 (Optimal linear threshold policy) *For belief states $\pi \in \Pi(X)$, the linear threshold policy $\mu_\theta(\pi)$ defined in (12.18) is*
(i) MLR increasing on lines $\mathcal{L}(e_X, \bar{\pi})$ iff $\theta(X-2) \geq 1$ and $\theta(i) \leq \theta(X-2)$ for $i < X-2$.
(ii) MLR increasing on lines $\mathcal{L}(e_1, \bar{\pi})$ iff $\theta(i) \geq 0$, for $i < X-2$. □

Proof Given any $\pi_1, \pi_2 \in \mathcal{L}(e_X, \bar{\pi})$ with $\pi_2 \geq_{L_X} \pi_1$, we need to prove: $\mu_\theta(\pi_1) \leq \mu_\theta(\pi_2)$ iff $\theta(X-2) \geq 1, \theta(i) \leq \theta(X-2)$ for $i < X-2$. But from the structure of (12.18), obviously $\mu_\theta(\pi_1) \leq \mu_\theta(\pi_2)$ is equivalent to $\begin{bmatrix} 0 & 1 & \theta' \end{bmatrix}' \begin{bmatrix} \pi_1 \\ -1 \end{bmatrix} \leq \begin{bmatrix} 0 & 1 & \theta' \end{bmatrix}' \begin{bmatrix} \pi_2 \\ -1 \end{bmatrix}$, or equivalently,

$$\begin{bmatrix} 0 & 1 & \theta(1) & \cdots & \theta(X-2) \end{bmatrix} (\pi_1 - \pi_2) \leq 0.$$

Now from Lemma 12.B.2(i), $\pi_2 \geq_{L_X} \pi_1$ implies that $\pi_1 = \epsilon_1 e_X + (1-\epsilon_1)\bar{\pi}$, $\pi_2 = \epsilon_2 e_X + (1-\epsilon_2)\bar{\pi}$ and $\epsilon_1 \leq \epsilon_2$. Substituting these into the above expression, we need to prove

$$(\epsilon_1 - \epsilon_2)\left(\theta(X-2) - \begin{bmatrix} 0 & 1 & \theta(1) & \cdots & \theta(X-2) \end{bmatrix}' \bar{\pi}\right) \leq 0, \quad \forall \bar{\pi} \in \mathcal{H}_X$$

iff $\theta(X-2) \geq 1, \theta(i) \leq \theta(X-2), i < X-2$. This is obviously true.

A similar proof shows that on lines $\mathcal{L}(e_1, \bar{\pi})$ the linear threshold policy satisfies $\mu_\theta(\pi_1) \leq \mu_\theta(\pi_2)$ iff $\theta(i) \geq 0$ for $i < X-2$. □

As a consequence of Theorem 12.4.1, the optimal linear threshold approximation to switching curve Γ of Theorem 12.3.4 is the solution of the following constrained optimization problem:

$$\theta^* = \arg\min_{\theta \in \mathbb{R}^X} J_{\mu_\theta}(\pi), \quad \text{subject to } 0 \leq \theta(i) \leq \theta(X-2),\ \theta(X-2) \geq 1 \text{ and } \theta(X-1) > 0 \tag{12.19}$$

where the cumulative cost $J_{\mu_\theta}(\pi)$ is obtained as in (11.1) by applying threshold policy μ_θ in (12.18).

Remark: The constraints in (12.19) are necessary *and* sufficient for the linear threshold policy (12.18) to be MLR increasing on lines $\mathcal{L}(e_X, \bar{\pi})$ an $\mathcal{L}(e_1, \bar{\pi})$. That is, (12.19) defines the set of all MLR increasing linear threshold policies – it does not leave out any MLR increasing polices; nor does it include any non-MLR increasing policies. Therefore optimizing over the space of MLR increasing linear threshold policies yields the optimal linear approximation to threshold curve Γ.

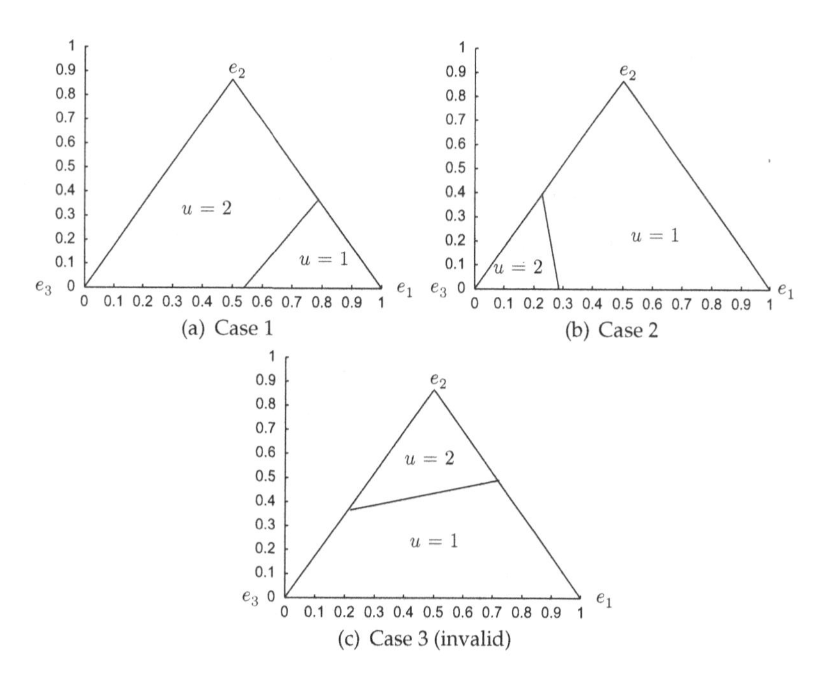

Figure 12.5 Examples of valid MLR increasing linear threshold policies for a stopping time POMDP on belief space $\Pi(X)$ for $X = 3$ (Case 1 and Case 2). Case 3 is invalid.

Intuition: Consider $\mathcal{X} = \{1, 2, 3\}$ so that the belief space $\Pi(X)$ is an equilateral triangle. Then with $(\omega(1), \omega(2))$ denoting Cartesian coordinates in the equilateral triangle, clearly $\pi(2) = 2\omega(2)/\sqrt{3}$, $\pi(1) = \omega(1) - \omega(2)/\sqrt{3}$ and the linear threshold satisfies

$$\omega(2) = \frac{\sqrt{3}\theta(1)}{2 - \theta(1)}\omega(1) + \big(\theta(2) - \theta(1)\big)\frac{\sqrt{3}}{2 - \theta(1)}. \tag{12.20}$$

So the conclusion of Theorem 12.4.1 that $\theta(1) \geq 1$ implies that the linear MLR increasing threshold has slope of 60° or larger. For $\theta(1) > 2$, it follows from (12.20) that the slope of the linear threshold becomes negative, i.e. more than 90°. For a non-degenerate threshold, the $\omega(1)$ intercept of the line should lie in $[0, 1]$ implying $\theta(1) > \theta(2)$ and $\theta(2) > 0$. Figure 12.5 illustrates these results. Figure 12.5(a) and (b) illustrate valid linear thresholds. In either case the conditions of Theorem 12.4.1 hold (the slope is larger than 60° and $\omega(1)$ intercept is in $[0, 1]$). Figure 12.5(c) shows an invalid threshold (since the slope is smaller than 60° and $\omega(1)$ intercept lies outside $[0, 1]$). In other words, Figure 12.5(c) shows an invalid threshold since it violates the requirement that $\mu_\theta(\pi)$ is decreasing on lines toward e_3 on $\Pi(3)$. A line segment $\mathcal{L}(e_3, \bar{\pi})$ starting from some point $\bar{\pi}$ on facet (e_2, e_1) and connected to e_3 would start in the region $u = 2$ and then go to region $u = 1$. This violates the requirement that $\mu_\theta(\pi)$ is increasing on lines toward e_3.

12.4.2 Algorithm to compute the optimal linear threshold policy

In this section a stochastic approximation algorithm is presented to estimate the threshold vector θ^* in (12.19). Because the cumulative cost $J_{\mu_\theta}(\pi)$ in (12.19) of the linear threshold policy μ_θ cannot be computed in closed form, we resort to simulation-based stochastic optimization. Let $n = 1, 2 \dots,$ denote iterations of the algorithm. The aim is to solve the following linearly constrained stochastic optimization problem:

$$\text{Compute } \theta^* = \arg\min_{\theta \in \Theta} \mathbb{E}\{J_n(\mu_\theta)\}$$
$$\text{subject to } 0 \leq \theta(i) \leq \theta(X-2), \theta(X-2) \geq 1 \text{ and } \theta(X-1) > 0. \tag{12.21}$$

Here the sample path cumulative cost $J_n(\mu_\theta)$ is evaluated as

$$J_n(\mu_\theta) = \sum_{k=0}^{\infty} \rho^k C(\pi_k, u_k), \quad \text{where } u_k = \mu_\theta(\pi_k) \text{ is computed via (12.18)} \tag{12.22}$$

with prior π_0 sampled uniformly from $\Pi(X)$. A convenient way of sampling uniformly from $\Pi(X)$ is to use the Dirichlet distribution (i.e. $\pi_0(i) = x_i / \sum_i x_i$, where $x_i \sim$ unit exponential distribution).

The above constrained stochastic optimization problem can be solved by a variety of methods.[7] One method is to convert it into an equivalent unconstrained problem via the following parametrization: Let $\phi = (\phi(1), \ldots \phi(X-1))' \in \mathbb{R}^{X-1}$ and parametrize θ as $\theta^\phi = \left[\theta^\phi(1), \ldots, \theta^\phi(X-1)\right]'$ where

$$\theta^\phi(i) = \begin{cases} \phi^2(X-1) & i = X-1 \\ 1 + \phi^2(X-2) & i = X-2 \\ (1 + \phi^2(X-2)) \sin^2(\phi(i)) & i = 1, \ldots, X-3 \end{cases}. \tag{12.23}$$

Then θ^ϕ trivially satisfies constraints in (12.21). So (12.21) is equivalent to the following unconstrained stochastic optimization problem:

$$\text{Compute } \mu_{\phi^*}(\pi) \text{ where } \phi^* = \arg\min_{\phi \in \mathbb{R}^{X-1}} \mathbb{E}\{J_n(\phi)\} \text{ and}$$
$$J_n(\phi) \text{ is computed using (12.22) with policy } \mu_{\theta^\phi}(\pi) \text{ in (12.23).} \tag{12.24}$$

Algorithm 14 uses the SPSA Algorithm 17 on page 350 of Chapter 15 to generate a sequence of estimates $\hat{\phi}_n$, $n = 1, 2, \ldots$, that converges to a local minimum of the optimal linear threshold ϕ^* with policy $\mu_{\phi^*}(\pi)$.

Algorithm 14 Policy gradient SPSA algorithm for computing optimal linear threshold policy

Assume (C), (F1), (F2), (S) hold so that the optimal policy is characterized by a switching curve in Theorem 12.3.4.
Step 1: Choose initial threshold coefficients $\hat{\phi}_0$ and linear threshold policy $\mu_{\hat{\phi}_0}$.
Step 2: For iterations $n = 0, 1, 2, \ldots$

- Evaluate sample cumulative cost $J_n(\hat{\phi}_n)$ using (12.24).
- Update threshold coefficients $\hat{\phi}_n$ via SPSA Algorithm 17 on page 350 as

$$\hat{\phi}_{n+1} = \hat{\phi}_n - \epsilon_{n+1} \widehat{\nabla}_\phi J_n(\hat{\phi}_n) \tag{12.25}$$

[7] More sophisticated gradient estimation methods can be used instead of the SPSA finite difference algorithm. Chapter 15 discuses the score function and weak derivative approaches for estimating the gradient of a Markov process with respect to a policy.

The stochastic gradient algorithm (12.25) converges to a local optimum. So it is necessary to try several initial conditions $\hat{\phi}_0$. The computational cost at each iteration is linear in the dimension of $\hat{\phi}$.

Summary

We have discussed three main structural results for stopping time POMDPs, namely convexity of the stopping set for linear costs (Theorem 12.2.1), monotone value function (Theorem 11.2.1) and monotone optimal policy (Theorem 12.3.4). Then a stochastic approximation algorithm was given to estimate the optimal linear MLR threshold policy. Using this policy, the real-time POMDP controller can be implemented according to Algorithm 11 on page 152. The structural results of this chapter are "class"-type results; that is, for parameters belonging to a set, the results hold. Hence there is an inherent robustness in these results since even if the underlying parameters are not exactly specified but still belong to the appropriate sets, the results still hold.

12.5 Example: Machine replacement POMDP

We continue here with the machine replacement example discussed in §6.2.3 and §7.3. We show that the conditions of Theorem 12.3.4 hold and so the optimal policy for machine replacement is characterized by a threshold switching curve.

Recall the state space is $\mathcal{X} = \{1, 2, \ldots, X\}$ where state X denotes the best state (brand new machine) while state 1 denotes the worst state, and the action space is $\mathcal{U} = \{1 \text{ (replace)}, 2 \text{ (continue)}\}$. Consider an infinite horizon discounted cost version of the problem. Bellman's equation reads

$$\mu^*(\pi) = \operatorname*{argmin}_{u \in \mathcal{U}} Q(\pi, u), \quad V(\pi) = \min_{u \in \mathcal{U}} Q(\pi, u)$$
$$Q(\pi, 1) = R + \rho\, V(e_X), \quad Q(\pi, 2) = c_2'\pi + \rho \sum_{y \in \mathcal{Y}} V(T(\pi, y))\sigma(\pi, y). \qquad (12.26)$$

Since every time action $u = 1$ (replace) is chosen, the belief state switches to e_X, Bellman's equation (12.26) is similar to that of a stopping POMDP. The cost of operating the machine $c(x, u = 2)$ is decreasing in state x since the smaller x is, the higher the cost incurred due to loss of productivity. So (C) holds. The transition matrix $P(2)$ defined in (6.13) satisfies (F2). Assume that the observation matrix B satisfies (F1). Finally, since $c(x, 2)$ is decreasing in x and $c(x, 1) = R$ is independent of x, it follows that $c(x, u)$ is submodular and so (S) holds. Then from Theorem 12.3.4, the optimal policy is MLR increasing and characterized by a threshold switching curve. Also from Theorem 12.2.1, the set of beliefs $\mathcal{R}_1$ where it is optimal to replace the machine is convex. Since the optimal policy is MLR increasing, if $\mathcal{R}_1$ is non-empty, then $e_1 \in \mathcal{R}_1$. Algorithm 14 can be used to estimate the optimal linear parametrized policy that is MLR increasing.

12.6 Multivariate stopping time POMDPs

This section discusses how the structural results obtained thus far in this chapter can be extended to POMDPs where the observation distribution is multivariate. The results are conceptually similar identical to the univariate case except that a multivariate version of the MLR ordering called the TP2 order is used.

Consider L possibly dependent Markov chains each with state space $\{1, 2, \ldots, S\}$. The composite S^L state Markov chain is denoted as

$$x_k = (x_k^{(1)}, \ldots, x_k^{(L)}) \text{ with state space } \mathcal{X} = S \times \cdots \times S. \tag{12.27}$$

We index the states of x_k by the vector indices $\mathbf{i}$ or $\mathbf{j}$, where

$$\mathbf{i} = (i_1,, \ldots, i_L) \in \mathcal{X} \text{ with generic element } i_l \in S,\ l = 1, 2, \ldots, L.$$

Assume the composite process x_k evolves with transition matrix

$$P = [P_{\mathbf{ij}}]_{S^L \times S^L},\ P_{\mathbf{ij}} = \mathbb{P}(x_k = \mathbf{j} | x_{k-1} = \mathbf{i}); \quad \pi_0 = [\pi_0(\mathbf{i})]_{S^L \times 1}, \\ \text{where } \pi_0(\mathbf{i}) = \mathbb{P}(x_0 = \mathbf{i}). \tag{12.28}$$

The observation vector y_k is recorded at time k is

$$y_k = (y_k^{(1),} \ldots, y_k^{(L)}) \text{ from } L\text{-variate distribution } \mathbb{P}(y_k | x_k, u_k). \tag{12.29}$$

Such multivariate observations with dependent individual Markov chains can model correlated targets moving in a convoy or flying in a formation.

The aim is to characterize the structure of the optimal policy of Bellman's equation (12.4) where the L-variate belief is defined as

$$\pi_k(\mathbf{i}) = \mathbb{P}(x_{k+1} = \mathbf{i} | y_{1:k}, u_{0,k-1}), \quad \mathbf{i} \in \mathcal{X}.$$

Note that π lives in the $S^L - 1$ dimension unit simplex denoted as $\Pi(X)$.

Multivariate definitions

We say that $\mathbf{i} \preceq \mathbf{j}$ (equivalently, $\mathbf{j} \succeq \mathbf{i}$), if $i_l \leq j_l$ for all $l = 1, \ldots, L$. (So $\preceq$ denotes the componentwise partial order.) Denote the element-wise minimum $\mathbf{i} \wedge \mathbf{j} = [\min(i_1, j_1), \ldots, \min(i_L, j_L)]$ and maximum $\mathbf{i} \vee \mathbf{j} = [\max(i_1, j_1), \ldots, \max(i_L, j_L)]$.

Define a function ϕ to be increasing if $i \preceq \mathbf{j}$ implies $\phi(\mathbf{i}) \leq \phi(\mathbf{j})$.

Let π and $\bar{\pi}$ be L-variate beliefs. Then we have the following multivariate counterparts of univariate stochastic dominance.

DEFINITION 12.6.1 *1. A belief π first-order multivariate stochastically dominates another belief $\bar{\pi}$ if for all increasing functions ϕ, $\sum_{\mathbf{i}} \phi(\mathbf{i})\pi(i) \geq \sum_{\mathbf{i}} \phi(\mathbf{i})\bar{\pi}(\mathbf{i})$.*

2. *π TP2 dominates $\bar{\pi}$, denoted as $\pi \underset{TP2}{\geq} \bar{\pi}$, if for $\mathbf{i}, \mathbf{j} \in \mathcal{X}$, $\pi(\mathbf{i})\,\bar{\pi}(\mathbf{j}) \leq \pi(\mathbf{i} \wedge j)\bar{\pi}(\mathbf{i} \vee \mathbf{j})$.*
3. *In general the TP2 order is not reflexive, i.e. $\pi \underset{TP2}{\geq} \pi$ does not necessarily hold. A multivariate distribution π is said to be multivariate TP2 (MTP2) if $\pi \underset{TP2}{\geq} \pi$ holds, i.e. $\pi(\mathbf{i})\pi(\mathbf{j}) \leq \pi(\mathbf{i} \vee \mathbf{j})\pi(\mathbf{i} \wedge \mathbf{j})$. This definition of reflexivity also applies to stochastic matrices and kernels.*
4. Result*: [249] MTP2 dominance implies first-order multivariate dominance.*

We used the TP2 order in §10.6.3 for monotone properties of filters.

Main result

For multivariate POMDPs, the assumptions (C), (F1), (F2) are modified as follows: in (C), $\geq_s$ is modified to multivariate first-order dominance. In (F1) and (F2), TP2 is replaced by MTP2. In particular, (F1) assumes that the L-variate observation probabilities $\mathbb{P}(y|x,u)$ are MTP2 in (y,x); see Definition 12.6.1(3). Also we assume that the expected costs are linear in the belief state.

THEOREM 12.6.2 *Consider a stopping time POMDP with L-variate observations.*

1. *The stopping region $\mathcal{R}_1$ is convex. (This follows from Theorem 12.2.1 since the costs are linear in the belief state.)*
2. *Under the multivariate versions of (C), (F1), (F2), the value function $V(\pi)$ in Bellman's equation is MTP2 decreasing in π. (The proof is similar to Theorem 11.2.1.)*
3. *Furthermore if (S) holds with $X = S^L$, then Statement 2 of Theorem 12.3.4 holds.*

Similar to §12.4, linear parametrized threshold policies that are TP2 increasing can be computed.

Remark. Why univariate orders will not work: Consider the case $L = 2$, $S = \{1,2\}$, $y_1, y_2 \in \{1,2\}$. Suppose the Markov chains $x_k^{(1)}$ and $x_k^{(2)}$ have observation probabilities

$$B^{(1)} = \begin{bmatrix} 0.8 & 0.2 \\ 0.3 & 0.7 \end{bmatrix}, \quad B^{(2)} = \begin{bmatrix} 0.9 & 0.1 \\ 0.4 & 0.6 \end{bmatrix}.$$

Clearly $B^{(1)}$ and $B^{(2)}$ are TP2 matrices. One might think that by defining the scalar observation process $y = \{1,2,3,4\}$ corresponding to the composite observations $y^{(1)}, y^{(2)}$ of $\{1,1\},\{1,2\},\{2,1\},\{2,2\}$, the resulting scalar observation matrix is TP2. However, even in the simple case where $y^{(1)}$ and $y^{(2)}$ are independent, the 4×4 matrix $\mathbb{P}(y^{(1)}, y^{(2)}|x^{(1)}, x^{(2)})$ is not TP2, but is MTP2 in $(y^{(1)}, y^{(2)}, x^{(1)}, x^{(2)})$. This follows from the classic result in [169, Proposition 3.5]. Therefore the multivariate version of (F1) holds.

To summarize for multivariate observations, similar structural results hold for POMDPs except that the MLR order is replaced with the TP2 order.

12.7 Radar scheduling with mutual information cost

So far this chapter has discussed POMDP stopping time problems involving HMM filters. In Chapter 13 we will present several examples of such stopping time POMDPs in the context of quickest change detection. However, in this section we take a small diversion by discussing structural results for a stopping time problem involving *Kalman filters*. The result is conceptually similar to stopping time POMDPs; the optimal policy is monotone (with respect to positive definite covariance matrices) and the parametrized policy can be computed via simulation-based optimization.

This section builds on the material in §8.3 where measurement control of Kalman filter tracking algorithms was described. However, the formulation differs from §8.3 in two ways: the cost used here is the mutual information which involves the determinant of the covariance matrix (instead of the trace used in §8.3). Second, the model used allows for missing observations which affects the Kalman filter covariance update.

Consider an adaptive radar which tracks L moving targets indexed by $l \in \{1, \ldots, L\}$. The costs involved are the mutual information (stochastic observability) of the targets. How much time should the radar spend observing each target so as to maintain the mutual information of all targets within a specified bound? The main result of this section is that the radar should deploy a policy which is a monotone function on the partially ordered set of positive definite covariance matrices.[8] This structure of the optimal policy facilitates developing numerically efficient algorithms to estimate and implement such policies.

The radar scheduler consists of a *macro-manager* that operates on a slow timescale and a *micro-manager* that operates on a fast timescale. The radar macro-manager allocates each target l a priority $\nu^l \in [0, 1]$ with $\sum_{l=1}^{L} \nu^l = 1$. Once the target priorities are assigned, the radar micro-manager is initiated. The L targets are then tracked using Kalman filters for time $k = 0, 1 \ldots, \tau$ where τ is a stopping time chosen by the micro-manager. The priority ν^l determines the fraction of the time the radar devotes to target l. The more time devoted to a target, the smaller the variance of its state estimate. Since the aim is keep all estimates within a specified bound, the question we seek to answer is: *Until what time τ should the micro-manager track the L targets with existing priority vector $\nu = (\nu^1, \ldots, \nu^L)$ before returning control to the macro-manager to pick a new priority vector*? This is formulated as a stopping time control problem.

12.7.1 Target kinematic model and tracker

The dynamics and measurement equations for each moving target l are modeled as a linear time invariant Gaussian state space system[9]

$$
\begin{aligned}
x_{k+1}^l &= A x_k^l + f w_k^l, \quad x_0 \sim N(\hat{x}_0, \Sigma_0), \\
y_k^l &= \begin{cases} H x_k^l + \frac{1}{\sqrt{\nu^l \Delta}} v_k^l, & \text{with probability } p_d^l, \\ \emptyset, & \text{with probability } 1 - p_d^l. \end{cases}
\end{aligned}
\tag{12.30}
$$

The state process $\{x_k^l\}$ consists of the position and velocity of target l. The matrices A and f are specified as in (2.39) of §2.6.

The noise processes $\{w_k^l\}$ and $\{v_k^l\}$ in (12.30) are mutually independent, i.i.d. zero-mean Gaussian vectors with covariance matrices Σ_w^l and Σ_v^l, respectively.

Regarding the observation process $\{y_k\}$ in (12.30), p_d^l denotes the probability of detection of target l. Also, $\emptyset$ represents a missed observation that contains no information about the state. Each tick at the k time scale consists of $\nu^l \Delta$ measurements of target l,

[8] As in §8.3, the belief state is a Gaussian with mean and covariance obtained from the Kalman filter. So the structural result that the optimal policy is monotone with respect to the covariance (belief) is consistent with the theme of this chapter.

[9] To avoid a notational clash with costs denoted as C, we use H for the observation matrix.

$l = 1, 2, \ldots, L$. These measurements are integrated[10] to yield a single observation y_k^l. Therefore, for target l with priority ν^l, the observation noise is $v_k^l/\sqrt{\nu^l \Delta}$ with covariance matrix is $\Sigma_v^l/(\nu^l \Delta)$.

In summary, the state noise and observation noise covariances in (12.30) are

$$Q^l = f\sigma_w^l f', \quad R^l(\nu^l) = \Sigma_v^l/(\nu^l \Delta).$$

The L targets are tracked using Kalman filters. The covariance matrix of the state estimate of target l at time k given observations $y_{1:k-1}$ is

$$\Sigma_{k|k-1}^l = \mathbb{E}\{(x_k^l - \mathbb{E}\{x_k^l|y_{1:k-1}^l\})(x_k^l - \mathbb{E}\{x_k^l|y_{1:k-1}^l\})'\}.$$

Based on (12.30), this is computed via the following measurement dependent Riccati equation (which follows from the Riccati equation (3.39) of Chapter 3):

$$\begin{aligned}\Sigma_{k+1|k}^l &= \mathcal{R}(\Sigma_{k|k-1}^l, y_k^l) \qquad (12.31)\\ &= A\Sigma_{k|k-1}^l A' + Q^l - I(y_k^l \neq \emptyset) A\Sigma_k^l H' \big(H\Sigma_{k|k-1}^l H' + R^l(\nu^l)\big)^{-1} H\Sigma_{k|k-1}^l A'.\end{aligned}$$

Here $I(\cdot)$ denotes the indicator function. In the special case when a target l is allocated zero priority ($\nu^{(l)} = 0$), or when there is a missing observation ($y_k^l = \emptyset$), then (12.31) becomes the Kalman predictor updated via the Lyapunov equation

$$\Sigma_{k+1|k}^l = \mathcal{L}(\Sigma_{k|k-1}^l) \stackrel{\text{def}}{=} A\Sigma_{k|k-1}^l A' + Q^l. \qquad (12.32)$$

12.7.2 Actions and costs

Denote the highest priority target as

$$a = \operatorname*{argmax}_l \nu^l.$$

At each time slot k, the radar micro-manager chooses action

$$u_k \in \{1 \text{ (stop)}, 2 \text{ (continue)}\}.$$

The costs incurred by these actions are:
Operating cost: Action $u_k = 2$ (continue) incurs the instantaneous cost C_o. This can be regarded as the threat rate of the highest priority target a.
Stopping cost: Action $u_k = 1$ (stop) incurs the following stopping cost:

$$C(x_k, y_{1:k}) = -I(x_k^a, y_{1:k-1}^a) + \max_{l \neq a} I(x_k^l, y_{1:k-1}^l). \qquad (12.33)$$

Here, $I(x_k^l; y_{1:k-1}^l)$ denotes the mutual information which is the average reduction in uncertainty of the target's coordinates x_k^l given measurements $y_{1:k-1}^l$. It is defined as

$$I(x_k^l; y_{1:k-1}^l) = h(x_k^l) - h(x_k^l|y_{1:k-1}^l), \qquad (12.34)$$

[10] In modern GMTI (ground moving target indicator radar) radars, individual measurements are obtained once every millisecond. These are averaged over a window $\Delta \approx 100$. implying that the k-time scale ticks at 0.1 of a second. The scheduling interval over which the micromanager operates before reverting back to the macro-manager is 10 to 50 epochs – in absolute time this corresponds to a range of 1 to 5 seconds. A ground target moving at 50 km per hour moves approximately 14 to 70 meters in this time period.

where $h(x_k^l)$ is the differential entropy of target l at time k, and $h(x_k^l|y_{1:k-1}^l)$ denotes the conditional differential entropy of target l at time k given the observation history $y_{1:k-1}^l$. Information theoretic criteria such as mutual information (stochastic observability) in target tracking have been used in [243, 220].

The stopping cost (12.33) is the difference in mutual information between the target with highest mutual information and the target with highest priority. If target a has a significantly higher priority than all other targets, this stopping cost discourages stopping too soon thereby allocating more resources to target a. One could replace $\max_{l \neq a}$ in (12.33) with $\min_{l \neq a}$ which can be viewed as a conservative stopping cost that gives preference to stop sooner.

By elementary information theory [90], for the Gaussian model (12.30), the mutual information-based stopping cost (12.33) becomes

$$C(\Sigma_k) = -\log|\bar{\Sigma}_{k|0}^a| + \log|\Sigma_{k|k-1}^a| + \max_{l \neq a}\left[\log|\bar{\Sigma}_{k|0}^l| - \log|\Sigma_{k|k-1}^l|\right]. \qquad (12.35)$$

Here, $|\cdot|$ denotes determinant. Also $\bar{\Sigma}_{k|0}^l = \mathbb{E}\{(x_k^l - \mathbb{E}\{x_k^l\})(x_k^l - \mathbb{E}\{x_k^l\})'\}$ is the predicted (a priori) covariance of target l at epoch k given no observations. It is computed using the Kalman predictor covariance update (12.32) for k iterations. Recall that $\Sigma_{k|k-1}^l$ is computed via the Riccati equation (12.31). On the left-hand side of (12.35), Σ_k denotes the covariances of all the targets:

$$\Sigma_k = (\Sigma_{k|k-1}^a, \bar{\Sigma}_{k|0}^a, \Sigma_{k|k-1}^{-a}, \bar{\Sigma}_{k|0}^{-a}), \qquad (12.36)$$

where Σ^a denotes the covariance of the high priority target a, and Σ^{-a} denotes the set of covariance matrices of the remaining $L-1$ targets.

12.7.3 Stopping time problem

With the above stopping and continuing costs, the radar stopping time control problem can now be formulated. Let μ denote a stationary policy

$$\mu : \Sigma_k \to u_k \in \{1 \text{ (stop)}, 2 \text{ (continue)}\}. \qquad (12.37)$$

For any prior 4-tuple Σ_0 and policy μ chosen by the micro-manager, define the stopping time $\tau = \min\{k : u_k = 1\}$. The cumulative cost incurred is

$$J_\mu(\Sigma) = \mathbb{E}_\mu\{\tau C_o + C(\Sigma_{\tau+1})|\Sigma_0 = \Sigma\}. \qquad (12.38)$$

This cost is a trade off between the threat of the highest priority target a versus the increasing uncertainty (mutual information) of the remaining targets. In (12.38), $\mathbb{E}_\mu$ denotes expectation with respect to stopping time τ.

The goal is to determine the optimal policy μ^* to minimize (12.38). Denote

$$J_{\mu^*}(\Sigma) = \min_\mu J_\mu(\Sigma). \qquad (12.39)$$

The optimal stationary policy $\mu^* \in \boldsymbol{\mu}$ and associated value function $V(\Sigma) = J_{\mu^*}(\Sigma)$ satisfy Bellman's equation (recall $\Sigma = (\Sigma^a, \bar{\Sigma}^a, \Sigma^{-a}, \bar{\Sigma}^{-a})$)

$$V(\Sigma) = \min\{Q(\Sigma,1), Q(\Sigma,2)\}, \quad \mu^*(\Sigma) = \operatorname{argmin}\{Q(\Sigma,1), Q(\Sigma,2)\},$$
$$Q(\Sigma,2) = C_o + \mathbb{E}_y\left[V\big(\mathcal{R}(\Sigma^a,y^a), \mathcal{L}(\bar{\Sigma}^a), \mathcal{R}(\Sigma^{-a},y^{-a}), \mathcal{L}(\bar{\Sigma}^{-a})\big)\right], \tag{12.40}$$
$$Q(\Sigma,1) = C(\Sigma)$$

where $\mathcal{R}$ and $\mathcal{L}$ are the Riccati and Lyapunov updates specified in (12.31) and (12.32). Here $\mathcal{R}(\Sigma^{-a},y^{-a})$ denotes the Kalman filter covariance update for the $L-1$ lower priority targets according to (12.31).

Since Σ belongs to an uncountable set it is not possible to compute the optimal policy in closed form. Hence the motivation to develop structural results.

12.7.4 Main structural result: monotone optimal decision policy

The result below shows that the optimal decision policy μ^* satisfying (12.40) is a monotone function of the covariance matrices of the targets.

Let $\mathcal{M}$ denote the set real-valued, symmetric positive semi-definite matrices. For $P, Q \in \mathcal{M}$ define the positive definite partial ordering $\succeq$ as $P \succeq Q$ if $x'Px \geq x'Qx$ for all $x \neq 0$, and $P \succ Q$ if $x'Px > x'Qx$ for $x \neq 0$. Define $\preceq$ with the inequalities reversed. Note that $[\mathcal{M}, \succeq]$ is a partially ordered set (poset).

We say that a scalar-valued function f is increasing if $P \preceq Q$ implies $f(P) \leq f(Q)$. Also $f(\Sigma^{-a})$ is increasing in Σ^{-a} if $f(\cdot)$ is increasing in each component Σ^l of Σ^{-a}, $l \neq a$.

The following is the main result. The proof in the appendix uses monotone properties of the Riccati and Lyapunov equations of the Kalman filter.

THEOREM 12.7.1 *Consider the stopping time problem (12.38) with stochastic observability cost (12.35). The optimal policy $\mu^*(\Sigma^a, \bar{\Sigma}^a, \Sigma^{-a}, \bar{\Sigma}^{-a})$ which satisfies (12.40) is increasing in Σ^a and $\bar{\Sigma}^{-a}$, and decreasing in $\bar{\Sigma}^a$ and Σ^{-a}, on the poset $[\mathcal{M}, \succeq]$.*

As a simple illustration of Theorem 12.7.1, suppose that there are two targets and the covariance of each target is a nonnegative scalar. Then Theorem 12.7.1 says that the optimal policy is a threshold curve of the form illustrated in Figure 12.6. This policy is intuitive since if the covariance of the high priority target increases, one needs to continue giving it high priority; if the covariance of the other target increases, then the micro-manager should return control to the macro-manager to reassign the target priorities.

Theorem 12.7.1 asserts that a monotone optimal policy $\mu^*(\Sigma)$ exists for the radar stopping time problem. How can this structure be exploited to implement a practical radar resource management algorithm? The monotone policy can be suitably parametrized by class of monotone functions and then the optimal parameter estimated via simulated-based stochastic optimization.

For example, define Θ as the set of unit-norm vectors in $\mathbb{R}^X$. Let θ^l and $\underline{\theta}^l \in \Theta$ denote unit-norm vectors, $\theta^{l\prime}\theta^l = 1$, $\underline{\theta}^{l\prime}\underline{\theta}^l = 1$ for $l = 1, \ldots, L$. Define the parametrized policy μ_θ, $\theta \in \Theta$ as

$$\mu_\theta(\Sigma) = \begin{cases} 1 & \text{if } -\theta^{a\prime}\Sigma^a\theta^a + \underline{\theta}^{a\prime}\bar{\Sigma}^a\underline{\theta}^a + \sum_{l\neq a}\theta^{l\prime}\Sigma^l\theta^l - \underline{\theta}^{l\prime}\bar{\Sigma}^l\underline{\theta}^l \geq 1, \\ 2 & \text{otherwise.} \end{cases} \tag{12.41}$$

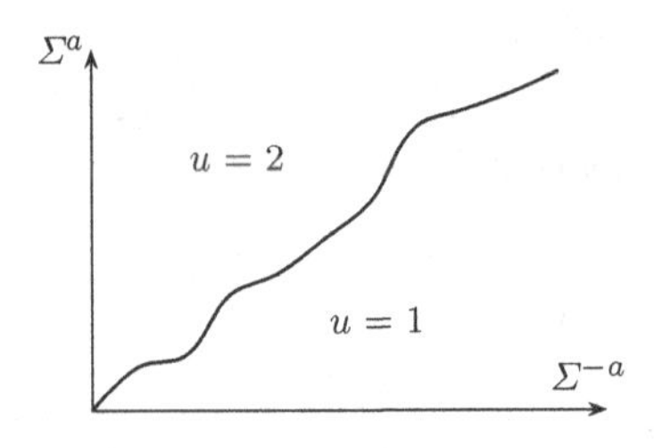

Figure 12.6 Threshold structure of optimal policy $\mu^*(\Sigma^a, \Sigma^{-a})$ for scalar covariances.

It is easily verified that this parametrized policy satisfies the monotone property of Theorem 12.7.1, namely $\mu_\theta(\Sigma)$ is increasing in $\Sigma^a, \bar{\Sigma}^{-a}$, and decreasing in $\bar{\Sigma}^a, \Sigma^{-a}$. The optimal parameter θ^* (which optimizes (12.38)) can be estimated via simulated-based stochastic optimization using for example the SPSA Algorithm 17 in Chapter 15. In [182], numerical examples are presented to illustrate these parametrized policies in persistent surveillance applications.

To summarize, the stopping time problem involving Kalman filters discussed in this section is similar to the POMDP stopping time problems considered in this chapter. In both cases:

1. The main structural result is that the optimal policy is monotone in the belief state with respect to a partial order (MLR in the POMDP case, positive definite ordering of covariance matrices in the Kalman filter case).
2. A parametrized version of the monotone policy can be estimated via simulated-based stochastic optimization.

12.8 Complements and sources

The convexity of the stopping region has been shown in several papers. Lovejoy [223] presents the clean four-line proof that we use in Theorem 12.2.1. The instruction problem of §12.2.3 is discussed in [49].

In [225] it is shown that the optimal policy for a stopping time POMDP is monotone increasing with respect to the MLR order. Theorem 12.3.4 extends [225] to nonlinear costs and shows the existence of a threshold switching curve by considering line segments in the simplex where the MLR order is a complete order. This result together with the characterization of the optimal linear threshold policy appears in [183, 179].

For multivariate POMDPs discussed in §12.6, the TP2 order and MTP2 kernels are detailed in the classic paper [169]. The TP2 order for multivariate POMDPs was used in [277]. [184] illustrates its use in radar resource management.

The radar resource allocation problem with mutual information cost (§12.7) is discussed in [182] where numerical examples are presented to illustrate fly-by and persistent surveillance applications of GMTI radar systems. In the appendix we derive monotone properties of the Kalman filter covariance updates.

Appendix 12.A Lattices and submodularity

Definition 12.3.2 on page 263 on submodularity suffices for our treatment of POMDPs. Here we outline a more abstract definition; see [322] for details.

(i) Poset: Let Ω denote a nonempty set and $\preceq$ denote a binary relation. Then $(\Omega, \preceq)$ is a partially ordered set (poset) if for any elements $a, b, c \in \Omega$, the following hold:

1. $a \preceq a$ (reflexivity)
2. if $a \preceq b$ and $b \preceq a$, then $a = b$ (anti-symmetry)
3. if $a \preceq b$ and $b \preceq c$, then $a \preceq c$ (transitivity).

For a POMDP, clearly $(\Pi(X), \leq_r)$ is a poset, where

$$\Pi(X) = \left\{\pi \in \mathbb{R}^X : \mathbf{1}'\pi = 1, \quad 0 \leq \pi(i) \leq 1 \text{ for all } i \in \mathcal{X}\right\}$$

is the belief space and $\leq_r$ is the MLR order defined in (10.2).

(ii) Lattice: A poset $(\Omega, \preceq)$ is called a lattice if the following property holds: $a, b \in \Omega$, then $a \vee b \stackrel{\text{defn}}{=} \max\{a, b\} \in \Omega$ and $a \wedge b \stackrel{\text{defn}}{=} \min\{a, b\} \in \Omega$. (Here min and max are with respect to partial order $\preceq$.)

Clearly, $(\Pi(X), \leq_r)$ is a lattice. Indeed if two beliefs $\pi_1, \pi_2 \in \Pi(X)$, then if $\pi_1 \leq_r \pi_2$, obviously $\pi_1 \vee \pi_2 = \pi_2$ and $\pi_1 \wedge \pi_2 = \pi_1$ belong to $\Pi(X)$. If π_1 and π_2 are not MLR comparable, then $\pi_1 \vee \pi_2 = e_X$ and $\pi_1 \wedge \pi_2 = e_1$, where e_i is the unit vector with 1 in the i-th position.

Note that $\Omega = \{e_1, e_2, e_3\}$ is not a lattice if one uses the natural element wise ordering. Clearly, $e_1 \vee e_2 = (1, 1, 0) \notin \Omega$ and $e_1 \wedge e_2 = (0, 0, 0) \notin \Omega$.

Finally, $\Pi(X) \times \{1, 2, \ldots, \mathcal{U}\}$ is also a lattice. This is what we use in our POMDP structural results.

(iii) Submodular function: Let $(\Omega, \preceq)$ be a lattice and $f : \Omega \to \mathbb{R}$. Then f is submodular if for all $a, b \in \Omega$,

$$f(a) + f(b) \geq f(a \vee b) + f(a \wedge b). \tag{12.42}$$

For the two component case, namely $a = (\pi_1, u+1)$, $b = (\pi_2, u)$, with $\pi_1 \leq_r \pi_2$, clearly Definition (12.42) is equivalent to Definition 12.3.2 on page 263; and this suffices for our purposes. When each of a and b consist of more than two components, the proof needs more work; see [322].

Appendix 12.B MLR dominance and submodularity on lines

Recall the definition of $\mathcal{L}(e_i, \bar{\pi})$ in (12.13).

LEMMA 12.B.1 *The following properties hold on* $[\Pi(X), \geq_r]$, $[\mathcal{L}(e_X, \bar{\pi}), \geq_{L_X}]$.
(i) On $[\Pi(X), \geq_r]$, e_1 *is the least and* e_X *is the greatest element. On* $[\mathcal{L}(e_X, \bar{\pi}), \geq_{L_X}]$, $\bar{\pi}$ *is the least and* e_X *is the greatest element and all points are MLR orderable.*
(ii) Convex combinations of MLR comparable belief states form a chain. For any $\gamma \in [0, 1]$, $\pi \leq_r \pi^0 \implies \pi \leq_r \gamma\pi + (1-\gamma)\pi^0 \leq_r \pi^0$.

LEMMA 12.B.2 *(i) For $i \in \{1, X\}$, $\pi_1 \geq_{L_i} \pi_2$ is equivalent to $\pi_j = (1 - \epsilon_j)\bar{\pi} + \epsilon_j e_X$ and $\epsilon_1 \geq \epsilon_2$ for $\bar{\pi} \in \mathcal{H}_i$ where $\mathcal{H}_i$ is defined in (12.12).*
(ii) So submodularity on $\mathcal{L}(e_i, \bar{\pi})$, $i \in \{1, X\}$, is equivalent to showing

$$\pi^\epsilon = (1-\epsilon)\bar{\pi} + \epsilon e_i \implies C(\pi^\epsilon, 2) - C(\pi^\epsilon, 1) \text{ decreasing w.r.t. } \epsilon. \tag{12.43}$$

The proof of Lemma 12.B.2 follows from Lemma 12.B.1 and is omitted.

As an example motivated by controlled sensing, consider costs that are quadratic in the belief. Suppose $C(\pi, 2) - C(\pi, 1)$ is of the form $\phi'\pi + \alpha(h'\pi)^2$. Then from (12.43), sufficient conditions for submodularity on $\mathcal{L}(e_X, \bar{\pi})$ and $\mathcal{L}(e_1, \bar{\pi})$ are for $\bar{\pi} \in \mathcal{H}_X$ and $\mathcal{H}_1$, respectively,

$$\phi_X - \phi'\bar{\pi} + 2\alpha h'\pi^\epsilon(h_X - h'\bar{\pi}) \leq 0, \quad \phi_1 - \phi'\bar{\pi} + 2\alpha h'\pi^\epsilon(h_1 - h'\bar{\pi}) \geq 0. \tag{12.44}$$

If $h_i \geq 0$ and increasing or decreasing in i, then (12.44) is equivalent to

$$\phi_X - \phi'\bar{\pi} + 2\alpha h_X(h_X - h'\bar{\pi}) \leq 0, \quad \phi_1 - \phi'\bar{\pi} + 2\alpha h_X(h_1 - h'\bar{\pi}) \geq 0 \tag{12.45}$$

where $\bar{\pi} \in \mathcal{H}_X$ and $\bar{\pi} \in \mathcal{H}_1$, respectively.

Appendix 12.C Proof of Theorem 12.3.4

Part 1: Establishing that $Q(\pi, u)$ is submodular, requires showing that $Q(\pi, 1) - Q(\pi, 2)$ is $\geq_{L_i}$ on lines $\mathcal{L}(e_X, \bar{\pi})$ for $i = 1$ and X. Theorem 11.2.1 shows by induction that for each k, $V_k(\pi)$ is $\geq_r$ decreasing on $\Pi(X)$ if (C), (F1), (F2) hold. This implies that $V_k(\pi)$ is $\geq_{L_i}$ decreasing on lines $\mathcal{L}(e_X, \bar{\pi})$ and $\mathcal{L}(e_1, \bar{\pi})$. So to prove $Q_k(\pi, u)$ in (12.9) is submodular, we only need to show that $C(\pi, 1) - C(\pi, 2)$ is $\geq_{L_i}$ decreasing on $\mathcal{L}(e_i, \bar{\pi})$, $i = 1, X$. But this is implied by (S). Since submodularity is closed under pointwise limits [322, Lemma 2.6.1 and Corollary 2.6.1], it follows that $Q(\pi, u)$ is submodular on $\geq_{L_i}$, $i = 1, X$ Having established $Q(\pi, u)$ is submodular on $\geq_{L_i}$, $i = 1, X$, Theorem 12.3.3 implies that the optimal policy $\mu^*(\pi)$ is $\geq_{L_1}$ and $\geq_{L_X}$ increasing on lines.

Part 2(a) Characterization of switching curve Γ. For each $\bar{\pi} \in \mathcal{H}_X$ (12.13), construct the line segment $\mathcal{L}(e_X, \bar{\pi})$ connecting $\mathcal{H}_X$ to e_X as in (12.13). By Lemma 12.B.1, on the line segment connecting $(1-\epsilon)\underline{\pi} + \epsilon e_X$, all belief states are MLR orderable. Since $\mu^*(\pi)$ is monotone increasing for $\pi \in \mathcal{L}(e_X, \bar{\pi})$, moving along this line segment toward e_X, pick the largest ϵ for which $\mu^*(\pi) = 1$. (Since $\mu^*(e_X) = 1$, such an ϵ always exists.) The belief state corresponding to this ϵ is the threshold belief state. Denote it by $\Gamma(\bar{\pi}) = \pi^{\epsilon^*, \bar{\pi}} \in \mathcal{L}(e_X, \bar{\pi})$ where $\epsilon^* = \sup\{\epsilon \in [0,1] : \mu^*(\pi^{\epsilon, \bar{\pi}}) = 1\}$.

The above construction implies that on $\mathcal{L}(e_X, \bar{\pi})$, there is a unique threshold point $\Gamma(\bar{\pi})$. Note that the entire simplex can be covered by considering all pairs of lines $\mathcal{L}(e_X, \bar{\pi})$, for $\bar{\pi} \in \mathcal{H}_X$, i.e. $\Pi(X) = \cup_{\bar{\pi} \in \mathcal{H}} \mathcal{L}(e_X, \bar{\pi})$. Combining all points $\Gamma(\bar{\pi})$ for all pairs of lines $\mathcal{L}(e_X, \bar{\pi})$, $\bar{\pi} \in \mathcal{H}_X$, yields a unique threshold curve in $\Pi(X)$ denoted $\Gamma = \cup_{\bar{\pi} \in \mathcal{H}} \Gamma(\bar{\pi})$.

Part 2(b) *Connectedness of* $\mathcal{R}_1$: Since $e_1 \in \mathcal{R}_1$, call $\mathcal{R}_{1a}$ the subset of $\mathcal{R}_1$ that contains e_1. Suppose $\mathcal{R}_{1b}$ was a subset of $\mathcal{R}_1$ that was disconnected from R_{1a}. Recall that every point in $\Pi(X)$ lies on a line segment $\mathcal{L}(e_1, \bar{\pi})$ for some $\bar{\pi}$. Then such a line segment starting from $e_1 \in \mathcal{R}_{1a}$ would leave the region $\mathcal{R}_{1a}$, pass through a region where action 2 was optimal, and then intersect the region $\mathcal{R}_{1b}$ where action 1 is optimal. But this violates the requirement that $\mu(\pi)$ is increasing on $\mathcal{L}(e_1, \bar{\pi})$. Hence $\mathcal{R}_{1a}$ and $\mathcal{R}_{1b}$ have to be connected.

Connectedness of $\mathcal{R}_2$: Assume $e_X \in \mathcal{R}_2$, otherwise $\mathcal{R}_2 = \emptyset$ and there is nothing to prove. Call the region $\mathcal{R}_2$ that contains e_X as $\mathcal{R}_{2a}$. Suppose $\mathcal{R}_{2b} \subset \mathcal{R}_2$ is disconnected from R_{2a}. Since every point in $\Pi(X)$ can be joined by the line segment $\mathcal{L}(e_X, \bar{\pi})$ to e_X. Then such a line segment starting from $e_X \in \mathcal{R}_{2a}$ would leave the region $\mathcal{R}_{2a}$, pass through a region where action 1 was optimal, and then intersect the region $\mathcal{R}_{2b}$ (where action 2 is optimal). This violates the property that $\mu(\pi)$ is increasing on $\mathcal{L}(e_X, \bar{\pi})$. Hence $\mathcal{R}_{2a}$ and $\mathcal{R}_{2b}$ are connected.

Part 3: Suppose $e_i \in \mathcal{R}_1$. Then considering lines $\mathcal{L}(e_i, \bar{\pi})$ and ordering $\geq_{L_i}$, it follows that $e_{i-1} \in \mathcal{R}_1$. Similarly if $e_i \in \mathcal{R}_2$, then considering lines $\mathcal{L}(e_{i+1}, \bar{\pi})$ and ordering $\geq_{L_{i+1}}$, it follows that $e_{i+1} \in \mathcal{R}_2$.

Part 4 follows trivially since for $X = 2$, $\Pi(X)$ is a one-dimensional simplex.

12.C.1 Proof of Theorem 12.7.1

It is convenient to rewrite Bellman's equation (12.40) as follows: define

$$\bar{C}(\Sigma) = C_o - C(\Sigma) + \sum_{y^a, y^{-a}} C(\mathcal{R}(\Sigma^a, y^a), \mathcal{L}(\Sigma^a), \mathcal{R}(\Sigma^{-a}, y^{-a}), \mathcal{L}(\Sigma^{-a})) q_{y^a} q_{y^{-a}},$$

$$\bar{V}(\Sigma) = V(\Sigma) - C(\Sigma), \quad \text{where } \Sigma = (\Sigma^a, \bar{\Sigma}^a, \Sigma^{-a}, \bar{\Sigma}^{-a}), \tag{12.46}$$

$$q_{y^l} = \begin{cases} p_d^l, & \text{if } y^l \neq \emptyset, \\ 1 - p_d^l, & \text{otherwise}, \end{cases}, \quad l = 1, \ldots, L.$$

In (12.46) we have assumed that missed observation events in (13.4) are statistically independent between targets, and so $q_{y^{-a}} = \prod_{l \neq a} q_{y^l}$. Then $\bar{V}(\cdot)$ and optimal policy $\mu^*(\cdot)$ satisfy Bellman's equation

$$\bar{V}(\Sigma) = \min\{\mathcal{Q}(\Sigma, 1), \mathcal{Q}(\Sigma, 2)\}, \mu^*(\Sigma) = \underset{u \in \{1,2\}}{\operatorname{argmin}} \{\mathcal{Q}(\Sigma, u)\}, \tag{12.47}$$

$$\mathcal{Q}(\Sigma, 1) = 0,$$

$$\mathcal{Q}(\Sigma, 2) = \bar{C}(\Sigma) + \sum_{y^a, y^{-a}} \bar{V}\big((\mathcal{R}(\Sigma^a, y), \mathcal{L}(\Sigma^a), \mathcal{R}(\Sigma^{-a}, y^{-a}), \mathcal{L}(\Sigma^{-a})\big) q_{y^a} q_{y^{-a}}.$$

The proof of Theorem 12.7.1 then follows from the following result.

THEOREM 12.C.1 *$\mathcal{Q}(\Sigma^a, \bar{\Sigma}^a, \Sigma^{-a}, \bar{\Sigma}^{-a}, u)$ is decreasing in Σ^a, decreasing in $\bar{\Sigma}^{-a}$, increasing in Σ^{-a} and increasing in $\bar{\Sigma}^a$. Hence there exists a version of the optimal policy $\mu^*(\Sigma^a, \bar{\Sigma}^a, \Sigma^{-a}, \bar{\Sigma}^{-a}) = \operatorname{argmin}_{u \in \{1,2\}} \mathcal{Q}(\Sigma^a, \bar{\Sigma}^a, \Sigma^{-a}, \bar{\Sigma}^{-a}, u)$, that is increasing in Σ^a, decreasing in $\bar{\Sigma}^a$, increasing in Σ^{-a} and decreasing in Σ^{-a}.*

The proof of Theorem 12.C.1 is by induction on the value iteration algorithm applied to (12.46). The value iteration algorithm reads:

$$\bar{V}_k(\Sigma) = \min\{\mathcal{Q}_k(\Sigma,1), \mathcal{Q}_k(\Sigma,2)\}, \quad \mu^*(\Sigma) = \underset{u\in\{1,2\}}{\operatorname{argmin}}\{\mathcal{Q}_k(\Sigma,u)\}, \tag{12.48}$$

$$\mathcal{Q}_k(\Sigma,2) = \bar{C}(\Sigma) + \sum_{y^a,y^{-a}} \bar{V}_{k-1}\left(\mathcal{R}(\Sigma^a,y), \mathcal{L}(\Sigma^a), \mathcal{R}(\Sigma^{-a},y^{-a}), \mathcal{L}(\Sigma^{-a})\right) q_{y^a} q_{y^{-a}}.$$

and $\mathcal{Q}_k(\Sigma,1) = 0$. Note $\bar{V}_0(\Sigma) = -C(\Sigma)$, is decreasing in $\Sigma^a, \bar{\Sigma}^{-a}$ and increasing in $\Sigma^{-a}, \bar{\Sigma}^a$ via Lemma 12.C.3 below.

Next assume $\bar{V}_k(\Sigma)$ is decreasing in $\Sigma^a, \bar{\Sigma}^{-a}$ and increasing in $\Sigma^{-a}, \bar{\Sigma}^a$. Since $\mathcal{R}(\Sigma^a,y)$, $\mathcal{L}(\bar{\Sigma}^a)$, $\mathcal{R}(\Sigma^{-a},y^{-a})$ and $\mathcal{L}(\bar{\Sigma}^{-a})$ are monotone increasing in Σ^a (see Lemma 12.C.2 below), $\bar{\Sigma}^a$, Σ^{-a} and $\bar{\Sigma}^{-a}$, it follows that the term

$$\bar{V}_k\left(\mathcal{R}(\Sigma^a,y^a), \mathcal{L}(\bar{\Sigma}^a), \mathcal{R}(\Sigma^{-a},y^{-a}), \mathcal{L}(\bar{\Sigma}^{-a})\right) q_{y^a} q_{y^{-a}}$$

is decreasing in $\Sigma^a, \bar{\Sigma}^{-a}$ and increasing in $\Sigma^{-a}, \bar{\Sigma}^a$. Next, it follows from Lemma 12.C.3 below that $\bar{C}(\Sigma^a,, \bar{\Sigma}^a, \Sigma^{-a}, \bar{\Sigma}^{-a})$ is decreasing in $\Sigma^a, \bar{\Sigma}^{-a}$ and increasing in $\Sigma^{-a}, \bar{\Sigma}^a$. Therefore from (12.48), $\mathcal{Q}_{k+1}(\Sigma,2)$ inherits this property. Hence $\bar{V}_{k+1}(\Sigma^a, \Sigma^{-a})$ is decreasing in $\Sigma^a, \bar{\Sigma}^{-a}$ and increasing in $\Sigma^{-a}, \bar{\Sigma}^a$. Since value iteration converges pointwise, i.e, $\bar{V}_k(\Sigma)$ pointwise $\bar{V}(\Sigma)$, it follows that $\bar{V}(\Sigma)$ is decreasing in Σ^a, $\bar{\Sigma}^{-a}$ and increasing in Σ^{-a}, $\bar{\Sigma}^a$.

Therefore, $\mathcal{Q}(\Sigma,2)$ is decreasing in Σ^a, $\bar{\Sigma}^{-a}$ and increasing in Σ^{-a}, $\bar{\Sigma}^a$. Therefore, from Bellman's equation $\mu^*(\Sigma) = \operatorname{argmin}\{0, Q(\Sigma,2)\}$, there exists a version of μ^* that is increasing in Σ^a, $\bar{\Sigma}^{-a}$ and decreasing in Σ^{-a}, $\bar{\Sigma}^a$. □

The two lemmas used in the above proof are now discussed.

LEMMA 12.C.2 ([17]) *The Riccati and Lyapunov updates $\mathcal{R}(\cdot)$ and $\mathcal{L}(\cdot)$ satisfy: $\Sigma_1 \succeq \Sigma_2$ (positive definite ordering), then $\mathcal{L}(\Sigma_1) \succeq \mathcal{L}(\Sigma_2)$ and $\mathcal{R}(\Sigma_1,y) \succeq \mathcal{R}(\Sigma_2,y)$.*

LEMMA 12.C.3 *The cost $\bar{C}(\Sigma^a, \bar{\Sigma}^a, \Sigma^{-a}, \bar{\Sigma}^{-a})$ defined in (12.46) is decreasing in Σ^a, $\bar{\Sigma}^{-a}$, and increasing in Σ^{-a}, $\bar{\Sigma}^a$.*

Proof Let $l^* = \arg\max_{l\neq a}\left[\log|\bar{\Sigma}_k^l| - \log|\Sigma_k^l|\right]$. From (12.46)

$$\begin{aligned}\bar{C}(\Sigma) = C_o - \log\frac{|\mathcal{L}(\bar{\Sigma}^a)|}{|\bar{\Sigma}^a|} + \sum_{y^a}\log\frac{|\mathcal{L}(\Sigma^a,y^a)|}{|\Sigma^a|}q(y^a)\\ + \log\frac{|\mathcal{L}(\bar{\Sigma}^{l^*})|}{|\bar{\Sigma}^{l^*}|} - \log\frac{|\mathcal{R}(\Sigma^{l^*},y^{l^*})|}{|\Sigma^{l^*}|}q(y^{l^*})\end{aligned} \tag{12.49}$$

where $|\cdot|$ denotes determinant. Theorem 12.C.4 below asserts that $\frac{|\mathcal{L}(\bar{\Sigma}^l)|}{|\bar{\Sigma}^l|}$ and $\frac{|\mathcal{R}(\Sigma^l,y^l)|}{|\Sigma^l|}$ are decreasing in $\bar{\Sigma}^l$ and Σ^l for all l, thereby proving the lemma. □

THEOREM 12.C.4 *Consider the Kalman filter Riccati and Lyapunov covariance updates, $\mathcal{R}(\Sigma,y)$ and $\mathcal{L}(\Sigma)$ defined in (12.31), (12.32). With* det *denoting determinant,*

$$\textit{(a):}\ \frac{\det(\mathcal{L}(\Sigma))}{\det(\Sigma)} \quad \textit{(b):}\ \frac{\det(\mathcal{R}(\Sigma,y))}{\det(\Sigma)} \ \textit{are decreasing in } \Sigma \textit{ (positive definite ordering)} \tag{12.50}$$

To prove Theorem 12.C.4, we start with the following lemma.

LEMMA 12.C.5 *For invertible matrices X, Z and conformable matrices W and Y,*

$$\det(Z)\ \det(X + YZ^{-1}W) = \det(X)\ \det(Z + WX^{-1}Y). \tag{12.51}$$

Proof The Schur complement formulae applied to $\begin{bmatrix} X & Y \\ -W & Z \end{bmatrix}$ yields

$$\begin{bmatrix} I & YZ^{-1} \\ 0 & I \end{bmatrix} \begin{bmatrix} X + YZ^{-1}W & 0 \\ 0 & Z \end{bmatrix} \begin{bmatrix} I & 0 \\ -Z^{-1}W & I \end{bmatrix}$$
$$= \begin{bmatrix} I & 0 \\ -WX^{-1} & I \end{bmatrix} \begin{bmatrix} X & 0 \\ 0 & Z + WX^{-1}Y \end{bmatrix} \begin{bmatrix} I & X^{-1}Y \\ 0 & I \end{bmatrix}.$$

Taking determinants yields (12.51). □

Proof of 12.50(a): Applying (12.51) with $[X, Y, W, Z] = [Q, A, A', \Sigma^{-1}]$ yields

$$\frac{\det(\mathcal{L}(\Sigma))}{\det(\Sigma)} = \frac{\det(A\Sigma A' + Q)}{\det(\Sigma)} = \det(\Sigma^{-1} + A'Q^{-1}A)\det(Q). \tag{12.52}$$

Since $\Sigma_1 \succ \Sigma_2 \succ 0$, then $0 \prec \Sigma_1^{-1} \prec \Sigma_2^{-1}$ and thus $0 \prec \Sigma_1^{-1} + A'Q^{-1}A \prec \Sigma_2^{-1} + A'Q^{-1}A$. Therefore $\det(\Sigma_1^{-1} + A'Q^{-1}A) < \det(\Sigma_2^{-1} + A'Q^{-1}A)$. Using (12.52), it then follows that $\det(\mathcal{L}(\Sigma))/\det(\Sigma)$ is increasing in Σ.

Proof of 12.50(b): Suppose $\Sigma_1 \succ \Sigma_2 \succ 0$. Then $\Sigma_1^{-1} \prec \Sigma_2^{-1}$ and so

$$\det(\Sigma_1^{-1} + H'R^{-1}H + A'Q^{-1}A) < \det(\Sigma_2^{-1} + H'R^{-1}H + A'Q^{-1}A)$$
$$\det(R + H\Sigma_1 H') > \det(R + H\Sigma_2 H')$$
$$\Longrightarrow \quad \frac{\det(Q)\det(\Sigma^{-1} + H'R^{-1}H + A'Q^{-1}A)\det(R)}{\det(R + H\Sigma H')} \text{ is decreasing in } \Sigma.$$

It only remains to show that the above expression is equal to $\det(\mathcal{R}(\Sigma, y))/\det(\Sigma)$. Via the matrix inversion lemma $(A + BCD)^{-1} = A^{-1} - A^{-1}B(C^{-1} + DA^{-1}B)^{-1}DA^{-1}$, it follows that $(\Sigma^{-1} + H'R^{-1}H)^{-1} = \Sigma - \Sigma H'(H\Sigma H' + R)^{-1}H\Sigma$. Therefore,

$$\det(\mathcal{R}(\Sigma, y)) = \det(A\Sigma A' - A\Sigma H'(H\Sigma H' + R)^{-1}H\Sigma A' + Q)$$
$$= \det(Q + A(\Sigma^{-1} + H'R^{-1}H)^{-1}A'). \tag{12.53}$$

Applying (12.51) with $[X, Y, W, Z] = [Q, A, A', \Sigma^{-1} + H'R^{-1}H]$ yields

$$\det(\Sigma^{-1} + H'R^{-1}H)\det(Q + A(\Sigma^{-1} + H'R^{-1}H)^{-1}A')$$
$$= \det(Q)\det(\Sigma^{-1} + H'R^{-1}H + A'Q^{-1}A). \tag{12.54}$$

Further, using (12.51) with $[X, Y, W, Z] = [\Sigma^{-1}, H', H, R]$, we have,

$$\det(\Sigma^{-1} + H'R^{-1}H) = \det(\Sigma^{-1})\det(R + H\Sigma H')/\det(R). \tag{12.55}$$

Substituting (12.55) into (12.54) yields

$$\frac{\det(\mathcal{R}(\Sigma, y))}{\det(\Sigma)} = \frac{\det(Q)\det(\Sigma^{-1} + H'R^{-1}H + A'Q^{-1}A)\det(R)}{\det(R + H\Sigma H')}. \tag{12.56}$$

13 Stopping time POMDPs for quickest change detection

Contents

Chapter 12 presented three structural results for stopping time POMDPs: convexity of the stopping region (for linear costs), the existence of a threshold switching curve for the optimal policy (under suitable conditions) and characterization of the optimal linear threshold policy. This chapter discusses several examples of stopping time POMDPs in quickest change detection. We will show that for these examples, convexity of the stopping set and threshold optimal policies arise naturally. Therefore, the structural results of Chapter 12 serve as a unifying theme and give substantial insight into what might otherwise be considered as a collection of sequential detection methods.

This chapter considers the following extensions of quickest change detection:

- Example 1: Quickest change detection with phase-distributed change time: classical quickest detection is equivalent to a stopping time POMDP where the underlying Markov chain jumps only once into an absorbing state (therefore the jump time is geometric distributed). How should quickest change detection be performed when the change time is phase-distributed and the stopping cost is quadratic in the belief state to penalize the variance in the state estimate?
- Example 2: Quickest transient detection: if the state of nature jumps into a state and then jumps out of the state, how should quickest detection of this transient detection be performed? The problem is equivalent to a stopping time POMDP where the Markov chain jumps only twice.

- Example 3: Risk-sensitive quickest detection: how to perform quickest detection with an exponential penalty.
- Example 4: Quickest detection with social learning: if individual agents learn an underlying state by performing social learning, how can the quickest change detection be applied by a global decision-maker? As will be shown, this interaction of local and global decision-makers results in interesting non-monotone behavior and the stopping set is not necessarily convex.
- Example 5: Quickest time herding with social learning: how should a decision-maker estimate an underlying state of nature when agents herd while performing social learning?
- Example 6: How should a monopoly optimally price a product when customers perform social learning? Each time a customer buys the product, the monopoly makes money and also gets publicity due to social learning. It is shown that it is optimal to start at a high price and then decrease the price over time.
- Example 7: Quickest detection with controlled sampling: if making measurements are expensive, when should a decision-maker look at a Markov chain to decide if the state has changed?

13.1 Example 1: Quickest detection with phase-distributed change time and variance penalty

Here we formulate quickest detection of a phase-distributed change time as a stopping time POMDP and analyze the structure of the optimal policy. The reader should review §12.2.2 where the classical quickest detection problem with geometric change times was discussed. We will consider two generalizations of the classical quickest detection problem: phase-type (PH) distributed change times and variance stopping penalty.

PH-distributions include geometric distributions as a special case and are used widely in modeling discrete event systems [250]. The optimal detection of a PH-distributed change point is useful since the family of PH-distributions forms a dense subset for the set of all distributions. As described in [250], a PH-distributed change time can be modeled as a multi-state Markov chain with an absorbing state. So the space of public belief states π now is a multidimensional simplex of probability mass functions. We will formulate the problem as a stopping time POMDP and characterize the optimal decision policy.

The second generalization we consider is a stopping penalty comprising of the false alarm and a variance penalty. The variance penalty is essential in stopping problems where one is interested in ultimately estimating the state x. It penalizes stopping too soon if the uncertainty of the state estimate is large.[1] The variance penalty results in a stopping cost that is quadratic in the belief state π. Recall that in Chapter 8.4 we discussed such nonlinear costs in the context of controlled sensing with the goal of penalizing uncertainty in the state estimate.

[1] In [33], a continuous-time stochastic control problem is formulated with a quadratic stopping cost, and the existence of the solution to the resulting quasi-variational inequality is proved.

13.1.1 Formulation of quickest detection as a stopping time POMDP

Below we formulate the quickest detection problem with PH-distributed change time and variance penalty as a stopping time POMDP. We can then use the structural results of Chapter 12 to characterize the optimal policy.

Transition matrix for PH distributed change time

The change point τ^0 is modeled by a *phase type (PH) distribution*. The family of all PH-distributions forms a dense subset for the set of all distributions [250] i.e., for any given distribution function F such that $F(0) = 0$, one can find a sequence of PH-distributions $\{F_n, n \geq 1\}$ to approximate F uniformly over $[0, \infty)$. Thus PH-distributions can be used to approximate change points with an arbitrary distribution. This is done by constructing a multi-state Markov chain as follows: assume state "1" (corresponding to belief e_1) is an absorbing state and denotes the state after the jump change. The states $2, \ldots, X$ (corresponding to beliefs $e_2, \ldots, e_X$) can be viewed as a single composite state that x resides in before the jump. To avoid trivialities, assume that the change occurs after at least one measurement. So the initial distribution π_0 satisfies $\pi_0(1) = 0$. The transition probability matrix is of the form

$$P = \begin{bmatrix} 1 & 0 \\ \underline{P}_{(X-1)\times 1} & \bar{P}_{(X-1)\times(X-1)} \end{bmatrix}. \tag{13.1}$$

The "change time" τ^0 denotes the time at which x_k enters the absorbing state 1:

$$\tau^0 = \min\{k : x_k = 1\}. \tag{13.2}$$

The distribution of τ^0 is determined by choosing the transition probabilities $\underline{P}, \bar{P}$ in (13.1). To ensure that τ^0 is finite, assume states $2, 3, \ldots X$ are transient. This is equivalent to $\bar{P}$ satisfying $\sum_{n=1}^{\infty} \bar{P}^n_{ii} < \infty$ for $i = 1, \ldots, X-1$ (where $\bar{P}^n_{ii}$ denotes the (i, i) element of the n-th power of matrix $\bar{P}$). The distribution of the absorption time to state 1 is

$$\nu_0 = \pi_0(1), \quad \nu_k = \bar{\pi}_0' \bar{P}^{k-1} \underline{P}, \quad k \geq 1, \tag{13.3}$$

where $\bar{\pi}_0 = [\pi_0(2), \ldots, \pi_0(X)]'$. The key idea is that by appropriately choosing the pair (π_0, P) and the associated state space dimension X, one can approximate any given discrete distribution on $[0, \infty)$ by the distribution $\{\nu_k, k \geq 0\}$; see [250, pp. 240–3]. The event $\{x_k = 1\}$ means the change point has occurred at time k according to PH-distribution (13.3). In the special case when x is a two-state Markov chain, the change time τ^0 is geometrically distributed.

Observations

The observation $y_k \in \mathcal{Y}$ given state x_k has conditional probability pdf or pmf

$$B_{xy} = p(y_k = y | x_k = x), \quad x \in \mathcal{X}, y \in \mathcal{Y}. \tag{13.4}$$

where $\mathcal{Y} \subset \mathbb{R}$ (in which case B_{xy} is a pdf) or $\mathcal{Y} = \{1, 2, \ldots, Y\}$ (in which case B_{xy} is a pmf). In quickest detection, states $2, 3, \ldots, X$ are fictitious and are defined to generate the PH-distributed change time τ^0 in (13.2). So states $2, 3, \dots, X$ are indistinguishable in terms of the observation y. That is, the observation probabilities B in (13.4) satisfy

$$B_{2y} = B_{3y} = \cdots = B_{Xy} \text{ for all } y \in \mathcal{Y}. \tag{13.5}$$

Actions

At each time k, a decision u_k is taken where

$$u_k = \mu(\pi_k) \in \mathcal{U} = \{1 \text{ (announce change and stop)}, 2 \text{ (continue) }\}. \tag{13.6}$$

In (13.6), the policy μ belongs to the class of stationary decision policies.

Stopping cost

If decision $u_k = 1$ is chosen, then the decision-maker announces that a change has occurred and the problem terminates. If $u_k = 1$ is chosen before the change point τ^0, then a false alarm and variance penalty is paid. If $u_k = 1$ is chosen at or after the change point τ^0, then only a variance penalty is paid. Below these costs are formulated.

Let $h = (h_1, \dots, h_X)'$ specify the physical state levels associated with states $1, 2, \dots, X$ of the Markov chain $x \in \{e_1, e_2, \dots, e_X\}$. The *variance penalty* is

$$\begin{aligned} &\mathbb{E}\{\|(x_k - \pi_k)'h\|^2 \mid \mathcal{I}_k\} = H'\pi_k(i) - (h'\pi_k)^2, \\ &\text{where } H_i = h_i^2 \text{ and } H = (H_1, H_2, \dots, H_X), \\ &\mathcal{I}_k = (y_1, \dots, y_k, u_0, \dots, u_{k-1}). \end{aligned} \tag{13.7}$$

This conditional variance penalizes choosing the stop action if the uncertainty in the state estimate is large. Recall we discussed POMDP with nonlinear costs in the context of controlled sensing in §8.4.

Next, the false alarm event $\cup_{i\geq 2}\{x_k = e_i\} \cap \{u_k = 1\} = \{x_k \neq e_1\} \cap \{u_k = 1\}$ represents the event that a change is announced before the change happens at time τ^0. To evaluate the *false alarm penalty*, let $f_i I(x_k = e_i, u_k = 1)$ denote the cost of a false alarm in state e_i, $i \in \mathcal{X}$, where $f_i \geq 0$. Of course, $f_1 = 0$ since a false alarm is only incurred if the stop action is picked in states $2, \dots, X$. The expected false alarm penalty is

$$\sum_{i\in\mathcal{X}} f_i \mathbb{E}\{I(x_k = e_i, u_k = 1)|\mathcal{I}_k\} = \psi'\pi_k I(u_k = 1),$$
$$\text{where } \psi = (f_1, \dots, f_X)',\ f_1 = 0. \tag{13.8}$$

The false alarm vector f is chosen with increasing elements so that states further from state 1 incur larger penalties.

Then with α, β denoting nonnegative constants that weight the relative importance of these costs, the stopping cost expressed in terms of the belief state at time k is

$$\bar{C}(\pi_k, u_k = 1) = \alpha(H'\pi_k - (h'\pi_k)^2) + \beta\, \psi'\pi_k. \tag{13.9}$$

One can view α as a Lagrange multiplier in a stopping time problem that seeks to minimize a cumulative cost subject to a variance stopping constraint.

Delay cost of continuing

We allow two possible choices for the delay costs for action $u_k = 2$:
(a) *Predicted delay*: If action $u_k = 2$ is taken then $\{x_{k+1} = e_1, u_k = 2\}$ is the event that no change is declared at time k even though the state has changed at time $k+1$. So with d denoting a nonnegative constant, $d\, I(x_{k+1} = e_1, u_k = 2)$ depicts a *delay cost*. The expected delay cost for decision $u_k = 2$ is

$$\bar{C}(\pi_k, u_k = 2) = d\,\mathbb{E}\{I(x_{k+1} = e_1, u_k = 2)|\mathcal{F}_k\} = d e_1' P' \pi_k. \tag{13.10}$$

The above cost is motivated by applications (e.g. sensor networks) where if the decision-maker chooses $u_k = 2$, then it needs to gather observation y_{k+1} thereby incurring an additional operational cost denoted as C_o. Strictly speaking, $\bar{C}(\pi, 2) = d e_1' P' \pi + C_o$. Without loss of generality set the constant C_o to zero, as it does not affect our structural results. The penalty $d\, I(x_{k+1} = e_1, u_k = 2)$ gives incentive for the decision-maker to predict the state x_{k+1} accurately.
(b) *Classical delay*: Instead of (13.10), a more "classical" formulation is that a delay cost is incurred when the event $\{x_k = e_1, u_k = 2\}$ occurs. The expected delay cost is

$$\bar{C}(\pi_k, u_k = 2) = d\,\mathbb{E}\{I(x_k = e_1, u_k = 2)|\mathcal{I}_k\} = d e_1' \pi_k. \tag{13.11}$$

Remark: Due to the variance penalty, the cost $\bar{C}(\pi, 1)$ in (13.9) is quadratic in the belief state π. Therefore, the formulation cannot be reduced to a standard stopping problem with linear costs in the belief state.
Summary: It is clear from the above formulation that quickest detection is simply a stopping-time POMDP of the form (12.3) with transition matrix (13.1) and costs (13.8), (13.10) or (13.11). In the special case of geometric change time ($X = 2$), and delay cost (13.11), the cost function assumes the classical *Kolmogorov–Shiryayev criterion* for detection of disorder (12.7).

13.1.2 Main result: threshold optimal policy for quickest detection

Note first that for $\alpha = 0$ in (13.9), the stopping cost is linear. Then Theorem 12.2.1 applies implying that the stopping set $\mathcal{R}_1$ is convex. Below we focus on establishing Theorem 12.3.4 to show the existence of a threshold curve for the optimal policy. As discussed at the end of §12.3.5, such a result goes well beyond establishing convexity of the stopping region.

We consider the predicted cost and delay cost cases separately below:

Quickest detection with predicted delay penalty

First consider the quickest detection problem with the predicted delay cost (13.10). For the stopping cost $\bar{C}(\pi, 1)$ in (13.9), choose $f = [0, 1, \cdots, 1]' = \mathbf{1}_X - e_1$. This weighs the states $2, \ldots, X$ equally in the false alarm penalty. With assumption (13.5), the variance penalty (13.7) becomes $\alpha(e_1'\pi - (e_1'\pi)^2)$. The delay cost $\bar{C}(\pi, 2)$ is chosen as (13.10). To summarize

$$\bar{C}(\pi, 1) = \alpha\left(e_1'\pi - (e_1'\pi)^2\right) + \beta(1 - e_1'\pi), \quad \bar{C}(\pi, 2) = d e_1' P' \pi. \tag{13.12}$$

Theorem 12.3.4 is now illustrated for the costs (13.12). Before proceeding there is one issue we need to fix. Even for linear costs in (13.12) ($\alpha = 0$), denoting $\bar{C}(\pi, u) = c_u'\pi$, it is seen that the elements of c_1 are increasing while the elements of c_2 are decreasing. So at first sight, it is not possible to apply Theorem 12.3.4 since assumption (C) of Theorem 12.3.4 requires that the elements of the cost vectors are decreasing. But the following novel transformation can be applied, which is nicely described in [144, pp. 389–90] (for fully observed MDPs).

Define

$$V(\pi) = \bar{V}(\pi) - (\alpha + \beta)f'\pi, \quad C(\pi, 1) = \alpha(H'\pi - (h'\pi)^2) - \alpha f'\pi$$
$$C(\pi, 2) = \bar{C}(\pi, 2) - (\alpha + \beta)\psi'\pi + \rho(\alpha + \beta)f'P'\pi. \tag{13.13}$$

Then clearly $V(\pi)$ satisfies Bellman's dynamic programming equation

$$\mu^*(\pi) = \arg\min_{u \in \mathcal{U}} Q(\pi, u), \ J_{\mu^*}(\pi) = V(\pi) = \min_{u \in \{1,2\}} Q(\pi, u), \tag{13.14}$$
$$\text{where } Q(\pi, 2) = C(\pi, 2) + \rho \sum_{y \in \mathcal{Y}} V(T(\pi, y))\, \sigma(\pi, y), \quad Q(\pi, 1) = C(\pi, 1).$$

Even though the value function is now changed, the optimal policy $\mu^*(\pi)$ and hence stopping set $\mathcal{R}_1$ remain unchanged with this coordinate transformation. The nice thing is that the new costs $C(\pi, u)$ in (13.13) can be chosen to be decreasing under suitable conditions. For example, if $\alpha = 0$, then clearly $C(\pi, 1) = 0$ (and so is decreasing by definition) and it is easily checked that if $d \geq \rho\beta$ then $C(\pi, 2)$ is also decreasing.

With the above transformation, we are now ready to apply Theorem 12.3.4. Assumptions (C) and (S) of Theorem 12.3.4 specialize to the following assumptions on the transformed costs in (13.13):

(C-Ex1) $d \geq \rho(\alpha + \beta)$
(S-Ex1) $(d - \rho(\alpha + \beta))(1 - P_{21}) \geq \alpha - \beta$

THEOREM 13.1.1 *Under (C-Ex1), (F1), (F2) and (S-Ex1), Theorem 12.3.4 holds implying that the optimal policy for quickest detection with PH-distributed change times has a threshold structure. Thus Algorithm 14 estimates the optimal linear threshold.*

Proof The proof follows from Theorem 12.3.4. We only need to show that (C-Ex1) and (S-Ex1) hold. These are specialized versions of conditions (C) and (S) arising in Theorem 12.3.4.

First consider (C-Ex1): Consider Lemma 11.1.1 with the choice of $\phi = 2\alpha e_1$, $h = e_1$ in (11.5). This yields $2\alpha \geq 0$ and $2\alpha \geq 2\alpha$ which always hold. So $C(\pi, 1)$ is $\geq_r$ decreasing in π for any nonnegative α. It is easily verified that (C-Ex1) is sufficient for the linear cost $C(\pi, 2)$ to be decreasing.

Next consider (S-Ex1). Set $\phi_i = (d - \rho(\alpha + \beta))P'e_i + (\beta - \alpha)e_1$, $h = e_1$ in (12.45). The first inequality is equivalent to: (i) $(d - \rho(\alpha + \beta))(P_{X1} - P_{i1}) \leq 0$ for $i \geq 2$ and (ii) $(d - \rho(\alpha + \beta))(1 - P_{X1}) \geq \alpha - \beta$. Note that (i) holds if $d \geq \rho(\alpha + \beta)$. The second inequality in (12.45) is equivalent to $(d - \rho(\alpha + \beta))(1 - P_{i1}) \geq \alpha - \beta$. Since P is

TP2, from Lemma 10.5.1 it follows that (S-Ex1) is sufficient for these inequalities to hold. □

Quickest detection with classical delay penalty

Finally, consider the "classical" delay cost $\bar{C}(\pi, 2)$ in (13.11) and stopping cost $\bar{C}(\pi, 1)$ in (13.9) with h in (13.5). Then

$$\bar{C}(\pi, 1) = \alpha\left(e_1'\pi - (e_1'\pi)^2\right) + \beta\,\psi'\pi_k, \quad \bar{C}(\pi, 2) = de_1'\pi. \tag{13.15}$$

Assume that the decision-maker designs the false alarm vector ψ to satisfy the following linear constraints:

(AS-Ex1) (i) $f_i \geq \max\{1, \rho\frac{\alpha+\beta}{\beta}\psi'P'e_i + \frac{\alpha-d}{\beta}\}, i \geq 2.$
(ii) $f_j - f_i \geq \rho\psi'P'(e_j - e_i), j \geq i, i \in \{2, \ldots, X-2\}$
(iii) $f_X - f_i \geq \frac{\rho(\alpha+\beta)}{\beta}\psi'P'(e_X - e_i), i \in \{2, \ldots, X-1\}.$

Feasible choices of f are easily obtained by a linear programming solver.

Then Theorem 12.3.4 continues to hold, under conditions (AS-Ex1), (F1),(F2).

Summary: We modeled quickest detection with PH-distributed change time as a multi-state POMDP. We then gave sufficient conditions for the optimal policy to have a threshold structure. The optimal linear parametrized policy can be estimated via the policy gradient Algorithm 12.4.2.

13.2 Example 2: Quickest transient detection

Our second example deals with Bayesian quickest transient detection; see [271] for a nice description of the problem with various cost functions. In quickest transient detection, a Markov chain state jumps from a starting state to a transient state at geometric distributed time τ, and then jumps out of the state to an absorbing state at another geometric distributed time. We show below that under similar assumptions to quickest time detection, the threshold switching curve of Theorem 12.3.4 holds. Therefore the linear threshold results and Algorithm 14 hold.

The setup comprises a stopping time POMDP with state space $\mathcal{X} = \{1, 2, 3\}$. The transition probability matrix and initial distribution are

$$P = \begin{bmatrix} 1 & 0 & 0 \\ p_{21} & p_{22} & 0 \\ 0 & p_{32} & p_{33} \end{bmatrix}, \quad \pi_0 = e_3. \tag{13.16}$$

So the Markov chain starts in state 3. After some geometrically distributed time it jumps to the transient state 2. Finally after residing in state 2 for some geometrically distributed time, it then jumps to the absorbing state 1.

In quickest transient detection, we are interested in detecting transition to state 2 with minimum cost. The action space is $\mathcal{U} = \{1 \text{ (stop)}, 2 \text{ (continue)}\}$. The stop action $u = 1$ declares that transient state 2 was visited.

We choose the following costs (see [271] for other choices). Similar to (13.11), let $d_i I(x_k = e_i, u_k = 2)$ denote the delay cost in state e_i, $i \in \mathcal{X}$. Of course $d_3 = 0$ since $x_k = e_3$ implies that the transient state has not yet been visited. So the expected delay cost is (where $\mathcal{I}_k$ is defined in (13.7))

$$\sum_{i \in \mathcal{X}} d_i \mathbb{E}\{x_k = e_i, u_k = 2 | \mathcal{I}_k\} = d'\pi_k \text{ where } d = (d_1, d_2, d_3)', \; d_3 = 0. \tag{13.17}$$

Typically the elements of the delay vector d are chosen as $d_1 \geq d_2 > 0$ so that state 1 (final state) accrues a larger delay than the transient state. This gives incentive to declare that transient state 2 was visited when the current state is 2, rather than wait until the process reaches state 1.

The false alarm cost for declaring $u = 1$ (transient state 2 was visited) when $x = 1$ is zero since the final state 1 could only have been reached after visiting transient state 2. So the false alarm penalty is $\mathbb{E}\{I(x_k = e_3, u_k = 1) | \mathcal{I}_k\} = 1 - (e_1 + e_2)'\pi_k$ for action $u_k = 1$. For convenience, in the variance penalty (13.7), we choose $h = [0, 0, 1]'$. So from (13.9), (13.11), the expected stopping cost and continuing costs are

$$\bar{C}(\pi, 1) = \alpha\big(h'\pi - (h'\pi)^2\big) + \beta(1 - (e_1 + e_2)'\pi), \quad \bar{C}(\pi, 2) = d'\pi. \tag{13.18}$$

The transformed costs and optimal decision policy $\mu^*(\pi)$ are as in (13.13) and (13.14).

Main result: The following assumptions are similar to those in quickest time detection. Note that due to its structure, P in (13.16) is always TP2 (i.e. (F2) of Theorem 12.3.4 holds).

(S-Ex2) The scaling factor for the variance penalty satisfies $\alpha \leq \frac{d_2 + \beta - \rho\beta P_{33}}{1 + \rho P_{33}}$.

For $\alpha = 0$ (zero variance penalty) (S-Ex2) holds trivially.

THEOREM 13.2.1 *Consider the quickest transient detection problem with delay and stopping costs in (13.18). Then under (F1), (13.16), (S-Ex2), the conclusions of Theorem 12.3.4 hold. Thus Algorithm 14 estimates the optimal linear threshold.*

Proof To show S-Ex2) holds we proceed as follows. Recall that the variance constraint $\alpha(e_3'\pi - (e_3'\pi)^2) = \alpha((\pi_1 + \pi_2) - (\pi_1 + \pi_2)^2)$. Set $\phi_1 = d_1 + \beta - \alpha - \rho(\alpha + \beta)$, $\phi_2 = d_2 + \beta - \alpha - \rho(\alpha + \beta)$, $f_3 = -\rho(\alpha + \beta)P_{32}$ in (12.45). The first inequality is equivalent to $\rho(\alpha + \beta)(1 - P_{32}) \leq \beta + d_i - \alpha$, $i = 1, 2$. The second inequality is equivalent to $d_1 \geq 0$ and $d_1 + \beta - \alpha \geq \rho(\alpha + \beta)(1 - P_{32})$. A sufficient condition for these is (S-Ex2). □

13.3 Example 3: Risk-sensitive quickest detection with exponential delay penalty

In this example, we generalize the results of [267], which deals with exponential delay penalty and geometric change times. We consider exponential delay penalty with PH-distributed change time. Our formulation involves risk-sensitive partially observed stochastic control. We first show that the exponential penalty cost function in [267] is a special case of risk-sensitive stochastic control cost function when the state space dimension $X = 2$. We then use the risk-sensitive stochastic control formulation to derive structural results for PH-distributed change time. In particular, the main result below (Theorem 13.3.1) shows that the threshold switching curve still characterizes the optimal stopping region $\mathcal{R}_1$. The assumptions and main results are conceptually similar to Theorem 12.3.4.

We discussed the general risk averse formulation of MDPs and POMDPs in §8.6. Here we deal with the special case of exponential costs. Risk-sensitive stochastic control with exponential cost has been studied extensively [95, 158, 37]; see the discussion in §8.7 for additional perspective and references. In simple terms, quickest time detection seeks to optimize the objective $\mathbb{E}\{J^0\}$ where J^0 is the accumulated sample path cost until some stopping time τ. In risk-sensitive control, one seeks to optimize $J = \mathbb{E}\{\exp(\epsilon J^0\}$. For $\epsilon > 0$, J penalizes heavily large sample path costs due to the presence of second-order moments. This is termed a risk-averse control. Risk-sensitive control provides a nice formalization of the exponential penalty delay cost and allows us to generalize the results in [267] to phase-distributed change times.

Below, we will use $c(e_i, u = 1)$ to denote false alarm costs and $c(e_i, u = 2)$ to denote delay costs, where $i \in \mathcal{X}$. Risk-sensitive control [46] considers the exponential cumulative cost function

$$J_\mu(\pi) = \mathbb{E}_\mu\left\{\exp\left(\epsilon \sum_{k=0}^{\tau-1} c(x_k, u_k = 2) + \epsilon\, c(x_\tau, u_\tau = 1)\right)\right\} \tag{13.19}$$

where $\epsilon > 0$ is the risk-sensitive parameter.

Let us first show that the exponential penalty cost in [267] is a special case of (13.19) for consider the case $X = 2$ (geometric distributed change time). For the state $x \in \{e_1, e_2\}$, choose $c(x, u = 1) = \beta I(x \neq e_1, u = 1) = \beta(1 - e_1'x)$ (false alarm cost), $c(x, u = 2) = d I(x = e_1, u = 2) = de_1'x$ (delay cost). Then it is easily seen that $\sum_{k=0}^{\tau-1} c(x_k, u_k = 2) + c(x_\tau, u_\tau = 1) = d\,|\tau - \tau^0|^+ + \beta I(\tau < \tau^0)$. Therefore (recall τ^0 is defined in (13.2) and τ is defined in (12.2)),

$$\begin{aligned}
J_\mu(\pi) &= \mathbb{E}_\mu\left\{\exp\left(\epsilon d\,|\tau - \tau^0|^+ + \epsilon\beta I(\tau < \tau^0)\right)\right\}\left[I(\tau < \tau^0) + I(\tau = \tau^0) + I(\tau > \tau^0)\right] \\
&= \mathbb{E}_\mu\left\{\exp(\epsilon\beta)I(\tau < \tau^0) + \exp(\epsilon d\,|\tau - \tau^0|^+)I(\tau > \tau^0) + 1\right\} \\
&= \mathbb{E}_\mu\left\{(e^{\epsilon\beta} - 1)I(\tau < \tau^0) + e^{\epsilon d|\tau - \tau^0|^+}\right\} \\
&= (e^{\epsilon\beta} - 1)\mathbb{P}_\mu(\tau < \tau^0) + \mathbb{E}_\mu\{e^{\epsilon d|\tau - \tau^0|^+}\}
\end{aligned} \tag{13.20}$$

which is identical to exponential delay cost function in [267, Eq. 40]. Thus the Bayesian quickest time detection with exponential delay penalty in [267] is a special case of a risk-sensitive stochastic control problem.

We consider the delay cost as in (13.10); so for state $x \in \{e_1, \dots, e_X\}$, $c(x, u_k = 2) = de_1'P'x$. To get an intuitive feel for this modified delay cost function, for the case $X = 2$,

$$\sum_{k=0}^{\tau-1} c(x_k, u_k = 2) + c(x_\tau, u_\tau = 1) = d|\tau - \tau^0|^+ + \beta I(\tau < \tau^0) + dP_{21}(\tau^0 - 1)I(\tau^0 < \tau).$$

Therefore, for $X = 2$, the exponential delay cumulative cost function is

$$J_\mu(\pi) = (e^{\epsilon\beta} - 1)\mathbb{P}_\mu(\tau < \tau^0) + \mathbb{E}_\mu\{e^{\epsilon d[|\tau-\tau^0|^+ + P_{21}(\tau_0-1)I(\tau_0<\tau)]}\}. \tag{13.21}$$

This is similar to (13.20) except for the additional term $P_{21}(\tau_0 - 1)I(\tau_0 < \tau)$ in the exponential.

With the above motivation, we consider risk-sensitive quickest detection for PH-distributed change time, i.e. $X \geq 2$. Let π denote the risk-sensitive belief state; see [104, 158] for extensive descriptions of the risk-sensitive belief state and verification theorems for dynamic programming in risk-sensitive control. From Theorem 8.6.2 in §8.6, Bellman's equation reads

$$\bar{V}(\pi) = \min\{\bar{C}(\pi,1), \sum_{y\in\mathcal{Y}} \bar{V}(T(\pi,y))\sigma(\pi,y)\} \quad \text{where} \tag{13.22}$$

$$\bar{C}(\pi,1) = R_1'\pi, \quad T(\pi,y) = \frac{B_y P' \text{diag}(R_2)\pi}{\sigma(\pi,y)}, \quad \sigma(\pi,y) = \mathbf{1}' B_y P' \text{diag}(R_2)\pi$$

$$R_1 = (1, e^{\epsilon\beta}, \ldots, e^{\epsilon\beta})', \; R_2 = (e^{\epsilon d}, e^{\epsilon d P_{21}}, \ldots, e^{\epsilon d P_{X1}})', \; B_y = \text{diag}(B_{1y}, \ldots B_{Xy}).$$

Similar to the transformation used in (13.13), define $V(\pi) = \bar{V}(\pi) - \bar{C}(\pi,1)$. Then $V(\pi)$ satisfies Bellman's equation (13.14) with

$$C(\pi,1) = 0, \quad C(\pi,2) = R_1'(P'\text{diag}(R_2) - I)\pi. \tag{13.23}$$

Assume the following condition holds

(C-Ex3) The elements of $R_1'(P'\text{diag}(R_2) - I)$ are decreasing w.r.t. $i = 1, 2, \ldots, X$.

Evaluating $C(\pi,2) = R_1'(P'\text{diag}(R_2) - I)\pi$, then (C-Ex3) is equivalent to

$$e^{\epsilon d} - 1 \geq e^{\epsilon d P_{21}}(P_{21} + e^{\epsilon\beta}(1 - P_{21})) - e^{\epsilon\beta} \text{ and } e^{\epsilon d P_{i1}}(P_{i1} + e^{\epsilon\beta}(1 - P_{i1}))$$

decreasing in $i \in \{2, \ldots, X\}$. For example, if $d = \epsilon = 1$, then for $\beta \geq 1$, the following are verified by elementary calculus:
(i) (C-Ex3) always holds for $\beta \geq 1$ when $X = 2$ (geometric distributed change time).
(ii) For PH-distributed change time, if (F2) holds, then (C-Ex3) always holds providing $P_{21} < 1/(e^\beta - 1)$.

THEOREM 13.3.1 *The stopping region $\mathcal{R}_1$ is a convex subset of $\Pi(X)$. Under (C-Ex3), (F1), (F2), Theorem 12.3.4 holds. Thus Algorithm 14 estimates the optimal linear threshold.* □

Proof The only difference compared to a standard stopping time POMDP is the update of the belief state (13.22) which now includes the term $\text{diag}(R_u)$. The elements of R_u are nonnegative and functionally independent of the observation y. Therefore the three main requirements that $T(\pi,y)$ is MLR increasing in π, $T(\pi,y)$ is MLR increasing in Y, and $\sigma(\pi,:)$ is $\geq_r$ increasing in π continue to hold. Then the rest of the proof is identical to Theorem 12.3.4. □

Remarks: (i) Delay formulation in [267]: Consider the formulation in [267] which is equivalent to (13.20). Then for the geometric distributed case $X = 2$, the convexity of

$\mathcal{R}_1$ holds using a similar proof to above. Since $\Pi(X)$ is a 1-dimensional simplex and $e_1 \in \mathcal{R}_1$, convexity implies there exists (a possible degenerate) threshold point π^* that characterizes $\mathcal{R}_1$ such that the optimal policy is of the form (12.8). As a sanity check, the analogous condition to (C-Ex3) reads $e^{\epsilon d} - 1 > P_{21}(1 - e^{\epsilon \beta})$. This always holds for $\epsilon \geq 0$. Therefore, assuming (F1) holds, the above theorem holds for the exponential delay penalty case under (F1). (Recall (F2) holds trivially when $X = 2$.) Finally, for $X > 2$, using a similar proof, the conclusions of Theorem 13.3.1 hold.

13.4 Examples 4, 5 and 6: Stopping time POMDPs in multi-agent social learning

This section discusses three examples of stopping time problems in multi-agent social learning.

(i) §13.4.2 (Example 4), deals with a multi-agent Bayesian stopping time problem where agents perform greedy social learning and reveal their actions to subsequent agents. Given such a protocol of local decisions, how can the multi-agent system make a global decision when to stop? We show that the optimal decision policy of the stopping time problem has multiple thresholds. The motivation for such problems arise in automated decision systems (e.g. sensor networks) where agents make local decisions and reveal these local decisions to subsequent agents. The multiple threshold behavior of the optimal global decision shows that making global decisions based on local decision involves non-monotone policies.

(ii) §13.4.3 (Example 5) deals with *constrained optimal* social learning which is formulated as a stopping time problem. We show that the optimal policy has a threshold switching curve similar to Theorem 12.3.4.

(iii) §13.4.4 (Example 6) discusses how a monopolist should optimally price a product when customers perform social learning to evaluate the quality of the product. It is shown that the monopolist should start at a high price and then gradually decrease its price.

All three examples, involve the interaction of local decision-makers (performing social learning) with a global decision-maker that chooses decisions by solving a POMDP. This is illustrated in Figure 13.1.

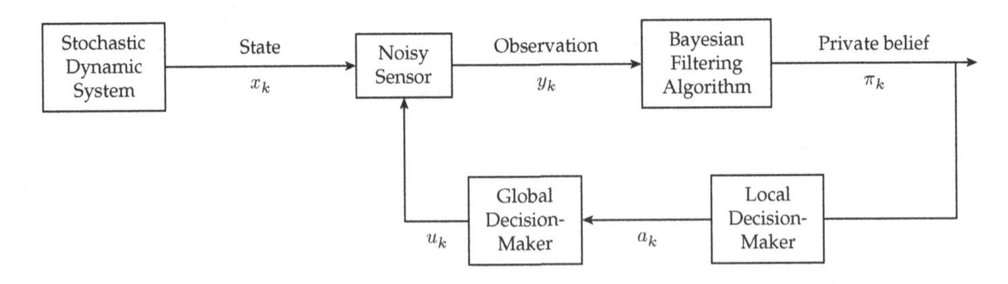

Figure 13.1 Interaction of local and global decision-makers.

13.4.1 Motivation: social learning amongst myopic agents

We have already described the social learning protocol in Chapter 5. Also in Chapter 8.5, we showed that a POMDP where agents perform social learning to update the belief results in a non-concave value function.

Consider a multi-agent system with agents indexed as $k = 1, 2, \ldots$ performing social learning to estimate an underlying random state x with prior π_0. (We assume in this section that x is a random variable and not a Markov chain.) Let $y_k \in \mathcal{Y} = \{1, 2, \ldots, Y\}$ denote the private observation of agent k and $a_k \in \mathbb{A} = \{1, 2, , \ldots, A\}$ denote the local action agent k takes. Define:

$$\mathcal{H}_k = (a_1, \ldots, a_{k-1}, y_k), \quad \mathcal{G}_k = (a_1, \ldots, a_{k-1}, a_k). \tag{13.24}$$

The social learning protocol was described in Chapter 5. Instead of repeating it, let us highlight the key Bayesian update equations:
A time k, based on its private observation y_k and public belief π_{k-1}, agent k:
1. Updates its private belief $\eta_k = \mathbb{E}\{x|\mathcal{H}_k\}$ as

$$\eta_k = \frac{B_{y_k}\pi_{k-1}}{\mathbf{1}'B_{y_k}\pi_{k-1}}. \tag{13.25}$$

2. Takes local myopic action $a_k = \arg\min_{a\in\mathbb{A}}\{c_a'\eta_k\}$ where $\mathbb{A} = \{1, 2, , \ldots, A\}$ denotes the set of local actions.
3. Based on a_k, the public belief $\pi_k = \mathbb{E}\{x_k|\mathcal{G}_k\}$ is updated (by subsequent agents) via the social learning filter (initialized with π_0)

$$\pi_k = T(\pi_{k-1}, a_k), \text{ where } T(\pi, a) = \frac{R_a^\pi \pi}{\sigma(\pi, a)}, \; \sigma(\pi, a) = \mathbf{1}_X' R_a^\pi \pi. \tag{13.26}$$

In (13.26), $R_a^\pi = \text{diag}(P(a|x = e_i, \pi), i \in \mathcal{X})$ with elements

$$P(a_k = a|x = e_i, \pi_{k-1} = \pi) = \sum_{y\in\mathcal{Y}} P(a_k = a|y, \pi)P(y|x = e_i) \tag{13.27}$$

$$= \sum_{y\in\mathcal{Y}} \prod_{\tilde{u}\in\mathbb{A}-\{a\}} I(c_a'B_y\pi < c_{\tilde{u}}'B_y\pi)P(y|x = e_i).$$

Here $I(\cdot)$ is the indicator function and $B_y = \text{diag}(\mathbb{P}(y|x = e_1), \ldots, \mathbb{P}(y|x = e_X)$. The procedure then repeats at time $k + 1$ and so on.

Recall from Theorem 5.3.1 that after some finite time $\bar{k}$, all agents choose the same action and the public belief freezes resulting in an information cascade.

13.4.2 Example 4: Stopping time POMDP with social learning: interaction of local and global decision-makers

Suppose a multi-agent system makes local decisions and performs social learning as described above. How can the multi-agent system make a global decision when to stop? Such problems are motivated in decision systems where a global decision needs to be made based on local decisions of agents. Figure 13.1 on the preceding page shows the setup with interacting local and global decision-makers.

We consider a Bayesian sequential detection problem for state $x = e_1$. Our goal below is to derive structural results for the optimal stopping policy. The main result below (Theorem 13.4.1) is that the global decision of when to stop is a multi-threshold function of the belief state.

Consider $\mathcal{X} = \mathcal{Y} = \{1, 2\}$ and $\mathbb{A} = \{1, 2\}$ and the social learning model of §13.4.1, where the costs $c(e_i, a)$ satisfy

$$c(e_1, 1) < c(e_1, 2), \quad c(e_2, 2) < c(e_2, 1). \tag{13.28}$$

Otherwise one action will always dominate the other action and the problem is uninteresting.

Let τ denote a stopping time adapted to $\mathcal{G}_k$, $k \geq 1$ (see (13.24)). In words, each agent has only the public belief obtained via social learning to make the global decision of whether to continue or stop. The goal is to solve the following stopping time POMDP to detect state e_1: Choose stopping time τ to minimize

$$J_\mu(\pi) = \mathbb{E}_\mu\{\sum_{k=0}^{\tau-1} \rho^k \mathbb{E}\{dI(x = e_1)| \mathcal{G}_k\} + \rho^\tau \beta \, \mathbb{E}\{I(x \neq e_1)|\mathcal{G}_\tau\}\}. \tag{13.29}$$

The first term is the delay cost and penalizes the decision of choosing $u_k = 2$ (continue) when the state is e_1 by the nonnegative constant d. The second term is the stopping cost incurred by choosing $u_\tau = 1$ (stop and declare state 1) at time $k = \tau$. It is the error probability of declaring state e_1 when the actual state is e_2. β is a positive scaling constant. In terms of the public belief, (13.29) is

$$J_\mu(\pi) = \mathbb{E}_\mu\{\sum_{k=0}^{\tau-1} \rho^k \bar{C}(\pi_{k-1}, u_k = 2) + \rho^\tau \bar{C}(\pi_{\tau-1}, u_\tau = 1)\} \tag{13.30}$$

$$\bar{C}(\pi, 2) = d e_1' \pi, \quad \bar{C}(\pi, 1) = \beta e_2' \pi.$$

The global decision $u_k = \mu(\pi_{k-1}) \in \{1 \text{ (stop)}, 2 \text{ (continue)}\}$ is a function of the public belief π_{k-1} updated according to the social learning protocol (13.25), (13.26). The optimal policy $\mu^*(\pi)$ and value function $V(\pi)$ satisfy Bellman's equation (13.14) with

$$Q(\pi, 2) = C(\pi, 2) + \rho \sum_{a \in \mathcal{U}} V(T(\pi, a)) \, \sigma(\pi, a) \text{ where} \tag{13.31}$$

$$C(\pi, 2) = \bar{C}(\pi, 2) - (1 - \rho)\bar{C}(\pi, 1), \quad Q(\pi, 1) = C(\pi, 1) = 0.$$

Here $T(\pi, a)$ and $\sigma(\pi, a)$ are obtained from the social learning filter (13.26). The above stopping time problem can be viewed as a macro-manager that operates on the public belief generated by micro-manager decisions. Clearly the micro- and macro-managers interact – the local decisions a_k taken by the micro-manager determine π_k and hence determines decision u_{k+1} of the macro-manager.

Since $\mathcal{X} = \{1, 2\}$, the public belief state $\pi = [1 - \pi(2), \ \pi(2)]'$ is parametrized by the scalar $\pi(2) \in [0, 1]$, and the belief space is the interval $[0, 1]$. Define the following intervals which form a partition of the interval [0,1]:

$$
\begin{aligned}
&\mathcal{P}_l = \{\pi(2) : \kappa_l < \pi(2) \leq \kappa_{l-1}\}, \quad l = 1,\dots,4 \text{ where} \\
&\kappa_0 = 1,\ \kappa_1 = \frac{(c(e_1,2) - c(e_1,1))B_{11}}{(c(e_1,2) - c(e_1,1))B_{11} + (c(e_2,1) - c(e_2,2))B_{21}} \\
&\kappa_2 = \frac{(c(e_1,2) - c(e_1,1))}{(c(e_1,2) - c(e_1,1)) + (c(e_2,1) - c(e_2,2))} \\
&\kappa_3 = \frac{(c(e_1,2) - c(e_1,1))B_{12}}{(c(e_1,2) - c(e_1,1))B_{12} + (c(e_2,1) - c(e_2,2))B_{22}}, \quad \kappa_4 = 0.
\end{aligned}
\tag{13.32}
$$

κ_0 corresponds to belief state e_2, and κ_4 corresponds to belief state e_1. (See discussion at the end of this section for more intuition about the intervals $\mathcal{P}_i$.)

It is readily verified that if the observation matrix B is TP2, then $\kappa_3 \leq \kappa_2 \leq \kappa_1$. The following is the main result.

THEOREM 13.4.1 *Consider the stopping time problem (13.30) where agents perform social learning using the social learning Bayesian filter (13.26). Assume (13.28) and B is symmetric and satisfies (F1). Then the optimal stopping policy $\mu^*(\pi)$ has the following structure: the stopping set $\mathcal{R}_1$ is the union of at most three intervals. That is $\mathcal{R}_1 = \mathcal{R}_1^a \cup \mathcal{R}_1^b \cup \mathcal{R}_1^c$ where $\mathcal{R}_1^a$, $\mathcal{R}_1^b$, $\mathcal{R}_1^c$ are possibly empty intervals. Here*

(i) The stopping interval $\mathcal{R}_1^a \subseteq \mathcal{P}_1 \cup \mathcal{P}_4$ and is characterized by a threshold point. That is, if $\mathcal{P}_1$ has a threshold point π^*, then $\mu^*(\pi) = 1$ for all $\pi(2) \in \mathcal{P}_4$ and

$$
\mu^*(\pi) = \begin{cases} 2 & \text{if } \pi(2) \geq \pi^* \\ 1 & \text{otherwise} \end{cases}, \quad \pi(2) \in \mathcal{P}_1. \tag{13.33}
$$

Similarly, if $\mathcal{P}_4$ has a threshold point π_4^*, then $\mu^*(\pi) = 2$ for all $\pi(2) \in \mathcal{P}_1$.

(ii) The stopping intervals $\mathcal{R}_1^b \subseteq \mathcal{P}_2$ and $\mathcal{R}_1^c \subseteq \mathcal{P}_3$.

(iii) The intervals $\mathcal{P}_1$ and $\mathcal{P}_4$ are regions of information cascades. That is, if $\pi_k \in \mathcal{P}_1 \cup \mathcal{P}_4$, then social learning ceases and $\pi_{k+1} = \pi_k$ (see Theorem 5.3.1 for definition of information cascade). □

The proof of Theorem 13.4.1 is in Appendix 13.A. The proof depends on properties of the social learning filter and these are summarized in Lemma 13.A.1 in Appendix 13.A. The proof is more complex than that of Theorem 12.2.1 since now $V(\pi)$ in is not necessarily concave over $\Pi(X)$, since $T(\cdot)$ and $\sigma(\cdot)$ are functions of R_a^π (13.27) which itself is an explicit (and in general non-concave) function of π.

Example: To illustrate the multiple threshold structure of the above theorem, consider the stopping time problem (13.30) with the following parameters:

$$
\rho = 0.9, \quad d = 1.8, \quad B = \begin{bmatrix} 0.9 & 0.1 \\ 0.1 & 0.9 \end{bmatrix}, \ c(e_i, a) = \begin{bmatrix} 4.57 & 5.57 \\ 2.57 & 0 \end{bmatrix}, \quad \beta = 2. \tag{13.34}
$$

Figure 13.2(a) and (b) show the optimal policy and value function. These were computed by constructing a grid of 500 values for $\Pi = [0, 1]$. The double threshold behavior of the stopping time problem when agents perform social learning is due to the discontinuous dynamics of the social learning filter (13.26).

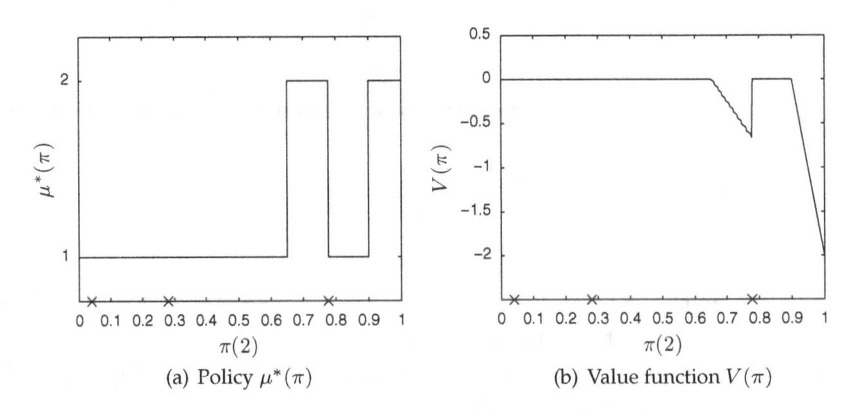

Figure 13.2 Double threshold policy in stopping time problem involving social learning. The parameters are specified in (13.34).

Discussion

The multiple threshold behavior (non-convex stopping set $\mathcal{R}_1$) of Theorem 13.4.1 is unusual. One would have thought that if it was optimal to "continue" for a particular belief $\pi^*(2)$, then it should be optimal to continue for all beliefs $\pi(2)$ larger than $\pi^*(2)$. The multiple threshold optimal policy shows that this is not true. Figure 13.2(a) shows that as the public belief $\pi(2)$ of state 2 decreases, the optimal decision switches from "continue" to "stop" to "continue" and finally "stop". Thus the global decision (stop or continue) is a non-monotone function of public beliefs obtained from local decisions.

The main reason for this unusual behavior is the dependence of the action likelihood R_a^π on the belief state π. This causes the social learning Bayesian filter to have a discontinuous update. The value function is no longer concave on $\Pi(X)$ and the optimal policy is not necessarily monotone. As shown in the proof of Theorem 13.4.1, the value function $V(\pi)$ is concave on each of the intervals $\mathcal{P}_l$, $l = 1, \dots, 4$.

To explain the final claim of the theorem, we characterize the intervals $\mathcal{P}_1$ and $\mathcal{P}_4$ defined in (13.32) more explicitly as follows:

$$\mathcal{P}_1 = \{\pi : \min_a c_a' B_y \pi = 2, \ \forall y \in \mathcal{Y}\} \tag{13.35}$$
$$\mathcal{P}_4 = \{\pi : \min_a c_a' B_y \pi = 1, \ \forall y \in \mathcal{Y}\}.$$

For public belief $\pi \in \mathcal{P}_1$, the optimal local action is $a = 2$ irrespective of the observation y; similarly for $\pi \in \mathcal{P}_4$, the optimal local action is $a = 1$ irrespective of the observation y. Therefore, on intervals $\mathcal{P}_1$ and $\mathcal{P}_4$, there is no social learning since the local action a reveals nothing about the observation y to subsequent agents. Social learning only takes place when the public belief is in $\mathcal{P}_2$ and $\mathcal{P}_3$.

Finally, we comment on the intervals $\mathcal{P}_l$, $l = 1, \dots, 4$. They form a partition of $\Pi(X)$ such that if $\pi \in \mathcal{P}_l$, then $T(\pi, 1) \in \mathcal{P}_{l+1}$ and $T(\pi, 2) \in \mathcal{P}_{l-1}$ (with obvious modifications for $l = 1$ and $l = 4$); see Lemma 13.A.1 in the appendix for details. In fact, κ_1 and κ_3 are fixed points of the composition Bayesian maps: $\kappa_1 = T(T(\kappa_1, 1), 2)$ and $\kappa_3 = T(T(\kappa_3, 2), 1)$. Given that the updates of the social Bayesian filter can be localized to specific intervals, we can then inductively prove that the value function is concave on each such interval. This is the main idea behind Theorem 13.4.1.

13.4.3 Example 5: Constrained social optimum and quickest time herding

In this subsection, we consider the constrained social optimum formulation in Chamley [84, Chapter 4.5]. We show in Theorem 13.4.2 that the resulting stopping time problem has a threshold switching curve. Thus optimal social learning can be implemented efficiently in a multi-agent system. This is in contrast to the multi-threshold behavior of the stopping time problem in §13.4.2 when agents were selfish in choosing their local actions.

The constrained social optimum formulation in [84] is motivated by the following question: how can agents aid social learning by acting benevolently and choosing their action to sacrifice their local cost but optimize a social welfare cost? In §13.4.2, agents ignore the information benefit their action provides to others resulting in information cascades where social learning stops. By constraining the choice by which agents pick their local action to two specific decision rules, the optimal choice between the two rules becomes a sequential decision problem.

Assume $\mathbb{A} = \mathcal{Y}$. As in the social learning model above, let c_a denote the cost vector for picking local action a and η_k (see (13.25)) denote the private belief of agent k.

Let τ denote a stopping time chosen by the decision-maker based on the information (more technically, adapted to the sequence of sigma-algebras $\mathcal{G}_k$, $k \geq 1$)

$$\mathcal{G}_k = (a_1, \ldots, a_{k-1}, a_k, u_1, \ldots, u_k).$$

As in §13.4.2, the goal is to solve the following sequential detection problem to detect state e_1: Pick the stopping time τ to minimize

$$\begin{aligned} J_\mu(\pi) = \mathbb{E}_\mu \Bigg\{ &\sum_{k=1}^{\tau-1} \rho^{k-1} \mathbb{E}\{c(x, a_k)|\mathcal{G}_{k-1}\} + \mathbb{E}_\mu\{\sum_{k=1}^{\tau-1} \rho^{k-1} d I(x = e_1)|\mathcal{G}_{k-1}\} \\ &+ \rho^{\tau-1}\mathbb{E}_\mu\{\beta I(x \neq e_1)|\mathcal{G}_{\tau-1}\} + \frac{\rho^{\tau-1}}{1-\rho} \min_{a \in \mathcal{Y}} \mathbb{E}\{c(x,a)|\mathcal{G}_{\tau-1}\} \Bigg\}. \end{aligned} \tag{13.36}$$

The second and third terms are the delay cost and error probability in stopping and announcing state e_1. The first and last terms model the cumulative social welfare cost involving all agents based on their local action. Let us explain these two terms. The agents pick their local action according to the decision rule $a(\pi, y, \mu(\pi)))$ (instead of myopically) as follows. As in [84], we constrain decision rule $a(\pi, y, \mu(\pi)))$ to two possible modes:

$$a_k = a(\pi_{k-1}, y_k, u_k) = \begin{cases} y_k & \text{if } u_k = \mu(\pi_{k-1}) = 2 \text{ (reveal observation)} \\ \arg\min_a c_a' \pi_{k-1} & \text{if } u_k = \mu(\pi_{k-1}) = 1 \text{ (stop)}\,. \end{cases} \tag{13.37}$$

Here the stationary policy $\mu : \pi_{k-1} \to u_k$ specifies which one of the two modes the benevolent agent k chooses. In mode $u_k = 2$, the agent k sacrifices is immediate cost

$c(x, a_k)$ and picks action $a_k = y_k$ to reveal full information to subsequent agents, thereby enhancing social learning.

In mode $u_k = 1$ the agent "stops and announces state 1". Equivalently, using the terminology of [84], the agent "herds" in mode $u_k = 1$. It ignores its private observation y_k, and chooses its action selfishly to minimizes its cost given the public belief π_{k-1}. So agent k chooses $a_k = \arg\min_a c_a'\pi_{k-1} = a_{k-1}$. Then clearly from (13.27), $P(a|e_x, \pi)$ is functionally independent of x since $P(y|x = e_i)$ is independent of i. Therefore from (13.26), if agent k herds, then $\pi_k = T(\pi_{k-1}, a_k) = \pi_{k-1}$, i.e. the public belief remains frozen. The cumulative cost incurred in herding is then equivalent to final term in (13.36):

$$\sum_{k=\tau}^{\infty} \rho^{k-1} \min_{a \in \mathcal{Y}} \mathbb{E}\{c(x,a)|\mathcal{G}_{k-1}\} = \sum_{k=\tau}^{\infty} \rho^{k-1} \min_a c_a'\pi_{\tau-1} = \frac{\rho^{\tau-1}}{1-\rho} \min_a c_a'\pi_{\tau-1}.$$

Define the *constrained social optimal policy* μ^* such that $J_{\mu^*}(\pi) = \min_\mu J_\mu(\pi)$. The sequential stopping problem (13.36) seeks to determine the optimal policy μ^* to achieve the optimal tradeoff between stopping and announcing state 1 and the cost incurred by agents that are acting benevolently. In analogy to Theorem 12.3.4, we show that $\mu^*(\pi)$ is characterized by a threshold curve.

Similar to (13.13) define the costs in terms of the belief state as

$$C(\pi, 1) = \frac{1}{1-\rho} \min_{a \in \mathcal{Y}} c_a'\pi, \quad C(\pi, 2) = \sum_{y \in \mathcal{Y}} c_y' B_y \pi + (d + (1-\rho)\beta) e_1'\pi - (1-\rho)\beta. \tag{13.38}$$

Below we list the assumptions and main structural result which is similar to Theorem 12.3.4. These assumptions involve the social learning cost $c(e_i, a)$ and are not required if these costs are zero.

(C-Ex5) $c(e_i, a) - c(e_{i+1}, a) \geq 0$

(S-Ex5) (i) $c(e_X, a) - c(e_i, a) \geq (1-\rho) \sum_y \big(c(e_X, y) B_{Xy} - c(e_i, y) B_{iy}\big)$

(ii)$(1-\rho) \sum_y \big(c(e_1, y) B_{1y} - c(e_i, y) B_{iy}\big) \geq c(e_1, a) - c(e_i, a)$.

(C-Ex5) is sufficient for $C(\pi, 1)$ and $C(\pi, 2)$ to be $\geq_r$ decreasing in $\pi \in \Pi(X)$. This implies that the costs $C(e_i, u)$ are decreasing in i, i.e. state 1 is the most costly state.

(S-Ex5) is sufficient for $C(\pi, u)$ to be submodular. It implies that $C(e_i, 2) - C(e_i, 1)$ is decreasing in i. This gives economic incentive for agents to herd when approaching the state e_1, since the differential cost between continuing and stopping is largest for e_1. Intuitively, the decision to stop (herd) should be made when the state estimate is sufficiently accurate so that revealing private observations is no longer required.

THEOREM 13.4.2 *Consider the sequential detection problem for state e_1 with social welfare cost in (13.36) and constrained decision rule (13.37). Then:*

(i) Under (C-Ex5), (F1), (S-Ex5), constrained social optimal policy $\mu^*(\pi)$ satisfies the structural properties of Theorem 12.3.4. (Thus a threshold switching curve exists).

(ii) The stopping set $\mathcal{R}_1$ is the union of $|\mathcal{Y}|$ convex sets (where $|\mathcal{Y}|$ denotes cardinality of $\mathcal{Y}$). Note also that $\mathcal{R}_1$ is a connected set by Statement (i). (Recall $\mathcal{Y} = \mathbb{A}$.) □

Proof Part (i) follows directly from the proof of Theorem 12.3.4. For Part (ii), define the convex polytopes $\mathcal{P}_a = \{\pi : c'_a\pi < c'_{\bar{a}}\pi, \quad \bar{a} \neq a\}$. Then on each convex polytope $\mathcal{P}_a$, since $C(\pi, 1) = c'_a\pi$ (recall $\alpha = 0$), we can apply the argument of Theorem 12.2.1 which yields that $\mathcal{R}_1 \cap \mathcal{P}_a$ is a convex region. Thus $\mathcal{R}_1$ is the union of A convex regions and is in general non-convex. However, it is still a connected set by part (i) of the theorem. □

The main implication of Theorem 13.4.2 is that the constrained optimal social learning scheme has a monotone structure. This is in contrast to the multi-threshold behavior of the stopping time problem in §13.4.2 when agents were selfish in picking their local actions. In [84, Chapter 4.5], the above formulation is used for pricing information externalities in social learning. From an implementation point of view, the existence of a threshold switching curve implies that the protocol only needs individual agents to store the optimal linear MLR policy (computed, for example, using Algorithm 14). Finally, $\mathcal{R}_1$ is the union of $|\mathcal{Y}|$ convex sets and is non-convex in general. This is different to standard stopping problems where the stopping set is convex.

13.4.4 Example 6: How to price a product when customers perform social learning

Suppose that a monopoly produces a product at fixed cost $c_o \in (0, 1)$. Let $x \in \{0, 1\}$ denote the intrinsic value of the product where 0 denotes a poor quality product and 1 denotes a good quality product.

At each time period $k = 1, 2, \ldots$, the monopoly receives a customer. The monopoly offers a take-it-or-leave-it price $p_k \in [0, 1]$ to customer k. The customer obtains a private observation $y_k \in \{0, 1\}$ of x with observation probabilities B_{xy_k}. The customer then decides whether to buy the product (action $a_k = 1$) or not to buy the product (action $a_k = 0$) by minimizing its expected cost:

$$a_k = \min_{a \in \{0,1\}} \mathbb{E}\{c_p(x, a)|a_1, \ldots, a_{k-1}, y_k\} = \min_a c'_{p,a}\eta_k$$

where $c_{p,0} = [0,\ 0]'$ and $c_{p,1} = [p_k,\ p_k - 1]'$. Here η_k is the private belief computed via (13.25) using the public belief π_{k-1}, which is evaluated via the social learning filter (13.26).

Notice that each time a customer buys from the monopoly, two things happen: the monopoly makes money due to the sale if $p_k > c_o$; also its gets publicity. The action a_k of agent k is recorded so that subsequent agents can use it to perform social learning to estimate x. (In a more sophisticated setting, the agent's action space is: buy and give a good review on social media, buy and give a bad review, don't buy; subsequent agents are influenced by these reviews.)

The monopolist chooses its price $p_k = \mu(\pi_{k-1})$ based on the public belief π_{k-1}. Here μ denotes a stationary policy. The monopoly aims to determine the optimal pricing policy μ^* to maximize its cumulative discounted reward:

$$J_\mu(\pi) = \mathbb{E}_\mu\{\sum_{k=1}^{\infty} \rho^{k-1}(p_k - c_o)I(a_k = 1) \mid \pi_0 = \pi\}, \qquad \text{where } p_k = \mu(\pi_{k-1}).$$

As in previous examples, the global decision-maker (monopolist) interacts with the local decision-maker (customers). The price (global action) determines the sales (local action) and social learning, which then determines the next price.

Optimal policy

The optimal pricing policy satisfies Bellman's equation

$$\begin{aligned} V(\pi) &= \max_{p\in[0,1]} Q(\pi,p), \quad \mu^*(\pi) = \operatorname{argmax}_{p\in[0,1]} Q(\pi,p) \\ \text{where } Q(\pi,p) &= r_p'\pi + \rho\sum_{a\in A} V(T(\pi,p,a))\,\sigma(\pi,p,a) \\ \text{and } r_p'\pi &= \sum_{y\in Y}\sum_{x\in X}(p-c_o)P(a=1|y,\pi)P(y|x)\pi(x) \\ &= \sum_{y\in Y}\sum_{x\in X}(p-c_o)I(c_1'\,B_y\,\pi_{k-1} < c_0'\,B_y\,\pi_{k-1})B_{xy}\pi(x). \end{aligned} \tag{13.39}$$

Due to the structure of the social learning filter (see Figure 5.3 on page 100),

$$Q(\pi,p) = \begin{cases} (p-c_o)+\rho V(\pi) & \text{if } p \le u_H(\pi); \\ \{\mathbf{1}'B_{y=1}\pi \times (p-c_o)+\rho\mathbb{E}\{V(\pi)\} & \text{if } u_H(\pi) < p \le u_S(\pi); \\ 0 & p > u_S(\pi) \end{cases} \tag{13.40}$$

$$\text{where } \mathbb{E}\{V(\pi)\} = \mathbf{1}'R^{\pi}_{a=0}\pi \times V(u_H(\pi)) + \mathbf{1}'R^{\pi}_{a=1}\pi \times V(u_S(\pi)),$$
$$u_H(\pi) = \frac{B_{y=0}\pi}{\mathbf{1}'B_{y=0}\pi}, \quad u_S(\pi) = \frac{B_{y=1}\pi}{\mathbf{1}'B_{y=1}\pi}.$$

Recall R^{π}_a are the action likelihoods defined in (5.4). As a result of (13.40), the optimal pricing policy can be characterized as follows:

THEOREM 13.4.3 *Consider the monopoly pricing problem where customers perform social learning. Then:*

1. *The optimal pricing policy $\mu^*(\pi)$ in (13.39) has the following threshold structure: There exist threshold beliefs $\pi^*, \pi^{**} \in (0,1)$ such that*

$$p = \mu^*(\pi) = \begin{cases} u_S(\pi)+\epsilon & \pi(2) \in [0,\pi^*) \\ u_S(\pi) & \pi(2) \in [\pi^*,\pi^{**}) \\ u_H(\pi) & \pi(2) \in [\pi^{**},1]. \end{cases} \tag{13.41}$$

   Here $u_S(\pi)$ and $u_H(\pi)$ defined in (13.40) are increasing in $\pi(2)$ and $u_H(\pi) \le u_S(\pi)$ for $\pi(2) \in [0,1]$ with equality holding at $\pi(2) = 0$ and $\pi(2) = 1$. Also ϵ is an arbitrary positive number.
2. *The price sequence $\{p_k = \mu^*(\pi_{k-1})\}$ generated by the optimal policy is a super-martingale. That is, $\mathbb{E}[p_{k+1}|\pi_k] \le p_k$.*

Before presenting the proof, we discuss some implications of Theorem 13.4.3. It is helpful to keep Figure 5.3 on page 100 in mind.

1. From the social learning filter we know that for beliefs $\pi(2) < \pi^*$ or $\pi(2) > \pi^{**}$, social learning freezes (information cascades occur). The optimal policy dictates that the monopoly should permanently exit the market if $\pi(2) < \pi^*$. Hence, $u_S(\pi) + \epsilon$ is called the *exit price*. $u_S(\pi)$ is called the *separating price* since it separates agents that have private observations $y = 0$ (and so choose $a = 0$) to those with $y = 1$ (and so choose $a = 1$). On the other hand, the monopoly induces a herd if $\pi(2) > \pi^{**}$ and makes a profit forever. Therefore, $u_H(\pi)$ is called the *pooling price* since it pools together all customers.
2. π^* and π^{**} can be viewed as the exit and capture thresholds, respectively. Learning is stopped eventually by the monopolist as one of the thresholds (exit or capture) is hit with probability one as the belief is a martingale.
3. If the monopoly is confident that its product is high quality, then it should start with a high price $p_k = u_S(\pi_{k-1})$ and then gradually reduce the price to capture additional sales. In particular, the optimal policy dictates that the monopoly should not start with a low promotional price; see [64]. Note also that since p_k is a super-martingale, $\mathbb{E}\{p_{k+1}\} \leq \mathbb{E}\{p_k\}$ meaning that the expected price drops with time.
4. [84] gives an interesting interpretation in terms of a restaurant owner. The owner should set the price sufficiently high initially to establish a flow of patrons. Once the reputation of the restaurant has been established in this elitist phase (by turning away poorer customers), the owner goes into mass production by lowering the price to make a profit on a larger number of customers.

Figure 13.3 illustrates the optimal pricing policy for the following parameters:

$$B = \begin{bmatrix} 0.6 & 0.4 \\ 0.4 & 0.6 \end{bmatrix}, c_o = 0.4, \rho = 0.6. \tag{13.42}$$

Proof of Theorem 13.4.3. Consider $Q(\pi, p)$ in (13.40). 1. For $p \leq u_H(\pi)$, clearly $Q(\pi, p)$ is maximized at $p = u_H(\pi)$. Next for $p \in (u_H(\pi), u_S(\pi)]$, the social learning filter update $T(\pi, a, p)$ is a constant with respect to p. Therefore, $\mathbb{E}\{V(\pi)\}$ is a constant with respect to p. Hence, $Q(\pi, p)$ is maximized at $p = u_S(\pi)$. Finally for $\pi < \pi^*$, since the $Q(\pi, p) = 0$ any price greater than $u_S(\pi)$ suffices. (Note that with this choice, the monopoly permanently leaves the market.)

2. From Theorem 5.3.1, π_k is a martingale. Since $u_S(\pi)$ is a concave function of π, it follows that if π_k and $\pi_{k+1} \in [\pi^*, \pi^{**})$, then $p_{k+1} = u_S(\pi_k)$ satisfies $\mathbb{E}[p_{k+1}|\pi_k] \leq p_k$ by Jensen's inequality. Next if $\pi_k(2) \in [\pi^{**}, 1]$, then an information cascade occurs and $\pi_k(2)$ freezes. But $u_S(\pi) \leq u_H(\pi)$ and so $p_{k+1} \leq p_k$. Finally, if $\pi_k \in [0, \pi^*)$, then π_k also freezes due to an information cascade. □

13.4.5 Summary

We considered three examples of POMDP stopping time problems in this section where the belief state is updated via the social learning filter (instead of the standard HMM filter). Social learning models multi-agent behavior where agents learn from their own observations and local decisions of previous agents. So POMDPs involving social

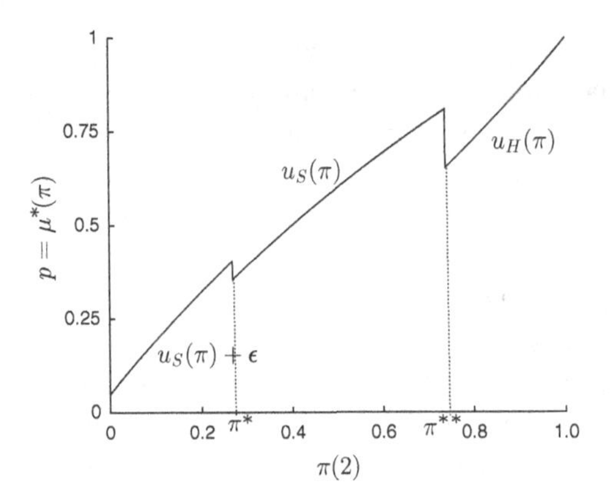

Figure 13.3 Optimal pricing policy $\mu^*(\pi)$ for monopolist with parameters specified in (13.42). The separating price $u_S(\pi)$ and pooling price $u_H(\pi)$ characterize the optimal policy (13.41). ϵ denotes an arbitrary positive number. π^* and π^{**} are threshold beliefs.

learning involves the interaction of local and global decision makers. We showed that stopping time problems with social learning results in unusual behavior; the stopping set in quickest detection is non-convex. We also studied constrained optimal social learning and how to optimally price a product when agents perform social learning.

13.5 Example 7: Quickest detection with controlled sampling

This section discusses quickest change detection when the decision-maker controls how often it observes (samples) a noisy Markov chain. The aim is to detect when a nosily observed Markov chain hits a target state by minimizing a combination of false alarm, delay cost and measurement sampling cost. There is an inherent trade-off between these costs: taking more frequent measurements yields accurate estimates but incurs a higher measurement cost. Making an erroneous decision too soon incurs a false alarm penalty. Waiting too long to declare the target state incurs a delay penalty. Since there are multiple "continue" actions, the problem is not a standard stopping time POMDP. For the two-state case, we show that under reasonable conditions, the optimal policy has the following intuitive structure: if the belief is away from the target state, look less frequently; if the belief is close to the target state, look more frequently.

13.5.1 Controlled sampling problem

Let $t = 0, 1, \ldots$ denote discrete-time and $\{x_t\}$ denote a Markov chain on the finite state space $\mathcal{X} = \{e_1, \ldots, e_X\}$ with transition matrix P.

Let $\tau_0, \tau_1, \ldots, \tau_{k-1}$ denote discrete time instants at which measurement samples have been taken, where by convention $\tau_0 = 0$. Let τ_k denote the current time-instant at which a measurement is taken. The measurement sampling protocol proceeds according to the following steps:

Step 1: Observation: A noisy measurement $y_k \in \mathcal{Y}$ at time $t = \tau_k$ of the Markov chain is obtained with conditional pdf or pmf $B_{xy} = p(y_k = y|x_{\tau_k} = x)$.
Step 2: Sequential decision-making: Let $\mathcal{I}_k = \{y_1, \dots, y_k, u_0, u_1, \dots, u_{k-1}\}$ denote the history of past decisions and available observations. At times τ_k, an action u_k is chosen according to the stationary policy μ, where

$$u_k = \mu(\mathcal{I}_k) \in \mathcal{U} = \{0 \text{ (announce change)}, 1, 2, \dots, L\}. \qquad (13.43)$$

Here, $u_k = l$ denotes take the next measurement after D_l time points, $l \in \{1, 2, \dots, L\}$. The initial decision at time $\tau_0 = 0$ is $u_0 = \mu(\pi_0)$ where π_0 is the initial distribution. Also, $D_1 < D_2 < \cdots < D_L$ are L distinct positive integers that denote the set of possible sampling time intervals. Thus the decision u_k specifies the next time τ_{k+1} to make a measurement as follows:

$$\tau_{k+1} = \tau_k + D_{u_k}, \quad u_k \in \{1, 2, \dots, L\}, \quad \tau_0 = 0. \qquad (13.44)$$

Step 3: Costs: If decision $u_k \in \{1, 2, \dots, L\}$ is chosen, a *decision cost* $c(x_t, u_k)$ is incurred by the decision-maker at each time $t \in [\tau_k, \dots, \tau_{k+1} - 1]$ until the next measurement is taken at time τ_{k+1}. Also at each time τ_k, $k = 0, 1, \dots, k^* - 1$, the decision-maker pays a nonnegative *measurement cost* $\bar{m}(x_{\tau_k}, x_{\tau_{k+1}}, y_{k+1}, u_k)$ to observe the noisy Markov chain at time $\tau_{k+1} = \tau_k + D_{u_k}$. In terms of $\mathcal{I}_k$, this is equivalent to choosing the measurement cost as (see (7.8))

$$r(x_{\tau_k} = e_i, u_k) = \sum_j P^{D_{u_k}}|_{ij} B_{jy}\, \bar{m}(x_{\tau_k} = e_i, x_{\tau_{k+1}} = e_j, y_{k+1} = y, u_k) \qquad (13.45)$$

where $P^{D_u}|_{ij}$ denotes the (i,j) element of matrix P^{D_u}.
Step 4: If at time $t = \tau_{k^*}$ the decision $u_{k^*} = 0$ is chosen, then a terminal cost $c(x_{\tau_{k^*}}, 0)$ is incurred and the problem terminates.
If decision $u_k \in \{1, 2, \dots, L\}$, set k to $k + 1$ and go to Step 1. □

In terms of the belief state, the objective to be minimized can be expressed as

$$J_\mu(\pi) = \mathbb{E}_\mu \left\{ \sum_{k=0}^{k^*-1} C(\pi_k, u_k) + C(\pi_{k^*}, u_{k^*} = 0) \right\} \qquad (13.46)$$

where $C(\pi, u) = C_u' \pi$ for $u \in \mathcal{U}$

$$C_u = \begin{cases} r_u + (I + P + \cdots + P^{D_u - 1}) c_u & u \in \{1, 2, \dots, L\} \\ c_0 & u = 0 \end{cases}$$
$$c_u = \left[c(e_1, u), \dots, c(e_X, u)\right]', \quad r_u = \left[r(e_1, u), \dots, r(e_X, u)\right]'. \qquad (13.47)$$

Define the stopping set $\mathcal{R}_1$ as

$$\mathcal{R}_1 = \{\pi \in \Pi(X) : \mu^*(\pi) = 0\} = \{\pi \in \Pi(X) : Q(\pi, 0) \leq Q(\pi, u),\ u \in \{1, 2, \dots, L\}\}. \qquad (13.48)$$

Bellman's dynamic programming equation reads

$$\mu^*(\pi) = \arg\min_{u \in \mathcal{U}} Q(\pi, u), \quad J_{\mu^*}(\pi) = V(\pi) = \min_{u \in \mathcal{U}} Q(\pi, u),$$
$$Q(\pi, u) = C(\pi, u) + \sum_{y \in \mathbf{Y}} V\left(T(\pi, y, u)\right) \sigma(\pi, y, u), \quad u = 1, \ldots, L,$$
$$Q(\pi, 0) = C(\pi, 0). \tag{13.49}$$

13.5.2 Example: Quickest change detection with optimal sampling

We now formulate the quickest detection problem with optimal sampling which serves as an example to illustrate the above model. Recall that decisions (whether to stop, or continue and take next observation sample after D_l time points) are made at times $\tau_1, \tau_2, \ldots$. In contrast, the state of the Markov chain (which models the change we want to detect) can change at any time t. We need to construct the delay and false alarm penalties to take this into account.

1. *Phase-distributed (PH) change time*: As in Example 1, in quickest detection, the target state (labeled as state 1) is absorbing. The transition matrix P is specified in (13.1). Denote the time at which the Markov chain hits the target state as

$$\tau^0 = \min\{t : x_t = 1\}. \tag{13.50}$$

2. *Observations*: As in (13.5) of Example 1, $B_{2y} = B_{3y} = \cdots = B_{Xy}$.

3. *Costs*: Associated with the quickest detection problem are the following costs.

(i) False alarm: Let τ_{k^*} denote the time at which decision $u_{k^*} = 0$ (stop and announce target state) is chosen, so that the problem terminates. If the decision to stop is made before the Markov chain reaches the target state 1, i.e. $\tau_{k^*} < \tau^0$, then a unit false alarm penalty is paid. Choosing $\psi = \mathbf{1} - e_1$ in (13.8), in terms of the belief state, the false alarm penalty at epoch $k = k^*$ is

$$\sum_{i \neq 1} \mathbb{E}\{I(x_{\tau_k} = e_i, u_k = 0) | \mathcal{I}_k\} = (\mathbf{1} - e_1)' \pi_k I(u_k = 0). \tag{13.51}$$

(ii) *Delay cost of continuing*: Suppose decision $u_k \in \{1, 2, \ldots, L\}$ is taken at time τ_k. So the next sampling time is $\tau_{k+1} = \tau_k + D_{u_k}$. Then for any time $t \in [\tau_k, \tau_{k+1} - 1]$, the event $\{x_t = e_1, u_k\}$ signifies that a change has occurred but not been announced by the decision-maker. Since the decision-maker can make the next decision (to stop or continue) at τ_{k+1}, the delay cost incurred in the time interval $[\tau_k, \tau_{k+1} - 1]$ is $d \sum_{t=\tau_k}^{\tau_{k+1}-1} I(x_t = e_1, u_k)$ where d is a nonnegative constant. For $u_k \in \{1, 2, \ldots, L\}$, the expected delay cost in interval $[\tau_k, \tau_{k+1} - 1] = [\tau_k, \tau_k + D_{u_k} - 1]$ is

$$d \sum_{t=\tau_k}^{\tau_{k+1}-1} \mathbb{E}\{I(x_t = e_1, u_k) | \mathcal{F}_k\} = d e_1' (I + P + \cdots + P^{D_{u_k}-1})' \pi_k.$$

(iii) *Measurement sampling cost*: Suppose decision $u_k \in \{1, 2, \ldots, L\}$ is taken at time τ_k. As in (13.47) let $r_{u_k} = (r(x_{\tau_k} = e_i, u_k), i \in \mathcal{X})$ denote the nonnegative measurement cost vector for choosing to take a measurement. Next, since in quickest detection, states

$2, \dots, X$ are fictitious states that are indistinguishable in terms of cost, choose $r(e_2, u) = \dots = r(e_X, u)$.

Choosing a constant measurement cost at each time (i.e. $r(e_i, u)$ independent of state i and action u), still results in nontrivial global costs for the decision-maker. This is because choosing a smaller sampling interval will result in more measurements until the final decision to stop, thereby incurring a higher total measurement cost for the global decision-maker.

Remarks: (i) *Quickest state estimation*: The setup is identical to above, except that unlike (13.1), the transition matrix P no longer has an absorbing target state. Therefore the Markov chain can jump in and out of the target state. To avoid pathological cases, we assume P is irreducible. Also there is no requirement for the observation probabilities to satisfy $B_{2y} = B_{3y} = \cdots = B_{Xy}$.

(ii) *Summary*: In the notation of (13.47), the costs for quickest detection/estimation optimal sampling are $C(\pi, u) = C_u'\pi$ where $C_0 = c_0 = \mathbf{1} - e_1$ and

$$C_u = r_u + (I + P + \cdots + P^{D_u - 1})c_u, \quad c_u = de_1, \quad u \in \{1, 2, \dots, L\}. \tag{13.52}$$

(iii) *Structural results*: As mentioned earlier, since there are multiple "continue" actions $u \in \{1, 2, \dots, L\}$, the problem is not a standard stopping time POMDP. Of course, Theorem 12.2.1 applies and so the optimal policy for the "stop" action, i.e. stopping set $\mathcal{R}_1$ (13.48), is a convex set. Characterizing the structure of the policy for the actions $\{1, 2, \dots, L\}$ is more difficult. For the two-state case, we obtain structural results in §13.5.3. For the multi-state case, we will develop results in Chapter 14.

13.5.3 Threshold optimal policy for quickest detection with sampling

Consider quickest detection with optimal sampling for geometric distributed change time. The transition matrix is $P = \begin{bmatrix} 1 & 0 \\ 1 - P_{22} & P_{22} \end{bmatrix}$ and expected change time is $\mathbb{E}\{\tau^0\} = \frac{1}{1 - P_{22}}$ where τ^0 is defined in (13.50). For a two-state Markov chain since $\pi(1) + \pi(2) = 1$, it suffices to represent π by its second element $\pi(2) \in [0, 1]$. That is, the belief space $\Pi(X)$ is the interval $[0, 1]$.

THEOREM 13.5.1 *Consider the quickest detection optimal sampling problem of §13.5.2 with geometric-distributed change time and costs (13.52). Assume the measurement cost $r(e_i, u)$ satisfies (C), (S) and the observation distribution satisfies (F2). Then there exists an optimal policy $\mu^*(\pi)$ with the following monotone structure: there exist up to L thresholds denoted $\pi_1^*, \dots, \pi_L^*$ with $0 = \pi_0^* \le \pi_1^* \le \pi_L^* \le \pi_{L+1}^* = 1$ such that, for $\pi(2) \in [0, 1]$,*

$$\mu^*(\pi) = l \ \text{if} \ \pi(2) \in [\pi_l^*, \pi_{l+1}^*), \quad l = 0, 1, \dots, L. \tag{13.53}$$

Here the sampling intervals are ordered as $D_1 < D_2 < \dots < D_L$. So the optimal sampling policy (13.53) makes measurements less frequently when the posterior $\pi(2)$ is away from the target state and more frequently when closer to the target state. (Recall the target state is $\pi(2) = 0$.)

The proof follows from that of Theorem 11.3.1 on page 246. There are two main conclusions regarding Theorem 13.5.1. First, for constant measurement cost, (C) and (S) hold trivially. For the general measurement cost $\bar{m}(x_{\tau_{k+1}} = e_j, y_{k+1}, u_k)$ (see (13.45)) that depends on the state at epoch $k+1$, then $r(e_i, u)$ in (13.45) automatically satisfies (S) if P satisfies (F1) and $\bar{m}$ is decreasing in j. Second, the optimal policy $\mu^*(\pi)$ is monotone in posterior $\pi(1)$ and therefore has a finite dimensional characterization. To determine the optimal policy, one only needs to compute the values of the L thresholds $\pi_1^*, \ldots, \pi_L^*$. These can be estimated via a simulation-based stochastic optimization algorithm.

13.5.4 Optimality of threshold policy for sequential optimal sampling

We now consider the general optimal sampling problem where the two-state Markov chain can jump in and out of the target state 1. Quickest detection considered above is a special case where the target state is absorbing. We will give sufficient conditions for the optimal policy to be monotone on a subset denoted $\Pi_{\mathcal{M}}$ of the belief space $\Pi = [0, 1]$. Define

$$\Pi_{\mathcal{M}} = \{\pi(2) : \pi_{\mathcal{M}} \leq \pi(2) \leq 1\},$$
$$\text{where } \pi_{\mathcal{M}} = \frac{P_{11} - P^2|_{11}}{P_{22} - P^2|_{22} + P_{11} - P^2|_{11}}. \tag{13.54}$$

For transition matrices that satisfy (F2), it can be shown that $\pi_{\mathcal{M}} \in [0, 1]$ and so $\Pi_{\mathcal{M}}$ is nonempty. For the quickest detection problem since $P_{11} = P^2|_{11} = 1$, clearly $\pi_{\mathcal{M}} = 0$, and so $\Pi_{\mathcal{M}} = \Pi(X)$ meaning that the optimal policy is monotone on the entire belief space [0, 1]. (Thus Theorem 13.5.1 holds on $\Pi(X)$.)

The intuition for specifying $\Pi_{\mathcal{M}}$ is that for $\pi \in \Pi_{\mathcal{M}}$, the filtering update (5.1) satisfies the following property under (F2): the second element $\pi(2)$ of $T(\pi, y, u)$ always is smaller than that of $T(\pi, y, u+1)$ (equivalently, $T(\pi, y, u)$ first-order stochastically dominates $T(\pi, y, u+1)$). This property is crucial to prove that the optimal policy has a monotone structure.

The following main result includes Theorem 13.5.1 as a special case.

THEOREM 13.5.2 *Consider the optimal sampling problem with state dimension $X = 2$ and action space $\mathcal{U}$ in (13.43). Then the optimal policy $\mu^*(\pi)$ defined in (13.49) has the following structure:*

(i) The optimal stopping set $\mathcal{R}_1$ (13.48) is a convex subset of Π. Therefore, the stopping set is the interval $\mathcal{R}_1 = [0, \pi_1^*)$ where the threshold $\pi_1^* \in [0, 1]$.

(ii) Under (C), (F1), (F2), (S) in §11.3 for a two-state POMDP, there exists an optimal sampling policy $\mu^*(\pi)$ defined in (13.49) that is increasing in $\pi(2)$ for $\pi(2) \in \Pi_{\mathcal{M}} \cup \mathcal{R}_1$ where $\Pi_{\mathcal{M}}$ is defined in (13.54).

(iii) As a consequence of (i) and (ii), there exist up to L thresholds denoted $\pi_1^*, \ldots, \pi_L^*$ in $\Pi_{\mathcal{M}} \cup \mathcal{R}_1$ such that, for $\pi(2) \in \Pi_{\mathcal{M}} \cup \mathcal{R}_1$, the optimal policy has the monotone structure of (13.53).

The proof of (i) follows from the convexity of the stopping region shown in Theorem 12.2.1 on page 258. The proof of (ii) follows from the monotone structure of the optimal policy for a two-state POMDP proved in Theorem 11.3.1 on page 246.

Note that in Theorem 11.3.1, we also needed (F3′). It turns out that in the case of sampling control for two-states, (F2) automatically implies that (F3′) holds. In particular, consider the expression

$$\begin{bmatrix} P_{21} & P_{22} \\ 0 & 1 \end{bmatrix} P = \begin{bmatrix} e_2' P^2 \\ e_2' P \end{bmatrix}.$$

On the left-hand side, the first matrix is TP2 by construction and P is TP2 by (F2). The product of TP2 matrices is TP2 (Lemma 10.5.2 on page 228). Therefore, the right-hand side is TP2. This is equivalent to saying that on the right-hand side, the first row is MLR dominated by the second row (see Definition 10.2.1 on page 223). That is, $P'e_2 \geq_r P^{2\prime} e_2$. Finally, $P' e_2 \geq_r P^{2'} e_2$ implies $P^{2'} e_2 \geq_r P^{3'} e_2$ by Theorem 10.3.11 under (F2). Therefore $P' e_2 \geq_r P^{2'} e_2 \geq_r P^{3'} e_3$. Hence, $P^{D_u{}'} e_2 \geq_r P^{D_{u+1}{}'} e_2$ since $D_u < D_{u+1}$ by assumption. This implies that for each $q \in \{1, 2\}$, $\sum_{j \geq q} P^{D_u}|_{ij}$ is submodular in (i, u), i.e. (F3′) in §11.3 holds.

COROLLARY 13.5.3 *Consider the quickest state estimation problem with setup identical to the quickest detection problem of Theorem 13.5.1 except that the transition matrix P does not necessarily have an absorbing state. Assume P satisfies (F2). Then the conclusions of Theorem 13.5.1 hold on* $\Pi_{\mathcal{M}} \cup \mathcal{R}_1$ *where* $\Pi_{\mathcal{M}}$ *is defined in (13.54).* □

Summary: We considered quickest change detection with measurement control, i.e., the decision-maker also has the choice of when to observe (sample) the Markov chain. It was shown that the optimal policy has a monotone structure: sample less frequently initially when the state is away from the target state and sample more frequently later. If satisfaction is viewed as the number of times the decision-maker looks at the Markov chain, Theorems 13.5.1 and 13.5.2 say that "delayed satisfaction" is optimal.

13.6 Complements and sources

This chapter has considered several examples of stopping time POMDPs in sequential quickest detection. In each section of this chapter, several references to the literature were given. The book [268] is devoted to quickest detection problems and contains numerous references. PH-distributed change times are used widely to model discrete event systems [250] and are a natural candidate for modeling arrival/demand processes for services that have an expiration date [93]. It would be useful to do a performance analysis of the various optimal detectors proposed in this chapter – see [317, 275] and references therein. [31] considers a measurement control problem for geometric-distributed change times (two-state Markov chain with an absorbing state). [180] considers joint change detection and measurement control POMDPs with more than two states.

Appendix 13.A Proof of Theorem 13.4.1

The proof is more involved than that of Theorem 12.2.1 since now $V(\pi)$ is not necessarily concave over $\Pi(X)$, since $T(\cdot)$ and $\sigma(\cdot)$ are functions of R^π_a (13.27) which itself is an explicit (and in general non-concave) function of π.

Define the matrix $R^\pi = (R^\pi(i,a), i = \{1,2\}, a \in \{1,2\})$, where $R^\pi(i,a) = P(a|x = e_i, \pi)$. From (13.27) there are only three possible values for R^π, namely,

$$R^\pi = \begin{bmatrix} 0 & 1 \\ 0 & 1 \end{bmatrix}, \pi \in \mathcal{P}_1, \quad R^\pi = B, \pi \in \mathcal{P}_2 \cup \mathcal{P}_3, \quad R^\pi = \begin{bmatrix} 1 & 0 \\ 1 & 0 \end{bmatrix}, \pi \in \mathcal{P}_4. \tag{13.55}$$

Based on Bellman's equation (13.31), the value iteration algorithm is

$$\begin{aligned} V_{n+1}(\pi) = \min\{&C(\pi,2) + \rho V_n(\pi) I(\pi \in \mathcal{P}_1) \\ &+ \rho \sum_{a \in \mathbb{A}} V_n(T(\pi,a))\sigma(\pi,a)[I(\pi \in \mathcal{P}_2) + I(\pi \in \mathcal{P}_3)] + \rho V_n(\pi) I(\pi \in \mathcal{P}_4), 0\}. \end{aligned} \tag{13.56}$$

Assuming $V_n(\pi)$ is MLR decreasing on $\mathcal{P}_1 \cup \mathcal{P}_4$ implies $V_{n+1}(\pi)$ is MLR decreasing on $\mathcal{P}_1 \cup \mathcal{P}_4$ since $C(\pi,2)$ is MLR decreasing. This proves claim (i).

We now prove inductively that $V_n(\pi)$ is piecewise linear concave on each interval $\mathcal{P}_l$, $l = 1, \ldots, 4$. The proof of concavity on $\mathcal{P}_1$ and $\mathcal{P}_4$ follows straightforwardly since $C(\pi,2)$ is piecewise linear and concave. The proof for intervals $\mathcal{P}_2$ and $\mathcal{P}_3$ is more delicate.

We need the following property of the social learning Bayesian filter. Define the two dimensional vector $\kappa_i = (1 - \kappa_i, \quad \kappa_i)'$.

LEMMA 13.A.1 *Consider the social learning Bayesian filter (13.26). Then* $T(\kappa_1, 1) = \kappa_2$, $T(\kappa_3, 2) = \kappa_2$. *Furthermore if B is symmetric TP2, then* $T(\kappa_2, 2) = \kappa_1$, $T(\kappa_2, 1) = \kappa_3$ *and* $\kappa_3 \leq \kappa_2 \leq \kappa_1$. *So*

(i) $\pi \in \mathcal{P}_2$ *implies* $T(\pi, 2) \in \mathcal{P}_1$ *and* $T(\pi, 1) \in \mathcal{P}_3$.
(ii) $\pi \in \mathcal{P}_3$ *implies* $T(\pi, 2) \in \mathcal{P}_2$ *and* $T(\pi, 1) \in \mathcal{P}_4$. □

Proof Recall from (13.55) that on intervals $\mathcal{P}_2$ and $\mathcal{P}_3$, $R^\pi = B$. Then it is straightforwardly verified from (13.26) that $T(\kappa_1, 1) = T(\kappa_3, 2) = \kappa_2$. Next, using (13.26) it follows that $B_{12}B_{11} = B_{22}B_{21}$ is a sufficient condition for $T(\kappa_2, 2) = \kappa_1$ and $T(\kappa_2, 1) = \kappa_3$. Also, applying Theorem 11.2.1(2), B TP2 implies $\kappa_3 \leq \kappa_2 \leq \kappa_1$. So B symmetric TP2 is sufficient for the claims of the lemma to hold. Statements (i) and (ii) then follow straightforwardly. In particular, from Theorem 11.2.1(1), $\kappa_1 \geq_r \pi \geq_r \kappa_2$ implies $T(\kappa_1, 1) = \kappa_2 \geq_r T(\pi, 1) \geq_r T(\kappa_2, 1) = \kappa_3$, which implies Statement (i) of the Lemma. Statement (ii) follows similarly. □

Returning to the proof of Theorem 13.4.1. Assume now that $V_n(\pi)$ is piecewise linear and concave on each interval $\mathcal{P}_l$, $l = 1, \ldots, 4$. That is, for two dimensional vectors γ_{m_l} in the set Γ_l, $V_n(\pi) = \sum_l \min_{m_l \in \Gamma_l} \gamma'_{m_l} \pi \, I(\pi \in \mathcal{P}_l)$. Consider $\pi \in \mathcal{P}_2$. From (13.55), since $R_a^\pi = B_a$, $a = 1, 2$, Lemma 13.A.1 (i) together with the value iteration algorithm (13.56) yields

$$V_{n+1}(\pi) = \min\{C(\pi, 2) + \rho \left[\min_{m_3 \in \Gamma_3} \gamma'_{m_3} B_1 \pi + \min_{m_1 \in \Gamma_1} \gamma'_{m_1} B_2 \pi \right], 0\}.$$

Since each of the terms in the above equation are piecewise linear and concave, it follows that $V_{n+1}(\pi)$ is piecewise linear and concave on $\mathcal{P}_2$. A similar proof holds for $\mathcal{P}_3$ and this involves using Lemma 13.A.1(ii). As a result the stopping set on each interval $\mathcal{P}_l$ is a convex region, i.e. an interval. This proves claim (ii).

14 Myopic policy bounds for POMDPs and sensitivity to model parameters

Contents

Chapter 12 discussed stopping time POMDPs and gave sufficient conditions for the optimal policy to have a monotone structure. In this chapter we consider more general POMDPs (not necessarily with a stopping action) and present the following structural results:

1. *Upper and lower myopic policy bounds using copositivity dominance*: For general POMDPs it is difficult to provide sufficient conditions for monotone policies. Instead, we provide sufficient conditions so that the optimal policy can be upper and lower bounded by judiciously chosen myopic policies. These sufficient conditions involve the *copositive ordering* described in Chapter 10. The myopic policy bounds are constructed to maximize the volume of belief states where they coincide with the optimal policy. Numerical examples illustrate these myopic policies for continuous and discrete valued observations.

2. *Lower myopic policy bounds using Blackwell dominance*: Suppose the observation probabilities for actions 1 and 2 can be related via the following factorization: $B(1) = B(2)\,R$ where R is a stochastic matrix. We then say that $B(2)$ *Blackwell* dominates $B(1)$. If this Blackwell dominance holds, we will show that a myopic policy coincides with the optimal policy for all belief states where choosing action 2 yields a smaller instantaneous cost than choosing action 1. Thus, the myopic policy forms a lower bound to the optimal policy. We provide two examples: scheduling an optimal filter versus an optimal predictor, and scheduling with ultrametric observation matrices.
3. *Sensitivity to POMDP parameters*: The final result considered in this chapter is: how does the optimal cumulative cost of POMDP depend on the transition and observation probabilities? We provide two sets of results: ordinal and cardinal. The ordinal results use the copositive ordering of transition matrices and Blackwell dominance of observation matrices that yield an ordering of the achievable optimal costs of a POMDP. The cardinal results determine explicit formulas for the sensitivity of the POMDP optimal costs and policy to small variations of the transition and observation probabilities.

14.1 The partially observed Markov decision process

Throughout this chapter we will consider discounted cost infinite horizon POMDPs discussed in §7.6. Let us briefly review this model. A discrete-time Markov chain evolves on the state space $\mathcal{X} = \{e_1, e_2, \dots, e_X\}$ where e_i denotes the unit X-dimensional vector with 1 in the i-th position. Denote the action space as $\mathcal{U} = \{1, 2, \dots, U\}$ and observation space as $\mathcal{Y}$. For discrete-valued observations $\mathcal{Y} = \{1, 2, \dots, Y\}$ and for continuous observations $\mathcal{Y} \subset \mathbb{R}$.

Let $\Pi(X) = \left\{\pi : \pi(i) \in [0,1], \sum_{i=1}^{X} \pi(i) = 1\right\}$ denote the belief space of X-dimensional probability vectors. For stationary policy $\mu : \Pi(X) \to \mathcal{U}$, initial belief $\pi_0 \in \Pi(X)$, discount factor $\rho \in [0, 1)$, define the discounted cost:

$$J_\mu(\pi_0) = \mathbb{E}\left\{\sum_{k=0}^{\infty} \rho^k c'_{\mu(\pi_k)} \pi_k\right\}. \tag{14.1}$$

Here $c_u = [c(1,u), \dots, c(X,u)]'$, $u \in \mathcal{U}$ is the cost vector for each action, and the belief state evolves as $\pi_k = T(\pi_{k-1}, y_k, u_k)$ where

$$T(\pi, y, u) = \frac{B_y(u) P'(u) \pi}{\sigma(\pi, y, u)}, \quad \sigma(\pi, y, u) = \mathbf{1}' B_y(u) P'(u) \pi,$$
$$B_y(u) = \operatorname{diag}\big(B_{1,y}(u), \cdots, B_{X,y}(u)\big). \tag{14.2}$$

Recall $\mathbf{1}$ represents a X-dimensional vector of ones, $P(u) = \left[P_{ij}(u)\right]_{X \times X}$ $P_{ij}(u) = \mathbb{P}(x_{k+1} = e_j | x_k = e_i, u_k = u)$ denote the transition probabilities, $B_{xy}(u) = \mathbb{P}(y_{k+1} = y | x_{k+1} = e_x, u_k = u)$ when $\mathcal{Y}$ is finite, or $B_{xy}(u)$ is the conditional probability density function when $\mathcal{Y} \subset \mathbb{R}$.

The aim is to compute the optimal stationary policy $\mu^* : \Pi(X) \to \mathcal{U}$ such that $J_{\mu^*}(\pi_0) \leq J_\mu(\pi_0)$ for all $\pi_0 \in \Pi(X)$. Obtaining the optimal policy μ^* is equivalent to solving Bellman's dynamic programming equation: $\mu^*(\pi) = \operatorname{argmin}_{u \in \mathcal{U}} Q(\pi, u)$, $J_{\mu^*}(\pi_0) = V(\pi_0)$, where

$$V(\pi) = \min_{u \in \mathcal{U}} Q(\pi, u), \quad Q(\pi, u) = c_u' \pi + \rho \sum_{y \in Y} V(T(\pi, y, u))\, \sigma(\pi, y, u). \tag{14.3}$$

Since $\Pi(X)$ is continuum, Bellman's equation (14.3) does not translate into practical solution methodologies. This motivates the construction of judicious myopic policies that upper and lower bound $\mu^*(\pi)$.

14.2 Myopic policies using copositive dominance: insight

For stopping time POMDPs, in Chapter 12 we gave sufficient conditions for $Q(\pi, u)$ in Bellman's equation to be submodular, i.e. $Q(\pi, u+1) - Q(\pi, u)$ is decreasing in belief π with respect to the monotone likelihood ratio order. This implied that the optimal policy $\mu^*(\pi)$ was MLR increasing in belief π and had a threshold structure.

Unfortunately, for a general POMDP, giving sufficient conditions for $Q(\pi, u)$ to be submodular is still an open problem.[1] Instead of showing submodularity, in this chapter we will give sufficient conditions for $Q(\pi, u)$ to satisfy

$$Q(\pi, u+1) - Q(\pi, u) \leq C(\pi, u+1) - C(\pi, u) \tag{14.4}$$

where $C(\pi, u)$ is a cleverly chosen instantaneous cost in terms of the belief state π. A nice consequence of (14.4) is the following: Let $\overline{\mu}(\pi) = \operatorname{argmin}_u C(\pi, u)$ denote the myopic policy that minimizes the instantaneous cost. Then (14.4) implies that the optimal policy $\mu^*(\pi)$ satisfies

$$\mu^*(\pi) \leq \overline{\mu}(\pi), \qquad \text{for all } \pi \in \Pi(X).$$

In words: if (14.4) holds, then the myopic policy $\overline{\mu}(\pi)$ is provably an upper bound to the optimal policy $\mu^*(\pi)$. Since the myopic policy is trivially computed, this is a useful result. But there is more! As will be described below, for discounted cost POMDPs, the optimal policy remains unchanged for a family of costs $C(\pi, u)$. So by judiciously choosing these costs we can also construct myopic policies $\underline{\mu}(\pi)$ that *lower bound* the optimal policy. To summarize, for any belief state π, we will present sufficient conditions under which the optimal policy $\mu^*(\pi)$ of a POMDP can be upper and lower bounded by myopic policies[2] denoted by $\overline{\mu}(\pi)$ and $\underline{\mu}(\pi)$, respectively, i.e. (see Figure 14.1 on the next page for a visual display)

$$\underline{\mu}(\pi) \leq \mu^*(\pi) \leq \overline{\mu}(\pi) \quad \text{for all } \pi \in \Pi(X). \tag{14.5}$$

[1] For the two-state case, conditions for submodularity are given in §11.3, but these do not generalize to more than two states.

[2] Obviously $\underline{\mu}(\pi) = 1$ for all $\pi \in \Pi(X)$ is a trivial (but useless) lower bound to the optimal policy. We are interested in constructing non trivial lower and upper bounds for the optimal policy and maximizing the volume where they overlap with the optimal policy.

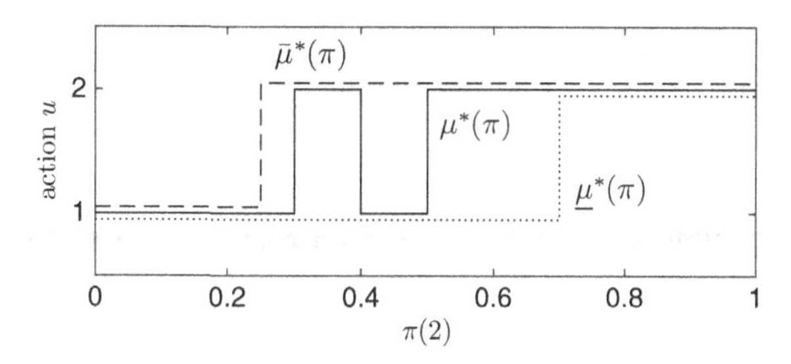

Figure 14.1 Illustration of main result of this chapter. The aim is to construct an upper bound $\bar{\mu}$ (dashed line) and lower bound $\underline{\mu}$ (dotted line), to the optimal policy μ^* (solid line) such that (14.5) holds for each belief state π. Thus the optimal policy is sandwiched between the judiciously chosen myopic policies $\underline{\mu}$ and $\bar{\mu}$ over the entire belief space $\Pi(X)$. Note for π where $\overline{\mu}(\pi) = \underline{\mu}(\pi)$, they coincide with the optimal policy $\mu^*(\pi)$. Maximizing the volume of beliefs where $\underline{\mu}(\pi) = \overline{\mu}(\pi)$ is achieved by solving a linear programming problem as described in §14.4.

Clearly, for belief states π where $\overline{\mu}(\pi) = \underline{\mu}(\pi)$, the optimal policy $\mu^*(\pi)$ is completely determined.

Interestingly, these judiciously constructed myopic policies are independent of the actual values of the observation probabilities (providing they satisfy a sufficient condition) which makes the structural results applicable to both discrete and continuous observations. Finally, we will construct the myopic policies, $\overline{\mu}(\pi)$ and $\underline{\mu}(\pi)$, to maximize the volume of the belief space where they coincide with the optimal policy $\mu^*(\pi)$.

As an extension of the above results, motivated by examples in controlled sensing [183, 339, 33], we show that similar myopic bounds hold for POMDPs with quadratic costs in the belief state. For such POMDPs, even for finite observation alphabet set and finite horizon, stochastic dynamic programming does not yield an explicit solution.

Numerical examples are presented to illustrate the performance of these myopic policies. To quantify how well the myopic policies perform we use two parameters: the volume of the belief space where the myopic policies coincide with the optimal policy, and an upper bound to the average percentage loss in optimality due to following this optimized myopic policy.

Context

The papers [225, 277, 278] give sufficient conditions for (14.4) so that the optimal policy of a POMDP can be upper bounded[3] by a myopic policy. Unfortunately, despite the enormous usefulness of such a result, the sufficient conditions given in [225] and [277] are not useful – it is impossible to generate nontrivial examples that satisfy the conditions (c), (e), (f) of [225, Proposition 2] and condition (i) of [277, Theorem 5.6]. In this chapter, we provide a fix to these sufficient conditions so that the results of [225, 277] hold for constructing a myopic policy that upper bounds the optimal policy. It turns out that Assumptions (F3′) and (F4) described in Chapter 10 are precisely the fix we

[3] Since [225] deals with maximization rather than minimization, the myopic policy constructed in [225] forms a lower bound.

need. We also show how this idea of constructing a upper bound myopic policy can be extended to constructing a *lower* bound myopic policy.

14.3 Constructing myopic policy bounds for optimal policy using copositive dominance

With the above motivation, we are now ready to construct myopic policies that provably sandwich the optimal policy for a POMDP.

Assumptions

(F3′) and (F4) below are the main copositivity assumptions.

(C1) There exists a vector $g \in \mathbb{R}^X$ such that the X-dimensional vector $\overline{C}_u \equiv c_u + (I - \rho P(u))\, g$ is strictly increasing elementwise for each action $u \in \mathcal{U}$.

(C2) There exists a vector $f \in \mathbb{R}^X$ such that the X-dimensional vector $\underline{C}_u \equiv c_u + (I - \rho P(u)) f$ is strictly decreasing elementwise for each action $u \in \mathcal{U}$.

(F1) $B(u)$, $u \in \mathcal{U}$ is totally positive of order 2 (TP2). That is, all second-order minors and nonnegative.

(F2) $P(u)$, $u \in \mathcal{U}$ is totally positive of order 2 (TP2).

(F3′) $\gamma_{mn}^{j,u,y} + \gamma_{nm}^{j,u,y} \geq 0 \; \forall m, n, j, u, y$ where

$$\gamma_{mn}^{j,u,y} = B_{j,y}(u) B_{j+1,y}(u+1) P_{m,j}(u) P_{n,j+1}(u+1) - B_{j+1,y}(u) B_{j,y}(u+1) P_{m,j+1}(u) P_{n,j}(u+1). \quad (14.6)$$

(F4) $\sum_{y \leq \bar{y}} \sum_{j=1}^{X} \left[P_{i,j}(u) B_{j,y}(u) - P_{i,j}(u+1) B_{j,y}(u+1) \right] \leq 0$ for $i \in \{1, 2, \ldots, X\}$ and $\bar{y} \in \mathcal{Y}$.

Discussion

Recall (F1), (F2), (F3′), (F4) were discussed in Chapter 10. As described in Chapter 10, (F3′) and (F4) are a relaxed version of Assumptions (c), (e), (f) of [225, Proposition 2] and Assumption (i) of [277, Theorem 5.6]. In particular, the assumptions (c), (e), (f) of [225] require that $P(u+1) \underset{\text{TP2}}{\geq} P(u)$ and $B(u+1) \underset{\text{TP2}}{\geq} B(u)$, where $\underset{\text{TP2}}{\geq}$ (TP2 stochastic ordering) is defined in [249], which is impossible for stochastic matrices, unless $P(u) = P(u+1)$, $B(u) = B(u+1)$ or the matrices $P(u), B(u)$ are rank 1 for all u meaning that the observations are non-informative.

Let us now discuss (C1) and (C2). If the elements of c_u are strictly increasing then (C1) holds trivially. Similarly, if the elements of c_u are strictly decreasing then (C2) holds; indeed then (C2) is equivalent to (C) on page 243.

(C1) and (C2) are easily verified by checking the feasibility of the following linear programs:

$$LP1 : \min_{g \in S_g} \mathbf{1}_X' g, \; LP2 : \min_{f \in S_f} \mathbf{1}_X' f. \tag{14.7}$$

$$S_g = \left\{ g : \overline{C}_u' e_i \leq \overline{C}_u' e_{i+1} \;\; \forall u \in \mathcal{U}, i \in \{1,2,\ldots,X\} \right\} \tag{14.8}$$

$$S_f = \left\{ f : \underline{C}_u' e_i \geq \underline{C}_u' e_{i+1} \;\; \forall u \in \mathcal{U}, i \in \{1,2,\ldots,X\} \right\} \tag{14.9}$$

where e_i is the unit X-dimensional vector with 1 at the i-th position.

14.3.1 Construction of myopic upper and lower bounds

We are interested in myopic policies of the form $\operatorname{argmin}_{u \in \mathcal{U}} C_u' \pi$ where cost vectors C_u are constructed so that when applied to Bellman's equation (14.3), they leave the optimal policy $\mu^*(\pi)$ unchanged. This is for several reasons: first, similar to [225], [277] it allows us to construct useful myopic policies that provide provable upper and lower bounds to the optimal policy. Second, these myopic policies can be straightforwardly extended to two-stage or multi-stage myopic costs. Third, such a choice precludes choosing useless myopic bounds such as $\overline{\mu}(\pi) = U$ for all $\pi \in \Pi(X)$.

Accordingly, for any two vectors g and $f \in \mathbb{R}^X$, define the myopic policies associated with the transformed costs $\overline{C}_u$ and $\underline{C}_u$ as follows:

$$\overline{\mu}(\pi) \equiv \operatorname*{argmin}_{u \in \mathcal{U}} \overline{C}_u' \pi, \quad \text{where } \overline{C}_u = c_u + (I - \rho P(u))\, g \tag{14.10}$$

$$\underline{\mu}(\pi) \equiv \operatorname*{argmin}_{u \in \mathcal{U}} \underline{C}_u' \pi, \quad \text{where } \underline{C}_u = c_u + (I - \rho P(u)) f. \tag{14.11}$$

It is easily seen that Bellman's equation (14.3) applied to optimize the objective (14.1) with transformed costs $\overline{C}_u$ and $\underline{C}_u$ yields the same optimal strategy $\mu^*(\pi)$ as the Bellman's equation with original costs c_u. The corresponding value functions are $\overline{V}(\pi) \equiv V(\pi) + g'\pi$ and $\underline{V}(\pi) \equiv V(\pi) + f'\pi$.

The following is the main result of this section.

THEOREM 14.3.1 *Consider a POMDP $(\mathcal{X}, \mathcal{U}, \mathcal{Y}, P(u), B(u), c, \rho)$ and assume (C1), (F1), (F2), (F3′), (F4) holds. Then the myopic policies, $\overline{\mu}(\pi)$ and $\underline{\mu}(\pi)$, defined in* (14.10), (14.11) *satisfy: $\underline{\mu}(\pi) \leq \mu^*(\pi) \leq \overline{\mu}(\pi)$ for all $\pi \in \Pi(X)$.*

The above result where the optimal policy $\mu^*(\pi)$ is sandwiched between $\underline{\mu}(\pi)$ and $\overline{\mu}(\pi)$ is illustrated in Figure 14.1 for $X = 2$.

Proof We show that under (C1),(F1), (F2), (F3′) and (F4), $\mu^*(\pi) \leq \overline{\mu}(\pi) \; \forall \pi \in \Pi(X)$. Let $\overline{V}$ and $\overline{Q}$ denote the variables in Bellman's equation (14.3) when using costs $\overline{C}_u$ defined in (14.10). Then from Theorem 11.2.1 in Chapter 12, $\overline{V}(T(\pi, y, u))$ is increasing in y. From Theorem 10.3.1(5), under (F4), $\sigma(\pi, u+1) \geq_s \sigma(\pi, u)$. Therefore,

$$\begin{aligned} \sum_{y \in \mathcal{Y}} \overline{V}(T(\pi, y, u))\sigma(\pi, y, u) &\underset{(a)}{\leq} \sum_{y \in \mathcal{Y}} \overline{V}(T(\pi, y, u))\sigma(\pi, y, u+1) \\ &\underset{(b)}{\leq} \sum_{y \in \mathcal{Y}} \overline{V}(T(\pi, y, u+1))\sigma(\pi, y, u+1). \end{aligned} \tag{14.12}$$

Inequality (b) holds since from Theorem 10.3.1(4) and Theorem 11.2.1,

$$\overline{V}(T(\pi, y, u+1)) \geq \overline{V}(T(\pi, y, u)) \quad \forall y \in \mathcal{Y}.$$

Equation (14.12) implies that $\sum_{y \in \mathcal{Y}} \overline{V}(T(\pi, y, u))\sigma(\pi, y, u)$ is increasing with regard to u or equivalently,

$$\overline{Q}(\pi, u) - \overline{C}_u' \pi \leq \overline{Q}(\pi, u+1) - \overline{C}_{u+1}' \pi. \tag{14.13}$$

It therefore follows that

$$\{\pi : \overline{C}_{u'}' \pi \geq \overline{C}_u' \pi\} \subseteq \{\pi : \overline{Q}(\pi, u') \geq \overline{Q}(\pi, u)\}, \; u' > u$$

which implies that $\overline{\mu}(\pi) \leq u \implies \mu^*(\pi) \leq u$. The proof that $\mu^*(\pi) \geq \underline{\mu}(\pi)$ is similar and omitted. (See [224, Lemma 1] for a more general statement.) □

14.4 Optimizing the myopic policy bounds to match the optimal policy

The aim of this section is to determine the vectors g and f, in (14.8) and (14.9), that maximize the volume of the simplex where the myopic upper and lower policy bounds, specified by (14.10) and (14.11), coincide with the optimal policy. That is, we wish to maximize the volume of the "overlapping region"

$$\text{vol}\,(\Pi_O), \text{ where } \Pi_O \equiv \{\pi : \overline{\mu}(\pi) = \underline{\mu}(\pi) = \mu^*(\pi)\}. \tag{14.14}$$

Notice that the myopic policies $\overline{\mu}$ and $\underline{\mu}$ defined in (14.10), (14.11) do not depend on the observation probabilities B_u and so neither does vol (Π_O). So $\overline{\mu}$ and $\underline{\mu}$ can be chosen to maximize vol (Π_O) independent of $B(u)$ and therefore work for discrete and continuous observation spaces. Of course, the proof of Theorem 14.3.1 requires conditions on $B(u)$.

Optimized myopic policy for two actions

For a two action POMDP, obviously for a belief π, if $\overline{\mu}(\pi) = 1$ then $\mu^*(\pi) = 1$. Similarly, if $\underline{\mu}(\pi) = 2$, then $\mu^*(\pi) = 2$. Denote the set of beliefs (convex polytopes) where $\overline{\mu}(\pi) = \mu^*(\pi) = 1$ and $\underline{\mu}(\pi) = \mu^*(\pi) = 2$ as

$$\begin{aligned} \Pi(X)_1^g &= \left\{\pi : \overline{C}_1' \pi \leq \overline{C}_2' \pi\right\} = \left\{\pi : (c_1 - c_2 - \rho(P(1) - P(2))g)'\pi \leq 0\right\}, \\ \Pi(X)_2^f &= \left\{\pi : \underline{C}_2' \pi \leq \underline{C}_1' \pi\right\} = \left\{\pi : (c_1 - c_2 - \rho(P(1) - P(2))f)'\pi \geq 0\right\}. \end{aligned} \tag{14.15}$$

Clearly $\Pi_O = \Pi(X)_1^g \cup \Pi(X)_2^f$. Our goal is to find $g^* \in S_g$ and $f^* \in S_f$ such that vol (Π_O) is maximized.

THEOREM 14.4.1 *Assume that there exists two fixed X-dimensional vectors g^* and f^* such that*

$$\begin{aligned} (P(2) - P(1))g^* &\preceq (P(2) - P(1))g, \; \forall g \in S_g \\ (P(1) - P(2))f^* &\preceq (P(1) - P(2))f, \; \forall f \in S_f \end{aligned} \tag{14.16}$$

where for X-dimensional vectors a and b, $a \preceq b \Rightarrow [a_1 \leq b_1, \cdots, a_X \leq b_X]$. If the myopic policies $\overline{\mu}$ and $\underline{\mu}$ are constructed using g^* and f^*, then vol(Π_O) is maximized.

Proof The sufficient conditions in (14.16) ensure that $\Pi(X)_1^{g^*} \supseteq \Pi(X)_1^g \ \forall g \in S_g$ and $\Pi(X)_2^{f^*} \supseteq \Pi(X)_2^f \ \forall f \in S_f$. Indeed, to establish that $\text{vol}\left(\Pi(X)_1^{g^*}\right) \geq \text{vol}\left(\Pi(X)_1^g\right)$ $\forall g \in S_g$:

$$\begin{aligned}
&(P(1) - P(2))\, g^* \succeq (P(1) - P(2))\, g \quad \forall g \in S_g \\
&\Rightarrow c_1 - c_2 - \rho\,(P(1) - P(2))\, g^* \preceq c_1 - c_2 - \rho\,(P(1) - P(2))\, g \ \ \forall g \in S_g \\
&\Rightarrow \Pi(X)_1^{g^*} \supseteq \Pi(X)_1^g \ \forall g \in S_g \Rightarrow \text{vol}\left(\Pi(X)_1^{g^*}\right) \geq \text{vol}\left(\Pi(X)_1^g\right) \forall g \in S_g
\end{aligned} \tag{14.17}$$

So $\text{vol}\left(\Pi(X)_1^{g^*}\right) \geq \text{vol}\left(\Pi(X)_1^g\right) \forall g \in S_g$ and $\text{vol}\left(\Pi(X)_2^{f^*}\right) \geq \text{vol}\left(\Pi(X)_2^f\right) \forall g \in S_g$. Since $\Pi_O = \Pi(X)_1^{g^*} \cup \Pi(X)_2^{f^*}$, the proof is complete. □

Theorem 14.4.1 asserts that myopic policies $\overline{\mu}$ and $\underline{\mu}$ characterized by two fixed vectors g^* and f^* maximize $\text{vol}(\Pi_O)$ over the entire belief space $\Pi(X)$. The existence and computation of these policies characterized by $g^* \in S_g$ and $f^* \in S_f$ are determined by Algorithm 15. Algorithm 15 solves X linear programs to obtain g^*. If no $g^* \in S_g$ satisfying (14.16) exists, then Algorithm 15 will terminate with no solution. The procedure for computing f^* is similar.

Algorithm 15 Compute g^*

1: **for all** $i \in X$ **do**
2: $\quad \alpha_i \leftarrow \min_{g \in S_g} e_i'(P(2) - P(1))g$
3: **end for**
4: $g^* \in S_{g^*}, S_{g^*} \equiv \left\{g^* : g^* \in S_g, e_i'(P(2) - P(1))g^* = \alpha_i, i = 1, \cdots, X\right\}$
5: $\overline{\mu}(\pi) = \operatorname{argmin}_{u \in \{1,2\}} \pi' \overline{C}_u^* \ \forall \pi \in \Pi(X)$, where $\overline{C}_u^* = c_u + (I - \rho P(u))\, g^*$
6: $\overline{\mu}(\pi) = \mu^*(\pi) = 1, \forall \pi \in \Pi(X)_1^{g^*}$.

Optimizing myopic policies for more than 2 actions

Unlike Theorem 14.4.1, for the case $U > 2$, we are unable to show that a single fixed choice of $\overline{\mu}$ and $\underline{\mu}$ maximizes $\text{vol}(\Pi_O)$. Instead at each time k, $\overline{\mu}$ and $\underline{\mu}$ are optimized depending on the belief state π_k. Suppose at time k, given observation y_k, the belief state, π_k, is computed by using (14.2). For this belief state π_k, the aim is to compute $g^* \in S_g$ (14.8) and $f^* \in S_f$ (14.9) such that the difference between myopic policy bounds, $\overline{\mu}(\pi_k) - \underline{\mu}(\pi_k)$, is minimized. That is,

$$\left(g^*, f^*\right) = \operatorname*{argmin}_{g \in S_g,\, f \in S_f} \overline{\mu}(\pi_k) - \underline{\mu}(\pi_k). \tag{14.18}$$

(14.18) can be decomposed into following two optimization problems,

$$g^* = \operatorname*{argmin}_{g \in S_g} \overline{\mu}(\pi_k), \ f^* = \operatorname*{argmax}_{f \in S_f} \underline{\mu}(\pi_k). \tag{14.19}$$

If assumptions (C1) and (C2) hold, then the optimizations in (14.19) are feasible. Then $\overline{\mu}(\pi_k)$ in (14.10) and g^*, in (14.19) can be computed as follows: starting with $\overline{\mu}(\pi_k) = 1$, successively solve a maximum of U feasibility LPs, where the i-th LP searches for a feasible $g \in S_g$ in (14.8) so that the myopic upper bound yields action i, i.e. $\overline{\mu}(\pi_k) = i$. The i-th feasibility LP can be written as

$$\begin{aligned} &\min_{g \in S_g} \mathbf{1}_X' g \\ &s.t., \quad \overline{C}_i' \pi_k \leq \overline{C}_u' \pi_k \; \forall u \in \mathcal{U}, u \neq i. \end{aligned} \tag{14.20}$$

The smallest i for which (14.20) is feasible yields the solution $(g^*, \overline{\mu}(\pi_k) = i)$ of the optimization in (14.19). The above procedure is straightforwardly modified to obtain f^* and the lower bound $\underline{\mu}(\pi_k)$ (14.11).

Summary

We have shown how to construct myopic policies that upper and lower bound the optimal policy of a POMDP. We also showed that the volume of the belief space where these bounds overlap with the optimal policy can be maximized by solving a linear programming problem. Numerical examples will be given in §14.6.

14.5 Controlled sensing POMDPs with quadratic costs

Thus far in this chapter, we considered POMDPs where the instantaneous cost is linear in the belief state. We now construct myopic policies for POMDPs where the instantaneous cost is quadratic in the belief state. The motivation for such POMDPs in controlled sensing was discussed extensively in §8.4. Consider the problem of estimating the state of a noisily observed Markov chain using a controlled sensor. At each time, the sensor can be controlled to either obtain high quality measurements by paying a high measurement cost or obtain low quality measurements by paying a lower measurement cost. The aim is to compute the optimal sensing policy which minimizes the infinite horizon discounted cost.

The POMDP is identical to §14.1, except that now the instantaneous cost incurred at time k is the sum of the measurement cost (linear in the belief) and the conditional variance of the estimate (quadratic function of the belief):

$$C(\pi_k, u) = c_u' \pi_k + D(\pi_k, u_k). \tag{14.21}$$

Here c_u denotes the state-dependent measurement cost vector accrued when choosing sensing mode (action) u. Also, D, which models the weighted conditional variance of Markov chain estimate, is quadratic of the form

$$\begin{aligned} D(\pi_k, u_k) &= \alpha(u_k) \mathbb{E}\left\{ \| (x_k - \pi_k)' h \|^2 | y_1, \cdots, y_k, u_0, \cdots, u_{k-1} \right\} \\ &= H' \pi_k - \left(h' \pi_k\right)^2, \text{where } H_i = h_i^2 \text{ and } H = \{H_1, \cdots, H_X\}. \end{aligned} \tag{14.22}$$

The nonnegative user-specified weights $\alpha(u)$ allow actions $u \in \mathcal{U}$ to be weighted differently. In (14.22), $h = (h_1, \cdots, h_X)$, represent the physical state levels associated with the states $1, 2, \cdots, X$ of the Markov chain, for example, the physical distances or threat levels of a target that is being sensed.

We now obtain myopic bounds for the optimal sensing policy. For $f, g \in \mathbb{R}^X$, define the myopic policies

$$\overline{\mu}(\pi) \equiv \underset{u \in \mathcal{U}}{\operatorname{argmin}}\, \overline{C}(\pi, u), \quad \text{where } \overline{C}(\pi, u) = C(\pi, u) - (g - \rho P(u) g)' \pi \tag{14.23}$$

$$\underline{\mu}(\pi) \equiv \underset{u \in \mathcal{U}}{\operatorname{argmin}}\, \underline{C}(\pi, u), \quad \text{where } \underline{C}(\pi, u) = C(\pi, u) - (f - \rho P(u) f)' \pi. \tag{14.24}$$

Consider the following assumptions:

(C1′) There exists a $g \in \mathbb{R}^X$ such that following set of linear inequalities hold

$$\overline{\phi}_i^u - \overline{\phi}_{i+1}^u \leq 2\alpha(u)\, h_X\, (h_i - h_{i+1})\, \forall u \in U, i = 1, \cdots, X-1,$$
$$\text{where } \overline{\phi}^u \equiv \alpha(u) H + c_u - (I - \rho P(u)) g. \tag{14.25}$$

(C2′) There exists a $f \in \mathbb{R}^X$ such that following set of linear inequalities hold

$$\underline{\phi}_i^u - \underline{\phi}_{i+1}^u \geq 2\alpha(u)\, h_1\, (h_i - h_{i+1})\, \forall u \in U, i = 1, \cdots, X-1,$$
$$\text{where } \overline{\phi}^u \equiv \alpha(u) H + c_u - (I - \rho P(u)) f. \tag{14.26}$$

The following result is identical to Theorem 14.3.1, except that (C1), (C2) are replaced with (C1′), (C2′) to handle the quadratic costs. Recall that Lemma 11.1.1 established the sufficiency of (C1′), (C2′).

THEOREM 14.5.1 *For a quadratic cost POMDP, Theorem 14.3.1 holds if (C1) and (C2) are replaced by (C1′) and (C2′), respectively.*

The algorithms in §14.4 are straightforwardly modified to optimize the myopic bounds $\overline{\mu}$ and $\underline{\mu}$ for POMDPs with quadratic costs. The polytopes S_g and S_f used for computing g and f, are redefined as,

$$S_g = \left\{ g : \overline{\phi}_i^u - \overline{\phi}_{i+1}^u \leq 2\alpha(u)\, h_X\, (h_i - h_{i+1})\ \forall u \in U, i = 1, \cdots, X-1 \right\},$$
$$S_f = \left\{ f : \underline{\phi}_i^u - \underline{\phi}_{i+1}^u \leq 2\alpha(u)\, h_1\, (h_i - h_{i+1})\ \forall u \in U, i = 1, \cdots, X-1 \right\}.$$

where $\overline{\phi}$ and $\underline{\phi}$ are defined in (14.25) and (14.26) respectively.

14.6 Numerical examples

Recall that on the set Π_O (14.14), the upper and lower myopic bounds coincide with the optimal policy $\mu^*(\pi)$. What is the performance loss outside the set Π_O? To quantify this, define the policy

$$\tilde{\mu}(\pi) = \begin{cases} \mu^*(\pi) & \forall \pi \in \Pi_O \\ \text{arbitrary action (e.g. 1)} & \forall \pi \notin \Pi_O \end{cases}.$$

Let $J_{\tilde{\mu}}(\pi_0)$ denote the discounted cost associated with $\tilde{\mu}(\pi_0)$. Also denote

$$\tilde{J}_{\mu^*}(\pi_0) = \mathbb{E}\left\{\sum_{k=0}^{\infty} \rho^k \tilde{c}'_{\mu^*(\pi_k)} \pi_k\right\},$$

where

$$\tilde{c}_{\mu^*(\pi)} = \begin{cases} c_{\mu^*(\pi)} & \pi \in \Pi_O \\ \left[\min_{u \in \mathcal{U}} c(1,u), \cdots, \min_{u \in \mathcal{U}} c(X,u)\right]' & \pi \notin \Pi_O \end{cases}$$

Clearly an upper bound for the percentage loss in optimality due to using policy $\tilde{\mu}$ instead of optimal policy μ^* is

$$\epsilon = \frac{J_{\tilde{\mu}}(\pi_0) - \tilde{J}_{\mu^*}(\pi_0)}{\tilde{J}_{\mu^*}(\pi_0)}. \tag{14.27}$$

In the numerical examples below, to evaluate ϵ, 1000 Monte Carlo simulations were run to estimate the discounted costs $J_{\tilde{\mu}}(\pi_0)$ and $\tilde{J}_{\mu^*}(\pi_0)$ over a horizon of 100 time units. The parameters ϵ and vol (Π_O) are used to evaluate the performance of the optimized myopic policy bounds constructed according to §14.4. Note that ϵ depends on the choice of observation distribution B, unlike vol (Π_O); see discussion below (14.14) and also Example 2 below.

Example 1: Sampling and measurement control with two actions: In this problem discussed in §13.5, at every decision epoch, the decision-maker has the option of either recording a noisy observation (of a Markov chain) instantly (action $u = 2$) or waiting for one time unit and then recording an observation using a better sensor (action $u = 1$). Should one record observations more frequently and less accurately or more accurately but less frequently?

We chose $X = 3$, $U = 2$ and $Y = 3$. Both transition and observation probabilities are action dependent (parameters specified in the Appendix). The percentage loss in optimality is evaluated by simulation for different values of the discount factor ρ. Table 14.1(a) displays vol (Π_O), ϵ_1 and ϵ_2. For each ρ, ϵ_1 is obtained by assuming $\pi_0 = e_3$ (myopic bounds overlap at e_3) and ϵ_2 is obtained by uniformly sampling $\pi_0 \notin \Pi_O$. Observe that vol (Π_O) is large and ϵ_1, ϵ_2 are small, which indicates the usefulness of the proposed myopic policies.

Example 2: 10-state POMDP: Consider a POMDP with $X = 10$, $U = 2$. Consider two sub-examples: the first with discrete observations $Y = 10$ (parameters in Appendix), the second with continuous observations obtained using the additive Gaussian noise model, i.e. $y_k = x_k + n_k$ where $n_k \sim N(0, 1)$. The percentage loss in optimality is evaluated by simulation for these two sub examples and denoted by $\epsilon_1^d, \epsilon_2^d$ (discrete observations) and $\epsilon_1^c, \epsilon_2^c$ (Gaussian observations) in Table 14.1(b).

ϵ_1^d and ϵ_1^c are obtained by assuming $\pi_0 = e_5$ (myopic bounds overlap at e_5). ϵ_2^d and ϵ_2^c are obtained by sampling $\pi_0 \notin \Pi_O$. Observe from Table 14.1(b) that vol (Π_O) decreases with ρ.

Example 3: 8-state and 8-action POMDP: Consider a POMDP with $X = 8$, $U = 8$ and $Y = 8$ (parameters in Appendix). Table 14.1(c) displays vol (Π_O), ϵ_1 and ϵ_2. For

Table 14.1 Performance of optimized myopic policies versus discount factor ρ for five numerical examples. The performance metrics vol (Π_O) and ϵ are defined in (14.14) and (14.27).

(a) Example 1

ρ	vol(Π_O)	ϵ_1	ϵ_2
0.4	95.3%	0.30%	16.6%
0.5	94.2%	0.61%	13.9%
0.6	92.4%	1.56%	11.8%
0.7	90.2%	1.63%	9.1%
0.8	87.4%	1.44%	6.3%
0.9	84.1%	1.00%	3.2%

(b) Example 2

vol(Π_O)	ϵ_1^d	ϵ_2^d	ϵ_1^c	ϵ_2^c
64.27%	7.73%	12.88%	6.92%	454.31%
55.27%	8.58%	12.36%	8.99%	298.51%
46.97%	8.97%	11.91%	12.4%	205.50%
39.87%	8.93%	11.26%	14.4%	136.31%
34.51%	10.9%	12.49%	17.7%	88.19%
29.62%	11.2%	12.24%	20.5%	52.16%

(c) Example 3

vol(Π_O)	ϵ_1	ϵ_2
61.4%	2.5%	10.1%
56.2%	2.3%	6.9%
47.8%	1.7%	4.9%
40.7%	1.4%	3.5%
34.7%	1.1%	2.3%
31.8%	0.7%	1.4%

(d) Example 4

ϵ_1	ϵ_2
0.13%	8.64%
0.28%	8.11%
0.48%	7.44 %
0.86%	6.94%
1.48%	6.59%
2.22%	6.18%

(e) Example 5

$\overline{\text{vol}}\,(\Pi_O)$	$\underline{\text{vol}}\,(\Pi_O)$	$\overline{\epsilon}_1$	$\underline{\epsilon}_1$	$\overline{\epsilon}_2$	$\underline{\epsilon}_2$
98.9%	84.5%	0.10%	6.17%	1.45%	1.71%
98.6%	80.0%	0.18%	7.75%	1.22%	1.50%
98.4%	75.0%	0.23%	11.62%	1.00%	1.31%
98.1%	68.9%	0.26%	14.82%	0.75%	1.10%
97.8%	61.5%	0.27%	19.74%	0.51%	0.89%
97.6%	52.8%	0.25%	24.08%	0.26%	0.61%

each ρ, ϵ_1 is obtained by assuming $\pi_0 = e_1$ (myopic bounds overlap at e_1) and ϵ_2 is obtained by uniformly sampling $\pi_0 \notin \Pi_O$. The results indicate that the myopic policy bounds are still useful for some values of ρ.

Example 4: Sampling and measurement control with quadratic costs: We extend Example 1 to the controlled sensing problem with quadratic costs (14.21), (14.22); parameters specified in appendix. The costs are chosen to tradeoff accuracy of the sensing action (conditional variance) versus the measurement cost: $\alpha(1) \leq \alpha(2)$ to penalize using the less accurate sensor and $c_1 \succeq c_2$, since the accurate sensing incurs a higher measurement cost. Table 14.1(d) displays ϵ_1 and ϵ_2 for different values of discount factor ρ. ϵ_1 is obtained by assuming $\pi_0 = e_3$ (myopic bounds overlap at e_3) and ϵ_2 is obtained by sampling $\pi_0 \notin \Pi_O$.

Example 5: Myopic bounds versus transition matrix: The aim here is to illustrate the performance of the optimized myopic bounds over a range of transition probabilities.

Consider a POMDP with $X = 3$, $U = 2$, additive Gaussian noise model of Example 2, and transition matrices

$$P(2) = \begin{pmatrix} 1 & 0 & 0 \\ 1-2\theta_1 & \theta_1 & \theta_1 \\ 1-2\theta_2 & \theta_2 & \theta_2 \end{pmatrix}, \quad P(1) = P^2(2).$$

It is straightforward to show that $\forall\ \theta_1, \theta_2$ such that $\theta_1 + \theta_2 \leq 1, \theta_2 \geq \theta_1$, $P(1)$ and $P(2)$ satisfy (F2) and (F3$'$). The costs are $c_1 = [1, 1.1, 1.2]'$ and $c_2 = [1.2, 1.1, 1.1]'$. Table 14.1(e) displays the worst case and best case values for performance metrics $(\text{vol}\,(\Pi_O)\,, \epsilon_1, \epsilon_2)$ versus discount factor ρ by sweeping over the entire range of (θ_1, θ_2). The worst case performance is denoted by $\underline{\text{vol}}\,(\Pi_O)$, $\underline{\epsilon}_1$, $\underline{\epsilon}_2$ and the best case by $\overline{\text{vol}}\,(\Pi_O), \overline{\epsilon}_1, \overline{\epsilon}_2$.

Discussion

(i) Recall in the above numerical examples, the percentage loss in optimality when starting with $\pi \in \Pi_O$ (respectively, $\pi \notin \Pi_O$) is denoted by $\epsilon_1(\epsilon_2)$. The dependence of ϵ on ρ is a tradeoff between two factors: vol(Π_O) and the fact that for small ρ, the first few decisions determine most of the cumulative cost. For example, if $\pi_0 \notin \Pi_O$, then even if the belief recovers and enters Π_O after a few time steps, the effect on the overall cost is minimal. The combination of these two factors can result in non-monotone behavior of ϵ_1 and ϵ_2 versus ρ (Examples 2 and 3).

(ii) *Construction of numerical examples*: Numerous additional examples can be constructed as follows: first consider the case where the observation probabilities are independent of the action, i.e. $B(u) = B\ \forall u \in \mathcal{U}$. (F4) hold. Such transition matrices are generated as follows: construct UX probability vectors each of dimension X that are MLR ordered in ascending order (obtained by solving a sequence of LPs). Assign the first X probability vectors to $P(1)$, the next X vectors to $P(2)$, and so on. Since the rows of each $P(a)$ are MLR increasing,(F2) holds. Actually, a larger class of transition matrices can be constructed by considering (F3$'$) directly. As discussed in §14.3, (F3$'$) ensures that $\Gamma^{j,u}$ is a nonnegative matrix and thus copositive. Since the sum of a positive semi-definite and nonnegative matrix is copositive, construction of $P(u)$ can be achieved via semidefinite programming [324]. Finally, for finite observation alphabet size, $B(u)$ can be constructed similarly via a sequence of LPs.

14.7 Blackwell dominance of observation distributions and optimality of myopic policies

In previous sections of this chapter, we used copositive dominance to construct upper and lower myopic bounds to the optimal policy of a POMDP. In this section we will use another concept, called Blackwell dominance, to construct lower myopic bounds to the optimal policy for a POMDP.

14.7.1 Myopic policy bound to optimal decision policy

Motivated by controlled sensing applications, consider the following POMDPs where based on the current belief state π_{k-1}, agent k chooses sensing mode

$$u_k \in \{1 \text{ (low resolution sensor)}, 2 \text{ (high resolution sensor)}\}.$$

Depending on its mode u_k, the sensor views the world according to this mode – that is, it obtains observation from a distribution that depends on u_k. Assume that for mode $u \in \{1,2\}$, the observation $y^{(u)} \in \mathcal{Y}^{(u)} = \{1,\dots,Y^{(u)}\}$ is obtained from the matrix of conditional probabilities

$$B(u) = \left(B_{iy^{(u)}}(u), i \in \{1,2,\dots,X\}, y^{(u)} \in \mathcal{Y}^{(u)}\right)$$
$$\text{where } B_{iy^{(u)}}(u) = \mathbb{P}(y^{(u)}|x = e_i, u).$$

The notation $\mathcal{Y}^{(u)}$ allows for mode dependent observation spaces. In sensor scheduling [178], the tradeoff is as follows: mode $u = 2$ yields more accurate observations of the state than mode $u = 1$, but the cost of choosing mode $u = 2$ is higher than mode $u = 1$. Thus there is an tradeoff between the cost of acquiring information and the value of the information.

The assumption that mode $u = 2$ yields more accurate observations than mode $u = 1$ is modeled as follows: we say mode 2 *Blackwell dominates* mode 1, denoted as

$$B(2) \succeq_{\mathcal{B}} B(1) \quad \text{if} \quad B(1) = B(2)R. \tag{14.28}$$

Here R is a $Y^{(2)} \times Y^{(1)}$ stochastic matrix. R can be viewed as a *confusion matrix* that maps $\mathcal{Y}^{(2)}$ probabilistically to $\mathcal{Y}^{(1)}$. (In a communications context, one can view R as a noisy discrete memoryless channel with input $y^{(2)}$ and output $y^{(1)}$.) Intuitively (14.28) means that $B(2)$ is more accurate than $B(1)$.

The goal is to compute the optimal policy $\mu^*(\pi) \in \{1,2\}$ to minimize the expected cumulative cost incurred by all the agents

$$J_\mu(\pi) = \mathbb{E}_\mu\{\sum_{k=0}^{\infty} \rho^k C(\pi_k, u_k)\}, \tag{14.29}$$

where $\rho \in [0,1)$ is the discount factor. Even though solving the above POMDP is computationally intractable in general, using Blackwell dominance, we show below that a myopic policy forms a lower bound for the optimal policy.

The value function $V(\pi)$ and optimal policy $\mu^*(\pi)$ satisfy Bellman's equation

$$\begin{aligned} V(\pi) &= \min_{u\in\mathcal{U}} Q(\pi,u), \quad \mu^*(\pi) = \arg\min_{u\in\mathcal{U}} Q(\pi,u),\ J_{\mu^*}(\pi) = V(\pi) \\ Q(\pi,u) &= C(\pi,u) + \rho \sum_{y^{(u)}\in\mathcal{Y}^{(u)}} V\left(T(\pi,y,u)\right)\sigma(\pi,y,u), \\ T(\pi,y,u) &= \frac{B_{y^{(u)}}(u)P'\pi}{\sigma(\pi,y,u)},\ \sigma(\pi,y,u) = \mathbf{1}_X' B_{y^{(u)}}(u)P'\pi. \end{aligned} \tag{14.30}$$

We now present the main structural result which uses Blackwell dominance. Let $\Pi^s \subset \Pi$ denote the set of belief states for which $C(\pi,2) < C(\pi,1)$. Define the myopic policy

$$\underline{\mu}(\pi) = \begin{cases} 2 & \pi \in \Pi^s \\ 1 & \text{otherwise} \end{cases}$$

THEOREM 14.7.1 *Consider a POMDP with cost $C(\pi,u)$ that is concave in the belief $\pi \in \Pi(X)$ for each action u. Suppose $B(2) \succeq_{\mathcal{B}} B(1)$, i.e. $B(1) = B(2)R$ holds where R is a stochastic matrix. Assume that the transition probabilities are action independent, i.e., $P(1) = P(2)$. Then the myopic policy $\underline{\mu}(\pi)$ is a lower bound to the optimal policy $\mu^*(\pi)$, i.e. $\mu^*(\pi) \geq \underline{\mu}(\pi)$ for all $\pi \in \Pi$. In particular, for $\pi \in \Pi^s$, $\mu^*(\pi) = \bar{\mu}(\pi)$, i.e. it is optimal to choose action 2 when the belief is in Π^s.* □

Remark: If $B(1) \succeq_{\mathcal{B}} B(2)$, then the myopic policy constitutes an upper bound to the optimal policy.

Theorem 14.7.1 is proved below. The proof exploits the fact that the value function is concave and uses Jensen's inequality. The usefulness of Theorem 14.7.1 stems from the fact that $\bar{\mu}(\pi)$ is trivial to compute. It forms a provable lower bound to the computationally intractable optimal policy $\mu^*(\pi)$. Since $\underline{\mu}$ is sub-optimal, it incurs a higher cumulative cost. This cumulative cost can be evaluated via simulation and is an upper bound to the achievable optimal cost.

Theorem 14.7.1 is nontrivial. The instantaneous costs satisfying $C(\pi,2) < C(\pi,1)$, does not trivially imply that the myopic policy $\underline{\mu}(\pi)$ coincides with the optimal policy $\mu^*(\pi)$, since the optimal policy applies to a cumulative cost function involving an infinite horizon trajectory of the dynamical system.

It is instructive to compare Theorem 14.7.1 with Theorem 14.3.1 on page 317. Theorem 14.3.1 used copositive dominance (with several assumptions on the POMDP model) to construct both upper and lower bounds to the optimal policy. In comparison, (Theorem 14.7.1) needs no assumptions on the POMDP model apart from the Blackwell dominance condition $B(1) = B(2)R$ and concavity of costs with respect to the belief; but only yields an upper bound.

14.7.2 Example 1: Optimal filter vs predictor scheduling

Suppose $u = 2$ is an active sensor (filter) which obtains measurements of the underlying Markov chain and uses the optimal HMM filter on these measurements to compute the belief and therefore the state estimate. So the usage cost of sensor 2 is high (since obtaining observations is expensive and can also result in increased threat of being discovered), but its performance cost is low (performance quality is high).

Suppose sensor $u = 1$ is a predictor which needs no measurement. So its usage cost is low (no measurement is required). However its performance cost is high since it is more inaccurate compared to sensor 2.

Since the predictor has non-informative observation probabilities, its observation probability matrix is $B(1) = \frac{1}{Y}\mathbf{1}_{X\times Y}$. So clearly $B(1) = B(2)B(1)$ meaning that the filter (sensor 2) Blackwell dominates the predictor (sensor 1). Theorem 14.7.1 says that if the current belief is π_k, then if $C(\pi_k, 2) < C(\pi_k, 1)$, it is always optimal to deploy the filter (sensor 2).

14.7.3 Example 2: Ultrametric matrices and Blackwell dominance

An $X \times X$ square matrix B is a symmetric stochastic ultrametric matrix if

1. B is symmetric and stochastic.
2. $B_{ij} \geq \min\{B_{ik}, B_{kj}\}$ for all $i, j, k \in \{1, 2, \ldots, X\}$.
3. $B_{ii} > \max\{B_{ik}\}, k \in \{1, 2, \ldots, X\} - \{i\}$ (diagonally dominant).

It is shown in [145] that if B is a symmetric stochastic ultrametric matrix, then the U-th root, namely $B^{1/U}$, is also a stochastic matrix[4] for any positive integer U. Then with $\succeq_{\mathcal{B}}$ denoting Blackwell dominance (14.28), clearly

$$B^{1/U} \succeq_{\mathcal{B}} B^{2/(U)} \succeq_{\mathcal{B}} \cdots \succeq_{\mathcal{B}} B^{(U-1)/U} \succeq_{\mathcal{B}} B.$$

Consider a social network where the reputations of agents are denoted as $u \in \{1, 2, \ldots, U\}$. An agent with reputation u has observation probability matrix $B^{(U-u+1)/U}$. So an agent with reputation 1 (lowest reputation) is U degrees of separation from the source signal while an agent with reputation U (highest reputation) is 1 degree of separation from the source signal. The underlying source (state) could be a news event, sentiment or corporate policy that evolves with time. A marketing agency can sample these agents – it can sample high reputation agents that have accurate observations but this costs more than sampling low reputation agents that have less accurate observations. Then Theorem 14.7.1 gives a suboptimal policy that provably lower bounds the optimal sampling policy.

14.7.4 Proof of Theorem 14.7.1

Recall from Theorem 8.4.1 that $C(\pi, u)$ concave implies that $V(\pi)$ is concave on $\Pi(X)$. We then use the Blackwell dominance condition (14.28). In particular,

$$T(\pi, y^{(1)}, 1) = \sum_{y^{(2)} \in \mathcal{Y}^{(2)}} T(\pi, y^{(2)}, 2) \frac{\sigma(\pi, y^{(2)}, 2)}{\sigma(\pi, y^{(1)}, 1)} P(y^{(1)} | y^{(2)})$$

$$\sigma(\pi, y^{(1)}, 1) = \sum_{y^{(2)} \in \mathcal{Y}^{(2)}} \sigma(\pi, y^{(2)}, 2) P(y^{(1)} | y^{(2)}).$$

Therefore $\frac{\sigma(\pi, y^{(2)}, 2)}{\sigma(\pi, y^{(1)}, 1)} P(y^{(1)} | y^{(2)})$ is a probability measure w.r.t. $y^{(2)}$ (since the denominator is the sum of the numerator over all $y^{(2)}$). Since $V(\cdot)$ is concave, using Jensen's inequality it follows that

[4] Although we do not pursue it here, conditions that ensure that the U-th root of a transition matrix is a valid stochastic matrix is important in interpolating Markov chains. For example, transition matrices for credit ratings on a yearly time scale can be obtained from rating agencies such as Standard & Poor's. Determining the transition matrix for periods of six months involves the square root of the yearly transition matrix [145].

$$
\begin{aligned}
V(T(\pi, y^{(1)}, 1)) &= V\left(\sum_{y^{(2)} \in \mathcal{Y}^{(2)}} T(\pi, y^{(2)}, 2) \frac{\sigma(\pi, y^{(2)}, 2)}{\sigma(\pi, y^{(1)}, 1)} P(y^{(1)}|y^{(2)}) \right) \\
&\geq \sum_{y^{(2)} \in \mathcal{Y}^{(2)}} V(T(\pi, y^{(2)}, 2)) \frac{\sigma(\pi, y^{(2)}, 2)}{\sigma(\pi, y^{(1)}, 1)} P(y^{(1)}|y^{(2)}) \\
&\implies \sum_{y^{(1)}} V(T(\pi, y^{(1)}, 1)) \sigma(\pi, y^{(1)}, 1) \geq \sum_{y^{(2)}} V(T(\pi, y^{(2)}, 2) \sigma(\pi, y^{(2)}, 2). \qquad (14.31)
\end{aligned}
$$

Therefore for $\pi \in \Pi^s$,

$$
C(\pi, 2) + \rho \sum_{y^{(2)}} V(T(\pi, y^{(2)}, 2) \sigma(\pi, y^{(2)}, 2) \leq C(\pi, 1) + \rho \sum_{y^{(1)}} V(T(\pi, y^{(1)}), 1) \sigma(\pi, y^{(1)}, 1).
$$

So for $\pi \in \Pi^s$, the optimal policy $\mu^*(\pi) = \arg\min_{u \in \mathcal{U}} Q(\pi, u) = 2$. So $\underline{\mu}(\pi) = \mu^*(\pi) = 2$ for $\pi \in \Pi^s$ and $\underline{\mu}(\pi) = 1$ otherwise, implying that $\underline{\mu}(\pi)$ is a lower bound for $\mu^*(\pi)$.

14.7.5 Combining Blackwell Dominance and Copositive Dominance. Accuracy vs Delay Tradeoff in Controlled Sensing

So far we have shown that copositive dominance can be used to construct myopic policies that upper and lower bound the optimal policy of a POMDP. Also for POMDPs where only the observation probabilities are action dependent, Blackwell dominance can be used to construct myopic policies that lower bound the optimal policy. Can copositive dominance and Blackwell dominance be combined to construct bounds for the optimal policy? The following theorem shows that this indeed is possible.

THEOREM 14.7.2 *Consider a POMDP with actions $u \in \{1, 2\}$. Suppose (C), (F1), (F2) hold. Also assume $B(2) \succeq_{\mathcal{B}} B(1)$ (Blackwell dominance (14.28)) and $P(2) \succeq P(1)$ (copositive dominance (Assumption (F3) on page 225)). Then Theorem 14.7.1 holds.*

Theorem 14.7.2 gives an alternative set of conditions (compared to Theorem 14.3.1) for a myopic policy to lower bound the optimal policy of a POMDP. The linear programming formulation of Theorem 14.4.1 can be used to maximize the volume where the myopic policy coincides with the optimal policy.

As an example, consider the controlled sensing problem of accuracy versus delay. Suppose sensor 2 is more accurate than sensor 1 but measures the state with a delay of 1 unit. The corresponding POMDP parameters are $P(2) = P^2(1)$ and $B(1) = B(2)R$ where R is a stochastic matrix. Then under the conditions of Theorem 14.7.2, the myopic policy is a lower bound to the optimal policy and when the instantaneous cost $C(\pi, 2) < C(\pi, 1)$ it is optimal to use sensor 2.

As a numerical example choose $B(2)$ and R as arbitrary TP2 matrices. Then their product is TP2. Suppose $P(1) = \begin{bmatrix} \theta_2 & \theta_2 & 1 - 2\theta_2 \\ \theta_1 & \theta_1 & 1 - 2\theta_1 \\ 0 & 0 & 1, \end{bmatrix}$, $\theta_1 + \theta_2 \leq 1, \theta_2 \geq \theta_1$. Then the conditions of Theorem 14.7.2 hold.

Proof (Theorem 14.7.2) Let $V(\pi)$ denote the value function of the POMDP. Recall the action dependent POMDP parameters are: For action $u = 1$, $(P(1), B(1))$, for action $u = 2$, $(P(2), B(2))$. Define the fictitious action $u = a$ with parameters $(P(2), B(1))$. By Blackwell dominance, Theorem 14.7.1 implies that

$$\sum_{y^{(2)}} V(T(\pi, y^{(2)}, 2)\sigma(\pi, y^{(2)}, 2) \leq \sum_{y^{(1)}} V(T(\pi, y^{(1)}), a)\sigma(\pi, y^{(1)}, a) \tag{14.32}$$

since actions 2 and a have the same transition matrices. Also under copositive assumption (F3), $T(\pi, y, a) \geq_r T(\pi, y, 1)$. From (C), (F1), (F2), $V(\pi)$ is MLR decreasing implying that $V(T(\pi, y, a) \leq V(T(\pi, y, 1)$. From (F3), (F1), (F2) it follows that $\sigma(\pi, a) \geq_s \sigma(\pi, 1)$. Therefore

$$\begin{aligned}\sum_{y^{(1)}} V(T(\pi, y^{(1)}), a)\sigma(\pi, y^{(1)}, a) &\leq \sum_{y^{(1)}} V(T(\pi, y^{(1)}), 1)\sigma(\pi, y^{(1)}, a) \\ &\leq \sum_{y^{(1)}} V(T(\pi, y^{(1)}), 1)\sigma(\pi, y^{(1)}, 1). \end{aligned}\tag{14.33}$$

Combining (14.32) and (14.33) yields (14.31). The rest of the proof follows as in Theorem 14.7.1. □

Summary

An observation matrix $B(2)$ Blackwell dominates $B(1)$ if $B(1) = B(2)R$ for some stochastic matrix R. So action 2 has a more accurate observation matrix than action 1. The main result discussed was that for a POMDP with Blackwell dominance, if the instantaneous cost $C(\pi_k, 1) < C(\pi_k, 2)$ at time k, then it is optimal to choose action 1, i..e, the myopic policy is optimal. The standard Blackwell dominance theorem requires that the transition matrices $P(u)$ are not action dependent. However, we showed that the result can be extended to action dependent transition matrices if $P(2) \preceq P(1)$ (copositive dominance).

14.8 Ordinal sensitivity: how does optimal POMDP cost vary with state and observation dynamics?

This and the next section focus on *achievable costs* attained by the optimal policy. This section presents gives bounds on the achievable performance of the optimal policies by the decision-maker. This is done by introducing a partial ordering of the transition and observation probabilities – the larger these parameters with respect to this order, the larger the optimal cumulative cost incurred.

How does the optimal expected cumulative cost J_{μ^*} of a POMDP vary with transition matrix P and observation distribution B? Can the transition matrices and observation distributions be ordered so that the larger they are, the larger the optimal cumulative cost? Such a result is very useful – it allows us to compare the optimal performance of different POMDP models, even though computing these is intractable. Recall that the transition matrix specifies the mobility of the state and the observation matrix specifies the noise distribution; so understanding how these affect the achievable optimal cost is important.

Consider two distinct POMDPs with transition matrices $\theta = P$ and $\bar{\theta} = \bar{P}$, respectively. Alternatively, consider two distinct POMDPs with observation distributions $\theta = B$ and $\bar{\theta} = \bar{B}$, respectively. Assume that the instantaneous costs $C(\pi, u)$ and discount factors ρ for both POMDPs are identical.

Let $\mu^*(\theta)$ and $\mu^*(\bar{\theta})$ denote, respectively, the optimal policies for the two POMDPs. Let $J_{\mu^*(\theta)}(\pi;\theta) = V(\pi;\theta)$ and $J_{\mu^*(\bar{\theta})}(\pi;\bar{\theta}) = V(\pi;\bar{\theta})$ denote the optimal value functions corresponding to applying the respective optimal policies.

Consider two arbitrary transition matrices P and $\bar{P}$. Recalling Definition 10.2.3 for $\succeq$ and (F3′), assume the copositive ordering

$$P \succeq \bar{P}. \tag{14.34}$$

$\bar{B}$ Blackwell dominates B denoted as

$$\bar{B} \succeq_{\mathcal{B}} B \text{ if } B = \bar{B}R \tag{14.35}$$

where $R = (R_{lm})$ is a stochastic kernel, i.e. $\sum_m R_{lm} = 1$.

The question we pose is: how does the optimal cumulative cost $J_{\mu^*(\theta)}(\pi;\theta)$ vary with transition matrix P or observation distribution B? For example, in quickest change detection, do certain phase-type distributions for the change time result in larger optimal cumulative cost compared to other phase-type distributions? In controlled sensing, do certain noise distributions incur a larger optimal cumulative cost than other noise distributions?

THEOREM 14.8.1 *1. Consider two distinct POMDP problems with transition matrices P and $\bar{P}$, respectively, where $P \succeq \bar{P}$ with respect to copositive ordering (14.34). If (C), (F1), (F2) hold, then the optimal cumulative costs satisfy*

$$J_{\mu^*(P)}(\pi;P) \leq J_{\mu^*(\bar{P})}(\pi;\bar{P}).$$

2. Consider two distinct POMDP problems with observation distributions B and $\bar{B}$, respectively, where $\bar{B} \succeq_{\mathcal{B}} B$ with respect to Blackwell ordering (14.35). Then

$$J_{\mu^*(B)}(\pi;B) \geq J_{\mu^*(\bar{B})}(\pi;\bar{B}).$$

The proof is in Appendix 14.B. Computing the optimal policy and associated cumulative cost of a POMDP is intractable. Yet, the above theorem facilitates comparison of these optimal costs for different transition and observation probabilities.

It is instructive to compare Theorem 14.8.1(1) with Theorem 9.4.1 of §9.4, which dealt with the optimal costs of two *fully* observed MDPs. Comparing the assumptions of Theorem 9.4.1 with Theorem 14.8.1(1), we see that the assumption on the costs (A1) is identical to (C). Assumption (A2) in Theorem 9.4.1 is replaced by (F1), (F2) which are conditions on the transition and observation probabilities. The first-order dominance condition (A2) in Theorem 9.4.1 is a weaker condition than the TP2 condition (F1). In particular, (F1) implies (A2). Finally, (A5) in Theorem 9.4.1 is replaced by the stronger assumption of copositivity ($\underline{\text{F3}}'$). Indeed, ($\underline{\text{F3}}'$) implies (A5).

Remark: An obvious consequence of Theorem 14.8.1(1) is that a Markov chain with transition probabilities $P_{iX} = 1$ for each state i incurs the lowest cumulative cost. After one transition such a Markov chain always remains in state X. Since the instantaneous costs are decreasing with state (C), clearly, such a transition matrix incurs the lowest cumulative cost. Similarly if $P_{i1} = 1$ for each state i, then the highest cumulative cost is incurred. A consequence of Theorem 14.8.1(2) is that the optimal cumulative cost incurred with perfect measurements is smaller than that with noisy measurements.

14.9 Cardinal sensitivity of POMDP

Here we give explicit bounds on the sensitivity of the discounted cumulative cost of a POMDP with respect to mis-specified model and mis-specified policy – these bounds can be expressed in terms of the Kullback–Leibler divergence. Such robustness is useful since even if a model violates the assumptions of the previous section, as long as the model is sufficiently close to a model that satisfies the conditions, then the optimal policy is close to a monotone policy.

Main result

Consider two distinct discounted cost POMDPs $\theta = (P, B)$ and $\bar{\theta} = (\bar{P}, \bar{B})$ with identical costs $c(x, u)$ and identical discount factor ρ. Let $\mu^*(\theta)$ and $\mu^*(\bar{\theta})$, respectively denote the optimal policies of these two POMDPs. Let $J_{\mu^*(\theta)}(\pi;\theta)$ and $J_{\mu^*(\theta)}(\pi;\bar{\theta})$ denote the discounted cumulative costs incurred by these POMDPs when using policy $\mu^*(\theta)$. Similarly, $J_{\mu^*(\bar{\theta})}(\pi;\theta)$ and $J_{\mu^*(\bar{\theta})}(\pi;\bar{\theta})$ denote the discounted cumulative costs (13.47) incurred by these POMDPs when using policy $\mu^*(\bar{\theta})$.

THEOREM 14.9.1 *Consider two distinct POMDPs with models $\theta = (P, B)$ and $\bar{\theta} = (\bar{P}, \bar{B})$, respectively. Then for mis-specified model and mis-specified policy, the following sensitivity bounds hold:*

$$\textit{Mis-specified Model:} \sup_{\pi \in \Pi} |J_{\mu^*(\theta)}(\pi;\theta) - J_{\mu^*(\theta)}(\pi;\bar{\theta})| \leq K\|\theta - \bar{\theta}\|. \tag{14.36}$$

$$\textit{Mis-specified policy:} J_{\mu^*(\bar{\theta})}(\pi, \theta) \leq J_{\mu^*(\theta)}(\pi, \theta) + 2K\|\theta - \bar{\theta}\|. \tag{14.37}$$

$$K = \frac{\max_{i,u} C(e_i, u)}{1 - \rho}, \quad \textit{and} \quad \|\theta - \bar{\theta}\| = \max_{i,u} \sum_{j,y} \left| B_{jy}(u)P_{ij}(u) - \bar{B}_{jy}(u)\bar{P}_{ij}(u) \right|.$$

If $P(u) = \bar{P}(u)$, and $D(B_j\|\bar{B}_j) = \sum_y B_{jy} \log(B_{jy}/\bar{B}_{jy})$ denotes the Kullback–Leibler divergence, then the following bound holds:

$$\|\theta - \bar{\theta}\| \leq \sqrt{2} \max_{i,u} \sum_j P_{ij}(u) \left[D(B_j \| \bar{B}_j)\right]^{1/2}. \tag{14.38}$$

In particular, for Gaussian observation distributions with variance σ^2, $\bar{\sigma}^2$, respectively,

$$\|\theta - \bar{\theta}\| \leq \left(\frac{\sigma}{\bar{\sigma}} - \log \frac{\sigma}{\bar{\sigma}} - 1\right)^{1/2}.$$

The proof of Theorem 14.9.1 is in Appendix 14.C. The bounds (14.36), (14.37) are tight since $\|\theta - \bar{\theta}\| = 0$ implies that the performance degradation is zero. Also (14.38) follows from Pinsker's inequality [90] that bounds the total variation norm $\|\theta - \bar{\theta}\|$ by the Kullback–Leibler divergence.

It is instructive to compare Theorem 14.9.1 with Theorem 14.8.1. Theorem 14.8.1 compared optimal cumulative costs for different models – it showed that $\theta \succeq \bar{\theta} \implies J_{\mu^*(\theta)}(\pi;\theta) \geq J_{\mu^*(\bar{\theta})}(\pi;\bar{\theta})$, where $\mu^*(\theta)$ and $\mu^*(\bar{\theta})$ denote the optimal sampling policies for models θ and $\bar{\theta}$, respectively (where the ordering $\succeq$ is specified in §14.8). In comparison, (14.36) applies the optimal policy $\mu^*(\theta)$ for model θ to the decision problem with a different model $\bar{\theta}$. Also (14.37) is a lower bound for the cumulative cost of applying the optimal policy for a different model $\bar{\theta}$ to the true model θ – this bound is in terms of the cumulative cost of the optimal policy for true model $\bar{\theta}$. What Theorem 14.9.1 says is that if the "distance" between the two models $\theta, \bar{\theta}$ is small, then the sub-optimality is small, as described by (14.36), (14.37).

14.10 Complements and sources

This chapter is based on [187] and extends the structural results of [225, 277]. Constructing myopic policies using Blackwell dominance goes back to [334]. [277] uses the multivariate TP2 order for POMDPs with multivariate observations. [145] shows the elegant result that the p-th root of a stochastic matrix P is a stochastic matrix providing P^{-1} is an M-matrix. An M-matrix A is a square matrix with negative non-diagonal elements which can be expressed as $A = \lambda I - L$ where L is a nonnegative matrix and λ is greater than the maximum modulus eigenvalue of L. The inverse of an ultrametric matrix is an M-matrix.

Appendix 14.A POMDP numerical examples

Parameters of Example 1: For the first example the parameters are defined as,

$$c = \begin{pmatrix} 1.0000 & 1.5045 & 1.8341 \\ 1.5002 & 1.0000 & 1.0000 \end{pmatrix}', \quad P(2) = \begin{pmatrix} 1.0000 & 0.0000 & 0.0000 \\ 0.4677 & 0.4149 & 0.1174 \\ 0.3302 & 0.5220 & 0.1478 \end{pmatrix}, \quad P(1) = P^2(2)$$

$$B(1) = \begin{pmatrix} 0.6373 & 0.3405 & 0.0222 \\ 0.3118 & 0.6399 & 0.0483 \\ 0.0422 & 0.8844 & 0.0734 \end{pmatrix}, \quad B(2) = \begin{pmatrix} 0.5927 & 0.3829 & 0.0244 \\ 0.4986 & 0.4625 & 0.0389 \\ 0.1395 & 0.79 & 0.0705 \end{pmatrix}.$$

Parameters of Example 2: For discrete observations $B(u) = B \; \forall u \in \mathcal{U}$,

$$
B = \begin{pmatrix}
0.0297 & 0.1334 & 0.1731 & 0.0482 & 0.1329 & 0.1095 & 0.0926 & 0.0348 & 0.1067 & 0.1391 \\
0.0030 & 0.0271 & 0.0558 & 0.0228 & 0.0845 & 0.0923 & 0.1029 & 0.0511 & 0.2001 & 0.3604 \\
0.0003 & 0.0054 & 0.0169 & 0.0094 & 0.0444 & 0.0599 & 0.0812 & 0.0487 & 0.2263 & 0.5075 \\
0 & 0.0011 & 0.0051 & 0.0038 & 0.0225 & 0.0368 & 0.0593 & 0.0418 & 0.2250 & 0.6046 \\
0 & 0.0002 & 0.0015 & 0.0015 & 0.0113 & 0.0223 & 0.0423 & 0.0345 & 0.2133 & 0.6731 \\
0 & 0 & 0.0005 & 0.0006 & 0.0056 & 0.0134 & 0.0298 & 0.0281 & 0.1977 & 0.7243 \\
0 & 0 & 0.0001 & 0.0002 & 0.0028 & 0.0081 & 0.0210 & 0.0227 & 0.1813 & 0.7638 \\
0 & 0 & 0 & 0.0001 & 0.0014 & 0.0048 & 0.0147 & 0.0183 & 0.1651 & 0.7956 \\
0 & 0 & 0 & 0 & 0.0007 & 0.0029 & 0.0103 & 0.0147 & 0.1497 & 0.8217 \\
0 & 0 & 0 & 0 & 0.0004 & 0.0017 & 0.0072 & 0.0118 & 0.1355 & 0.8434
\end{pmatrix}
$$

$$
P(1) = \begin{pmatrix}
0.9496 & 0.0056 & 0.0056 & 0.0056 & 0.0056 & 0.0056 & 0.0056 & 0.0056 & 0.0056 & 0.0056 \\
0.9023 & 0.0081 & 0.0112 & 0.0112 & 0.0112 & 0.0112 & 0.0112 & 0.0112 & 0.0112 & 0.0112 \\
0.8574 & 0.0097 & 0.0166 & 0.0166 & 0.0166 & 0.0166 & 0.0166 & 0.0166 & 0.0166 & 0.0167 \\
0.8145 & 0.0109 & 0.0218 & 0.0218 & 0.0218 & 0.0218 & 0.0218 & 0.0218 & 0.0218 & 0.0220 \\
0.7737 & 0.0119 & 0.0268 & 0.0268 & 0.0268 & 0.0268 & 0.0268 & 0.0268 & 0.0268 & 0.0268 \\
0.7351 & 0.0126 & 0.0315 & 0.0315 & 0.0315 & 0.0315 & 0.0315 & 0.0315 & 0.0315 & 0.0318 \\
0.6981 & 0.0131 & 0.0361 & 0.0361 & 0.0361 & 0.0361 & 0.0361 & 0.0361 & 0.0361 & 0.0361 \\
0.6632 & 0.0136 & 0.0404 & 0.0404 & 0.0404 & 0.0404 & 0.0404 & 0.0404 & 0.0404 & 0.0404 \\
0.6301 & 0.0139 & 0.0445 & 0.0445 & 0.0445 & 0.0445 & 0.0445 & 0.0445 & 0.0445 & 0.0445 \\
0.5987 & 0.0141 & 0.0484 & 0.0484 & 0.0484 & 0.0484 & 0.0484 & 0.0484 & 0.0484 & 0.0484
\end{pmatrix}
$$

$$
P(2) = \begin{pmatrix}
0.5688 & 0.0143 & 0.0521 & 0.0521 & 0.0521 & 0.0521 & 0.0521 & 0.0521 & 0.0521 & 0.0522 \\
0.5400 & 0.0144 & 0.0557 & 0.0557 & 0.0557 & 0.0557 & 0.0557 & 0.0557 & 0.0557 & 0.0557 \\
0.5133 & 0.0145 & 0.0590 & 0.0590 & 0.0590 & 0.0590 & 0.0590 & 0.0590 & 0.0590 & 0.0592 \\
0.4877 & 0.0145 & 0.0622 & 0.0622 & 0.0622 & 0.0622 & 0.0622 & 0.0622 & 0.0622 & 0.0624 \\
0.4631 & 0.0145 & 0.0653 & 0.0653 & 0.0653 & 0.0653 & 0.0653 & 0.0653 & 0.0653 & 0.0653 \\
0.4400 & 0.0144 & 0.0682 & 0.0682 & 0.0682 & 0.0682 & 0.0682 & 0.0682 & 0.0682 & 0.0682 \\
0.4181 & 0.0144 & 0.0709 & 0.0709 & 0.0709 & 0.0709 & 0.0709 & 0.0709 & 0.0709 & 0.0712 \\
0.3969 & 0.0143 & 0.0736 & 0.0736 & 0.0736 & 0.0736 & 0.0736 & 0.0736 & 0.0736 & 0.0736 \\
0.3771 & 0.0141 & 0.0761 & 0.0761 & 0.0761 & 0.0761 & 0.0761 & 0.0761 & 0.0761 & 0.0761 \\
0.3585 & 0.0140 & 0.0784 & 0.0784 & 0.0784 & 0.0784 & 0.0784 & 0.0784 & 0.0784 & 0.0787
\end{pmatrix}
$$

$$
c = \begin{pmatrix}
0.5986 & 0.5810 & 0.6116 & 0.6762 & 0.5664 & 0.6188 & 0.7107 & 0.4520 & 0.5986 & 0.7714 \\
0.6986 & 0.6727 & 0.7017 & 0.7649 & 0.6536 & 0.6005 & 0.6924 & 0.4324 & 0.5790 & 0.6714
\end{pmatrix}'
$$

Parameters of Example 3: $B(u) = \Upsilon_{0.7}\ \forall u \in \mathcal{U}$, where Υ_ε is a tridiagonal matrix defined as

$$\Upsilon_\varepsilon = \left[\varepsilon_{ij}\right]_{X\times X}, \varepsilon_{ij} = \begin{cases} \varepsilon & i=j \\ 1-\varepsilon & (i,j) = (1,2),(X-1,X) \\ \dfrac{1-\varepsilon}{2} & (i,j) = (i,i+1),(i,i-1), i \neq 1, X \\ 0 & \text{otherwise} \end{cases}$$

$$P(1) = \begin{pmatrix} 0.1851 & 0.1692 & 0.1630 & 0.1546 & 0.1324 & 0.0889 & 0.0546 & 0.0522 \\ 0.1538 & 0.1531 & 0.1601 & 0.1580 & 0.1395 & 0.0994 & 0.0667 & 0.0694 \\ 0.1307 & 0.1378 & 0.1489 & 0.1595 & 0.1472 & 0.1143 & 0.0769 & 0.0847 \\ 0.1157 & 0.1307 & 0.1437 & 0.1591 & 0.1496 & 0.1199 & 0.0840 & 0.0973 \\ 0.1053 & 0.1196 & 0.1388 & 0.1579 & 0.1520 & 0.1248 & 0.0888 & 0.1128 \\ 0.0850 & 0.1056 & 0.1326 & 0.1618 & 0.1585 & 0.1348 & 0.0977 & 0.1240 \\ 0.0707 & 0.0906 & 0.1217 & 0.1578 & 0.1629 & 0.1447 & 0.1078 & 0.1438 \\ 0.0549 & 0.0757 & 0.1095 & 0.1502 & 0.1666 & 0.1576 & 0.1189 & 0.1666 \end{pmatrix}$$

$$P(2) = \begin{pmatrix} 0.0488 & 0.0696 & 0.1016 & 0.1413 & 0.1599 & 0.1614 & 0.1270 & 0.1904 \\ 0.0413 & 0.0604 & 0.0882 & 0.1292 & 0.1503 & 0.1661 & 0.1425 & 0.2220 \\ 0.0329 & 0.0482 & 0.0752 & 0.1195 & 0.1525 & 0.1694 & 0.1519 & 0.2504 \\ 0.0248 & 0.0388 & 0.0649 & 0.1097 & 0.1503 & 0.1732 & 0.1643 & 0.2740 \\ 0.0196 & 0.0309 & 0.0566 & 0.0985 & 0.1429 & 0.1805 & 0.1745 & 0.2965 \\ 0.0158 & 0.0258 & 0.0517 & 0.0934 & 0.1392 & 0.1785 & 0.1794 & 0.3162 \\ 0.0134 & 0.0221 & 0.0463 & 0.0844 & 0.1335 & 0.1714 & 0.1822 & 0.3467 \\ 0.0110 & 0.0186 & 0.0406 & 0.0783 & 0.1246 & 0.1679 & 0.1899 & 0.3691 \end{pmatrix}$$

$$P(3) = \begin{pmatrix} 0.0077 & 0.0140 & 0.0337 & 0.0704 & 0.1178 & 0.1632 & 0.1983 & 0.3949 \\ 0.0058 & 0.0117 & 0.0297 & 0.0659 & 0.1122 & 0.1568 & 0.1954 & 0.4225 \\ 0.0041 & 0.0090 & 0.0244 & 0.0581 & 0.1011 & 0.1494 & 0.2013 & 0.4526 \\ 0.0032 & 0.0076 & 0.0210 & 0.0515 & 0.0941 & 0.1400 & 0.2023 & 0.4803 \\ 0.0022 & 0.0055 & 0.0165 & 0.0439 & 0.0865 & 0.1328 & 0.2006 & 0.5120 \\ 0.0017 & 0.0044 & 0.0132 & 0.0362 & 0.0751 & 0.1264 & 0.2046 & 0.5384 \\ 0.0012 & 0.0033 & 0.0106 & 0.0317 & 0.0702 & 0.1211 & 0.1977 & 0.5642 \\ 0.0009 & 0.0025 & 0.0091 & 0.0273 & 0.0638 & 0.1134 & 0.2004 & 0.5826 \end{pmatrix}$$

$$P(4) = \begin{pmatrix} 0.0007 & 0.0020 & 0.0075 & 0.0244 & 0.0609 & 0.1104 & 0.2013 & 0.5928 \\ 0.0005 & 0.0016 & 0.0063 & 0.0208 & 0.0527 & 0.1001 & 0.1991 & 0.6189 \\ 0.0004 & 0.0013 & 0.0049 & 0.0177 & 0.0468 & 0.0923 & 0.1981 & 0.6385 \\ 0.0003 & 0.0009 & 0.0038 & 0.0149 & 0.0407 & 0.0854 & 0.2010 & 0.6530 \\ 0.0002 & 0.0007 & 0.0031 & 0.0123 & 0.0346 & 0.0781 & 0.2022 & 0.6688 \\ 0.0001 & 0.0005 & 0.0023 & 0.0100 & 0.0303 & 0.0713 & 0.1980 & 0.6875 \\ 0.0001 & 0.0004 & 0.0019 & 0.0083 & 0.0266 & 0.0683 & 0.1935 & 0.7009 \\ 0.0001 & 0.0003 & 0.0014 & 0.0069 & 0.0240 & 0.0651 & 0.1878 & 0.7144 \end{pmatrix}$$

$$P(5) = \begin{pmatrix} 0.0000 & 0.0002 & 0.0010 & 0.0054 & 0.0204 & 0.0590 & 0.1772 & 0.7368 \\ 0.0000 & 0.0001 & 0.0008 & 0.0041 & 0.0168 & 0.0515 & 0.1663 & 0.7604 \\ 0.0000 & 0.0001 & 0.0006 & 0.0038 & 0.0156 & 0.0480 & 0.1596 & 0.7723 \\ 0.0000 & 0.0001 & 0.0005 & 0.0032 & 0.0139 & 0.0450 & 0.1603 & 0.777 \\ 0.0000 & 0.0001 & 0.0004 & 0.0028 & 0.0124 & 0.0418 & 0.1590 & 0.7835 \\ 0.0000 & 0.0001 & 0.0003 & 0.0023 & 0.0106 & 0.0389 & 0.1547 & 0.7931 \\ 0.0000 & 0.0000 & 0.0003 & 0.0018 & 0.0090 & 0.0351 & 0.1450 & 0.8088 \\ 0.0000 & 0.0000 & 0.0002 & 0.0015 & 0.0080 & 0.0325 & 0.1386 & 0.8192 \end{pmatrix}$$

$$P(6) = \begin{pmatrix} 0.0000 & 0.0000 & 0.0001 & 0.0012 & 0.0067 & 0.0296 & 0.1331 & 0.8293 \\ 0.0000 & 0.0000 & 0.0001 & 0.0010 & 0.0059 & 0.0275 & 0.1238 & 0.8417 \\ 0.0000 & 0.0000 & 0.0001 & 0.0009 & 0.0056 & 0.0272 & 0.1238 & 0.8424 \\ 0.0000 & 0.0000 & 0.0001 & 0.0009 & 0.0053 & 0.0269 & 0.1234 & 0.8434 \\ 0.0000 & 0.0000 & 0.0001 & 0.0006 & 0.0043 & 0.0237 & 0.1189 & 0.8524 \\ 0.0000 & 0.0000 & 0.0001 & 0.0005 & 0.0038 & 0.0215 & 0.1129 & 0.8612 \\ 0.0000 & 0.0000 & 0.0000 & 0.0004 & 0.0032 & 0.0191 & 0.1094 & 0.8679 \\ 0.0000 & 0.0000 & 0.0000 & 0.0003 & 0.0025 & 0.0161 & 0.1011 & 0.8800 \end{pmatrix}$$

$$P(7) = \begin{pmatrix} 0.0000 & 0.0000 & 0.0000 & 0.0003 & 0.0022 & 0.0143 & 0.0938 & 0.8894 \\ 0.0000 & 0.0000 & 0.0000 & 0.0002 & 0.0019 & 0.0136 & 0.0901 & 0.8942 \\ 0.0000 & 0.0000 & 0.0000 & 0.0002 & 0.0017 & 0.0126 & 0.0849 & 0.9006 \\ 0.0000 & 0.0000 & 0.0000 & 0.0002 & 0.0015 & 0.0118 & 0.0819 & 0.9046 \\ 0.0000 & 0.0000 & 0.0000 & 0.0001 & 0.0013 & 0.0108 & 0.0754 & 0.9124 \\ 0.0000 & 0.0000 & 0.0000 & 0.0001 & 0.0011 & 0.0098 & 0.0714 & 0.9176 \\ 0.0000 & 0.0000 & 0.0000 & 0.0001 & 0.0010 & 0.0090 & 0.0713 & 0.9186 \\ 0.0000 & 0.0000 & 0.0000 & 0.0001 & 0.0009 & 0.0084 & 0.0675 & 0.9231 \end{pmatrix}$$

$$P(8) = \begin{pmatrix} 0.0000 & 0.0000 & 0.0000 & 0.0001 & 0.0008 & 0.0078 & 0.0665 & 0.9248 \\ 0.0000 & 0.0000 & 0.0000 & 0.0000 & 0.0007 & 0.0068 & 0.0626 & 0.9299 \\ 0.0000 & 0.0000 & 0.0000 & 0.0000 & 0.0006 & 0.0061 & 0.0581 & 0.9352 \\ 0.0000 & 0.0000 & 0.0000 & 0.0000 & 0.0005 & 0.0057 & 0.0561 & 0.9377 \\ 0.0000 & 0.0000 & 0.0000 & 0.0000 & 0.0005 & 0.0053 & 0.0558 & 0.9384 \\ 0.0000 & 0.0000 & 0.0000 & 0.0000 & 0.0004 & 0.0051 & 0.0558 & 0.9387 \\ 0.0000 & 0.0000 & 0.0000 & 0.0000 & 0.0004 & 0.0045 & 0.0522 & 0.9429 \\ 0.0000 & 0.0000 & 0.0000 & 0.0000 & 0.0003 & 0.0040 & 0.0505 & 0.9452 \end{pmatrix}$$

$$c = \begin{pmatrix} 1.0000 & 2.2486 & 4.1862 & 6.9509 & 11.2709 & 15.9589 & 21.4617 & 27.6965 \\ 31.3230 & 8.8185 & 9.6669 & 11.4094 & 14.2352 & 17.8532 & 22.3155 & 27.5353 \\ 50.0039 & 26.3162 & 14.6326 & 15.3534 & 17.1427 & 19.7455 & 23.1064 & 27.3025 \\ 65.0359 & 40.2025 & 27.5380 & 19.5840 & 20.3017 & 21.8682 & 24.2022 & 27.4108 \\ 79.1544 & 53.1922 & 39.5408 & 30.5670 & 23.3697 & 23.9185 & 25.1941 & 27.4021 \\ 90.7494 & 63.6983 & 48.6593 & 38.6848 & 30.4868 & 25.7601 & 26.0012 & 27.1867 \\ 99.1985 & 71.1173 & 55.0183 & 44.0069 & 34.7860 & 29.0205 & 26.9721 & 27.1546 \\ 106.3851 & 77.2019 & 60.0885 & 47.8917 & 37.6330 & 30.8279 & 27.7274 & 26.4338 \end{pmatrix}$$

Parameters of Example 4: $\alpha(1) = 0.7, \alpha(2) = 1$,

$$c = \begin{pmatrix} 0.9023 & 1.1169 & 4.8158 \\ 0.1 & 1.1169 & 4.8158 \end{pmatrix}', \; P(2) = \begin{pmatrix} 1.0000 & 0.0000 & 0.0000 \\ 0.1641 & 0.1563 & 0.6796 \\ 0.1247 & 0.1639 & 0.7114 \end{pmatrix}, \; P(1) = (P(2))^2$$

$$B(1) = \begin{pmatrix} 0.49 & 0.3565 & 0.1535 \\ 0.2977 & 0.4604 & 0.2419 \\ 0.0035 & 0.5002 & 0.4963 \end{pmatrix}, \; B(2) = \begin{pmatrix} 0.4798 & 0.3634 & 0.1568 \\ 0.3473 & 0.4273 & 0.2254 \\ 0.0255 & 0.4901 & 0.4844 \end{pmatrix}.$$

Appendix 14.B Proof of Theorem 14.8.1

Part 1: We prove that dominance of transition matrices $P \succeq \bar{P}$ (with respect to (14.34)) results in dominance of optimal costs, i.e. $V(\pi; P) \leq V(\pi; \bar{P})$. The proof is by induction. For $n = 0$, $V_n(\pi; P) \leq V_n(\pi; \bar{P}) = 0$ by the initialization of the value iteration algorithm (7.29).

Next, to prove the inductive step assume that $V_n(\pi; P) \leq V_n(\pi; \bar{P})$ for $\pi \in \Pi$. By Theorem 11.2.1, under (C), (F1), (F2), $V_n(\pi; P)$ and $V_n(\pi; \bar{P})$ are MLR decreasing in $\pi \in \Pi$. From Theorem 10.3.1(10.3.1), under (F3′) it follows that $T(\pi, y, u; P) \geq_r T(\pi, y, u; \bar{P})$ where $\geq_r$ denotes MLR dominance. This implies

$$V_n(T(\pi, y, u; P); \bar{P}) \leq V_n(T(\pi, y, u; \bar{P}); \bar{P}), \quad P \succeq \bar{P}. \tag{14.39}$$

Since $V_n(\pi; P) \leq V_n(\pi; \bar{P})$ $\forall \pi \in \Pi$ by the induction hypothesis, clearly

$$V_n(T(\pi, y, u, P); P) \leq V_n(T(\pi, y, u, P); \bar{P}). \tag{14.40}$$

Combining (14.39) and (14.40) yields

$$V_n(T(\pi, y, u; P); P) \leq V_n(T(\pi, y, u, \bar{P}); \bar{P}), \quad P \succeq \bar{P}.$$

Under (F1), (F2), Statement 10.3.1 of Theorem 10.3.1 says that $T(\pi, y, u; P)$ is MLR increasing in y. Therefore, $V_n(T(\pi, y, u; P); P)$ is increasing in y. Also from Statement 10.3.1 of Theorem 10.3.1, $\sigma(\pi, \cdot, u; P) \geq_s \sigma((\pi, \cdot, u; \bar{P})$ for $P \succeq \bar{P}$ where $\geq_s$ denotes first order stochastic dominance. Therefore,

$$\sum_y V_n(T(\pi, y, u; P); P)\sigma(\pi, \cdot, u; P) \leq \sum_y V_n(T(\pi, y, u; \bar{P}); \bar{P})\sigma(\pi, \cdot, u; \bar{P}).$$

and so

$$C(\pi, u) + \sum_y V_n(T(\pi, y, u; P); P)\sigma(\pi, \cdot, u; P)$$
$$\leq C(\pi, u) + \sum_y V_n(T(\pi, y, u; \bar{P}); \bar{P})\sigma(\pi, \cdot, u; \bar{P}).$$

Minimizing both sides with respect to action u yields $V_{n+1}(\pi; P) \geq V_{n+1}(\pi; \bar{P})$ and concludes the induction argument.

Part 2: Next we show that dominance of observation distributions $\bar{B} \succeq_{\mathcal{B}} B$ (with respect to the Blackwell order (14.35)) results in dominance of the optimal costs, namely

$V(\pi;B) \geq V(\pi,\bar{B})$. In general B and $\bar{B}$ can have different observation spaces. Denote them by $y^{(1)} \in \mathcal{Y}^{(1)}$ and $y^{(2)} \in \mathcal{Y}^{(2)}$, respectively. Let $T(\pi, y^{(1)}, u)$ and $\bar{T}(\pi, y^{(1)}, u)$ denote the Bayesian filter update (14.30) with observation B and $\bar{B}$, respectively. Let $\sigma(\pi, y, u)$ and $\bar{\sigma}(\pi, y, u)$ denote the corresponding normalization measures. Then using an identical proof to (14.31), it follows that

$$\sum_{y^{(2)} \in \mathcal{Y}^{(2)}} V_n(T(\pi, y^{(2)}, u); \bar{B})\sigma(\pi, y^{(2)}, u) \geq \sum_{y^{(1)} \in \mathcal{Y}^{(1)}} V_n(\bar{T}(\pi, y^{(1)}, u); \bar{B})\bar{\sigma}(\pi, y^{(1)}, u). \tag{14.41}$$

With the above inequality, the proof of the theorem follows by mathematical induction using the value iteration algorithm (7.29). Assume $V_n(\pi;B) \geq V_n(\pi;\bar{B})$ for $\pi \in \Pi$. Then

$$\begin{aligned}
& C(\pi, u) + \sum_{y^{(2)}} V_n(T(\pi, y^{(2)}, u); B)\sigma(\pi, y^{(2)}, u) \\
& \geq C(\pi, u) + \sum_{y^{(2)}} V_n(T(\pi, y^{(2)}, u); \bar{B})\sigma(\pi, y^{(2)}, u) \\
& \geq C(\pi, u) + \sum_{y^{(1)}} V_n(\bar{T}(\pi, y^{(1)}, u); \bar{B})\bar{\sigma}(\pi, y^{(1)}, u)
\end{aligned}$$

where the second inequality follows from (14.41). Thus $V_{n+1}(\pi;B) \geq V_{n+1}(\pi;\bar{B})$. This completes the induction step.

Appendix 14.C Proof of Theorem 14.9.1

The aim is to prove (14.36), (14.37). Most of the efforts below are to prove (14.36). We will prove that (14.36) holds for any policy μ, not just the optimal policy. That is, for any policy μ, it will be shown that

$$\sup_{\pi \in \Pi} |J_\mu(\pi;\theta) - J_\mu(\pi;\bar{\theta})| \leq K\|\theta - \bar{\theta}\|. \tag{14.42}$$

We start this appendix with the proof of (14.37) since it follows straightforwardly from (14.42). To prove (14.37), note that trivially

$$\begin{aligned}
J_{\mu^*(\bar{\theta})}(\pi, \theta) &\leq J_{\mu^*(\bar{\theta})}(\pi, \bar{\theta}) + \sup_\pi |J_{\mu^*(\bar{\theta})}(\pi, \theta) - J_{\mu^*(\bar{\theta})}(\pi, \bar{\theta})| \\
J_{\mu^*(\theta)}(\pi, \bar{\theta}) &\leq J_{\mu^*(\theta)}(\pi, \theta) + \sup_\pi |J_{\mu^*(\theta)}(\pi, \theta) - J_{\mu^*(\theta)}(\pi, \bar{\theta})|.
\end{aligned}$$

Also by definition $J_{\mu^*(\bar{\theta})}(\pi, \bar{\theta}) \leq J_{\mu^*(\theta)}(\pi, \bar{\theta})$ since $\mu^*(\bar{\theta})$ is the optimal policy for model $\bar{\theta}$. Therefore,

$$\begin{aligned}
J_{\mu^*(\bar{\theta})}(\pi, \theta) &\leq J_{\mu^*(\theta)}(\pi, \theta) + \sup_\pi |J_{\mu^*(\bar{\theta})}(\pi, \theta) - J_{\mu^*(\bar{\theta})}(\pi, \bar{\theta})| \\
&\quad + \sup_\pi |J_{\mu^*(\theta)}(\pi, \theta) - J_{\mu^*(\theta)}(\pi, \bar{\theta})| \\
&\leq J_{\mu^*(\theta)}(\pi, \theta) + 2 \sup_\mu \sup_\pi |J_{\mu(\bar{\theta})}(\pi, \theta) - J_{\mu(\bar{\theta})}(\pi, \bar{\theta})|.
\end{aligned}$$

Then from (14.42), clearly (14.37) follows.

Proof of (14.36): We now present the proof of (14.42) and thus (14.36).

The cumulative cost incurred by applying policy $\mu(\pi)$ to model θ satisfies at time n

$$J_\mu^{(n)}(\pi;\theta) = C'_{\mu(\pi)}\pi + \rho \sum_y J_\mu^{(n-1)}(T(\pi,y,\mu(\pi);\theta)\sigma(\pi,y,\mu(\pi);\theta).$$

Therefore, the absolute difference in cumulative costs for models $\theta, \bar{\theta}$ satisfies

$$\begin{aligned}
&|J_\mu^{(n)}(\pi;\theta) - J_\mu^{(n)}(\pi;\bar{\theta})| \\
&\quad \leq \rho \sum_y \sigma(\pi,y,\mu(\pi);\theta) \left| J_\mu^{(n-1)}(T(\pi,y,\mu(\pi);\theta)) - J_\mu^{(n-1)}(T(\pi,y,\mu(\pi);\bar{\theta})) \right| \\
&\qquad + \rho \sum_y J_\mu^{(n-1)}(T(\pi,y,\mu(\pi);\bar{\theta})) \left| \sigma(\pi,y,\mu(\pi);\theta) - \sigma(\pi,y,\mu(\pi);\bar{\theta}) \right| \\
&\quad \leq \rho \sup_{\pi\in\Pi} |J_\mu^{(n-1)}(\pi;\theta) - J_\mu^{(n-1)}(\pi;\bar{\theta})| \sum_y \sigma(\pi,y,\mu(\pi);\theta) \\
&\qquad + \rho \sup_{\pi\in\Pi} J_\mu^{(n-1)}(\pi;\bar{\theta}) \sum_y \left| \sigma(\pi,y,\mu(\pi);\theta) - \sigma(\pi,y,\mu(\pi);\bar{\theta}) \right|. \qquad (14.43)
\end{aligned}$$

Note that $\sum_y \sigma(\pi,y,\mu(\pi);\theta) = 1$. Also evaluating $\sigma(\pi,y,u;\theta) = \mathbf{1}'_X B_y(u) P'(u)\pi$ yields

$$\begin{aligned}
&\sum_y \left| \sigma(\pi,y,\mu(\pi);\theta) - \sigma(\pi,y,\mu(\pi);\bar{\theta}) \right| \\
&\qquad \leq \max_u \sum_y \sum_i \sum_j \left| B_{jy}(u) P_{ij}(u) - \bar{B}_{jy}(u) \bar{P}_{ij}(u) \right| \pi(i) \\
&\qquad \leq \max_u \max_i \sum_y \sum_j \left| B_{jy}(u) P_{ij}(u) - \bar{B}_{jy} \bar{P}_{ij}(u) \right|. \qquad (14.44)
\end{aligned}$$

Finally, $\sup_{\pi\in\Pi} J_\mu^{(n-1)}(\pi;\bar{\theta}) \leq \frac{1}{1-\rho} \max_{i\in\{1,2,\dots,X\}} C(e_i,u)$. Using these bounds in (14.43) yields

$$\begin{aligned}
&\sup_{\pi\in\Pi} |J_\mu^{(n)}(\pi;\theta) - J_\mu^{(n)}(\pi;\bar{\theta})| \\
&\quad \leq \rho \sup_{\pi\in\Pi} |J_\mu^{(n-1)}(\pi;\theta) - J_\mu^{(n-1)}(\pi;\bar{\theta})| + \frac{\rho}{1-\rho} \max_{i\in\{1,2,\dots,X\}} C(e_i,u) \|\theta - \bar{\theta}\|. \qquad (14.45)
\end{aligned}$$

Then starting with $J_\mu^{(0)}(\pi;\theta) = J_\mu^{(0)}(\pi;\bar{\theta}) = 0$, unraveling (14.45) yields (14.42). In particular, choosing $\mu = \mu^*(\theta)$ yields (14.36).

Proof of (14.38): When θ and $\bar{\theta}$ have identical $P(u)$, then (14.44) becomes

$$\max_u \max_i \sum_j P_{ij}(u) \sum_y |B_{jy} - \bar{B}_{jy}|.$$

From Pinsker's inequality [90], the total variation norm is bounded by Kullback–Leibler distance D defined in (14.38) as

$$\sum_y |B_{jy} - \bar{B}_{jy}| \leq \sqrt{2\, D(B_j \| \bar{B}_j)}.$$

Part IV

Stochastic approximation and reinforcement learning

Parts II and III of the book discussed stochastic dynamic programming for POMDPs. The aim was to determine the globally optimal policy. Part IV deals with stochastic gradient algorithms that converge to a local optimum. Such gradient algorithms are computationally efficient unlike dynamic programming which can be intractable. Furthermore, stochastic gradient algorithms form the basis for reinforcement learning – that is, they facilitate estimating the optimal policy when one does not know the parameters of the MDP or POMDP.

Chapter 15 discusses gradient estimation for Markov processes via stochastic simulation. This forms the basis of gradient-based reinforcement learning.

Chapter 16 presents simulation-based stochastic approximation algorithms for estimating the optimal policy of MDPs when the transition probabilities are not known. These algorithms are also described in the context of POMDPs. The Q-learning algorithm and gradient-based reinforcement learning algorithms are presented.

Chapter 17 gives a brief description of convergence analysis of stochastic approximation algorithms. Examples given include recursive maximum likelihood estimation of HMM parameters, the least mean squares algorithm for estimating the state of an HMM (instead of a Bayesian HMM filter), discrete stochastic optimization algorithms, and mean field dynamics for approximating the dynamics of information flow in social networks.

Unlike Parts I to III, Part IV of the book is non-Bayesian (in the sense that Bayes' formula is not the main theme).

15 Stochastic optimization and gradient estimation

Contents

To motivate Part IV of the book, consider the discounted cost MDP problem:

$$\text{Compute } \mu^* = \operatorname*{argmin}_{\mu} \mathbb{E}_{\mu}\Big\{\sum_{k=0}^{\infty} \rho^k c(x_k, \mu(x_k))\Big\}.$$

where x_k is the controlled state of the system, $c(x_k, u_k)$ is the cost incurred at time k by choosing action $u_k = \mu(x_k)$ and $\rho < 1$ is a discount factor. In Parts II and III of the book we used stochastic dynamic programming to compute the *globally* optimal policy μ^* for such problems.

In comparison, Part IV deals with computing (estimating) *local* minima using stochastic gradient algorithms. Suppose the action at time k is chosen according to the following parameterized policy: $u_k = \mu_\theta(x)$ for some pre-specified function μ_θ parametrized by the vector $\theta \in \mathbb{R}^p$. Then the aim is:

$$\text{Compute } \theta^* = \operatorname*{argmin}_{\theta} C(\theta), \quad C(\theta) = \mathbb{E}_{\mu_\theta}\Big\{\sum_{k=0}^{\infty} \rho^k c(x_k, \mu_\theta(x_k)\Big\}. \tag{15.1}$$

This will be achieved using a stochastic gradient algorithm of the form

$$\theta_{k+1} = \theta_k - \epsilon_k \widehat{\nabla}_\theta C_k(\theta_k). \tag{15.2}$$

Here ϵ_k is a scalar step size and $\widehat{\nabla}_\theta C_k(\theta_k)$ denotes an estimate of gradient $\nabla_\theta C(\theta)$ evaluated at θ_k. These gradient estimates need to be computed using the observed realization of $\{x_k, \{c(x_k, u_k)\}$ since this is the only information available to the decision-maker. Since in general, $C(\theta)$ is non-convex, at best one can expect (15.2) to converge (in a sense to be made precise below) to a local stationary point of (15.1).

Even though stochastic gradient algorithms typically converge to a local stationary point, there are several advantages compared to stochastic dynamic programming. First, in many cases, (15.2) can operate without knowing the transition matrices of the MDP, whereas dynamic programming requires complete knowledge of these parameters. (This is the basis of reinforcement learning in Chapter 16.) Second, (15.2) is often substantially cheaper[1] to compute, especially for very large state and action spaces, where dynamic programming can be prohibitively expensive. We have already encountered stochastic gradient algorithms in §9.5 and §12.4.2 where the SPSA stochastic gradient algorithm was used to estimate monotone policies for MDPs and POMDPs.

This chapter focuses on gradient estimation for Markov processes. Several algorithms will be discussed including the SPSA algorithm, score function gradient estimator and weak derivative gradient estimator. Such gradient estimators form the basis for implementing stochastic gradient algorithms for MDPs and POMDPs (discussed in Chapter 16). Simulation-based gradient estimation is a mature area; our aim in this chapter is to present a few key algorithms that are useful for MDPs and POMDPs.

15.1 Stochastic gradient algorithm

The aim is to solve a continuous-valued stochastic optimization problem.[2] Suppose a parameter vector $\theta \in \mathbb{R}^p$ specifies a transition matrix (or more generally, kernel) P_θ and stationary distribution π_θ from which a Markov process $\{x_k\}$ is simulated. The aim is to compute $\theta^* \in \Theta$ that minimizes[3] the expected cost

$$C(\theta) = \mathbb{E}_{\pi_\theta}\{c(x,\theta)\} = \int_{\mathcal{X}} c(x,\theta)\pi_\theta(x)dx. \tag{15.3}$$

In stochastic optimization, the stationary distribution π_θ and/or the cost $c(x,\theta)$ are not known explicitly[4]; instead the sample path sequence $\{c(x_k,\theta), k = 1,2,\ldots\}$ of (noisy) costs can be observed for any choice of model θ. Therefore, it is not possible to compute the integral in (15.3) explicitly. This is in contrast to a deterministic optimization problem where $c(x,\theta)$ and π_θ are known so that the right-hand side of (15.3) can be evaluated explicitly.

[1] This relies on determining a parsimonious representation of μ_θ (called a "feature" in machine learning) parametrized by the feature vector θ.

[2] The term "continuous" stochastic optimization is used to distinguish from discrete stochastic optimization which is described in §17.4 where Θ is a finite set.

[3] Assume $\Theta \subset \mathbb{R}^p$ is compact and $C(\theta)$ is a continuous function so that the minimum exists.

[4] We assume that P_θ is known but computing π_θ in closed form is not tractable. More generally, P_θ itself may not be known, but for any choice of θ, samples of $c(x_k,\theta)$ can be simulated.

In stochastic optimization, a widely used algorithm to compute (estimate) the minimizer θ^* of (15.3) is the following stochastic gradient algorithm run over time $k = 0, 1, \ldots$

$$\theta_{k+1} = \theta_k - \epsilon_k \widehat{\nabla}_\theta C_k(\theta_k). \tag{15.4}$$

Here $\widehat{\nabla}_\theta C_k(\theta_k)$ denotes an estimate of gradient $\nabla_\theta C(\theta)$ evaluated at θ_k. Note that $\widehat{\nabla}_\theta C_k(\theta_k)$ needs to be computed using the observed realization $\{c(x_k, \theta)\}$ since this is the only information available.

In (15.4), ϵ_k is a positive scalar at each time k and denotes the step size of the algorithm. There are two general philosophies for choosing the step size.

1. Decreasing step size $\epsilon_k = 1/k^\alpha$, where $\alpha \in (0.5, 1]$. Under reasonable conditions, it can be proved that the stochastic gradient algorithm (15.4) converges with probability 1 to a local stationary point of $C(\theta)$.
2. Constant step size $\epsilon_k = \epsilon$ where ϵ is a small positive constant. Under reasonable conditions, it can be proved that the stochastic gradient algorithm (15.4) converges weakly to a local stationary point. Such constant step size algorithms are useful for estimating a time-varying minimum. Suppose $C(\theta)$ itself varies slowly with time so that its minimum evolves slowly with time. Then the constant step size stochastic gradient algorithm can *track* the time-varying minimum, i.e. estimate a local stationary point of the time varying cost $C(\theta)$.

Remark: Robbins–Monro algorithm: The stochastic gradient algorithm (15.4) can be viewed as a special case of the following Robbins–Monro algorithm. Suppose one wishes to solve numerically the following equation for θ:

$$\mathbb{E}_{\pi_\theta}\{c(x, \theta)\} = 0, \text{ or equivalently } \int_{\mathcal{X}} c(x, \theta)\, \pi_\theta(x) dx = 0. \tag{15.5}$$

Assume that for any choice of θ, one can obtain random (noisy) samples $c(x_k, \theta)$. The Robbins–Monro algorithm for solving (15.5) is the following stochastic approximation algorithm:

$$\theta_{k+1} = \theta_k + \epsilon_k\, c(x_k, \theta_k).$$

Notice that the stochastic gradient algorithm (15.4) can be viewed as an instance of the Robbins–Monro algorithm for solving $\mathbb{E}_{\pi_\theta}\{\nabla_\theta C_k(\theta)\} = 0$. Chapter 16 shows that the Q-learning algorithm (reinforcement learning algorithm) for solving MDPs is also an instance of the Robbins–Monro algorithm.

Examples

Examples of stochastic gradient algorithms include the following:

1. Least mean square (LMS) algorithm: The LMS algorithm belongs to the class of adaptive filtering algorithms and is used widely in adaptive signal processing. The objective to minimize is

$$C(\theta) = \mathbb{E}_\pi\{(y_k - \psi_k'\theta)^2\}$$

where y_k and ψ_k are observed at each time k. Then using (15.4) we obtain the LMS algorithm

$$\theta_{k+1} = \theta_k + \epsilon \psi_k (y_k - \psi_k' \theta_k). \tag{15.6}$$

Note that this case is a somewhat simple example of a stochastic optimization problem since we know $c(x, \theta) = (y_k - \psi_k' \theta)^2$ explicitly as a function of $x = (y, \psi)$ and θ. Also the measure π does not depend explicitly on θ. In §17.6.1, we will discuss the consensus LMS algorithm.

2. *Adaptive modulation*: Consider a wireless communication system where the packet error rate of the channel x_k evolves according to a Markov process. The modulation scheme is adapted over time depending on the observed (empirical) packet error rate y_k measured at the receiver, where $y_k \in [0, 1]$. Suppose at each time slot k, one of two modulation schemes can be chosen, namely,

$$u_k \in \{r_1 \text{ (low data rate)}, r_2 \text{ (high data rate)}\}.$$

The modulation scheme u_k affects the transition matrix (kernel) $P(u_k)$ of the error probability process x_k. Define the instantaneous throughput at time k as $r(x_k, u_k) = (1 - x_k) u_k$. The aim is to maximize the average throughput $\lim_{N\to\infty} \frac{1}{N+1} \mathbb{E}_\mu \{\sum_{k=0}^N r(x_k, u_k)\}$ where the policy μ maps the available information $\mathcal{I}_k = \{u_0, y_1, u_1, \ldots, u_{k-1}, y_k\}$ to action u_k.

The setup is depicted in Figure 15.1. The problem is an average cost POMDP and in general intractable to solve.

Consistent with the theme of Part IV of the book, consider the following "simple" parametrized policy for the adaptive modulation scheme:

$$u_{k+1} = \mu_\theta(y_k) = \begin{cases} r_1 & \text{if } y_k \geq \theta \\ r_2 & \text{if } y_k < \theta \end{cases}. \tag{15.7}$$

This family of policies μ_θ parametrized by θ is intuitive since the individual policies use a less aggressive modulation scheme (low data rate) when the channel quality is poor (large error rate y_k), and vice versa. The aim is to determine the value of θ in the adaptive modulation policy μ_θ which maximizes the average throughput.

Let π_θ denote the joint stationary probability distribution of the Markov process (x_k, y_{k-1}) induced by the policy μ_θ. The aim is to determine θ which maximizes the average throughput:

$$\text{Compute } \theta^* = \underset{\theta \in \Theta}{\operatorname{argmax}}\, R(\theta), \quad \text{where } R(\theta) = \mathbb{E}_{\pi_\theta}\{r(x_k, \mu(y_{k-1}))\} \tag{15.8}$$

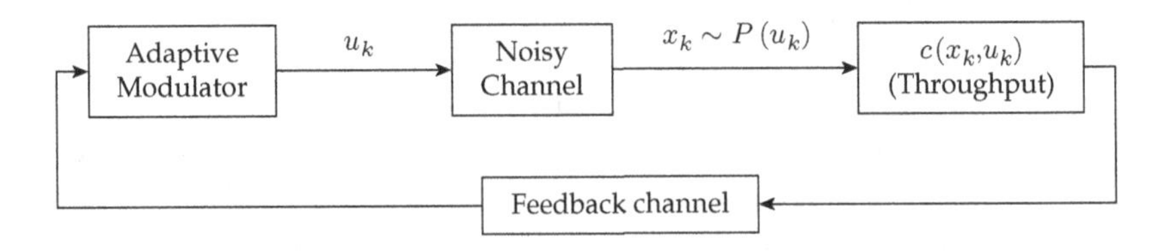

Figure 15.1 Schematic setup of adaptive modulation scheme.

given noisy measurements $r_k = (1 - y_k)u_k$ of the instantaneous throughput. An explicit formula for π_θ is not known since it is a complicated function of the modulation scheme, error correction codes, channel coding, medium access control protocol, etc. So (15.8) is not a deterministic optimization problem.

Assuming $\mathbb{E}\{y_k\} = x_k$ (the empirical observations of the channel are unbiased), then θ^* can be estimated using the stochastic gradient algorithm

$$\theta_{k+1} = \theta_k + \epsilon_k\,(\hat{\nabla}_\theta r_k).$$

Here $\hat{\nabla}_\theta r_k$ is an estimate of the gradient $\nabla_\theta R(\theta) = \int r(x, \mu(y)) \nabla_\theta \pi_\theta(x, y) dx dy$. In §15.5 and §15.6, two classes of simulated-based algorithms are given to estimate the gradient with respect to the stationary distribution of a Markov process. Alternatively, any of the finite difference methods in §15.4 can be used.

15.2 How long to simulate a Markov chain?

Before proceeding with estimating the gradient of a cost function, let us digress briefly to review how to estimate the expected cost function given the sample path of a Markov chain. Appendix A gives several algorithms for simulating a Markov chain with specified initial distribution π_0 and transition matrix P.

Suppose a Markov chain is simulated for n-time points. Let $\mathcal{X} = \{e_1, e_2, \ldots, e_X\}$ denote the state space, where e_i are the X-dimensional unit vectors. Given $h \in \mathbb{R}^X$, let

$$\phi_n = \frac{1}{n} \sum_{k=0}^{n-1} h' x_k$$

denote the time averaged estimate. By the strong law of large numbers, if P is regular (recall definition in §2.4.3), then as $n \to \infty$, $\phi_n \to h'\pi_\infty$ with probability 1 for any initial distribution π_0. However, in practice, for finite sample size simulations, one needs to determine the accuracy of estimate and guidelines for choosing the sample size n. For sample size n, define the bias, variance and mean square deviation of the estimate ϕ_n as

$$\begin{aligned}
\text{Bias}(\phi_n) &= \mathbb{E}\{\phi_n\} - h'\pi_\infty \\
\text{Var}(\phi_n) &= \mathbb{E}\{\phi_n - \mathbb{E}\{\phi_n\}\}^2 \\
\text{MSD}(\phi_n) &= \mathbb{E}\{\phi_n - h'\pi_\infty\}^2 = \text{Var}(\phi_n) + (\text{Bias}(\phi_n))^2.
\end{aligned}$$

The main result is as follows; the proof is in the appendix.

THEOREM 15.2.1 *Consider an n-point sample path of a Markov chain $\{x_k\}$ with regular transition matrix P. Then*

$$|\,\text{Bias}(\phi_n)| \leq \frac{\max_{i,j} |h_i - h_j|}{n(1-\rho)} \|\pi_0 - \pi_\infty\|_{TV} \tag{15.9}$$

$$\text{MSD}(\phi_n) \leq 2\,\frac{\max_{i,j} |h_i - h_j|^2}{n(1-\rho)} \sum_{i \in \mathcal{X}} (\|e_i - \pi_\infty\|_{TV})^2\, \pi_\infty(i) + O(\frac{1}{n^2}) \tag{15.10}$$

$$\mathbb{P}\left(|\phi_n - E\{\phi_n\}| > \epsilon\right) \leq 2\exp\left(-\frac{\epsilon^2(1-\rho)^2 n}{\max_{l,m}|h_l - h_m|^2}\right). \tag{15.11}$$

Here, for any two pmfs $\pi, \bar{\pi}$, $\|\pi - \bar{\pi}\|_{TV}$ denotes the variational distance

$$\|\pi - \bar{\pi}\|_{TV} = \frac{1}{2}\|\pi - \bar{\pi}\|_1 = \frac{1}{2}\sum_i |\pi(i) - \bar{\pi}(i)|$$

and ρ denotes the Dobrushin coefficient of transition matrix P (see § 2.7):

$$\rho = \frac{1}{2}\max_{i,j}\sum_{l\in\mathcal{X}} |P_{il} - P_{jl}|.$$

In summary, the bias is small if the initial distribution π_0 is chosen close to the stationary distribution π_∞, or the sample size n is chosen large, or the transition matrix has a small coefficient of ergodicity ρ. The mean square deviation on the other hand also depends on the sum of the squares of the variational distance between the stationary distribution and the unit state vectors. Finally, (15.11) is an elementary concentration of measure result for a finite state Markov chain – the time average concentrates with exponential probability around the mean. The proof of (15.11) in the appendix is similar to [177, Theorem 1.1] and uses a standard trick involving martingales with the Azuma–Hoeffding inequality.

15.3 Gradient estimators for Markov processes

Consider the stochastic optimization problem (15.3). To solve (15.3) using the stochastic gradient algorithm (15.4) requires estimating the gradient $\nabla_\theta C(\theta)$. We wish to compute the gradient estimate $\hat{\nabla}_\theta C_N(\theta)$ using the observed realization of costs $\{c(x_k, \theta)\}$, $k = 0, \ldots, N$. Here x_k is an ergodic Markov process with transition matrix or kernel P_θ. The rest of this chapter discusses two types of simulated-based gradient estimators for Markov processes:

- Finite difference gradient estimators such as the Kiefer–Wolfowitz and SPSA algorithms discussed in §15.4. They do not require knowledge of the transition kernel of the underlying Markov process. However, finite difference gradient estimators suffer from a bias variance tradeoff. Recall that the SPSA algorithm was used in §9.5 and §12.4.2 for MDPs and POMDPs.
- Gradient estimators that exploit knowledge of the Markov transition kernel. We discuss the score function gradient estimator in §15.5 and the weak derivative estimator below in §15.6. Unlike the finite difference estimators, these gradient estimators are unbiased for random variables.

§16 shows how these gradient estimators can be used for reinforcement learning of MDPs, constrained MDPs, and POMDPs.

15.4 Finite difference gradient estimators and SPSA

This section describes two finite difference gradient estimators that can be used with the stochastic gradient algorithm to minimize (15.3). The algorithms do not require the transition matrix of the Markov chain to be known.

15.4.1 Kiefer–Wolfowitz algorithm

This uses the two-sided numerical approximation of a derivative and is described in Algorithm 16. Here Δ_n is the gradient step while ϵ_n is the step size of the stochastic approximation algorithm. These need to be chosen so that

$$\Delta_n > 0,\ \epsilon_n > 0,\ \Delta_n \to 0,\ \epsilon_n \to 0,\ \sum_n \epsilon_n = \infty,\ \sum_n \epsilon_n^2/\Delta_n^2 < \infty. \tag{15.12}$$

A disadvantage of the Kiefer–Wolfowitz algorithm is that $2p$ independent simulations are required to evaluate the gradient along all the possible directions $i = 1, \ldots, p$.

Algorithm 16 Kiefer–Wolfowitz algorithm ($\theta_n \in \mathbb{R}^p$)

For iterations $n = 0, 1, 2, \ldots$

- Evaluate the $2p$ sampled costs $\hat{C}_n(\theta + \Delta_n e_i)$ and $\hat{C}_n(\theta - \Delta_n e_i)$, $i = 1, 2, \ldots, p$ where e_i is a unit vector with 1 in element i.
- Compute gradient estimate

$$\hat{\nabla}_{\theta(i)} C_n(\theta) = \frac{1}{2\Delta_n}\left[\hat{C}_n(\theta + \Delta_n e_i) - \hat{C}_n(\theta - \Delta_n e_i)\right], \quad i = 1, \ldots, p \tag{15.13}$$

  Here $\Delta_n = \frac{\Delta}{(n+1)^\gamma}$ denotes the gradient step size with $0.5 \leq \gamma \leq 1$ and $\Delta > 0$.
- Update model estimate θ_n via stochastic gradient algorithm

$$\theta_{n+1} = \theta_n - \epsilon_n \hat{\nabla} C(\theta_n), \quad \epsilon_n = \frac{\epsilon}{(n+1+s)^\zeta}, \quad 0.5 < \zeta \leq 1, \quad \text{and } \epsilon, s > 0.$$

15.4.2 Simultaneous perturbation stochastic approximation (SPSA)

The SPSA algorithm [309] has been pioneered by J. Spall (please see the website `www.jhuapl.edu/SPSA/`). It overcomes the problem of requiring $2p$ independent simulations in the Kiefer–Wolfowitz algorithm by choosing a single random direction in $\mathbb{R}^p$ along which to evaluate the finite difference gradient. So only two simulations are required. The SPSA algorithm is described in Algorithm 17. The step sizes Δ_n and ϵ_n are chosen as in (15.12).

In Algorithm 17 the random directions have been chosen from a Bernoulli distribution. The random directions d_n need to be chosen from a distribution such that the inverse moments $\mathbb{E}\{|1/d_n(i)|^{2+2\alpha}\}$ are finite for some $\alpha > 0$. Suitable choices (apart from Bernoulli) include segmented uniform and U-shaped densities; see [309, p.185] for a detailed description.

Spall shows that SPSA is asymptotically as efficient as Kiefer–Wolfowitz.

Algorithm 17 SPSA algorithm ($\theta_n \in \mathbb{R}^p$)

For iterations $n = 0, 1, 2, \ldots$

- Simulate the p dimensional vector d_n with random elements

$$d_n(i) = \begin{cases} -1 & \text{with probability } 0.5 \\ +1 & \text{with probability } 0.5. \end{cases}$$

- Evaluate sample costs $\hat{C}_n(\theta_n + \Delta_n d_n)$ and $\hat{C}_n(\theta_n - \Delta_n d_n)$
- Compute gradient estimate

$$\hat{\nabla} C(\theta_n) = \frac{\hat{C}_n(\theta_n + \Delta_n d_n) - \hat{C}_n(\theta_n - \Delta_n d_n)}{2\Delta_n} d_n.$$

  Here $\Delta_n = \frac{\Delta}{(n+1)^\gamma}$ denotes the gradient step size with $0.5 \leq \gamma \leq 1$ and $\Delta > 0$.
- Update model estimate θ_n via stochastic gradient algorithm

$$\theta_{n+1} = \theta_n - \epsilon_n \hat{\nabla} C(\theta_n), \quad \epsilon_n = \frac{\epsilon}{(n+1+s)^\zeta}, \quad 0.5 < \zeta \leq 1, \text{ and } \epsilon, s > 0.$$

Finite difference methods such as SPSA suffer from the bias-variance trade-off. The bias[5] in the gradient estimate is proportional to Δ^2. On the other hand, if $\hat{C}_n(\theta - \Delta e_i)$ and $\hat{C}_n(\theta + \Delta e_i)$ are sampled independently, then the variance is proportional to $1/\Delta^2$. To decrease the bias, one needs a small Δ, but this results in an increase in the variance.

For small Δ, by using common random numbers (see Appendix A), one would expect that $\hat{C}_n(\theta - \Delta e_i)$ and $\hat{C}_n(\theta + \Delta e_i)$ are highly correlated and significant variance reduction in $\hat{\nabla} C(\theta_n)$ can be obtained.

15.5 Score function gradient estimator

This section describes the score function gradient estimator. Unlike the SPSA algorithm, no finite difference approximation is used. We assume that the transition kernel P_θ of the Markov process x_k is known but computing the stationary distribution π_θ is intractable. The aim is to compute the gradient estimate $\hat{\nabla}_\theta C_n(\theta_n)$ in the stochastic gradient algorithm (15.4) given the realization of costs $\{c(x_k, \theta)\}$, $k \in \iota_n$ for any choice of θ.

[5] This follows from the Taylor series expansion (assuming sufficient differentiability)

$$\hat{C}_n(\theta_n + \Delta e_i) = \hat{C}_n(\theta_n) + \Delta \frac{d\hat{C}_n(\theta_n)}{d\theta} + \frac{1}{2}\Delta^2 \frac{d^2\hat{C}_n(\theta_n)}{d\theta^2} + O(\Delta^3)$$

$$\hat{C}_n(\theta_n - \Delta e_i) = \hat{C}_n(\theta_n) - \Delta \frac{d\hat{C}_n(\theta_n)}{d\theta} + \frac{1}{2}\Delta^2 \frac{d^2\hat{C}_n(\theta_n)}{d\theta^2} + O(\Delta^3).$$

Then evaluate $\hat{\nabla}_\theta C_n(\theta_n)$ as in (15.13) and take the expected value.

15.5.1 Score function gradient estimator for RVs

To highlight the main ideas, the score function gradient estimator for random variables is discussed first.

Assuming sufficient regularity[6] to swap the order of integration and gradient, we have

$$\nabla_\theta C(\theta) = \int c(X) \nabla_\theta \pi_\theta(X) dX = \int c(X) \frac{\nabla_\theta \pi_\theta(X)}{\pi_\theta(X)} \pi_\theta(X) dX.$$

The score function algorithm proceeds as follows: simulate $X_k \sim \pi_\theta$ and compute for any N

$$\hat{\nabla}_\theta C_N = \frac{1}{N} \sum_{k=1}^{N} c(X_k) \frac{\nabla_\theta \pi_\theta(X_k)}{\pi_\theta(X_k)}.$$

The term "score function" stems from the fact that $\frac{\nabla_\theta \pi_\theta(X)}{\pi_\theta(X)} = \nabla_\theta \log \pi_\theta$. For any N, this is an unbiased estimator of $\nabla_\theta C(\theta)$. The derivative of the log of density function is often called the "score function" in statistics.

Example: If $\pi_\theta(x) = \theta e^{-\theta x}$, then $\frac{\nabla_\theta \pi_\theta(X)}{\pi_\theta(X)} = \nabla_\theta \log \pi_\theta = \frac{1}{\theta} - X$.

15.5.2 Score function gradient estimator for Markov process

We now describe the score function simulated-based gradient estimator for a Markov process. The eventual goal is to solve finite state MDPs and POMDPs via stochastic gradient algorithms. So to avoid technicalities, we consider a finite state Markov chain x and assume that transition matrix P_θ is regular so that there exists a unique stationary distribution π_θ. Therefore, for a cost $c(x)$, the expected cost is

$$\mathbb{E}_{\pi_\theta}\{c(x)\} = \sum_{i=1}^{X} c(i)\, \pi_\theta(i) = c' \pi_\theta.$$

Suppose that for any choice of θ, one can observe via simulation a sample path of the costs $c(x_k)$, $k = 1, 2, \ldots$ where the Markov chain $\{x_k\}$ evolves with transition matrix P_θ. Given such simulated sample paths, the aim is to estimate the gradient

$$\nabla_\theta \mathbb{E}_{\pi_\theta}\{c(x)\} = \nabla_\theta \left(c' \pi_\theta \right).$$

This is what we mean by simulated-based gradient estimation. We assume that the transition matrix P_θ is known but the stationary distribution π_θ is not known explicitly.[7] Also the cost[8] $c(x)$ may not be known explicitly and the simulated sampled path $c(x_k)$ may be observed in zero mean noise.

[6] More precisely, suppose $\theta \in \Theta, x \in \mathcal{X}$ where Θ is an open set. Then from the dominated convergence theorem [287], $\nabla_\theta \int_{\mathcal{X}} f(x,\theta) dx = \int_{\mathcal{X}} \nabla_\theta f(x,\theta) dx$ if the following conditions hold: (i) f and $\nabla_\theta f$ are continuous on $\mathcal{X} \times \Theta$. (ii) $|f(x,\theta)| \leq g_0(x)$ and $\|\nabla_\theta f(x,\theta)\| \leq g_1(x)$ where g_0 and g_1 are integrable functions , that is, $\int_{\mathcal{X}} g_0(x) dx < \infty$ and $\int_{\mathcal{X}} g_1(x) dx < \infty$.

[7] Even though P_θ is known, it is intractable to compute the stationary distribution π_θ in closed form. So $\nabla_\theta \left(c' \pi_\theta \right)$ cannot be evaluated explicitly. Actually for the reinforcement learning algorithms for MDPs in Chapter 16, the gradient estimator does not need explicit knowledge of P_θ.

[8] We assume that $c(x)$ is not an explicit function of θ. In the MDP case considered in Chapter 16, the state will be chosen as $z_k = (x_k, u_k)$ where x_k is the Markov state and u_k is the action. Then the cost $c(z_k)$ is not an explicit function of θ, where θ will denote a parametrized policy that determines the transition matrix P_θ of the process $\{z_k\}$.

Clearly as $N \to \infty$, for any initial distribution π_0, $\lim_{N\to\infty} c' P_\theta'^N \pi_0 \to c'\pi_\theta$. So the finite sample approximation is (write P_θ as P to simplify notation)

$$\nabla_\theta \left(c'\pi_\theta\right) \approx \pi_0' \sum_{k=0}^{N-1} P^{N-k-1}\, \nabla_\theta P\, P^k\, c \tag{15.14}$$

$$= \sum_{x_{0:N}} c(x_N) \Big[(\nabla_\theta P_{x_0 x_1}) P_{x_1 x_2} \cdots P_{x_{N-1}x_N} + P_{x_0 x_1} (\nabla_\theta P_{x_1 x_2}) \cdots P_{x_{N-1}x_N}$$

$$+ \cdots + P_{x_0 x_1} P_{x_1 x_2} \cdots (\nabla_\theta P_{x_{N-1}x_N}) \Big] \pi_{x_0}$$

$$= \sum_{x_{0:N}} c(x_N) \left[\frac{\nabla_\theta P_{x_0 x_1}}{P_{x_0 x_1}} + \cdots + \frac{\nabla_\theta P_{x_{N-1}x_N}}{P_{x_{N-1}x_N}} \right] P_{x_0 x_1} P_{x_1 x_2} \cdots P_{x_{N-1}x_N} \pi_{x_0}$$

$$= \mathbb{E}_{x_{0:N}} \left\{ c(x_N) \sum_{k=1}^{N} \frac{\nabla_\theta P_{\theta\, x_{k-1}x_k}}{P_{\theta\, x_{k-1}x_k}} \right\}.$$

This leads to the score function gradient estimator in Algorithm 18.

Algorithm 18 Score function gradient estimation algorithm

Step 1: Simulate Markov chain $x_0, \ldots, x_N$ with transition matrix P_θ.

Step 2: Compute $S_N^\theta = \sum_{k=1}^N \frac{\nabla_\theta P_{\theta\, x_{k-1}x_k}}{P_{\theta\, x_{k-1}x_k}}$. This is evaluated recursively as

$$S_k^\theta = \frac{\nabla_\theta P_{\theta\, x_{k-1}x_k}}{P_{\theta\, x_{k-1}x_k}} + S_{k-1}^\theta, \quad k = 1, \ldots, N.$$

Step 3: Evaluate the score function gradient estimate via simulation as

$$\widehat{\nabla_\theta} C_N(\theta) = \frac{1}{N} \sum_{k=1}^{N} c(x_k) S_k^\theta.$$

(Note that this is the average over N samples.)

15.6 Weak derivative gradient estimator

This section describes the weak derivative gradient estimator. Like the score function estimator, for random variables it provides an unbiased estimator of the gradient.

A probability distribution F_θ is weakly differentiable at a point θ (on an open subset of $\mathbb{R}$) if there exists a signed measure[9] denoted as $\nu_\theta = \nabla_\theta F$ such that

[9] Let Ω be the set of outcomes and $\mathcal{A}$ denote a sigma-algebra defined on Ω. Then a signed measure ν is a real valued function on $\mathcal{A}$ which satisfies: (i) σ-additive – meaning that if $A_i, i = 1, 2, \ldots$ are disjoint sets (events) in $\mathcal{A}$, then $\nu(\cup_i A_i) = \sum_i \nu(A_i)$. (ii) $\nu(\emptyset) = 0$.

A signed measure is finite if $\nu(\Omega) < \infty$. Note a nonnegative signed measure is called a measure. Finally, ν is a probability measure if $\nu(\Omega) = 1$.

$$\lim_{s \to 0} \frac{1}{s} \left[\int c(x) dF_{\theta+s}(x) - \int c(x) dF_\theta(x) \right] = \int c(x) d\nu_\theta(x) \tag{15.15}$$

for any bounded continuous function $c(x)$. The term "weak derivative" is used since $\nabla_\theta F(x)$ may not be a function in the classical sense; it could be a generalized function (e.g. Dirac delta function). Hence the above definition involves integration with a test function $c(x)$.

A well known result in measure theory is that any finite signed measure ν can be decomposed as

$$\nu = g_1 \mu_1 - g_2 \mu_2 \tag{15.16}$$

where g_1 and g_2 are constants and μ_1 and μ_2 are probability measures. In our case, since the signed measure is obtained as the derivative of a probability measure, i.e. since $\int dF_\theta(x) = 1$, therefore $\int d\nu_\theta(x) = 0$, implying that $g_1 = g_2$ in (15.16). So the definition of weak derivative of a probability distribution can be re-expressed as:

DEFINITION 15.6.1 *A probability distribution F_θ is weakly differentiable at a point θ (on an open subset of $\mathbb{R}$) if there exist probability distributions $\dot{F}_\theta$ and $\ddot{F}_\theta$ and a constant g_θ such that for any bounded continuous function $c(x)$, the following holds:*

$$\nabla_\theta \int c(x) dF_\theta(x) = g_\theta \left[\int c(x) d\dot{F}_\theta(x) - \int c(x) d\ddot{F}_\theta(x) \right].$$

In more familiar engineering notation using probability density functions, the definition of the weak derivative is

$$\nabla_\theta \int c(x) p_\theta(x) = g_\theta \left[\int c(x) \dot{p}_\theta(x) dx - \int c(x) \ddot{p}_\theta(x) dx) \right] \tag{15.17}$$

where $\dot{p}_\theta$ and $\ddot{p}_\theta$ are probability density functions, or equivalently,

$$\nabla p_\theta = g_\theta (\dot{p}_\theta - \ddot{p}_\theta). \tag{15.18}$$

The weak derivative is specified by the triplet $(g_\theta, \dot{p}_\theta, \ddot{p}_\theta)$. A similar characterization holds in terms of probability mass functions.

In general the representation (15.18) of the weak derivative is not unique. One specific representation of interest is obtained via the so-called Hahn–Jordan decomposition. This is a deep result in measure theory – for our practical needs, the following simplistic version suffices: a signed measure can be decomposed as in (15.17) such that densities $\dot{p}_\theta$ and $\ddot{p}_\theta$ are orthogonal. This means that the set of x where density $\dot{p}_\theta(x)$ is nonzero coincides with the set of x where density $\ddot{p}_\theta(x)$ is zero, and vice versa.

15.6.1 Weak derivative of random variables

Based on (15.18), the weak derivative gradient estimation for a random variable is as follows: simulate N samples $\dot{X}_k \sim \dot{p}_\theta$ and $\ddot{X}_k \sim \ddot{p}_\theta$, $k = 1, 2, \ldots, N$. Then

$$\hat{\nabla}_\theta C_N = \frac{1}{N} \sum_{k=1}^{N} g_\theta \left[c(\dot{X}_k) - c(\ddot{X}_k) \right].$$

Examples: Here are examples of the weak derivative of random variables:
1. *Exponential*: $\pi_\theta(x) = \theta e^{-\theta x}$. Then the Hahn–Jordan decomposition is

$$\nabla_\theta \pi_\theta(x) = e^{-\theta x}(1 - x\theta) I(x > \frac{1}{\theta}) - e^{-\theta x}(x\theta - 1) I(x \leq \frac{1}{\theta})$$

So $g_\theta = \theta e$, $\dot{p}_\theta(x) = \theta e^{-\theta x}(1 - x\theta) I(x > \frac{1}{\theta})$, $\ddot{p}_\theta(x) = \theta e^{-\theta x}(x\theta - 1) I(x \leq \frac{1}{\theta})$.
2. *Poisson*: For a Poisson random variable, the probability mass function is

$$p_\theta(x) = \frac{e^\theta \theta^x}{x!}, \quad x = 0, 1, \ldots .$$

So clearly, $\nabla_\theta p_\theta(x) = p_\theta(x-1) - p_\theta(x)$. So one possible weak derivative implementation is

$$g_\theta = 1, \; \dot{p}_\theta = p_\theta(x-1), \; \ddot{p}_\theta = p_\theta(x).$$

15.6.2 Weak derivative gradient estimator for Markov process

Let π_θ denote the stationary distribution of a Markov chain with regular transition matrix P_θ. Suppose that for any choice of θ, one can observe via simulation sample paths $\{c(x_k)\}$ of the costs where Markov chain x_k evolves with transition matrix P_θ. The aim is to estimate via simulation the gradient $\nabla_\theta \mathbb{E}_{\pi_\theta}\{c(x)\} = \nabla_\theta \left(\pi_\theta' c\right)$. It is assumed that P_θ is known but π_θ is not known explicitly. Also $c(x)$ may not be known explicitly and the simulated sampled path $c(x_k)$ may be observed in zero mean noise.

Analogous to random variable case (15.17), define the weak derivative of the transition matrix P_θ as follows: a transition probability matrix P_θ is weakly differentiable at a point θ (on an open subset of $\mathbb{R}$) if there exists transition probability matrices $\dot{P}_\theta$ and $\ddot{P}_\theta$ and a diagonal matrix g_θ such that:

$$\begin{aligned} &\nabla_\theta P_\theta = g_\theta(\dot{P}_\theta - \ddot{P}_\theta), \\ &\text{equivalently, } \nabla_\theta P_{\theta\, ij} = g_{\theta,i}(\dot{P}_{\theta\, ij} - \ddot{P}_{\theta\, ij}), \quad i,j \in \{1,2,\ldots,X\}. \end{aligned} \tag{15.19}$$

So the weak derivative of transition matrix P_θ is specified by the triplet $(\dot{P}_\theta, \ddot{P}_\theta, g_\theta)$. Obviously, $\nabla_\theta P_\theta \mathbf{1} = \nabla_\theta \mathbf{1} = 0$ implying that each row of $\nabla_\theta P_\theta$ adds to zero.

THEOREM 15.6.2 *For regular transition matrix P_θ with stationary distribution π_θ and cost vector $c \in \mathbb{R}^X$,*

$$\nabla_\theta \left(\pi_\theta' c\right) = \pi_\theta'(\nabla_\theta P_\theta) \sum_{k=0}^{\infty} {P_\theta}^k c. \tag{15.20}$$

Proof

$$\begin{aligned} \nabla_\theta \left(\pi_\theta'(P_\theta)^N c\right) &= (\nabla_\theta \pi_\theta')(P_\theta)^N c + \sum_{k=0}^{N-1} \pi_\theta'(P_\theta)^k (\nabla_\theta P_\theta)(P_\theta)^{N-k-1} c \\ &= (\nabla_\theta \pi_\theta')(P_\theta)^N c + \pi_\theta'(\nabla_\theta P_\theta) \sum_{k=0}^{N-1} (P_\theta)^k c. \end{aligned} \tag{15.21}$$

Next assume that the transition matrix satisfies $\rho(P_\theta) < 1$ where ρ denotes the Dobrushin coefficient defined in (2.45). (Recall that for an irreducible aperiodic matrix P, $\rho(P^m) < 1$ for some positive integer m. For notational simplicity, assume $m = 1$.) We now show that $\rho(P_\theta) < 1$ implies that the first term in (15.21) goes to zero as $N \to \infty$. From the definition of weak derivative in (15.15),

$$s(\nabla_\theta \pi'_\theta)(P_\theta)^N = \left(\pi_{\theta+s}(P_\theta)^N - \pi_\theta(P_\theta)^N\right) + o(s)$$

But since $\rho(P_\theta) < 1$, it follows from Theorem 2.7.2 that as $N \to \infty$

$$\| \left(\pi_{\theta+s}(P_\theta)^N - \pi_\theta(P_\theta)^N\right) \| \leq (\rho(P_\theta))^N \|\pi_{\theta+s} - \pi_\theta\|_{\text{TV}} = o(s)$$

Therefore, $\lim_{N\to\infty}(\nabla_\theta \pi'_\theta)(P_\theta)^N c = 0$.

□

The above theorem leads to the following finite sample estimate:

$$\nabla_\theta \left(\pi'_\theta c\right) \approx \pi'_0 (P_\theta)^m (\nabla_\theta P_\theta) \sum_{k=0}^{N} (P_\theta)^k c \tag{15.22}$$

for any initial distribution π_0. For sufficiently large m and N (15.22) will approach (15.20). The sample path interpretation of (15.22) leads to the weak derivative estimation Algorithm 19. Step 1 simulates $\pi'_0(P_\theta)^m$, Step 2 implements the weak derivative $(\nabla_\theta P_\theta)$ and propagates the two chains $\dot{x}$ and $\ddot{x}$ for N steps. Finally Step 3 estimates the right-hand side of (15.22). So Step 3 yields that

$$\widehat{\nabla}_\theta C_N(\theta) = g_{\theta,x_m} \sum_{k=m}^{N} c(\dot{x}_k) - c(\ddot{x}_k).$$

Note if the Markov chains $\dot{x}$ and $\ddot{x}$ are simulated with common random numbers (see Appendix A.2.4 on page 437 for its use of common random numbers for variance reduction), then at some time point τ, $\dot{x}_\tau = \ddot{x}_\tau$, and the processes $\dot{x}$, $\ddot{x}$ merge and evolve identically after time τ. This yields Step 3 of Algorithm 19.

Example: Consider the transition matrix

$$P_\theta = \frac{1}{8}\begin{bmatrix} 2+3\theta & 5-5\theta & 1+2\theta \\ 2+5\theta & 5+\theta & 1-6\theta \\ 2-\theta & 3-3\theta & 3+4\theta \end{bmatrix}$$

Then the weak derivative triplet is:

$$\dot{P}_\theta = \begin{bmatrix} 3/5 & 0 & 2/5 \\ 5/6 & 1/6 & 0 \\ 0 & 0 & 1 \end{bmatrix}, \quad \ddot{P}_\theta = \begin{bmatrix} 0 & 1 & 0 \\ 0 & 0 & 1 \\ 1/4 & 3/4 & 0 \end{bmatrix}, \quad g_\theta = \text{diag}\begin{bmatrix} 2/3 \\ 5/6 \\ 1/2 \end{bmatrix}.$$

15.7 Bias and variance of gradient estimators

Here we characterize the statistical properties of the score function and weak derivative gradient estimators discussed above.

Algorithm 19 Weak derivative-based gradient estimation algorithm

---

Evaluate triplet $(\dot{P}_\theta, \ddot{P}_\theta, g_\theta)$ using formula $\nabla_\theta P_{\theta ij} = g_{\theta,i}(\dot{P}_{\theta ij} - \ddot{P}_{\theta ij})$.

Step 1: Simulate the Markov chain $x_0, \dots x_{m-1}$ with transition matrix P_θ.

Step 2a: Starting with state x_{m-1}, simulate at time m the states $\dot{x}_m$ and $\ddot{x}_m$ with transition matrix $\dot{P}_\theta$ and $\ddot{P}_\theta$, respectively.

Step 2b: Starting with states $\dot{x}_m$ and $\ddot{x}_m$, respectively, simulate the two Markov chains $\dot{x}_{m+1}, \dot{x}_{m+2}, \dots, \dot{x}_N$ and $\ddot{x}_{m+1}, \ddot{x}_{m+2}, \dots, \ddot{x}_N$ with transition matrix P_θ.
Use the same random numbers to simulate these Markov chains (for example, according to the algorithm in Appendix A.1.4 on page 432).

Step 3: Evaluate the weak derivative estimate via simulation as

$$\widehat{\nabla}_\theta C_{m,N}(\theta) = g_{\theta, x_m} \sum_{k=m}^{m+\tau} c(\dot{x}_k) - c(\ddot{x}_k), \text{ where } \tau = \min\{k : \dot{x}_k = \ddot{x}_k, k \le N\}. \tag{15.23}$$

THEOREM 15.7.1 *For a Markov chain with initial distribution π_0, regular transition matrix P_θ with coefficient of ergodicity ρ (see §2.7) and stationary distribution π_θ:*

1. *The score function gradient estimator of Algorithm 18 has:*
   (a) *Bias:* $\mathbb{E}\{\widehat{\nabla_\theta C_N}\} - \nabla_\theta c' \pi_\theta = O(1/N)$.
   (b) *Variance:* $\text{Var}\{\widehat{\nabla_\theta C_N}\} = O(N)$.
2. *The weak derivative gradient estimator of Algorithm 19 has:*
   (a) *Bias:* $\mathbb{E}\{\widehat{\nabla_\theta C_{m,N}}\} - \nabla_\theta c' \pi_\theta = O(\rho^m)\, \|\pi_0 - \pi_\theta\|_{TV} + O(\rho^N)$.
   (b) *Variance:* $\text{Var}\{\widehat{\nabla_\theta C_N}\} = O(1)$.

The proof is in the appendix. The result shows that despite the apparent simplicity of the score function gradient estimator (and its widespread use), the weak derivative estimator performs better in both bias and variance. The variance of the score function estimator actually grows with sample size! We will show in numerical examples for reinforcement learning of MDPs in §16.5, that the weak derivative estimator has a substantially smaller variance than the score function gradient estimator.

Why is the variance of the score function gradient estimator $O(N)$ while the variance of the weak derivative estimator is $O(1)$? The weak derivative estimator uses the difference of two sample paths, $\{\dot{x}_m\}$ and $\{\ddot{x}_m\}$, $m = 1, 2 \dots, N$. Its variance is dominated by a term of the form $\sum_{m=1}^N g'(P_\theta{}^m)'(\pi_0 - \bar{\pi}_0)$ where π_0 and $\bar{\pi}_0$ are the distributions of $\dot{x}_0$ and $\ddot{x}_0$. From §2.7.2, this sum is bounded by constant $\times \sum_{m=1}^N \rho^m$ which is $O(1)$ since $\rho < 1$ for a regular transition matrix P_θ. In comparison, the score function estimator uses a single sample path and its variance is dominated by a term of the form $\sum_{m=1}^N g'(P_\theta{}^m)'\pi_0$. This sum grows as $O(N)$. The proof in the appendix formalizes this argument.

To gain additional insight consider the following simplistic examples.

1. I.I.D. process. Suppose $P_\theta = 1\pi_\theta'$. If we naively use the score function estimator for a Markov chain, the variance is $O(N)$. In comparison, the weak derivative estimator

is identical to that of a random variable since P_θ has identical rows. In particular, Step 2b of Algorithm 19 simulates exactly the same value of $\dot{x}$ and $\ddot{x}$ after one time step, and so the estimator is $g_{\theta,x_m} c(\dot{x}_m) - c(\ddot{x}_m)$. So the variance of the weak derivative estimator is substantially smaller.

2. Constant cost $c(x) = 1$. In this trivial case, $\nabla_\theta \mathbb{E}_{\pi_\theta}\{c(x)\} = 0$. If we naively use the score function estimator for a Markov chain, the variance is $O(N)$. In comparison the weak derivative estimator yields 0 implying that the variance is zero. So for nearly constant costs, the weak derivative estimator is substantially better than the score function derivative estimator.

15.8 Complements and sources

This chapter presented an introductory description of simulation-based gradient estimation and is not meant to be a comprehensive account. We have discussed only *scalar* step size stochastic approximation algorithms. The reason is that *matrix* step size algorithms incur quadratic computational cost (in the parameter dimension) at each time compared to linear cost for scalar step size algorithms. Also in terms of gradient estimation, the estimated Hessian typically has larger variance than the estimated first-order derivatives. We should point out, however, that in cases where the Hessian of the cost can be computed explicitly, then matrix step size algorithms can be used. For example, in adaptive filtering settings where the scalar step size LMS algorithm is used, the matrix step size recursive least squares (RLS) algorithm is also widely used.

Our use of Azuma–Hoeffding's inequality for a finite state Markov chain in Theorem 15.2.1 is elementary; we simply express the Markov chain in terms of a martingale difference process. In §17.5, we will use similar ideas to characterize the mean field dynamics of population processes. [65] contains a comprehensive treatment of much more sophisticated concentration of measure results.

The books [139, 290, 300] present comprehensive accounts of adaptive filtering. Adaptive filtering constitutes an important class of stochastic approximation algorithms – [48, 200] are *the* books in the analysis of stochastic approximation algorithms. The SPSA algorithm was pioneered by Spall; see [309] and the website `www.jhuapl.edu/spsa/` for repository of code and references. [261] is an excellent book for coverage of simulation-based gradient estimation; indeed, §15.5 and §15.6 are based on [261].

In the discussion of gradient estimators we have omitted the important topic of infinitesimal perturbation analysis (IPA) and process derivatives. This method stems from the inverse transform method of simulation described in Appendix A which says that a random variable with distribution function F_θ can be generated as $X \sim F_\theta^{-1}(U)$ where U is a uniform random number in $[0, 1]$. So $\frac{dX}{d\theta} = \frac{d}{d\theta} F_\theta^{-1}(U)$. For example, if $F_\theta(x) = 1 - e^{-\theta x}$ is the exponential distribution, then $X = -\frac{1}{\theta}\log(1 - U)$ is an exponentially distributed random number. Then the process derivative is $\frac{dX}{d\theta} = \frac{1}{\theta^2}\log(1 - U)$. Developing IPA algorithms for controlled Markov chains can be tricky and needs careful construction; please refer to [82] for a highly accessible textbook treatment.

Appendix 15.A Proof of Theorem 15.2.1

To prove (15.9), recall $\pi_k = \mathbb{E}\{x_k\} = P'^k \pi_0$ and $\pi_\infty = P'^k \pi_\infty$. Then

$$\begin{aligned}
\left|\mathbb{E}\left\{\frac{1}{n}\sum_{k=0}^{n-1} h' x_k\right\} - h'\pi_\infty\right| &= \frac{1}{n}\left|\sum_{k=0}^{n-1} h'(\pi_k - \pi_\infty)\right| \\
&\leq \frac{1}{n}\sum_{k=0}^{n-1} \max_{i,j}|h_i - h_j|\,\|\pi_k - \pi_\infty\|_{\text{TV}} \\
&\quad \text{(Lemma 2.7.3 on page 29)} \\
&= \frac{1}{n}\max_{i,j}|h_i - h_j| \sum_{k=0}^{n-1} \|P'^k\pi_0 - P'^k\pi_\infty\|_{\text{TV}} \\
&\leq \frac{1}{n}\max_{i,j}|h_i - h_j|\,\|\pi_0 - \pi_\infty\|_{\text{TV}} \sum_{k=0}^{n-1}\rho^k \\
&\quad \text{(Theorem 2.7.2, property 6)} \\
&\leq \frac{1}{n}\max_{i,j}|h_i - h_j|\,\|\pi_0 - \pi_\infty\|_{\text{TV}} \frac{1}{1-\rho}
\end{aligned}$$

where the last inequality follows because $\rho < 1$ since P is regular.

The second assertion (15.10) is proved as follows.

$$\begin{aligned}
\text{Var}\{\phi_n\} &= \frac{1}{n^2}\mathbb{E}\{(\sum_{k=0}^{n-1} h'(x_k - \mathbb{E}\{x_k\}))^2\} \leq \frac{1}{n^2}\mathbb{E}\{(\sum_{k=0}^{n-1} h'(x_k - \pi_\infty))^2\} \\
&= \frac{1}{n^2}\sum_{l=0}^{n-1}\sum_{m=0}^{n-1}\mathbb{E}\{h'(x_l - \pi_\infty)h'(x_m - \pi_\infty)\} \\
&\overset{(a)}{=} \frac{2}{n^2}\sum_{l=0}^{n-1}\sum_{m\geq l}\mathbb{E}\big\{h'(x_l - \pi_\infty)\mathbb{E}\{h'(x_m - \pi_\infty)|x_l\}\big\} \\
&= \frac{2}{n^2}\sum_{l}\sum_{m\geq l}\mathbb{E}\left\{h'(x_l - \pi_\infty)\,h'P'^{m-l}(x_l - \pi_\infty)\right\} \\
&\overset{(b)}{\leq} \frac{2}{n^2}\max_{i,j}|h_i - h_j|^2 \sum_{l}\sum_{m\geq l}\rho^{m-l}\,\mathbb{E}\{\|x_l - \pi_\infty\|_{\text{TV}}^2\} \\
&\leq \frac{2}{n^2}\max_{i,j}|h_i - h_j|^2 \frac{1}{1-\rho}\sum_{l=0}^{n-1}\mathbb{E}\{\|x_l - \pi_\infty\|_{\text{TV}}^2\} \\
&= \frac{2}{n}\max_{i,j}|h_i - h_j|^2 \frac{1}{1-\rho}\sum_{i=1}^{X}\|e_i - \pi_\infty\|_{\text{TV}}^2\pi_\infty(i) + O(\frac{1}{n^2})
\end{aligned}$$

(a) follows from the smoothing property of conditional expectations (3.11); (b) follows from Lemma 2.7.3 on page 29 and property 6 of Theorem 2.7.2 on page 28.

To prove (15.11), we start with stating Azuma–Hoeffding's inequality.

THEOREM 15.A.1 (Azuma–Hoeffding inequality) *Suppose* $S_n = \sum_{k=1}^n M_k + S_0$ *where* $\{M_k\}$ *is a martingale difference process*[10] *with bounded differences satisfying* $|M_k| \leq \Delta_k$ *almost surely where* Δ_k *are finite constants. Then for any* $\epsilon > 0$

$$\mathbb{P}(S_n - S_0 \geq \epsilon) \leq \left(-\frac{\epsilon^2}{\sum_{k=1}^n \Delta_k^2}\right) \text{ and } \mathbb{P}(S_n - S_0 \leq -\epsilon) \leq \left(-\frac{\epsilon^2}{\sum_{k=1}^n \Delta_k^2}\right). \quad (15.24)$$

Therefore

$$\mathbb{P}(|S_n - S_0| \geq \epsilon) \leq 2\exp\left(-\frac{\epsilon^2}{\sum_{k=1}^n \Delta_k^2}\right). \quad (15.25)$$

We will apply the Azuma–Hoeffding inequality to $L_n = \sum_{k=0}^{n-1} h' x_k$. The idea is to express L_n as the sum of a martingale difference process by defining

$$M_k = \mathbb{E}\{L_n|\mathcal{F}_k\} - \mathbb{E}\{L_n|\mathcal{F}_{k-1}\}, \quad \text{where } \mathcal{F}_k = \{x_0, \ldots, x_{k-1}\}.$$

Clearly, $\mathbb{E}\{M_k|\mathcal{F}_{k-1}\} = 0$ from the smoothing property of conditional expectations and so $\{M_k\}$ is a martingale difference process. Also

$$\begin{aligned}\sum_{k=1}^n M_k &= \mathbb{E}\{L_n|\mathcal{F}_n\} - \mathbb{E}\{L_n|\mathcal{F}_{n-1}\} + \cdots + \mathbb{E}\{L_n|\mathcal{F}_1\} - \mathbb{E}\{L_{n-1}|\mathcal{F}_0\} \\ &= L_n - \mathbb{E}\{L_n\} \end{aligned} \quad (15.26)$$

since $\mathbb{E}\{L_n|\mathcal{F}_0\} = \mathbb{E}\{L_n\}$ (unconditional expectation) as $\mathcal{F}_0$ contains no information ($\mathcal{F}_0$ is the trivial sigma algebra).

To use the Azuma–Hoeffding inequality, we first need to bound $|M_k|$.

$$\begin{aligned} M_k &= \mathbb{E}\{L_n|\mathcal{F}_k\} - \sum_{j=1}^X \mathbb{E}\{L_n|j, \mathcal{F}_{k-1}\} P_{x_{k-1}j} \\ &= \sum_{j=1}^X P_{x_{k-1}j} \left[\mathbb{E}\{L_n|x_k, \mathcal{F}_{k-1}\} - \mathbb{E}\{L_n|j, \mathcal{F}_{k-1}\}\right]. \end{aligned}$$

Therefore,

$$\begin{aligned} |M_k| &\leq \max_{i,j} |\mathbb{E}\{L_n|i, \mathcal{F}_{k-1}\} - \mathbb{E}\{L_n|j, \mathcal{F}_{k-1}\}| \\ &= \max_{i,j} |h'(I + P + \cdots + P^{n-k})'(e_i - e_j)| \\ &\overset{(a)}{\leq} \max_{i,j} \max_{l,m} |h_l - h_m|(1 + \rho + \cdots + \rho^{n-k})\|e_i - e_j\|_{\text{TV}} \leq \frac{1}{1-\rho} \max_{l,m} |h_l - h_m| \end{aligned}$$

$$|M_k|^2 \leq \frac{1}{(1-\rho)^2} \max_{l,m} |h_l - h_m|^2 \quad (15.27)$$

[10] Let $\{z_k\}$ be a sequence of random variables. Suppose $\mathcal{F}_k = \{z_1, \ldots, z_k\}$ and $\{M_k\}$ is a $\mathcal{F}_k$ measurable process, meaning that $M_k = \phi(z_1, \ldots, z_k)$ for some (Borel) function ϕ. Then $\{M_k\}$ is called a martingale difference process if $\mathbb{E}\{M_k|\mathcal{F}_{k-1}\} = 0$ for $k = 1, 2, \ldots$. Note that

1. If $\{M_k\}$ is a martingale difference process then $S_n = \sum_{k=1}^n M_k + S_0$ is a martingale since
$$\mathbb{E}\{S_n|\mathcal{F}_{n-1}\} = \mathbb{E}\{M_n|\mathcal{F}_{n-1}\} + \mathbb{E}\{S_{n-1}|\mathcal{F}_{n-1}\} = S_{n-1}.$$
2. Conversely, if S_k, $k = 0, 1, \ldots$ is a martingale, then clearly $M_k = S_k - S_{k-1}$, $k = 1, 2, \ldots$ is a martingale difference process.

where (a) follows from Lemma 2.7.3 on page 29 and property 6 of Theorem 2.7.2 on page 28. Denoting the right-hand side of (15.27) as Δ^2, and applying Azuma–Hoeffding's inequality (15.25) to $L_n - \mathbb{E}\{L_n\}$ in (15.26) yields (15.11).

Appendix 15.B Proof of Theorem 15.7.1

We start with the following useful lemma.

LEMMA 15.B.1 *For $c \in \mathbb{R}^X$, regular transition matrix P_θ with Dobrushin coefficient ρ and stationary distribution π_θ,*

$$\big|\pi_\theta' \nabla_\theta P_\theta \sum_{k=n}^{\infty} {P_\theta}^k c\big| = O(\rho^n) \tag{15.28}$$

Also, for any two X-dimensional pmfs α and β, and $c \in \mathbb{R}^X$,

$$\big|(\alpha - \beta)' {P_\theta}^k c\big| \le C\rho^k \text{ for some constant } C \ge 0. \tag{15.29}$$

Proof From (15.19), we have

$$\begin{aligned}
|\pi_\theta' \nabla_\theta P_\theta \sum_{k=n}^{\infty} {P_\theta}^k c| &= |\pi_\theta' g(\dot{P}_\theta - \ddot{P}_\theta) \sum_{k=n}^{\infty} {P_\theta}^k c| \le \max_i |g_i|\, |\pi_\theta'(\dot{P}_\theta - \ddot{P}_\theta) \sum_{k=n}^{\infty} {P_\theta}^k c| \\
&\overset{(a)}{\le} \max_i |g_i|\, \|\dot{P}_\theta{}' \pi_\theta - \ddot{P}_\theta{}' \pi_\theta\|_{\text{TV}} \max_{ij} |(e_i - e_j)' \sum_{k=n}^{\infty} {P_\theta}^k c| \\
&\overset{(b)}{\le} C \sum_{k=n}^{\infty} \rho^k = O(\rho^n).
\end{aligned}$$

where (a) follows from Lemma 2.7.3 on page 29 and property 6 of Theorem 2.7.2 on page 28, and (b) follows from (15.29) where C is a non-negative constant. □

1(a) *Bias of score function estimator*. From (15.14), the bias of $c(x_n)S_n^\theta$ is

$$\begin{aligned}
&|\pi_0' \sum_{k=0}^{n-1} {P_\theta}^{n-k-1} (\nabla_\theta P_\theta) {P_\theta}^k c - \nabla_\theta \pi_\theta' c| \\
&= |\pi_0' \sum_{k=0}^{n-1} {P_\theta}^{n-k-1} \nabla_\theta P_\theta P^k c - \pi_\theta'(\nabla_\theta P_\theta) \sum_{k=0}^{\infty} {P_\theta}^k c| \\
&\le \sum_{k=0}^{n-1} |(\pi_0' {P_\theta}^{n-k-1} - \pi_\theta')(\nabla_\theta P_\theta) {P_\theta}^k c| + |\pi_\theta'(\nabla_\theta P_\theta) \sum_{k=n}^{\infty} {P_\theta}^k c| \\
&\le \|\pi_0 - \pi_\theta\|_{\text{TV}} O(n\rho^n) + O(\rho^n)
\end{aligned}$$

where the last inequality follows from Lemma 15.B.1. So the bias of the averaged score $\frac{1}{n}\sum_{k=1}^{n} c(x_k)S_k^\theta$ in Step 3 of Algorithm 18 is

$$\|\pi_0 - \pi_\theta\|_{\text{TV}}\, O(\frac{1}{n}\sum_{k=1}^{n} k\rho^k) + O(\frac{1}{n}\sum_{k=1}^{n}\rho^k) = O(\frac{1}{n})$$

1(b) *Variance of score function estimator.* The variance is

$$\text{Var}\left(\widehat{\nabla}_\theta C_N(\theta)\right) = \mathbb{E}\{\left(\widehat{\nabla}_\theta C_N(\theta)\right)^2\} - \mathbb{E}^2\{\widehat{\nabla}_\theta C_N(\theta)\}. \tag{15.30}$$

Recall the bias is $O(1/N)$ implying that $\mathbb{E}^2\{\widehat{\nabla}_\theta C_N(\theta)\} = O(1)$. We now show that $\mathbb{E}\{\left(\widehat{\nabla}_\theta C_N(\theta)\right)^2\} = O(N)$ thereby implying that the variance grows as $O(N)$.

From Step 3 of of Algorithm 18, $\widehat{\nabla_\theta} C_N(\theta) = \frac{1}{N}\sum_{k=1}^{N} c(x_k)S_k^\theta$. So

$$\mathbb{E}\{\left(\widehat{\nabla}_\theta C_N(\theta)\right)^2\} = \mathbb{E}\{\left(\frac{1}{N}\sum_{k=1}^{N} c(x_k)S_k^\theta\right)^2\} \in [\underline{a}, \bar{a}]$$

where $\underline{a} = \mathbb{E}\{\left(\frac{\min_i c(i)}{N}\sum_{k=1}^{N} S_k^\theta\right)^2\}$ and $\bar{a} = \mathbb{E}\{\left(\frac{\max_i c(i)}{N}\sum_{k=1}^{N} S_k^\theta\right)^2\}$. So it suffices to bound

$$\mathbb{E}\{\left(\frac{1}{N}\sum_{k=1}^{N} S_k^\theta\right)^2\} = \mathbb{E}\{\left(\frac{1}{N}\sum_{k=1}^{N}(N-k+1)\frac{\nabla_\theta P_{\theta x_{k-1}x_k}}{P_{\theta x_{k-1}x_k}}\right)^2\}.$$

The key step is that $S_n^\theta = \sum_{k=1}^{n} \frac{\nabla_\theta P_{\theta x_{k-1}x_k}}{P_{\theta x_{k-1}x_k}}$ in Algorithm 18 is a zero mean martingale.[11] Therefore, for initial distribution π_0, stationary distribution π_θ and defining $\Phi_\theta \in \mathbb{R}^x$ with elements $\Phi_\theta(j) = \sum_{l=1}^{X}\left(\frac{\nabla_\theta P_{\theta jl}}{P_{\theta jl}}\right)^2 P_{\theta jl}$, it follows that

$$\begin{aligned}
\mathbb{E}\{\left(\frac{1}{N}\sum_{k=1}^{N} S_k^\theta\right)^2\} &= \frac{1}{N^2}\sum_{k=1}^{N}(N-k+1)^2\,\mathbb{E}\{\left(\frac{\nabla_\theta P_{\theta x_{k-1}x_k}}{P_{\theta x_{k-1}x_k}}\right)^2\} \\
&= \frac{1}{N^2}\sum_{k=1}^{N}(N-k+1)^2\,\pi_0' {P_\theta}^k \Phi_\theta \\
&= \frac{1}{N^2}\left[\sum_{k=1}^{N}(N-k+1)^2\,(\pi_0-\pi_\theta)'{P_\theta}^k\Phi_\theta + \pi_\theta'\Phi_\theta\sum_{k=1}^{N}(N-k+1)^2\right] \\
&\leq \frac{1}{N^2}\left[N^2\sum_{k=1}^{N}(\pi_0-\pi_\theta)'{P_\theta}^k\Phi_\theta + \pi_\theta'\Phi_\theta\sum_{k=1}^{N}(N-k+1)^2\right] \\
&\overset{a}{=} \sum_{k=1}^{N} C\rho^k + \frac{1}{N^2}\pi_\theta'\Phi_\theta\sum_{k=1}^{N}(N-k+1)^2 = O(1) + O(N).
\end{aligned}$$

(a) follows from (15.29) where C is a non-negative constant. So the variance of the score function estimator grows as $O(N)$.

[11] The reader should verify this and the fact that if $S_n = \sum_{k=1}^{n} M_k$ is a zero mean martingale then $\mathbb{E}\{S_n^2\} = \text{Var}\{S_n\} = \sum_{k=1}^{n}\mathbb{E}\{M_k^2\}$.

2(a) *Bias of weak derivative estimator.* For the weak derivative estimator, from (15.20) and (15.22), the bias is

$$
\begin{aligned}
&|\pi_0'(P_\theta)^m(\nabla_\theta P_\theta)\sum_{k=0}^{n}(P_\theta)^k c - \pi_\theta'(\nabla_\theta P_\theta)\sum_{k=0}^{\infty}(P_\theta)^k c| \\
&\leq |(\pi_0'(P_\theta)^m - \pi_\theta')(\nabla_\theta P_\theta)\sum_{k=0}^{n}(P_\theta)^k c| + |\pi_\theta'(\nabla_\theta P_\theta)\sum_{k=n+1}^{\infty}(P_\theta)^k c| \\
&\overset{(a)}{\leq} |(\pi_0'(P_\theta)^m - \pi_\theta')(\nabla_\theta P_\theta)\sum_{k=0}^{n}\left[(P_\theta)^k - \mathbf{1}\pi_\theta'\right] c| + |\pi_\theta'(\nabla_\theta P_\theta)\sum_{k=n+1}^{\infty}(P_\theta)^k c| \\
&\overset{(b)}{\leq} O(\rho^m)\|\pi_0 - \pi_\theta\|_{\text{TV}} + O(\rho^n).
\end{aligned}
$$

(a) follows because $\nabla_\theta P_\theta \mathbf{1} = \nabla_\theta(P_\theta \mathbf{1}) = \nabla_\theta \mathbf{1} = 0$. (b) follows since for regular transition matrix P_θ, $Z_n \overset{\text{defn}}{=} \sum_{k=0}^{n}\left[(P_\theta)^k - \mathbf{1}\pi_\theta'\right]$ is well defined (bounded) and converges to the fundamental matrix $Z = (I - P_\theta + \mathbf{1}\pi_\theta')^{-1}$.

2(b) *Variance of weak derivative estimator.* Since the bias is bounded, it follows that $\mathbb{E}^2\{\widehat{\nabla}_\theta C_N(\theta)\}$ is $O(1)$. So in (15.30) we only need to show that

$$
\mathbb{E}\{\big(\widehat{\nabla}_\theta C_N(\theta)\big)^2\} = \mathbb{E}\{\big(\sum_{k=0}^{n} c(\dot{x}_k) - c(\ddot{x}_k)\big)^2\} = O(1).
$$

Expanding we get

$$
\sum_{k=0}^{n}\mathbb{E}\{\big(c(\dot{x}_k) - c(\ddot{x}_k)\big)^2\} + 2\sum_{k}\sum_{j=0}^{k-1}\mathbb{E}\{\big(c(\dot{x}_k) - c(\ddot{x}_k)\big)\big(c(\dot{x}_j) - c(\ddot{x}_j)\big)\} \tag{15.31}
$$

The first term in (15.31) is bounded as (C_1, C_2 denote non-negative constants):

$$
\begin{aligned}
\sum_{k}\mathbb{E}\{\big(c(\dot{x}_k) - c(\ddot{x}_k)\big)^2\} &\leq \max_{i,j}|c(i) - c(j)|\sum_{k}\mathbb{E}\{\big(c(\dot{x}_k) - c(\ddot{x}_k)\big)\} \\
&\leq C_1\sum_{k}(\pi_0 - \bar{\pi}_0)'P^k c \leq C_2\sum_{k}\rho^k = O(1).
\end{aligned}
$$

where π_0 and $\bar{\pi}_0$ are the distributions of $\dot{x}_0$ and $\ddot{x}_0$. The second term in (15.31) is

$$
\begin{aligned}
&2\sum_{k=0}^{n}\sum_{j=0}^{k-1}\mathbb{E}\{\big(c(\dot{x}_k) - c(\ddot{x}_k)\big)\big(c(\dot{x}_j) - c(\ddot{x}_j)\big)\} \\
&= 2\sum_{k=0}^{n}\sum_{j=0}^{k-1}\mathbb{E}\big\{\,\mathbb{E}\{\big(c(\dot{x}_k) - c(\ddot{x}_k)\big)|\dot{x}_j\ddot{x}_j\}\big(c(\dot{x}_j) - c(\ddot{x}_j)\big)\,\big\} \\
&\leq 2\sum_{k=0}^{n}\sum_{j=0}^{k-1}\mathbb{E}\{c'(P^{k-j})'(e_{\dot{x}_j} - e_{\ddot{x}_j})\,\big(c(\dot{x}_j) - c(\ddot{x}_j)\big)\} \\
&\leq C_1\sum_{k=0}^{n}\sum_{j=0}^{k-1}\rho^{k-j}\|e_{\dot{x}_j} - e_{\ddot{x}_j}\|_{\text{TV}}\,\big(c(\dot{x}_j) - c(\ddot{x}_j)\big)\}
\end{aligned}
$$

$$
\begin{aligned}
&\le C_2 \sum_{k=0}^{n} \sum_{j=0}^{k-1} \rho^{k-j} \left(c(\dot{x}_j) - c(\ddot{x}_j)\right)\} \\
&= C_2 \sum_{k=0}^{n} \sum_{j=0}^{k-1} \rho^{k-j} c'(P^j)'(\pi - \bar{\pi}) \le C_3 \sum_{k=0}^{n} \sum_{j=0}^{k-1} \rho^{k-j} \rho^j = O(1).
\end{aligned}
$$

for constants C_1, C_2, C_3, where π and $\bar{\pi}$ are the distributions of $\dot{x}_j$ and $\ddot{x}_j$.

16 Reinforcement learning

Contents

Parts II and III of the book discussed dynamic programming algorithms for solving MDPs and POMDPs. Dynamic programming assumes that the MDP or POMDP model is completely specified. This chapter presents simulation-based stochastic approximation algorithms for estimating the optimal policy of MDPs when the transition probabilities are not known. *Simulation-based* means that although the transition probabilities are unknown, the decision-maker can observe the system trajectory under any choice of control actions. The simulation-based algorithms given in this chapter also apply as suboptimal methods for solving POMDPs.

The following algorithms are presented in this chapter:

1. The Q-learning algorithm is described in §16.1. It uses the Robbins–Monro algorithm (described in Chapter 15) to estimate the value function for an unconstrained MDP. It is also shown how a primal-dual Q-learning algorithm can be used for MDPs with monotone optimal policies. The Q-learning algorithm also applies as a suboptimal method for POMDPs.
2. Policy gradient algorithms are presented in §16.2 to §16.7. These use gradient estimation (described in Chapter 15) of the cost function together with a stochastic gradient algorithm to estimate the optimal policy. The policy gradient algorithms apply to MDPs and constrained MDPs. They also yield a suboptimal policy search method for POMDPs.

Some terminology: Determining the optimal policy of an MDP or POMDP when the parameters are not known falls under the class of stochastic *adaptive* control problems.

Stochastic adaptive control algorithms are of two types: *direct methods*, where the unknown transition probabilities $P_{ij}(u)$ are estimated simultaneously while updating the control policy, and *implicit methods* (such as simulation-based methods), where the transition probabilities are not directly estimated in order to compute the control policy. In this chapter, we focus on implicit simulation-based algorithms for solving MDPs and POMDPs. These are also called *reinforcement learning algorithms*.[1] One motivation for such implicit methods is that, since they are simulation based, only regions of the state space visited by the simulated sample path are used to determine the controller. Effort is not wasted on determining parameters for low probability regions which are rarely or never visited.

16.1 Q-learning algorithm

The Q-learning algorithm is a widely used reinforcement learning algorithm. As described below, the Q-learning algorithm is the simply the Robbins–Monro stochastic approximation algorithm (15.5) of Chapter 15 applied to estimate the value function of Bellman's dynamic programming equation.

16.1.1 Discounted cost MDP

From (6.21) in Chapter 6, Bellman's dynamic programming equation for a discounted cost MDP (with discount factor ρ) reads

$$\begin{aligned} V(i) &= \min_{u \in \mathcal{U}} \left(c(i,u) + \rho \sum_j P_{ij}(u) V(j) \right) \\ &= \min_{u \in \mathcal{U}} \left(c(i,u) + \rho \, \mathbb{E}\{V(x_{k+1}) | x_k = i, u_k = u\} \right). \end{aligned} \tag{16.1}$$

For each state action pair (i, u) define the Q-factors as

$$Q(i,u) = c(i,u) + \rho \sum_j P_{ij}(u) V(j), \quad i \in \mathcal{X}, u \in \mathcal{U}.$$

We see from (16.1) that the Q-factors $Q(i, u)$ can be expressed as

$$Q(i,u) = c(i,u) + \rho \sum_j P_{ij}(u) \min_{u'} Q(j,u') \tag{16.2}$$

$$= c(i,u) + \rho \, \mathbb{E}\{\min_{u'} Q(x_{k+1}, u') | x_k = i, u_k = u)\}. \tag{16.3}$$

[1] To quote the highly influential book [313] "Reinforcement learning is learning what to do–how to map situations to actions–so as to maximize a numerical reward signal. The learner is not told which actions to take, . . . but instead must discover which actions yield the most reward by trying them. . . Actions may affect not only the immediate reward but also the next situation and, through that, all subsequent rewards. These two characteristics–trial-and-error search and delayed reward–are the two most important distinguishing features of reinforcement learning."

Equation (16.1) has an expectation $\mathbb{E}$ inside the minimization, whereas (16.3) has the expectation *outside* the minimization. It is this crucial observation that forms the basis for using stochastic approximation algorithms to estimate the Q-factors. Indeed (16.3) is simply of the form

$$\mathbb{E}\{f(Q)\} = 0 \tag{16.4}$$

where the random variables $f(Q)$ are defined as

$$f(Q) = c(x_k, u_k) + \rho \min_{u'} Q(x_{k+1}, u') - Q(x_k, u_k). \tag{16.5}$$

The Robbins–Monro algorithm ((15.5) in §15.1) can be used to estimate the solution Q^* of (16.4) as follows: generate a sequence of estimates $\hat{Q}_k$ as

$$\hat{Q}_{k+1}(x_k, u_k) = \hat{Q}_k(x_k, u_k) + \epsilon_k f(\hat{Q}_k). \tag{16.6}$$

The step size ϵ_k is chosen[2] as

$$\epsilon_k = \frac{\epsilon}{\text{Visit}(i, u, k)} \tag{16.7}$$

where ϵ is a positive constant and $\text{Visit}(i, u, k)$ is the number of times the state-action pair (i, u) has been visited until time k by the algorithm.

Algorithm 20 Q-learning algorithm

For $n = 0, 1, \ldots,$ (slow time scale):

  Update policy as $\mu_n(i) = \min_{u \in \mathcal{U}} \hat{Q}_{n\Delta}(i, u)$ for $i = 1, 2, \ldots, X$.

    For $k = n\Delta, n\Delta + 1, \ldots, (n+1)\Delta - 1$ (fast timescale)

      Given state x_k, choose action $u_k = \mu_n(x_k)$.

      Simulate next state $x_{k+1} \sim P_{x_k, x_{k+1}}(u_k)$.

      Update Q-factors as (step size ϵ_k is chosen according to (16.7))

$$\hat{Q}_{k+1}(x_k, u_k) = \hat{Q}_k(x_k, u_k) + \epsilon_k \left[c(x_k, u_k) + \rho \min_{u'} \hat{Q}_k(x_{k+1}, u') - \hat{Q}_k(x_k, u_k) \right]$$

For average cost MDP, update Q-factors using (16.8) instead of above.

Algorithm 20 summarizes the entire procedure as a two-timescale stochastic approximation algorithm. On the fast timescale, the Q factors are updated applying the same policy for a fixed period of time slots referred to as *update interval*, denoted as Δ in Algorithm 20. After that n-th interval, the new policy $\mu_{n+1}(i)$ is chosen based on current Q-factors as $\mu_{n+1}(i) = \min_{u \in \mathcal{U}} \hat{Q}_{(n+1)\Delta}(i, u)$. This update is done on the slow time scale.

Note that Q-learning does not require explicit knowledge of the transition probabilities – all that is needed is access to the controlled system so as to measure its next state x_{k+1} when an action u_k is applied. For a finite state MDP, the Q-learning algorithm converges with probability one to the optimal solution of Bellman's equation; see [200, 53]

[2] In general, the decreasing step size of a stochastic approximation algorithm needs to satisfy $\sum_k \epsilon_k = \infty$ and $\sum_k \epsilon_k^2 < \infty$

for conditions and proof. Please refer to [54] for novel variations of the Q-learning algorithm for discounted cost MDPs.

16.1.2 Average cost MDP

For average cost MDPs recall ACOE (6.30) and the relative value iteration algorithm (6.31) of §6.5.2. In analogy to §16.1.1, define the relative Q-factors

$$Q(i,u) = c(i,u) + \sum_j P_{ij}(u) \min_{u'} Q(j,u') - \min_u Q(1,u), \quad i > 1$$

$$Q(1,u) = c(1,u) + \sum_j P_{1j}(u) \min_{u'} Q(j,u')$$

so that $g = \min_u Q(1,u)$. Then the Q-learning algorithm for average cost MDPs proceeds as in Algorithm 20 with Q-factor update:

$$\hat{Q}_{k+1}(x_k,u_k) = \hat{Q}_k(x_k,u_k) + \epsilon_k \Big[c(x_k,u_k) + \min_{u'} \hat{Q}_k(x_{k+1},u') - \hat{Q}_k(x_k,u_k) - \min_{u'} \hat{Q}_k(1,u') I(x_k \neq 1) \Big]. \tag{16.8}$$

Please see [2] for several novel Q-learning algorithms for average cost MDPs.

16.1.3 Primal-dual Q-learning for submodular MDPs

If we know the optimal policy of an MDP is monotone, how can Q-learning exploit this structure? It was shown in Chapter 9 that under suitable conditions on the transition matrix and cost, the Q-factors of a MDP are submodular. This resulted in the optimal policy having a monotone structure. Suppose we do not have explicit knowledge of the transition matrix, but we know that the costs and transition matrix satisfy the conditions (A1), (A2), (A3) and (A4) of §9.3 implying that the Q-factors are submodular. This implies from Theorem 9.3.1 that the optimal policy has a threshold structure. *How can the Q-learning algorithm be designed to exploit this submodular property of the Q-factors?*

The submodularity condition is a linear inequality constraint on Q:

$$Q(i,u+1) - Q(i,u) \leq Q(i+1,u+1) - Q(i+,u) \tag{16.9}$$

Write this as the inequality constraint

$$QM \geq 0 \tag{16.10}$$

where $\geq$ is element-wise, and the definition of M is obvious from (16.9).

In order to incorporate the constraint (16.10), it is convenient to interpret Q-learning as a stochastic gradient algorithm that minimizes an objective function. Accordingly, define $g(Q)$ so that $\nabla_Q g(Q) = -f(Q)$ where $f(\cdot)$ is defined in (16.5). Then we can write

(16.4) as $\mathbb{E}\{f(Q)\} = \nabla_Q \mathbb{E}\{g(Q)\} = 0$. Then Q-learning can be interpreted as a stochastic approximation algorithm to find

$$Q^* = \operatorname*{argmin}_Q \mathbb{E}\{g(Q)\}.$$

From (16.2) and (16.5), $\nabla_Q^2 f(Q) = -\nabla_Q g(Q)$ is a diagonal matrix with non-negative elements and hence positive semidefinite. Therefore $g(Q)$ is convex.

As the objective is convex and the constraint set (linear inequality) is convex, we can use the primal-dual stochastic approximation algorithm to estimate Q^*:

$$\hat{Q}_{k+1} = \hat{Q}_k + \epsilon_k^{(1)} \left[f(\hat{Q}_k) + \lambda_k M\right]$$
$$\lambda_{k+1} = \max[\lambda_k - \epsilon_k^{(1)} Q_k M,\ 0].$$

Here $\lambda_k \geq 0$ are interpreted as Lagrange multipliers for the constraint (16.9). The step sizes $\epsilon_k^{(1)}$ and $\epsilon_k^{(1)}$ are evaluated as in (16.7).

16.1.4 Q-learning for POMDP

So far we have discussed Q-learning for MDPs. For POMDPs, one can use the following sub-optimal approach [208]: Compute or estimate the Q-factors for the underlying MDP as in §16.1.1. Then given the belief state π, compute the POMDP policy as

$$\mu(\pi) = \operatorname*{argmax}_u \sum_x \pi(x) Q(x, u). \tag{16.11}$$

Although this is effective in some applications, the policy $\mu(\pi)$ does not take actions to gain information since $Q(x, u)$ is computed for the fully observed MDP, which assumes that uncertainty regarding the state disappears after taking one action. We refer the reader to [258] for examples where this approach can fail. The Q-learning algorithm (16.11) for POMDPs requires complete knowledge of the POMDP parameters since the belief state π needs to be evaluated at each time instant. A simpler heuristic (which can also lead to poor performance) is to choose the POMDP policy as

$$\mu(\pi) = \mu^*_{\text{MDP}}(\hat{x}_k)$$

where μ^*_{MDP} is computed via Q-learning and the clean state estimate $\hat{x}_k$ is computed via the stochastic approximation algorithm described in Chapter 17. Then no knowledge of the underlying POMDP parameters are required.

16.2 Policy gradient reinforcement learning for MDP

The Q-learning algorithms described in §16.1 operate in the value space and aim to estimate the value function. The rest of this chapter focuses on solving MDPs and POMDPs using reinforcement learning algorithms that operate in the *policy* space. That is, with

μ_θ denoting a policy parametrized by θ, the aim is to minimize the expected cumulative cost $\mathbb{E}\{C_n(\mu_\theta)\}$ with respect to θ by using a stochastic gradient algorithm of the form[3]

$$\theta_{n+1} = \theta_n - \epsilon_n \hat{\nabla}_\theta C_n(\mu_{\theta_n}). \tag{16.12}$$

Here $C_n(\mu_{\theta_n})$ denotes the observed cumulative cost by the decision-maker when using policy μ_{θ_n} and $\hat{\nabla}_\theta C_n(\mu_{\theta_n})$ denotes the estimated gradient of the cost $\mathbb{E}\{C_n(\mu_\theta)\}$ evaluated at μ_{θ_n}. The phrase "policy gradient algorithm" applies to algorithm (16.12) since it moves along the gradient of the cost in parametrized policy space θ to determine the optimal parametrized policy.

One way of implementing the policy gradient algorithm (16.12) is to use the finite difference SPSA Algorithm 17. Indeed in Part III of the book, we used it for estimating optimal monotone policies for MDPs and POMDPs. In this section we focus on using the more sophisticated score function and weak derivative gradient estimators of Chapter 15 to design policy gradient algorithms for solving MDPs and constrained MDPs.

Consider the average cost unichain MDP formulated in §6.5. Recall $x_k \in \mathcal{X} = \{1, 2, \ldots, X\}$ is the state and $u_k \in \mathcal{U} = \{1, 2, \ldots, U\}$ is the action. Then from (6.37), the Markov process

$$z_k = (x_k, u_k)$$

has transition matrix given by

$$\begin{aligned} P_{i,u,j,\bar{u}}(\theta) &= \mathbb{P}(x_{k+1} = j, u_{k+1} = \bar{u} \mid x_k = i, u_k = u) = \theta_{j\bar{u}} P_{ij}(u) \\ \text{where} \quad \theta_{j\bar{u}} &= \mathbb{P}(u_{k+1} = \bar{u} | x_{k+1} = j). \end{aligned} \tag{16.13}$$

The action probabilities θ defined in (16.13) specify the policy for the MDP. Let $\pi_\theta(i, a)$ denote the stationary distribution of the Markov process z_k. In §6.5 we solved for $\pi_\theta(i, a)$ as a linear program (6.34) and then computed the action probabilities as $\theta_{ia} = \pi_\theta(i, a) / \sum_{i=1}^X \pi_\theta(i, a)$.

Unfortunately, if the transition probabilities $P_{ij}(u)$ are not known, then the LP (6.34) cannot be solved. Instead, here we consider the following equivalent formulation for the optimal policy θ^*:

$$\begin{aligned} &\theta^* = \underset{\theta \in \Theta}{\operatorname{argmin}}\, C(\theta), \quad C(\theta) = \mathbb{E}_{\pi_\theta}\{c(x, u)\} = \sum_{i \in \mathcal{X}} \sum_{u \in \mathcal{U}} \pi_\theta(i, a) c(i, a). \\ &\Theta = \{\theta_{ia} \geq 0, \sum_{a \in \mathcal{U}} \theta_{ia} = 1\}. \end{aligned} \tag{16.14}$$

Note from (16.14) that the optimal policy specified by θ^* depends on the stationary distribution π_θ rather than the unknown transition probabilities $P(u)$.

16.2.1 Parametrizations of policy

To estimate the optimal θ^* in (16.14), we need to ensure that $\theta^* \in \Theta$. To this end, it is convenient to parametrize the action probabilities θ by some judiciously chosen parameter vector ψ so that

[3] We use n to denote the batch time index. The policy gradient algorithm operates on batches of data where each batch comprises of N time points.

$$\theta_{ia}(\psi) = \mathbb{P}(u_n = a | x_n = i), \quad a \in \mathcal{U} = \{1, 2 \dots, U\},\ i \in \mathcal{X} = \{1, 2, \dots, X\}$$

The optimization problem is then

$$\min_{\psi \in \Psi} C(\psi), \qquad \text{where } C(\psi) = \mathbb{E}_{\pi_\psi}\{c(x, u)\} \tag{16.15}$$

Note that the instantaneous cost is independent of ψ but the expectation is with respect to a measure parametrized by ψ.

With the above parametrization, we will use the stochastic gradient algorithm

$$\psi_{n+1} = \psi_n - \epsilon_n \hat{\nabla}_\psi C_n(\psi_n) \tag{16.16}$$

to estimate the minimum $\theta(\psi^*)$. Here $\hat{\nabla}_\psi C_n(\theta(\psi_n))$ denotes an estimate of the gradient $\nabla_\psi C(\theta(\psi))$ evaluated at ψ_n. The aim is obtain a gradient estimator which does not require explicitly knowing the transition matrices $P(u)$ of the MDP. The algorithm (16.16) needs to operate recursively on batches of the observed system trajectory $\{(x_k, u_k), k \in \{nN, (n+1)N - 1\}\}$ to yield a sequence of estimates $\{\psi_n\}$ of the optimal solution ψ^*.

Before proceeding with algorithm (16.16) and the gradient estimator, we first introduce two useful parametrizations ψ that automatically encode the constraints (16.14) on the action probabilities θ.

1. Exponential parametrization: The exponential parameterization for θ is

$$\theta_{ia}(\psi) = \frac{e^{\psi_{ia}}}{\sum_{u \in \mathcal{U}} e^{\psi_{iu}}}, \qquad \psi_{ia} \in \mathbb{R}, i \in \mathcal{X},\ a \in \mathcal{U} = \{1, 2, \dots, U\}. \tag{16.17}$$

In (16.17), the $\psi_{ia} \in \mathbb{R}$ are unconstrained, yet clearly the constraint (16.14) holds. Note ψ has dimension UX.

2. Spherical coordinate parametrization: To each value θ_{iu} associate the values $\lambda_{iu} = \sqrt{\theta_{iu}}$. Then (16.14) yields $\sum_{u \in \mathcal{U}} \lambda_{iu}^2 = 1$, and λ_{iu} can be interpreted as the coordinates of a vector that lies on the surface of the unit sphere in $\mathbb{R}^U$. In spherical coordinates, the angles are $\psi_{ia}, a = 1, \dots U - 1$, and the radius is unity. For $U \geq 2$, the spherical coordinates parameterization ψ is defined as:

$$\theta_{iu}(\psi) = \begin{cases} \cos^2(\psi_{i,1}) & \text{if } u = 1 \\ \cos^2(\psi_{i,u}) \prod_{p=1}^{u-1} \sin^2(\psi_{i,p}) & 2 \leq u \leq U - 1\,. \\ \sin^2(\psi_{i,U-1})) \prod_{p=1}^{U-2} \sin^2(\psi_{i,p}) & u = U \end{cases} \tag{16.18}$$

To summarize, the spherical coordinate parameters are

$$\psi_{ia} \in \mathbb{R}, \quad i = 1, 2, \dots, X,\ a = 1, 2, \dots, U - 1.$$

The ψ_{ia} are unconstrained, yet the constraint (16.14) holds.

Note ψ has dimension $(U - 1)X$. For example, if $U = 2$, $\theta_{i1} = \cos^2 \psi_{i,1}$ and $\theta_{i2} = \sin^2 \psi_{i,1}$ where ψ_{i1} is unconstrained; clearly $\theta_{i1}^2 + \theta_{i2}^2 = 1$.

16.3 Score function policy gradient algorithm for MDP

This section uses the score function gradient estimator of Algorithm 18 on page 352 to estimate $\hat{\nabla}_\psi C_n(\psi_n)$. Together with the stochastic gradient algorithm (16.16), it constitutes a reinforcement learning algorithm for estimating the optimal policy ψ^* for the MDP without requiring knowledge of the transition matrices.

We now describe the score function estimator for an MDP. Consider the augmented Markov process $z_k = (x_k, u_k)$. From the transition probabilities (16.13), it follows that[4]

$$\nabla_\psi P_{i,u,x,\bar{u}}(\theta(\psi)) = P_{ix}(u)\, \nabla_\psi \theta_{x\bar{u}}(\psi). \tag{16.19}$$

The aim is to estimate the gradient with respect to each component ψ_{xa} for the exponential and spherical parametrizations defined above.

For the exponential parametrization (16.17), $\nabla_{\psi_{xa}}\theta_{xa}(\psi) = \theta_{xa} - \theta_{xa}^2$ and $\nabla_{\psi_{xa}}\theta_{x\bar{u}}(\psi) = -\theta_{xa}\theta_{x\bar{u}}$ for $\bar{u} \neq a$. So for $a = 1, 2, \ldots, U$, Step 2 of Algorithm 18 for the n-th batch comprising of times $k \in \{nN+1, \ldots, (n+1)N\}$ is

$$S_k^{\psi_{xa}} = \frac{\nabla_{\psi\, xa} P_{z_{k-1}, z_k}(\theta)}{P_{z_{k-1} z_k}(\theta)} + S_{k-1}^{\psi_{xa}} \tag{16.20}$$

$$\text{where } \frac{\nabla_{\psi\, x,a} P_{z_{k-1}, z_k}(\theta)}{P_{z_{k-1} z_k}(\theta)} = \begin{cases} 1 - \theta_{x_k, u_k} & \text{if } a = u_k, x = x_k \\ -\theta_{x_k, a} & \text{if } a \in \mathcal{U} - \{u_k\}, x = x_k \\ 0 & \text{otherwise.} \end{cases}$$

If instead we use the spherical coordinates (16.18), then Step 2 of Algorithm 18 for $a = 1, 2, \ldots, U-1$ is:

$$S_k^{\psi_{xa}} = \frac{\nabla_{\psi\, xa} P_{z_{k-1}, z_k}(\theta)}{P_{z_{k-1} z_k}(\theta)} + S_{k-1}^{\psi_{xa}} \tag{16.21}$$

$$\frac{\nabla_{\psi\, xa} P_{z_{k-1}, z_k}(\theta)}{P_{z_{k-1} z_k}(\theta)} = \begin{cases} \frac{2}{\tan \psi_{xa}} & a < u_k, x = x_k \\ -2 \tan \psi_{xa} & a = u_k, x = x_k \\ 0 & a > u_{k+1} \end{cases}.$$

Finally, for either parametrization, Step 3 of Algorithm 18 for the n-th batch reads: $\widehat{\nabla}_{\psi_{xa}} C_n(\theta) = \frac{1}{N} \sum_{k=1}^N c(x_k, u_k) S_k^{\psi_{xa}}$.

The stochastic gradient algorithm (16.16) together with score S_k^ψ constitute a parameter-free reinforcement learning algorithm for solving a MDP. As can be seen from (16.20), explicit knowledge of the transition probabilities $P(u)$ or costs $c(x, u)$ are not required; all that is required is that the cost and next state can be obtained (simulated) given the current state by action.

[4] This is the key reason why the score function gradient estimator for an MDP does not require knowledge of $P(u)$. One sees that score function $\frac{\cancel{P_{ix}(u)} \nabla_\psi \theta_{x\bar{u}}}{\cancel{P_{ix}(u)} \theta_{x\bar{u}}}$ is independent of $P_{ix}(u)$ and therefore only depends on the action probabilities.

The main issue with the score function gradient estimator is its large variance as described in Theorem 15.7.1. To reduce the variance, [35] replaces Step 2 with

$$S_k^{\psi} = \frac{\nabla_{\psi} P_{x_{k-1}x_k}(\theta)}{P_{x_{k-1}x_k}(\theta)} + \beta\, S_{k-1}^{\psi} \tag{16.22}$$

where $\beta \in (0,1)$ is a forgetting factor. Other variance reduction techniques [261] include regenerative estimation, finite horizon approximations; Appendix A discusses variance reduction in simulation.

16.4 Weak derivative gradient estimator for MDP

This section uses the weak derivative gradient estimator of Algorithm 19 on page 356 to estimate $\hat{\nabla}_{\psi} C_k(\theta_k)$. Together with the stochastic gradient algorithm (16.16), it constitutes a reinforcement learning algorithm for estimating the optimal policy ψ^* for an MDP.

Consider the augmented Markov process $z_k = (x_k, u_k)$. From the transition probabilities (16.13), it follows that

$$\nabla_{\psi} P_{i,u,x,\bar{u}}(\theta) = P_{ix}(u)\, \nabla_{\psi} \theta_{x,\bar{u}}. \tag{16.23}$$

Our plan is as follows: recall from Algorithm 19 on page 356 that the weak derivative estimator generates two Markov process $\dot{z}$ and $\ddot{z}$. The weak derivative representations we will choose below imply that $\dot{z}_k = z_k$ for all time k, where z_k is the sample path of the MDP. So we only need to worry about simulating the process $\ddot{z}$. It is shown later in this section that $\ddot{z}$ can also be obtained from the MDP sample path z by using cut-and-paste arguments.

For the exponential parameterization (16.17), one obtains from (16.23) the following derivative with respect to each component ψ_{xa}:

$$\nabla_{\psi_{xa}} P(i,u,x,\bar{u}) = \theta_{xa}(1-\theta_{xa}) \left[P_{ix}(u)\, I(\bar{u}=a) - P_{ix}(u) \frac{\theta_{x\bar{u}}}{1-\theta_{xa}} I(\bar{u} \neq a) \right].$$

For the spherical coordinate parameterization (16.18), elementary calculus yields that with respect to each component ψ_{xa}, $a = 1, 2, \ldots, U-1$

$$\nabla_{\psi_{xa}} P(i,u,x,\bar{u}) = -2\theta_{xa} \tan \psi_{xa} \left[P_{ix}(u)\, I(\bar{u}=a) - P_{ix}(u) \frac{\theta_{x\bar{u}}}{\theta_{xa} \tan^2 \psi_{xa}} I(\bar{u} > a) \right].$$

Comparing this with the weak derivative decomposition (15.19), we can use the weak derivative gradient estimator of Algorithm 19. This yields Algorithm 21, which is the weak derivative estimator of an MDP.

Finally, the stochastic gradient algorithm (16.16) together with Algorithm 21 result in a policy gradient algorithm for solving an MDP.

Algorithm 21 Weak derivative estimator for MDP

Let $k = 0, 1, \ldots,$ denote local time within the n-th batch.

Step 1: Simulate $z_0, \ldots, z_{m-1}$ with transition probability $P_{i,u,j,\bar{u}}(\theta(\psi))$ defined in (16.13).

Step 2a: Starting with z_{m-1}, choose $\dot{z}_m = z_m = (x_m, u_m)$. Choose $\ddot{x}_m = x_m$. Choose $\ddot{z}_m = (\ddot{x}_m, \ddot{u}_m)$ where $\ddot{u}_m$ is simulated with

$$\mathbb{P}(\ddot{u}_m = \bar{u}) = \frac{\theta_{x_m \bar{u}}}{1 - \theta_{x_m a}}, \quad \bar{u} \in \mathcal{U} - \{a\} \text{ (exponential parameterization)}$$

$$\mathbb{P}(\ddot{u}_m = \bar{u}) = \frac{\theta_{x_m \bar{u}}}{\theta_{x_m a} \tan^2 \psi_{x_m a}}, \quad \bar{u} \in \{a+1, \ldots, U\} \text{ spherical coordinates)}$$

Step 2b: Starting with $\dot{z}_m$ and $\ddot{z}_m$, simulate the two Markov chains $\dot{z}_{m+1}, \dot{z}_{m+2}, \ldots$ and $\ddot{z}_{m+1}, \ddot{z}_{m+2}, \ldots$ with transition probabilities as in Step 1. (Note $\dot{z}_k = z_k$ for all time k by construction.)

Step 3: Evaluate the weak derivative estimate for the n-th batch as

$$\widehat{\nabla}_{\psi_{xa}} C_n(\theta) = g_{x_m a} \sum_{k=m}^{m+\tau} c(z_k) - c(\ddot{z}_k), \text{ where } \tau = \min\{k : z_k = \ddot{z}_k\}, \tag{16.24}$$

$$g_{xa} = \begin{cases} \theta_{xa}(1 - \theta_{xa}) & \text{exponential parametrization} \\ -2\theta_{xa} \tan \psi_{xa} & \text{spherical coordinates} \end{cases}$$

Parameter-free weak derivative estimator

Unlike the score function gradient estimator (16.20), the weak derivative estimator requires explicit knowledge of the transition probabilities $P(u)$ in order to propagate the process $\{\ddot{z}_k\}$ in the evaluation of (16.24) in Algorithm 21.

How can the weak derivative estimator in Algorithm 21 be modified to work without knowledge of the transition matrix? We need to propagate $\ddot{z}_k$ in Step 2b. This is done by a cut and paste technique originally proposed by [146]. Given $\ddot{z}_m = (x_m, \ddot{u}_m)$ at time m, define

$$\nu = \min\{k > 0 : z_{m+k} = (x_m, \ddot{u}_m)\}.$$

Since z is unichain, it follows that ν is finite with probability 1. Then $\ddot{z}$ is constructed as follows:

Step (i): Choose $\ddot{z}_{m+k} = z_{m+\nu+k}$ for $k = 1, 2, \ldots, N$ where N denotes some pre-specified batch size.

Step (ii): Compute the cost differences in (16.24) as

$$\sum_{k=m}^{N-\nu} c(z_k) - c(\ddot{z}_k) = \sum_{k=m}^{m+\nu-1} c(z_k) + \cancel{\sum_{k=m+\nu}^{N-\nu} c(z_k)} - \cancel{\sum_{k=m+\nu}^{N-\nu} c(z_k)} - \sum_{k=N-\nu+1}^{N} c(z_k)$$

Another possible implementation is

$$\sum_{k=m}^{N} c(z_k) - \sum_{k=m+\nu}^{N} c(\ddot{z}_k) - \nu\,\hat{c} = \sum_{k=m}^{N-\nu-1} c(z_k) - \nu\,\hat{c}. \tag{16.25}$$

Here $\hat{c}$ denotes any estimator that converges as $N \to \infty$ to $C(\psi)$ a.s., where $C(\psi)$ is defined in (16.14). For example, $\hat{c} = \frac{1}{N}\sum_{m=1}^{N} c(z_m)$. Either of the above implementations in (16.24) together with Algorithm 21 results in a policy gradient algorithm (16.16) for solving the MDP without explicit knowledge of the transition matrices.

16.5 Numerical comparison of gradient estimators

This section compares the score function gradient estimator (used in the influential paper [35]) with the parameter-free weak derivative estimator of §16.4. It is shown that the weak derivative estimator has substantially smaller variance.

The following MDP was simulated: $\mathcal{X} = \{1, 2\}$ (two states), $U = 3$ (three actions),

$$P(1) = \begin{pmatrix} 0.9 & 0.1 \\ 0.2 & 0.8 \end{pmatrix}, \quad P(2) = \begin{pmatrix} 0.3 & 0.7 \\ 0.6 & 0.4 \end{pmatrix}, \quad P(3) = \begin{pmatrix} 0.5 & 0.5 \\ 0.1 & 0.9 \end{pmatrix}.$$

The action probability matrix $(\theta(i,u))$ and cost matrix $(c(i,u))$ were chosen as:

$$(\theta(i,a)) = \begin{bmatrix} 0.2 & 0.6 & 0.2 \\ 0.4 & 0.4 & 0.2 \end{bmatrix}, \quad (c(i,a)) = -\begin{bmatrix} 50.0 & 200.0 & 10.0 \\ 3.0 & 500.0 & 0.0 \end{bmatrix}.$$

We work with the exponential parametrization (16.17). First we compute the ground-truth. By solving the linear program (6.34) for the optimal parameter ψ^*, we obtain the true derivative at ψ^* as

$$\nabla_\psi[C(\theta(\psi))] = \begin{pmatrix} -9.010 & 18.680 & -9.670 \\ -45.947 & 68.323 & -22.377 \end{pmatrix}. \tag{16.26}$$

We simulated the parameter-free weak derivative estimator using (16.25) in Algorithm 21 for the exponential parametrization. For batch sizes $N = 100$ and 1000 respectively, the weak derivative gradient estimates are

$$\widehat{\nabla C}_{100}^{\mathrm{WD}} = \begin{pmatrix} -7.851 \pm 0.618 & 17.275 \pm 0.664 & -9.425 \pm 0.594 \\ -44.586 \pm 1.661 & 66.751 \pm 1.657 & -22.164 \pm 1.732 \end{pmatrix}$$
$$\widehat{\nabla C}_{1000}^{\mathrm{WD}} = \begin{pmatrix} -8.361 \pm 0.215 & 17.928 \pm 0.240 & -9.566 \pm 0.211 \\ -46.164 \pm 0.468 & 68.969 \pm 0.472 & -22.805 \pm 0.539 \end{pmatrix}.$$

The numbers after $\pm$ above, denote the confidence intervals at level 0.05 with 100 batches. The variance of the gradient estimator is shown in Table 16.1, together with the corresponding CPU time.

Table 16.1 Variance of weak derivative estimator with exponential parametrization.

$N = 1000$		$\mathrm{Var}[\widehat{\nabla C}_N^{\mathrm{WD}}]$	
$i = 1$	1.180	1.506	1.159
$i = 2$	5.700	5.800	7.565
CPU		1 unit	

Table 16.2 Variance of score function estimator.

$N = 1000$		$\mathrm{Var}[\widehat{\nabla C}_N^{\mathrm{Score}}]$	
$i = 1$	89083	135860	89500
$i = 2$	584012	593443	393015
CPU		687 units	
$N = 10000$		$\mathrm{Var}[\widehat{\nabla C}_N^{\mathrm{Score}}]$	
$i = 1$	876523	1310900	880255
$i = 2$	5841196	5906325	3882805
CPU		6746 units	

We implemented the score function gradient estimator of [35] with the following parameters: forgetting factor 1, batch sizes of $N = 1000$ and 10000. In both cases a total number of 10,000 batches were simulated. The score function gradient estimates are

$$\widehat{\nabla C}_{10000}^{\mathrm{Score}} = \begin{pmatrix} -3.49 \pm 5.83 & 16.91 \pm 7.17 & -13.42 \pm 5.83 \\ -41.20 \pm 14.96 & 53.24 \pm 15.0 & -12.12 \pm 12.24 \end{pmatrix}$$

$$\widehat{\nabla C}_{1000}^{\mathrm{Score}} = \begin{pmatrix} -6.73 \pm 1.84 & 19.67 \pm 2.26 & -12.93 \pm 1.85 \\ -31.49 \pm 4.77 & 46.05 \pm 4.75 & -14.55 \pm 3.88 \end{pmatrix}$$

The variance of the score function gradient estimates are given Table 16.2.

Notice that even with substantially larger batch sizes and number of batches (and hence computational time), the variance of the score function estimator is orders of magnitude larger than the weak derivative estimator. For further numerical results please see [181].

16.6 Policy gradient reinforcement learning for constrained MDP (CMDP)

This section describes how the policy gradient algorithms for MDPs described in §16.2–16.4 can be extended to solve constrained MDPs (CMDPs). Recall that for CMDPs (see Theorem 6.6.1 on page 137), the optimal policy is randomized. Assuming a unichain CMDP (see Theorem 6.6.1 on page 137), our aim is to obtain the optimal parametrized policy:

$$\text{Compute } \theta^* = \underset{\theta}{\operatorname{argmin}}\, C(\theta) = \mathbb{E}_{\pi_\theta}\{c(x,u)\} = \sum_{i\in\mathcal{X}}\sum_{u\in\mathcal{U}} \pi_\theta(i,u)c(i,u) \tag{16.27}$$

$$\text{subject to } B_l(\theta) = \sum_i \sum_u \pi_\theta(i,u)\,\beta_l(i,u) \leq \gamma_l, \quad l = 1,2,\ldots,L. \tag{16.28}$$

The key difference compared to the unconstrained MDP are the L constraints (16.28). As in the unconstrained MDP case (previous section), $\pi_\theta(i,a)$ is the stationary distribution of the controlled Markov process $z_k = (x_k, u_k)$. The optimal policy depends on the action probabilities $\theta(\psi^*)$ where $\theta_{ia}(\psi) = \mathbb{P}(u_n = a|x_n = i)$ and ψ denotes a suitably chosen parametrization of $\theta(\psi)$.

As described in Theorem 6.6.1, the optimal policy of a CMDP is randomized for up to L states. Our aim is to devise stochastic approximation algorithms to estimate this randomized policy, i.e. optimize (16.27) subject to (16.28).

Assumption (O): The minima ψ^* of (16.27), (16.28) are regular, i.e. $\nabla_\psi B_l(\psi^*)$, $l = 1,\ldots,L$ are linearly independent. Then ψ^* belongs to the set of Kuhn–Tucker points

$$\left\{\psi^* \in \Psi : \exists \mu_l \geq 0,\ l = 1,\ldots,L \text{ such that } \quad \nabla_\psi C + \nabla_\psi B\mu = 0, \quad B'\mu = 0\right\}$$

where $\mu = (\mu_1 \ldots, \mu_L)'$. Also, ψ^* satisfies the second-order sufficiency condition $\nabla^2_\psi C(\psi^*) + \nabla^2_\psi B(\psi^*)\mu > 0$ (positive definite) on the subspace $\{y \in \mathbb{R}^L : \nabla_\psi B_l(\psi^*)y = 0\}$ for all $l : B_l(\psi^*) = 0, \mu_l > 0$.

Algorithm 22 Primal-dual reinforcement learning algorithm for CMDP

Parameters: Cost matrix $(c(i,a))$, constraint matrix $(\beta(i,a))$, batch size N, step size $\epsilon > 0$.

Step 0: Initialize: Set $n = 0$, initialize $\psi(n)$ and vector $\lambda(n) \in \mathbb{R}^L_+$.

Step 1: System trajectory observation: Observe MDP over batch $I_n \stackrel{\text{defn}}{=} \{k \in [nN, (n+1)N - 1]\}$ using randomized policy $\theta(\psi(n))$ of (16.18) and compute estimate $\hat{B}(n)$ of the constraints as

$$\hat{B}^\epsilon_l(n+1) = \hat{B}^\epsilon_l(n) + \sqrt{\epsilon}\left(\frac{1}{N}\sum_{k\in I_n}\beta_l(Z_k) - \hat{B}^\epsilon_l(n)\right), \quad l = 1,\ldots,L. \tag{16.29}$$

Step 2: Gradient estimation: Compute $\widehat{\nabla_\psi C}(n)$, $\widehat{\nabla_\psi B}(n)$ over the batch I_n using a gradient estimator (such as weak derivative or score function).

Step 3: Update policy $\theta(\psi(n))$: Use a penalty function primal-dual-based stochastic approximation algorithm to update ψ as follows:

$$\psi_{n+1} = \psi_n - \epsilon\left(\widehat{\nabla_\psi C}(n) + \widehat{\nabla_\psi B}(n)\max\left[0, \lambda_n + \Delta\,\widehat{B}(n)\right]\right) \tag{16.30}$$

$$\lambda_{n+1} = \max\left[\left(1 - \frac{\epsilon}{\Delta}\right)\lambda_n, \lambda_n + \epsilon\widehat{B}_n\right]. \tag{16.31}$$

The "penalization" Δ is a suitably large positive constant and $\max[\cdot,\cdot]$ above is taken element wise.

Step 4: Set $n = n + 1$ and go to Step 1.

The policy gradient algorithm for solving the CMDP is described in Algorithm 22. Below we elaborate on the primal-dual algorithm (16.30), (16.31). The constrained optimization problem (16.27), (16.28) is in general non-convex in the parameter ψ. One solution methodology is to minimize the augmented Lagrangian via a primal-dual stochastic gradient algorithm. Such multiplier algorithms are widely used in deterministic optimization [50, p. 446] with extension to stochastic approximation in [199]. First, convert the inequality MDP constraints (16.28) to equality constraints by introducing the variables $z = (z_1, \dots, z_L) \in \mathbb{R}^L$, so that $B_l(\psi) + z_l^2 = 0$, $l = 1, \dots, L$. Define the augmented Lagrangian,

$$\mathcal{L}_\Delta(\psi, z, \lambda) \stackrel{\text{defn}}{=} C(\psi) + \sum_{l=1}^{L} \lambda_l (B_l(\psi) + z_l^2) + \frac{\Delta}{2} \sum_{l=1}^{L} \left(B_l(\psi) + z_l^2 \right)^2 . \tag{16.32}$$

Here Δ denotes a large positive constant. After some further calculations detailed in [50, pg.396 and 397], the primal-dual algorithm operating on the augmented Lagrangian reads

$$\begin{aligned} \psi_{n+1}^\epsilon &= \psi_n^\epsilon - \epsilon \left(\nabla_\psi C(\psi_n^\epsilon) + \nabla_\psi B(\psi_n^\epsilon) \max\left[0, \lambda_n^\epsilon + \Delta B(\psi_n^\epsilon) \right] \right) \\ \lambda_{n+1}^\epsilon &= \max\left[\left(1 - \frac{\epsilon}{\Delta}\right) \lambda_n^\epsilon, \lambda_n^\epsilon + \epsilon B(\psi_n^\epsilon) \right] \end{aligned} \tag{16.33}$$

where $\epsilon > 0$ denotes the step size and $\max[\cdot, \cdot]$ is taken elementwise.

LEMMA 16.6.1 *Under Assumption (O), for sufficiently large $\Delta > 0$, there exists $\bar{\epsilon} > 0$, such that for all $\epsilon \in (0, \bar{\epsilon}]$, the sequence $\{\psi^\epsilon(n), \lambda^\epsilon(n)\}$ generated by the primal-dual algorithm (16.33) is attracted to a local Kuhn–Tucker pair (ψ^*, λ^*).*

Proof The proof follows from Proposition 4.4.2 in [50]. □

16.7 Policy gradient algorithm for POMDPs

This section discusses how the policy gradient algorithm yields a suboptimal reinforcement learning algorithm for solving POMDPs. Consider the discounted cost POMDP formulated in §7.6 with parameters

$$(\mathcal{X}, \mathcal{U}, \mathcal{Y}, P(u), B(u), c(u), \rho).$$

Recall that in the dynamic programming formulation for POMDPs, the information available to the decision-maker at each time k is $\mathcal{I}_k = \{\pi_0, u_0, y_1, \dots, u_{k-1}, y_k\}$.

In this section on reinforcement learning, it is assumed that the POMDP parameters $P(u), B(u), c(u)$ are not known to the decision-maker. Also different to the dynamic programming formulation, the information available to the decision-maker at each time k is $(\mathcal{I}_k, c(x_k, u_k))$. Even though the instantaneous cost incurred, namely, $c(x_k, u_k)$ is available to the decision-maker at each time, since the function $c(x, u)$ is not known (and is not necessarily invertible), the decision-maker does not know the underlying state x_k.

Since the transition and observation probabilities of the POMDP are not known to the decision-maker, it is not possible to compute the belief state (posterior state distribution) π_k at each time k. Recall that in Chapter 7, we showed in (7.9) that the belief π_k is a sufficient statistic for $\mathcal{I}_k$.

Consider the following suboptimal approach: approximate $\mathcal{I}_k$ by a sliding window of Δ past actions and observations denoted as

$$\mathcal{I}_k^\Delta = \{u_{k-\Delta}, y_{k-\Delta+1}, \dots, u_{k-1}, y_k\}$$

Consider the augmented Markov process $z_k = (x_k, u_k, \mathcal{I}_k^\Delta)$. Its transition probability is

$$\mathbb{P}(x_{k+1}, u_{k+1}, \mathcal{I}_{k+1}^\Delta | x_k, u_k, \mathcal{I}_k^\Delta) = \mathbb{P}(x_{k+1}|x_k, u_k)\, \mathbb{P}(y_{k+1}|x_{k+1}, u_k)\, \mathbb{P}(u_{k+1}|\mathcal{I}_k^\Delta)$$

Here the POMDP policy $\mathbb{P}(u_{k+1}|\mathcal{I}_k^\Delta)$ is parametrized by ψ as

$$\theta(\psi_{\mathcal{I}_k^\Delta, a}) = \mathbb{P}(u_k = a|\mathcal{I}_k^\Delta).$$

for example using the exponential parametrization or spherical coordinates.

In analogy to the MDP case (16.14), the aim is to determine the minimizer

$$\psi^* = \underset{\theta}{\operatorname{argmin}}\, C(\theta(\psi)), \quad \text{where } C(\theta) = \mathbb{E}_{\pi_\theta}\{c(x,u)\} = \sum_{i \in \mathcal{X}} \sum_{u \in \mathcal{U}} \sum_{\mathcal{I}^\Delta} \pi_\theta(i, a, \mathcal{I}^\Delta) c(i, a)$$

where π_θ denotes the stationary distribution of $z_k = (x_k, u_k, \mathcal{I}_k^\Delta)$.

The stochastic gradient algorithm (16.16) can be used to estimate the minimizer ψ^*. The gradient estimate in (16.16) can be determined by the score function or weak derivative method as described in previous sections. From the transition probabilities of the augmented process z_k, it follows that

$$\nabla_\psi \mathbb{P}(x_{k+1}, u_{k+1}, \mathcal{I}_{k+1}^\Delta | x_k, u_k, \mathcal{I}_k^\Delta) = P_{x_k, x_{k+1}}(u_k)\, B_{x_{k+1}, y_{k+1}}(u_k)\, \nabla_\psi \theta_{\mathcal{I}_{k+1}^\Delta, u_{k+1}}.$$

Similar to §16.3, in the score function estimator, the transition and observation probabilities cancel out in the score function:

$$\frac{\nabla_\psi \mathbb{P}(x_{k+1}, u_{k+1}, \mathcal{I}_{k+1}^\Delta | x_k, u_k, \mathcal{I}_k^\Delta)}{\mathbb{P}(x_{k+1}, u_{k+1}, \mathcal{I}_{k+1}^\Delta | x_k, u_k, \mathcal{I}_k^\Delta)} = \frac{\cancel{P_{x_k, x_{k+1}}(u_k) B_{x_{k+1}, y_{k+1}}(u_k)} \nabla_\psi \theta_{\mathcal{I}_{k+1}^\Delta, u_{k+1}}}{\cancel{P_{x_k, x_{k+1}}(u_k) B_{x_{k+1}, y_{k+1}}(u_k)} \theta_{\mathcal{I}_{k+1}^\Delta, u_{k+1}}}$$

So knowledge of the transition and observation probabilities is not required.

To simplify notation consider the case where $\Delta = 0$, so that $\mathcal{I}_k^\Delta = y_k$. (The paper [35] uses this choice.) Algorithm 23 describes the score function algorithm for estimating the gradient of the POMDP parametrized policy with the exponential parametrization (16.20). This is called the GPOMDP (gradient POMDP) algorithm in [35]. Instead of the exponential parametrization, the spherical coordinates (16.21) can be used.

The discount factor β in (16.34) is to reduce the variance of the score function estimator. As shown in §15.7 and §16.5, the weak derivative estimator has significantly smaller variance than the score function for Markovian dynamics. In particular, Algorithm 21 can be used with the parameter free cut-and-paste argument to obtain estimates with substantially smaller variances.

We also refer the reader to [251] for the so-called PEGASUS method and [1] for other policy gradient-based policy search algorithms for POMDPs.

Algorithm 23 Score function gradient algorithm for POMDP with $\mathcal{I}^{\Delta}_{k+1} = y_{k+1}$

For each batch $n = 1, 2, \ldots,$ evaluate

For $k = 1, \ldots, N$:

$$S^{\psi_{xa}}_{k+1} = \frac{\nabla_{\psi}\theta_{\mathcal{I}^{\Delta}_{k+1}, u_{k+1}}}{\theta_{\mathcal{I}^{\Delta}_{k+1}, u_{k+1}}} + \beta\, S^{\psi_{xa}}_{k}, \quad \beta \in (0,1) \tag{16.34}$$

$$\text{where } \frac{\nabla_{\psi}\theta_{\mathcal{I}^{\Delta}_{k+1}, u_{k+1}}}{\theta_{\mathcal{I}^{\Delta}_{k+1}, u_{k+1}}} = \begin{cases} 1 - \theta_{y_{k+1}, u_{k+1}} & \text{if } a = u_{k+1}, y = y_{k+1} \\ -\theta_{y_{k+1}, a} & \text{if } a \in \mathcal{U} - \{u_{k+1}\}, y = y_{k+1} \\ 0 & \text{otherwise.} \end{cases}$$

Evaluate the gradient estimate

$$\hat{\nabla}_{\psi_{ya}} C_n(\theta) = \frac{1}{n} \sum_{k=1}^{n} c(x_k, u_k) S^{\psi_{ya}}_{k}$$

Update parametrized policy vector with stochastic gradient algorithm

$$\psi_{n+1} = \psi_n - \epsilon_n \hat{\nabla}_{\psi} C_n(\theta_n)$$

16.8 Complements and sources

The book [313] is the "bible" of reinforcement learning. [53] (though published in the mid 1990s) is a remarkably clear exposition on reinforcement learning algorithms. [200, 53] present convergence analyses proofs for several types of reinforcement learning algorithms. [54] has several novel variations of the Q-learning algorithm for discounted cost MDPs. [2] has novel Q-learning-type algorithms for average cost MDPs. [14] uses ant-colony optimization for Q-learning. [35] is an influential paper that uses the score function gradient estimator in a policy gradient algorithm for POMDPs.

The area of reinforcement learning is evolving with rapid dynamics. The proceedings of the Neural Information Processing Systems (NIPS) conference and International Conference on Machine Learning (ICML) have numerous recent advances in reinforcement learning algorithms.

17 Stochastic approximation algorithms: examples

Contents

This final chapter, presents four case studies of stochastic approximation algorithms in state/parameter estimation and modeling in the context of POMDPs.

Example 1 discusses online estimation of the parameters of an HMM using the recursive maximum likelihood estimation algorithm. The motivation stems from classical adaptive control: the parameter estimation algorithm can be used to estimate the parameters of the POMDP for a fixed policy; then the policy can be updated using dynamic programming (or approximation) based on the parameters and so on.

Example 2 shows that for an HMM comprised of a slow Markov chain, the least mean squares algorithm can provide satisfactory state estimates of the Markov chain without any knowledge of the underlying parameters. In the context of POMDPs, once the state estimates are known, a variety of suboptimal algorithms can be used to synthesize a reasonable policy.

Example 3 shows how discrete stochastic optimization problems can be solved via stochastic approximation algorithms. In controlled sensing, such algorithms can be used to compute the optimal sensing strategy from a finite set of policies.

Example 4 shows how large-scale Markov chains can be approximated by a system of ordinary differential equations. This mean field analysis is illustrated in the context of information diffusion in a social network. As a result, a tractable model can be obtained for state estimation via Bayesian filtering.

We also show how consensus stochastic approximation algorithms can be analyzed using standard stochastic approximation methods.

17.1 A primer on stochastic approximation algorithms

This section presents a rapid summary of the convergence analysis of stochastic approximation algorithms. Analyzing the convergence of stochastic approximation algorithms is a highly technical area. The books [48, 305, 200] are seminal works that study the convergence of stochastic approximation algorithms under general conditions. Our objective here is much more modest. We merely wish to point out the final outcome of the analysis and then illustrate how this analysis can be applied to the four case studies relating to POMDPs.

Consider a constant step size stochastic approximation algorithms of the form

$$\theta_{k+1} = \theta_k + \epsilon\, H(\theta_k, x_k), \quad k = 0, 1, \ldots \tag{17.1}$$

where $\{\theta_k\}$ is a sequence of parameter estimates generated by the algorithm, ϵ is small positive fixed step size, and x_k is a discrete-time geometrically ergodic Markov process (continuous or discrete state) with transition kernel $P(\theta_k)$ and stationary distribution π_{θ_k}.

Such algorithms are useful for tracking time varying parameters and are widely studied in adaptive filtering. Because of the constant step size, convergence with probability one of the sequence $\{\theta_k\}$ is ruled out. Instead, under reasonable conditions, $\{\theta_k\}$ converges weakly (in distribution) as will be formalized below.

Analysis of stochastic approximation algorithms is typically of three types:

1. Mean square error analysis
2. Ordinary differential equation (ODE) analysis
3. Diffusion limit for tracking error

The mean square analysis seeks to show that for large time k, $\mathbb{E}\|\theta_k - \theta^*\|^2 = O(\varepsilon)$ where θ^* is the true parameter. §17.3.2 provides a detailed example.

In comparison, the ODE analysis and diffusion limit deal with suitable scaled sequences of iterates that are treated as *stochastic processes* rather than random variables. These two analysis methods seek to characterize the behavior of the entire trajectory (random process) rather than just at a specific time k (random variable) that the mean square error analysis does. The price to pay for the ODE and diffusion limit analysis is the highly technical machinery of weak convergence analysis. Below we will provide a heuristic incomplete treatment that only scratches the surface of this elegant analysis tool.

17.1.1 Weak convergence

For constant step size algorithms, under reasonable conditions one can show that the estimates generated by the algorithm converge weakly. For a comprehensive treatment of weak convergence of Markov processes please see [108].

Weak convergence is a function space generalization of convergence in distribution of random variables.[1] Consider a continuous-time random process $X(t), t \in [0, T]$ which we will denote as X. A sequence of random processes $\{X^{(n)}\}$ (indexed by $n = 1, 2, \ldots$) converges weakly to X if for each bounded continuous real-valued functional ϕ,

$$\lim_{n\to\infty} \mathbb{E}\{\phi(X^{(n)})\} = \mathbb{E}\{\phi(X)\}.$$

Equivalently, a sequence of probability measures $\{P^{(n)}\}$ converges weakly to P if $\int \phi\, dP^{(n)} \to \int \phi\, dP$ as $n \to \infty$. Note that the functional ϕ maps the entire trajectory of $X^{(n)}(t), 0 \le t \le T$ of the random process to a real number. (The definition specializes to the classical convergence in distribution if $X^{(n)}$ is a random variable and ϕ is a function mapping $X^{(n)}$ to a real number.)

In the above definition, the trajectories of the random processes lie in the function space $C[0, T]$ (the class of continuous functions on the interval $[0, T]$), or more generally, $D[0, T]$ which is the space of piecewise constant functions that are continuous from the right with limit on the left – these are called "cadlag" functions ("continue à droite, limite à gauche" in French). Appendix D gives some basic definitions and an example of weak convergence (Donsker's functional central limit theorem).

17.1.2 Ordinary differential equation (ODE) analysis of stochastic approximation algorithms

Consider a generic stochastic approximation algorithm

$$\theta_{k+1} = \theta_k + \epsilon\, H(\theta_k, x_k), \quad k = 0, 1, \ldots, \tag{17.2}$$

where $\{x_k\}$ is a random process and $\theta_k \in \mathbb{R}^p$ is the estimate generated by the algorithm at time $k = 1, 2, \ldots$.

The ODE analysis for stochastic approximation algorithms was pioneered by Ljung (see [216, 218]) and subsequently by Kushner and co-workers [198, 200]. It aims to show that the sample path $\{\theta_k\}$ generated by the stochastic approximation algorithm (17.2) behaves asymptotically according to a deterministic ordinary differential equation (ODE). Let us make this more precise. First, we represent the sequence $\{\theta_k\}$ generated by the algorithm (17.2) as a continuous-time process since we want to show that its limiting behavior is the continuous-time ODE. This is done by constructing the continuous-time trajectory via piecewise constant interpolation of $\{\theta_k\}$ as

$$\theta^\epsilon(t) = \theta_k \quad \text{for } t \in [k\epsilon, k\epsilon + \epsilon), \quad k = 0, 1, \ldots. \tag{17.3}$$

In electrical engineering, the terminology "zero-order hold" is used to describe this piecewise constant interpolation. Then we are interested in studying the limit[2] of the continuous time interpolated process $\theta^\epsilon(t)$ as $\epsilon \to 0$ over the time interval $[0, T]$.

[1] A sequence of random variables $\{X_n\}$ converges in distribution if $\lim_{n\to\infty} F_n(x) = F(x)$ for all x for which F is continuous. Here F_n and F are the cumulative distribution functions of X_n and X, respectively. An equivalent statement is that $\mathbb{E}\{\phi(X_n)\} \to \mathbb{E}\{\phi(X)\}$ for every bounded continuous function ϕ.

[2] The intuition is as follows: take a step size of $1/n$ and run the algorithm for n times as many iterations. In the limit, take a step size of ϵ that is vanishingly small and run the algorithm for $1/\epsilon$ iterations. We are interested in this asymptotic behavior as $\epsilon \to 0$.

By using stochastic averaging theory, it can be shown that the following weak convergence result holds; see [200] for proof.[3]

THEOREM 17.1.1 (ODE analysis) *Consider the stochastic approximation algorithm (17.2) with constant step size ϵ. Assume*

(SA1) *$H(\theta, x)$ is uniformly bounded for all $\theta \in \mathbb{R}^p$ and $x \in \mathbb{R}^q$.*
(SA2) *For any $\ell \geq 0$, there exists $h(\theta)$ such that*

$$\frac{1}{N} \sum_{k=\ell}^{N+\ell-1} \mathbb{E}_l\{H(\theta, x_k)\} \to h(\theta) \text{ in probability as } N \to \infty. \tag{17.4}$$

where $\mathbb{E}_l$ denotes expectation with respect to the σ-algebra generated by $\{x_k, k < l\}$.
(SA3) *The ordinary differential equation (ODE)*

$$\frac{d\theta(t)}{dt} = h\big(\theta(t)\big), \quad \theta(0) = \theta_0 \tag{17.5}$$

has a unique solution for every initial condition.

Then the interpolated estimates $\theta^\epsilon(t)$ defined in (17.3) satisfies

$$\lim_{\epsilon \to 0} \mathbb{P}\big(\sup_{0 \leq t \leq T} |\theta^\epsilon(t) - \theta(t)| \geq \eta \big) = 0 \quad \text{for all } T > 0, \eta > 0$$

where $\theta(t)$ is the solution of the ODE (17.5).

Numerous variants of Theorem 17.1.1 exist with weaker conditions. Even though the proof of Theorem 17.1.1 is highly technical, the theorem gives remarkable insight. It says that for sufficiently small step size, the entire interpolated trajectory of estimates generated by the stochastic approximation algorithm is captured by the trajectory of a deterministic ODE. Put differently, if the ODE is designed to follow a specific deterministic trajectory, then the stochastic approximation algorithm will converge weakly to this trajectory.

Assumption (SA1) is a tightness condition; see Appendix D. It can be weakened considerably as shown in the books [48, 200]. More generally, (SA1) can be replaced by uniform integrability of $H(\theta_k, x_k)$, namely that for some $\alpha > 0$,

$$\sup_{k \leq T/\epsilon} \mathbb{E}\|H(\theta_k, x_k)\|^{1+\alpha} < \infty.$$

Assumption (SA2) is a weak law of large numbers. It allows us to work with correlated sequences whose remote past and distant future are asymptotically independent. Examples include sequences of i.i.d. random variables with bounded variance, martingale difference sequences with finite second moments, moving average processes driven by a martingale difference sequence, mixing sequences in which remote past and distant future are asymptotically independent, and certain non-stationary sequences such as functions of Markov chains.

[3] For random variables, convergence in distribution to a constant implies convergence in probability. The generalization to weak convergence is: weak convergence to a deterministic trajectory (specified in our case by an ODE) implies convergence in probability to this deterministic trajectory. The statement in the theorem is a consequence of this.

Some intuition: The intuition behind Theorem 17.1.1 stems from *stochastic averaging theory*: for systems comprised of fast and slow timescales, at the slow timescale, the fast dynamics can be approximated by their average. In the stochastic approximation algorithm (17.2), $\{\theta_k\}$ evolves on a slow timescale due to the small step size ϵ. The signal $\{x_k\}$ evolves on a fast timescale. So on the slow timescale, the system can be approximated by averaging the fast dynamics $\{x_k\}$, that is, replacing x_k with its expected value. This yields the deterministic difference equation (where $h(\cdot)$ below is defined in (17.4))

$$\theta_{k+1} = \theta_k + \epsilon h(\theta_k), \quad k \geq 0.$$

With $\theta^\epsilon(t)$ defined in (17.3), we can write this difference equation as

$$\theta^\epsilon(t) = \theta^\epsilon(0) + \int_0^t h\big(\theta^\epsilon(t)\big) + o(\epsilon)$$

which in turn yields the ODE (17.5) as $\epsilon \to 0$.

Example: Suppose $\{x_k\}$ is a geometrical ergodic Markov process with transition probability matrix (kernel) $P(\theta_k)$ and stationary distribution π_{θ_k}. Then by Theorem 17.1.1, the stochastic approximation converges to the ODE

$$\frac{d\theta(t)}{dt} = h(\theta(t))$$

$$\text{where } h(\theta(t) = \int_{\mathcal{X}} H(\theta(t), x)\pi_{\theta(t)}(x)dx = \mathbb{E}_{\pi_{\theta(t)}}\{H(\theta(t), x)\}. \tag{17.6}$$

17.1.3 Diffusion limit for tracking error

The ODE analysis of Theorem 17.1.1 says that the interpolated trajectory of estimates generated by the algorithm converges weakly to the trajectory of the ODE (17.5) over a time interval $[0, T]$. What is the rate of convergence? This is addressed by the diffusion limit analysis that we now describe.

Define the scaled tracking error as the continuous-time process

$$\tilde{\theta}^\epsilon(t) = \frac{1}{\sqrt{\epsilon}}(\theta^\epsilon(t) - \theta(t)) \tag{17.7}$$

where $\theta(t)$ evolves according to the ODE (17.5) and $\theta^\epsilon(t)$ is the interpolated trajectory of the stochastic approximation algorithm.

Define the following continuous-time diffusion process:

$$\tilde{\theta}(t) = \int_0^t \nabla_\theta h(\theta(s))\, \tilde{\theta}(s)\, ds + \int_0^t R^{1/2}(\theta(s))\, dw(s) \tag{17.8}$$

where $h(\theta(t))$ is defined in the ODE (17.5), the covariance matrix

$$R(\theta) = \sum_{n=-\infty}^{\infty} \operatorname{cov}_\theta \left(H(\theta, x_n), H(\theta, x_0)\right), \tag{17.9}$$

and $w(t)$ is standard Brownian motion (defined in Appendix D). The last term in (17.8) is interpreted as a stochastic (Itô) integral; see, for example, [167].

Then the main result regarding the limit of the tracking error is the following:

THEOREM 17.1.2 (Diffusion limit of tracking error) *As $\epsilon \to 0$, the scaled error process $\tilde{\theta}^\epsilon$ defined in (17.7) converges weakly to the diffusion process $\tilde{\theta}$ defined in (17.8).*

17.1.4 Infinite horizon asymptotics and convergence rate

The ODE and diffusion analyses outlined above apply to a finite time interval $[0, T]$. Often we are interested in the asymptotic behavior as $T \to \infty$. In particular, we want an expression for the asymptotic rate of convergence of a stochastic approximation algorithm. In stochastic analysis, the rate of convergence refers to the asymptotic variance of the normalized errors about the limit point.

Under stability conditions so that $\lim_{T\to\infty} \lim_{\epsilon\to 0} \theta^\epsilon(t) = \lim_{\epsilon\to 0} \lim_{T\to\infty} \theta^\epsilon(t)$, it can be shown [192] that the stable fixed points of the ordinary differential equation (17.5) coincide with the attractors of the stochastic approximation algorithm. Suppose that this limit exists and denote it as θ^*. Then for large t, the diffusion approximation (17.8) becomes the linear (Gaussian) diffusion

$$\tilde{\theta}(t) = \int_0^t \nabla_\theta h(\theta)|_{\theta=\theta^*} \tilde{\theta}(s)\, ds + \int_0^t R^{1/2}(\theta^*)\, dw(s). \tag{17.10}$$

In other words, the scaled error $\tilde{\theta}^\epsilon(t)$ converges to a Gaussian process – this can be viewed as a functional central limit theorem.

Suppose that the matrix $\nabla_\theta h(\theta^*)$ is stable, that is, all its eigenvalues have strictly negative real parts. Then the diffusion process $\tilde{\theta}(t)$ has a stationary Gaussian distribution, that is for large t,

$$\tilde{\theta}(t) \sim N(0, \Sigma) \tag{17.11}$$

where the positive definite matrix Σ satisfies the algebraic Lyapunov equation[4]

$$\nabla_\theta h(\theta)\, \Sigma + \Sigma\, \nabla'_\theta h(\theta) + R(\theta)\big|_{\theta=\theta^*} = 0. \tag{17.12}$$

where R is defined in (17.9). The covariance matrix Σ, which is the solution of (17.12), is interpreted as the *asymptotic rate of convergence* of the stochastic approximation algorithm. For large n and small step size ϵ, (17.11) says that the tracking error of the stochastic approximation algorithm (17.2) behaves as

$$\epsilon^{-1/2}(\theta_n - \theta^*) \sim N(0, \Sigma),$$

where Σ is the solution of the algebraic Lyapunov equation (17.12).

[4] Recall we discussed the algebraic Lyapunov equation for discrete-time systems in §2.3.3. In comparison (17.12) is the continuous-time analog of the algebraic Lyapunov equation.

17.2 Example 1: Recursive maximum likelihood parameter estimation of HMMs

Chapter 4 presented offline algorithms for computing the maximum likelihood estimate (MLE) of an HMM given a batch of observations. This section discusses recursive (online) estimation of the parameters of hidden Markov models and presents stochastic approximation algorithms to carry out the estimation task.

Let x_k, $k = 0, 1, \ldots$ denote a Markov chain on the state space $\mathcal{X} = \{1, 2, \ldots, X\}$ where X is fixed and known. The HMM parameters $P(\theta)$ (transition probability matrix) and $B(\theta)$ (observation probabilities) are functions of the parameter vector θ in a compact subset Θ of Euclidean space. Assume that the HMM observations y_k, $k = 1, 2, \ldots$ are generated by a true parameter vector $\theta^o \in \Theta$ which is not known. The aim is to design a recursive algorithm to estimate the HMM parameter vector θ^o.

Example: Consider the Markov chain observed in Markov modulated Gaussian noise via the observation process

$$y_k = x_k + \sigma(x_k) v_k$$

where $v_k \sim N(0, 1)$ is i.i.d. and $\sigma(1), \ldots, \sigma(X)$ are fixed positive scalars in the interval $[\underline{\sigma}, \bar{\sigma}]$. One possible parametrization of this Gaussian noise HMM is

$$\theta = [P_{11}, \ldots, P_{XX}, \sigma(1), \sigma(2), \ldots, \sigma(X)]'.$$

So $\Theta = \{\theta : P_{ij} \geq 0, \ \sum_j P_{ij} = 1, \ \sigma(i) \in [\underline{\sigma}, \bar{\sigma}]\}$. A more useful parametrization that automatically encodes these constraints on the transition matrix is to use spherical coordinates or the exponential parametrization discussed in §16.2.1.

The normalized log likelihood of the HMM parameter θ based on the observations $y_{1:n} = (y_1, \ldots, y_n)$ is $l_n(\theta) = \frac{1}{n} \log p(y_{1:n}|\theta)$; see (4.44) in §4.6.2. As described in (4.46), it can be expressed as the arithmetic mean of terms involving the observations and the HMM prediction filter as follows:

$$l_n(\theta) = \frac{1}{n} \sum_{k=1}^{n} \log \left[\mathbf{1}' B_{y_k}(\theta) \pi^{\theta}_{k|k-1} \right] \tag{17.13}$$

where $B_y(\theta) = \text{diag}(p(y|x = 1, \theta), \ldots, p(y|x = X, \theta))$. Here $\pi^{\theta}_{k|k-1}$ denotes the HMM predictor assuming the HMM parameter is θ:

$$\pi^{\theta}_{k|k-1} = [\pi^{\theta}_{k|k-1}(1), \ldots, \pi^{\theta}_{k|k-1}(X)]', \ \pi^{\theta}_{k|k-1}(i) = \mathbb{P}_{\theta}(x_k = i|y_1, \ldots . y_{k-1}).$$

Recall from §3.5, the HMM predictor is computed via the recursion

$$\pi^{\theta}_{k+1|k} = \frac{P'(\theta) B_{y_k}(\theta) \pi^{\theta}_{k|k-1}}{\mathbf{1}' B_{y_k}(\theta) \pi^{\theta}_{k|k-1}}. \tag{17.14}$$

Define the incremental score vector as

$$S(\pi^{\theta}_{k|k-1}, y_k, \theta) = \nabla_{\theta} \log \left[\mathbf{1}' B_{y_k}(\theta) \pi^{\theta}_{k|k-1} \right]. \tag{17.15}$$

Then the recursive MLE (RMLE) algorithm for online estimation of the HMM parameters is a stochastic approximation algorithm of the form

$$\theta_{k+1} = \Pi_\Theta\left(\theta_k + \epsilon\, S(\pi^{\theta_k}_{k|k-1}, y_k, \theta_k)\right) \tag{17.16}$$

where ϵ denotes a positive step size, and Π_Θ denotes the projection of the estimate to the set Θ.

17.2.1 Computation of score vector

The score vector in the RMLE algorithm (17.16) can be computed recursively by differentiating the terms within the summation in (17.13) with respect to each element $\theta(l)$ of θ, $l = 1, \ldots, p$. This yields the l-th component of S as

$$S^{(l)}(\pi^\theta_{k|k-1}, y_k, \theta) = \frac{\mathbf{1}' B_{y_k}(\theta)\, w^{(l)}_k(\theta)}{\mathbf{1}' B_{y_k}(\theta) \pi^\theta_{k|k-1}} + \frac{\mathbf{1}'[\nabla_{\theta(l)} B_{y_k}(\theta)]\pi^\theta_{k|k-1}}{\mathbf{1}' B_{y_k}(\theta)\pi^\theta_{k|k-1}} \tag{17.17}$$

where $w^{(l)}_k(\theta) = \nabla_{\theta(l)} \pi^\theta_{k|k-1}$ denotes the partial derivative of $\pi^\theta_{k|k-1}$ in (17.14) with respect to the lth component of the parameter vector θ. Define the $X \times p$ matrix $w_k(\theta) = ((w^{(1)}_k(\theta), \ldots, w^{(p)}_k(\theta))$. Clearly $w_k(\theta)$ belongs to Ξ defined by $\Xi = \{w \in \mathbb{R}^{X \times p} : \mathbf{1}' w = \mathbf{0}\}$ since $\mathbf{1}' \nabla_{\theta(l)} \pi^\theta_{k|k-1} = \nabla_{\theta(l)} \mathbf{1}' \pi^\theta_{k|k-1} = \nabla_{\theta(l)} 1 = 0$.

We need a recursion for evaluating $w^{(l)}_k(\theta)$ in (17.17). Differentiating $\pi^\theta_{k+1|k}$ with respect to $\theta(l)$ yields

$$w^{(l)}_{k+1}(\theta) = \nabla_{\theta(l)} \pi^\theta_{k+1|k} = R_1(y_k, \pi^\theta_{k|k-1}, \theta) w^{(l)}_k(\theta) + R^{(l)}_2(y_k, \pi^\theta_{k|k-1}, \theta) \tag{17.18}$$

where

$$R_1(y_k, \pi^\theta_{k|k-1}, \theta) = P'(\theta)\left[I - \frac{B_{y_k}(\theta)\, \pi^\theta_{k|k-1} \mathbf{1}'}{\mathbf{1}' B_{y_k}(\theta)\, \pi^\theta_{k|k-1}}\right] \frac{B_{y_k}(\theta)}{\mathbf{1}' B_{y_k}(\theta)\, \pi^\theta_{k|k-1}}$$

$$R^{(l)}_2(z_k, \pi^\theta_{k|k-1}, \theta) = P'(\theta)\left[I - \frac{B_{y_k}(\theta)\, \pi^\theta_{k|k-1} \mathbf{1}'}{\mathbf{1}' B_{y_k}(\theta)\, \pi^\theta_{k|k-1}}\right] \frac{\mathbf{1}' \nabla_{\theta(l)} B_{y_k}(\theta)\, \pi^\theta_{k|k-1}}{\mathbf{1}' B_{y_k}(\theta)\, \pi^\theta_{k|k-1}} + \frac{[\nabla_{\theta(l)} P'(\theta)]\, B_{y_k}(\theta)\, \pi^\theta_{k|k-1}}{\mathbf{1}' B_{y_k}(\theta)\, \pi^\theta_{k|k-1}}.$$

To summarize (17.16), (17.17) and (17.18) constitute the RMLE algorithm for online parameter estimation of an HMM.

17.2.2 ODE analysis of RMLE algorithm

This section analyzes the convergence of the RMLE algorithm (17.16) using the ODE method of Theorem 17.1.1. The assumptions used here are identical to §4.6.2 where we discussed the consistency of the MLE. Assume the conditions (L1) to (L5) in §4.6.2 hold. Define the augmented Markov process $z_k = (x_k, y_k, \pi^\theta_{k|k-1}, w^\theta_k)$. Similar to the proof of Theorem 3.7.1, it can be shown that z_k is geometrically ergodic and has an unique invariant distribution, denoted as $\nu_{\theta,\theta^o}(x, y, \pi, w)$ where θ^o denotes the true model that generates the observations. As a consequence, the following strong law of large numbers holds; see (4.47): $l_n(\theta, \theta^o) \to l(\theta)$, P_{θ^o} w.p.1 as $n \to \infty$, where

$$l(\theta, \theta^o) = \mathbb{E}_{\nu_{\theta,\theta^o}}\{l_n(\theta)\} = \int_{\mathcal{Y}\times\Pi(X)} \log\left[\mathbf{1}'B_y(\theta)\,\pi(\theta)\right]\,\nu_{\theta,\theta^o}(dy, d\pi),$$

and ν_{θ,θ^o} denotes the marginal of the invariant distribution on $\mathcal{Y}\times\Pi(X)$. Here $\mathcal{Y}$ denotes the observation space and $\Pi(X) = \left\{\pi : \pi(i) \in [0,1], \sum_{i=1}^{X}\pi(i) = 1\right\}$. denotes the belief space (unit simplex).

We are now ready to apply Theorem 17.1.1 to specify the ODE that determines the limiting behavior of the RMLE algorithm (17.16). In particular, from (17.6) and the definition (17.15) it follows that the ODE is

$$\frac{d\theta}{dt} = \mathbb{E}_{\nu_{\theta,\theta^o}}\{S(\pi_{k|k-1}^{\theta_k}, y_k, \theta_k)\} + \tilde{m}(t) = \nabla_\theta l(\theta, \theta^o) + \tilde{m}(t).$$

where $\tilde{m}(t)$ is the minimum force (projection) needed to keep $\theta \in \Theta$. Recall that the Kullback–Leibler information is

$$D(\theta^o, \theta) = l(\theta^o, \theta^o) - l(\theta, \theta^o) \geq 0.$$

So we can rewrite the ODE associated with the RMLE algorithm (17.16) as

$$\frac{d\theta}{dt} = -\nabla_\theta D(\theta^o, \theta) + \tilde{m}(t). \tag{17.19}$$

For simplicity suppose θ_n is bounded with probability 1 without the projection. Then the RMLE algorithm (17.16) converges to a local stationary point of $D(\theta^o, \theta)$, i.e. to a point in the set $\{\theta : \nabla_\theta D(\theta^o, \theta) = 0\}$.

Let $\mathcal{M}$ denote the set of global minimizers of $D(\theta^o, \theta)$. It was shown via Jensen's inequality in Step 2 below (4.48) that the true model θ^o belongs to $\mathcal{M}$ and $D(\theta^o, \theta^o) = 0$. Therefore $D(\theta^o, \theta)$ serves as a Lyapunov function in θ for the ODE (17.19). Indeed, from the ODE (17.19), for $\theta \notin \mathcal{M}$,

$$\frac{d}{dt}D(\theta^o, \theta) = (\nabla_\theta D(\theta^o, \theta))'\,\frac{d\theta}{dt} = -(\nabla_\theta D(\theta^o, \theta))^2 < 0,$$

implying that the global minimizers of $D(\theta^o, \theta)$ are asymptotic stable equilibria[5] of the ODE and thus of the RMLE algorithm (17.16). Of course, this does not mean that there are no other asymptotic stable equilibria; in general, the RMLE converges to a local stationary point of $D(\theta^o, \theta)$.

[5] More precisely, a Lyapunov function $V(x)$ is defined as follows [173]. Let x^* be an equilibrium point for the ODE $\dot{x} = f(x)$, i.e. $f(x^*) = 0$. Let $D \in \mathbb{R}^p$ be a domain containing x^*. Let $V : D \to \mathbb{R}$ be a continuously differentiable function such that

$$V(x^*) = 0, \text{ and } V(x) > 0 \text{ for } x \in D - \{x^*\}.$$

1. If $\dot{V}(x) \leq 0$ for $x \in D$, then x^* is stable. That is, for each $\epsilon > 0$, there is $\delta(\epsilon) > 0$ such that $\|x^* - x(0)\| < \delta(\epsilon)$ implies $\|x^* - x(t)\} < \epsilon$ for all $t \geq 0$.
2. If $\dot{V}(x) < 0$ for $x \in D - \{x^*\}$, then x^* is asymptotically stable. That is $\delta(\epsilon)$ can be chosen such that $\|(x^* - x(0)\| < \delta(\epsilon)$, implies $\lim_{t\to\infty} x(t) = x^*$.

17.3 Example 2: HMM state estimation via LMS algorithm

This section discusses how the least mean squares (LMS) stochastic gradient algorithm can be used to estimate the underlying state of a slow Markov chain given noisy observations. Recall in Chapter 3, a Bayesian HMM filter was used for this purpose. Unlike the HMM filter, the LMS algorithm does not require exact knowledge of the underlying transition matrix or observation probabilities. In the context of POMDPs, this implies that for slow Markov chains, the underlying state can be estimated (with provable performance bounds) and then an MDP controller can be run. This approach lies within the class of MDP-based heuristics for solving POMDPs outlined in Chapter 18.

17.3.1 Formulation

Let $\{y_n\}$ be a sequence of real-valued signals representing the observations obtained at time n, and $\{x_n\}$ be the time-varying true parameter, an $\mathbb{R}^r$-valued random process. Suppose that

$$y_n = \varphi_n' x_n + v_n, \; n = 0, 1, \ldots, \tag{17.20}$$

where $\varphi_n \in \mathbb{R}^r$ is the regression vector and $\{v_n\} \in \mathbb{R}$ is a zero mean sequence. Note that (17.20) is a variant of the usual linear regression model, in which a time-varying stochastic process x_n is in place of a fixed parameter. We assume that x_n is a slow discrete-time Markov chain.

(A1) Suppose that there is a small parameter $\varepsilon > 0$ and that $\{x_n\}$ is a Markov chain with states and transition probability matrix given by

$$\mathcal{X} = \{1, 2, \ldots, X\}, \quad \text{and} \quad P^\varepsilon = I + \varepsilon Q. \tag{17.21}$$

Here I denotes the $X \times X$ identity matrix and $Q = (Q_{ij})$ is a $X \times X$ generator matrix of a continuous-time Markov chain (i.e. Q satisfies $Q_{ij} \geq 0$ for $i \neq j$ and $\sum_{j=1}^X Q_{ij} = 0$ for each $i = 1, \ldots, X$). For simplicity, assume the initial distribution $\mathbb{P}(x_0 = g_i) = \pi_0(i)$ to be independent of ε for each $i = 1, \ldots, X$.

Note that the small parameter $\varepsilon > 0$ in (A1) ensures that the identity matrix I dominates. In fact, $Q_{ij} \geq 0$ for $i \neq j$ thus the small parameter $\varepsilon > 0$ ensures the entries of the transition matrix to be positive since $P^\varepsilon_{ij} = \delta_{ij} + \varepsilon Q_{ij} \geq 0$ for $\varepsilon > 0$ small enough, where $\delta_{ij} = 1$ if $i = j$ and is 0 otherwise. The use of the generator Q makes the row sum of the matrix P be one since $\sum_{j=1}^X P^\varepsilon_{ij} = 1 + \varepsilon \sum_{j=1}^X Q_{ij} = 1$. The essence is that although the true parameter is time varying, it is piecewise constant. In addition, the process does not change too frequently due to the dominating identity matrix in the transition matrix (17.20). It remains as a constant most of the time and jumps into another state at random instants. Hence the terminology "slow" Markov chain.

LMS algorithm

The LMS algorithm will be used to track the Markov chain $\{x_n\}$. Recall from (15.6) that the LMS algorithm generates estimates $\{\widehat{\theta}_n\}$ according to

$$\widehat{\theta}_{n+1} = \widehat{\theta}_n + \mu\varphi_n(y_n - \varphi_n'\widehat{\theta}_n), \qquad n = 0, 1, \ldots, \tag{17.22}$$

where $\mu > 0$ is a small constant step size for the algorithm. By using (17.20) with $\widetilde{\theta}_n = \widehat{\theta}_n - x_n$, the tracking error satisfies

$$\widetilde{\theta}_{n+1} = \widetilde{\theta}_n - \mu\varphi_n\varphi_n'\widetilde{\theta}_n + \mu\varphi_n v_n + (x_n - x_{n+1}). \tag{17.23}$$

The aim is to determine bounds on the deviation $\widetilde{\theta}_n = \widehat{\theta}_n - x_n$. This goal is accomplished by the following four steps:

1. Obtain mean square error bounds for $\mathbb{E}|\widehat{\theta}_n - x_n|^2$.
2. Obtain a limit ODE of centered process.
3. Obtain a weak convergence result of a suitably scaled sequence.
4. Obtain probabilistic bounds on $P(|\widehat{\theta}_n - x_n| > \alpha)$ for $\alpha > 0$, and hence probability of error bounds based on the result obtained in part 2 above.

The Markov chain x_n is called a *hypermodel* in [48]. It is important to note that the Markovian dynamics of the hypermodel x_n are used only in our analysis, it is not used in the implementation of the LMS algorithm (17.22).

Assumptions on the signals

Let $\mathcal{F}_n$ be the σ-algebra generated by $\{(\varphi_j, v_j),\ j < n, x_j,\ j \le n\}$, and denote the conditional expectation with respect to $\mathcal{F}_n$ by $\mathbb{E}_n$. We will use the following conditions on the signals.

(A2) The signal $\{\varphi_n, v_n\}$ is independent of $\{x_n\}$. Either $\{\varphi_n, v_n\}$ is a sequence of bounded signals such that there is a symmetric and positive definite matrix $B \in \mathbb{R}^{r\times r}$ such that $\mathbb{E}\{\varphi_n\varphi_n'\} = B$

$$\left|\sum_{j=n}^{\infty}\mathbb{E}_n\{\varphi_j\varphi_j' - B\}\right| \le K, \quad \text{and} \left|\sum_{j=n}^{\infty}\mathbb{E}_n\{\varphi_n v_j\}\right| \le K, \tag{17.24}$$

or $\{\varphi_n, v_n\}$ is a martingale difference satisfying $\sup_n E\{|\varphi_n|^{4+\Delta}\} < \infty$ and $\sup_n E\{|\varphi_n v_n|^{2+\Delta}\} < \infty$ for some $\Delta > 0$.

Remark: Inequalities (17.24) are modeled after mixing processes and are in the almost sure (a.s.) sense with the constant K independent of ω, the sample point. [Note, however, we use the same kind of notation as in, for example, the mixing inequalities [58, p. 166, Eq. (20.4)] and [198, p. 82, Eqs. (6.6) and (6.7)].] This allows us to work with correlated signals whose remote past and distant future are asymptotically independent. To obtain the desired result, the distribution of the signal need not be known. The boundedness is a mild restriction, for example, one may consider truncated Gaussian processes etc.

Moreover, dealing with recursive procedures in practice, in lieu of (17.22), one often uses a projection or truncation algorithm. For instance, one may use

$$\widehat{\theta}_{n+1} = \pi_H[\widehat{\theta}_n + \mu\varphi_n(y_n - \varphi_n'\widehat{\theta}_n)], \tag{17.25}$$

where π_H is a projection operator and H is a bounded set. When the iterates are outside H, it will be projected back to the constrained set H. Extensive discussions for such projection algorithms can be found in [200].

17.3.2 Analysis of tracking error

Mean square error bounds

This section establishes a mean square error estimate for $\mathbb{E}|\widehat{\theta}_n - x_n|^2$. It is important to note that the mean square error analysis below holds for small positive but fixed μ and ε. Indeed, let $\lambda_{\min} > 0$ denote the smallest eigenvalue of the symmetric positive definite matrix B defined in (A2). Then in the following theorem, it is sufficient to pick out μ and ε small enough so that

$$\lambda_{\min}\mu > O(\mu^2) + O(\varepsilon\mu); \tag{17.26}$$

see (17.99) in the proof below. The phrase "for sufficiently large n" in what follows means that there is an $n_0 = n_0(\varepsilon, \mu)$ such that (17.27) holds for $n \geq n_0$. In fact, (17.27) holds uniformly for $n \geq n_0$.

THEOREM 17.3.1 *Under conditions* (A1) *and* (A2), *for sufficiently large n, as* $\varepsilon \to 0$ *and* $\mu \to 0$,

$$\mathbb{E}|\widetilde{\theta}_n|^2 = \mathbb{E}|\widehat{\theta}_n - x_n|^2 = O\left(\mu + \varepsilon/\mu\right)\exp(\mu + \varepsilon/\mu). \tag{17.27}$$

The theorem is proved in the appendix to this chapter.

In view of Theorem 17.3.1, it is clear that in order for the adaptive algorithm to track the time-varying parameter, due to the presence of the term $\exp(\varepsilon/\mu)$, we need to have at least $\varepsilon/\mu = O(1)$. Thus, the ratio ε/μ must not be large. A glance of the order of magnitude estimate $O(\mu + \varepsilon/\mu)$, to balance the two terms μ and ε/μ, we need to choose $\varepsilon = O(\mu^2)$. Therefore, we arrive at the following corollary.

COROLLARY 17.3.2 *Under the conditions of Theorem 17.3.1, if* $\varepsilon = O(\mu^2)$, *then for sufficiently large n,* $\mathbb{E}|\theta_n|^2 = O(\mu)$.

Mean ODE and diffusion approximation

Due to the form of the transition matrix given by (17.21), the underlying Markov chain belongs to the category of two-time-scale Markov chains. For some of the recent work on this subject, we refer the reader to [349] and the references therein. It is assumed that $\varepsilon = O(\mu^2)$ (see Corollary 17.3.2), i.e. the adaptation speed of the LMS algorithm (17.22) is faster than the Markov chain dynamics.

Recall that the mean square error analysis in §17.3.2 deals with the mean square behavior of the random variable $\widetilde{\theta}_n = \widehat{\theta}_n - x_n$ as $n \to \infty$, for small but fixed μ and ε. In contrast, the mean ODE and diffusion approximation analysis of this section deal with how the entire discrete-time trajectory (stochastic process) $\{\widetilde{\theta}_n : n = 0, 1, 2 \dots, \}$ converges (weakly) to such a limiting continuous-time process (on a suitable function space) as $\mu \to 0$ on a timescale $O(1/\mu)$. Since the underlying true parameter x_n evolves according to a Markov chain (unlike standard stochastic approximation proofs where the parameter is assumed constant), the proofs of the ODE and diffusion limit are nonstandard and require use of the so-called "martingale problem" formulation for stochastic diffusions; please see [347, 348] for details.

Mean ordinary differential equation (ODE)

Define $\widetilde{\theta}_n = \widehat{\theta}_n - x_n$ and the interpolated process

$$\widetilde{\theta}^\mu(t) = \widetilde{\theta}_n \quad \text{for } t \in [n\mu, n\mu + \mu). \tag{17.28}$$

We will need another condition.

(A3) As $n \to \infty$,

$$\frac{1}{n}\sum_{j=n_1}^{n_1+n} \mathbb{E}_{n_1} \varphi_j v_j \to 0, \text{ in probability,} \qquad \frac{1}{n}\sum_{j=n_1}^{n_1+n} \mathbb{E}_{n_1} \varphi_j \varphi_j' \to B, \text{ in probability.}$$

THEOREM 17.3.3 *Under* (A1)–(A3) *and assuming that* $\widetilde{\theta}_0 = \widetilde{\theta}_0^\mu$ *converges weakly to* $\widetilde{\theta}^0$, *then* $\widetilde{\theta}^\mu(\cdot)$ *defined in* (17.28) *converges weakly to* $\widetilde{\theta}(\cdot)$, *which is a solution of the ODE*

$$\frac{d}{dt}\widetilde{\theta}(t) = -B\widetilde{\theta}(t), \; t \geq 0, \; \widetilde{\theta}(0) = \widetilde{\theta}^0. \tag{17.29}$$

This theorem provides us with the evolution of the tracking errors. It shows that $\widehat{\theta}_n - x_n$ evolves dynamically so that its trajectories follows a deterministic ODE. Since the ODE is asymptotically stable, the errors decay exponentially fast to 0 as time grows.

Diffusion limit

The aim here is to characterize the tracking error of the LMS algorithm when estimating the state of an HMM.

Define the scaled tracking error $u_n = (\widehat{\theta}_n - x_n)/\sqrt{\mu}$. Then

$$u_{n+1} = u_n - \mu\varphi_n\varphi_n' u_n + \sqrt{\mu}\varphi_n v_n + \frac{x_n - x_{n+1}}{\sqrt{\mu}}. \tag{17.30}$$

Define the continuous-time interpolation $u^\mu(\cdot)$ as

$$u^\mu(t) = u_n \text{ for } t \in [\mu(n - N_\mu), \mu(n - N_\mu) + \mu). \tag{17.31}$$

(A4) $\sqrt{\mu}\sum_{j=n}^{n+t/\mu} \mathbb{E}_n \varphi_j v_j$ converges weakly to a Brownian motion with a covariance Σt for a positive definite Σ. Moreover,

$$\left| \sum_{j=n}^{n+t/\mu} \sum_{k=n}^{n+t/\mu} \mathbb{E}_n \varphi_j \varphi_k' v_j v_k \right| \leq K.$$

Remark: Note that (A4) assumes convergence to a Brownian motion. Sufficient conditions guaranteeing the convergence of the scaled sequence to the Brownian motion in (A4) are readily available in the literature; see, for example, [58, 108] among others. For instance, if $\{\varphi_n v_n\}$ is a uniform mixing sequence with mixing rate ψ_n satisfying $\sum_k \psi_k^{1/2} < \infty$, the classical result in functional central limit theorem (see [58, 108]) implies that $\sqrt{\mu}\sum_j^{t/\mu-1} \varphi_j v_j$ converges weakly to a Brownian motion process $\widetilde{w}(t)$ whose covariance is given by Rt, where

$$R = \mathbb{E}\varphi_0\varphi_0' v_0^2 + \sum_{j=1}^{\infty} \mathbb{E}\varphi_j\varphi_0' v_j v_0 + \sum_{j=1}^{\infty} \mathbb{E}\varphi_0\varphi_j' v_0 v_j. \tag{17.32}$$

THEOREM 17.3.4 *Assume that* (A1)–(A4) *hold, and that* $u^\mu(0)$ *converges weakly to* u^0. *Then the interpolated process* $u^\mu(\cdot)$ *defined in* (17.31) *converges weakly to* $u(\cdot)$, *which is the solution of the stochastic differential equation*

$$du(t) = -Bu(t)dt + R^{1/2}dw(t), \; u(0) = u^0, \tag{17.33}$$

where $w(\cdot)$ *is a standard r-dimensional Brownian motion, B and R are given in* (A2), §17.3.1 *and* (A3), §17.3.2, *and* $R^{1/2}$ *denotes the square root of R, (i.e.* $R = R^{1/2}(R^{1/2})'$).

Since B is symmetric positive definite, the matrix $-B$ is negative definite and hence Hurwitz (i.e. all of its eigenvalue being negative). It follows that

$$\Sigma = \int_0^\infty \exp(-Bt) R \exp(-Bt) dt, \tag{17.34}$$

is well defined. In fact, Σ is the stationary covariance of the diffusion process and can be obtained as the solution of the Lyapunov equation

$$\Sigma B + B\Sigma = R. \tag{17.35}$$

In view of Theorem 17.3.4, $\widehat{\theta}_n - x_n$ is asymptotically normally distributed with mean 0 and covariance $\mu\Sigma$. This covariance is identical to that of the constant step size LMS algorithm estimating a constant parameter (i.e. $\varepsilon = 0$). The expression for the asymptotic covariance is not surprising. Since x_n is a slowly varying Markov chain, its structure of transition probability matrix P^ε makes the chain acts almost like a constant parameter, with infrequent jumps. As a result, the process with suitable scaling leads to a diffusion limit, whose stationary covariance is the solution of the Lyapunov equation (17.35).

Some intuition: Consider the first-order time discretization of the stochastic differential equation (17.33) with discretization interval μ (see (17.31)) equal to the step size of the LMS algorithm. The resulting discretized system is

$$u_{n+1} = u_n - \mu B u_n + \sqrt{\mu} R^{1/2} v_n \tag{17.36}$$

where $v_n = w_{n+1} - w_n$ is a discrete-time white Gaussian noise. By comparing (17.36) with (17.33), it is intuitively clear that they are equivalent in distribution. In particular, by stochastic averaging principle, the fast variable $\phi_n\phi_n'$ behaves as its average B yielding the second term in (17.36). The equivalence in distribution of the third terms in (17.36) and (17.33) can be seen similarly. Thus discretizing the continuous-time process (17.33) yields (at least intuitively) the discrete-time process (17.36).

The proof Theorem 17.3.4 given in [347, 348] goes in the reverse direction, i.e. it shows that as the discretization interval $\mu \to 0$, the discrete-time process (17.33) converges weakly (in distribution) to the continuous-time process (17.36) under suitable technical assumptions. The key technical assumption for this weak convergence of a discrete-time process to a continuous-time process is that between discrete-time sample points, the process should be well behaved in distribution. This well-behavedness is captured by the tightness assumption which roughly speaking states that the between sample points, the process is bounded in probability; see Appendix D. The main tool used to prove the weak convergence is the "martingale problem" formulation; see [347, 348] for details.

17.3.3 Example: tracking slow HMMs

This subsection analyzes the usefulness of the LMS algorithm (17.22) for estimating the state of a hidden Markov model with slow dynamics. As in §17.3.2, assume that $\varepsilon = O(\mu^2)$ or $\mu = O(\sqrt{\varepsilon})$, i.e. the adaptation speed of the LMS algorithm (17.22) is faster than the Markov chain dynamics. As mentioned in Section 1, given the computationally efficiency of the LMS algorithm for estimating the state of an HMM, there is strong motivation to analyze the performance of the algorithm. Given that the underlying signal x_n is a finite state Markov chain, the probability of error of the estimate quantized to the nearest state value is a more meaningful performance measure than the asymptotic covariance.

A conventional HMM [107, 274] comprising of a finite state Markov chain observed in noise is of the form (17.20) where $\varphi_n = 1$ for all n and the states of the Markov chain g_i, $1 \leq i \leq m$ are real-valued scalars. For this HMM case, the LMS algorithm (17.22) has computational cost $O(1)$, i.e. independent of m.

Let $\widehat{\theta}_n^H$ denote the estimate $\widehat{\theta}_n$ of (17.22) quantized to the nearest Markov state:

$$\widehat{\theta}_n^H = g_{i^*} \quad \text{where } i^* = \arg\min_{1 \leq i \leq m} |g_i - \widehat{\theta}_n|. \tag{17.37}$$

For notational convenience assume that the states of the above HMM are ordered in ascending order and are equally spaced, i.e. $g_1 < g_2 < \cdots < g_X$, and $d = g_{i+1} - g_i$ is a positive constant.

Eq. (17.35) implies that $\Sigma = \sigma_e^2/2$. The probability of error can be computed as follows:

$$\begin{aligned} P(\widehat{\theta}_n^H \neq x_n) &= P(\widehat{\theta}_n - x_n > d/2 \mid x_n = g_1)P(x_n = g_1) \\ &\quad + \sum_{i=2}^{m-1} P(|\widehat{\theta}_n - x_n| > d/2 \mid x_n = g_i)P(x_n = g_i) \\ &\quad + P(x_n - \widehat{\theta}_n > d/2 \mid x_n \neq g_m)P(x_n = g_m). \end{aligned} \tag{17.38}$$

We summarize this in the following result.

THEOREM 17.3.5 *Suppose that the HMM satisfies* (A1), (A2) *and* $\{v_n\}$ *is a sequence of zero mean i.i.d. random variables satisfying* $\mathbb{E}X_n = 0$, $\mathbb{E}X_n^2 = \sigma^2 > 0$, *and* $\rho = \mathbb{E}|X_n|^3 < \infty$. *Then the probability of error of the LMS algorithm estimate* θ_n^H *(17.37) in estimating the state* x_n *of the HMM is*

$$\begin{aligned}P(\widehat{\theta}_n^H \neq x_n) &= \pi_1(\varepsilon n)\Phi^c\Big(\frac{d}{2\sqrt{\Sigma\mu}}\Big) + 2(\pi_2(\varepsilon n) + \cdots \\ &\quad + \pi_{m-1}(\varepsilon n))\Phi^c\Big(\frac{d}{2\sqrt{\Sigma\mu}}\Big) + \pi_m(\varepsilon n)\Phi^c\Big(\frac{d}{2\sqrt{\Sigma\mu}}\Big) \\ &= (2 - \pi_1(\varepsilon n) - \pi_m(\varepsilon n))\Phi^c\Big(\frac{d}{2\sqrt{\Sigma\mu}}\Big)\end{aligned} \tag{17.39}$$

where $\Phi^c(\cdot)$ *is the complementary Gaussian cumulative distribution function.*

The above result (based on weak convergence) is to be contrasted with the following computation of the error probability using the mean square convergence in Corollary 17.3.2 above. Using Chebyshev's inequality that $P(|\widehat{\theta}_n - x_n| > d/2 | x_n = g_i) \leq K\mu/d^2$ (where K is a positive constant independent of μ and d) together with (17.38) and Corollary 17.3.2 yields

$$P(\widehat{\theta}_n^H \neq x_n) \leq K\mu/d^2 \tag{17.40}$$

The above expression (17.40) is less useful than Theorem 17.3.5. However, it serves as a consistency check. As μ and $\varepsilon \to 0$, the probability of error of the tracking algorithm goes to zero.

Comparison of error probabilities

It is instructive to compare the probability of error expression (17.39) for the LMS algorithm (17.22) tracking a slow Markov chain parameter with that of the asymptotic HMM filter derived in [127]. By virtue of [330, Eq.3.35], the following upper bound holds for Φ^c in (17.39):

$$\Phi^c\left(\frac{d/2}{\sqrt{\mu\Sigma}}\right) \leq \frac{1}{\sqrt{2\pi}}\frac{\sqrt{\mu\Sigma}}{d/2}\exp\left(-\frac{(d/2)^2}{2\mu\Sigma}\right). \tag{17.41}$$

As described in [330], the above is an excellent approximation for small μ. In [127], it is shown that the steady state asymptotic HMM filter with $O(m)$ computational complexity has error probability $K\mu^2\log(1/\mu^2)$ where the constant K depends on the steady state probabilities of the Markov chain. These error probabilities converge to zero much faster than that of the $O(1)$ complexity LMS algorithm with error probability upper bound in (17.41).

17.3.4 Closing remark: tracking fast Markov chains

So far in this section, we have assumed that the Markov chain with transition matrix $P = I + \epsilon Q$ evolves on a slower timescale than the LMS algorithm with step size μ, i.e. $\epsilon = O(\mu^2)$. What happens if $\epsilon = O(\mu)$ so that the Markov chain evolves on the same timescale as the LMS algorithm? The analysis is more complex [348] and the main results are as follows:

- The mean square tracking error is (compare with Theorem 17.3.2)

$$\mathbb{E}|\widetilde{\theta}_n|^2 = \mathbb{E}|\widehat{\theta}_n - x_n|^2 = O(\mu). \tag{17.42}$$

- The ODE becomes a Markov modulated ODE (compare with Theorem 17.3.3)

$$\frac{d}{dt}\widetilde{\theta}(t) = -B\widetilde{\theta}(t) + Qx(t), \quad t \geq 0,\ \widetilde{\theta}(0) = \widetilde{\theta}^0.$$

  where $x(t)$ is the interpolated Markov process (constructed as in (17.28)) with transition rate matrix Q. So stochastic averaging no longer results in a deterministic ODE. Instead the limiting behavior is specified by a differential equation driven by a continuous-time Markov chain with transition rate matrix Q; see [348, 346].

17.4 Example 3: Discrete stochastic optimization for policy search

In this section we consider a POMDP modeled as a back box: a decision-maker can choose one of a finite number of policies and can observe the noisy costs incurred over a time horizon. The aim is to solve the following discrete stochastic optimization problem:

$$\text{Compute } \theta^* = \operatorname*{argmin}_{\theta \in \Theta} C(\theta), \quad \text{where } C(\theta) = \mathbb{E}\{c_n(\theta)\}. \tag{17.43}$$

Here

- $\Theta = \{1, 2, \dots, S\}$ denotes the finite discrete space of possible policies.
- θ is a parameter that specifies a policy – for example, it could specify how long and in which order to use specific sensors (sensing modes) in a controlled sensing problem.
- $c_n(\theta)$ is the observed cost incurred when using strategy θ for sensing. Typically, this cost is evaluated by running the POMDP over a specified horizon. So n can be viewed as an index for the n-th run of the POMDP.
- $\theta^* \in \mathcal{G}$ where $\mathcal{G} \subset \Theta$ denotes the set of global minimizers of (17.43).

It is assumed that the probability distribution of the noisy cost c_n is not known. Therefore, $C(\theta)$ in (17.43) cannot be evaluated at each time n; otherwise the problem is a deterministic integer optimization problem. We can only observe noisy samples $c_n(\theta)$ of the system performance $C(\theta)$ via simulation for any choice of $\theta \in \Theta$. Given this noisy performance $\{c_n(\theta)\}$, $n = 1, 2, \dots$, the aim is to estimate θ^*. The problem is shown schematically in Figure 17.1.

Context

An obvious brute force method for solving (17.43) involves an exhaustive enumeration: for each $\theta \in \Theta$, compute the empirical average

$$\widehat{c}_N(\theta) = \frac{1}{N}\sum_{n=1}^{N} c_n(\theta),$$

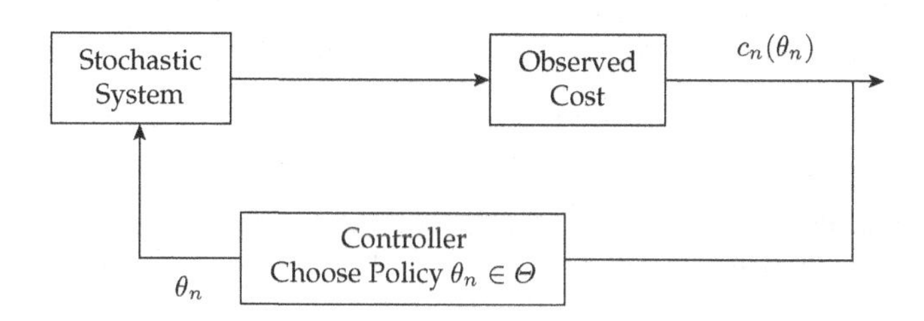

Figure 17.1 Discrete stochastic optimization problem. For any choice of $\theta_n \in \Theta = \{1, 2, \ldots, S\}$, noisy observations of the incurred cost $c_n(\theta_n)$ can be obtained. The aim is to determine the global minimizer $\theta^* = \operatorname{argmin}_{\theta \in \Theta} \mathbb{E}\{c_n(\theta)\}$. In the process of estimating θ^*, we want the controller to choose θ^* more often than any other value of $\theta \in \Theta$ so as not to waste resources making observations at non-optimal values of θ.

via simulation for large N. Then pick $\widehat{\theta} = \operatorname{argmin}_{\theta \in \Theta} \widehat{c}_N(\theta)$. Since for any fixed $\theta \in \Theta$, $\{c_n(\theta)\}$ is an i.i.d. sequence of random variables, by virtue of Kolmogorov's strong law of large numbers, $\widehat{c}_N(\theta) \to \mathbb{E}\{c_n(\theta)\}$ w.p.1, as $N \to \infty$. This and the finiteness of Θ imply that

$$\operatorname*{argmin}_{\theta \in \Theta} \widehat{c}_N(\theta) \to \operatorname*{argmin}_{\theta \in \Theta} \mathbb{E}\{c_n(\theta)\} \text{ w.p.1 as } N \to \infty. \tag{17.44}$$

In principle, the above brute force simulation method can solve the discrete stochastic optimization problem (17.43) and the estimate is *consistent*, i.e. (17.44) holds. However, the method is highly inefficient since $\widehat{c}_N(\theta)$ needs to be evaluated for each $\theta \in \Theta$. The evaluations of $\widehat{c}_N(\theta)$ for $\theta \notin \mathcal{G}$ are wasted because they contribute nothing to the estimation of $\widehat{c}_N(\theta)$, $\theta \in \mathcal{G}$.

The idea of discrete stochastic approximation is to design an algorithm that is both *consistent* and *attracted* to the minimum. That is, the algorithm should spend more time obtaining observations $c_n(\theta)$ in areas of the state space near $\mathcal{G}$ and less in other areas. The discrete stochastic optimization problem (17.43) is similar in spirit to the stochastic bandit problem which will be discussed briefly in §17.6 with the key difference that (17.43) aims to determine the minimizer θ^* while bandits aim to minimize an average observed function over a period of time (called the regret).

In the remainder of this section we provide two examples of discrete stochastic optimization algorithms:

- The smooth best-response adaptive search algorithm;
- Discrete stochastic search algorithm.

A numerical example is then given to illustrate the performance of these algorithms. Also the performance is compared with the *upper confidence bound (UCB) algorithm* which is a popular multi-armed bandit algorithm.

We are interested in situations where the objective function $C(\theta)$ in (17.43) evolves randomly over a slow timescale and hence the global minimizers evolve slowly with time. So the stochastic optimization algorithms described below have constant step sizes to facilitate tracking time varying global minimizers.

17.4.1 Algorithm 1: Smooth best-response adaptive search

Algorithm 24 describes the smooth best-response adaptive search algorithm for solving the discrete stochastic optimization problem (17.43). In this algorithm, $\gamma \in (0, D]$ denotes the exploration weight where $D \geq \max_\theta C(\theta) - \min_\theta C(\theta)$ upper bounds the maximum difference in objective function among the feasible solutions.

Algorithm 24 Smooth best-response adaptive search

Step 0: **Initialization:** Set exploration parameter $\gamma \in (0, D]$. Initialize $\psi_0 = \mathbf{0}_S$.

Step 1: **Sampling:** Sample $\theta_n \in \Theta$ according to S-dimensional probability vector

$$b^\gamma(\psi_n) = \left[b_1^\gamma(\psi_n), \ldots, b_S^\gamma(\psi_n)\right], \quad \text{where } b_i^\gamma(\psi) = \frac{\exp\left(-\psi_i/\gamma\right)}{\sum_{j\in\Theta}\exp\left(-\psi_j/\gamma\right)}.$$

Step 2: **Evaluation:** Perform simulation to obtain $c_n(\theta_n)$.

Step 3: **Belief update:** Update $\psi_{n+1} \in \mathbb{R}^S$ as

$$\psi_{n+1} = \psi_n + \mu\left[f\left(\theta_n, \psi_n, c_n(\theta_n)\right) - \psi_n\right], \tag{17.45}$$

$$\text{where } f = (f_1, \ldots, f_S)', \quad f_i\left(\theta_n, \psi_n, c_n(\theta_n)\right) = \frac{c_n(\theta_n)}{b_i^\gamma(\psi_n)} I(\theta_n = i).$$

Step 4: **Recursion:** Set $n \leftarrow n + 1$ and go to Step 1.

The S-dimensional probability vector $b^\gamma(\psi_n)$ in Step 1 of Algorithm 24 is actually a special case of the following general framework: let Π denote the unit S-dimensional simplex, and int(Π) denote the interior of this simplex. Given the vector $\psi = [\psi_1, \ldots, \psi_S]' \in \mathbb{R}^S$ of beliefs about objective values at different candidate solutions, the *smooth best-response sampling* strategy $b^\gamma \in \Pi$ is

$$b^\gamma(\psi) := \underset{\sigma\in\Pi}{\operatorname{argmin}}\left(\sigma'\psi - \gamma\rho(\sigma)\right), \quad 0 < \gamma < D, \tag{17.46}$$

where $\sigma \in \Pi$ and the perturbation function $\rho(\sigma) : \text{int}(\Pi) \to \mathbb{R}$ satisfy

i) $\rho(\cdot)$ is continuously differentiable, strictly concave, and $|\rho| \leq 1$;
ii) $\|\nabla\rho(\sigma)\| \to \infty$ as σ approaches the boundary of Π where $\|\cdot\|$ denotes the Euclidean norm.

In Algorithm 24, we chose $\rho(\cdot)$ as the *entropy function* [119] $\rho(\sigma) = -\sum_{i\in\Theta}\sigma_i \log(\sigma_i)$. Then applying (17.46) yields Step 1 which can be viewed as a Boltzmann exploration strategy with constant temperature

$$b_i^\gamma(\psi) = \frac{\exp\left(-\psi_i/\gamma\right)}{\sum_{j\in\Theta}\exp\left(-\psi_j/\gamma\right)}. \tag{17.47}$$

Such an exploration strategy is used widely in the context of game-theoretic learning and is called logistic fictitious-play [120] or logit choice function [147].

ODE analysis

The aim is to show that the sequence $\{\theta_n\}$ generated by Algorithm 24 spends more time in the set of global maximizers $\mathcal{G}$ than any other point in Θ. We first define two performance measures: regret and empirical sampling distribution.

The *regret* is defined as the "opportunity loss" and compares the performance of an algorithm, selecting among S alternatives, to the performance of the best of those alternatives in hindsight. Accordingly, define the regret as

$$r_n = r_{n-1} + \mu\left[c_n(\theta_n) - C(\theta^*)\right]. \tag{17.48}$$

Recall $C(\theta^*)$ defined in (17.43) is the expected cost of the global minimizer.

The *empirical sampling distribution* z_n is an S-dimensional probability vector with the θ-th element denoting the fraction of time that the algorithm samples a particular candidate $\theta \in \Theta$ until time n:

$$z_{n+1} = z_n + \mu\left[e_{\theta_n} - z_n\right]. \tag{17.49}$$

Here e_θ is the unit S-dimensional indicator vector with 1 in the θ-th position.

For both performance measures defined above, the step size $\mu \in (0,1)$ introduces an exponential forgetting and facilitates tracking a slowly time varying optimum θ^*. Choosing $\mu = 1/(n+1)$ gives the average regret and empirical sampling distribution; however motivated by tracking slowly time varying optimum, we are interested in fixed step size μ.

With the notation in (17.45), (17.48) and (17.49), define

$$Y_n = [\hat{\psi}_n', r_n, z_n']', \quad C = \left[C(1), \ldots, C(S)\right]'$$

$$\hat{\psi}_n = \psi_n - C, \; A_n\big(\theta_n, \hat{\psi}_n, c_n(\theta_n)\big) = \begin{bmatrix} f\left(\theta_n, \hat{\psi}_n + C, c_n(\theta_n)\right) - C \\ c_n(\theta_n) - C(\theta^*) \\ e_{\theta_n} \end{bmatrix}.$$

Then from (17.45), (17.48) and (17.49), Y_n satisfies the recursion

$$Y_{n+1} = Y_n + \mu\left(A_n\big(\theta_n, \hat{\psi}_n, c_n(\theta_n)\big) - Y_n\right). \tag{17.50}$$

We can now use the ODE analysis on the stochastic approximation (17.50) to analyze Algorithm 24. Define the piecewise constant continuous-time interpolated processes $Y^\mu(t) = Y_n$ for $t \in [n\mu, (n+1)\mu)$. Then Theorem 17.1.1 implies that the interpolated process Y^μ converges weakly to Y as $\mu \to 0$ such that the limit $Y = [\hat{\psi}', r, z']'$ is a solution to the ODE

$$\frac{dY}{dt} = F(Y) - Y, \quad \text{where } F(Y) = \begin{bmatrix} \mathbf{0}_S \\ C'\, b^\gamma\left(\hat{\psi} + C\right) - C(\theta^*) \\ b^\gamma\left(\hat{\psi} + C\right) \end{bmatrix}. \tag{17.51}$$

Let us explain the right hand side of (17.51). $F(Y)$ is obtained by averaging $A_n\big(\theta_n, \hat{\psi}_n, c_n(\theta_n)\big)$ in (17.50) with respect to the distribution $b^\gamma\left(\psi\right)$ of θ (recall Step 1 of Algorithm 24). In particular, with $\hat{\psi} = \psi - C$ we have

$$\begin{aligned}
\mathbb{E}\{f_i\left(\theta_n, \hat{\psi}_n + C, c_n(\theta_n)\right) - C(i)\} &= \mathbb{E}\{\frac{c_n(\theta_n)}{b_i^\gamma(\psi_n)} I(\theta_n = i)\} - C(i) \\
&= \frac{C(i)}{b_i^\gamma(\psi)} b_i^\gamma(\psi) - C(i)) = 0 \\
\mathbb{E}\{c_n(\theta_n) - C(\theta^*)\} &= \sum_{i=1}^{S} C(i) b_i^\gamma(\psi) - C(\theta^*) \\
\mathbb{E}\{e_{\theta_n}\} &= b^\gamma(\psi)
\end{aligned}$$

thereby yielding $F(Y)$ in (17.51).

Next, we investigate the stability and global attractor of the ODE (17.51). Notice that $F(Y)$ only depends on $\hat{\psi}$ and so there is no explicit interconnection between the dynamics of r and z in (17.51). Therefore, we start by examining

$$\frac{d}{dt}\begin{bmatrix} \hat{\psi} \\ r \end{bmatrix} = \begin{bmatrix} \mathbf{0}_S \\ C'\, b^\gamma(\hat{\psi} + C) - C(\theta^*) \end{bmatrix} - \begin{bmatrix} \hat{\psi} \\ r \end{bmatrix}. \tag{17.52}$$

The first component is asymptotically stable, and any trajectory $\hat{\psi}(t)$ decays exponentially fast to $\mathbf{0}_S$ as $t \to \infty$. We now show that the r, is globally asymptotically stable with global attracting set $\mathbb{R}_{[0,\eta)}$. Substituting the global attractor $\hat{\psi} = \mathbf{0}_S$ in (17.52) yields

$$\frac{dr}{dt} = C'\, b^\gamma(C) - C(\theta^*) - r.$$

Choosing the Lyapunov function $V(r) = r^2$, yields

$$\frac{d}{dt}V(r) = 2r\left[b^\gamma(C) \cdot C - C(\theta^*) - r\right] \le 2r\left[K(\gamma) - r\right]$$

for some constant $K(\gamma)$ which is monotone increasing in γ. So for $\eta > 0$, $\widehat{\gamma}$ can be chosen small enough such that, if $\gamma \le \widehat{\gamma}$ and $r \ge \eta$, then $\frac{d}{dt}V(r) \le -V(r)$. Therefore, for $\gamma \le \widehat{\gamma}$, it follows that $\lim_{t\to\infty} d(r(t), \mathbb{R}_{[0,\eta)}) = 0$ where the distance $d(r(t), \mathbb{R}_{[0,\eta)}) = \inf_{x\in\mathbb{R}_{[0,\eta)}} |x - r(t)|$.

It remains to analyze stability of

$$\frac{d}{dt}\begin{bmatrix} \hat{\psi} \\ z \end{bmatrix} = \begin{bmatrix} \mathbf{0}_S \\ b^\gamma(\hat{\psi} + C) \end{bmatrix} - \begin{bmatrix} \hat{\psi} \\ z \end{bmatrix}.$$

The first component is asymptotically stable. Substituting the global attractor $\hat{\psi} = \mathbf{0}_S$ yields $\frac{dz}{dt} = b^\gamma(C) - z$ which is globally asymptotically stable. Hence any trajectory $z(t)$ converges exponentially fast to $b^\gamma(C)$ as $t \to \infty$.

To summarize: The ODE analysis shows that for sufficiently small μ, Algorithm 24 spends approximately $b_\theta^\gamma(C(\theta))$ fraction of time at candidate $\theta \in \Theta$ where b^γ is defined in Step 1. So it spends most time at the global minimizer θ^*.

17.4.2 Algorithm 2: Discrete stochastic search

The second discrete stochastic optimization algorithm that we discuss is displayed in Algorithm 25. This random search algorithm was proposed by Andradottir [19, 20] for computing the global minimizer in (17.43). Recall $\mathcal{G}$ denotes the set of global minimizers of (17.43) and $\Theta = \{1, 2, \ldots, S\}$ is the search space. The following assumption is required:

(O) For each $\theta, \tilde{\theta} \in \Theta - \mathcal{G}$ and $\theta^* \in \mathcal{G}$,

$$\mathbb{P}(c_n(\theta^*) < c_n(\theta)) \geq \mathbb{P}(c_n(\theta) > c_n(\theta^*))$$
$$\mathbb{P}(c_n(\tilde{\theta}) > c_n(\theta^*)) \geq \mathbb{P}(c_n(\tilde{\theta}) > c_n(\theta))$$

Assumption (O) ensures that the algorithm is more likely to jump toward a global minimum than away from it. Suppose $c_n(\theta) = \theta + w_n(\theta)$ in (17.43) for each $\theta \in \Theta$, where $\{w_n(\theta)\}$ has a symmetric probability density function or probability mass function with zero mean. Then assumption (O) holds.

Algorithm 25 Random search (RS) algorithm

Step 0: **Initialization:** Select $\theta_0 \in \Theta$. Set $\pi_{i,0} = 1$ if $i = \theta_0$, and 0 otherwise. Set $\theta_0^* = \theta_0$ and $n = 1$.

Step 1: **Sampling:** Sample a candidate solution $\tilde{\theta}_n$ uniformly from the set $\Theta - \theta_{n-1}$.

Step 2: **Evaluation:** Simulate samples $c_n(\theta_{n-1})$ and $c_n(\tilde{\theta}_n)$.
If $c_n(\tilde{\theta}_n) < c_n(\theta_{n-1})$, set $\theta_n = \tilde{\theta}_n$, else, set $\theta_n = \theta_{n-1}$.

Step 3: **Belief (occupation probability) update:**

$$\pi_n = \pi_{n-1} + \mu \left[e_{\theta_n} - \pi_{n-1} \right], \quad 0 < \mu < 1, \tag{17.53}$$

where e_{θ_n} is the unit vector with the θ_n-th element being equal to one.

Step 4: **Global optimum estimate:** $\theta_n^* \in \operatorname{argmax}_{i \in \Theta} \pi_{i,n}$.

Step 5: **Recursion:** Set $n \leftarrow n + 1$ and go to Step 1.

Convergence analysis of Algorithm 25

We now show that Algorithm 25 is attracted to the global minimizer set $\mathcal{G}$ in the sense that it spends more time in $\mathcal{G}$ than at any other candidate in $\Theta - \mathcal{G}$. The sequence $\{\theta_n\}$ generated by Algorithm 25 is a homogeneous Markov chain on the state space Θ - this follows directly from Steps 1 and 2 of Algorithm 25. Let P denote the $S \times S$ transition probability matrix of this Markov chain. Providing P is aperiodic irreducible, then P has a stationary distribution π_∞ which satisfies $P'\pi_\infty = \pi_\infty$, $\sum_{i=1}^{S} \pi_\infty[i] = 1$.

Let us now use apply the ODE analysis to (17.53). Define the interpolated continuous-time process $\pi^\mu(t) = \pi_n$ for $t \in [n\mu, (n+1)\mu)$. Then by Theorem 17.1.1, the process $\pi^\mu(\cdot)$ converges weakly at $\mu \to 0$ to $\pi(\cdot)$ which satisfies the ODE $\frac{d\pi}{dt} = \pi_\infty - \pi_t$ where π_∞ is the stationary distribution of transition matrix P. Thus clearly π_t converges exponentially fast to π_∞.

It only remains to prove that the elements of the stationary distribution π_∞ are maximized at the global minimizer – this implies that the algorithm spends more time at the global minimizer $\theta^* \in \mathcal{G}$ than any other candidate.

THEOREM 17.4.1 *Consider the discrete stochastic optimization problem (17.43). Then under assumption (O), the Markov chain $\{\theta_n\}$ generated by Algorithm 25 has the following property for its stationary distribution π_∞:*

$$\pi_\infty(i) > \pi_\infty(j), \text{ for } i \in \mathcal{G},\ j \in \Theta - \mathcal{G}.$$

The proof is in the appendix and was originally given in [19]. The proof follows from Assumption (O) which shapes the transition probability matrix P and hence invariant distribution π_∞ by imposing the following constraints: $P_{ij} < P_{ji}$ for $i \in \mathcal{G}, j \in \Theta - \mathcal{G}$, i.e. it is more probable to jump from a state outside $\mathcal{G}$ to a state in $\mathcal{G}$ than the reverse; $P_{ij} < P_{lj}$ for $i \in \mathcal{G}, j, l \in \Theta - \mathcal{G}$, i.e. it is less probable to jump out of the global optimum i to another state j compared to any other state l. Intuitively, one would expect that such a transition probability matrix would generate a Markov chain that is attracted to the set $\mathcal{G}$ (spends more time in $\mathcal{G}$ than other states), i.e. $\pi_\infty[i] > \pi_\infty[j]$, $i \in \mathcal{G}, j \notin \mathcal{G}$.

17.4.3 Algorithm 3: Upper confidence bound (UCB) algorithm

The final algorithm that we discuss is the UCB algorithm. UCB [27] belongs to the family of "follow the perturbed leader" algorithms and is widely used in the context of multi-armed bandits (see §17.6.2). Since bandits are typically formulated in terms of maximization (rather than minimization), we consider here *maximization* of $C(\theta)$ in (17.43) and denote the global maximizers as θ^*. Let B denote an upper bound on the objective function and $\xi > 0$ be a constant. The UCB algorithm is summarized in Algorithm 26. For a static discrete stochastic optimization problem, we set $\mu = 1$; otherwise, the discount factor μ has to be chosen in the interval $(0, 1)$. Each iteration of UCB requires $O(S)$ arithmetic operations, one maximization and one simulation of the objective function.

Algorithm 26 Upper confidence bound (UCB) algorithm for maximization of objective $C(\theta)$, $\theta \in \Theta = \{1, 2, \ldots, S\}$ with discount factor $\mu \in (0, 1]$ and exploration constant $\xi > 0$. B is an upper bound on the objective function.

Step 0. **Initialization.** Simulate each $\theta \in \Theta = \{1, 2, \ldots, S\}$ once to obtain $c_1(\theta)$. Set $\widehat{c}_{\theta,S} = c_1(\theta)$ and $m_{\theta,S} = 1$. Set $n = S + 1$.

Step 1a. **Sampling.** At time n sample candidate solution

$$\theta_n = \operatorname*{argmax}_{\theta \in \Theta} \left[\widehat{c}_{\theta,n-1} + B \sqrt{\frac{\xi \log(M_{n-1} + 1)}{m_{\theta,n-1}}} \right]$$

where $\widehat{c}_{\theta,n-1} = \frac{1}{m_{\theta,n-1}} \sum_{\tau=1}^{n-1} \mu^{n-\tau-1} c_\tau(\theta_\tau) I(\theta_\tau = \theta)$,

$$m_{\theta,n-1} = \sum_{\tau=1}^{n-1} \mu^{n-\tau-1} I(\theta_\tau = \theta), \quad M_{n-1} = \sum_{i=1}^{S} m_{i,n-1}.$$

Step 1b. **Evaluation.** Simulate to obtain $c_n(\theta_n)$.

Step 2. **Global Optimum Estimate.** $\theta_n^* \in \operatorname{argmax}_{i \in \Theta} \widehat{c}_{i,n}$.

Step 3. **Recursion.** Set $n \leftarrow n + 1$ and go to Step 1.

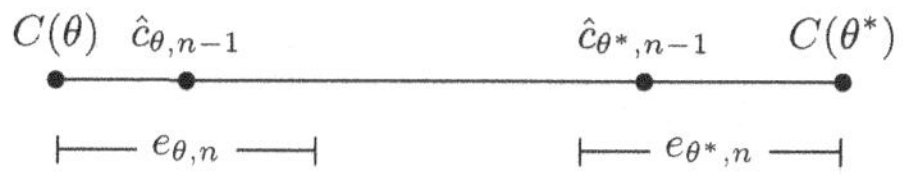

$$
\begin{array}{cccc}
C(\theta) \ \hat{c}_{\theta,n-1} & & \hat{c}_{\theta^*,n-1} & C(\theta^*) \\
\vdash e_{\theta,n} \dashv & & \vdash e_{\theta^*,n} \dashv &
\end{array}
$$

Figure 17.2 Illustration of confidence bounds in UCB algorithm.

Analysis of UCB Algorithm

Consider the UCB Algorithm 26 with discount factor $\mu = 1$. So $M_{n-1} + 1 = n$ and m_{in} is the number of times candidate i has been sampled until time n.

THEOREM 17.4.2 *Consider the UCB Algorithm 26 with discount factor $\mu = 1$ and exploration constant $\xi > 1$. Then the expected number of times candidate θ is sampled in n time points satisfies*

$$
\mathbb{E}\{m_{\theta n}\} \leq \frac{4B^2}{(C(\theta^*) - C(\theta))^2} \xi \log n + \frac{3}{2} + \frac{1}{2(\xi - 1)}. \tag{17.54}
$$

The proof is in the appendix. The intuition behind UCB when $\mu = 1$ is as follows. Let $e_{\theta,n} = B\sqrt{\frac{\xi \log(n)}{m_{i,n}}}$. Then $\hat{c}_{\theta,n-1} + e_{\theta,n}$ in Step 1a can be viewed as the upper bound of a confidence interval. If a particular candidate θ has been sampled sufficiently many times, then with high probability

$$
\hat{c}_{\theta^*,n-1} \geq C(\theta^*) - e_{\theta^*,n} \quad \text{and } \hat{c}_{\theta,n-1} \leq C(\theta) + e_{\theta,n}. \tag{17.55}
$$

as illustrated in the Figure 17.2. Clearly from Step 1 of the UCB algorithm, candidate θ is no longer sampled if

$$
\hat{c}_{\theta,n-1} + e_{\theta,n} < \hat{c}_{\theta^*,n-1} + e_{\theta^*,n}. \tag{17.56}
$$

From (17.55), it follows that $\hat{c}_{\theta,n-1} + e_{\theta,n} \leq C(\theta) + 2e_{\theta,n}$ and $\hat{c}_{\theta^*,n-1} + e_{\theta^*,n} > C(\theta^*)$. So a sufficient condition for (17.56) is that $C(\theta) + 2e_{\theta,n} < C(\theta^*)$, i.e.,

$$
m_{\theta,n-1} \geq \frac{4B^2}{(C(\theta^*) - C(\theta))^2} \xi \log n
$$

That is, in n time points, each suboptimal candidate θ in expectation is not sampled more often than $\log(n)/(C(\theta^*) - C(\theta))^2$ multiplied by a constant. So UCB achieves logarithmic regret, see §17.6.2 for a short discussion of regret.

The above analysis holds for the UCB algorithm with no discount ($\mu = 1$) and assumes that the global optimizer θ^* does not evolve with time. Suppose in a time horizon of N, the true optimizer θ^* jump changes Γ times within the feasible set Θ. Then to track this time varying optimum (denoted as θ_n^*), one chooses a discount factor $\mu < 1$ in Algorithm 26. It is of interest to determine the expected time the UCB algorithm spends in candidate θ when it is not optimal. It is shown in [122] that by choosing $\mu = 1 - (4B)^{-1}\sqrt{\Gamma/N}$,

$$
\sum_{n=1}^{N} I(\theta_n = \theta \neq \theta_n^*) = O(\sqrt{N\Gamma} \log N).
$$

This is to be compared with the standard UCB for fixed θ^* and $\mu = 1$ which spends $O(\log N)$ time at candidate θ.

17.4.4 Numerical examples

This section illustrates the performance of the above discrete stochastic optimization algorithms in estimating the mode of an unknown probability mass function. Let $\rho(\theta)$, $\theta \in \Theta = \{0, 1, \ldots, S\}$ denote the degree distribution[6] of a social network (random graph). Suppose that a pollster aims to estimate the mode of the degree distribution, namely

$$\theta^* = \underset{\theta \in \{0,1,\ldots,S\}}{\operatorname{argmax}} \ \rho(\theta). \tag{17.57}$$

The pollster does not know $\rho(\theta)$. It uses the following protocol to estimate the mode. At each time n, the pollster chooses a specific $\theta_n \in \Theta$, and then asks a randomly sampled individual in the social network: Is your degree θ_n? The individual replies "yes" (1) or "no" (0). Given these responses $\{I(\theta_n), n = 1, 2, \ldots\}$ where $I(\cdot)$ denotes the indicator function, the pollster aims to solve the discrete stochastic optimization problem: compute

$$\theta^* = \underset{\theta \in \{0,1,\ldots,S\}}{\operatorname{argmin}} \ -\mathbb{E}\left\{I(\theta_n)\right\}. \tag{17.58}$$

Clearly the global optimizers of (17.57) and (17.58) coincide. Below, we illustrate the performance of discrete stochastic optimization Algorithm 24 (abbreviated as AS), Algorithm 25 (abbreviated as RS), and Algorithm 26 (abbreviated as UCB) for estimating θ^* in (17.58).

In the numerical examples below, we simulated the degree distribution as a Poisson pmf with parameter λ:

$$\rho(\theta) = \frac{\lambda^\theta \exp(-\lambda)}{\theta!}, \quad \theta \in \{0, 1, 2, \ldots, S\}.$$

It is straightforward to show that the mode of the Poisson distribution is

$$\theta^* = \underset{\theta}{\operatorname{argmax}} \, \rho(\theta) = \lfloor \lambda \rfloor. \tag{17.59}$$

If λ is integer-valued, then both λ and $\lambda - 1$ are modes of the Poisson pmf. Since the ground truth θ^* is explicitly given by (17.59) we have a simple benchmark numerical study. (The algorithms do not use the knowledge that ρ is Poisson.)

Example 1: Static discrete stochastic optimization

First consider the case where the Poisson rate λ is a constant. We consider two examples: i) $\lambda = 1$, which implies that the set of global optimizers is $\mathcal{G} = \{0, 1\}$, and ii) $\lambda = 10$, in which case $\mathcal{G} = \{9, 10\}$. For each case, we further study the effect of the size of search

[6] The degree of a node is the number of connections or edges the node has to other nodes. The degree distribution $\rho(\theta)$ is the fraction of nodes in the network with degree θ, where $\theta \in \Theta = \{0, 1, \ldots\}$. Note that $\sum_\theta \rho(\theta) = 1$ and so the degree distribution is a pmf with support on Θ.

Table 17.1 Example 1: Percentage of independent runs of algorithms that converged to the global optimum Set in n iterations.

(a) $\lambda - 1$

Iteration	$S = 10$			$S = 100$		
n	AS	RS	UCB	AS	RS	UCB
10	55	39	86	11	6	43
50	98	72	90	30	18	79
100	100	82	95	48	29	83
500	100	96	100	79	66	89
1000	100	100	100	93	80	91
5000	100	100	100	100	96	99
10000	100	100	100	100	100	100

(b) $\lambda = 10$

Iteration	$S = 10$			$S = 100$		
n	AS	RS	UCB	AS	RS	UCB
10	29	14	15	7	3	2
100	45	30	41	16	9	13
500	54	43	58	28	21	25
1000	69	59	74	34	26	30
5000	86	75	86	60	44	44
10000	94	84	94	68	49	59
20000	100	88	100	81	61	74
50000	100	95	100	90	65	81

space on the performance of algorithms by considering two instances: i) $S = 10$, and ii) $S = 100$. Since the problem is static in the sense that $\mathcal{G}$ is fixed for each case, one can use the results of [41] to show that if the exploration factor γ in (17.46) decreases to zero sufficiently slowly, the sequence of samples $\{\theta_n\}$ converges almost surely to the global minimum. We consider the following modifications to AS Algorithm 24:

(i) The constant step-size μ in (17.45) is replaced by decreasing step-size $\mu(n) = \frac{1}{n}$;
(ii) The exploration factor γ in (17.46) is replaced by $\frac{1}{n^\beta}$, $0 < \beta < 1$.

We chose $\beta = 0.2$ and $\gamma = 0.01$. Also in Algorithm 26, we set $B = 1$, and $\xi = 2$.

Table 17.1 displays the performance of the algorithms AS, RS and UCB. To give a fair comparison, the performance of the algorithms are compared based on number of simulation experiments performed by each algorithm. Observe the following from Table 17.1: in all three algorithms, the speed of convergence decreases when either S or λ (or both) increases. However, the effect of increasing λ is more substantial since the objective function values of the worst and best states are closer when $\lambda = 10$. Given equal numbers of simulation experiments, a higher percentage of cases that a particular method has converged to the global optima indicates faster convergence rate.

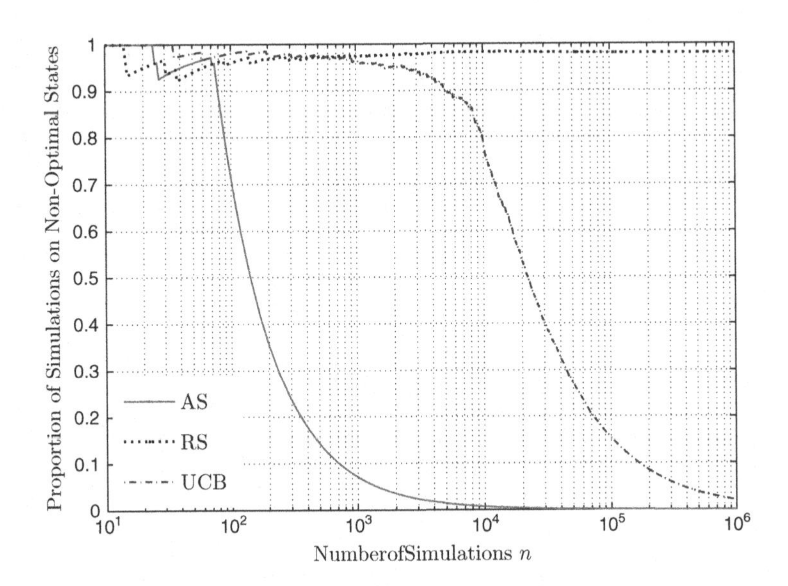

Figure 17.3 Example 1: Proportion of simulation effort expended on states outside the global optima set ($\lambda = 1, S = 100$).

To evaluate and compare efficiency of the algorithms, the sample path of the number of simulation experiments performed on non-optimal feasible solutions is displayed in Figure 17.3, when $\lambda = 1$ and $S = 100$. As can be seen, since the RS method randomizes among all (except the previously sampled) feasible solutions at each iteration, it performs approximately 98% of the simulations on non-optimal elements. The UCB algorithm switches to its exploitation phase after a longer period of exploration as compared to the AS algorithm.

Example 2: Regime-switching discrete stochastic optimization.

Consider now the case where the Poisson rate $\lambda(\theta_n)$ jump changes between the values of 1 and 10 according to a slow Markov chain $\{\theta_n\}$ with state space $\mathcal{M} = \{1, 2\}$, and transition probability matrix

$$P^\epsilon = I + \epsilon Q, \quad Q = \begin{bmatrix} -0.5 & 0.5 \\ 0.5 & -0.5 \end{bmatrix}. \tag{17.60}$$

Since the modes of a Poisson pmf with integer valued rate λ are λ and $\lambda - 1$, the sets of global optimizers are $\mathcal{G}(1) = \{0, 1\}$ and $\mathcal{G}(2) = \{9, 10\}$, respectively. We chose $\gamma = 0.1$, and $\mu = \epsilon = 0.01$ in Algorithm 24. Such Markov modulated social networks arise in friendship networks where the dynamics of link formation change for example with the occurrence of a large festival.

Figure 17.4 compares efficiency of algorithms for different values of ε in (17.60), which determines the speed of evolution of the hyper-model. Each point on the graph is an average over 100 independent runs of 10^6 iterations of the algorithms. As expected, the estimate of the global optimum spends more time in the global optimum for all

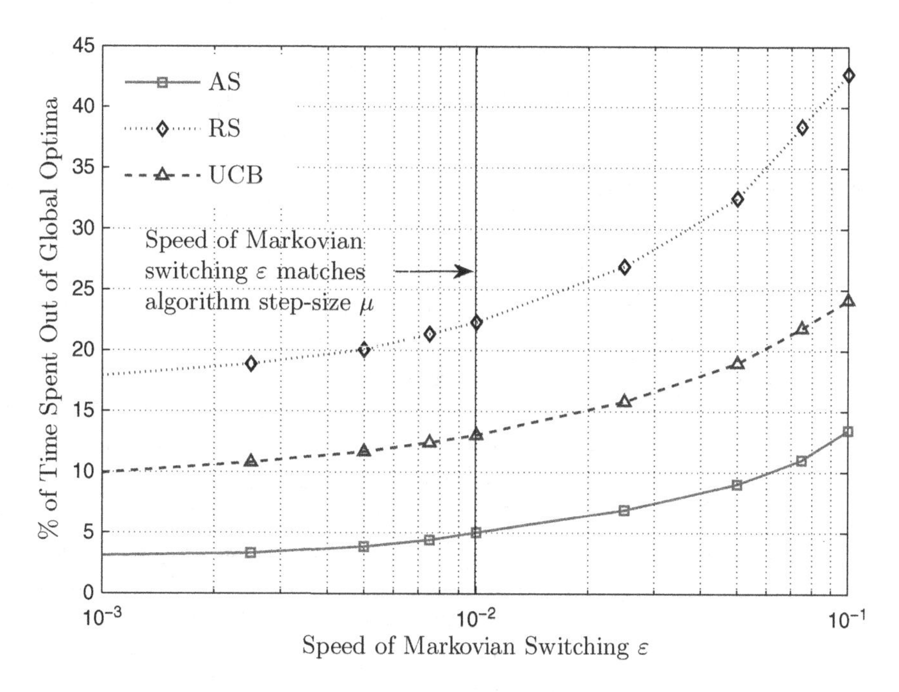

Figure 17.4 Example 2: Proportion of time the global optimum estimate does not match the true global optimum versus the switching speed of the global optimizers set after $n = 10^6$ iterations.

methods as the speed of time variations decreases. The estimate provided by AS differs from the true global optimum less frequently compared with the RS and UCB.

17.5 Example 4: Mean field population dynamics models for social sensing

So far in this chapter, the ODE analysis has been used to analyze stochastic approximation algorithms. This section shows that the ODE analysis can be used as a modeling tool to obtain a tractable model for Markov chains with large state spaces that model population dynamics. We are interested in constructing tractable models for the diffusion of information over a social network comprising of a population of interacting agents. As described in [221], such models arise in a wide range of social phenomena such as diffusion of technological innovations, ideas, behaviors, trends [129], cultural fads and economic conventions [84] where individual decisions are influenced by the decisions of others. Once a tractable model has been constructed, Bayesian filters can be used to estimate the underlying state of nature (such as sentiment).

Consider a social network comprised of individuals (agents) who may or may not adopt a specific new technology (such as a particular brand of smartphone). We are interested in modeling how adoption of this technology diffuses in the social network. Figure 17.5 shows the schematic setup. The underlying state of nature $\{x_k\}$ can be viewed as the market conditions or competing technologies that evolve with time and affect the

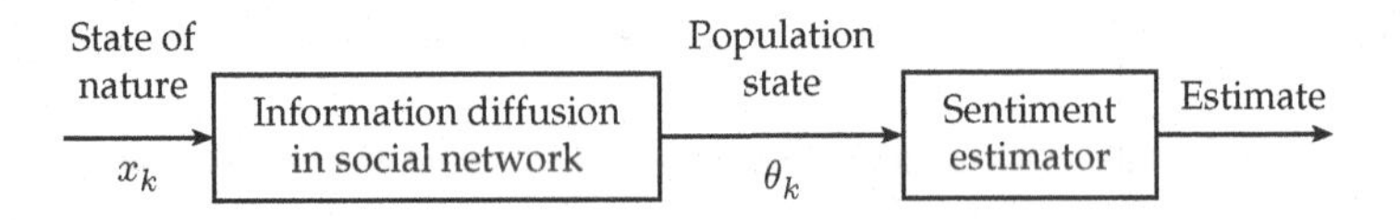

Figure 17.5 Social sensing of sentiment.

adoption of the technology. The information (adoption of technology) diffuses over the network – the states of individual nodes (adopt or don't adopt) evolve over time as a probabilistic function of the states of their neighbors[7] and the state of nature $\{x_k\}$. Let θ_k denote the population state vector at time k; as explained below the l-th element of this vector, denoted $\theta_k(l)$, is the fraction of the population with k neighbors that has adopted the technology. As the adoption of the new technology diffuses through the network, its effect is observed by social sensing – such as user sentiment[8] on a micro-blogging platform like Twitter. The nodes that tweet their sentiments act as social sensors. Suppose the state of nature x_k changes suddenly due to a sudden market shock or presence of a new competitor. The goal for a market analyst or product manufacturer is to estimate the state of nature x_k so as to detect the market shock or new competitor.

As a signal processing problem, the state of nature x_k can be viewed as the signal (Markov chain), and the social network can be viewed as a sensor. The observed sentiment can be viewed as an HMM: noisy measurement of θ_k (population state) which in turn depends on x_k (state of nature). The key difference compared to classical signal processing is that the social network (sensor) has dynamics due to the information diffusion over a graph. Estimating x_k can be formulated as a filtering problem, while detecting sudden changes in x_k can be formulated as a quickest detection problem.

Dealing with large populations results in an combinatorial explosion in the size of the state space for θ_k. The aim of this section is to construct tractable models of the population dynamics by using the mean field dynamics approximation.

17.5.1 Population model dynamics

Let x_k denote the state of nature at time k. Consider a population consisting of M agents indexed by $m = 1, 2, \dots .M$. Each agent m at time k has state $s_k^{(m)} \in \{1, 2 \dots, L\}$. Let the L dimensional vector θ_k denote the fraction of agents in the L different states at time k. We call θ_k the *population state*:

$$\theta_k(l) = \frac{1}{M} \sum_{m=1}^{M} I(s_k^{(m)} = l), \quad l = 1, 2, \dots, L. \tag{17.61}$$

where $I(\cdot)$ denotes the indicator function.

[7] In a study conducted by social networking site myYearbook, 81% of respondents said they had received advice from friends and followers relating to a product purchase through a social site; 74% of those who received such advice found it to be influential in their decision (ClickZ, January 2010).

[8] To quote from Wikipedia: "Sentiment analysis refers to the use of natural language processing, text analysis and computational linguistics to identify and extract subjective information in source materials."

Given the states $s_k^{(m)}$, $m = 1,\dots,M$ and hence population state θ_k (17.61), the population evolves as a Markov process in the following 2-steps: At each time k

Step 1. An agent m_k is chosen uniformly from the M agents. The state $s_{k+1}^{(m_k)}$ at time $k+1$ of this agent is simulated with transition probability

$$P_{ij}(x_k,\theta_k) = \mathbb{P}(s_{k+1} = j | s_k = i, x_k, \theta_k)$$

which depends on both the state of nature and the population state.

Step 2. Then the population state θ_{k+1} reflects this updated state of agent m_k:

$$\theta_{k+1} = \theta_k + \frac{1}{M}\left[e_{s_{k+1}^{(m_k)}} - e_{s_k^{(m_k)}} \right] \tag{17.62}$$

$$\mathbb{P}\big(\theta_{k+1} = \theta_k + \frac{1}{M}[e_j - e_i] \mid \theta_k, x_k\big) = \frac{1}{M} P_{ij}(x_k,\theta_k).$$

Note that $\{\theta_k\}$ is an $\binom{M+L-1}{L-1}$ state Markov chain with state space

$$\Theta = \{\theta : \sum_{l=1}^{L} \theta(l) = 1,\ \theta(l) = n/L \text{ for some integer } n \geq 0\}.$$

Note also that Θ is a subset of the $L-1$ dimensional simplex denoted as $\Pi(L)$.
Example: If $M = 3$ and $L = 2$, then there are four population states:

$$\theta_k \in \Theta = \{[1,0],\ [2/3,1/3],\ [1/3,2/3],\ [0,1]\}.$$

Suppose the current population state is $\theta_k = [2/3, 1/3]$. If the chosen agent m_k was in state 1 at time k, then it can either remain in state 1, in which case $\theta_{k+1} = [2/3,1/3]$, or jump to state 2 in which case $\theta_{k+1} = [1/3,2/3]$ according to (17.62).

We are interested in modeling the evolution of the population process $\{\theta_k\}$ in (17.62) when the number of agents M is large. The finite state model θ_k becomes intractable due to the combinatorial explosion in the number of states. For large population M, it turns out that the evolution of the population process θ_k resembles a stochastic approximation algorithm.

THEOREM 17.5.1 *For a population size of M agents, where each agent has L possible states, the population distribution process θ_k evolves in $\Theta \subset \Pi(L)$ as*

$$\theta_{k+1} = \theta_k + \frac{1}{M} H(x_k,\theta_k) + \nu_k, \quad \theta_0 \in \Theta, \tag{17.63}$$

$$\textit{where } H(x_k,\theta_k) = \sum_{i=1}^{L}\sum_{j=1}^{L} (e_j - e_i) P_{ij}(x_k,\theta_k).$$

Here ν_k is an L dimensional finite-state martingale increment process with $\|\nu_k\|_2 \leq \Gamma/M$ for some positive constant Γ.

Theorem 17.5.1 is the martingale representation of the Markov chain $\{\theta_k\}$ and was proved in §2.4.2. In particular, (17.63) implies that the population dynamics resemble a stochastic approximation algorithm: the new state is the old state plus a noisy update (the "noise" being a martingale difference process) that is weighed by a step size $1/M$ which is small when the population M is large.

17.5.2 Mean field dynamics

The main result below shows that for large population M, the population process θ_k in (17.63) converges to a deterministic difference equation (or equivalently, an ODE) called the mean field dynamics. Thus the ODE method serves the purpose of constructing a tractable model for the population dynamics.

THEOREM 17.5.2 (Mean field dynamics) *Consider the deterministic mean field dynamics process with state $\bar{\theta}_k \in \Pi(L)$ (the $L-1$ dimensional unit simplex):*

$$\bar{\theta}_{k+1} = \bar{\theta}_k + \frac{1}{M} H(x_k, \bar{\theta}_k), \quad \bar{\theta}_0 = \theta_0. \tag{17.64}$$

Assume that $H(\theta, x)$ is Lipschitz continuous[9] *in θ. Then for a time horizon of N points, the deviation between the mean field dynamics $\bar{\theta}_k$ in (17.64) and actual population distribution θ_k in (17.63) satisfies*

$$\mathbb{P}\{\max_{0 \leq k \leq N} \left\|\bar{\theta}_k - \theta_k\right\|_\infty \geq \epsilon\} \leq C_1 \exp(-C_2 \epsilon^2 M) \tag{17.65}$$

providing $N = O(M)$.

The proof is in the appendix. Equivalently, we can express the above result in continuous time as an ODE. Define the sampling period as $1/M$. Define the continuous-time linear interpolated process[10] $\theta(t)$, $t \in [0, N/M]$ associated with θ_k, $k = 0, 1, \ldots, N-1$ as

$$\theta(t) = \theta_k + \frac{t - k/M}{1/M} (\theta_{k+1} - \theta_k), \ t \in \left[\frac{k}{M}, \frac{(k+1)}{M}\right). \tag{17.66}$$

Define the continuous-time process $x(t)$ for the state of nature similarly. Define the continuous-time mean field dynamics process $\bar{\theta}(t)$ via the ODE

$$\frac{d\bar{\theta}}{dt} = H(x, \bar{\theta}). \tag{17.67}$$

Then the main assertion of Theorem 17.5.2, namely (17.65), is equivalent to saying that the interpolated population process $\theta(t)$ defined in (17.66) converges to the mean field dynamics process $\bar{\theta}(t)$ (17.67) in the following sense:

$$\mathbb{P}\left\{\max_{0 \leq t \leq T} \left\|\bar{\theta}(t) - \theta(t)\right\|_\infty \geq \epsilon\right\} \leq C_1 \exp(-C_2 \epsilon^2 M) \tag{17.68}$$

where C_1 and C_2 are positive constants and $T = N/M$ is a finite time horizon.

Unlike the mean square error analysis of Theorem 17.3.1, Theorem 17.5.2 gives an exponential bound to the probability of maximum deviation. As explained in [44], since the exponential bound $\sum_M \exp(-CM)$ is summable, the Borel–Cantelli lemma[11]

[9] For any $\alpha, \beta \in \Pi(L)$, $\|H(x,\alpha) - H(x,\beta)\|_\infty \leq \lambda \|\alpha - \beta\|_\infty$ for some positive constant λ.

[10] One can also consider a piecewise constant interpolation as done earlier in this chapter.

[11] If a sequence of events $\{A_n\}$ satisfies $\sum_n \mathbb{P}(\mathbb{A}_n) < \infty$, then $\mathbb{P}(A_n$ infinitely often$) = 0$.

applies. This facilities concluding important results on the exit times of the population process. In particular, consider the continuous-time processes $\theta(t)$ and $\bar{\theta}(t)$. Given an open set U in the $L-1$ unit simplex $\Pi(L)$, define the exit time as the random variable

$$\tau^M(U) = \inf\{t \geq 0 : \theta(t) \notin U\}. \tag{17.69}$$

Let $\gamma^+(\theta)$ denote the closure of the set of states visited by the mean field trajectory $\theta(t)$, $t \geq 0$. Then using Theorem 17.5.2, [44, Proposition 1] shows that

$$\mathbb{P}\left\{\lim_{M\to\infty} \tau^M(U) = \infty\right\} = 1. \tag{17.70}$$

In words, the exit time of the population process from any neighborhood of the mean field dynamics trajectory is probabilistically very large. Indeed, [44] concludes that attractors of the mean field dynamics are good predictors of the stochastic process when the initial conditions are close.

17.5.3 Example: Diffusion of information in social network

The aim of this section is to illustrate the population model and mean field dynamics of §17.5.1 and §17.5.2 for the diffusion of contagion (or technology) in a social network. The model we present below for the diffusion of information is called the *Susceptible-Infected-Susceptible* (SIS) model [259, 329] and we follow the setup in [221]. The social network that we consider is an undirected graph:

$$G = (V, E), \text{ where } V = \{1, 2, \ldots, M\}, \text{ and } E \subseteq V \times V. \tag{17.71}$$

Here, V denotes the set of M nodes (vertices), and E denotes the set of links (edges). The degree of a node m is its number of neighbors:

$$D^{(m)} = |\{n \in V : m, n \in E\}|,$$

where $|\cdot|$ denotes cardinality. Let $M(l)$ denote the number of nodes in the social network G with degree l, and let the degree distribution $\rho(l)$ specify the fraction of nodes with degree l. That is, for $l = 0, 1, \ldots, L$,

$$M(l) = \sum_{m\in V} I\big(D^{(m)} = l\big), \quad \rho(l) = \frac{M(l)}{M}.$$

Here, $I(\cdot)$ denotes the indicator function. Note that $\sum_l \rho(l) = 1$. The degree distribution can be viewed as the probability that a node selected randomly with uniform distribution on V has a connectivity l. We assume that the degree distribution of the social network is arbitrary but known – allowing an arbitrary degree distribution facilities modeling complex networks.

Assume each agent m has two possible states ($L = 2$),

$$s_k^{(m)} \in \{1 \text{ (infected)}, 2 \text{ (succeptible)}\}.$$

Let $\theta_k(l)$ denote the fraction of nodes with degree l at time k that are infected. We call the L-dimensional vector θ_k as the *infected population state* at time k. So

$$\theta_k(l) = \frac{1}{M(l)} \sum_m I\big(D^{(m)} = l, s_k^{(m)} = 1\big), \quad l = 1, \ldots, L. \tag{17.72}$$

Similar to §17.5.1, the infected population evolves as follows: if node m has degree $D^{(m)} = l$, then it jumps from state i to j with probabilities

$$P_{ij}(l, a, x) = \mathbb{P}\left(s_{k+1}^{(m)} = j | s_k^{(m)} = i, D^{(m)} = l, A_k^{(m)} = a, x_k = x\right) \quad i, j \in \{1, 2\}. \tag{17.73}$$

Here, $A_k^{(m)}$ denotes the number of infected neighbors of node m at time k. In words, the transition probability of an agent depends on its degree and the number of infected neighbors. The infected population state is updated as

$$\theta_{k+1}(l) = \theta_k(l) + \frac{1}{M(l)}[I(s_{k+1}^{(l)} = 1, s_k^{(l)} = 2) - I(s_{k+1}^{(l)} = 2, s_k^{(l)} = 1)].$$

We are interested in modeling the evolution of infected population using mean field dynamics. The following statistic forms a convenient parametrization of the transition probabilities of θ. Define $\alpha(\theta_k)$ as the probability that a uniformly sampled link in the network at time k has at least one node that is infected. We call $\alpha(\theta_k)$ as the *infected link probability*. Clearly

$$\begin{aligned} \alpha(\theta_k) &= \frac{\sum_{l=1}^{L} (\text{\# of links from infected node of degree } l)}{\sum_{l=1}^{L} (\text{\# of links of degree } l)} \\ &= \frac{\sum_{l=1}^{L} l\, \rho(l)\, \theta_k(l)}{\sum_l^L l\, \rho(l)}. \end{aligned} \tag{17.74}$$

In terms of the infected link probability α, we can now specify the scaled transition probabilities[12] of the process θ:

$$\begin{aligned} \bar{P}_{12}(l, \alpha_k, x) &\overset{\text{defn}}{=} \frac{1}{\rho(l)} \mathbb{P}\left(\theta_{k+1}(l) = \theta_k(l) + \frac{1}{M(l)} \Big| x_k = x\right) \\ &= (1 - \theta_k(l)) \sum_{a=0}^{l} P_{12}(l, a, x)\, \mathbb{P}(a \text{ out of } l \text{ neighbors infected}) \\ &= (1 - \theta_k(l)) \sum_{a=0}^{l} P_{12}(l, a, x) \binom{l}{a} \alpha_k^a (1 - \alpha_k)^{l-a}, \\ \bar{P}_{21}(l, \alpha_k, x) &\overset{\text{defn}}{=} \frac{1}{\rho(l)} \mathbb{P}\left(\theta_{k+1}(l) = \theta_k(l) - \frac{1}{M(l)} \Big| x_k = x\right) \\ &= \theta_k \sum_{a=0}^{l} P_{21}(l, a, x) \binom{l}{a} \alpha_k(a) (1 - \alpha_k)^{l-a}. \end{aligned} \tag{17.75}$$

[12] The transition probabilities are scaled by the degree distribution ρl for notational convenience. Indeed, since $M(l) = M\rho(l)$, by using these scaled probabilities we can express the dynamics of the process θ in terms of the same-step size $1/M$ as described in Theorem 17.5.1. We assume that the degree distribution $\rho(l)$, $l \in \{1, 2, \ldots, L\}$, is uniformly bounded away from zero.

where P_{12}, P_{21} are defined in (17.73). In the above, the notation α_k is the short form for $\alpha(\theta_k)$. Using these transition probabilities, Theorem 17.5.2 then yields the following mean field dynamics for the infected population process θ:

$$\bar{\theta}_{k+1}(l) = \bar{\theta}_k(l) + \frac{1}{M}\left[\bar{P}_{12}(l, \alpha(\bar{\theta}_k), x_k) - \bar{P}_{21}(l, \alpha(\bar{\theta}_k), x_k)\right]. \tag{17.76}$$

For a time horizon of N points, the deviation between the mean field dynamics $\bar{\theta}_k$ in (17.76) and actual infected distribution in θ_k satisfies the exponential bound (17.65).

Bayesian filtering of sentiment: Typically two kinds of observations are available from the social network. Extrinsic measurements z_k are noisy observations of the state of nature x_k. For example, if x_k denotes the price/availability of products offered by competitors or the current market, then economic indicators yield noisy measurements z_k. Also sentiment-based observations y_k are obtained by sampling the social network. Given these observations

$$z_k \sim p(\cdot|x_k), \quad y_k \sim (p(\cdot|\bar{\theta}_k) \tag{17.77}$$

estimating the underlying state of nature x_k given the dynamics (17.76) is a filtering problem. If x_k is modeled as a finite state Markov chain, then (17.76) together with (17.77) is a partially observed nonlinear jump Markov system. The state can be estimated using the particle filters of Chapter 3. In a more sophisticated setting, the infected population dynamics for $\bar{\theta}_k$ (fraction of population that buys a product) can be controlled by advertising. Then determining the optimal advertising strategy given the observations in (17.77) is a partially observed control problem.

17.6 Complements and sources

The ODE analysis for stochastic approximation algorithms was pioneered by Ljung (see [216, 218]) and subsequently by Kushner and co-workers [198, 200]. In this chapter we have only scratched the surface of this remarkably powerful analysis tool. Apart from the books listed above, [48, 305, 63] are also excellent references for analysis of such algorithms. The papers [42, 43] illustrate the power of ODE method (and generalizations to differential inclusions) for analyzing the dynamics of game-theoretic learning.

§17.3 uses the LMS algorithm for tracking parameters that jump infrequent but by possibly large amounts. In most traditional analyses, the parameter changes by small amounts over small intervals of time. As mentioned in §17.3.4, one can also analyze the tracking capability for Markov chains that jump by large amounts on short intervals of time [348, 346]. In such cases, stochastic averaging leads to a Markov modulated ODE (instead of a deterministic ODE).

It would be remiss of us not to mention the substantial literature in the analysis of adaptive filtering algorithms [290, 139]. The proof of Theorem 17.3.1 uses perturbed Lyapunov functions. Solo [305] was influential in developing discounted perturbed Lyapunov function methods for analysis of adaptive filtering algorithms. The mean field dynamics proof of Theorem 17.5.2 is based on [44] and uses Azuma–Hoeffding's inequality. It requires far less mathematical machinery. The monograph [195] contains several deep results in Markov chain approximations of population processes and dynamics.

Survey papers in discrete stochastic optimization include [55]. There has also been much recent work in using the ODE analysis for ant-colony optimization algorithms and the use of ant-colony optimization in Q-learning [14].

Due to lack of space, we have omitted several important aspects of stochastic approximation algorithms. One particular intriguing result is Polyak's iterate averaging [266]. By choosing a larger scalar step size together with an averaging step, it can be shown [200] that one can achieve the asymptotic convergence rate of a matrix step size.

17.6.1 Consensus stochastic approximation algorithms

There has been much recent work on *diffusion* and *consensus* stochastic approximation algorithms, where multiple stochastic approximation algorithms communicate with each other over a graph. This area has been pioneered by Sayed (see [291] and references therein) and shows remarkable potential in a variety of distributed processing applications.

In this section we briefly analyze the consensus stochastic approximation algorithm. The consensus stochastic approximation algorithm is of the form

$$\theta_{k+1} = A\,\theta_k + \epsilon H(\theta_k, x_k), \tag{17.78}$$

where A is a symmetric positive definite stochastic matrix. For $A = I$, (17.78) becomes the standard stochastic approximation algorithm.

One can analyze the consensus algorithm as follows. Define the matrix Q such that $A = \exp(Q\epsilon)$ where $\exp(\cdot)$ denotes the matrix exponential. So Q is proportional to the matrix logarithm of A. (Since A is positive definite, the real-valued matrix logarithm always exists.) Indeed Q is a generator matrix with $Q_{ii} < 0$ and $Q\mathbf{1} = 0$. Then as ϵ becomes sufficiently small (recall the ODE analysis applies for the interpolated process with $\epsilon \to 0$) since $\exp(Q\epsilon) \approx I + \epsilon Q + o(\epsilon)$, one can express (17.78) as

$$\theta_{k+1} = \theta_k + \epsilon\big(Q\theta_k + H(\theta_k, x_k)\big). \tag{17.79}$$

Therefore, the consensus ODE associated with (17.78) is

$$\frac{d\theta}{dt} = Q\theta + \mathbb{E}_{\pi_\theta}\{H(\theta, x)\}. \tag{17.80}$$

Typically Q is chosen such that at the optimum value θ^*, $Q\theta^* = 0$ implying that the consensus ODE and original ODE have identical attractors.

To summarize, despite at first sight looking different to a standard stochastic approximation algorithm, the consensus stochastic approximation algorithm (17.78) in fact is equivalent to a standard stochastic approximation algorithm (17.79) by taking the matrix logarithm of the consensus matrix A.

Example: *Consensus LMS*: Given that the consensus algorithm can be analyzed using the ODE method, below we determine its asymptotic convergence rate using the formula

(17.12). For notational convenience, consider two *scalar* LMS algorithms that aim to estimate a common scalar parameter θ^*:

$$\theta_{k+1}^{(l)} = \theta_k^{(l)} + \epsilon\psi_k^{(l)}(y_k^{(l)} - \psi_k^{(l)}\theta_k^{(l)}), \quad l = 1, 2. \tag{17.81}$$

Here the observations are $y_k^{(l)} = \psi_k^{(l)}\theta^* + v_k^{(l)}$. Assume that the i.i.d. observation noise $v_k^{(l)}$ for $l = 1, 2$ has the same variance σ^2, and also $\mathbb{E}\{(\psi_k^{(l)})^2\} = B$.

The ODEs for the two individual LMS algorithms are

$$\frac{d}{dt}\theta^{(l)} = B\big(\theta^* - \theta^{(l)}\big), \quad l = 1, 2.$$

Clearly both these ODEs have stable attractors at θ^*.

Next, we consider the following consensus LMS algorithm:

$$\begin{bmatrix} \theta_{k+1}^{(1)} \\ \theta_{k+1}^{(2)} \end{bmatrix} = \begin{bmatrix} a & 1-a \\ 1-a & a \end{bmatrix} \begin{bmatrix} \theta_k^{(1)} \\ \theta_k^{(2)} \end{bmatrix} + \epsilon \begin{bmatrix} \psi_k^{(1)}(y_k^{(1)} - \psi_k^{(1)}\theta_k^{(1)}) \\ \psi_k^{(2)}(y_k^{(2)} - \psi_k^{(2)}\theta_k^{(2)}) \end{bmatrix}, \quad a \in (0.5, 1]. \tag{17.82}$$

Let us analyze the consensus LMS algorithm (17.82) using the ODE approach. Define the generator matrix Q such that $A = \exp(Q\epsilon)$. Then Q is of the form $\begin{bmatrix} -q & q \\ q & -q \end{bmatrix}$. As in (17.80), the ODE for the consensus LMS algorithm (17.82) is

$$\frac{d}{dt}\begin{bmatrix} \theta^{(1)} \\ \theta^{(2)} \end{bmatrix} = Q \begin{bmatrix} \theta^{(1)} \\ \theta^{(2)} \end{bmatrix} + \begin{bmatrix} B(\theta^* - \theta^{(1)}) \\ B(\theta^* - \theta^{(2)}) \end{bmatrix}.$$

Clearly $Q\begin{bmatrix} \theta^* \\ \theta^* \end{bmatrix} = 0$. So each component of the consensus LMS converges to θ^*.

We can now compare the asymptotic rate of convergence of LMS versus consensus LMS. Recall that the asymptotic rate of convergence is given by the Lyapunov equation (17.12). The asymptotic convergence rate for LMS is

$$B\Sigma_{\text{LMS}} + \Sigma_{\text{LMS}}B = R \implies \Sigma_{\text{LMS}} = \frac{\sigma^2}{2}$$

since $R = B\sigma^2$. From (17.12), the asymptotic covariance matrix Σ_C (this is a 2×2 matrix) of the consensus LMS algorithm satisfies

$$\begin{aligned} &\big(\text{diag}(B, B) - Q\big)\Sigma_C + \Sigma_C\big(\text{diag}(B, B) - Q\big) = \text{diag}(R, R) \\ &\implies \Sigma_C = \frac{\sigma^2}{2}\begin{bmatrix} \frac{B+q}{B+2q} & \frac{q}{B+2q} \\ \frac{q}{B+2q} & \frac{B+q}{B+2q} \end{bmatrix}. \end{aligned} \tag{17.83}$$

Note that $\Sigma_C(1,1)$ and $\Sigma_C(2,2)$, which are the asymptotic variances at node 1 and node 2 of consensus LMS, are always smaller than Σ_{LMS} by the multiplicative factor $\frac{B+q}{B+2q}$. Indeed, for sufficiently large q, $\Sigma_C(1,1) \to \Sigma_{\text{LMS}}/2$.

17.6.2 Multi-armed bandits

The discrete stochastic optimization problem considered in §17.4 is similar in spirit to the stochastic bandit problem which we now discuss briefly. The bandit problem comprises of S probability density functions $p_1, p_2, \ldots, p_S$ with associated expected values $\mu_1, \mu_2, \ldots, \mu_S$. These distributions are unknown.[13] If player selects "arm" (distribution) $\theta_n \in \{1, 2, \ldots, S\}$ at time n, she gains a reward $r_n \sim p_{\theta_n}$. The player aims to determine which distribution yields the highest expected reward while gaining as much reward while playing. Bandit algorithms specify a strategy that a player can use to choose the arm θ_n at each time n. The performance measure for the bandit problem is the expected cumulative regret over a horizon N:

$$R_N = N \max_i \mu_i - \sum_{n=1}^{N} \mu_{\theta_n} = N \max_i \mu_i - N \sum_{\theta=1}^{S} \mu_\theta \, \mathbb{E}\{\pi_N(\theta)\} \tag{17.84}$$

where $\pi_N(\theta)$ is the fraction of time the arm θ is played during the time interval N. The objective (17.84) seeks to achieve a trade-off between exploration and exploitation: one needs to try all possible arms to determine which is the best one; on the other hand playing suboptimal arms incurs a regret. Too little exploration can wrongly make a suboptimal arm appear better that the optimal arm due to the random fluctuations; too much exploration prevents the algorithm form playing the optimal arm enough which incurs a larger regret.

A key result regarding the multi-armed bandit problem was given by Lai and Robbins [201]. They gave policies for the bandit problem that satisfy

$$\mathbb{E}\{\pi_N(\theta)\} \leq \left[\frac{1}{D(p_\theta || p^*)} + o(1)\right] \frac{\log N}{N}$$

where $o(1) \to 0$ and $N \to \infty$ and $D(p_\theta || p^*)$ is the Kullback–Leibler divergence between the density p_θ of a suboptimal arm and the reward density p^* of the optimal arm, that is,

$$D(p_\theta || p^*) = \int p_\theta(x) \log \frac{p_\theta(x)}{p^*(x)} \, dx.$$

Therefore, under these policies, for large N, the optimal arm is played exponentially more often than any other arm and the regret grows at least logarithmically with N. Lai and Robbins also proved that this regret is the best possible. In other words, for any policy and suboptimal arm θ,

$$\mathbb{E}\{\pi_N(\theta)\} \geq \left[\frac{1}{D(p_\theta || p^*)}\right] \frac{\log N}{N}.$$

That is, one needs to spend at least $\log N$ time in each of the suboptimal arms in order to estimate the optimal arm. An algorithm is said to solve the multi-armed bandit problem if it has regret $R_N = O(\log N)$. Clearly by Theorem 17.4.2, the UCB Algorithm 26 achieves logarithmic expected regret over a finite time horizon. This was shown in the

[13] This non-Bayesian framework is different to the Bayesian bandit formulation given in §11.4 where we discussed POMDP multi-armed bandits with known transition and observation probabilities. Here we are interested in reinforcement learning algorithms where the underlying distributions are not known.

influential paper [27]. The UCB algorithm has since become a benchmark to compare new algorithms. The monograph [71] is an excellent survey of the bandit problem and summarizes several state of the art results. Recently, Thompson sampling [6] is an area that has witnessed much interest due to its excellent empirical performance in many situations.

Note that the objective of discrete stochastic optimization, namely to estimate the parameter θ^* which optimizes (17.43), is different to that of the multi-armed bandit problem which seeks to minimize the average observed function (17.84) over a period of time.

Appendix 17.A Proof of Theorem 17.3.1

It suffices to prove (17.27) for the Euclidian norm $|x| = (x'x)^{1/2}$. Define $V(x) = (x'x)/2$. Direct calculation leads to

$$\begin{aligned}
&\mathbb{E}_n V(\widetilde{\theta}_{n+1}) - V(\widetilde{\theta}_n) \\
&= \mathbb{E}_n \left\{ \widetilde{\theta}_n' \left[-\mu\varphi_n\varphi_n'\widetilde{\theta}_n + \mu\varphi_n v_n + (x_n - x_{n+1}) \right] \right\} \\
&\quad + \mathbb{E}_n | -\mu\varphi_n\varphi_n'\widetilde{\theta}_n + \mu\varphi_n v_n + (x_n - x_{n+1})|^2.
\end{aligned} \tag{17.85}$$

In view of the Markovian assumption, the independence of the Markov chain with the signals $\{(\varphi_n, v_n)\}$, and the structure of the transition probability matrix given by (17.21),

$$\begin{aligned}
\mathbb{E}_n(x_n - x_{n+1}) &= \sum_{i=1}^{X} \mathbb{E}(g_i - x_{n+1} | x_n = g_i) I(x_n = g_i) \\
&= \sum_{i=1}^{X} \left[g_i - \sum_{j=1}^{X} g_j p_{ij}^{\varepsilon} \right] I(x_n = g_i) \\
&= -\varepsilon \sum_{i=1}^{X} \sum_{j=1}^{X} g_j Q_{ij} I(x_n = g_i) = O(\varepsilon).
\end{aligned} \tag{17.86}$$

Using an elementary inequality $ab \le (a^2 + b^2)/2$ for two real numbers a and b, we have

$$\begin{aligned}
|\widetilde{\theta}_n| = |\widetilde{\theta}_n| \cdot 1 \le (|\widetilde{\theta}_n|^2 + 1)/2, \text{ so} \\
O(\varepsilon)|\widetilde{\theta}_n| \le O(\varepsilon)(V(\widetilde{\theta}_n) + 1).
\end{aligned}$$

By virtue of the boundedness of the signal $\{(\varphi_n, v_n)\}$,

$$\mathbb{E}_n | -\mu\varphi_n\varphi_n'\widetilde{\theta}_n + \mu\varphi_n v_n + (x_n - x_{n+1})|^2 = O(\mu^2 + \mu\varepsilon + \varepsilon)(V(\widetilde{\theta}_n) + 1).$$

Using the above two inequalities in (17.85) together with (17.86) yields

$$\begin{aligned}
&\mathbb{E}_n V(\widetilde{\theta}_{n+1}) - V(\widetilde{\theta}_n) \\
&= \mathbb{E}_n \left\{ \widetilde{\theta}_n' \left[-\mu\varphi_n\varphi_n'\widetilde{\theta}_n + \mu\varphi_n v_n \right] \right\} + O(\varepsilon)(V(\widetilde{\theta}_n) + 1) + O(\mu^2 + \mu\varepsilon)(V(\widetilde{\theta}_n) + 1).
\end{aligned} \tag{17.87}$$

To obtain the desired estimate, we need to "average out" the terms inside the curly bracket {} in (17.87). Roughly speaking, if φ_n, v_n were independent and also independent of $\widetilde{\theta}_n$, then if $\mathbb{E}_n\{\varphi_n\varphi_n'\} = B$, the expectation of the term in the curly brackets would be

$-\mu\widetilde{\theta}_n B x_n$. The idea below is to extend this rigorously to correlated noise (of mixing type) $\{\varphi_n, v_n\}$ that satisfies (A2) by using perturbed Lyapunov functions.

To do so, define two perturbations of the Lyapunov function by

$$V_1^\varepsilon(\widetilde{\theta}, n) = -\mu \sum_{j=n}^{\infty} \mathbb{E}_n \widetilde{\theta}'(\varphi_j \varphi_j' - B)\widetilde{\theta},$$

$$V_2^\varepsilon(\widetilde{\theta}, n) = \mu \sum_{j=n}^{\infty} \widetilde{\theta}' \mathbb{E}_n \varphi_j e_j.$$

For each $\widetilde{\theta}$, by virtue of (A2), it is easily verified that

$$\mu \Big| \sum_{j=n}^{\infty} [\mathbb{E}_n \varphi_j \varphi_j' - B] \Big| |\widetilde{\theta}|^2 \leq O(\mu)(V(\widetilde{\theta}) + 1),$$

so

$$|V_1^\varepsilon(\widetilde{\theta}, n)| \leq O(\mu)(V(\widetilde{\theta}) + 1). \tag{17.88}$$

Similarly, for each $\widetilde{\theta}$,

$$|V_2^\varepsilon(\widetilde{\theta}, n)| \leq O(\mu)(V(\widetilde{\theta}) + 1). \tag{17.89}$$

Note that

$$\begin{aligned} &\mathbb{E}_n V_1^\varepsilon(\widetilde{\theta}_{n+1}, n+1) - V_1^\varepsilon(\widetilde{\theta}_n, n) \\ &\quad = \mathbb{E}_n V_1^\varepsilon(\widetilde{\theta}_{n+1}, n+1) - \mathbb{E}_n V_1^\varepsilon(\widetilde{\theta}_n, n+1) + \mathbb{E}_n V_1^\varepsilon(\widetilde{\theta}_n, n+1) - V_1^\varepsilon(\widetilde{\theta}_n, n). \end{aligned} \tag{17.90}$$

It follows that

$$\mathbb{E}_n V_1^\varepsilon(\widetilde{\theta}_n, n+1) - V_1^\varepsilon(\widetilde{\theta}_n, n) = \mu \mathbb{E}_n \widetilde{\theta}_n'(\varphi_n \varphi_n' - B)\widetilde{\theta}_n, \tag{17.91}$$

by virtue of (A2). In addition,

$$\begin{aligned} &\mathbb{E}_n V_1^\varepsilon(\widetilde{\theta}_{n+1}, n+1) - \mathbb{E}_n V_1^\varepsilon(\widetilde{\theta}_n, n+1) \\ &\quad = -\mu \sum_{j=n+1}^{\infty} \mathbb{E}_n(\widetilde{\theta}_{n+1} - \widetilde{\theta}_n)'[\mathbb{E}_{n+1}\varphi_j\varphi_j' - B]\widetilde{\theta}_{n+1} \\ &\quad\quad -\mu \sum_{j=n+1}^{\infty} \mathbb{E}_n \widetilde{\theta}_n'[\mathbb{E}_{n+1}\varphi_j\varphi_j' - B](\widetilde{\theta}_{n+1} - \widetilde{\theta}_n). \end{aligned} \tag{17.92}$$

Using (17.23), similar estimates as that of (17.86) yields

$$\begin{aligned} \mathbb{E}_n|\widetilde{\theta}_{n+1} - \widetilde{\theta}_n| &\leq \mu \mathbb{E}_n |\varphi_n \varphi_n'||\widetilde{\theta}_n| + \mu \mathbb{E}_n|\varphi_n v_n| + O(\varepsilon) \\ &= O(\mu)(V(\widetilde{\theta}_n) + 1) + O(\varepsilon). \end{aligned} \tag{17.93}$$

Moreover,

$$\begin{aligned}
&\Big|\mu \sum_{j=n+1}^{\infty} \mathbb{E}_n \widetilde{\theta}_n' [\mathbb{E}_{n+1}\varphi_j\varphi_j' - B](\widetilde{\theta}_{n+1} - \widetilde{\theta}_n)\Big| \\
&\leq K\mu\mathbb{E}_n\Big[\Big|\sum_{j=n+1}^{\infty} [\mathbb{E}_{n+1}\varphi_j\varphi_j' - B]\Big| |\widetilde{\theta}_n||\widetilde{\theta}_{n+1} - \widetilde{\theta}_n|\Big] \\
&\leq K\mu|\widetilde{\theta}_n|\mathbb{E}_n|\widetilde{\theta}_{n+1} - \widetilde{\theta}_n| \\
&\leq O(\mu^2 + \mu\varepsilon)(V(\widetilde{\theta}_n) + 1).
\end{aligned} \tag{17.94}$$

Likewise,

$$\begin{aligned}
&\Big| - \mu \sum_{j=n+1}^{\infty} \mathbb{E}_n(\widetilde{\theta}_{n+1} - \widetilde{\theta}_n)'[\mathbb{E}_{n+1}\varphi_j\varphi_j' - B]\widetilde{\theta}_{n+1}\Big| \\
&\leq O(\mu^2 + \mu\varepsilon)(V(\widetilde{\theta}_n) + 1).
\end{aligned} \tag{17.95}$$

Thus, we arrive at

$$\begin{aligned}
&\mathbb{E}_n V_1^\varepsilon(\widetilde{\theta}_{n+1}, n+1) - V_1^\varepsilon(\widetilde{\theta}_n, n) \\
&= \mu\mathbb{E}_n\widetilde{\theta}_n'(\varphi_n\varphi_n' - B)\widetilde{\theta}_n + O(\mu^2 + \mu\varepsilon)(V(\widetilde{\theta}_n) + 1).
\end{aligned} \tag{17.96}$$

A similar computation for $V_2^\varepsilon(\widetilde{\theta}_n, n)$ yields that

$$\begin{aligned}
&\mathbb{E}_n V_2^\varepsilon(\widetilde{\theta}_{n+1}, n+1) - V_2^\varepsilon(\widetilde{\theta}_n, n) \\
&= O(\mu^2 + \mu\varepsilon)(V(\widetilde{\theta}_n) + 1).
\end{aligned} \tag{17.97}$$

Define

$$W(\widetilde{\theta}, n) = V(\widetilde{\theta}) + V_1^\varepsilon(\widetilde{\theta}, n) + V_2^\varepsilon(\widetilde{\theta}, n).$$

Then using (17.85), (17.96), and (17.97), we obtain

$$\begin{aligned}
&\mathbb{E}_n W(\widetilde{\theta}_{n+1}, n+1) - W(\widetilde{\theta}_n, n) \\
&\leq -\mu\widetilde{\theta}_n' B\widetilde{\theta}_n + O(\mu^2 + \mu\varepsilon + \varepsilon)(V(\widetilde{\theta}_n) + 1).
\end{aligned} \tag{17.98}$$

Since B is positive definite, there is a $\lambda > 0$ such that $\widetilde{\theta}' B\widetilde{\theta} \geq \lambda V(\widetilde{\theta})$. For example, from the Rayleigh–Ritz theorem, $\lambda = \lambda_{\min}$ satisfies $\widetilde{\theta}' B\widetilde{\theta} \geq \lambda V(\widetilde{\theta})$, where $\lambda_{\min} > 0$ is the smallest eigenvalue of B. This together with (17.88) and (17.89) implies

$$\begin{aligned}
&\mathbb{E}_n W(\widetilde{\theta}_{n+1}, n+1) - W(\widetilde{\theta}_n, n) \\
&\leq -\lambda\mu W(\widetilde{\theta}_n, n) + O(\mu^2 + \mu\varepsilon + \varepsilon)(W(\widetilde{\theta}_n, n) + 1).
\end{aligned}$$

Choose μ and ε small enough so that there is a $\lambda_0 > 0$ satisfying $\lambda_0 \leq \lambda$ and

$$-\lambda\mu + O(\mu^2) + O(\mu\varepsilon) \leq -\lambda_0\mu. \tag{17.99}$$

Note that this is equivalent to (17.26). Then for such a small fixed μ and ε, we obtain (using $(O(\mu^2 + \varepsilon\mu + \varepsilon) \leq O(\mu^2 + \varepsilon))$,

$$\mathbb{E}_n W(\widetilde{\theta}_{n+1}, n+1) \leq (1 - \lambda_0\mu)W(\widetilde{\theta}_n, n) + O(\varepsilon)W(\widetilde{\theta}_n, n) + O(\mu^2 + \varepsilon).$$

Taking expectation above and iterating on the resulting inequality yield

$$\mathbb{E}W(\widetilde{\theta}_{n+1}, n+1) \le (1-\lambda_0\mu)^n \mathbb{E}W(\widetilde{\theta}_0, 0) + \varepsilon \sum_{j=0}^{n} (1-\lambda_0\mu)^{n-j} \mathbb{E}W(\widetilde{\theta}_j, j) + O(\mu + \varepsilon/\mu)\,.$$

By taking n large enough, we can make $(1-\lambda_0\mu)^n \le O(\mu)$. Thus an application of the Gronwall's inequality leads to

$$\begin{aligned}
\mathbb{E}W(\widetilde{\theta}_{n+1}, n+1) &\le O(\mu + \varepsilon/\mu)\exp\left(\varepsilon \sum_{j=0}^{n} (1-\lambda_0\mu)^{n-j}\right) \\
&\le O(\mu + \varepsilon/\mu)\exp(\varepsilon/\mu).
\end{aligned}$$

Finally, applying (17.88) and (17.89) again, we also obtain

$$\mathbb{E}V(\widetilde{\theta}_{n+1}) \le O(\mu + \varepsilon/\mu)\exp(\varepsilon/\mu).$$

Thus the desired result follows. □

Appendix 17.B Proof of Theorem 17.4.1

Define $\gamma_n^{ij} = c_n(i) - c_n(j)$. For convenience we omit the subscript n in γ_n^{ij} since the samples are i.i.d. Then the transition probabilities can be expressed as

$$P_{ij} = \begin{cases} \frac{1}{S-1}\mathbb{P}(\gamma^{ij} > 0), & i,j \in \Theta, i \ne j \\ \frac{1}{S-1}\sum_{j\in\Theta-\{i\}} \mathbb{P}(\gamma^{ij} \le 0) & i = j \end{cases}.$$

So the stationary distribution satisfies

$$\pi_\infty(j) = \sum_{i=1}^{S} P_{ij}\pi_\infty(i) = \frac{1}{S-1}\sum_{i\in\Theta-\{j\}} \mathbb{P}(\gamma^{ij} > 0)\pi_\infty(i) + \frac{\pi_\infty(j)}{S-1}\sum_{i\in\Theta-\{j\}} \mathbb{P}(\gamma^{ji} \le 0).$$

Suppose that $i \in \mathcal{G}$ and $j \notin \mathcal{G}$. Then by (O) it follows that

$$\begin{aligned}
\pi_\infty(j) \le \frac{1}{S-1}\sum_{l\in\Theta-\{i,j\}} \mathbb{P}(\gamma^{li} > 0)\pi_\infty(l) &+ \frac{\pi_\infty(j)}{S-1}\sum_{l\in\Theta-\{i,j\}} \mathbb{P}(\gamma^{il} \le 0) \\
&+ \frac{\pi_\infty(i)}{S-1}\mathbb{P}(\gamma^{ij} > 0) + \frac{\pi_\infty(j)}{S-1}\mathbb{P}(\gamma^{ji} \le 0).
\end{aligned}$$

Therefore,

$$\begin{aligned}
\pi_\infty(i) - \pi_\infty(j) \ge \frac{\pi_\infty(i) - \pi_\infty(j)}{S-1} &\sum_{l\in\Theta-\{i,j\}} \mathbb{P}(\gamma^{il} \le 0) + \frac{\pi_\infty(j)}{S-1}\left[\mathbb{P}(\gamma^{ji} > 0) - \mathbb{P}(\gamma^{ji} \le 0)\right] \\
&+ \frac{\pi_\infty(i)}{S-1}\left[\mathbb{P}(\gamma^{ij} \le 0) - \mathbb{P}(\gamma^{ij} > 0)\right]
\end{aligned}$$

$$
\begin{aligned}
&= \frac{\pi_\infty(i) - \pi_\infty(j)}{S-1} \sum_{l \in \Theta - \{i,j\}} \mathbb{P}(\gamma^{il} \le 0) + \frac{\pi_\infty(j)}{S-1}\left[2\,\mathbb{P}(\gamma^{ji} > 0) - 1\right] \\
&\quad + \frac{\pi_\infty(i)}{S-1}\left[1 - 2\,\mathbb{P}(\gamma^{ij} > 0)\right] \\
&= \frac{K}{S-1}\,(\pi_\infty(i) - \pi_\infty(j))\,.
\end{aligned}
$$

Here $K = \sum_{l \in \Theta - \{i,j\}} \mathbb{P}(\gamma^{il} \le 0) + 1 - 2\mathbb{P}(\gamma^{ij} > 0) \le S - 1$. Therefore, $\pi_\infty(i) > \pi_\infty(j)$ for $i \in \mathcal{G}, j \in \Theta - \mathcal{G}$.

Appendix 17.C Proof of Theorem 17.4.2

We start with the following lemma that upper bounds m_{in}. For notational convenience, denote the right hand side of Step 1 of the UCB algorithm as

$$
H_{\theta st} = \widehat{c}_{\theta,t} + B\sqrt{\frac{\xi \log t}{s}}
$$

LEMMA 17.C.1 *Suppose each of the S candidates has been sampled once. Then for any $u > 1$ and $\tau \in \mathbb{R}$, the number of times candidate θ is sampled is upper bounded as*

$$
\begin{aligned}
m_{\theta n} \le u &+ \sum_{t=u+S-1}^{n} I(\exists s : s \in \{u, \dots, t-1\},\ \textit{such that } H_{\theta st} > \tau) \\
&+ \sum_{t=u+S-1}^{n} I(\exists s^* : s^* \in \{u, \dots, t-1\},\ \textit{such that } H_{\theta^* s^* t} \le \tau). \quad (17.100)
\end{aligned}
$$

Note that Lemma 17.C.1 holds independent of the form of $H_{\theta st}$. Using Lemma 17.C.1, it follows from the union bound that

$$
\mathbb{E}\{m_{\theta n}\} \le u + \sum_{t=u+S-1}^{n} \sum_{s=u}^{t-1} \mathbb{P}(H_{\theta st} > \tau) + \sum_{t=u+S-1}^{n} \sum_{s^*=u}^{t-1} \mathbb{P}(H_{\theta^* s^* t} \le \tau). \qquad (17.101)
$$

It only remains to choose τ and u to upper bound the right hand side. Choose

$$
\tau = C(\theta^*) = \mathbb{E}\{c_n(\theta^*)\}, \quad \text{and} \quad u = \frac{B^2}{(C(\theta^*) - C(\theta))^2}\,\xi\,\frac{\log t}{s}.
$$

Then from Hoeffding's inequality (15.24)

$$
\begin{aligned}
&\mathbb{P}\big(H_{\theta st} > C(\theta^*)\big) \le \exp\Big(-\frac{u}{2B^2}(C(\theta^*) - C(\theta))^2\Big) = u^{-2\xi} \le n^{-2\xi} \\
&\mathbb{P}\big(H_{\theta^* st} \le C(\theta^*)\big) \le t^{-2\xi}.
\end{aligned}
$$

Therefore we can bound the right hand side of (17.101) as

$$\sum_{t=u+S-1}^{n} \sum_{s=u}^{t-1} \mathbb{P}(H_{\theta st} > \tau) \leq \frac{n^{2(1-\xi)}}{2} < \frac{1}{2} \quad n \geq 1.$$

$$\sum_{t=u+S-1}^{n} \sum_{s^*=u}^{t-1} \mathbb{P}(H_{\theta^* s^* t} \leq \tau) \leq \sum_{t=u+1}^{n} t^{1-2\xi} \leq \int_{u}^{\infty} t^{1-2\xi}\, dt = \frac{u^{-2(\xi-1)}}{2(\xi-1)}, \quad \xi > 1.$$

($\sum_{t=1}^{\infty} \frac{1}{t^{2\xi-1}}$ is the Riemann zeta function and there are several other bounds.)
Proof of Lemma 17.C.1 For any $u \in \{2, 3, \ldots\}$,

$$\begin{aligned}
m_{\theta n} &= \sum_{t=1}^{u+S-2} I(\theta_t = \theta) + \sum_{t=u+S-1}^{n} I(\theta_t = \theta) \\
&\leq u + \sum_{t=u+S-1}^{n} I(\theta_t = \theta) I(m_{\theta,t} > u) \\
&\overset{(a)}{=} u + \sum_{t=u+S-1}^{n} I\big(\theta_t = \theta, u \leq m_{\theta,t-1}, H_{\theta,m_\theta(t-1),t} > \max_{j \neq \theta} H_{j,m_j(t-1),t}\big) \\
&\leq u + \sum_{t=u+S-1}^{n} I\big(\theta_t = \theta, u \leq m_{\theta,t-1}, H_{\theta,m_{\theta,t-1},t} > H_{\theta^*,m_{\theta^*,t-1},t}\big) \\
&\leq u + \sum_{t=u+S-1}^{n} I(\exists s \in \{u, t-1\} \text{ s. t. } H_{\theta,s,t} > \tau) \\
&\qquad + \sum_{t=u+S-1}^{n} I(\exists s^* \in \{u, t-1\} \text{ s. t. } H_{\theta^*,s^*,t} \leq \tau)
\end{aligned}$$

where $\tau \in \mathbb{R}$. Note (a) follows from Step 1 of Algorithm 26. The other inequalities follow since the event within the indicator function before the inequality is a subset of the event within the indicator function after the inequality.

Appendix 17.D Proof of Theorem 17.5.2

Recalling the notation of Theorem 17.5.1, define

$$\tilde{\theta}_n = \theta_n - \bar{\theta}_n, \quad S_N = \max_{1 \leq n \leq N} \| \sum_{k=1}^{n} \nu_k \|_\infty.$$

LEMMA 17.D.1 *(Recall from Footnote 9 that λ below is the Lipschitz constant.)*

$$\|\tilde{\theta}_{n+1}\|_\infty \leq \|\tilde{\theta}_0\|_\infty + \frac{\lambda}{M} \sum_{k=1}^{n} \|\tilde{\theta}_k\|_\infty + S_N.$$

LEMMA 17.D.2 *(Recall the notation in Theorem 17.5.1 for Γ.)*

$$\mathbb{P}(S_N \geq \epsilon) \leq 2\exp\left(-\frac{\epsilon^2 M^2}{2\Gamma N}\right)$$

Using Lemmas 17.D.1 and 17.D.2 the proof of Theorem 17.5.2 is as follows. Applying Gronwall's inequality[14] to Lemma 17.D.1 yields $\|\tilde{\theta}_n\|_\infty \leq S_N \exp\left[\frac{\lambda n}{M}\right]$, which in turn implies that

$$\max_{1\leq n\leq N} \|\tilde{\theta}_n\|_\infty \leq S_N \exp\left[\frac{\lambda N}{M}\right].$$

As a result

$$\mathbb{P}(\max_{1\leq n\leq N} \|\tilde{\theta}_n\|_\infty > \epsilon) \leq \mathbb{P}(S_N \exp\left[\frac{\lambda N}{M}\right] > \epsilon) = \mathbb{P}\big(S_N > \exp\left[-\frac{\lambda N}{M}\right]\epsilon\big)$$

Next applying Lemma 17.D.2 to the right-hand side yields

$$\mathbb{P}(\max_{1\leq n\leq N} \|\tilde{\theta}_n\|_\infty > \epsilon) \leq 2\exp\left(-\exp(\frac{-2\lambda N}{M})\,\epsilon^2 \frac{M^2}{2\Gamma N}\right).$$

Finally choosing $N = c_1 M$, for some positive constant c_1 yields

$$\mathbb{P}(\max_{1\leq n\leq N} \|\tilde{\theta}_n\|_\infty > \epsilon) \leq 2\exp(-C_2\epsilon^2 M), \quad \text{where } C_2 = \exp(-2\lambda c_1)\frac{1}{2\Gamma c_1}.$$

This completes the proof of Theorem 17.5.2.

Proof of Lemma 17.D.1: Recall $\tilde{\theta}_n = \theta_n - \bar{\theta}_n$. It satisfies

$$\begin{aligned}
\tilde{\theta}_{n+1} &= \tilde{\theta}_n + \frac{1}{M}[H(x_n, \theta_n) - H(x_n, \bar{\theta}_n)] + \nu_n \\
&= \tilde{\theta}_0 + \frac{1}{M}\sum_{k=1}^{n}[H(x_k, \theta_k) - H(x_k, \bar{\theta}_k)] + \sum_{k=1}^{n} \nu_k \\
\|\tilde{\theta}_{n+1}\|_\infty &\leq \|\tilde{\theta}_0\|_\infty + \frac{1}{M}\sum_{k=1}^{n} \|[H(x_k, \theta_k) - H(x_k, \bar{\theta}_k)]\|_\infty + \|\sum_{k=1}^{n} \nu_k\|_\infty \\
&\leq \|\tilde{\theta}_0\|_\infty + \frac{\lambda}{M}\sum_{k=1}^{n} \|\tilde{\theta}_k\|_\infty + S_N
\end{aligned}$$

since $\|[H(x_k, \theta_k) - H(x_k, \bar{\theta}_k)]\|_\infty \leq \lambda\|\theta_k - \bar{\theta}_k\|_\infty$ where λ is the Lipschitz constant (see Footnote 9 on page 410).

[14] Gronwall's inequality: if $\{x_k\}$ and $\{b_k\}$ are nonnegative sequences and $a \geq 0$, then

$$x_n \leq a + \sum_{k=1}^{n-1} x_k b_k \implies x_n \leq a\exp(\sum_{k=1}^{n-1} b_k)$$

Proof of Lemma 17.D.2: $\|\sum_{k=1}^n \nu_k\|_\infty = \max_i |\sum_{k=1}^n e_i' \nu_k| = \sum_{k=1}^n e_{i^*}' \nu_k|$ for some i^*. Since $e_{i^*}' \nu_k$ is a martingale difference process with $|e_{i^*}' \nu_k| \leq \sqrt{\Gamma}/M$ applying the Azuma–Hoeffding inequality (Theorem 15.A.1) yields

$$\mathbb{P}(\|\sum_{k=1}^n \nu_k\|_\infty \geq \epsilon) = \mathbb{P}(|\sum_{k=1}^n e_{i^*}' \nu_k| \geq \epsilon) \leq 2 \exp\left[-\frac{\epsilon^2 M^2}{2\Gamma n}\right]$$

The right-hand side is increasing with n. So clearly

$$\mathbb{P}(\max_{1\leq n \leq N} \|\sum_{k=1}^n \nu_k\|_\infty \geq \epsilon) \leq 2 \exp\left[-\frac{\epsilon^2 M^2}{2\Gamma N}\right]$$

18 Summary of algorithms for solving POMDPs

For convenience, we summarize various algorithms for solving POMDPs that appear in Parts II, III and IV of the book. The algorithms can be classified into:

1. Optimal algorithms

For finite horizon POMDPs, optimal algorithms are based on exact value iteration (which are computationally intractable apart from small dimensional examples). These are described in §7.5 and include the Incremental Pruning algorithm, Monahan's Algorithm and Witness algorithm. Code for running these algorithms is freely downloadable from `www.pomdp.org`

For discounted cost infinite horizon POMDPs with discount factor ρ, based on Theorem 7.6.3, running value iteration for N iterations yields an accuracy

$$|V_N(\pi) - V(\pi)| \leq \frac{\rho^{N+1}}{1-\rho} \max_{x,u} |c(x,u)|.$$

2. Suboptimal algorithms based on value iteration

Lovejoy's algorithm in §s 7.5.2 provably sandwiches the optimal value function between low complexity lower and upper bounds. Lovejoy's algorithms is easily implementable in Matlab and can be run in conjunction with the POMDP solvers in `www.pomdp.org`

Point-based value iteration algorithms outlined in §7.5.3 include SARSOP. Details and code can be obtained from `http://bigbird.comp.nus.edu.sg /pmwiki/farm/appl/`

The belief compression algorithm is outlined in §7.5.4. The open loop feedback control algorithm is described in §7.5.5.

§8.4 gives similar algorithms for POMDPs that have costs which are nonlinear in the belief state. Such POMDPs arise in controlled sensing where one wishes to penalize uncertainty of the estimates.

3. Algorithms that exploit POMDP structural results

§12.7 gives structural results for a stopping time problem involving Kalman filters in radar scheduling. The costs involved are the stochastic observability (mutual information) of the targets. The optimal policy is a monotone function of the covariance matrices with respect to the positive definite ordering.

Chapter 12 gives structural results for stopping time POMDPs. For a variety of examples, sufficient conditions are given for the optimal policy to have a monotone structure on the belief space. In §12.4, this monotone structure is exploited to devise gradient algorithms that estimate the optimal linear monotone policy. Algorithm 14 uses the SPSA algorithm to estimate the optimal linear monotone policy.

For more general POMDPs, Chapter 14 presents structural results that provably upper and lower bound the optimal policy for a POMDP under certain assumptions. In this case, the upper and lower bounds are judiciously chosen myopic policies that can be computed via linear programming. The upper and lower bound myopic policies that optimize the overlap with the optimal POMDP policy are obtained using linear programming in Algorithm 15. Naturally, once the myopic policies are obtained, their implementation is trivial.

4. Heuristics based on MDP solutions

The aim here is to use the value function or policy on the underlying MDP, and then combine this with the belief state in various heuristic ways.

§16.1.4 gives a Q-learning for POMDPs that is based on computing the Q-factors of the underlying MDP. Given the belief state π, compute the POMDP policy as

$$\mu(\pi) = \operatorname*{argmax}_u \sum_x \pi(x) Q(x, u).$$

This is called the Q-MDP algorithm in [78]. For the above Q-MDP algorithm, if the underlying MDP has the structure of Chapter 9 so that the value function (and hence Q-factors) are monotone, then this can be exploited. Similarly, if using the structural results of Chapter 9, if the Q-factors are submodular, this property can also be exploited as a linear constraint as described in §16.1.3 and a primal-dual Q-learning algorithm can be used.

A simpler heuristic is to choose the POMDP policy as

$$\mu(\pi) = \mu^*_{\text{MDP}}(\hat{x}_k)$$

where μ^*_{MDP} is computed via Q-learning on the underlying MDP and the clean state estimate $\hat{x}_k$ is computed via the stochastic approximation algorithm described in Chapter 17. Then no knowledge of the underlying POMDP parameters is required. In particular, if by using the structural results in Chapter 9 it is established that the underlying MDP has a threshold policy, then $\mu(\pi)$ has a simple implementation.

5. Gradient-based policy search

Chapter 16 presents gradient-based policy search algorithms for constrained MDP and POMDPs. For the POMDP case (§16.7), the algorithm proposed is model-free in the sense that the parameters of the POMDP are not required to be known. Define the state as a sliding window Δ of past actions and observations:

$$x_k = (y_k, y_{k-1}, u_{k-1}, y_{k-2}, u_{k-2}, \dots, y_{k-\Delta}, u_{k-\Delta})$$

where Δ is a positive integer chosen by the user. The aim is to determine the optimal parameterized policy μ that maps x_k to u_k. A stochastic gradient algorithm is used to achieve this. The algorithm requires a gradient estimator to estimate the gradient of the POMDP cost with respect to the policy. Both the score function and weak derivative algorithms are discussed in §16.3 and §16.4, respectively. The score function algorithm was used in [35]. The weak derivative method has significantly smaller variance as demonstrated in §16.5.

6. Discrete stochastic optimization-based search

§17.4 discusses several stochastic search algorithms for estimating the best policy from a finite set of possible policies. Algorithm 24, 25 and 26 are examples.

7. Benchmark problems

In the computational POMDP community, there are now several benchmark POMDP problems motivated by applications such as path planning, and human–robot interaction. The following table from [297] lists some benchmark POMDP problems.

Domain	X	U	Y	Transitions	Observations	Horizon
Hallway [208]	92	5	17	stochastic	stochastic	29
Tag [262]	870	5	30	stochastic	deterministic	30
Rocksample [304]	12545	13	2	deterministic	stochastic	33
Navigation [194]	2653	6	102	deterministic	deterministic	62
Wumpus 7 [9]	1568	4	2	deterministic	deterministic	16

Appendix A Short primer on stochastic simulation

The main use of stochastic simulation in this book was in Parts II, III and IV for simulation-based gradient estimation and stochastic optimization. Also simulation was used in Chapter 3 for particle filters. This appendix presents some elementary background material in stochastic simulation that is of relevance to filtering, POMDPs and reinforcement learning. Our coverage is necessarily incomplete and only scratches surface of a vast and growing area. The books [284, 280] are accessible treatments of stochastic simulation to an engineering audience.

Assume that uniformly distributed pseudo random numbers $u \sim U[0,1]$ can be generated efficiently, where $U[0,1]$ denotes the uniform probability density function with support from 0 to 1 ("pseudo random" since a computer is a deterministic device). For example, the Matlab command `rand(n)` generates an $n \times n$ matrix where each element is $U[0,1]$ and statistically independent of other elements. Starting with $U[0,1]$ random numbers, the aim is to generate samples of random variables and random processes with specified distributions.

An important motivation for stochastic simulation stems from computing multidimensional integrals efficiently via Monte Carlo methods. Given a function $\phi : \mathbb{R}^X \to \mathbb{R}$, then if $p(\cdot)$ denotes a pdf having support over $\mathbb{R}^X$, the multi-dimensional integral can be expressed as

$$\int_{\mathbb{R}^X} \phi(x)dx = \int_{\mathbb{R}^X} \frac{\phi(x)}{p(x)} p(x)dx = \mathbb{E}_p\left\{\frac{\phi(x)}{p(x)}\right\}$$

where $\mathbb{E}_p$ denotes expectation with respect to p. By simulating independently and identically distributed (i.i.d.) samples $\{x_k\}$, $k = 1, \ldots, N$, from the pdf $p(\cdot)$, classical Monte Carlo methods compute the above integral approximately as

$$\frac{1}{N}\sum_{k=1}^{N} \frac{\phi(x_k)}{p(x_k)} \quad \text{where } x_k \sim p(x)\,.$$

By the strong law of large numbers (see Appendix D), for large N, one would expect the approximation to be accurate. The logic is that direct computation of the integral via deterministic methods can be difficult, whereas the Monte Carlo method can be implemented efficiently by proper choice of $p(\cdot)$.

In classical Monte Carlo methods, the samples $\{x_k\}$ are generated i.i.d. In the last 25 years, there have been significant advances in Markov chain Monte Carlo (MCMC) methods where, in order to evaluate the above integral, $\{x_k\}$ is generated according to a geometrically ergodic Markov chain whose stationary distribution is $p(\cdot)$.

A.1 Simulation of random variables

Assuming an algorithm is available for generating uniform $U[0, 1]$ random numbers, we describe below three elementary methods for simulating random variables, namely, the inverse transform method, the acceptance rejection method and the composition method.

A.1.1 Inverse transform method

Suppose we wish to generate a random variable x with cumulative distribution function F. The inverse transform method proceeds as follows:

Step 1: Generate $u \sim U[0, 1]$.
Step 2: Generate $x = F^{-1}(u)$.

Define $F^{-1}(u) = \inf\{x : F(x) = u\}$ if $F^{-1}(\cdot)$ is not unique.

Therefore, if p_i, $i = 1, \ldots, m$ is a probability mass function, then the inverse transform method generates $x \sim p$ as

1. Generate $u \sim U[0, 1]$.
2. Generate $x = l^* = \min\{l : u \leq \sum_{k=1}^{l} p_i\}$.

The inverse transform works due to the following argument: let $\bar{F}$ denote the distribution function of x generated by the algorithm. By definition for $\zeta \in \mathbb{R}$,

$$\bar{F}(\zeta) = \mathbb{P}\{x \leq \zeta\} = \mathbb{P}\{F^{-1}(u) \leq \zeta\} = \mathbb{P}\{F(F^{-1}(u)) \leq F(\zeta)\}.$$

The last equality follows since F is a monotone non-decreasing function of ζ and so for any two real numbers α, β, $\alpha \leq \beta$ is equivalent to $F(\alpha) \leq F(\beta)$. Thus, $\bar{F}(\zeta) = \mathbb{P}\{u \leq F(\zeta)\} = F(\zeta)$ where the last equality follows since u is uniformly distributed in $[0, 1]$.

Example 1: For generating a discrete valued random variable, Step 2 comprises of $m-1$ `if` statements in a computer program and can be inefficient in runtime execution if m is large. One special case where the inverse transform method for discrete valued random variables can be implemented efficiently is the *discrete uniform mass function*. In this case $p_i = 1/m$ for all $i = 1, 2, \ldots, m$. Then, the above method yields

$$x = l \text{ if } \frac{l-1}{m} \leq u < \frac{l}{m} \text{ or equivalently } x = \text{Int}(mu) + 1. \tag{A.1}$$

Example 2: Exponentially distributed random variables with distribution $F(x) = 1 - e^{-\lambda x}$, $x \geq 0$, $\lambda > 0$ can be generated as $x = -\frac{1}{\lambda}\log(1-u)$.

Example 3: Generate normal random variables as follows. If $\Theta \sim U[0, 2\pi]$ (uniform pdf) and $R \sim \lambda e^{-\lambda r}$ (exponential pdf) with $\lambda = 1/2$ are independent random variables, then it can be shown that $X = \sqrt{R}\cos\Theta$ and $Y = \sqrt{R}\sin\Theta$ are independent $N(0, 1)$ random variables. So if $u_1, u_2 \sim U[0, 1]$ are independent, then $x = \sqrt{-2\log u_1}\cos(2\pi u_2)$ and $Y = \sqrt{-2\log u_1}\sin(2\pi u_2)$ are independent $N(0, 1)$ random variables.

Example 4 (Skorohod representation theorem): A standard result in probability theory is that almost sure convergence implies convergence in distribution on a probability space $(\Omega, \mathcal{F}, P)$. The Skorohod representation theorem says the following: Suppose $Y_n \sim F_n$, converges in distribution (weakly) to $Y \sim F$. Then there exists random variables $\bar{Y}_n \sim F_n$ and $Y \sim F$ on another probability space $(\bar{\Omega}, \bar{\mathcal{F}}, \bar{P})$ such that $\bar{Y}_n \to \bar{Y}$ almost surely.

An explicit construction of the probability space for $\bar{Y}_n$ and $\bar{Y}$ is based on inverse transform method: define $\bar{Y}_n = F_n^{-1}(u)$, $\bar{Y} = F^{-1}(u)$, $u \sim U[0,1]$. That is $\bar{P}$ is the uniform distribution[1] on $[0,1]$. Then the inverse transform method implies $\bar{Y}_n \sim F_n$ and $\bar{Y} \sim F$. Also since $F_n \to F$ pointwise monotonically, $F_n^{-1} \to F^{-1}$ pointwise. This means that $\bar{Y}_n \to \bar{Y}$ almost surely.

A.1.2 Acceptance Rejection method

Suppose one can generate samples (relatively easily) from pdf q. How can random samples be simulated from pdf p? Assuming that $\sup_\zeta \frac{p(\zeta)}{q(\zeta)} < \infty$, the acceptance rejection algorithm described in Algorithm 27 can be used.

Algorithm 27 Acceptance Rejection algorithm

Let c denote a constant such that $c \geq \sup_\zeta \frac{p(\zeta)}{q(\zeta)}$. Then:

Step 1: Generate $y \sim q$.
Step 2: Generate $u \sim U[0,1]$.
Step 3: If $u < \frac{p(y)}{c\,q(y)}$, set $x = y$.
Otherwise go back to step 1.

The acceptance rejection method works since

$$\mathbb{P}(x \leq \zeta) = \mathbb{P}(y \leq \zeta \,|\, u \leq \frac{p(y)}{c\,q(y)}) = \frac{\mathbb{P}\left(y \leq \zeta, u \leq \frac{p(y)}{c\,q(y)}\right)}{\mathbb{P}\left(u \leq \frac{p(y)}{cq(y)}\right)}$$

$$= \frac{\text{Probability of } y \leq \zeta \text{ and accept}}{\text{Probability of accept}} = \frac{\int_{-\infty}^{\zeta} \int_0^{\frac{p(y)}{cq(y)}} du\, q(y) dy}{\int_{-\infty}^{\infty} \int_0^{\frac{p(y)}{cq(y)}} du\, q(y) dy} = \frac{\frac{1}{c}\int_{-\infty}^{\zeta} p(y) dy}{\frac{1}{c}\int_{-\infty}^{\infty} p(y) dy}.$$

Remarks: Note that the acceptance rejection method operates on the density functions while the inverse transform method operates on the distribution function. The acceptance rejection algorithm has several nice properties:

(i) The algorithm is self-normalizing in the sense that the pdf $p(\cdot)$ does not need to be normalized.

(ii) Since $c \geq \sup_\zeta \frac{p(\zeta)}{q(\zeta)}$, obviously[2] $c \geq 1$. The expected number of iterations for the algorithm is c. To see this, note from the above derivation that each iteration

[1] More technically, $\bar{P}$ is the Lebesgue measure on $[0,1]$.

[2] Clearly $\frac{p(\zeta)}{q(\zeta)} \leq c$ and so $p(\zeta) \leq cq(\zeta)$ implying $\int_{-\infty}^{\infty} p(\zeta) d\zeta \leq c \int_{-\infty}^{\infty} q(\zeta) d\zeta$, i.e. $1 \leq c$.

independently yields a probability of acceptance of $\frac{1}{c}$. So the number of iterations to accept is a geometric random variable[3] with mean c and variance $c(c-1)$.
(iii) The same approach applies to generate discrete valued random variables where $p_1, \dots, p_m$ and $q_1, \dots, q_m$ are now probability mass functions. In this discrete case, a particularly simple choice of q is the discrete uniform distribution: $q_i = 1/m$, $i = 1, \dots, m$ which can be implemented as

$$y = \text{Int}(m\, u) + 1, \qquad \text{where } u \sim U[0, 1].$$

$\text{Int}(x)$ denotes the integer part of real number x; see (A.1).
Example: A unit normal $\boldsymbol{N}(0, 1)$ can be generated from an exponentially distributed random variable with pdf $q(x) = e^{-x}, x \geq 0$ as follows. First generate a random variable with pdf

$$p(x) = \frac{2}{\sqrt{2\pi}} e^{-x^2/2}, \quad x \geq 0.$$

This corresponds to the absolute value of a $\boldsymbol{N}(0, 1)$ random variable. Note

$$\sup_{\zeta} \frac{p(\zeta)}{q(\zeta)} = \sup_{\zeta \in \mathbb{R}} \sqrt{\frac{2}{\pi}} e^{\zeta - \zeta^2/2} = \sqrt{\frac{2e}{\pi}}.$$

So choosing $c = \sqrt{2e/\pi}$, Step 1 generates an exponentially distributed random variable y (using for example the inverse transform method). Step 2 generates a uniform random variable u. Finally, Step 3 sets $x = y$ if $u \leq e^{-(y-1)^2/2}$. Finally to generate a $\boldsymbol{N}(0, 1)$ random variable, we multiply x by ± 1 where 1 and -1 are chosen with probability 1/2 each.

A.1.3 Composition method

The composition method simulates a random variable from a probability distribution that is a convex combination of probability distributions. Suppose it is possible to simulate random samples from each of the n distributions $F_i(\zeta)$, $i = 1, 2, \dots, n$. The aim is to simulate samples from the convex combination

$$F(\zeta) = \sum_{i=1}^{n} p_i F_i(\zeta)$$

where $p_i \geq 0$, $i = 1, 2 \dots, n$ and $\sum_{i=1}^{N} p_i = 1$.
The composition method proceeds as follows:

1. Generate the integer random sample $i^* \in \{1, \dots, n\}$ with probability mass function $p_1, \dots, p_n$, using for example the inverse transform or acceptance rejection method.
2. Then generate a random sample x from distribution F_{i^*}.

[3] Recall that a geometric distribution models number of trials to the first success when trials are independent and identically distributed with success probability p: So for $n \geq 1$, the probability of n trials to the first success is $p(1-p)^{n-1}$. The expected value and variance of a geometric random variable are $1/p$ and $\frac{1-p}{p^2}$, respectively.

A similar composition method holds for a "continuum" convex combination of probability density functions. That is, suppose one can simulate samples from $x \sim p_{x|y}(x|y)$ and $y \sim p_y(y)$ where $y \in \mathbb{R}^m$. Then samples from

$$p_x(\zeta) = \int_{\mathbb{R}^m} p_{x|y}(\zeta|y) p_y(y) dy \tag{A.2}$$

can be simulated via the following composition method:

Step 1: Simulate $y^* \sim p_y(\cdot)$.
Step 2: Simulate $x \sim p_{x|y}(\cdot|y^*)$.

The nice property of the above algorithm is that we do not need to compute the integral in (A.2) in order to simulate from $p_x(\zeta)$. §2.5.2 uses the composition method for optimal state prediction of a jump Markov linear system.

A.1.4 Simulating a Markov chain

The sample path of an i.i.d. discrete-time process can be simulated by repeated use of the inverse transform or acceptance rejection algorithms as long as the uniform numbers generated at each step are independent. Since for a Markov chain, given x_k, state x_{k+1} is conditionally independent of the past, the following procedure for simulating a Markov chain is obvious: Let P_i denote the i-th row of transition matrix P.

1. Generate $x_0 \sim \pi_0$.
2. For $k = 1, \ldots,$ generate $x_k \sim P_{x_{k-1}}$.

Here Step 1 and 2 can be implemented using, for example, the inverse transform or acceptance rejection method discussed earlier.

A.2 Variance reduction in simulation

In classical Monte Carlo evaluation of an integral, one simulates N i.i.d. samples of $x_k; k = 1, 2, \ldots, N$ from some pdf p. Then as $N \to \infty$,

$$\frac{1}{N} \sum_{k=1}^{N} c(x_k) \to \mathbb{E}_p\{c(x)\} = \int_{\mathcal{X}} c(x) p(x) dx \quad \text{with probability 1.}$$

Note that $\frac{1}{N} \sum_{k=1}^{N} c(x_k)$ is an unbiased estimate of $\mathbb{E}_p\{c(x)\}$ for any N; however, the variance of the estimate can be large. Below we outline some popular strategies for reducing the variance of the estimate.

A.2.1 Importance sampling

Let $p(x)$ denote a target distribution. Then clearly for any density $q(x)$

$$\mathbb{E}_p\{c(x)\} = \int_{\mathcal{X}} c(x) \frac{p(x)}{q(x)} q(x) dx$$

as long as $q(x)$ is chosen so that $p(x)/q(x)$ is finite for all x. In importance sampling, to estimate $\mathbb{E}_p\{c(x)\}$, one samples $x_k; k = 1, 2, \dots, N$ from the importance distribution (also called "instrumental distribution") $q(x)$ which is chosen so that $p(x)/q(x)$ is finite for all x. Denote the importance sampling estimate as

$$\hat{c}_N = \frac{1}{N}\sum_{k=1}^{N} c(x_k)\frac{p(x_k)}{q(x_k)}, \quad x_k \sim q. \tag{A.3}$$

If the sequence $\{x_k\}$ is i.i.d., then via the strong law of large numbers, as $N \to \infty$, the importance sampled estimate $\hat{c}_N \to \mathbb{E}_p\{c(x)\}$ almost surely. Also, by the central limit theorem

$$\lim_{N\to\infty} \sqrt{N}\left(\hat{c}_N - \mathbb{E}_p\{c(x)\}\right) \stackrel{\mathcal{L}}{=} \mathbf{N}(0, \operatorname{Var}_q(c(x))),$$

$$\text{where } \operatorname{Var}_q(c(x)) = \int c^2(x)\frac{p^2(x)}{q(x)}dx - \mathbb{E}^2\{c(x)\}. \tag{A.4}$$

Self-normalized importance sampling

Often times $p(x)$ is only known up to a normalization constant and evaluating the normalization constant can be computationally intractable. In such cases, self-normalized importance sampling can be used. Define the importance weight $w(x) = \frac{p(x)}{q(x)}$. Then the self normalized importance sampling estimate is

$$\hat{c}_N = \frac{\sum_{k=1}^{N} c(x_k)w(x_k)}{\sum_{k=1}^{N} w(x_k)}, \quad x_k \sim q.$$

The self-normalized estimate is biased for any finite N. For an i.i.d. sequence $\{x_k\}$, it follows via the strong law of large numbers that

$$\frac{1}{N}\sum_{k=1}^{N} c(x_k)w(x_k) \to \mathbb{E}_p\{c(x)\}, \quad \frac{1}{N}\sum_{k=1}^{N} w(x_k) \to 1$$

implying that $\hat{c}_N \to \mathbb{E}_p\{c(x)\}$ with probability one. By the central limit theorem,

$$\lim_{N\to\infty} \sqrt{N}\left(\hat{c}_N - \mathbb{E}_p\{c(x)\}\right) \stackrel{\mathcal{L}}{=} \mathbf{N}(0, \sigma^2),$$

$$\text{where } \sigma^2 = \int (c(x) - \mathbb{E}_p\{c(x)\})^2 \frac{p^2(x)}{q(x)}dx \tag{A.5}$$

which is to be compared with (A.4). This follows since

$$\sqrt{N}\left(\hat{c}_N - \mathbb{E}_p\{c(x)\}\right) = \frac{\frac{1}{\sqrt{N}}\sum_{k=1}^{N}\left(c(x_k)w(x_k) - \mathbb{E}_p\{c(x)\}\right)}{\frac{1}{N}\sum_{k=1}^{N} w(x_k)}.$$

The numerator converges in distribution to $\mathbf{N}(0, \sigma^2)$. The denominator converges with probability one to 1. By Slutsky's theorem,[4] the ratio converges to $\mathbf{N}(0, \sigma^2)$.

[4] One of the consequences of Slutsky's theorem is: if X_n converges in distribution to X and Y_n converges in distribution to a nonzero constant C, then X_n/Y_n converges in distribution to X/C. Recall convergence with probability one implies convergence in distribution.

Optimal importance density

Returning to the standard importance sampling estimator (A.3): what is the best choice of importance distribution $q(\cdot)$ to minimize the variance of the estimate? Since

$$\mathrm{Var}_q(c(x)) = \int c^2(x)\frac{p^2(x)}{q^2(x)}q(x)dx - \mathbb{E}^2\{c(x)\}$$
$$\mathrm{Var}_p(c(x)) = \int c^2(x)p(x)dx - \mathbb{E}^2\{c(x)\}.$$

So to estimate $\mathbb{E}\{c(x)\}$, in order to get a variance reduction using importance sampling, namely, $\mathrm{Var}_q(c(x)) < \mathrm{Var}_p(c(x))$ requires

$$\int c^2(x)\left(1 - \frac{p(x)}{q(x)}\right)p(x)dx > 0.$$

Assuming $c(x)$ to be nonnegative, the above formula yields the following insight into choosing the importance density $q(x)$: in regions of x where $c(x)p(x)$ is large choose $p(x)/q(x) < 1$; equivalently choose $q(x)$ to be large. Thus $q(x)$ needs to be chosen large for important values of x where $c(x)p(x)$ is large.

Regarding the optimal importance sampling density, assuming $c(x) > 0$, it is straightforwardly verified that choosing

$$q(x) = \frac{c(x)p(x)}{\int c(x)p(x)dx}$$

in (A.4) results in $\mathrm{Var}_q(c(x)) = 0$. Of course the above optimal importance sampling density has no practical value since determining it requires evaluating $\int c(x)p(x)dx = \mathbb{E}\{c(x)\}$; if we can evaluate this expression then there is no need for a Monte Carlo estimator.

One useful class of importance sampling densities are the so-called tilted densities obtained Esscher transform of the original pdf $p(x)$:

$$q(x) = \frac{e^{tx}p(x)}{\int e^{tx}p(x)dx}, \quad t \in \mathbb{R}. \tag{A.6}$$

For example, if p is Gaussian $\mathbf{N}(\mu, \sigma^2)$, then the tilted density q is $\mathbf{N}(\mu + \sigma^2 t, \sigma^2)$. If p is exponential with rate λ, then the tilted density q is exponential with rate $\lambda - t$.

Example 1: Suppose for a fixed real number α, the aim is to evaluate

$$c = \mathbb{P}(x > \alpha) \quad \text{where } x \sim \mathbf{N}(0, 1).$$

The standard Monte Carlo estimator based on N i.i.d. samples is

$$\hat{c} = \frac{1}{N}\sum_{k=1}^{N} I(x_k > \alpha), \quad x_k \sim \mathbf{N}(0, 1) \text{ i.i.d.} \tag{A.7}$$

Choosing the importance density $q = \mathbf{N}(\mu, 1)$, yields the estimator

$$\hat{c} = \frac{1}{N}\sum_{k=1}^{N} I(x_k > \alpha)\exp\left(-\frac{\mu^2}{2} - \mu x_k\right), \quad x_k \sim \mathbf{N}(0, \mu) \text{ i.i.d.} \tag{A.8}$$

This importance density q is a titled version of $N(0, 1)$ by choosing $t = \mu$ in (A.6). While it is difficult to compute the optimal choice of μ for the importance sampling density q, it can be shown that the optimal choice of μ satisfies $\mu > \alpha$. So choosing $\mu = \alpha$ is a reasonable choice.

To illustrate the above importance sampling estimator, consider estimating $\mathbb{P}(x > \alpha)$ where $\alpha = 8, x \sim N(0, 1)$. A Matlab simulation for $N = 50000$ points using the standard Monte Carlo estimate (A.7) yields $\hat{c} = 0$ (in double precision) which is useless. In comparison, a Matlab simulation for $N = 50000$ points using the importance sampling estimator (A.8) with $\mu = \alpha = 8$, yields the estimate $\hat{c} = 6.25 \times 10^{-16}$.

Example 2: This example is from [280]. Consider a two-state Markov chain with transition matrix

$$P = \begin{bmatrix} P_1 & 1 - P_1 \\ 1 - P_2 & P_2 \end{bmatrix}, \quad P_1, P_2 \in (0, 1).$$

We wish to also impose the constraint that $P_1 + P_2 < 1$. More specifically, assume that the prior distribution of transition probabilities P_1, P_2 is uniformly distributed on the simplex $P_1 + P_2 < 1$; so the prior density function is

$$p(P_1, P_2) = 2I(P_1 + P_2 < 1).$$

Denote the number of jumps from state i to j as J_{ij}, $i, j \in \{1, 2\}$. It is easily shown that the posterior is

$$p(P_1, P_2 | J_{11}, J_{12}, J_{21}, J_{22}) \propto P_1^{J_{11}} (1 - P_1)^{J_{12}} P_2^{J_{22}} (1 - P_2)^{J_{21}} I(P_1 + P_2 < 1).$$

The aim is to estimate the conditional expectation $\mathbb{E}\{P_1 | J_{11}, J_{12}, J_{21}, J_{22}\}$. Chapter 3 shows that such conditional mean estimates are optimal in the mean square error sense.

The above posterior is the product of two Bernoulli distributions restricted to the simplex $P_1 + P_2 < 1$. The standard Monte Carlo estimator simulates samples P_1 and P_2 independently from these two Bernoulli distributions until the sum of the two samples $P_1 + P_2 < 1$. This can be inefficient. For example, as pointed out in [280], suppose the data is $J_{11}, J_{12}, J_{21}, J_{22} = (68, 28, 17, 4)$, then

$$\mathbb{P}(P_1 + P_2 < 1 | J_{11}, J_{12}, J_{21}, J_{22}) = 0.21.$$

Consider now importance sampling with importance density

$$q(P_1, P_2) \propto N(P_1; \hat{P}_1, \hat{P}_1 \frac{(1 - \hat{P}_1)}{J_{12} + J_{11}}) \times N(P_2; \hat{P}_2, \hat{P}_2 \frac{(1 - \hat{P}_2)}{J_{21} + J_{22}}).$$

Here $\hat{P}_1 = J_{11}/(J_{11} + J_{12})$ and $\hat{P}_2 = J_{22}/(J_{21} + J_{22})$ are the maximum likelihood estimates of the transition probabilities (this will be shown in Chapter 4). Sampling from q can be done in two steps: Simulate P_1 from the first normal distribution above restricted to $[0, 1]$. Then simulate P_2 from the second normal distribution restricted to $[0, 1 - P_1]$.

It is shown in [280], on a 10,000-sample simulation experiment, that the importance sampling estimator yields an error of approximately 1% compared to the true value; in contrast, the standard Monte Carlo estimator error is more than 7%.

A.2.2 Variance reduction by conditioning

The smoothing property of conditional expectations implies that $\mathbb{E}\{\mathbb{E}\{X|Z\}\} = \mathbb{E}\{X\}$ (see (3.11) in Chapter 3). Then using the conditional variance formula

$$\text{Var}(X) = \mathbb{E}\{\text{Var}(X|Z)\} + \text{Var}(\mathbb{E}\{X|Z\})$$

it follows that for any two random variables X and Z:

$$\text{Var}(X) \geq \text{Var}(\mathbb{E}\{X|Z\}).$$

Thus the conditional mean estimate $\mathbb{E}\{X|Z\}$ always has a smaller variance than the raw estimate X, unless of course, X and Z are independent, in which case equality holds in the above equation. Variance reduction by conditioning is also called Rao–Blackwellization. This is illustrated in the context of particle filtering in Chapter 3.

A.2.3 Variance reduction by stratified sampling

Variance reduction by conditioning used the fact that $\text{Var}(X) \geq \text{Var}(\mathbb{E}\{X|Z\})$. Variance reduction by stratified sampling uses the fact that variances satisfy

$$\text{Var}(X) \geq \mathbb{E}\{\text{Var}(X|Z)\}. \tag{A.9}$$

In stratified sampling, to estimate $\mathbb{E}\{c(x)\}$ the sample space of x is partitioned into L disjoint regions (strata) and the mean of $c(x)$ is estimated in each strata. Finally $\mathbb{E}\{c(x)\}$ is estimated as a weighted average of these strata means.

For $k = 1, \ldots, N$:

1. Simulate the finite valued random variable $z_k \in \{1, 2, \ldots, L\}$ with $\mathbb{P}(z_k = l) = p_l$ where $\sum_{l=1}^{L} p_l = 1$.
2. Simulate x_k to be dependent on z_k such that

$$\hat{c}_l = \frac{\sum_{k=1}^{N} I(z_k = l) c(x_k)}{\sum_{k=1}^{N} I(z_k = l)}, \quad l = 1, 2, \ldots, L$$

3. Finally compute the estimate $\hat{c} = \sum_{l=1}^{L} p_l \hat{c}_l$.

From (A.9), it follows that

$$\mathbb{E}\{\text{Var}(c(x)|z)\} = \sum_{l=1}^{L} p_l \,\text{Var}(c(x)|z = l) \leq \text{Var}(c(x)).$$

Stratified sampling can also be viewed as a special case of importance sampling. Partition the sample space of x into L disjoint regions $\mathcal{R}_1, \ldots, \mathcal{R}_L$ and choose the importance density as $q(x) = p(x)/p_l$, for $x \in \mathcal{R}_l$, where $p_l = \int I(x \in \mathcal{R}_l) p(x) dx$. Then applying (A.4) yields

$$\text{Var}_q(c(x)) = \sum_{l} \text{Var}_p(c(x)|x \in \mathcal{R}_i) p_i \leq \text{Var}_p(c(x)).$$

Example: Suppose one wishes to estimate the integral $\int_0^1 c(x)dx$. The standard Monte Carlo estimator based on N samples is $\frac{1}{N} \sum_{k=1}^{N} c(x_k)$ where $x_k \sim U[0, 1]$. The stratified sampling estimator based on N samples is as follows: Partition $[0,1]$ into the L intervals $[0,\frac{1}{L}),[\frac{1}{L},\frac{2}{L}),\ldots,[\frac{L-1}{L},1]$. Then for each interval l generate N/L samples and compute

$$\hat{c}_l = \frac{1}{N/L}\sum_{k=1}^{N/L} c(x_k^{(l)}), \quad x_k^{(l)} \sim U[\frac{l-1}{L},\frac{l}{L}].$$

Finally compute $\hat{c} = \frac{1}{L}\sum_{l=1}^{L}\hat{c}_l$.

A.2.4 Variance reduction by common random numbers

Using the formula that

$$\mathrm{Var}(\frac{X+Y}{2}) = \frac{1}{4}(\mathrm{Var}(X) + \mathrm{Var}(Y) + 2\,\mathrm{cov}(X,Y))$$

it is clear that if X and Y were negatively correlated so that $\mathrm{cov}(X,Y) < 0$, then the variance is smaller than the case if X and Y are independent (in which case $\mathrm{cov}(X,Y) = 0$).

Accordingly, the estimator $\frac{1}{N}\sum_{k=1}^{N/2} c(x_k) + c(y_k)$ where $c(x_k), c(y_k)$ are negatively correlated has smaller variance compared to the standard Monte Carlo estimator $\frac{1}{N}\sum_{k=1}^{N} c(x_k)$. In particular, if $c(\cdot)$ is a monotone function, then $c(u)$ and $c(1-u)$ are negatively correlated for uniform $u \sim U[0,1]$.

A.3 Markov chain Monte Carlo (MCMC) methods

Recall, the stationary distribution of the transition probability matrix (or more generally transition kernel) satisfies $\pi_\infty = P'\pi_\infty$. In MCMC, the aim is to simulate a Markov process with specified stationary distribution π_∞ which is called the "target distribution". So the aim is to simulate a Markov chain whose transition matrix (kernel) admits the target distribution as a stationary distribution. The main question is how to construct a transition matrix P that admits π_∞ as a stationary distribution.

Why sample from a Markov chain rather than simulate i.i.d. samples from π_∞? It is often impossibly difficult to simulate i.i.d. samples from an arbitrary multivariate distribution. Also often π_∞ might only be known up to a normalizing constant. For continuous or large state spaces, computing the normalizing constant is intractable. This occurs in implementations of Bayes' rule where the denominator is the integral of the numerator and can be difficult to compute.

A.3.1 Reversible Markov chains and the Metropolis–Hastings algorithm

The Metropolis–Hastings algorithm is one of the earliest MCMC methods. It constructs a *reversible* Markov chain that admits a pre-specified π_∞ as the stationary distribution. We first recall a few facts about reversible Markov chains.

Reversible Markov chain

A reversible Markov chain is one where the transition matrix P and stationary distribution π_∞ satisfy

$$P_{ij}\,\pi_\infty(i) = P_{ji}\,\pi_\infty(j). \tag{A.10}$$

Denoting the diagonal matrix $\Pi_\infty \stackrel{\text{defn}}{=} \operatorname{diag}(\pi_\infty(1),\cdots,\pi_\infty(X))$, the reversibility condition reads:

$$\Pi_\infty P = P'\Pi_\infty.$$

(A.10) is called the "detailed balance equation". If there exists a probability vector π_∞ that satisfies (A.10), then π_∞ has to be a stationary distribution (as can be confirmed by summing the equation over i). The name "reversible" arises since for Markov chains satisfying (A.10), $(x_{k_1}, x_{k_2}, \ldots, x_{k_n})$ has the same distribution as $(x_{\tau-k_1}, x_{\tau-k_2}, \ldots, x_{\tau-k_n})$ for any times $k_1 < k_2 < \cdots < k_n$, τ and any n. That is, running the process forward in time is statistically indistinguishable from running the process backwards in time.

Why reversible Markov chains? For the purposes of simulation, given π_∞, it is substantially easier to a construct a reversible transition matrix that satisfies (A.10) which is a componentwise relationship, than to construct a general transition matrix that admits π_∞ as an eigenvector. Also, optimizing the second largest eigenvalue modulus of a reversible transition matrix can be formulated as a convex optimization problem. There are also useful bounds available for the second largest eigenvalue modulus of reversible Markov chains.[5]

The following theorem summarizes useful properties of reversible Markov chains.

THEOREM A.3.1 *Suppose P and π_∞ are the transition matrix and stationary distribution of a reversible Markov chain. Then*

1. *P is similar to a symmetric matrix. In particular, the similarity transformation $P^* = \Pi_\infty^{1/2} P \Pi_\infty^{-1/2}$ yields a symmetric matrix. Thus all eigenvalues of P are real.*
2. *Minimizing the second largest eigenvalue modulus of P is equivalent to the following convex optimization problem:*

$$\begin{aligned} &\text{Minimize } \lambda_2(P) = \|\Pi_\infty^{1/2} P \Pi_\infty^{-1/2} - qq'\|_2 \\ &\text{subject to } P_{ij} \geq 0, \quad P\mathbf{1} = \mathbf{1}, \quad \Pi_\infty P = P'\Pi_\infty \end{aligned}$$

where $q = \begin{bmatrix} \sqrt{\pi_\infty(1)} & \cdots & \sqrt{\pi_\infty(X)} \end{bmatrix}'$.

Thus, a nice property of the transition probability matrix of a reversible Markov chain is that all its eigenvalues are real. The second largest eigenvalue modulus, which determines the rate of convergence of the state probabilities of the Markov chain to the stationary distribution, can be minimized as a convex optimization problem; see [66] for a tutorial on maximizing the convergence rate of a reversible Markov chains via convex optimization.

[5] Reversibility also results in simpler conditions for a central limit theorem to hold for Markov processes on general state spaces (which we will not consider in this book). Reversibility can be imposed in the construction of a MCMC algorithm by additional simulation steps [320].

Proof Statement 1: The reversibility equation states $\Pi_\infty^{1/2}\Pi_\infty^{1/2}P = P'\Pi_\infty^{1/2}\Pi_\infty^{1/2}$ which is equivalent to $\Pi_\infty^{1/2}P\Pi_\infty^{-1/2} = (\Pi_\infty^{1/2}P\Pi_\infty^{-1/2})'$.
Statement 2: (From [66].) Since they are real-valued, denote the eigenvalues of P in descending order as $\{1, \lambda_2(P), \ldots, \lambda_X(P)\}$. So the second largest eigenvalue modulus is $\max\{\lambda_2(P), -\lambda_X(P)\}$. Now,

$$\lambda_2(P) = \sup\{u'Pu : \|u\|_2 \leq 1, \mathbf{1}'u = 0\}.$$

Note $u'Pu$ is linear in P and so $\lambda_2(P)$ is the pointwise supremum of linear functions of P. Thus $\lambda_2(P)$ is convex in P. Similarly, $-\lambda_n(P) = \sup\{-u'Pu : \|u\}_2 \leq 1\}$ is also convex. So the second largest eigenvalue modulus which is $\max\{\lambda_2(P), -\lambda_X(P)\}$ is convex in P.
□

Metropolis–Hastings algorithm

Suppose π_∞ is a target distribution and we wish to construct a reversible Markov chain that has π_∞ as its stationary distribution. Choose any irreducible transition matrix Q with $Q_{ii} = 0$, from which one can easily simulate a Markov chain. Next, construct a transition matrix P with elements P_{ij} as follows:

$$P_{ij} = \alpha_{ij}Q_{ij} + \left(1 - \sum_{l \neq i} \alpha_{il}Q_{il}\right) I(i = j), \quad i,j \in \mathcal{X}. \tag{A.11}$$

Here $\alpha_{ij} \in [0, 1]$ are called acceptance probabilities with $\alpha_{ii} = 0$. The above construction can be interpreted as the following two stage procedure:

1. Given the current state i, simulate the next state j with transition probabilities Q_{ij}, $j \in \{1, 2 \ldots, X\} - \{i\}$.
2. Accept the new state j with probability α_{ij}; the combination of the two events (generate and accept) has probability $P_{ij} = Q_{ij}\alpha_{ij}$. If not accepted, remain in the previous state i; this rejection event has probability $\left(1 - \sum_{l \neq i} \alpha_{il}Q_{il}\right)$ (Note $Q_{ii} = 0$).

It only remains to specify the acceptance probabilities α_{ij}, $i \neq j$ in (A.11). We choose them as follows:

$$\alpha_{ij} = \frac{S_{ij}}{1 + \frac{\pi_\infty(i)Q_{ij}}{\pi_\infty(j)Q_{ji}}}, \tag{A.12}$$

where S is a symmetric $X \times X$ matrix with nonnegative elements. Then as long as $\alpha_{ij} \leq 1$, clearly P is a stochastic matrix. To ensure that $\alpha_{ij} \leq 1$, one needs to choose S_{ij} in (A.12) so that

$$S_{ij} \leq 1 + \frac{\pi_\infty(i)Q_{ij}}{\pi_\infty(j)Q_{ji}}, \quad \text{and } S_{ij} \leq 1 + \frac{\pi_\infty(j)Q_{ji}}{\pi_\infty(i)Q_{ij}}$$

since $S_{ij} = S_{ji}$. This is equivalent to choosing

$$S_{ij} \leq 1 + \min\{\frac{\pi_\infty(i)Q_{ij}}{\pi_\infty(j)Q_{ji}}, \frac{\pi_\infty(j)Q_{ji}}{\pi_\infty(i)Q_{ij}}\}. \tag{A.13}$$

Finally, it is straightforward to check that with the choice (A.12) for the acceptance probabilities, (P, π_∞) satisfies the reversibility equation (A.10). Indeed, since S is symmetric,

$$\pi_\infty(i)\, Q_{ij} \frac{\cancel{S_{ij}}}{1 + \frac{\pi_\infty(i) Q_{ij}}{\pi_\infty(j) Q_{ji}}} = \pi_\infty(j)\, Q_{ji} \frac{\cancel{S_{ji}}}{1 + \frac{\pi_\infty(j) Q_{ji}}{\pi_\infty(i) Q_{ij}}}.$$

Summary: Given a target distribution π_∞, we constructed a reversible Markov chain with transition matrix P according to (A.11) where:

1. Q is an arbitrary irreducible stochastic matrix with $Q_{ii} = 0$.
2. The acceptance probabilities α_{ij} are chosen as (A.12) where the symmetric matrix S is chosen to satisfy the constraints (A.13).

Various algorithms can be obtained for different choices of Q and S. Consider the special case when the inequality (A.13) is chosen with equality. Then the acceptance probabilities (A.12) become

$$\alpha_{ij} = \min\left(1, \frac{\pi_\infty(j) Q_{ji}}{\pi_\infty(i) Q_{ij}}\right).$$

This corresponds to the standard Metropolis Hastings Algorithm 28.

Algorithm 28 Metropolis–Hastings

Choose any irreducible transition matrix Q with zero diagonal elements.
Given state $x_k = i$ at time k:

1. Simulate next state of the Markov chain $j \sim Q_{ij}$.
2. Generate $u \sim U[0, 1]$.
3. If $u < \frac{\pi_\infty(j) Q_{ji}}{\pi_\infty(i) Q_{ij}}$ then set $x_{k+1} = j$; otherwise set $x_{k+1} = x_k = i$.

Note step 3 of Algorithm 28 accepts $x_{k+1} = j$ with probability α_{ij}; and remains in the state $x_k = i$ with probability $1 - \alpha_{ij}$. A useful property of the Metropolis Hastings algorithm is that the target distribution π_∞ only needs to be known up to a normalizing constant, since the acceptance probabilities of step 3 depend on the ratio $\pi_\infty(j)/\pi_\infty(i)$.

Metropolis–Hastings for continuous-valued random variables

To simulate a continuous-valued random variable π_∞ in state space $\mathcal{X} \subseteq \mathbb{R}^m$, the approach is analogous to above. Choose a transition kernel (conditional density) $q(y|x)$ which is strictly positive for all $x, y \in \mathcal{X}$. (This is a sufficient condition for π_∞-irreducibility defined in the Appendix C.) Then implement Algorithm 28 with x replacing i, y replacing j, and $q(y|x)$ replacing Q_{ij}. Similar to (A.11), the Metropolis–Hastings algorithm will generate the sample path of a reversible Markov process on $\mathcal{X}$ with transition kernel

$$p(y|x) = \alpha_{xy} q(y|x) + (1 - \int_{\mathcal{X}} \alpha_{xy} q(y|x) dy)\, \delta_x(y)$$

where $\delta_x(y)$ denotes the Dirac delta function at x and $\alpha_{xy} = \min\left(1, \frac{\pi_\infty(y) q(x|y)}{\pi_\infty(x) q(y|x)}\right)$.

A.3.2 Gibbs sampling

Gibbs sampling is a special case of the Metropolis–Hastings algorithm. We want to simulate samples from the multivariate distribution $p(x_1, x_2, \ldots, x_L)$. Suppose that the L conditional distributions, $p(x_1|x_2, \ldots, x_L)$, $p(x_2|x_1, x_3, \ldots, x_L)$, etc can be computed relatively easily. Then the Gibbs sampling algorithm is given in Algorithm 29. Notice that each iteration has L stages.

Algorithm 29 Gibbs sampling

Given samples $x_1^{(n)}, x_2^{(n)}, \ldots, x_L^{(n)}$ from iteration n.
At $(n+1)$th iteration generate

1. $x_1^{(n+1)} \sim p(x_1|x_2^{(n)}, x_3^{(n)}, \ldots, x_L^{(n)})$
2. $x_2^{(n+1)} \sim p(x_2|x_1^{(n)}, x_3^{(n)}, \ldots, x_L^{(n)})$

$\vdots$

L. $x_L^{(n+1)} \sim p(x_L|x_1^{(n)}, x_2^{(n)}, \ldots, x_{L-1}^{(n)})$

Clearly $(x_1^{(n)}, \ldots, x_L^{(n)})$ is a Markov process evolving over iterations n. Gibbs sampling is equivalent to the composition of L Metropolis–Hastings algorithms with acceptance probability of 1. To see this, denoting $x = (x_1, \ldots, x_L)$ and $x' = (x'_1, \ldots, x'_L)$, note that in each of the steps $l = 1, \ldots, L$ in any iteration, the instrumental distribution in step l of the Gibbs sampling algorithm is

$$Q_{x,x'} = I(x_1 = x'_1, x_2 = x'_2, \ldots, x_{l-1} = x'_{l-1}, x_{l+1} = x'_{l+1}, \ldots, x_L = x'_L) \\ \times p(x'_l|x_1, \ldots, x_{k-1}, x_{l+1}, \ldots, x_L).$$

Therefore the acceptance probabilities in the Metropolis–Hastings algorithm are

$$\frac{\pi_\infty(x')Q_{x',x}}{\pi_\infty(x)Q_{x,x'}} = \frac{p(x')Q_{x',x}}{p(x)Q_{x,x'}} = \frac{p(x')p(x_l|x_1, \ldots, x_{l-1}, x_{l+1}, \ldots, x_L)}{p(x)p(x'_l|x_1, \ldots, x_{l-1}, x_{l+1}, \ldots, x_L)} = 1.$$

Remark: Data augmentation algorithm: The data augmentation algorithm was introduced by Tanner and Wong [316, 315]; see [323] for a tutorial description. Suppose we wish to sample from $p(x_1|y)$. Sometimes it is easier to sample from the augmented distribution $p(x_1, x_2|y)$ via Gibbs sampling where random variable x_2 and joint distribution $p(x_1, x_2|y)$ are carefully chosen.

The data augmentation algorithm is simply a two-stage Gibbs sampling algorithm, i.e. Algorithm 29 with $L = 2$. Stage 1 samples $x_1^{(n+1)}$ from $p(x_1|x_2^{(n)}, y)$ and stage 2 samples $x_2^{(n+1)}$ from $p(x_2|x_1^{(n+1)}, y)$. Given the samples $(x_1^{(n)}, x_2^{(n)})$, $n = 1, 2, \ldots$ generated from $p(x_1, x_2|y)$ by Gibbs sampling, clearly the samples $\{x_1^{(n)}, n = 1, 2, \ldots\}$ are from the marginal distribution $p(x_1|y)$.

Appendix B Continuous-time HMM filters

B.1 Introduction

Continuous-time filtering involves advanced concepts such as Itô calculus and martingale theory which are not covered in this book. This appendix presents the continuous-time HMM filter for three reasons: first, it will be shown that the continuous-time HMM filter can be time-discretized to obtain exactly the discrete-time HMM filter. From a structural point of view, this means that the monotone results for filtering developed in Part III of the book apply to the continuous-time HMM filter. From an algorithm point of view, the practitioner who wishes to numerically implement a continuous-time HMM filter can use the discrete-time HMM filter as an excellent approximation. Second, this discretization procedure reveals an interesting link with adaptive filtering: for sufficiently small discretization intervals, the HMM filter behaves as a stochastic approximation algorithm. Finally, optimal filters for Markov modulated Poisson processes can be obtained from the continuous-time HMM filter.

B.2 Continuous time Markov processes

Let $t \in [0,T]$ denotes continuous time. Consider a continuous-time random process $\{x_t\}$ where at each time t, x_t is a random variable in the finite state space $\mathcal{X} = \{e_1, e_2, \ldots, e_X\}$. Here e_i denotes the unit X-vector with 1 in the i-th position. Then the process $\{x_t\}$ is a continuous-time Markov process if the following property holds: given any n time instants, $0 \leq t_1 < t_2 < \ldots < t_n < t_{n+1}$, the following Markov property holds: $\mathbb{P}(x_{t_{n+1}} \in S | x_{t_1}, x_{t_2}, \ldots, x_{t_n}) = \mathbb{P}(x_{t_{n+1}} \in S | x_{t_n})$ where $S \subset \mathcal{X}$.

A continuous-time finite-state Markov chain is characterized by the transition rate matrix (infinitesimal generator) denoted as Q: here Q is a $X \times X$ matrix and is interpreted in terms of transition probabilities as follows:

$$Q_{ij} = \begin{cases} \lim_{\Delta \to 0} \frac{\mathbb{P}(x_\Delta = e_j | x_0 = e_i)}{\Delta} & i \neq j \\ \lim_{\Delta \to 0} \frac{\mathbb{P}(x_\Delta = e_i | x_0 = e_i) - 1}{\Delta} & i = j \end{cases} \tag{B.1}$$

$$\text{where } Q_{ii} < 0, \quad \sum_{j=1}^{X} Q_{ij} = 0, Q_{ij} \geq 0, i \neq j, \quad \text{for } 1 \leq i,j \leq X.$$

Optimal predictor: Fokker–Planck equation: Recall from Chapter 2.2 that in discrete time, the Chapman–Kolmogorov equation characterizes the optimal state predictor. In continuous time, the Fokker–Planck equation (Kolmogorov forward equation) characterizes the optimal predictor. For an X-state homogeneous Markov chain with generator Q and initial distribution π_0, define the X-dimensional vector state probability vector $\pi_t = [\pi_t(1), \ldots, \pi_t(X)]'$ where $\pi_t(i) = \mathbb{P}(x_t = i)$. The Fokker–Planck equation reads

$$\frac{d\pi_t}{dt} = Q'\pi_t, \quad \text{initialized with } \pi_0.$$

It can be solved explicitly as $\pi_t = \exp(Q't)\pi_0$. Discretizing this over time with discretization interval Δ and sampling times denoted as t_k, $k = 1, 2, \ldots$ yields: $\pi_{t_k} = P'\pi_{t_{k-1}}$ where $P = \exp(Q'\Delta)$ is the transition probability matrix. This is the discrete-time HMM predictor of Chapter 2.4.3.

B.3 Continuous-time hidden Markov model (HMM) filter

On the state space $\mathcal{X} = \{e_1, e_2, \ldots, e_X\}$, consider a continuous-time Markov chain $\{x_t\}$ with transition rate matrix Q. Assume the observation process is

$$y_t = \int_0^t C'x_s ds + v_t$$

where the X-dimensional vector $C = [C(1), \ldots, C(X)]$ denotes the physical state levels of the Markov chain corresponding to states $e_1, \ldots, e_X$. $\{v_t\}$ denotes standard Brownian motion defined in Appendix D. Then $\{y_t\}$ denotes the observation trajectory of a continuous-time HMM. The aim is to estimate the underlying Markov chain given the observations.

Given the observation trajectory $y_{0:t}$, define the un-normalized filtered density q_t and its normalized counterpart as

$$q_t(x) = p(x_t = x, y_{0:t}), \quad \pi_t(x) = \mathbb{P}(x_t = x | y_{0:t}). \tag{B.2}$$

In continuous time, there two ways for expressing the optimal filter:

1. Duncan–Mortensson–Zakai (DMZ) filter: This is a stochastic differential equation that gives the time evolution of the un-normalized filtered density q_t. The resulting stochastic differential equation is linear in q_t.
2. Innovations filter: This uses the innovations representation, to yield directly a stochastic differential equation for the evolution of the normalized filtered density π_t. However, in general this equation is nonlinear in π_t.

Define the X-vector of the normalized and un-normalized filtered densities

$$q_t = [q_t(1), q_t(2), \cdots, q_t(X)]', \quad \pi_t = [\pi_t(1), \pi_t(2), \cdots, \pi_t(X)]'. \tag{B.3}$$

Define the diagonal $X \times X$ observation matrix

$$B = \operatorname{diag}(C(1), \ldots, C(X)) \tag{B.4}$$

The DMZ continuous-time HMM filter reads

$$q_t = q_0 + \int_0^t Q'q_s ds + \int_0^t Bq_s dy_s. \tag{B.5}$$

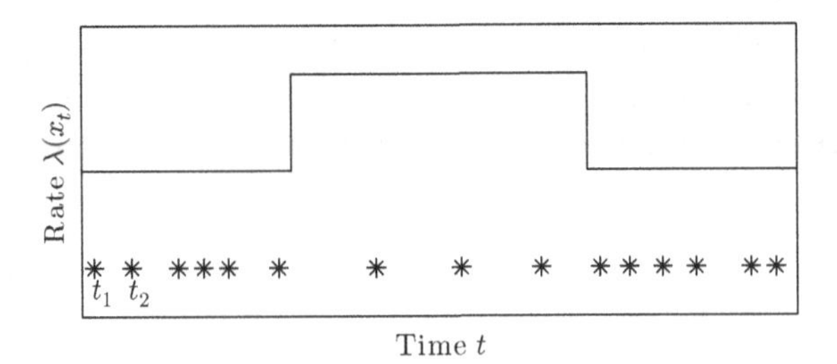

Figure B.1 Markov modulated Poisson process trajectory. The rate $\lambda(x_t)$ evolves according to the realization of a two state Markov chain. The Poisson events occur according to the time sequence $\{t_n, n \geq 1\}$. The figure illustrates the case when the rate for state 1 is larger than that for state 2. Given these Poisson events, the aim is to estimate the underlying Markov chain x_t.

The innovations HMM filter equation reads:

$$\pi_t = \pi_0 + \int_0^t Q'\pi_s ds + \int_0^t \big[B\pi_s - C'\pi_s\pi_s\big]\big[dy_s - C'\pi_s ds\big], \tag{B.6}$$

The conditional mean estimate is (recall the state space comprises unit vectors)

$$\hat{x}_t = \pi_t = q_t / \mathbf{1}' q_t.$$

The innovations form of the continuous-time HMM filter is also called the *Wonham* filter. The HMM predictor is obtained by setting $B = 0$ in the Zakai equation or setting $B = 0$ and $C = 0$ in the innovations equation. The HMM predictor reads: $\pi_t = \pi_0 + \int_0^t Q'\pi_s ds$ or equivalently $\pi_t = \exp(Q't)\pi_0$.

B.3.1 Markov modulated Poisson process (MMPP) filter

A MMPP is an inhomogeneous Poisson process where the Poisson rate is modulated by a Markov chain. Figure B.1 illustrates the sample path of a MMPP.

Given observation of the Poisson process, the aim is to estimate the underlying Markov chain The observation model is

$$N_t^{(l)} = \int_0^t \lambda^{(l)} x_s ds + \nu_t^{(l)}, \quad l = 1, \ldots, L.$$

The L-variate observation process N_t represents an inhomogeneous (Markov modulated) Poisson process with nonnegative rates $\lambda^{(l)}(x_t)$ that evolve in time according to the realization of the Markov process x_t. Each $N_t^{(l)}$ is an integer valued process representing the number of Poisson events of mark l that have occurred in the interval $[0, t]$. That is

$$\mathbb{P}(N_t^{(l)} = k) = \frac{\left(m_t^{(l)}\right)^k}{k!} \exp(-m_t^{(l)}), \quad \text{where } m_t^{(l)} = \int_0^t \lambda^{(l)} x_s ds.$$

The process $\nu_t^{(l)}$ is a zero mean Poisson martingale. Note $dN_t^{(l)} = 1$ if an event occurs at time t, and is zero otherwise.

Let $B^{(l)} \stackrel{\text{defn}}{=} \text{diag}(\lambda^{(l)})$ denote a diagonal $X \times X$ matrix. Then the Zakai form of the MMPP filter reads (where q_t is defined in (B.3)):

$$q_t = q_0 + \int_0^t Q' q_s ds + \int_0^t \sum_{l=1}^L (B^{(l)} - I) q_s \left[dN_s^{(l)} - ds\right]. \tag{B.7}$$

The innovations MMPP filter equation reads

$$\pi_t = \pi_0 + \int_0^t Q'\pi_s ds + \int_0^t \sum_{l=1}^{L} \left[\frac{B^{(l)}\pi_{s-}}{\lambda^{(l)}\pi_{s-}} - \pi_{s-} \right] \left[dN_s^{(l)} - \lambda^{(l)}\pi_{s-} ds \right]. \tag{B.8}$$

Since N_t is a counting process, the integral involving $dN_s^{(l)}$ is just a Stieltjes integral. In fact, the above MMPP filter can be implemented exactly. Let $\tau_k, k = 1, 2, \ldots$ denote the event times of the MMPP $\{N_t\}$. Then the Zakai equation (B.7) can be written as

$$q_t = q_0 + \int_0^t \left[Q' - \left(\sum_{l=1}^{L} (B^{(l)} - I) \right) \right] q_s ds + \sum_{l=1}^{L} (B^{(l)} - I) \sum_{\tau_k < t} q_{\tau_{k-}}. \tag{B.9}$$

This leads to the following exact implementation algorithm for an MMPP filter: For $t \in (\tau_k, \tau_{k+1})$, i.e. between event times,

$$q_t = \exp\left[\left(Q' - \left(\sum_{l=1}^{L} (B^{(l)} - I) \right) \right) (t - \tau_k) \right] q_{\tau_k}. \tag{B.10}$$

That is, q_t evolves deterministically between Poisson events. At the event time $t = \tau_{k+1}$, update $q_{\tau_{k+1}}$ as

$$q_{\tau_{k+1}} = q_{\tau_{k+1}^-} + \sum_{l=1}^{L} (B^{(l)} - I) q_{\tau_{k+1}^-}.$$

B.4 Robust nonlinear filters and numerical implementation

This section presents a *robust* version of the HMM Zakai filtering equation (B.5). Robust means that the resulting filtering equation is locally Lipschitz continuous in the observations – i.e. the equation depends continuously on the observation path. Indeed, the equations turn out to be non-stochastic ordinary differential equations whose coefficients depend on the observations. These robust equations are useful from a practical point of view – their numerical solution via time discretization can be performed without worrying about the Itô terms.

The idea of robust filtering – i.e. re-expressing the stochastic differential equation as non-stochastic differential equation with random coefficients has been used extensively in the context of nonlinear filtering; see, for example, [87], [197], [92], [257] or Chapter 4 of [46]. In [157] versions of these robust filters, probabilistic interpretations and implicit and explicit discretization schemes were developed for continuous-time hidden Markov models.

Define the robust filtered state distribution as

$$\bar{q}_t = \Phi_t^{-1} q_t, \qquad \text{where } \Phi_t = \exp(By_t - \frac{1}{2}B^2 t).$$

Recall q_t satisfies the HMM filter Zakai equation (B.5) and B is defined in (B.4).

THEOREM B.4.1 *[157] (Robust HMM filter) Consider the continuous-time HMM of §B.3. The robust forward filters evolve as*

$$\frac{d\bar{q}_t}{dt} = \Phi_t^{-1} Q' \Phi_t \bar{q}_t, \qquad \bar{q}_0 = q_0 \tag{B.11}$$

Furthermore, the robust filtered state estimate

$$\hat{x}_{t|t} = \frac{C'\Phi_t\bar{q}_t}{\mathbf{1}'\Phi_t\bar{q}_t}$$

defines a locally Lipschitz version of $\mathbb{E}\{x_t|\mathbf{Y}_t\}$ *in the sense that for any two observation trajectories* $y^{(1)}, y^{(2)} \in C(\mathbb{R}^n \times [0,T])$ *and for some constant* K *depending on* $\|y^{(1)}\|$ *and* $\|y^{(2)}\|$, $|\hat{x}_t(y^{(1)}) - \hat{x}_t(y^{(2)})| \leq K\|y^{(1)} - y^{(2)}\|$.

The proof of (B.11) involves Itô calculus (see §B.5). From a numerical implementation point of view, (B.11) is exceedingly useful as we now discuss.

B.4.1 Time discretization of robust HMM filter

The purpose of this section is to provide a computable approximation of the continuous-time HMM filter. Consider a regular partition $0 = t_0 < t_1 < \cdots < t_{k-1} < t_k < \cdots$ with constant time step $\Delta = t_k - t_{k-1}$. Write $M = [T/\Delta]$ for the largest integer such that $M\Delta \leq T$.

Two approaches are available to obtain discrete-time filtering equations.

1. *Discretization of model*: One approach is to sample the continuous-time observations $\{y_t, t \geq 0\}$ and approximate the original continuous-time HMM. The filtering equations for the discrete-time HMM would then provide an approximation of the filtering equations for the continuous-time HMM.
2. *Discretization of filter*: The other approach is to directly discretize the filtering equations, or their robust versions obtained above.

It turns out that an explicit time discretization of the robust filtering equation yields exactly the discrete-time HMM equation thus linking the two above mentioned approaches. This is an important consistency property in that it links discretized continuous-time results with discrete-time results. The discretization is as follows: between sampling times t_{k-1} and t_k,

$$\begin{aligned}\bar{q}_{t_k} &= \bar{q}_{t_{k-1}} + \int_{t_{k-1}}^{t_k} \Phi_s^{-1}\, Q'\, \Phi_s\, \bar{q}_s\, ds \\ &\simeq \bar{q}_{t_{k-1}} + \Phi_{t'_k}^{-1}\, Q'\, \Phi_{t'_k}\, \bar{q}_{t''_k}\, \Delta\end{aligned}$$

for some $t_{k-1} \leq t'_k, t''_k \leq t_k$. Various approximations can be obtained, for different choices of t'_k and t''_k. Choosing $t'_k = t''_k = t_{k-1}$ gives

$$\bar{q}_{t_k} \simeq [I + \Delta\, \Phi_{t_{k-1}}^{-1}\, Q'\, \Phi_{t_{k-1}}]\, \bar{q}_{t_{k-1}}\,,$$

which results in the following explicit discrete-time approximation

$$\bar{q}_k = [I + \Delta\, \Phi_{t_{k-1}}^{-1}\, Q'\, \Phi_{t_{k-1}}]\, \bar{q}_{k-1}\,. \tag{B.12}$$

Note that (B.12) looks like a stochastic approximation algorithm with step size Δ:

$$\bar{q}_k = \bar{q}_{k-1} + \Delta\, \Phi_{t_{k-1}}^{-1}\, Q'\, \Phi_{t_{k-1}} \bar{q}_{k-1}\,.$$

Multiplying both sides of (B.12) by Φ_{t_k} gives the following approximation for the HMM Zakai equation (B.5)

$$q_k = \Phi_{t_k}\, \Phi_{t_{k-1}}^{-1}\, [I + \Delta\, Q']\, q_{k-1} = \Psi_k\, [I + \Delta\, Q']\, q_{k-1}\,, \tag{B.13}$$
$$\text{where } \Psi_t^s = \Phi_t\, \Phi_s^{-1} = \exp\left\{ B\,[y_t - y_s] - \frac{1}{2}\, B^2\, [t-s] \right\} \text{ and } \Psi_k = \Psi_{t_k}^{t_{k-1}}\,.$$

Here $\Psi_k = \Phi_{t_k}\, \Phi_{t_{k-1}}^{-1} = \operatorname{diag}(\psi_k^1, \cdots, \psi_k^N)$ with $\psi_k^i = \phi_{t_k}^i / \phi_{t_{k-1}}^i$. Equation (B.13) is exactly the discrete-time un-normalized HMM filter of (3.59) in Chapter 3 for Gaussian observation noise.

Summary: (B.13) is a useful numerical implementation for a continuous-time HMM filter. This time discretized robust version of the continuous-time HMM filter coincides with the discrete-time HMM filter. Therefore the monotone properties of the discrete-time filter in Chapter 10 apply to continuous-time HMM filters. These monotone properties were the basis of structural results for POMDPs.

B.4.1 Remark: Continuous-time stochastic control

The analog of Bellman's equation in continuous time is the Hamilton–Jacobi–Bellman (HJB) equation which is a nonlinear partial differential equation. It can involve significant mathematical technicalities; for example, stopping time problems in continuous time involve quasi-variational inequalities [47].

From a numerical implementation point of view, HJB can be solved via finite difference methods. For HJBs arising in quantitative finance, the Barles–Souganidis framework [34] is a very useful finite difference-based method. This framework shows that under suitable conditions, the discretized finite difference scheme converges to the HJB as the discretization intervals go to zero. (The solution of the HJB is interpreted in terms of "viscosity solutions" [117].) Suppose Bellman's equation for an MDP or POMDP obtained from by the finite difference approximation of the HJB equation, satisfies the various monotone properties described in Part III (e.g. submodularity, monotone optimal policy). Then under suitable conditions, the resulting continuous-time stochastic control problem inherits these monotone properties. Therefore, despite the formidable technicalities, the algorithms in Part III can be used for constructing parametrized monotone policies.

B.5 Itô's formula

Consider the following continuous-time diffusion in $\mathbb{R}^X$ where w_t denotes vector-valued Brownian motion with independent components:

$$x_t = x_0 + \int_0^t A_s(x_s)ds + \int_0^t \Gamma_s(x_s)\, dw_s, \quad x_0 \sim \pi_0(\cdot). \tag{B.14}$$

The term $\int_0^t \Gamma_s(x_s)\, dw_s$ is interpreted as an Itô integral and is a zero mean martingale. Then Itô's rule [167] says that for any twice differentiable function ϕ,

$$\begin{aligned} \phi(x_t) &= \phi(x_0) + \int_0^t (\mathcal{L}\phi)(x_s)\, ds + \int_0^t (\nabla\phi(x_s))\Gamma(x_s)dw_s \\ \text{where } \mathcal{L}_t\phi &= \frac{1}{2}\operatorname{trace}[Q_t(x)\nabla^2\phi] + A_t'(x)\nabla\phi, \quad Q_t(x) = \Gamma_t(x)\Gamma_t'(x). \end{aligned} \tag{B.15}$$

Example 1: Choose $x_t = w_t$, $\phi(x) = x^2$, then Itô's formula (B.15) implies

$$w_t^2 = 2\int_0^t w_s dw_s + t,$$

which is different to classical integration due to the present of the term t.

Example 2: Geometric Brownian motion evolves according to the scalar process

$$x_t = x_0 + a\int_0^t x_s ds + \sigma \int_0^t x_s dw_s.$$

In the famous Black-Scholes model in finance, x_t models the price of a stock at time t, a is called the drift and σ is called the volatility. Using Itô's formula,

$$\log x_t = \log x_0 + (a - \sigma^2/2)t + \sigma w_t$$

which in turn implies that

$$x_t = \exp\left((a - \sigma^2/2)t + \sigma w_t\right) x_0.$$

Since w_t is Gaussian, the exponential term has a log-normal distribution. Note that x_t is non-negative if x_0 is non-negative; hence its use in modeling prices.

Next, Itô's integration by parts formula says that

$$x_t y_t = x_0 y_0 + \int_0^t x_s dy_s + \int_0^t dx_s y_s + [x, y]_t. \tag{B.16}$$

Here $[x, y]_t = \sum_{0 \le s \le t} \Delta x_s \Delta y_s$ denotes the quadratic variation where $\Delta x_s = x_s - x_{s-}$. For our purposes it suffices that for Brownian motion, the quadratic variation $[w, w]_t = t$. (Intuitively, Brownian motion satisfies $(dw_t)^2 = dt$.)

Proof of (B.11) in Theorem B.4.1. We can now establish (B.11) for the robust HMM filter. Denote Φ_t^{-1} as $\bar{\Phi}_t$. Since $\bar{q}_t = \bar{\Phi}_t q_t$, applying formula (B.16) yields

$$\bar{q}_t = \bar{q}_0 + \int_0^t \bar{\Phi}_s dq_s + \int_0^t d\bar{\Phi}_s q_s + [\bar{\Phi}, \bar{q}]_t.$$

Next using Itô's rule (B.15) yields $d\bar{\Phi}_s = \bar{\Phi}_s(-B dy_s + \frac{1}{2}B^2 ds + \frac{1}{2}B^2 ds)$. Therefore

$$\bar{q}_t = \bar{q}_0 + \int_0^t \bar{\Phi}_s(Q' q_s ds + \cancel{B q_s dy_s}) + \int_0^t \bar{\Phi}_s(\cancel{-B q_s dy_s} + B^2 q_s ds) + [\bar{\Phi}, \bar{q}]_t.$$

The quadratic variation term is

$$\begin{aligned}
[\bar{\Phi}, \bar{q}]_t &= \sum_{0 \le s \le t} \Delta\bar{\Phi}_s \Delta\bar{q}_s = \sum_{0 \le s \le t} \bar{\Phi}_s\left(-B\Delta y_s + \frac{1}{2}B^2 \Delta s\right)(Q' q_s \Delta s + B q_s \Delta y_s) \\
&= -\int_0^t \bar{\Phi}_s B^2 q_s ds \quad \text{(since only the } \Delta y_s \text{ terms contain Brownian motion)}
\end{aligned}$$

and cancels with the corresponding term in $\bar{q}_t$ yielding

$$\bar{q}_t = \bar{q}_0 + \int_0^t \bar{\Phi}_s Q' q_s ds = \bar{q}_0 + \int_0^t \bar{\Phi}_s Q' \Phi_s \bar{q}_s ds. \qquad \square$$

Appendix C Markov processes

Countable and finite state

1. Recurrence: For a countable state Markov chain $\{x_n\}$ with state space $\mathcal{X}$ as the set of integers, denote the first return time as $T_i = \inf\{n > 0 : x_n = i | x_0 = i\}$.

- A state i is recurrent if $P(T_i < \infty) = 1$ or equivalently[1],

$$\sum_{n=1}^{\infty} P_{ii}^n = \mathbb{E}\{\sum_{n=1}^{\infty} I(x_n = i) | x_0 = i\} = \infty.$$

  - A recurrent state is called positive recurrent if $\mathbb{E}\{T_i\} < \infty$
  - A recurrent state is called null recurrent if $\mathbb{E}\{T_i\} = \infty$.
- A state i is called transient if $P(T_i < \infty) < 1$ or equivalently if $\sum_{n=1}^{\infty} P_{ii}^n$ is finite.

For finite state Markov chains null recurrence cannot exist. Also for a finite state Markov chain, all states cannot be transient. Since this book mainly considers finite state Markov chains, null recurrence does not play any role.

Null recurrence can arise in countable state Markov chains. For example, consider a 1-dimensional random walk on the set of integers: $x_{k+1} = x_k + w_k$ where $P(w_k = -1) = p$, $P(w_k = 1) = 1 - p$. Then this random walk is null recurrent if $p = 0.5$ (symmetric random walk). If $p > 0.5$ or $p < 0.5$, then the process is transient. The 2-dimensional symmetric random walk is also null-recurrent. Symmetric random walks for dimensions larger than 2 are always transient.

2. Aperiodic: A state i has a period n if any return to state i must occur at multiples of n time steps. That is with gcd denoting greatest common divisor,

$$n = \gcd\{k : \mathbb{P}(x_k = i | x_0 = i) > 0\}.$$

If $n = 1$, then the state is said to be *aperiodic*. Put differently, a state is aperiodic if there exists a positive n such that for all $m \geq n$, $P_{ii}^m = \mathbb{P}(x_m = i | x_0 = i) > 0$.

A Markov chain is called aperiodic if every state is aperiodic.

3. Communicating Classes: Two states i and j are said to communicate, if for some positive m and n, $P_{ij}^m > 0$ and $P_{ji}^n > 0$.

THEOREM C.0.1 *Suppose i and j are communicating states. Then if state i is "P" then so is state j; where property "P" denotes either positive recurrent, null recurrent, transient, or aperiodic.*

[1] P_{ii}^n denotes the (i, i) element of P^n.

Recall, a Markov chain is irreducible if every state communicates with every other state. As a consequence of Theorem C.0.1, an irreducible Markov chain requires only one aperiodic state to imply that all states are aperiodic.

THEOREM C.0.2 *For a finite state Markov chain, if $\{x_n\}$ is aperiodic and irreducible, then the transition matrix is regular. That is, there exists a positive integer n such that all elements of P^n are nonzero.*

Markov processes on general state space

We summarize some basic definitions for Markov processes on a general state space $\mathcal{X} \subseteq \mathbb{R}^X$; please refer to [240] for a seminal exposition.

We have already defined transition densities $p(x_{n+1}|x_n)$ for such Markov process in Chapter 2. Assume for notational convenience that these densities are time homogeneous. For $S \subseteq \mathcal{X}$, define the transition kernel

$$P(x, S) = \mathbb{P}(x_{k+1} \in S | x_k = x) = \int_S p(x_{k+1} = \zeta | x_k = x)\, d\zeta.$$

Define the n-step transition kernel as

$$P^n(x, S) = \int_{\mathcal{X}} P(x, d\zeta)\, P^{n-1}(\zeta, S).$$

In terms of the transition density $P^n(x, S) = \int_S p(x_n = \zeta | x_0 = x)\, d\zeta$.
ϕ-irreducibility: Let ϕ be a probability measure on $\mathcal{X}$. Then the transition kernel P is said to be ϕ-irreducible if $\phi(S) > 0$ implies that for every $x \in \mathcal{X}$, there exists a positive integer n such that $P^n(x, S) > 0$.
Example 1: Consider the time invariant linear Gaussian state space model

$$x_{k+1} = A x_k + w_k, \quad x_0 \sim N(\hat{x}_0, \Sigma_0)$$

where $x_k \in \mathbb{R}^X$ and $w_k \sim N(0, Q)$ is an iid process. Then from (2.21),

$$p(x_k|x_0) = N(\hat{x}_k, \Sigma_k), \quad \text{where } \hat{x}_k = A^k \hat{x}_0, \quad \Sigma_k = A^k \Sigma_0 A'^k + \sum_{n=0}^{k-1} A^n Q A'^n.$$

Define M such that the state noise covariance $Q = MM'$. Assume that $[A, M]$ is completely reachable; see (2.24). Then for all $k \geq X$, it follows that $\sum_{n=0}^{k-1} A^n Q A'^n$ is positive definite implying Σ_k is positive definite. Therefore, $p(x_k|x_0)$ has a density which is positive everywhere. Hence, the process $\{x_k\}$ is irreducible with respect to the Lebesgue measure.
Example 2: As an example of a Markov process that is not irreducible consider the scalar process $x_{n+1} = A x_n + w_n$ where $A > 1$ and w_n is iid uniformly distributed on $[-1, 1]$. Then

$$x_{n+1} - x_n = (A-1)x_n + w_n \geq (A-1)x_n - 1 > 0, \quad \text{if } x_n > 1/(A-1).$$

So choosing $S = [0, x_n]$, it follows that $P^n(x_n, S) = 0$ for all $n > 0$ if $x_n > 1/(A-1)$. Hence the Markov process is not irreducible. (The process is monotone increasing and never visits previous states.)

Appendix D Some limit theorems

Strong law of large numbers (SLLN)

The aim is to characterize the sample average of a sequence of random variables $\{X_k\}$. Define the sample average as

$$\mu_n = \frac{1}{n}\sum_{k=1}^{n} X_k.$$

Then the SLLN states

THEOREM D.0.3 *(IID) Suppose $\{X_k\}$ is an independent and identically distributed (i.i.d.) sequence of random variables. Then $\lim_{n\to\infty}\mu_n = \mathbb{E}\{X_1\}$ almost surely iff $\mathbb{E}\{|X_1|\} < \infty$ and $\mathbb{E}\{X_1\} = \mu$.*

(Finite-state Markov) Suppose $\{X_n\}$ is an X-state Markov chain with state space comprising of X-dimensional unit vectors. Assume that the transition matrix P is regular (i.e. for some positive integer k, P^k has strictly positive elements) implying a unique stationary distribution π_∞ exists. Then $\lim_{n\to\infty}\mu_n = \pi_\infty$ almost surely.

The i.i.d. version of the above SLLN is also called Kolmogorov's SLLN. In this i.i.d. case, the condition $\mathbb{E}\{|X_1|\} < \infty$ and $\mathbb{E}\{X_1\} = \mu$ is necessary and sufficient for the SLLN to hold. Also since $\{X_k\}$ are i.i.d. clearly $\mathbb{E}\{\mu_n\} = \mu$.

An important application of the law of large numbers is the estimation of a cumulative distribution from random samples. Suppose $\{X_n\}$ is an i.i.d. sequence simulated from an unknown cumulative distribution function F. The empirical distribution function is defined as (where $I(\cdot)$ below denotes the indicator function)

$$F_n(x) = \frac{1}{n}\sum_{k=1}^{n} I(X_k \le x) \tag{D.1}$$

$F_n(x)$ is a natural estimator of F when F is not known. Clearly by the law of large numbers $\lim_{n\to\infty} F_n(x) = F(x)$ almost surely for each x. Actually for estimation of cumulative distributions one has a stronger result, namely, uniform almost sure convergence.

THEOREM D.0.4 (Glivenko–Cantelli theorem) *Suppose $\{X_n\}$ is an i.i.d. sequence with cumulative distribution function F on the real line. Then*

$$\lim_{n\to\infty}\sup_{x\in\mathbb{R}} |F_n(x) - F(x)| = 0 \text{ almost surely.}$$

The theorem says that a large number of random samples uniformly approximates the distribution itself.

Central limit theorem

THEOREM D.0.5 *With* $\mu_n = \frac{1}{n}\sum_{k=1}^{n} X_k$ *and* $\stackrel{\mathcal{L}}{=}$ *denoting convergence in distribution, the following hold:*

(IID) Suppose $\{X_k\}$ *is an independent and identically distributed sequence of random variables with zero mean and unit variance. Then* $\lim_{n\to\infty} \sqrt{n}\mu_n \stackrel{\mathcal{L}}{=} N(0,1)$.

(Finite-state Markov) Suppose $\{X_n\}$ *is an X-state Markov chain with state space comprising of X-dimensional unit vectors. Suppose the transition matrix P is regular implying a unique stationary distribution* π_∞ *exists. Then for any vector* $g \in \mathbb{R}^X$,

$$\lim_{n\to\infty} \sqrt{n} g'(\mu_n - \pi_\infty) \stackrel{\mathcal{L}}{=} N(0,\sigma^2)$$

where the variance

$$\sigma^2 = 2g'\text{diag}(\pi_\infty)Zg - g'\text{diag}(\pi_\infty)(I + \mathbf{1}\pi'_\infty)g.$$

$Z = (I - (P - \mathbf{1}\pi'_\infty))^{-1}$ *is called the fundamental matrix of the regular Markov chain.*

The above central limit theorems have been stated for sequences of scalars and the limiting distributions are univariate Gaussians. It suffices to consider sequences of scalars. This is a consequence of the famous Cramér-Wold device.

THEOREM D.0.6 (Cramér–Wold device) *Suppose* $\{X_n\}$ *is a sequence of random X-dimensional vectors. Suppose for each unit norm vector* $g \in \mathbb{R}^X$, $g'X_n$ *converges in distribution to* $g'X_\infty$ *where* X_∞ *is an X-dimensional random vector with joint (multivariate) distribution F. Then* $\{X_n\}$ *converges in distribution to F.*

Continuing with the example involving empirical distribution functions defined in (D.1), it follows from the central limit theorem[1] that $\lim_{n\to\infty}\sqrt{n}(F_n(x) - F(x)) \stackrel{\mathcal{L}}{=} N\big(0, F(x)(1 - F(x))\big)$ for each x. In analogy to the Glivenko–Cantelli theorem, the central limit theorem can be extended to a uniform or functional central limit theorem; see Donsker's theorem below. This involves more advanced concepts in weak convergence.

Weak convergence

Consider a continuous-time random process $X(t), t \in [0,T]$ which we will denote as X. A sequence of random processes $X^{(n)}$ (indexed by $n = 1, 2, \ldots$) converges weakly to X if for each bounded continuous real-valued functional ϕ,

$$\lim_{n\to\infty} \mathbb{E}\{\phi(X^{(n)})\} = \mathbb{E}\{\phi(X)\}.$$

Equivalently, a sequence of probability measures $P^{(n)}$ converges weakly to P if $\int \phi \, dP^{(n)} \to \int \phi \, dP$ as $n \to \infty$. Note that the functional ϕ maps the entire trajectory of $X^{(n)}(t), 0 \le t \le T$ of the random process to a real number.

[1] The Dvoretzky–Kiefer–Wolfowitz inequality says that $\mathbb{P}(\sup_{x\in\mathbb{R}} |F_n(x) - F(x)| > \epsilon) \le 2e^{-2n\epsilon^2}$ for any $\epsilon > 0$ and finite n.

In the above definition, the trajectories of the random processes lie in the function space $C[0, T]$ (the class of continuous functions on the interval $[0, T]$), or more generally, $D[0, T]$ which is the space of piecewise constant functions that are continuous from the right with limit on the left – these are called "cadlag" functions (continue à droite, limite à gauche) in French.

To give an example of weak convergence, we first define Brownian motion.

Brownian motion

A continuous-time random process $w(t)$, $t \in [0, T]$ is called Brownian motion (Wiener process) if:

1. $\mathbb{P}(w(0) = 0) = 1$.
2. $w(t)$ has continuous trajectories with probability one.
3. $w(t) \sim \mathbf{N}(0, t)$, i.e. normally distributed marginals with zero mean and variance t.
4. $w(t)$ has independent increments. That is, for every partition $[t_1, \ldots, t_k]$ of $[0, T]$, the increments $w(t_1) - w(t_0), w(t_2) - w(t_1), \ldots, w(t_k) - w(t_{k-1})$ are statistically independent.

Actually, due to a remarkable theorem by Billingsley [58, Theorem 19.1], if $w(t)$ is considered to lie in $D[0, T]$, then the Gaussian property (3) follows automatically from the other three properties.

From an engineering point of view Brownian motion can be viewed as the integral of continuous-time white noise – however, this definition is mathematically imprecise. More specifically, from the above definition, $(w(t + h) - w(t))/h \sim \mathbf{N}(0, 1/h)$. The limit $h \to 0$ does not exist – indeed Brownian motion is continuous everywhere but differentiable nowhere.

Example: Donsker's functional central limit theorem.

The classical central limit theorem D.0.5 deals with a sequence of random variables converging in distribution to a Gaussian random variable. It states: Suppose $\{X_k\}$ is an independent and identically distributed sequence of random variables with zero mean and unit variance. Define $\mu_n = \frac{1}{n}\sum_{k=1}^{n} X_k$. Then $\lim_{n\to\infty} \sqrt{n}\mu_n \stackrel{\mathcal{L}}{=} \mathbf{N}(0, 1)$ where $\stackrel{\mathcal{L}}{=}$ denotes convergence in distribution and $\mathbf{N}(0, 1)$ is the unit variance zero mean Gaussian density.

The functional central limit theorem deals with a sequence of random processes indexed by ϵ converging weakly to a Gaussian random process as $\varepsilon \to 0$. Let $\lfloor x \rfloor$ denote the integer part of x. With the same assumptions as the above classical central limit theorem, define the continuous-time random process

$$\mu^{\varepsilon}(t) = \frac{\sqrt{\varepsilon}}{t} \sum_{k=1}^{\lfloor t/\varepsilon \rfloor} X_k = \varepsilon^{-1/2} \mu_{\lfloor t/\varepsilon \rfloor}, \quad t \geq 0. \tag{D.2}$$

Donsker's functional central limit theorem states that as $\varepsilon \to 0$, the process μ^{ε} defined in (D.2) converges weakly to standard Brownian motion.

By the definition of weak convergence, any continuous bounded functional ϕ of process μ^ε should therefore converge weakly to the corresponding functional of Brownian motion. Choosing for example $\phi(\mu^\varepsilon) = \max_{t\leq T} |\mu^\varepsilon(t)|$ (which is continuous on the function space $C[0,T]$), Donsker's theorem implies $\max_{t\leq T} |\mu^\varepsilon(t)|$ converges weakly to $\max_{t\leq T} |w(t)|$.

Tightness

To prove weak convergence of a stochastic approximation algorithm (see Theorem 17.1.1), a constructive approach is required that starts with finite dimensional distributions. If a random process $X^{(n)}$ converges weakly to X, then clearly all finite dimensional distributions of $X^{(n)}$ converge to X. But the reverse is not true in general. In order to go from convergence of finite dimensional distributions to weak convergence (and therefore establish a constructive proof for weak convergence), one needs to ensure that the process is well behaved between the time points considered in the finite dimensional distribution. Tightness of the sequence of measures $\{P^{(n)}\}$ is the property required for convergence of finite dimensional distributions to imply weak convergence.

We use the following tightness criterion from [198, p. 47]. A sequence of random process $\{X^{(n)}\}$ indexed by n is tight if for any $\delta > 0$ and $0 < s \leq \delta$,

$$\lim_{\delta\to 0} \limsup_{n\to\infty} \mathbb{E}\{|X^{(n)}(t+s) - X^{(n)}(t)|^2\} = 0. \tag{D.3}$$

LEMMA D.0.7 *Consider the stochastic approximation algorithm*

$$\theta_{k+1} = \theta_k + \epsilon\, H(\theta_k, x_k), \quad k = 0, 1, \ldots$$

and interpolated process $\theta^\epsilon(t) = \theta_k$ *for* $t \in [k\epsilon, k\epsilon + \epsilon)$, $k = 0, 1, \ldots$. *Then if* θ_k *is bounded for all time* k *and* $H(\cdot, x_k)$ *is continuous, then the sequence of processes* $\{\theta^\epsilon\}$ *indexed by* ϵ *is tight.*

Assumption **(SA1)** in Theorem 17.1.1 on page 383 is a tightness condition.

Proof

$$\theta^\epsilon(t+s) - \theta^\epsilon(t) = \epsilon \sum_{k=t/\epsilon}^{(t+s)/\epsilon - 1} H(\theta_k, x_k)$$

Since θ_k is bounded and $H(\cdot, x_k)$ is continuous, therefore $H(\theta_k, x_k)$ is bounded. So

$$\begin{aligned}\mathbb{E}\{|\theta^\epsilon(t+s) - \theta^\epsilon(t)|^2\} &= \mathbb{E}\Big\{\Big(\epsilon \sum_{k=t/\epsilon}^{(t+s)/\epsilon - 1} H(\theta_k, x_k)\Big)' \Big(\epsilon \sum_{k=t/\epsilon}^{(t+s)/\epsilon - 1} H(\theta_k, x_k)\Big)\Big\} \\ &\leq \text{ constant } \epsilon^2 \left(\frac{t+s}{\epsilon} - \frac{t}{\epsilon}\right)^2 = O(s^2).\end{aligned}$$

Taking $\limsup_{\epsilon\to 0}$ followed by $\lim_{\delta\to 0}$, (D.3) is verified. Thus $\{\theta^\epsilon\}$ is tight. □

References

[1] D. Aberdeen and J. Baxter, Scaling internal-state policy-gradient methods for POMDPs. In *International Conference on Machine Learning*, pp. 3–10, 2002.

[2] J. Abounadi, D. P. Bertsekas and V. Borkar, Learning algorithms for Markov decision processes with average cost. *SIAM Journal on Control and Optimization*, 40(3):681–98, 2001.

[3] D. Acemoglu and A. Ozdaglar, Opinion dynamics and learning in social networks. *Dynamic Games and Applications*, 1(1):3–49, 2011.

[4] S. Afriat, The construction of utility functions from expenditure data. *International Economic Review*, 8(1):67–77, 1967.

[5] S. Afriat, *Logic of Choice and Economic Theory*. (Oxford: Clarendon Press, 1987).

[6] S. Agrawal and N. Goyal, Analysis of Thompson sampling for the multi-armed bandit problem. In *Proceedings 25th Annual Conference Learning Theory*, volume 23, 2012.

[7] R. Ahuja and J. Orlin, Inverse optimization. *Operations Research*, 49(5):771–83, 2001.

[8] I. F. Akyildiz, W. Su, Y. Sankarasubramaniam and E. Cayirci, Wireless sensor networks: A survey. *Computer Networks*, 38(4):393–422, 2002.

[9] A. Albore, H. Palacios and H. Geffner, A translation-based approach to contingent planning. In *International Joint Conference on Artificial Intelligence*, pp. 1623–28, 2009.

[10] S. C. Albright, Structural results for partially observed Markov decision processes. *Operations Research*, 27(5):1041–53, Sept.–Oct. 1979.

[11] E. Altman, *Constrained Markov Decision Processes*. (London: Chapman and Hall, 1999).

[12] E. Altman, B. Gaujal and A. Hordijk, *Discrete-Event Control of Stochastic Networks: Multimodularity and Regularity*. (Springer-Verlag, 2004).

[13] T. Ben-Zvi and A. Grosfeld-Nir, Partially observed Markov decision processes with binomial observations. *Operations Research Letters*, 41(2):201–6, 2013.

[14] M. Dorigo and M. Gambardella, Ant-q: A reinforcement learning approach to the traveling salesman problem. In *Proceedings of the 12th International Conference on Machine Learning*, pp. 252–60, 2014.

[15] R. Amir, Supermodularity and complementarity in economics: An elementary survey. *Southern Economic Journal*, 71(3):636–60, 2005.

[16] M. S. Andersland and D. Teneketzis, Measurement scheduling for recursive team estimation. *Journal of Optimization Theory and Applications*, 89(3):615–36, June 1996.

[17] B. D. O. Anderson and J. B. Moore, *Optimal Filtering*. (Englewood Cliffs, NJ: Prentice Hall, 1979).

[18] B. D. O. Anderson and J. B. Moore, *Optimal Control: Linear Quadratic Methods*. (Englewood Cliffs, NJ: Prentice Hall, 1989).

[19] S. Andradottir, A global search method for discrete stochastic optimization. *SIAM Journal on Optimization*, 6(2):513–30, May 1996.

[20] S. Andradottir, Accelerating the convergence of random search methods for discrete stochastic optimization. *ACM Transactions on Modelling and Computer Simulation*, 9(4):349–80, Oct. 1999.

[21] A. Arapostathis, V. Borkar, E. Fernández-Gaucherand, M. K. Ghosh and S. I. Marcus, Discrete-time controlled Markov processes with average cost criterion: A survey. *SIAM Journal on Control and Optimization*, 31(2):282–344, 1993.

[22] P. Artzner, F. Delbaen, J. Eber and D. Heath, Coherent measures of risk. *Mathematical Finance*, 9(3):203–28, July 1999.

[23] P. Artzner, F. Delbaen, J. Eber, D. Heath and H. Ku, Coherent multiperiod risk adjusted values and bellmans principle. *Annals of Operations Research*, 152(1):5–22, 2007.

[24] K. J. Åström, Optimal control of Markov processes with incomplete state information. *Journal of Mathematical Analysis and Applications*, 10(1):174–205, 1965.

[25] R. Atar and O. Zeitouni, Lyapunov exponents for finite state nonlinear filtering. *SIAM Journal on Control and Optimization*, 35(1):36–55, 1997.

[26] S. Athey, Monotone comparative statics under uncertainty. *The Quarterly Journal of Economics*, 117(1):187–223, 2002.

[27] P. Auer, N. Cesa-Bianchi and P. Fischer, Finite-time analysis of the multiarmed bandit problem. *Machine Learning*, 47(2-3):235–56, 2002.

[28] A.Young and S. Russell, Algorithms for inverse reinforcement learning. In *Proceedings of the 17th International Conference on Machine Learning*, pp. 663–70, 2000.

[29] A. Banerjee, A simple model of herd behavior. *Quaterly Journal of Economics*, 107(3):797–817, August 1992.

[30] A. Banerjee, X. Guo and H. Wang, On the optimality of conditional expectation as a Bregman predictor. *IEEE Transactions on Information Theory*, 51(7):2664–9, 2005.

[31] T. Banerjee and V. Veeravalli, Data-efficient quickest change detection with on-off observation control. *Sequential Analysis*, 31:40–77, 2012.

[32] Y. Bar-Shalom, X. R. Li and T. Kirubarajan, *Estimation with Applications to Tracking and Navigation*. John Wiley, New York, 2008.

[33] J. S. Baras and A. Bensoussan, Optimal sensor scheduling in nonlinear filtering of diffusion processes. *SIAM Journal Control and Optimization*, 27(4):786–813, July 1989.

[34] G. Barles and P. E. Souganidis, Convergence of approximation schemes for fully nonlinear second order equations. In *Asymptotic Analysis*, number 4, pp. 2347–9, 1991.

[35] P. Bartlett and J. Baxter, Estimation and approximation bounds for gradient-based reinforcement learning. *Journal of Computer and System Sciences*, 64(1):133–50, 2002.

[36] M. Basseville and I.V. Nikiforov, *Detection of Abrupt Changes — Theory and Applications*. Information and System Sciences Series. (Englewood Cliffs, NJ: Prentice Hall, 1993).

[37] N. Bäuerle and U. Rieder, More risk-sensitive Markov decision processes. *Mathematics of Operations Research*, 39(1):105–20, 2013.

[38] L. E. Baum and T. Petrie, Statistical inference for probabilistic functions of finite state Markov chains. *Annals of Mathematical Statistics*, 37:1554–63, 1966.

[39] L. E. Baum, T. Petrie, G. Soules and N. Weiss, A maximisation technique occurring in the statistical analysis of probabilistic functions of Markov chains. *Annals of Mathematical Statistics*, 41(1):164–71, 1970.

[40] R. Bellman, *Dynamic Programming*. 1st edition (Princeton, NJ: Princeton University Press, 1957).

[41] M. Benaim and M. Faure, Consistency of vanishingly smooth fictitious play. *Mathematics of Operations Research*, 38(3):437–50, Aug. 2013.

[42] M. Benaim, J. Hofbauer and S. Sorin, Stochastic approximations and differential inclusions. *SIAM Journal on Control and Optimization*, 44(1):328–48, 2005.

[43] M. Benaim, J. Hofbauer and S. Sorin, Stochastic approximations and differential inclusions, Part II: Applications. *Mathematics of Operations Research*, 31(3):673–95, 2006.

[44] M. Benaim and J. Weibull, Deterministic approximation of stochastic evolution in games. *Econometrica*, 71(3):873–903, 2003.

[45] V. E. Beneš, Exact finite-dimensional filters for certain diffusions with nonlinear drift. *Stochastics*, 5:65–92, 1981.

[46] A. Bensoussan, *Stochastic Control of Partially Observable Systems*. (Cambridge University Press, 1992).

[47] A. Bensoussan and J. Lions, *Impulsive Control and Quasi-Variational Inequalities*. (Paris: Gauthier-Villars, 1984).

[48] A. Benveniste, M. Metivier and P. Priouret, *Adaptive Algorithms and Stochastic Approximations*, volume 22 of *Applications of Mathematics*. (Springer-Verlag, 1990).

[49] D. P. Bertsekas, *Dynamic Programming and Optimal Control*, volume 1 and 2. (Belmont, MA: Athena Scientific, 2000).

[50] D. P. Bertsekas, *Nonlinear Programming*. (Belmont, MA: Athena Scientific, 2000).

[51] D. P. Bertsekas, Dynamic programming and suboptimal control: A survey from ADP to MPC. *European Journal of Control*, 11(4):310–34, 2005.

[52] D. P. Bertsekas and S. E. Shreve, *Stochastic Optimal Control: The Discrete-Time Case*. (New York, NY: Academic Press, 1978).

[53] D. P. Bertsekas and J. N. Tsitsiklis, *Neuro-Dynamic Programming*. (Belmont, MA: Athena Scientific, 1996).

[54] D. P. Bertsekas and H. Yu, Q-learning and enhanced policy iteration in discounted dynamic programming. *Mathematics of Operations Research*, 37(1):66–94, 2012.

[55] L. Bianchi, M. Dorigo, L. Gambardella and W. Gutjahr, A survey on metaheuristics for stochastic combinatorial optimization. *Natural Computing: An International Journal*, 8(2):239–87, 2009.

[56] S. Bikchandani, D. Hirshleifer and I. Welch, A theory of fads, fashion, custom, and cultural change as information cascades. *Journal of Political Economy*, 100(5):992–1026, October 1992.

[57] P. Billingsley, *Statistical inference for Markov processes*, volume 2. (University of Chicago Press, 1961).

[58] P. Billingsley, *Convergence of Probability Measures*. (New York, NY: John Wiley, 1968).

[59] P. Billingsley, *Probability and Measure*. (New York, NY: John Wiley, 1986).

[60] S. Blackman and R. Popoli, *Design and Analysis of Modern Tracking Systems*. (Artech House, 1999).

[61] R. Bond, C. Fariss, J. Jones, A. Kramer, C. Marlow, J. Settle and J. Fowler, A 61-million-person experiment in social influence and political mobilization. *Nature*, 489:295–8, September 2012.

[62] J. G. Booth and J. P. Hobert, Maximizing generalized linear mixed model likelihoods with an automated monte carlo em algorithm. *Journal Royal Statistical Society*, 61:265–85, 1999.

[63] V. S. Borkar, *Stochastic Approximation. A Dynamical Systems Viewpoint*. (Cambridge University Press, 2008).

[64] S. Bose, G. Orosel, M. Ottaviani and L. Vesterlund, Dynamic monopoly pricing and herding. *The RAND Journal of Economics*, 37(4):910–28, 2006.

[65] S. Boucheron, G. Lugosi and P. Massart, *Concentration Inequalities: A Nonasymptotic Theory of Independence*. (Oxford University Press, 2013).

[66] S. Boyd, P. Diaconis and L. Xiao, Fastest mixing Markov chain on a graph. *SIAM Review*, 46(4):667–89, 2004.

[67] S. Boyd, N. Parikh, E. Chu, B. Peleato and J. Eckstein, Distributed optimization and statistical learning via the alternating direction method of multipliers. *Foundations and Trends in Machine Learning*, 3(1):1–122, 2011.

[68] S. Boyd and L. Vandenberghe, *Convex Optimization*. (Cambridge University Press, 2004).

[69] P. Bremaud. *Markov Chains: Gibbs Fields, Monte Carlo Simulation, and Queues*. (Springer-Verlag, 1999).

[70] R. W. Brockett and J. M. C. Clarke. The geometry of the conditional density equation. In O. L. R. Jacobs et al., editor, *Analysis and Optimization of Stochastic Systems*, pp. 299–309 (New York, 1980).

[71] S. Bubeck and N. Cesa-Bianchi, Regret analysis of stochastic and nonstochastic multi-armed bandit problems. *arXiv preprint arXiv:1204.5721*, 2012.

[72] S. Bundfuss and M. Dür, Algorithmic copositivity detection by simplicial partition. *Linear Algebra and Its Applications*, 428(7):1511–23, 2008.

[73] S. Bundfuss and M. Dür, An adaptive linear approximation algorithm for copositive programs. *SIAM Journal on Optimization*, 20(1):30–53, 2009.

[74] P. E. Caines. *Linear Stochastic Systems*. (John Wiley, 1988).

[75] E. J. Candès and T. Tao, The power of convex relaxation: Near-optimal matrix completion. *IEEE Transactions on Information Theory*, 56(5):2053–80, May 2009.

[76] O. Cappe, E. Moulines and T. Ryden, *Inference in Hidden Markov Models*. (Springer-Verlag, 2005).

[77] A. R. Cassandra, Tony's POMDP page. www.cs.brown.edu/research/ai/pomdp/

[78] A. R. Cassandra, *Exact and Approximate Algorithms for Partially Observed Markov Decision Process*. PhD thesis, Dept. Computer Science, Brown University, 1998.

[79] A. R. Cassandra, A survey of POMDP applications. In *Working Notes of AAAI 1998 Fall Symposium on Planning with Partially Observable Markov Decision Processes*, pp. 17–24, 1998.

[80] A. R. Cassandra, L. Kaelbling and M. L. Littman, Acting optimally in partially observable stochastic domains. In *AAAI*, volume 94, pp. 1023–8, 1994.

[81] A. R. Cassandra, M. L. Littman and N. L. Zhang, Incremental pruning: A simple fast exact method for partially observed Markov decision processes. In *Proceedings of the 13th Annual Conference on Uncertainty in Artificial Intelligence (UAI-97)*. (Providence, RI 1997).

[82] C. G. Cassandras and S. Lafortune, *Introduction to Discrete Event Systems*. (Springer-Verlag, 2008).

[83] O. Cavus and A. Ruszczynski, Risk-averse control of undiscounted transient Markov models. *SIAM Journal on Control and Optimization*, 52(6):3935–66, 2014.

[84] C. Chamley, *Rational Herds: Economic Models of Social Learning*. (Cambridge University Press, 2004).

[85] C. Chamley, A. Scaglione and L. Li, Models for the diffusion of beliefs in social networks: An overview. *IEEE Signal Processing Magazine*, 30(3):16–29, 2013.

[86] W. Chiou, A note on estimation algebras on nonlinear filtering theory. *Systems and Control Letters*, 28:55–63, 1996.

[87] J. M. C. Clark, The design of robust approximations to the stochastic differential equations of nonlinear filtering. In J. K. Skwirzynski, editor, *Communication Systems and Random Processes Theory, Darlington 1977*. (Alphen aan den Rijn: Sijthoff and Noordhoff, 1978).
[88] T. F. Coleman and Y. Li, An interior trust region approach for nonlinear minimization subject to bounds. *SIAM Journal on Optimization*, 6(2):418–45, 1996.
[89] T. M. Cover and M. E. Hellman, The two-armed-bandit problem with time-invariant finite memory. *IEEE Transactions on Information Theory*, 16(2):185–95, 1970.
[90] T. M. Cover and J. A. Thomas, *Elements of Information Theory*. (Wiley-Interscience, 2006).
[91] A. Dasgupta, R. Kumar and D. Sivakumar, Social sampling. In *Proceedings of the 18th ACM SIGKDD International Conference on Knowledge Discovery and Data mining*, 235–43, (Beijing, 2012). ACM.
[92] M. H. A. Davis, On a multiplicative functional transformation arising in nonlinear filtering theory. *Z. Wahrscheinlichkeitstheorie verw. Gebiete*, 54:125–39, 1980.
[93] S. Dayanik and C. Goulding, Detection and identification of an unobservable change in the distribution of a Markov-modulated random sequence. *IEEE Transactions on Information Theory*, 55(7):3323–45, 2009.
[94] A. P. Dempster, N. M. Laird and D. B. Rubin, Maximum likelihood from incomplete data via the EM algorithm. *Journal of the Royal Statistical Society, B*, 39:1–38, 1977.
[95] E. Denardo and U. Rothblum, Optimal stopping, exponential utility, and linear programming. *Mathematical Programming*, 16(1):228–44, 1979.
[96] C. Derman, G. J. Lieberman and S. M. Ross, Optimal system allocations with penalty cost. *Management Science*, 23(4):399–403, December 1976.
[97] R. Douc, E. Moulines and T. Ryden, Asymptotic properties of the maximum likelihood estimator in autoregressive models with Markov regime. *The Annals of Statistics*, 32(5):2254–304, 2004.
[98] A. Doucet, N. De Freitas and N. Gordon, editors, *Sequential Monte Carlo Methods in Practice*. (Springer-Verlag, 2001).
[99] A. Doucet, S. Godsill and C. Andrieu, On sequential Monte-Carlo sampling methods for Bayesian filtering. *Statistics and Computing*, 10:197–208, 2000.
[100] A. Doucet, N. Gordon and V. Krishnamurthy, Particle filters for state estimation of jump Markov linear systems. *IEEE Transactions on Signal Processing*, 49:613–24, 2001.
[101] A. Doucet and A. M. Johansen, A tutorial on particle filtering and smoothing: Fiteen years later. In D. Crisan and B. Rozovsky, editors, *Oxford Handbook on Nonlinear Filtering*. (Oxford University Press, 2011).
[102] E. Dynkin, Controlled random sequences. *Theory of Probability & Its Applications*, 10(1):1–14, 1965.
[103] J. N. Eagle, The optimal search for a moving target when the search path is constrained. *Operations Research*, 32:1107–15, 1984.
[104] R. J. Elliott, L. Aggoun and J. B. Moore, *Hidden Markov Models – Estimation and Control*. (New York, NY: Springer-Verlag, 1995).
[105] R. J. Elliott and V. Krishnamurthy, Exact finite-dimensional filters for maximum likelihood parameter estimation of continuous-time linear Gaussian systems. *SIAM Journal on Control and Optimization*, 35(6):1908–23, November 1997.
[106] R. J. Elliott and V. Krishnamurthy, New finite dimensional filters for estimation of discrete-time linear Gaussian models. *IEEE Transactions on Automatic Control*, 44(5):938–51, May 1999.

[107] Y. Ephraim and N. Merhav, Hidden Markov processes. *IEEE Transactions on Information Theory*, 48:1518–69, June 2002.

[108] S. N. Ethier and T. G. Kurtz, *Markov Processes—Characterization and Convergence*. (Wiley, 1986).

[109] J. Evans and V. Krishnamurthy, Hidden Markov model state estimation over a packet switched network. *IEEE Transactions on Signal Processing*, 42(8):2157–66, August 1999.

[110] R. Evans, V. Krishnamurthy and G. Nair, Networked sensor management and data rate control for tracking maneuvering targets. *IEEE Transactions on Signal Processing*, 53(6):1979–91, June 2005.

[111] M. Fanaswala and V. Krishnamurthy, Syntactic models for trajectory constrained track-before-detect. *IEEE Transactions on Signal Processing*, 62(23):6130–42, 2014.

[112] M. Fanaswalla and V. Krishnamurthy, Detection of anomalous trajectory patterns in target tracking via stochastic context-free grammars and reciprocal process models. *IEEE Journal on Selected Topics Signal Processing*, 7(1):76–90, Feb. 2013.

[113] M. Fazel, H. Hindi and S. P. Boyd, Log-det heuristic for matrix rank minimization with applications to Hankel and Euclidean distance matrices. In *Proceedings of the 2003 American Control Conference*, 2003.

[114] E. Feinberg and A. Shwartz, editors, *Handbook of Markov Decision Processes*. (Springer-Verlag, 2002).

[115] J. A. Fessler and A. O. Hero. Space–Alternating Generalized Expectation–Maximization algorithm. *IEEE Transactions on Signal Processing*, 42(10):2664–77, 1994.

[116] J. Filar, L. Kallenberg and H. Lee, Variance-penalized Markov decision processes. *Mathematics of Operations Research*, 14(1):147–61, 1989.

[117] W. H. Fleming and H. M. Soner, *Controlled Markov Processes and Viscosity Solutions*, volume 25. (Springer Science & Business Media, 2006).

[118] A. Fostel, H. Scarf and M. Todd. Two new proofs of Afriat's theorem. *Economic Theory*, 24(1):211–19, 2004.

[119] D. Fudenberg and D. K. Levine, *The Theory of Learning in Games*. (MIT Press, 1998).

[120] D. Fudenberg and D. K. Levine, Consistency and cautious fictitious play. *Journal of Economic Dynamics and Control*, 19(5-7):1065–89, 1995.

[121] F. R. Gantmacher, *Matrix Theory*, volume 2. (New York, NY: Chelsea Publishing Company, 1960).

[122] A. Garivier and E. Moulines. On upper-confidence bound policies for switching bandit problems. In *Algorithmic Learning Theory*, pages 174–188. Springer, 2011.

[123] E. Gassiat and S. Boucherone, Optimal error exponents in hidden Markov models order estimation. *IEEE Transactions on Information Theory*, 49(4):964–80, 2003.

[124] D. Ghosh, Maximum likelihood estimation of the dynamic shock-error model. *Journal of Econometrics*, 41(1):121–43, 1989.

[125] J. C. Gittins, *Multi–Armed Bandit Allocation Indices*. (Wiley, 1989).

[126] S. Goel and M. J. Salganik, Respondent-driven sampling as Markov chain Monte Carlo. *Statistics in Medicine*, 28:2209–29, 2009.

[127] G. Golubev and R. Khasminskii, Asymptotic optimal filtering for a hidden Markov model. *Math. Methods Statist.*, 7(2):192–208, 1998.

[128] N. J. Gordon, D. J. Salmond and A. F. M. Smith, Novel approach to nonlinear/non-Gaussian Bayesian state estimation. *IEE Proceedings-F*, 140(2):107–13, 1993.

[129] M. Granovetter, Threshold models of collective behavior. *American Journal of Sociology*, 83(6):1420–43, May 1978.

[130] A. Grosfeld-Nir, Control limits for two-state partially observable Markov decision processes. *European Journal of Operational Research*, 182(1):300–4, 2007.

[131] D. Guo, S. Shamai and S. Verdú, Mutual information and minimum mean-square error in Gaussian channels. *IEEE Transactions on Information Theory*, 51(4):1261–82, 2005.

[132] M. Hamdi, G. Solman, A. Kingstone and V. Krishnamurthy, Social learning in a human society: An experimental study. *arXiv preprint arXiv:1408.5378*, 2014.

[133] J. D. Hamilton and R. Susmel, Autoregressive conditional heteroskedasticity and changes in regime. *Journal of Econometrics*, 64(2):307–33, 1994.

[134] J. E. Handschin and D. Q. Mayne, Monte Carlo techniques to estimate the conditional expectation in multi-stage non-linear filtering. *International Journal Control*, 9(5):547–59, 1969.

[135] E. J. Hannan and M. Deistler, *The Statistical Theory of Linear Systems*. Wiley series in probability and mathematical statistics. Probability and mathematical statistics. (New York, NY: John Wiley, 1988).

[136] T. Hastie, R. Tibshirani and J. Friedman, *The Elements of Statistical Learning*. (Springer-Verlag, 2009).

[137] M. Hauskrecht, Value-function approximations for partially observable Markov decision processes. *Journal of Artificial Intelligence Research*, 13(1):33–94, 2000.

[138] S. Haykin, Cognitive radio: Brain-empowered wireless communications. *IEEE Journal on Selected Areas Communications*, 23(2):201–20, Feb. 2005.

[139] S. Haykin, *Adaptive Filter Theory* 5th edition. Information and System Sciences Series. (Prentice Hall, 2013).

[140] D. D. Heckathorn, Respondent-driven sampling: A new approach to the study of hidden populations. *Social Problems*, 44:174–99, 1997.

[141] D. D. Heckathorn, Respondent-driven sampling ii: Deriving valid population estimates from chain-referral samples of hidden populations. *Social Problems*, 49:11–34, 2002.

[142] M. E. Hellman and T. M. Cover, Learning with finite memory. *The Annals of Mathematical Statistics*, 41(3):765–82, 1970.

[143] O. Hernández-Lerma and J. Bernard Laserre, *Discrete-Time Markov Control Processes: Basic Optimality Criteria*. (New York, NY: Springer-Verlag, 1996).

[144] D. P. Heyman and M. J. Sobel, *Stochastic Models in Operations Research*, volume 2. (McGraw-Hill, 1984).

[145] N. Higham and L. Lin, On pth roots of stochastic matrices. *Linear Algebra and Its Applications*, 435(3):448–63, 2011.

[146] Y.-C. Ho and X.-R. Cao. *Discrete Event Dynamic Systems and Perturbation Analysis*. (Boston, MA: Kluwer Academic, 1991).

[147] J. Hofbauer and W. Sandholm, On the global convergence of stochastic fictitious play. *Econometrica*, 70(6):2265–94, November 2002.

[148] R. A. Horn and C. R. Johnson, *Matrix Analysis*. (Cambridge University Press, 2012).

[149] R. A. Howard, *Dynamic Probabilistic Systems*, volume 1: Markov Models. (New York: John Wiley, 1971).

[150] R. A. Howard, *Dynamic Probabilistic Systems*, volume 2: Semi-Markov and Decision Processes. (New York: John Wiley, 1971).

[151] D. Hsu, S. Kakade and T. Zhang, A spectral algorithm for learning hidden Markov models. *Journal of Computer and System Sciences*, 78(5):1460–80, 2012.

[152] S. Hsu,. Chuang and A. Arapostathis, On the existence of stationary optimal policies for partially observed mdps under the long-run average cost criterion. *Systems & Control Letters*, 55(2):165–73, 2006.

[153] M. Huang and S. Dey, Stability of Kalman filtering with Markovian packet losses. *Automatica*, 43(4):598–607, 2007.

[154] Ienkaran I. Arasaratnam and S. Haykin, Cubature Kalman filters. *IEEE Transactions on Automatic Control*, 54(6):1254–69, 2009.

[155] K. Iida, *Studies on the Optimal Search Plan*, volume 70 of *Lecture Notes in Statistics*. (Springer-Verlag, 1990).

[156] M. O. Jackson. *Social and Economic Networks*. (Princeton, NJ: Princeton University Press, 2010).

[157] M. R. James, V. Krishnamurthy and F. LeGland, Time discretization of continuous-time filters and smoothers for HMM parameter estimation. *IEEE Transactions on Information Theory*, 42(2):593–605, March 1996.

[158] M. R. James, J. S. Baras and R. J. Elliott, Risk-sensitive control and dynamic games for partially observed discrete-time nonlinear systems. *IEEE Transactions on Automatic Control*, 39(4):780–92, April 1994.

[159] B. Jamison, Reciprocal processes. *Probability Theory and Related Fields*, 30(1):65–86, 1974.

[160] A. H. Jazwinski, *Stochastic Processes and Filtering Theory*. (NJ: Academic Press, 1970).

[161] A. Jobert and L. C. G. Rogers, Valuations and dynamic convex risk measures. *Mathematical Finance*, 18(1):1–22, 2008.

[162] L. Johnston and V. Krishnamurthy, Opportunistic file transfer over a fading channel – a POMDP search theory formulation with optimal threshold policies. *IEEE Transactions on Wireless Commun.*, 5(2):394–405, Feb. 2006.

[163] T. Kailath, *Linear Systems*. (NJ: Prentice Hall, 1980).

[164] R. E. Kalman, A new approach to linear filtering and prediction problems. *Trans. ASME, Series D (J. Basic Engineering)*, 82:35–45, March 1960.

[165] R. E. Kalman, When is a linear control system optimal? *J. Basic Engineering*, 51–60, April 1964.

[166] R. E. Kalman and R. S. Bucy, New results in linear filtering and prediction theory. *Trans. ASME, Series D (J. Basic Engineering)*, 83:95–108, March 1961.

[167] I. Karatzas and S. Shreve, *Brownian Motion and Stochastic Calculus*, 2nd edition. (Springer, 1991).

[168] S. Karlin, *Total Positivity*, volume 1. (Stanford Univrsity, 1968).

[169] S. Karlin and Y. Rinott, Classes of orderings of measures and related correlation inequalities. I. Multivariate totally positive distributions. *Journal of Multivariate Analysis*, 10(4):467–98, December 1980.

[170] S. Karlin and H. M. Taylor, *A Second Course in Stochastic Processes*. (Academic Press, 1981).

[171] K. V. Katsikopoulos and S. E. Engelbrecht, Markov decision processes with delays and asynchronous cost collection. *IEEE Transactions on Automatic Control*, 48(4):568–74, 2003.

[172] J. Keilson and A. Kester, Monotone matrices and monotone Markov processes. *Stochastic Processes and Their Applications*, 5(3):231–41, 1977.

[173] H. K. Khalil, *Nonlinear Systems* 3rd edition. (Prentice Hall, 2002).

[174] M. Kijima, *Markov Processes for Stochastic Modelling*. (Chapman and Hall, 1997).

[175] A. N. Kolmogorov, Interpolation and extrapolation of stationary random sequences. *Bull. Acad. Sci. U.S.S.R, Ser. Math.*, 5:3–14, 1941.

[176] A. N. Kolmogorov, Stationary sequences in Hilbert space. *Bull. Math. Univ. Moscow*, 2(6), 1941.

[177] L. Kontorovich and K. Ramanan, Concentration inequalities for dependent random variables via the martingale method. *The Annals of Probability*, 36(6):2126–58, 2008.

[178] V. Krishnamurthy, Algorithms for optimal scheduling and management of hidden Markov model sensors. *IEEE Transactions on Signal Processing*, 50(6):1382–97, June 2002.

[179] V. Krishnamurthy, Bayesian sequential detection with phase-distributed change time and nonlinear penalty – A lattice programming POMDP approach. *IEEE Transactions on Information Theory*, 57(3):7096–124, October 2011.

[180] V. Krishnamurthy, How to schedule measurements of a noisy Markov chain in decision making? *IEEE Transactions on Information Theory*, 59(9):4440–61, July 2013.

[181] V. Krishnamurthy and F. Vazquez Abad, Gradient based policy optimization of constrained unichain Markov decision processes. In S. Cohen, D. Madan, and T. Siu, editors, *Stochastic Processes, Finance and Control: A Festschrift in Honor of Robert J. Elliott*. (World Scientific, 2012). http://arxiv.org/abs/1110.4946.

[182] V. Krishnamurthy, R. Bitmead, M. Gevers and E. Miehling, Sequential detection with mutual information stopping cost: Application in GMTI radar. *IEEE Transactions on Signal Processing*, 60(2):700–14, 2012.

[183] V. Krishnamurthy and D. Djonin, Structured threshold policies for dynamic sensor scheduling: A partially observed Markov decision process approach. *IEEE Transactions on Signal Processing*, 55(10):4938–57, Oct. 2007.

[184] V. Krishnamurthy and D.V. Djonin, Optimal threshold policies for multivariate POMDPs in radar resource management. *IEEE Transactions on Signal Processing*, 57(10), 2009.

[185] V. Krishnamurthy, O. Namvar Gharehshiran and M. Hamdi, Interactive sensing and decision making in social networks. *Foundations and Trends in Signal Processing*, 7(1-2):1–196, 2014.

[186] V. Krishnamurthy and W. Hoiles, Online reputation and polling systems: Data incest, social learning and revealed preferences. *IEEE Transactions Computational Social Systems*, 1(3):164–79, January 2015.

[187] V. Krishnamurthy and U. Pareek, Myopic bounds for optimal policy of POMDPs: An extension of Lovejoy's structural results. *Operations Research*, 62(2):428–34, 2015.

[188] V. Krishnamurthy and H. V. Poor, Social learning and Bayesian games in multiagent signal processing: How do local and global decision makers interact? *IEEE Signal Processing Magazine*, 30(3):43–57, 2013.

[189] V. Krishnamurthy and C. Rojas, Reduced complexity HMM filtering with stochastic dominance bounds: A convex optimization approach. *IEEE Transactions on Signal Processing*, 62(23):6309–22, 2014.

[190] V. Krishnamurthy, C. Rojas and B. Wahlberg, Computing monotone policies for Markov decision processes by exploiting sparsity. In *3rd Australian Control Conference (AUCC)*, 1–6. IEEE, 2013.

[191] V. Krishnamurthy and B. Wahlberg, POMDP multiarmed bandits – structural results. *Mathematics of Operations Research*, 34(2):287–302, May 2009.

[192] V. Krishnamurthy and G. Yin, Recursive algorithms for estimation of hidden Markov models and autoregressive models with Markov regime. *IEEE Transactions on Information Theory*, 48(2):458–76, February 2002.

[193] P. R. Kumar and P. Varaiya, *Stochastic Systems – Estimation, Identification and Adaptive Control*. (Prentice Hall, 1986).

[194] H. Kurniawati, D. Hsu and W. S. Lee, Sarsop: Efficient point-based POMDP planning by approximating optimally reachable belief spaces. In 2008 *Robotics: Science and Systems Conference*, Zurich, Switzerland, 2008.

[195] T. G. Kurtz, *Approximation of Population Processes*, volume 36. SIAM, 1981.

[196] H. J. Kushner, Dynamical equations for optimal nonlinear filtering. *Journal of Differential Equations*, 3:179–90, 1967.

[197] H. J. Kushner, A robust discrete state approximation to the optimal nonlinear filter for a diffusion. *Stochastics*, 3(2):75–83, 1979.

[198] H. J. Kushner, *Approximation and Weak Convergence Methods for Random Processes, with Applications to Stochastic Systems Theory*. (Cambridge, MA: MIT Press, 1984).

[199] H. J. Kushner and D. S. Clark, *Stochastic Approximation Methods for Constrained and Unconstrained Systems*. (Springer-Verlag, 1978).

[200] H. J. Kushner and G. Yin, *Stochastic Approximation Algorithms and Recursive Algorithms and Applications*, 2nd edition. (Springer-Verlag, 2003).

[201] T. Lai and H. Robbins, Asymptotically efficient adaptive allocation rules. *Advances in Applied Mathematics*, 6(1):4–22, 1985.

[202] A. Lansky, A. Abdul-Quader, M. Cribbin, T. Hall, T. J. Finlayson, R. Garffin, L. S. Lin and P. Sullivan, Developing an HIV behavioral surveillance system for injecting drug users: the National HIV Behavioral Surveillance System. *Public Health Reports*, 122(S1):48–55, 2007.

[203] S. Lee, Understanding respondent driven sampling from a total survey error perspective. *Survey Practice*, 2(6), 2009.

[204] F. LeGland and L. Mevel, Exponential forgetting and geometric ergodicity in hidden Markov models. *Mathematics of Controls, Signals and Systems*, 13(1):63–93, 2000.

[205] B. G. Leroux, Maximum-likelihood estimation for hidden Markov models. *Stochastic Processes and Its Applications*, 40:127–43, 1992.

[206] R. Levine and G. Casella, Implementations of the Monte Carlo EM algorithm. *Journal of Computational and Graphical Statistics*, 10(3):422–39, September 2001.

[207] T. Lindvall. *Lectures on the Coupling Method*. (Courier Dover Publications, 2002).

[208] M. Littman, A. R. Cassandra and L. Kaelbling, Learning policies for partially observable environments: Scaling up. In *ICML*, volume 95, pages 362–70. Citeseer, 1995.

[209] M. L. Littman, *Algorithms for Sequential Decision Making*. PhD thesis, Brown University, 1996.

[210] M. L. Littman, A tutorial on partially observable Markov decision processes. *Journal of Mathematical Psychology*, 53(3):119–25, 2009.

[211] C. Liu and D.B. Rubin, The ECME algorithm: A simple extension of EM and ECM with faster monotone convergence. *Biometrica*, 81(4):633–48, 1994.

[212] J. S. Liu, *Monte Carlo Strategies in Scientific Computing*. (Springer-Verlag, 2001).

[213] J. S. Liu and R. Chen, Sequential monte carlo methods for dynamic systems. *Journal American Statistical Association*, 93:1032–44, 1998.

[214] K. Liu and Q. Zhao, Indexability of restless bandit problems and optimality of Whittle index for dynamic multichannel access. *IEEE Transactions on Information Theory*, 56(11):5547–67, 2010.

[215] Z. Liu and L. Vandenberghe, Interior-point method for nuclear norm approximation with application to system identification. *SIAM Journal on Matrix Analysis and Applications*, 31(3):1235–56, 2009.

[216] L. Ljung, Analysis of recursive stochastic algorithms. *IEEE Transactions on Auto. Control*, AC-22(4):551–75, 1977.

[217] L. Ljung, *System Identification*, 2nd edition. (Prentice Hall, 1999).

[218] L. Ljung and T. Söderström, *Theory and Practice of Recursive Identification*. (Cambridge, MA: MIT Press, 1983).

[219] I. Lobel, D. Acemoglu, M. Dahleh and A. E. Ozdaglar, Preliminary results on social learning with partial observations. In *Proceedings of the 2nd International Conference on Performance Evaluation Methodolgies and Tools* (Nantes, France, 2007). ACM.

[220] A. Logothetis and A. Isaksson. On sensor scheduling via information theoretic criteria. In *Proc. American Control Conf.*, pages 2402–06, (San Diego, 1999).

[221] D. López-Pintado, Diffusion in complex social networks. *Games and Economic Behavior*, 62(2):573–90, 2008.

[222] T. A. Louis, Finding the observed information matrix when using the EM algorithm. *Journal of the Royal Statistical Society*, 44(B):226–33, 1982.

[223] W. S. Lovejoy, On the convexity of policy regions in partially observed systems. *Operations Research*, 35(4):619–21, July–August 1987.

[224] W. S. Lovejoy, Ordered solutions for dynamic programs. *Mathematics of Operations Research*, 12(2):269–76, 1987.

[225] W. S. Lovejoy, Some monotonicity results for partially observed Markov decision processes. *Operations Research*, 35(5):736–43, September–October 1987.

[226] W. S. Lovejoy, Computationally feasible bounds for partially observed Markov decision processes. *Operations Research*, 39(1):162–75, January–February 1991.

[227] W. S. Lovejoy, A survey of algorithmic methods for partially observed Markov decision processes. *Annals of Operations Research*, 28:47–66, 1991.

[228] M. Luca, *Reviews, Reputation, and Revenue: The Case of Yelp.com, Technical Report 12-016*. Harvard Business School, September 2011.

[229] D. G. Luenberger, *Optimization by Vector Space Methods*. (New York, NY: John Wiley, 1969).

[230] I. MacPhee and B. Jordan, Optimal search for a moving target. *Probability in the Engineering and Information Sciences*, 9:159–82, 1995.

[231] C. D. Manning and H. Schütze, *Foundations of Statistical Natural Language Processing*. (Cambridge, MA: The MIT Press, 1999).

[232] S. I. Marcus, Algebraic and geometric methods in nonlinear filtering. *SIAM Journal on Control and Optimization*, 22(6):817–44, November 1984.

[233] S. I. Marcus and A. S. Willsky, Algebraic structure and finite dimensional nonlinear estimation. *SIAM J. Math. Anal.*, 9(2):312–27, April 1978.

[234] H. Markowitz, Portfolio selection. *The Journal of Finance*, 7(1):77–91, 1952.

[235] D. Q. Mayne, J. B. Rawlings, C. V. Rao and P. Scokaert, Constrained model predictive control: Stability and optimality. *Automatica*, 36(6):789–814, 2000.

[236] G. J. McLachlan and T. Krishnan, *The EM Algorithm and Extensions*. Wiley series in probability and statistics. Applied probability and statistics. (New York, NY: John Wiley, 1996).

[237] L. Meier, J. Perschon and R. M. Dressler, Optimal control of measurement subsystems. *IEEE Transactions on Automatic Control*, 12(5):528–36, October 1967.

[238] J. M. Mendel, *Maximum-Likelihood Deconvolution: A Journey into Model-Based Signal Processing*. (Springer-Verlag, 1990).

[239] X. L. Meng, On the rate of convergence of the ecm algorithm. *The Annals of Statistics*, 22(1):326–39, 1994.

[240] S. P. Meyn and R. L. Tweedie, *Markov Chains and Stochastic Stability*. (Cambridge University Press, 2009).

[241] P. Milgrom, Good news and bad news: Representation theorems and applications. *Bell Journal of Economics*, 12(2):380–91, 1981.

[242] P. Milgrom and C. Shannon, Monotone comparative statics. *Econometrica*, 62(1):157–180, 1994.

[243] R. R. Mohler and C. S. Hwang, Nonlinear data observability and information. *Journal of Franklin Institute*, 325(4):443–64, 1988.

[244] G. E. Monahan, A survey of partially observable Markov decision processes: Theory, models and algorithms. *Management Science*, 28(1), January 1982.

[245] P. Del Moral, *Feynman-Kac Formulae – Genealogical and Interacting Particle Systems with Applications*. (Springer-Verlag, 2004).

[246] W. Moran, S. Suvorova and S. Howard, Application of sensor scheduling concepts to radar. In A. Hero, D. Castanon, D. Cochran and K. Kastella, editors, *Foundations and Applications for Sensor Management*, pages 221–56. (Springer-Verlag, 2006).

[247] G. B. Moustakides, Optimal stopping times for detecting changes in distributions. *Annals of Statistics*, 14:1379–87, 1986.

[248] A. Muller, How does the value function of a Markov decision process depend on the transition probabilities? *Mathematics of Operations Research*, 22:872–85, 1997.

[249] A. Muller and D. Stoyan, *Comparison Methods for Stochastic Models and Risk*. (Wiley, 2002).

[250] M. F. Neuts, *Structured Stochastic Matrices of M/G/1 Type and Their Applications*. (Marcel Dekker, 1989).

[251] A. Ng and M. Jordan, Pegasus: A policy search method for large MDPs and POMDPs. In *Proceedings of the Sixteenth Conference on Uncertainty in Artificial Intelligence*, pages 406–15. (Morgan Kaufmann Publishers Inc., 2000).

[252] M. H. Ngo and V. Krishnamurthy, Optimality of threshold policies for transmission scheduling in correlated fading channels. *IEEE Transactions on Communications*, 57(8):2474–83, 2009.

[253] M. H. Ngo and V. Krishnamurthy, Monotonicity of constrained optimal transmission policies in correlated fading channels with ARQ. *IEEE Transactions on Signal Processing*, 58(1):438–51, 2010.

[254] N. Noels, C. Herzet, A. Dejonghe, V. Lottici, H. Steendam, M. Moeneclaey, M. Luise and L. Vandendorpe, Turbo synchronization: an EM algorithm interpretation. In *Proceedings of IEEE International Conference on Communications ICC'03*, volume 4, 2933–7. IEEE, 2003.

[255] M. Ottaviani and P. Sørensen, Information aggregation in debate: Who should speak first? *Journal of Public Economics*, 81(3):393–421, 2001.

[256] C. H. Papadimitriou and J. N. Tsitsiklis, The complexity of Markov decision processes. *Mathematics of Operations Research*, 12(3):441–50, 1987.

[257] E. Pardoux, Equations du filtrage nonlineaire de la prediction et du lissage. *Stochastics*, 6:193–231, 1982.

[258] R. Parr and S. Russell, Approximating optimal policies for partially observable stochastic domains. In *IJCAI*, volume 95, pages 1088–94. (Citeseer, 1995).

[259] R. Pastor-Satorras and A. Vespignani, Epidemic spreading in scale-free networks. *Physical Review Letters*, 86(14):3200, 2001.

[260] S. Patek, On partially observed stochastic shortest path problems. In *Proceedings of 40th IEEE Conference on Decision and Control*, pages 5050–5, Orlando, Florida, 2001.

[261] G. Pflug, *Optimization of Stochastic Models: The Interface between Simulation and Optimization*. Kluwer Academic Publishers, 1996.

[262] J. Pineau, G. Gordon and T. Sebastian, Point-based value iteration: An anytime algorithm for POMDPs. In *IJCAI*, volume 3, 1025–32, 2003.

[263] M. L. Pinedo, *Scheduling: Theory, Algorithms, and Systems.* (Springer-Verlag, 2012).

[264] L. K. Platzman, Optimal infinite-horizon undiscounted control of finite probabilistic systems. *SIAM Journal on Control and Optimization*, 18:362–80, 1980.

[265] S. M. Pollock, A simple model of search for a moving target. *Operations Research*, 18:893–903, 1970.

[266] B. T. Polyak and A. B. Juditsky, Acceleration of stochastic approximation by averaging. *SIAM Journal of Control and Optimization*, 30(4):838–55, July 1992.

[267] H. V. Poor, Quickest detection with exponential penalty for delay. *Annals of Statistics*, 26(6):2179–205, 1998.

[268] H. V. Poor and O. Hadjiliadis, *Quickest Detection.* (Cambridge University Press, 2008).

[269] H.V. Poor, *An Introduction to Signal Detection and Estimation*, 2nd edition. (Springer-Verlag, 1993).

[270] B. M. Pötscher and I. R. Prucha, *Dynamic Nonlinear Econometric Models: Asymptotic Theory.* (Springer-Verlag, 1997).

[271] K. Premkumar, A. Kumar and V. V. Veeravalli, Bayesian Quickest Transient Change Detection. In Proceedings of International Workshop in Applied Probability, Madrid, 2010.

[272] M. Puterman, *Markov Decision Processes.* (John Wiley, 1994).

[273] J. Quah and B. Strulovici, Aggregating the single crossing property. *Econometrica*, 80(5):2333–48, 2012.

[274] L. R. Rabiner, A tutorial on hidden Markov models and selected applications in speech recognition. *Proceedings of the IEEE*, 77(2):257–85, 1989.

[275] V. Raghavan and V. Veeravalli, Bayesian quickest change process detection. In *ISIT*, 644–648, Seoul, 2009.

[276] F. Riedel, Dynamic coherent risk measures. *Stochastic Processes and Their Applications*, 112(2):185–200, 2004.

[277] U. Rieder, Structural results for partially observed control models. *Methods and Models of Operations Research*, 35(6):473–90, 1991.

[278] U. Rieder and R. Zagst, Monotonicity and bounds for convex stochastic control models. *Mathematical Methods of Operations Research*, 39(2):187–207, June 1994.

[279] B. Ristic, S. Arulampalam and N. Gordon, *Beyond the Kalman Filter: Particle Filters for Tracking Applications.* (Artech, 2004).

[280] C. P. Robert and G. Casella, *Monte Carlo Statistical Methods.* (Springer-Verlag, 2013).

[281] R. T. Rockafellar and S. Uryasev, Optimization of conditional value-at-risk. *Journal of Risk*, 2:21–42, 2000.

[282] S. Ross, Arbitrary state Markovian decision processes. *The Annals of Mathematical Statistics*, 2118–22, 1968.

[283] S. Ross, *Introduction to Stochastic Dynamic Programming.* (San Diego, CA: Academic Press, 1983).

[284] S. Ross, *Simulation*, 5th edition. (Academic Press, 2013).

[285] D. Rothschild and J. Wolfers, Forecasting elections: Voter intentions versus expectations, 2010.

[286] N. Roy, G. Gordon and S. Thrun, Finding approximate POMDP solutions through belief compression. *Journal of Artificial Intelligence Research*, 23:1–40, 2005.

[287] W. Rudin, *Principles of Mathematical Analysis.* (McGraw-Hill, 1976).

[288] A. Ruszczyński, Risk-averse dynamic programming for Markov decision processes. *Mathematical Programming*, 125(2):235–61, 2010.

[289] T. Sakaki, M. Okazaki and Y. Matsuo, Earthquake shakes Twitter users: Real-time event detection by social sensors. In *Proceedings of the 19th International Conference on World Wide Web*, pages 851–60. (New York, 2010). ACM.

[290] A. Sayed, *Adaptive Filters*. (Wiley, 2008).

[291] A. H. Sayed, Adaptation, learning, and optimization over networks. *Foundations and Trends in Machine Learning*, 7(4–5):311–801, 2014.

[292] M. Segal and E. Weinstein, A new method for evaluating the log-likelihood gradient, the hessian, and the Fisher information matrix for linear dynamic systems. *IEEE Transactions on Information Theory*, 35(3):682–7, May 1989.

[293] E. Seneta, *Non-Negative Matrices and Markov Chains*. (Springer-Verlag, 1981).

[294] L. I. Sennott, *Stochastic Dynamic Programming and the Control of Queueing Systems*. (Wiley, 1999).

[295] M. Shaked and J. G. Shanthikumar, *Stochastic Orders*. (Springer-Verlag, 2007).

[296] G. Shani, R. Brafman and S. Shimony, Forward search value iteration for POMDPs. In *IJCAI*, 2619–24, 2007.

[297] G. Shani, J. Pineau and R. Kaplow, A survey of point-based POMDP solvers. *Autonomous Agents and Multi-Agent Systems*, 27(1):1–51, 2013.

[298] A. N. Shiryaev, On optimum methods in quickest detection problems. *Theory of Probability and Its Applications*, 8(1):22–46, 1963.

[299] R. H. Shumway and D. S. Stoffer, An approach to time series smoothing and forecasting using the EM algorithm. *Journal of Time Series Analysis*, 253–64, 1982.

[300] R. Simmons and S. Konig, Probabilistic navigation in partially observable environments. In *Proceedings of 14th International Joint Conference on Artificial Intelligence*, 1080–87, (Montreal, CA: Morgan Kaufman).

[301] S. Singh and V. Krishnamurthy, The optimal search for a Markovian target when the search path is constrained: the infinite horizon case. *IEEE Transactions on Automatic Control*, 48(3):487–92, March 2003.

[302] R. D. Smallwood and E. J. Sondik, Optimal control of partially observable Markov processes over a finite horizon. *Operations Research*, 21:1071–88, 1973.

[303] J. E. Smith and K. F. McCardle, Structural properties of stochastic dynamic programs. *Operations Research*, 50(5):796–809, 2002.

[304] T. Smith and R. Simmons, Heuristic search value iteration for pomdps. In *Proceedings of the 20th Conference on Uncertainty in Artificial Intelligence*, 520–7. (AUAI Press, 2004).

[305] V. Solo and X. Kong, *Adaptive Signal Processing Algorithms – Stability and Performance*. (NJ: Prentice Hall, 1995).

[306] E. J. Sondik, *The Optimal Control of Partially Observed Markov Processes*. PhD thesis, Electrical Engineering, Stanford University, 1971.

[307] E. J. Sondik, The optimal control of partially observable Markov processes over the infinite horizon: discounted costs. *Operations Research*, 26(2):282–304, March–April 1978.

[308] M. Spaan and N. Vlassis, Perseus: Randomized point-based value iteration for POMDPs. *J. Artif. Intell. Res.(JAIR)*, 24:195–220, 2005.

[309] J. Spall, *Introduction to Stochastic Search and Optimization*. (Wiley, 2003).

[310] L. Stone, What's happened in search theory since the 1975 Lanchester prize. *Operations Research*, 37(3):501–06, May–June 1989.

[311] R. L. Stratonovich, Conditional Markov processes. *Theory of Probability and Its Applications*, 5(2):156–78, 1960.

[312] J. Surowiecki, *The Wisdom of Crowds*. (New York, NY: Anchor, 2005).

[313] R. Sutton and A. Barto, *Reinforcement Learning: An Introduction.* (Cambridge, MA: MIT Press, 1998).

[314] M. Taesup and T. Weissman, Universal Filtering Via Hidden Markov Modeling. *IEEE Transactions on Information Theory*, 54(2):692–708, 2008.

[315] M. A. Tanner, *Tools for Statistical Inference: Methods for the Exploration of Posterior Distributions and Likelihood Functions.* Springer series in statistics. (New York, NY: Springer-Verlag, 1993).

[316] M. A. Tanner and W. A. Wong, The calculation of posterior distributions by data augmentation. *J. Am. Statis. Assoc.*, 82:528–40, 1987.

[317] A. G. Tartakovsky and V. V. Veeravalli, General asymptotic Bayesian theory of quickest change detection. *Theory of Probability and Its Applications*, 49(3):458–97, 2005.

[318] R. Tibshirani. Regression shrinkage and selection via the lasso. *Journal of the Royal Statistical Society. Series B (Methodological)*, pages 267–288, 1996.

[319] P. Tichavsky, C. H. Muravchik and A. Nehorai, Posterior Cramér-Rao bounds for discrete-time nonlinear filtering. *IEEE Transactions on Signal Processing*, 46(5):1386–96, May 1998.

[320] L. Tierney, Markov chains for exploring posterior distributions. *The Annals of Statistics*, 1701–28, 1994.

[321] D. M. Topkis, Minimizing a submodular function on a lattice. *Operations Research*, 26:305–21, 1978.

[322] D. M. Topkis, *Supermodularity and Complementarity.* (Princeton, NJ: Princeton University Press, 1998).

[323] D. van Dyk and X. Meng, The art of data augmentation. *Journal of Computational and Graphical Statistics*, 10(1):1–50, 2001.

[324] L. Vandenberghe and S. Boyd, Semidefinite programming. *SIAM review*, 38(1):49–95, 1996.

[325] V. N. Vapnik, *Statistical Learning Theory.* (Wiley, 1998).

[326] H. Varian, The nonparametric approach to demand analysis. *Econometrica*, 50(1):945–73, 1982.

[327] H. Varian, Non-parametric tests of consumer behaviour. *The Review of Economic Studies*, 50(1):99–110, 1983.

[328] H. Varian, Revealed preference and its applications. *The Economic Journal*, 122(560):332–8, 2012.

[329] F. Vega-Redondo, *Complex Social Networks*, volume 44. (Cambridge University Press, 2007).

[330] S. Verdu, *Multiuser Detection.* (Cambridge University Press, 1998).

[331] B. Wahlberg, S. Boyd, M. Annergren and Y. Wang, An ADMM algorithm for a class of total variation regularized estimation problems. In *Proceedings 16th IFAC Symposium on System Identification*, July 2012.

[332] A. Wald, Note on the consistency of the maximum likelihood estimate. *The Annals of Mathematical Statistics*, 595–601, 1949.

[333] E. Wan and R. Van Der Merwe, The unscented Kalman filter for nonlinear estimation. In *Adaptive Systems for Signal Processing, Communications, and Control Symposium 2000. AS-SPCC. The IEEE 2000*, pages 153–8. IEEE, 2000.

[334] C. C. White and D. P. Harrington, Application of Jensen's inequality to adaptive suboptimal design. *Journal of Optimization Theory and Applications*, 32(1):89–99, 1980.

[335] L. B. White and H. X. Vu, Maximum likelihood sequence estimation for hidden reciprocal processes. *IEEE Transactions on Automatic Control*, 58(10):2670–74, 2013.

[336] W. Whitt, Multivariate monotone likelihood ratio and uniform conditional stochastic order. *Journal Applied Probability*, 19:695–701, 1982.

[337] P. Whittle, Multi-armed bandits and the Gittins index. *J. R. Statist. Soc. B*, 42(2):143–9, 1980.

[338] N. Wiener, *The Extrapolation, Interpolation and Smoothing of Stationary Time Series*. (New York, NY: John Wiley, 1949).

[339] J. Williams, J. Fisher, and A. Willsky, Approximate dynamic programming for communication-constrained sensor network management. *IEEE Transactions on Signal Processing*, 55(8):4300–11, 2007.

[340] E. Wong and B. Hajek. *Stochastic Processes in Engineering Systems*, 2nd edition. (Berlin: Springer-Verlag, 1985).

[341] W. M. Wonham, Some applications of stochastic differential equations to optimal nonlinear filtering. *SIAM J. Control*, 2(3):347–69, 1965.

[342] C. F. J. Wu, On the convergence properties of the EM algorithm. *Annals of Statistics*, 11(1):95–103, 1983.

[343] J. Xie, S. Sreenivasan, G. Kornis, W. Zhang, C. Lim and B. Szymanski, Social consensus through the influence of committed minorities. *Physical Review E*, 84(1):011130, 2011.

[344] B. Yakir, A. M. Krieger and M. Pollak, Detecting a change in regression: First-order optimality. *Annals of Statistics*, 27(6):1896–1913, 1999.

[345] D. Yao and P. Glasserman, *Monotone Structure in Discete-Event Systems*. (Wiley, 1st edition, 1994).

[346] G. Yin, C. Ion and V. Krishnamurthy, How does a stochastic optimization/approximation algorithm adapt to a randomly evolving optimum/root with jump Markov sample paths. *Mathematical Programming*, 120(1):67–99, 2009.

[347] G. Yin and V. Krishnamurthy, LMS algorithms for tracking slow Markov chains with applications to hidden Markov estimation and adaptive multiuser detection. *IEEE Transactions on Information Theory*, 51(7), July 2005.

[348] G. Yin, V. Krishnamurthy and C. Ion, Regime switching stochastic approximation algorithms with application to adaptive discrete stochastic optimization. *SIAM Journal on Optimization*, 14(4):117–1215, 2004.

[349] G. Yin and Q. Zhang, *Discrete-time Markov Chains: Two-Time-Scale Methods and Applications*, volume 55. (Springer, 2006).

[350] S. Young, M. Gasic, B. Thomson and J. Williams, POMDP-based statistical spoken dialog systems: A review. *Proceedings of the IEEE*, 101(5):1160–79, 2013.

[351] F. Yu and V. Krishnamurthy, Optimal joint session admission control in integrated WLAN and CDMA cellular network. *IEEE Transactions Mobile Computing*, 6(1):126–39, January 2007.

[352] M. Zakai, On the optimal filtering of diffusion processes. *Z. Wahrscheinlichkeitstheorie verw. Gebiete*, 11:230–43, 1969.

[353] Q. Zhao, L. Tong, A. Swami and Y. Chen, Decentralized cognitive MAC for opportunistic spectrum access in ad hoc networks: A POMDP framework. *IEEE Journal on Selected Areas Communications*, pages 589–600, 2007.

[354] K. Zhou, J. Doyle and K. Glover, *Robust and Optimal Control*, volume 40. (NJ: Prentice Hall, 1996).

Index